NUMERICAL METHODS AND ANALYSIS

International Series in Pure and Applied Mathematics

Also Available from McGraw-Hill

Schaum's Outline Series in Mathematics & Statistics

Most outlines include basic theory, definitions, and hundreds of solved problems and supplementary problems with answers.

Titles on the Current List Include:

Advanced Calculus
Advanced Mathematics
Analytic Geometry
Beginning Calculus
Boolean Algebra
Calculus, 3d edition
Calculus for Business, Economics, & the Social Sciences
Calculus of Finite Differences & Difference Equations
College Algebra
College Mathematics, 2d edition
Complex Variables
Descriptive Geometry
Differential Equations
Differential Geometry
Discrete Math
Elementary Algebra
Essential Computer Math
Finite Mathematics
Fourier Analysis
General Topology
Geometry, 2d edition
Group Theory
Laplace Transforms
Linear Algebra, 2d edition
Mathematical Handbook of Formulas & Tables
Matrices
Matrix Operations
Modern Algebra
Modern Elementary Algebra
Modern Introductory Differential Equations
Numerical Analysis, 2d edition
Partial Differential Equations
Probability
Probability & Statistics
Projective Geometry
Real Variables
Review of Elementary Mathematics
Set Theory & Related Topics
Statistics, 2d edition
Technical Mathematics
Tensor Calculus
Trigonometry, 2d edition
Vector Analysis

Schaum's Solved Problems Books

Each title in this series is a complete and expert source of solved problems containing thousands of problems with worked out solutions.

Titles on the Current List Include:

3000 Solved Problems in Calculus
2500 Solved Problems in College Algebra and Trigonometry
2500 Solved Problems in Differential Equations
2000 Solved Problems in Discrete Mathematics
3000 Solved Problems in Linear Algebra
2000 Solved Problems in Numerical Analysis
3000 Solved Problems in Precalculus

Available at your College Bookstore. A complete list of Schaum titles may be obtained by writing to: Schaum Division
McGraw-Hill, Inc.
Princeton Road S-1
Highstown, NJ 08520

NUMERICAL METHODS AND ANALYSIS

James L. Buchanan

U.S. Naval Academy

Peter R. Turner

U.S. Naval Academy

McGraw-Hill, Inc.

New York St. Louis San Francisco Auckland Bogotá
Caracas Lisbon London Madrid Mexico Milan Montreal
New Delhi Paris San Juan Singapore Sydney Tokyo Toronto

To Our Families

This book was set in Times Roman.
The editors were Richard Wallis and David A. Damstra;
the production supervisor was Leroy A. Young.
The cover was designed by Rafael Hernandez.
R. R. Donnelley & Sons Company was printer and binder.

NUMERICAL METHODS AND ANALYSIS

1 2 3 4 5 6 7 8 9 0 DOC DOC 9 0 9 8 7 6 5 4 3 2

ISBN 0-07-008717-2

Library of Congress Cataloging-in-Publication Data

Buchanan, James L.
    Numerical methods and analysis / James L. Buchanan and Peter R.
Turner.
    p.   cm.—(International series in pure and applied
mathematics)
    Includes bibliographical references and index.
    ISBN 0-07-008717-2
    1. Numerical analysis.   I. Turner, P. R. (Peter R.), (date).
II. Title.   III. Series.
QA297.B79   1992
519.4′0285—dc20   91-43465

ABOUT THE AUTHORS

James L. Buchanan is professor of mathematics at the U.S. Naval Academy. He received his Ph.D. from the University of Delaware in 1980 and has been at the Naval Academy since 1979. He is the coauthor of a book on elliptic systems of partial differential equations and the author of a number of journal articles on partial differential equations and the theory of elastic materials. He has also given several workshops on the use of the microcomputer in the teaching of differential equations. Dr. Buchanan is coauthor of the public domain computer programs *Mathematics Plotting Program* and the *Midshipman Differential Equations Program*.

Peter R. Turner has been a member of the faculty of the mathematics department at the U.S. Naval Academy since 1987. He received his B.Sc. and Ph.D. degrees from the University of Sheffield, England, in 1970 and 1973. He then joined the faculty of the applied mathematics department there as a junior research fellow for one year before moving on to the University of Lancaster as a lecturer in numerical analysis. He is the author of many publications in both technical journals and conference proceedings as well as an introductory text in numerical analysis; he also edited three volumes of lectures from the UK Science and Engineering Research Council summer schools in numerical analysis. In addition to his research in numerical analysis and computer arithmetic, Dr. Turner has maintained an active interest in the development of the mathematics and, especially, the scientific computing curriculum.

CONTENTS

PREFACE

FEATURES OF THE TEXT

Emphasis on the nature of computer arithmetic. The axioms of computer arithmetic differ in significant ways from those of real arithmetic that students learn in introductory mathematics courses. These differences cannot be ignored in formulating algorithms for accomplishing some mathematical task and assessing their limitations. Thus we discuss in detail in Chapter 1 how computers do arithmetic, and we return in subsequent chapters to show how this process affects the design and choice of algorithms and the ways in which they may fail.

Emphasis on the differences among computers. While we certainly do not neglect measures of the effectiveness of an algorithm (such as the order of its error term) that are independent of the computer hardware and software used in implementing the algorithm, such measures simply do not tell the entire story. Hence we discuss the various ways in which computers differ and how these differences can affect the performance of algorithms. The implications of computer architecture and the choice of algorithms to carry out the fundamental arithmetic operations and evaluation of standard mathematical functions will be all the more difficult to ignore as we move into the era of parallel rather than serial scientific computing.

Test results on algorithms. In many sections of the text we compare the times required to execute different algorithms on a microcomputer. In isolation such results might lead students to conclude that there is a clear winner among the different algorithms for accomplishing a specified task when in fact there is not; however, we take pains to emphasize how differences among computers can lead to different results. Students are often invited in the exercises to conduct the same tests on their computer systems and explain any differences in the results.

Use of graphics. In creating this text we have availed ourselves of the greatly enhanced possibilities for graphical illustration that microcomputers afford, including numerous figures in the text and many graphics oriented exercises. This helps students better understand algorithms that have a geometric rationale and dispels the impression that numerical considerations seldom affect anything to the left of the sixth decimal place.

Additional topics. We have included in this text two topics rarely treated in introductory texts. In Chapter 3 evaluation of standard mathematical functions is treated in great detail. In particular the CORDIC algorithms, which are almost universally used for evaluation of transcendental functions in hand-held calculators, are presented. We have found that this topic interests students and serves as a good illustration of the mathematical sophistication required to develop effective numerical algorithms. In Chapter 13 we have included a brief introduction to parallel scientific computing. Full discussion of numerical algorithms for even one of the several different parallel architectures that are currently employed is a book length topic; however, we feel that parallel computing is becoming so important that it warrants mention even in a first course. In Chapter 13 we give a general discussion of the various architectures in use and the implications of the choice of architecture on the efficiency and accuracy of numerical algorithms. A discussion of numerical linear algebra serves to illustrate and augment the presentation.

PREREQUISITES

In writing this text we have endeavored to make the early parts of most sections accessible to students who have had no more than a three- or four-semester introduction to calculus and differential equations. Indeed, as outlined below, a semester course can be constructed from the introductory material in a number of chapters. For courses that cover all the material in certain chapters in depth, familiarity with real analysis and linear algebra will be helpful. In Appendixes A and B we review relevant topics in these two areas.

PRESENTATION OF ALGORITHMS

Usually the algorithms in this text are presented in *pseudocode*. From this form they may be readily rendered by students in any of the standard computer languages (BASIC, Fortran, Pascal, or C). Sometimes a generic implementation of an algorithm was too clumsy, and thus the algorithm afforded a good opportunity to illustrate the use of constructs available in many but not all languages. In these cases we have instead presented the algorithm in Turbo Pascal. Translation to other languages should not be difficult, however, providing these languages contain constructs comparable to those used in the program presented.

POSSIBLE COURSE OUTLINES

The book covers a wide range of topics in numerical analysis and can be used for a variety of different courses ranging from a one-semester "broad-spectrum" *methods* course through a two- or even three-semester sequence in numerical *analysis*. There is great flexibility in the choice of topics to be used in introductory courses and in further study including reading courses and/or extended projects.

Three possible one-semester courses would consist of:

(a) Chapters 1, 2, (3), 4, 5, 6, (7)
(b) Chapters 1, 4, 9, 10, 12
(c) Chapters 1, (2), 6, (7), 8, 11, 13

where the parentheses indicate optional topics. The first of these is a fairly typical course covering iteration, interpolation, and linear equations; the second follows the polynomial interpolation, quadrature, and ordinary differential equations path; the third is largely a numerical linear algebra course. Clearly other possibilities abound.

For a broad methods-oriented course, elementary material from the first ten chapters could be used. For example such a course might consist of Sections 1.1–1.5, (1.7), 2.1–2.4, 3.1, 3.2, 4.1–4.3, 5.1, 5.2, (5.3), 6.1, 6.2, 6.4, 7.3, (7.4), 8.2, 8.4, 9.1–9.3, 10.1–10.4.

ACKNOWLEDGMENTS

We wish to acknowledge the help, support, and encouragement of family, friends, and colleagues here at the Naval Academy and elsewhere. The various contributions are too numerous to list individually but we thank all of them. Some of these people and some of our students have suffered (almost) as much as the authors from the growing pains of this work. The reviewers also provided several suggestions for improvements in the draft, and we thank them too: Sue Brown, Naval Postgraduate School; Bruce Edwards, University of Florida; Mostafa Ghandehari, Naval Postgraduate School; Murli Gupta, George Washington University; and James Hummel, University of Maryland. Among colleagues there is one who merits special mention: Lt. Dennis Frazier, USNR, who has read through almost all the text in several versions and made many useful comments and suggestions. In particular he has greatly enhanced our knowledge and understanding of methods for the eigenvalue problem. Students, mostly at the U.S. Naval Academy and the University of Lancaster, England, have over the years helped us develop and polish the material. Two deserve special credit for their efforts in finding many typographical and other errors in the various drafts: former midshipmen Robert Baker and Matt Heck.

James L. Buchanan
Peter R. Turner

GENERAL INTRODUCTION

Before beginning with the detailed description of the book and the specific material we will be studying, it is well worthwhile to answer the fundamental questions "What is numerical analysis?" and "Why do we need numerical methods?" It is probably easier to treat the second of these first.

There is a tendency in lower-level mathematical courses to teach recipes for solving particular problems which fit into very special forms. These can be solved by some convenient procedure which leads to the exact solution just by following prescribed rules. Unfortunately, these idealized problems are by no means typical of "real-life" mathematical, physical, or engineering problems. Many physical situations are described by differential equations, and these almost invariably must be solved approximately on a computer by numerical methods. If we think of the problem of solving just one equation in one unknown, we can write down a formula for its solutions in the case of a polynomial equation of degree not more than four. For any more complicated equations (and often for these simple ones too) the only techniques we have available are numerical methods such as Newton's method which many readers will recall from earlier calculus courses.

If you start to think about what mathematical functions we can evaluate *exactly*, and for what values of their argument this is true, you will come up with a very short list indeed—essentially consisting of rational functions with rational coefficients and rational arguments or, equivalently, a quotient of two polynomials with integer coefficients and integer values of the variable. For almost all functions it is therefore necessary to develop efficient numerical techniques for their *approximate* evaluation.

The foregoing is just a brief introduction to the vast area of mathematical problems for which we need numerical methods. We will see many further examples as the course proceeds. For any particular problem it is not enough simply to develop a method which we hope will give a reasonable approximation to the solution; we need to *know* that it is indeed giving us such an approximation, *and* we need to know how good that approximation is.

The role of the numerical analyst is to develop and analyze numerical techniques. This includes proving that an iterative method such as Newton's converges and that the limit is a solution of the original problem, and analyzing and *bounding* the errors introduced in the numerical process. There are many sources of these errors. The accuracy of the mathematical model of the physical situation, the arithmetic system of the computer, and conditions we use to stop a particular process are among the most important; all these must be taken into consideration to guarantee any particular accuracy in the final computed solution.

A significant portion of this book will concentrate on one area of numerical analysis, namely, the evaluation of a mathematical function. However, as we will soon see, this involves the use of many of the ideas mentioned above: the solution of a nonlinear equation, a system of linear equations, and many other aspects of interpolation and approximation. The following simple examples will serve to illustrate the different considerations and approaches which may be adopted in trying to solve the problem

Evaluate $f(x)$ to within a specified accuracy.

The techniques we are likely to adopt will depend not just on the function but, among other things, on whether one, a few, or many values of the argument x are needed, the accuracy desired, the machinery available, and the arithmetic system being used. This list takes no account of the type of information we have—an explicit formula, a power (or other) series expansion, a mathematical definition of the function (such as the integral of another function), or perhaps just some function values at other points which may themselves be subject to experimental error.

Examples

1. The square root function, $f(t) = \sqrt{t}$
   a. "By hand": bisection to solve $x^2 - t = 0$
   b. Using calculator/computer: Newton or fast iterative method to solve $x^2 - t = 0$
   c. (Binary) Computer library routine: Suitably modified Newton's method
   d. (Decimal) Hand-held calculator routine: CORDIC algorithm (or, possibly, Newton's method)
2. The exponential function, $f(x) = \exp x = e^x$
   a. Small x: Taylor series or power series
   b. Invert logarithm function by solving $x = \ln y$
   c. (Binary) Computer routine: Range reduction and power series for 2^x

3. The natural logarithm function, $f(x) = \ln x$
   a. $|1 - x| < 1$: Power series
   b. Integrate the reciprocal function
   c. (Binary) Computer routine:

$$\ln(f2^{E}) = \ln f + E \ln 2$$

   d. (Decimal) Calculator routine: CORDIC routine
4. The function $\sin\sqrt{x}$ given a table of values of

$$\sin \frac{n}{100} \qquad (n = 0, 1, \dots, 100)$$

   a. For a single value of x: Polynomial or piecewise polynomial interpolation
   b. For multiple values of x: Spline interpolation or an approximation method such as a least squares fit or discrete Fourier transform.

In this book we will consider all the possible approaches just mentioned and find out by computational experiment and experience as well as mathematical analyses of the methods how good or bad some of the proposed solutions really are. We will begin with a look at how computers store and manipulate numerical data and how this can affect our decisions as to the approach to be used and the accuracy which can be achieved by any particular algorithm. The next chapter deals with the numerical approaches to the solution of equations—mostly nonlinear equations in a single unknown. We consider the various iterative, that is, successive approximation, schemes and their convergence (or nonconvergence) to the desired solution. This also necessitates the study of *rates of convergence* of sequences and leads to the development of techniques for the acceleration of slowly convergent sequences. We also consider here the application of Newton's method to a system of two nonlinear equations and apply it to the problem of obtaining quadratic factors—and therefore the complex roots—of polynomial equations.

From there we will proceed to the approximate evaluation of elementary functions—starting with the most elementary, polynomials, then considering the use of the equation-solving techniques and concluding with a look at the design of computer and hand-held calculator algorithms, including the fascinating CORDIC algorithm which is used by nearly all hand-held calculators for almost all the elementary functions including multiplication and division. The classical interpolation and approximation methods discussed in the next two chapters will give you an introduction to a vast and still growing area of numerical analysis. It will also make evident the need to develop efficient numerical techniques for the solution of systems of linear equations.

Of course, the problem of evaluating functions includes much more than we have just outlined. One of the most important aspects is certainly the numerical solution of differential equations. This enormous field of study is the chief objective of the second half of the text and is itself dependent on the methods of the numerical calculus—numerical techniques for differentiation, integration, and the location of stationary points. These, too, are treated in considerable detail and raise their own interesting questions and solutions. Numerical integration will reintroduce many readers to the trapezoidal and Simpson's formulas for estimating areas. The solution of differential equations is so central to (almost) every area of applied mathematics—whether in the sciences and engineering or in operations research and economic applications— that it is probably still the biggest single user of computer time, certainly on the number-crunching supercomputers and parallel processors. We conclude the text with some consideration of the effect of possible vectorizing or paralleliza- tion of algorithms. This is by no means restricted to the efficient parallel implementation of the algorithms which are to be preferred for a serial machine but entails some new considerations which result in a different choice of fundamental algorithm to implement.

CHAPTER
1

COMPUTER ARITHMETIC AND ERRORS

In this chapter, we study the methods and properties of computer number representations and arithmetic. We will see that there are many aspects to this which have an important bearing on the choice and design of numerical procedures. There are pitfalls which must be avoided, errors which must be accounted for, and assumptions which must *not* be made. We will see, for example, that many of the standard axioms and properties of the familiar real number system no longer hold in the computer or calculator.

The other major area of interest for this chapter is the study of errors and how these can affect our calculations. There are many types of error (including plain old mistakes) which we must account for. There are the errors inherent in the mathematical model caused by making simplifying assumptions in the original development of the problem to be solved; there are errors caused by the abbreviation of the mathematical process—known as *truncation errors*—and there are errors resulting from the nonexactness of the computer representation and manipulation of numerical data—*rounding errors*.

1.1 INTEGER REPRESENTATION

We are, of course, familiar with the ordinary decimal representation of integers in which we identify $a_n a_{n-1} \cdots a_1 a_0$ with $a_n 10^n + a_{n-1} 10^{n-1} + \cdots + a_1 10^1 + a_0 10^0$, where each of the coefficients (or digits) is one of $0, 1, 2, 3, \ldots, 9$. Similarly, the binary representation would identify this same string with the integer $a_n 2^n + a_{n-1} 2^{n-1} + \cdots + a_1 2^1 + a_0 2^0$, where now each of the coefficients is a binary digit, or *bit*, 0 or 1. It is this binary system that is almost invariably used for the computer representation of positive integers.

Typically, an integer is stored in a computer word of a fixed length; that is, a fixed number of bits are used. (For integers this wordlength will usually be 8, 16, or 32.) The first bit will always be used to store the sign of the number; 0 is used for the positive and 1 for the negative sign. The reason for this choice will become apparent shortly.

For the purposes of illustration suppose that we are working with a computer which has a 4-bit wordlength for integers. Then the representable nonnegative integers and their representations are as given in Table 1.1.1.

Now the result of adding another 1 to the last of the entries in this table would be to force a carry all the way across to the left-hand, *most significant*, end of the word; this yields the representation 1000 which in standard binary—*but not in the computer*—would be interpreted as the positive integer 8. However, the first bit, in this case the 1, is used to represent the negative sign. So what does 1000 represent on our hypothetical computer?

In what may at first seem the most natural interpretation this would be −0, but that is obviously wasteful as −0 and 0 are identical integers. Nonetheless, we do want to retain the representation 0000 for the integer 0. The convenient solution is to store negative integers in what is called *complemented* form. There are two such forms available within the binary system, namely, *one's complement*, in which every bit is simply *negated* or reversed (1→0 and 0→1), and *two's complement*, which is the same as the one's complement with a unit added in the least significant position *without a carry beyond the most significant bit*. See Table 1.1.2.

The right-hand column in the table is used to represent the negatives of the corresponding numbers in the left-hand one. Thus, for example, we have −5 represented by 1011. Notice that there is one representation which has not yet been used; 1000 represents −8 in this 4-bit integer format we have described.

For a more realistic wordlength such as you might find on a genuine computer, the principles are identical: if n bits are allocated to the storage of integers, then the smallest representable integer $I_{\min}$ is -2^{n-1} (represented by

TABLE 1.1.1
Four-bit binary representation of nonnegative integers

n	4-bit binary
0	0000
1	0001
2	0010
3	0011
4	0100
5	0101
6	0110
7	0111

TABLE 1.1.2

Integer	4-bit binary	One's complement	Two's complement and value
0	0000	1111	$0000 = 0$
1	0001	1110	$1111 = -1$
2	0010	1101	$1110 = -2$
3	0011	1100	$1101 = -3$
4	0100	1011	$1100 = -4$
5	0101	1010	$1011 = -5$
6	0110	1001	$1010 = -6$
7	0111	1000	$1001 = -7$

$100\cdots0$), and adding 1 to this and its successors adds 1 to the value represented. This continues up to and through $0\ (00\cdots0)$, $1\ (00\cdots01)$, and on up to the largest representable integer $I_{max} = 2^{n-1} - 1\ (011\cdots1)$.

From the above description, what is the effect of running the following Pascal program?

Program 1.1.3*

```
program Integer_Overflow;
var   i: integer;
begin
    i:=1;
    while i<>0 do begin
        i:=i+1;
        writeln(i);
    end;
end.
```

At first you might be tempted to believe that the values of the variable i being printed to the screen will grow and grow indefinitely. But, what happens when the maximum representable integer I_{max} is reached?

The computer tries to make the assignment

$$i := I_{max} + 1 \qquad (1.1.1)$$

which *overflows* the allowable range. On many computer systems this will result in an "integer overflow" message and termination of the execution of the program. On other computer systems, including those running Turbo Pascal, however, the addition will be performed with the carry from the most significant end spilling over into the sign bit. This yields the binary representa-

---

* Note that numbered text elements (Tables, Examples, Programs, etc.) are triple-numbered by section *in sequence with each other*. Thus Program 1.1.3 here follows Tables 1.1.1 and 1.1.2. Equations, on the other hand, are triple-numbered by section in their own sequence.

tion $100\cdots0$ which, of course, represents the value I_{min} of the smallest representable integer -2^{n-1}.

On such a system, it follows that eventually i will assume negative values and then the repeated addition of unity will, again eventually, result in the assignment $i = 0$ which will terminate the program.

Before leaving the topic of integer representation, it is desirable to say a little more about the use of two's complement form for negative integers. The great benefit of this representation is that integer addition and subtraction become the same operation. The subtraction $p - q$ is achieved by forming the two's complement of q and *adding* the result to p. Thus, for example, on our hypothetical 4-bit machine of Table 1.1.2, the subtraction operation $m = 3 - 6$ is performed as the addition of (binary) 0011 and 1010:

$$\begin{array}{r} 0011 \\ +1010 \\ \hline 1101 \end{array}$$

which is the two's-complement-form binary representation of -3, as expected. The details of the computer implementation of the various arithmetic operations and data types are not central to the theme of this text, and we will not pursue them further here.

EXERCISES 1.1

1. Write down the 4-bit two's complement binary representations of the integers 2 and -3. Multiply these two by performing long multiplication of the binary strings, *ignoring any bits to the left of the sign bit*. Verify that the result is the correct representation of -6.

2. Form a table similar to Table 1.1.2 for 5-bit representation of integers. Verify that the two's complement form is suitable for multiplication by performing experiments similar to those of Exercise

1 with two positive factors, two negative factors, and one of each.

3. Write down I_{max} and I_{min} for the 4- and 5-bit representations of Exercises 1 and 2.

4. Find I_{max} and I_{min} for the Turbo Pascal types *integer* and *longint* which are, respectively, 8- and 16-bit two's complement binary representations. Multiply 17 by -5 in type *integer*.

5. Show that for 16-bit two's complement integers with integer wraparound, 8! appears to be negative!

1.2 FLOATING-POINT NUMBERS

The *floating-point* system of number representation and computer arithmetic is used throughout the world of scientific computing. It is the computer equivalent of the "scientific notation" used for many physical constants and by most hand-held calculators. Indeed, scientific notation is a decimal floating-point system.

For any arithmetic base β, a number X can be written as

$$X = \pm f \times \beta^{\pm E} \tag{1.2.1}$$

where E is a nonnegative integer and f a real number. This is a base β

floating-point representation for X with *exponent* $\pm E$ and *fraction*, or *mantissa*, f. (The base β is invariably a positive integer, though in principle this is not necessary.)

The representation (1.2.1) is very far from unique, since, for example, we also could write

$$X = \pm(f\beta) \times \beta^{\pm E - 1} \qquad (1.2.2)$$

To render the floating-point representation unique, most systems use a *normalized* floating-point format in which X is expressed as in (1.2.1) with

$$\frac{1}{\beta} \leq f < 1 \qquad (1.2.3)$$

(That normalized representations are unique is a simple exercise left to the reader.)[1]

In a normalized floating-point representation with k *significant βits*, we thus have

$$X \simeq \pm(0.f_1 f_2 \cdots f_k) \times \beta^{\pm E} \qquad (1.2.4)$$

where the base β digits, or βits, are integers satisfying

$$1 \leq f_1 < \beta \qquad \text{and} \qquad 0 \leq f_i < \beta \qquad (1.2.5)$$

and the base β fraction

$$0.f_1 f_2 \cdots f_k = f_1 \beta^{-1} + f_2 \beta^{-2} + \cdots + f_k \beta^{-k} \qquad (1.2.6)$$

Example 1.2.1. Write the number e in normalized floating-point form for (*a*) base 10, (*b*) base 2, and (*c*) base 16.

Solution
(*a*) For the decimal base, we have, to seven decimal places,

$$e \simeq 2.718\,281\,8 = (0.271\,828\,18)_{10} \times 10^1 \qquad (1.2.7)$$

which is the required normalized form with eight significant digits.
(*b*) For the binary system,

$$e = 1 \times 2^1 + 0 \times 2^0 + 1 \times 2^{-1} + 0 \times 2^{-2} + 1 \times 2^{-3} + 1 \times 2^{-4} + 0 \times 2^{-5} + 1 \times 2^{-6}$$

$$+ \cdots$$

$$\simeq (10.1011011111100001010100010110)_2$$

$$= (0.101011011111100001010100010110)_2 \times 2^2 \qquad (1.2.8)$$

which is the normalized binary representation with 30 significant bits.
(*c*) For the *hexadecimal*, or base 16, system, we can obtain the representation easily from (1.2.8). Using the fact that $16 = 2^4$, and so $2^2 = \frac{1}{4} \times 16^1$, we see that

---

[1] Some systems use instead the normalization $1 \leq f < \beta$. This is particularly common for the scientific notation of hand-held calculators.

$$(0.1010110111111000010101000010110)_2 \times 2^2$$

$$= \tfrac{1}{4} \times (0.1010110111111000010101000010110)_2 \times 16^1$$

$$= (0.0010101101111110000010101000010110)_2 \times 16^1$$

Again using the fact that $16 = 2^4$, it follows that each group of 4 bits represents a single hex (or hexadecimal) digit; thus the leading 0.0010 represents $\tfrac{1}{8}$, or $\tfrac{2}{16}$, the next 1011 represents $(8+2+1)/2^8$, or $11/16^2$, and so on. We obtain, therefore,

$$e \simeq \left(\tfrac{2}{16} + \tfrac{11}{16^2} + \tfrac{7}{16^3} + \tfrac{14}{16^4} + \tfrac{1}{16^5} + \tfrac{5}{16^6} + \tfrac{1}{16^7} + \tfrac{6}{16^8} \right) \times 16^1$$

and using the standard notation for the additional hex digits, where A, B, C, D, E, and F are used to represent 10, 11, 12, 13, 14, and 15, this is

$$e \simeq (0.2B7E1516)_{16} \times 16^1 \tag{1.2.9}$$

which is the normalized hexadecimal floating-point representation of e with eight significant hex digits.

One of the first things which must be settled in the design of any computer or calculator system is the choice of base for the arithmetic. In the case of a computer this will almost inevitably be a power of 2 and usually 2 itself. For a calculator the decimal system is normally to be favored because of the fact that the result of every intermediate calculation is displayed and the time penalty for conversion to and from the binary system would be too great.

We can run a simple test to ascertain the base of the floating-point arithmetic system within any particular machine as follows.

Example 1.2.2. The value $x = 1/n$ is a fixed point of the function given by

$$f(x) = (n+1)x - 1 \tag{1.2.10}$$

since

$$f\left(\frac{1}{n}\right) = \frac{n+1}{n} - 1 = \frac{1}{n} \tag{1.2.11}$$

It follows then that iterating this function for any particular value n using the following loop should, if the floating-point arithmetic of the computer were exact, simply result in $x = 1/n$:

```
x:=1/n
for i = 1 to 30
    x:=(n + 1)*x − 1
```

We use here the symbol := for computer assignment, so "x:=1/n", means "Assign the value $1/n$ to the stored variable x."

Table 1.2.3 shows the results of implementing such a loop in Turbo Pascal for different values of n. We see that for the various powers of 2 the arithmetic is indeed exact. By contrast, for other values of n the error grows steadily from approximately -3.5×10^5 for $n = 3$ through about -2.3×10^{18} for $n = 10$.

These errors are the direct result of the propagation of the rounding error made in the binary representation of $\tfrac{1}{3}$ and $\tfrac{1}{10}$, respectively. To see how this error

TABLE 1.2.3

n	Final x
1	1.000 000 000 000 00 E + 0000
2	5.000 000 000 000 00 E − 0001
3	1.747 630 000 000 00 E + 0005
4	2.500 000 000 000 00 E − 0001
5	−4.021 311 173 693 75 E + 0010
6	−1.952 324 816 734 00 E + 0012
7	−4.021 071 095 865 60 E + 0013
8	1.250 000 000 000 00 E − 0001
9	−1.616 879 469 807 53 E + 0017
10	−2.308 383 841 816 94 E + 0018
11	−1.962 659 088 425 60 E + 0019
12	−2.443 971 442 609 19 E + 0020
13	−1.935 039 516 698 14 E + 0021
14	−1.328 735 789 941 45 E + 0022
15	7.052 067 281 085 79 E + 0022
16	6.250 000 000 000 00 E − 0002

grows as the program proceeds, suppose that within the machine x is initially stored as $1/n + \delta$, where δ will be a small quantity—of the order of $\pm x 10^{-12}$ for the computation reported above. With each iteration of the loop this error will be affected by the assignment

$$x := (n+1)x - 1 \tag{1.2.12}$$

For this first step, with $x = 1/n + \delta$, this results in the new value

$$x = (n+1)\left(\frac{1}{n} + \delta\right) - 1 = \left[1 + \frac{1}{n} + (n+1)\delta\right] - 1 \tag{1.2.13}$$

from which we see that the error has grown by a factor $n + 1$. Each iteration will have this same effect. (There will also be a relatively small contribution from the rounding error of the arithmetic of the iteration itself, but we can safely neglect that for the time being.) The program runs through the loop 30 times, and so the final error (or, the major part of it) will be $(n+1)^{30}\delta$. For the case $n = 10$, this quantity has magnitude around 2×10^{18}.

We will make a more detailed study of *error analysis* (of which this was a simple example) later on.

The advantages of the binary base for computer arithmetic are many, and not least among them is that, for a given wordlength, the mean representation error resulting from the use of the binary system is smaller than for any other base. This situation is further improved by the fact that a normalized binary floating-point number necessarily has a 1 as the first bit of its mantissa. Since this is universally true, there is no need for it to be stored explicitly—of course, it must be accounted for in the arithmetic processing within this *hidden-bit*, or *implicit-bit*, binary system.

The fact that the representation errors for binary arithmetic are smaller than those of other bases is a consequence of a somewhat surprising mathematical fact, namely, that the real numbers are logarithmically distributed. What this means is that the proportion of (base β) numbers whose leading significant bit is n, say, is given by

$$F_\beta(n) = \log_\beta(n+1) - \log_\beta n = \frac{\ln(1 + 1/n)}{\ln \beta} \qquad (1.2.14)$$

In particular, this implies that 30 percent of decimally represented numbers have the leading significant digit 1. A relatively simple computer program will at least make this apparently ridiculous statement look reasonable. Algorithm 1.2.4 below multiplies 1000 pairs of random numbers between 1 and 10 and produces a table of the frequencies of the leading significant digit (lsd) of their products. The function "random" generates different uniformly distributed random numbers in $[0, 1)$ on each call so that $9*\text{random} + 1$ produces a random number in the interval $[1, 10)$. In Table 1.2.5, the results of running this experiment are listed together with those of similar trials where more factors were multiplied together. We will return to these considerations in Sec. 1.4.

Algorithm 1.2.4

Initialize	count(n) = 0, for n = 1, 2, . . . , 9;
Loop	for i = 1 to 1000
	x:=9*random + 1; y:=random*9 + 1;
	z:=x*y;
	if z < 10 then n:=[z] else n:=[z/10];
	count(n):=count(n) + 1;
Output	count(n), for n = 1, 2, . . . , 9 .

As usual $[z]$ denotes the integer part of z, that is, the largest integer not greater than z.

Of course, Table 1.2.5 presents the results of just one small experiment; more extensive testing would produce closer agreement with the logarithmic distribution. (See Exercises 6 and 7.)

Many other studies related to the distribution of errors in computer arithmetic are dependent on this logarithmic distribution, and all lead to the conclusion that 2 is the optimal base to choose. As we have already remarked, the decimal base is generally to be preferred for a hand-held calculator, and performing the steps of Example 1.2.2 on your calculator will convince you that this is indeed the base used there.

Just as there are maximum and minimum integers which can be represented within the machine, similarly the set of floating-point numbers is finite and bounded. The range of representable numbers is, of course, dependent on the wordlength used for their storage and on how the individual bits are allocated to the mantissa and exponent parts. This varies enormously from machine to machine and from operating system to operating system.

TABLE 1.2.5
Frequencies of leading significant digits in 1000 products with two, three, or four random factors together with those given by the logarithmic distribution

lsd	Two factors	Three factors	Four factors	Logarithmic distribution
1	0.236	0.313	0.324	0.301
2	0.183	0.182	0.164	0.176
3	0.156	0.142	0.139	0.125
4	0.104	0.098	0.090	0.097
5	0.113	0.080	0.072	0.079
6	0.076	0.060	0.063	0.067
7	0.052	0.041	0.055	0.058
8	0.045	0.039	0.049	0.051
9	0.035	0.045	0.044	0.046

Two standard implementations have been adopted by IEEE (the Institute of Electrical and Electronics Engineers) for single-length (using a 32-bit word) and double-length (64-bit word) floating-point formats, but the particular details are not important to the development of the principles we are concerned with here. We will discuss more of the details of the various IEEE and Turbo Pascal real variable types along with the study of floating-point representation and arithmetic errors in Sec. 1.4.

For the remainder of the discussion let us suppose that our hypothetical machine has a binary floating-point system using M bits for the mantissa and N bits for the exponent, each together with its sign. (In most practical implementations the sign of the exponent will be incorporated into the exponent itself which is then stored in what is termed *biased* form; similarly the mantissa and its sign are often stored together using a complemented form for this part of the word.) Again, for our purposes, it does not matter whether or not the leading bit is stored explicitly or implicitly; we simply regard it as one of the N.

To add two such floating-point numbers, they must first be *aligned* so that like powers of 2 appear in the same position.

Example 1.2.6. Consider the addition of the floating-point representation of

$$a = 123.45 \quad \text{and} \quad b = 1.2345$$

Now,

$$a \simeq 64 + 32 + 16 + 8 + 2 + 1 + \tfrac{1}{4} + \tfrac{1}{8} + \tfrac{1}{16} + \tfrac{1}{128} + \tfrac{1}{256}$$

$$= (0.111101101110011)_2 \times 2^7$$

and

$$b \simeq 1 + \tfrac{1}{8} + \tfrac{1}{16} + \tfrac{1}{32} + \tfrac{1}{64} + \left(\tfrac{1}{2}\right)^{13}$$

$$= (0.100111100000010)_2 \times 2^1$$

each to 15-significant-bit accuracy.

To align these two quantities we must first *shift b* six places to the right (6 being the difference in the exponents of the two quantities). Thus b is rewritten as

$$b \simeq (0.000000100111100000010)_2 \times 2^7$$

In most machines the extra bits (or at least some of them) needed for this representation of b would be retained in the accumulator and may affect the final rounding of the result. The sum $a + b$ is now given by 2^7 times

$$(0.111101101110011)_2 + (0.000000100111100000010)_2$$

$$= (0.111110010101111000010)_2$$

which must be rounded to 15 bits to yield

$$(0.111110010101111)_2 \times 2^7 = 64 + 32 + 16 + 8 + 4 + \tfrac{1}{2} + \tfrac{1}{8} + \tfrac{1}{32} + \tfrac{1}{64} + \tfrac{1}{128} + \tfrac{1}{256}$$

$$= 124.68$$

to the same accuracy as the original data.

In the shift operation at the start of the above example we saw how some of the information in the smaller of the two numbers was lost to the final answer. In the extreme case the whole of b could be shifted so far to the right that it has no effect on the sum at all. The number of significant bits stored in the floating-point format, that is, the length of the mantissa, determines the relative magnitudes of two numbers for this to happen. This information can be obtained from the value of the machine-dependent constant μ, called *machine epsilon* or the *machine unit*. This is defined to be the smallest positive number x for which

$$1 + x > 1 \tag{1.2.15}$$

This quantity can be found by running a very simple program such as Program 1.2.7 below. For type *real* in Turbo Pascal we find that this quantity is 2^{-39}.

Program 1.2.7

```
          var n: integer;
              nu,mu: real;
          begin
              mu:=1; nu:=2; n:=0;
              while nu > 1 do begin
                  mu:=mu/2; nu:=1 + mu;
                  n:=n + 1;
              end;
              writeln('Machine epsilon is 2^-', n - 1);
          end.
```

Machine epsilon should not be confused with the smallest positive representable number. This much smaller quantity can be found by a similar, but simpler, program to that above. For systems which use normalized binary floating-point arithmetic, it will be $2^{E_{min}-1}$. Here, E_{min} is the smallest possible value of the exponent. The corresponding maximum value is denoted E_{max}, which is $2^N - 1$ for our hypothetical machine with N bits for the exponent. The largest representable number is therefore $2^{E_{max}}(1 - 2^{-M})$, where M is the number of bits in the mantissa, so that the maximum value of this fraction is $1 - 2^{-M}$. These last two machine constants—or, more accurately, system constants—are sometimes denoted by ω for the smallest and Ω for the largest representable (positive) quantities.

In the above program, it may seem initially strange that two variables mu and nu are declared and used in the loop rather than the apparently simpler control

$$\text{while } 1 + \text{mu} > 1 \text{ do begin}$$

The reason is simply that we must test for the smallest positive number such that the *stored* result of the addition is greater than 1. The simpler version above would apply the test to the quantity *in the accumulator* which will typically have one or more additional bits called *guard bits* for use in performing correct rounding and normalization of results.

If the result of any arithmetic operation exceeds Ω, then it is said to cause *overflow*, and the program will terminate with an error message. In the case where the result is smaller, in absolute value, than ω, it is said to *underflow*. In many systems underflow will not result in termination of the program but will simply cause the computer to treat the result as zero.

In the next few sections we will see some of the anomalies that arise out of the finiteness of the computer's floating-point arithmetic—and some of the things we can do to alleviate the difficulties caused by them.

EXERCISES 1.2

1. Obtain the normalized floating-point representation of π using (a) six significant digits of decimal, (b) 20 binary digits, and (c) five hexadecimal digits.

2. Write down the normalized binary representations of the integers $1, 2, \ldots, 10$ and each of their reciprocals using 6 bits for the mantissa.

3. Find the arithmetic base of your computer and your calculator.

4. Write and run a program to find the machine unit μ for your computer. Extend this to find the smallest representable positive number ω and the largest, Ω. What are the numbers of bits allocated to the storage of the mantissa and exponent? Is your machine using normalized floating-point arithmetic? Is the arithmetic of your computer balanced?

5. Prove that the normalized floating-point representation of a positive number is unique.

6. Write a program to implement Algorithm 1.2.4, and compare the results with the logarithmic distribution.

7. Try to justify the claim that the distribution of lsd is logarithmic.
   (a) Write a program to implement an algorithm similar to Algorithm 1.2.4 which examines 10,000 products of two, three, four, and five randomly generated factors in the interval $[1, 10]$.
   (b) See if you can come up with a theoretical explanation.

1.3   ERRORS: THEIR SOURCES AND MEASUREMENT

By far the most frequent source of error—and often the most significant—is just plain human error, or blunders. We are going to proceed with the discussion of errors in a utopian dreamworld in which infallibility has spread to all mathematicians and computer programmers, and so such mistakes cannot occur. Even in our dream world, however, errors are an important aspect of computational life. They are everywhere and unavoidable.

However, by careful analysis of the errors in any numerical process, we can at least obtain bounds for these errors and, therefore, some measure of the accuracy of our final solution. This analysis must include a study of the sources and the propagation of the errors, paying heed to their interactions. (Frequently, by careful design of our algorithms we can restrict their effect quite significantly.)

In this section we consider the three major sources of error, often with reference forward to methods and topics which will, at this stage, be unfamiliar. It is the basic principles which are important here rather than the detail. We begin though with some of the fundamental definitions of error analysis.

Suppose that a number x is approximated by $\bar{x}$, then the *absolute error* in this approximation is defined to be $|x - \bar{x}|$; this corresponds to the idea of two numbers agreeing to a certain number of decimal places.

The *relative error* is defined as $|x - \bar{x}|/|x| = |1 - \bar{x}/x|$. This corresponds to the notion of agreement of a certain number of significant figures. Frequently, because $\bar{x}$ is the quantity that is actually available, the above definition is modified to $|x - \bar{x}|/|\bar{x}| = |1 - x/\bar{x}|$. Note that this definition is unsymmetric and, for this reason, is somewhat unsatisfactory. (This can be overcome by use of the very similar notion of relative precision which is symmetric and therefore yields a metric.)

Example 1.3.1
(*a*) Find the absolute and relative errors in approximating π by $\bar{\pi} = 3.1416$.
(*b*) Find the absolute and relative errors in approximating 100π by 314.16.

Solution
(*a*) Now the true value of π is $3.141\,592\,65$ to eight decimal places, and so

$$|\pi - \bar{\pi}| \simeq 0.000\,007\,35$$

is the absolute error, while

$$\frac{|\pi - \bar{\pi}|}{\pi} \simeq 2.34 \times 10^{-6}$$

is the relative error.

Since the first of these is less than 5×10^{-5} but greater than 5×10^{-6}, we say that the approximation is accurate to four decimal places; similarly, from the second of these we say we have five significant figures of accuracy.
(*b*) From the above, we find that the absolute error is now approximately $0.000\,735$, so the approximation is accurate to two decimal places. For the

relative error, we again obtain 2.34×10^{-6} and still have the same five significant figures of accuracy.

As a general and simple principle, we normally regard absolute error as appropriate for quantities of the order of unity, while relative error is more suited to both larger and smaller magnitudes. Therefore, when considering errors in numerical processes using the floating-point system, it is appropriate to consider relative error throughout.

For measuring the accuracy with which a function f is approximated by another function p, say, on an interval $[a, b]$, we use other measures which are usually based on absolute error but can readily be adapted to relative error versions. Three of the most commonly used metrics here are

$$\|f - p\|_\infty = \max_{a \le x \le b} |f(x) - p(x)| \tag{1.3.1}$$

$$\|f - p\|_1 = \int_a^b |f(x) - p(x)|\, dx \tag{1.3.2}$$

$$\|f - p\|_2 = \sqrt{\int_a^b |f(x) - p(x)|^2\, dx} \tag{1.3.3}$$

These are known, respectively, as the L_∞ (or *supremum*), L_1, and L_2 or (*continuous*) *least squares* metrics.

There are also discrete versions of these metrics which can be used in the situation where a function is only known by its values at certain points. These are used, for example, in assessing the accuracy of a least squares approximation to data.

At this point we turn our attention to some of the principal potential sources of error within a particular computation.

Rounding Error

Rounding, or roundoff, errors arise from the storage of numbers within a finite computer word. In particular, we see that the floating-point binary representation entails rounding the mantissa to a specified number of bits accuracy or, equivalently, within a specified relative error. We can get some idea of the effect of this rounding by considering the following simple example of a quadratic equation.

Example 1.3.2. The equation

$$x^2 - 5000.0002x + 1 = 0$$

has the exact solutions 5000 and 0.0002. If we try to compute these roots by the quadratic formula on a computer which has approximately 11 (decimal) significant figures of accuracy in all calculations, we obtain the results

$$5.000\,000\,000\,0\,E + 03 \quad \text{and} \quad 2.000\,033\,855\,4\,E - 04$$

Even in this simple program we can see the effect of accumulation of rounding error. If the program is modified to compute the smaller of the roots by using the fact that the product of the roots is 1, then we obtain full machine accuracy in this solution too. (Recall that the product of the roots of the quadratic equation $ax^2 + bx + c = 0$ is c/a.)

We see from this example that it is possible to reduce the effect of rounding error very significantly by efficient algorithm design. In the case of quadratic equations this is *always* the method that should be adopted: the larger of the roots (in absolute value) should be computed by the formula and the smaller one from the product of the roots.

The floating-point representation itself entails the introduction of round-off errors right at the outset of any computation. To estimate the absolute error we can obtain a bound as follows. We have seen in Sec. 1.2 that for a machine with an M-bit mantissa in its floating-point word, machine epsilon is given by

$$\mu = 2^{-M} \tag{1.3.4}$$

from which it follows that this is the difference between successive values of the mantissa. Therefore, if

$$\bar{x} = \bar{f} \times 2^E \tag{1.3.5}$$

the next larger representable number is given by

$$\bar{x} + \delta = (\bar{f} + \mu) \times 2^E \tag{1.3.6}$$

It follows that the rounding error in representing x by $\bar{x}$ by rounding to the nearest representable number is bounded by $\delta/2$ and therefore that the absolute error satisfies

$$|x - \bar{x}| \le \frac{\delta}{2} = 2^{E-M-1} = \mu \times 2^{E-1} \tag{1.3.7}$$

Furthermore, since the representation is normalized, it follows that its (exact) mantissa f is greater than $\frac{1}{2}$ and therefore that

$$|x - \bar{x}| \le \mu|x| \tag{1.3.8}$$

This last inequality establishes that the machine unit μ is an upper bound for the floating-point relative representation error.

Truncation Error

Truncation errors are the errors arising from the particular algorithm and approximation even with the assumption that all arithmetic is exact. Typically, it is therefore the error resulting from the truncation of the numerical process. Among the more obvious examples of sources of truncation error are the use of some finite number of terms to estimate the sum of an infinite series and the use of a number of discrete steps in the solution of a differential equation by, say, Euler's method (see Chap. 10).

We must be a little careful in describing such errors since some truncation errors could perhaps be better termed "truncation blunders." One simple example of this possibility would be the attempted summation of the harmonic series

$$1 + \tfrac{1}{2} + \tfrac{1}{3} + \cdots = \sum_{n=1}^{\infty} \frac{1}{n} \qquad (1.3.9)$$

which for any particular accuracy requirement will appear, on the computer, to converge. For example, working to two (base 10) significant figures, no term beyond $n = 20$ will contribute to the sum, while each term for $n > 2$ would be rounded to just one decimal place. We would be led by this very crude computer to the conclusion that the sum of the harmonic series is just 3.9.

Of course, we know that this series diverges.

The problem here is due partly to truncation and partly to rounding error. It is the effect of the rounding error that terms beyond the twentieth make no contribution to the sum.

To be certain that a finite partial sum of a series approximates the infinite sum to a specified accuracy ϵ, say, we must establish that the complete tail does not affect the sum to that precision. That is, the truncation error must be bounded by ϵ. Internally, such a sum must be computed using greater precision than is desired in the final sum in order to overcome the accumulated effect of rounding error. We will discuss the question of summation of series in greater detail in Secs. 1.6 and 1.7.

Modeling Error

In the development of mathematical models of physical (or other) situations, there are almost inevitably simplifying assumptions made. Such simplifications certainly result in errors in the output from such models. Ideally, we should analyze the assumptions and their effects, but the reason for their being made is (typically) because this analysis cannot be successfully completed. One way in which their effect can be analyzed is by perturbing the data slightly and measuring the relative change in the results.

If the mathematical model and the numerical methods are stable, then such small perturbations should induce comparable small changes in the output. This technique does not give any sure guide to the accuracy of the results and the model. However, stability is commonly regarded as an indicator of likely accuracy. A numerical process in which small perturbations in input induce wild fluctuations in the output are termed *ill-conditioned*.

One of the most famous examples of an ill-conditioned numerical problem is Wilkinson's polynomial equation

$$(x - 1)(x - 2) \cdots (x - 20) = 0$$

where the polynomial is given in the form

$$x^{20} + a_{19}x^{19} + \cdots + a_1 x + a_0$$

The coefficient a_{19} is -210. If this coefficient is changed by about 2^{-23}, then the roots are changed dramatically; there are now five complex conjugate pairs and just ten real roots, one of which is approximately 20.85. A change in one coefficient of less than 1 part in 10^9 has certainly produced significant changes in the solution to the problem.

With the possible exception of chaotic dynamical systems, it would be highly unlikely that a physical, economic, or engineering system would have such inherent instability. In cases where such ill-conditioning occurs in practice, the model, the mathematics, the numerical procedure, and the data must all be suspect. Only after careful reexamination of all these and the recomputation of stable numerical results can we have any confidence in those results.

Even in the case of chaotic systems, we must still be very skeptical about numerical results. Chaotic systems can be characterized by their extreme sensitivity to changes in initial conditions. The rounding error of the floating-point representation of those initial conditions imposes such a change, while further implicit changes are imposed by the accumulated roundoff errors in the subsequent calculations. Much analysis of the physical system under consideration is needed to ensure that computed chaotic behavior is a genuine physical phenomenon rather than the effect of rounding and truncation errors in a numerically unstable process.

EXERCISES 1.3

1. What is the absolute error in approximating $\frac{1}{3}$ by 0.33333? If 0.33333 is the correct representation of a quantity x to the five decimal places shown, what is the range of possible values of x?
2. If a real number x is approximated by $\bar{x} = 123.456$, with relative error bounded by 0.1, what are the endpoints of the interval of possible values for x?
3. Estimate the absolute and relative errors in the representations for e given in Example 1.2.1.
4. Estimate the absolute and relative errors in the representations of π found in Exercise 1 of Sec. 1.2.
5. Suppose the function $\cos x$ is approximated by $1 - x^2/2$ on $[0, \pi/2]$. Find the error in this approximation as measured by the L_1, L_2, and L_∞ metrics.
6. Compute the difference $\sqrt{10.1} - \sqrt{10}$, first, directly but rounding each step of the calculation to four significant figures. Note how there are now just two

significant figures in the result. Estimate the absolute and relative errors. Reorganize the calculation using

$$\sqrt{b} - \sqrt{a} = \frac{b - a}{\sqrt{b} + \sqrt{a}}$$

and we obtain four significant figures in the result. (Indeed, this latter calculation delivers the correct answer to this precision.)

7. Use the quadratic formula to solve the equation

$$x^2 - 10.1x + 1 = 0$$

(a) rounding and (b) chopping the results of every intermediate calculation to four (decimal) significant figures. Estimate the absolute and relative precisions in these results.

Compare the results with those obtained by finding the larger root from this formula and the smaller by dividing this into the product of the roots.

1.4 ERRORS IN FLOATING-POINT ARITHMETIC

We begin by comparing and contrasting some fundamental properties of the mathematicians' real number system with its precise axioms and laws of arithmetic and ordering with the "real-world" floating-point arithmetic of the scientific computer in which many of these rules break down—some of them in surprising ways. We have already seen earlier in Sec. 1.3 that computer arithmetic is not exact since rounding errors are committed every time any floating-point arithmetic is performed. Many of the differences highlighted here are direct consequences of the rounding errors in floating-point arithmetic.

To begin with, the set of real numbers, $\mathbb{R}$, forms a commutative group under addition. That is:

The system is closed under addition:	$a, b \in \mathbb{R} \Rightarrow a + b \in \mathbb{R}$
Addition is commutative:	$a + b = b + a$
Addition is associative:	$a + (b + c) = (a + b) + c$
There is a *unique* zero element, 0:	$a + 0 = 0 + a = a$
Each element a has a *unique* negative, $-a$:	$a + (-a) = 0$

Of these five laws, only the commutativity carries over to floating-point arithmetic. Similarly, some of the rules relating to multiplication and division break down within floating-point systems; in particular, there are floating-point numbers for which there is no reciprocal within the system, and this system is not closed under multiplication and division. Even some of the fundamental order relations of the real numbers fail in computer systems.

Most of these apparently fundamental flaws are far less critical than may at first seem likely.

To see that the system is not closed, it is sufficient to attempt the addition of two representable numbers, each of which is greater than $\Omega/2$. Most of the other failings in the floating-point arithmetic system stem from the fact that for any (positive) representable quantity x, the result of the computer floating-point addition denoted $x \oplus y$ will be returned as x whenever $y < \mu x$. We have already seen, for example, that $1 \oplus \mu > 1$, while for any $y < \mu$, the computer returns $1 \oplus y$ as 1. In this case, y acts exactly like 0, and so we see that 0 is not unique. (It should be noted that 0 is the only *universal* zero in the system; that is, $y = 0$ is the only representable quantity such that $x \oplus y = x$ for every representable number x.)

The loss of associativity can be seen by taking $a = 1$ and $b = c = \mu/2$, in which case we obtain

$$(a \oplus b) \oplus c = \left(1 \oplus \frac{\mu}{2}\right) \oplus \frac{\mu}{2} = 1 \oplus \frac{\mu}{2} = 1 \qquad (1.4.1)$$

and

$$a \oplus (b \oplus c) = 1 \oplus \left(\frac{\mu}{2} \oplus \frac{\mu}{2}\right) = 1 \oplus \mu \qquad (1.4.2)$$

The above example also demonstrates the breakdown of one of the fundamental order relations for real numbers, namely, that

$$x > 0 \rightarrow 1 + x > 1$$

The lack of closure under multiplication and division is easy to see since the multiplication of floating-point numbers entails the summation of their exponents. It follows that if x and y have exponents greater than $E_{max}/2$, then their product will overflow. What is somewhat more surprising is that in most floating-point systems there are representable numbers whose reciprocals cannot be represented. We are not referring here to the rounding error in the reciprocation process but to the fact that the computer assignment $1.0/x$ will result in floating-point overflow for some representable numbers x.

The reason for this is that exponents are commonly stored in biased form so that the range of exponent values is not symmetric, typically going from

$$E_{min} = -2^N \qquad \text{up to} \qquad E_{max} = 2^N - 1 \qquad (1.4.3)$$

where, as before, N is the number of bits available for storage of the exponent. For a normalized system, this yields

$$\omega = 2^{-2^{N-1}} \qquad \text{and} \qquad \Omega < 2^{2^{N-1}} \qquad (1.4.4)$$

and so we see that any quantity lying in the interval $[\omega, 4\omega]$ has reciprocal greater than Ω, which cannot be represented.

A floating-point system is said to be *balanced* if $\omega\Omega \approx 1$. This corresponds, for a normalized system, to the condition that $E_{max} = -E_{min} + 1$. A system, such as that described in (1.4.3) above, in which $E_{min} = -E_{max} - 1$ is not balanced. For unnormalized systems this lack of balance is more marked.

We turn now to the specific format of various floating-point representations. We begin with the IEEE *single-precision* format which uses a 32-bit word of which 23 are allocated to the storage of the mantissa and 8 to the biased exponent. It follows that

$$E_{max} = 127 \qquad \text{and} \qquad E_{min} = -128$$

Therefore

$$\Omega \approx 2^{127} \approx 1.7 \times 10^{38}$$

Similarly,

$$\omega = \tfrac{1}{2} \times 2^{-128} \approx 1.5 \times 10^{-39}$$

In this case $\omega\Omega \approx \tfrac{1}{4}$, and so the system is not balanced.

Both the IEEE floating-point representations use the implicit- or hidden-bit binary format. Consequently, the effective mantissa length is increased from 23 to 24 bits. With symmetric rounding and, at least, one *guard bit*, it follows that the machine unit for this system is given by $\mu = 2^{-24}$.

The other IEEE standard—*double precision*—is similar but uses a 64-bit word for the normalized binary floating-point representation in which 11 are allocated to the (biased) exponent and 52 to the (implicit-bit) mantissa. For this system it follows that

$$\Omega \approx 2^{1023} \approx 10^{308} \qquad \omega \approx 2^{-1025} \approx 10^{-308.5}$$

The machine unit $\mu = 2^{-53}$ which yields a relative accuracy in the representation of approximately 10^{-16}. That is, 15 or 16 significant figures will be correct *in the representation*—but, of course, not necessarily in the results of our computation.

Many machines today are equipped with special hardware units for performing floating-point arithmetic very fast. One commonly encountered example is the personal computer equipped with a *mathematics coprocessor*. Turbo Pascal uses such hardware when it is available or can emulate it if it is not.

The Turbo Pascal implementation of the IEEE data types uses *gradual underflow*; that is, unnormalized representations are used with the minimum exponent to allow the representation of quantities much closer to zero. The hidden bit is not used in the software emulation on machines without a mathematics coprocessor, and so the machine unit in this case is double the figure quoted above.

The default Turbo Pascal type *real* is a normalized 48-bit format with the same exponent range as for *single*. This leaves (taking account of the sign bit) 39 bits for the implicit-bit fraction, which therefore has a relative accuracy of 2^{-40} or about 12 significant (decimal) figures. Again in the non-coprocessor implementation, the hidden bit is lost, and the machine unit increases to $\mu = 2^{-39}$.

The remaining Turbo Pascal real data type is *extended* that uses 80 bits of which 64 are used for the mantissa. This yields a very high precision of about 19 significant figures and a much increased range of representable numbers with a decimal exponent range of -4951 up to 4932. Even in the situation where a mathematics coprocessor is available, there are no guard bits or hidden bit available for type extended, so rounding errors can build up more rapidly than for the other data types. It is probably better, therefore, to think of type extended as yielding a relative precision similar to that of type double. In the coprocessor environment, however, arithmetic in type extended is faster than any of the other real data types since the others are first converted into type extended and truncated back to the appropriate lengths before storing the result. The saving to be made from the shorter formats on a coprocessor machine is therefore in memory requirement.

At this stage it is worthwhile to discuss briefly the question of efficiency of numerical methods. There is no absolute and universal measure of efficiency, but the time taken to execute a program on a particular machine is one commonly used criterion. Unfortunately, even this simple measure is highly machine-, or system-, dependent. In Table 1.4.1 we illustrate some of this variability even within the limited environment of Turbo Pascal versions 5.0 or 5.5 running with or without a mathematics coprocessor.

In algorithms for linear algebraic problems, the solution of systems of linear equations, or the algebraic eigenvalue problem (see Chaps. 6 and 11), a commonly used indicator of computational efficiency is an arithmetic operation count. That is, we count the number of additions, subtractions, multiplications,

TABLE 1.4.1
Relative timings of arithmetic operations and function evaluations in Turbo Pascal

Operation	With coprocessor		Without coprocessor	
	TP5	TP5.5	TP5	TP5.5
Sign change	0.69	0.67	0.14	0.13
Addition	1.00	1.00	1.00	1.00
Subtraction	1.14	0.97	1.00	0.97
Multiplication	1.20	1.17	4.38	1.41
Division	1.51	1.39	6.14	5.31
Square root	1.20	1.17	38.46	35.38
Cosine	6.74	6.53	46.03	18.67
Arctangent	3.66	3.56	35.46	14.26
Natural log (ln)	4.54	4.44	56.76	24.59
Exponential function	4.34	4.19	45.08	20.56

and divisions required by a particular *implementation* of the method. Traditionally, the numbers of multiplications and divisions have been regarded as the important quantities on the basis that these two operations are more "expensive" (that is, take more time) than addition and subtraction. On many modern machines this distinction is much less appropriate. Frequently, the time required for multiplication is similar to that for addition, while division is more expensive. We see that in the case where a coprocessor is available, even this distinction is inappropriate unless the number of divisions is excessive. The only reasonable operation count for such linear algebra problems is thus an overall count of *all* arithmetic operations.

In all cases a long loop repeating the same operation is timed and compared with the time for a similar loop of additions after allowance is made for the loop control itself by timing an empty loop of the same length. The base times for the addition loop were almost identical in the four cases. (It should be noted that the relative timings in Table 1.4.1 will vary slightly with the details of the operations.)

Many numerical methods require the repeated evaluation of (perhaps complicated) functions—numerical integration, optimization, and differential equations are just three important examples. It is clear from Table 1.4.1 that almost any such function evaluation will be equivalent to many arithmetic operations and is therefore likely to dominate the overhead involved in the management of the algorithm. Clearly, such comparisons are highly dependent on the system in use. We see, for example, that the square root operation is comparable with a multiplication when a coprocessor is available but is many times more expensive otherwise.

Questions of the efficiency of the implementation of an algorithm will be addressed in context throughout the book. From the enormous variation in the relative timings, it is also plain that the development of efficient algorithms for the evaluation of the elementary functions is of great potential benefit. We discuss this question in some detail in Chap. 3.

We return now to the main purpose of this section, namely, the errors inherent in the use of floating-point arithmetic. It is necessary first to give a little more consideration to the distribution of floating-point numbers and the consequent distribution of the representation errors. These are, of course, the base errors from which all the others are propagated. It was commented in Sec. 1.2 that the real numbers are "logarithmically distributed," and some justification for this claim was provided by Program 1.2.4. There has been much work devoted to the explanation of this apparently outlandish claim based on many different approaches.

The original observation of the phenomenon was made by Benford in 1937, when he found that the early pages of a book of logarithm tables were much more heavily used than the later ones which related to the logarithms of numbers with the higher leading significant digits. (Benford remarked that this might have been expected if he had been looking at a cheap novel which might be discarded after a brief initial reading, but few of his students spent their free time reading through books of tables of logarithms.) Benford conjectured that scientists and engineers had significantly greater need for the logarithms of numbers whose leading significant digit (lsd) was 1, 2, or 3 than they did for those with lsd 8 or 9. This prompted him to a survey of real-life numbers which he took from every source he could, including physical constants, areas of rivers, and numbers appearing in newspapers and magazines. In all he collected some 20,000 such numbers and found remarkably close agreement with the logarithmic distribution which, for decimally represented numbers, states that the frequency $F_{10}(n)$ of lsd n is given, according to (1.2.14), by

$$F_{10}(n) = \log_{10}(n + 1) - \log_{10} n = \log_{10}\left(1 + \frac{1}{n}\right) \qquad (1.4.5)$$

While this claim is widely accepted as a fact of computational life, it is still by no means fully explained.

Among the approaches used, one of the most popular is the "invariance to scaling" argument which takes as an axiom that the distribution of the lsd of, say, lengths of rivers should be unchanged by a change of measurement units from, say, miles to kilometers. In essence this reduces to the observation that there are just as many numbers in the interval $[1, 2)$ as there are in $[2, 4)$ since any element of the second can be obtained by doubling an element of the first. The only continuous distribution consistent with this claim is the logarithmic one. Unfortunately, this assumption of invariance to scaling turns out to be equivalent to the original observation.

Perhaps the most convincing argument is that based on the continuous version of the experiment carried out using Algorithm 1.2.4. This was pursued successfully and independently by Barlow and Bareiss (1985) and Turner (1982). In its simplest form, this results in establishing the convergence of a sequence of functions which represent the density functions of the distribution for products of an increasing number of factors. It turns out that irrespective of the initial distribution from which the factors are drawn (save only for the assumption of its integrability), the sequence does converge uniformly and its limit does indeed yield the logarithmic distribution. (The details of the argument are not central to the present discussion, but the interested reader is referred to the papers of Barlow and Bareiss, and Turner.)

To see one of the immediate consequences of this for computer arithmetic, we simply consider the relative representation errors achievable for a fixed 32-bit wordlength and for different bases with a similar overall range of representable numbers.

Example 1.4.2. Consider the representation errors for the binary and hexadecimal systems in a 32-bit format.

To make the ranges comparable, we will assume that the binary system allocates 9 bits to the (signed or biased) exponent with 1 for the sign and the remaining 22 for the mantissa. Taking account of the implicit leading bit, this yields a relative representation accuracy of 2^{-23}. Now the maximum exponent E_{max} is $2^8 - 1$, or equivalently, $\Omega \approx 2^{255}$.

Now $2^{256} = (2^4)^{64}$ and $2^4 = 16$, so the hexadecimal representation requires a maximum exponent of about 64, or 2^6; that is, a total of 7 $(= 6 + 1)$ bits is needed for the hexadecimal exponent. Again, taking account of the sign, this leaves 24 bits for the hexadecimal mantissa which therefore has six hexadecimal digits. Since the leading hexadecimal digit can be any one of $1, 2, \ldots, 15$, there is no possibility of using an implicit leading bit. Thus six hexadecimal digits is the genuine accuracy available in this representation.

How do these representations compare? We must consider four separate cases, namely, those where the binary exponent is of the form $4N$, $4N - 1$, $4N - 2$, or $4N - 3$. In each case the hexadecimal exponent is N. The difference between successive representable values in the hexadecimal system is therefore 16^{N-6}, or 2^{4N-24}.

According to the logarithmic distribution, the frequency of the (hexadecimal) lsd being 1 is $\log_{16} 2 = \frac{1}{4}$, which is the same as the combined frequency for lsd 2 or 3, which in turn is the same as those for lsd 4, 5, 6, or 7 and for $8, 9, \ldots, 15$ since, for example,

$$F_{16}(4) + F_{16}(5) + F_{16}(6) + F_{16}(7) = \log_{16}\tfrac{5}{4} + \log_{16}\tfrac{6}{5} + \log_{16}\tfrac{7}{6} + \log_{16}\tfrac{8}{7}$$

$$= \log_{16}\tfrac{8}{4} = \tfrac{1}{4}$$

These four cases correspond to the binary exponent being $4N - 3$, $4N - 2$, $4N - 1$, and $4N$, respectively. For binary exponent k, the difference between representable numbers is 2^{k-23}. For these four cases, these differences are therefore 2^{4N-26}, 2^{4N-25}, 2^{4N-24}, and 2^{4N-23}. Thus, we see that 50 percent of the time the representation error will be smaller for the binary system than for the hexadecimal one. It will only be greater in 25 percent of the cases.

There are many studies of the distribution of floating-point arithmetic errors which also establish the superiority of the binary system over any other base. The advantage is always increased with the use of the hidden-bit binary system. We do not pursue these statistical analyses here but turn now to the propagation of error through floating-point arithmetic operations.

At this stage we treat this propagation algebraically. Let us suppose that true positive values x and y are approximated by $\bar{x}$ and $\bar{y}$, respectively, and consider the absolute error in the result of the various arithmetic operations. It is convenient to denote the absolute errors in the representations by δ_x and δ_y so that

$$x = \bar{x} \pm \delta_x \qquad \text{and} \qquad y = \bar{y} \pm \delta_y$$

It is easy to see that

$$|\bar{x} \pm \bar{y} - (x \pm y)| = |(\bar{x} - x) \pm (\bar{y} - y)|$$

$$\leq \delta_x + \delta_y \tag{1.4.6}$$

In a similar way we obtain for multiplication,

$$|\bar{x}\bar{y} - xy| \leq y\delta_x + x\delta_y + \delta_x\delta_y \tag{1.4.7}$$

and, for division,

$$\left| \frac{\bar{x}}{\bar{y}} - \frac{x}{y} \right| \leq \frac{y\delta_x + x\delta_y}{y\bar{y}} \tag{1.4.8}$$

We have commented already that the appropriate measure of precision for the floating-point arithmetic system is relative error. Neglecting any *second-order effects*, that is, neglecting terms involving products of more than one error such as $\delta_x\delta_y$ in (1.4.7), the last two of these equations reduce to the simple approximations

$$\rho_{xy} \simeq \rho_{x/y} \simeq \rho_x + \rho_y \tag{1.4.9}$$

where for each of the quantities in this approximate equation ρ_a denotes the relative error in the representation of a. In the next section we will introduce a slightly different notation for the absolute and relative errors and a more useful general technique for the estimation of propagated errors.

EXERCISES 1.4

1. Verify for your computer that

$$\left(1 \oplus \frac{\mu}{2}\right) \oplus \frac{\mu}{2} \neq 1 \oplus \left(\frac{\mu}{2} \oplus \frac{\mu}{2}\right)$$

where $\oplus$ represents the floating-point addition operation.

2. Find a machine-representable floating-point number whose reciprocal overflows.

3. For our hypothetical computer with M bits for the mantissa and N for the exponent, find an example of a floating-point number which does not have a unique negative.

4. Derive the error estimates (1.4.7) and (1.4.8). Deduce the relative error estimates (1.4.9).

1.5 PROPAGATION OF ERROR

In the analysis of numerical procedures it is not the individual representation errors which are vital but the overall effect of the rounding and truncation errors on the final result. For this reason it is important to spend some time on the arithmetic of error propagation. We will examine the effect of this in the next section where, by way of example, we consider the summation of a few simple series.

Of course, we cannot expect to be able to find exact expressions for the error in a particular computation—if we could, then we would eliminate the error altogether. The best we can do is to obtain *error bounds*, and, in general, we want these to be as sharp as possible. If we can obtain good bounds for a particular process, then we can decide on how much computational effort is necessary to achieve some specified accuracy. We begin with the error resulting from a single computational step.

Suppose we wish to evaluate—or, perhaps more precisely, approximate— $f(x)$ where f is a differentiable function and we are given the approximation $\bar{x}$ to x. Now, by the Mean Value Theorem (MVT, hereafter) we know that

$$f(x) - f(\bar{x}) = (x - \bar{x})f'(\xi) \tag{1.5.1}$$

for some ξ between x and $\bar{x}$. It follows that

$$|f(x) - f(\bar{x})| \le |x - \bar{x}| \max_{t \in I} |f'(t)| \tag{1.5.2}$$

where I is the interval with endpoints x and $\bar{x}$. Using $\delta(\cdot)$ to denote the absolute error in computed values, we may rewrite (1.5.2) as

$$\delta(f(\bar{x})) \le \delta(\bar{x}) \max_{t \in I} |f'(t)| \tag{1.5.3}$$

Usually, we anticipate that $\delta(\bar{x})$ will be small, and so we can make the first-order approximation

$$\delta(f(\bar{x})) \simeq \delta(\bar{x})|f'(\bar{x})| \tag{1.5.4}$$

which, though not a rigorous bound for the error, does give an often useful estimate. Note, however, that this estimate has been obtained without any account of error in the approximation and/or evaluation of the function f itself. We will consider such matters a little later. To obtain an estimate of the relative error, we could simply divide (1.5.4) through by $f(\bar{x})$.

Example 1.5.1. Use (1.5.4) to estimate the absolute and relative errors for the function $f(x) = x^{1/3}$ for $\bar{x} = 64.00$.

Solution. Here we assume that $\bar{x}$ is correct to the quoted precision of the representation, from which it follows that the absolute error in $\bar{x}$ is bounded by 5×10^{-3}. Also $f'(x) = x^{-2/3}/3$, and so (1.5.4) gives the approximation

$$\delta(\bar{x}^{1/3}) \simeq \frac{\delta(\bar{x})}{3\bar{x}^{2/3}} = \frac{5 \times 10^{-3}}{48} \simeq 10^{-4} \tag{1.5.5}$$

To test the validity of this estimate, suppose the true value of x had been at the extreme end of its possible range, $x = 63.995$, in which case $f(x) = 3.9999$ to four decimal places. The computed value $f(64.00)$ is 4.0, and so the estimated error 10^{-4} given by (1.5.5) is correct.

To estimate the relative error $\rho(f(\bar{x}))$, we recall from the definition that

$$\rho(f(\bar{x})) = \frac{\delta(f(\bar{x}))}{f(\bar{x})} \tag{1.5.6}$$

which therefore yields

$$\rho(f(\bar{x})) \simeq \frac{10^{-4}}{4} = 2.5 \times 10^{-5}$$

The fact that the error estimate (1.5.4) is dependent on the derivative $f'(\bar{x})$ may lead to concern that for cases, such as in Example 1.5.1 above, where this derivative becomes unbounded as x approaches 0, the estimate will be inappropriate.

However, if we assume the same *relative* accuracy in the representation $\bar{x} = 0.000\,001\,000$, then the estimate (1.5.4) yields

$$\delta(f(\bar{x})) \simeq \frac{5 \times 10^{-10}}{3 \times 10^{-4}} \simeq 1.67 \times 10^{-6}$$

Comparing this with the true error for the extreme case, $x = 0.000\,000\,999\,5$, we have $f(x) = 0.009\,998\,33$, whereas $f(\bar{x}) = 0.01$. The true error is thus

$$0.01 - 0.009\,998\,33 = 1.67 \times 10^{-6}$$

as predicted above. The relative error here is approximately 1.67×10^{-4}, so we observe that there is indeed some deterioration in the relative accuracy of the computation as $\bar{x}$ approaches the singular point 0.

The approach used to derive the error estimate (1.5.4) can be readily extended to functions of more than one variable. For a function f of two variables x, y we obtain

$$\delta(f(\bar{x}, \bar{y})) \simeq |f_x(\bar{x}, \bar{y})| \delta(\bar{x}) + |f_y(\bar{x}, \bar{y})| \delta(\bar{y}) \tag{1.5.7}$$

where, conventionally, f_x, f_y denote the partial derivatives $\partial f / \partial x$, $\partial f / \partial y$. We use the estimate (1.5.7) to derive expressions for the rounding errors of floating-point arithmetic operations.

For the cases of addition and subtraction, we get

$$\delta(\bar{x} \pm \bar{y}) \simeq \delta(\bar{x}) + \delta(\bar{y}) \tag{1.5.8}$$

since, for example, if $f(x, y) = x - y$, we have $f_x = 1$ and $f_y = -1$.

Similarly, for multiplication and division, we get

$$\delta(\overline{xy}) \simeq |\bar{y}| \delta(\bar{x}) + |\bar{x}| \delta(\bar{y}) \tag{1.5.9}$$

and

$$\delta\left(\frac{\bar{x}}{\bar{y}}\right) \simeq \left|\frac{1}{\bar{y}}\right| \delta(\bar{x}) + \left|\frac{\bar{x}}{\bar{y}^2}\right| \delta(\bar{y})$$

$$= \frac{\delta(\bar{x}) + |\bar{x}/\bar{y}| \delta(\bar{y})}{|\bar{y}|} \tag{1.5.10}$$

[Compare (1.4.6) to (1.4.8).] This last result demonstrates the reason why division by a smaller number can lead to severe rounding error as there is a factor $|1/\bar{y}|$ in this estimate.

Example 1.5.2. Estimate the error in evaluating $f(x, y, z) = x^2 + y/z$ for

$$x \simeq \bar{x} = 1.23 \qquad y \simeq \bar{y} = 2.34 \qquad z \simeq \bar{z} = 3.45$$

where each of these is assumed to be correctly rounded to the number of figures shown.

Solution. We may assume, therefore, that

$$\delta(\bar{x}), \delta(\bar{y}), \delta(\bar{z}) \le 0.005$$

From (1.5.9) and (1.5.10), we deduce

$$\delta(\bar{x}^2) \simeq 2\bar{x}\delta(\bar{x}) \le 2.46 \times 0.005 = 0.0123$$

$$\delta\left(\frac{\bar{y}}{\bar{z}}\right) \simeq \frac{\delta(\bar{y}) + |\bar{y}/\bar{z}|\delta(\bar{z})}{|\bar{z}|} \le 0.00243$$

and hence, using (1.5.8), we obtain the estimate

$$\delta\left(\bar{x}^2 + \frac{\bar{y}}{\bar{z}}\right) \simeq 0.0147$$

(You should verify this estimate from first principles and check its accuracy for specific values of x, y, z.)

The corresponding relations for the relative errors ρ in the arithmetic operations can easily be derived from those above. [Again compare with (1.4.9).] We obtain

$$\rho(\overline{xy}) \simeq \rho(\bar{x}) + \rho(\bar{y})$$

$$\rho\left(\frac{\bar{x}}{\bar{y}}\right) \simeq \rho(\bar{x}) + \rho(\bar{y}) \qquad (1.5.11)$$

$$\rho(\bar{x} \pm \bar{y}) \simeq \frac{\delta(\bar{x}) + \delta(\bar{y})}{|\bar{x} \pm \bar{y}|}$$

which, in the special case of addition with $\bar{x}$ and $\bar{y}$ having the same sign, yields

$$\rho(\bar{x} + \bar{y}) \simeq \max(\rho(\bar{x}), \rho(\bar{y})) \qquad (1.5.12)$$

These relations can be readily established by the reader using the corresponding results for absolute errors and the definition of relative error.

We now turn to the question of how errors are propagated through a computation. The major drawback of the error bounds above lies in the fact that we assumed that the function f—and, in particular, the arithmetic operations—can be computed exactly, with the only source of error being the representation error of the arguments of the functions. We have already observed that there are further errors introduced by the arithmetic itself. There is then likely to be another roundoff error resulting from the need to store the result in our standard floating-point representation.

Thus, for example, we should really be estimating the error in the rounded representation of $\bar{f}(\bar{x})$, which we will denote by $\overline{\bar{f}(\bar{x})}$. Similar new errors are introduced at every stage of a computation, and so a full analysis of a major program, taking account of every rounding error, would be exceedingly tedious and time consuming. More importantly, its result would almost certainly be a gross overestimate of the true error committed.

Example 1.5.3. Estimate the final relative error in computing $f(x, y, z) = x^2 + y/z$ from floating-point data $\bar{x}$, $\bar{y}$, $\bar{z}$.

Solution. First the two quantities $\bar{x}^2$ and $\bar{y}/\bar{z}$ are computed, and each is stored after rounding to yield $\overline{\bar{x}^2}$ and $\overline{\bar{y}/\bar{z}}$. In each case the rounding introduces a relative error μ which must be combined with the propagated error. This yields

$$\rho(\overline{\bar{x}^2}) \simeq 2\rho(\bar{x}) + \mu$$

$$\rho\left(\overline{\frac{\bar{y}}{\bar{z}}}\right) \simeq \rho(\bar{y}) + \rho(\bar{z}) + \mu$$

The final addition entails a further rounding error. On the assumption that the quantities $\overline{\bar{x}^2}$ and $\overline{\bar{y}/\bar{z}}$ are both positive, we may deduce

$$\rho(\overline{\overline{\bar{x}^2} + \overline{\bar{y}/\bar{z}}}) \simeq \max\left(\rho(\overline{\bar{x}^2}), \rho\left(\overline{\frac{\bar{y}}{\bar{z}}}\right)\right) + \mu$$

$$\simeq \max\left(2\rho(\bar{x}) + \mu, \rho\left(\frac{\bar{y}}{\bar{z}}\right) + \mu\right) + \mu \qquad (1.5.13)$$

Note that for the case where $\bar{x}$, $\bar{y}$, $\bar{z}$ are the floating-point representations of *exact* data, the relative errors $\rho(\bar{x})$, $\rho(\bar{y})$, $\rho(\bar{z})$ are bounded by μ, and, therefore, the final error estimate above is just 4μ.

For the same data of Example 1.5.2, namely, $\bar{x} = 1.23$, $\bar{y} = 2.34$, $\bar{z} = 3.45$, the relative errors were bounded by 5×10^{-3}, and so the estimate (1.5.13) gives a relative error of 2×10^{-2}. (The actual relative error observed here was approximately 6.7×10^{-3}, so even for this small computation the estimate obtained is very pessimistic.)

We saw, in the case of solving a quadratic equation, that reorganizing the computational algorithm can have a very beneficial (or detrimental!) effect on the accuracy of the results. The benefit derives from eliminating severe cancellation or reducing the effect of accumulated roundoff errors. It is similarly the case that the ordering of the arithmetic steps can have a significant effect on this accumulated error, because the addition of a small number to a large one may have no effect—as we observed for the case of $1 + \epsilon$ where $\epsilon < \mu$.

It should be noted here, too, that the repeated addition of a small quantity to itself can lead to severe errors. This particular operation is very common in numerical computing wherever a continuous range is *discretized* into a finite number of (equal) small steps; examples occur in numerical integration, interpolation, and the solution of differential equations.

Example 1.5.4. Consider the result of running the following simple Turbo Pascal program:

```
var
      step, left, right:real;
      nsteps:integer;
begin
      write('Input number of steps'); readln(nsteps);
      step:=1.0/nsteps;
      left:=0; right:=1;
      repeat
            left:=left + step;
            writeln(left);
      until left > =right;
end.
```

With "nsteps" $= 1000$, for example, the program stops with "left" $=$ 1.000 999 999 6. This is caused by the rounding error in the initial floating-point representation of 0.001. The effect can be more marked with other numbers of steps. If such a loop were used to generate the points for, say, the trapezium rule to estimate the integral of a function (see Chap. 9), then the actual right-hand endpoint used would result in one extra step being taken and even the breakdown of the program if the function ceased to be defined outside the range of integration. (With the apparently sufficient test "until left $=$ right" instead, the program would run indefinitely for almost all input values of nsteps.)

This difficulty can be easily overcome by using an (exact) integer-controlled loop:

$$\text{for i:=1 to nsteps do left:=left + step;}$$

The final value of "left" for "nsteps" $= 1000$ is now 0.999 999 999 6, so the error has been reduced to just 4×10^{-10}.

The summation of a series of decreasing terms is another operation which can be greatly affected by how the calculation is organized. We will consider this and other aspects of this problem in the next section. The effect of the order of summation is illustrated by the summation of the first n terms of a geometric series with *common ratio* x,

$$1 + x + x^2 + \cdots + x^{n-1}$$

first from left to right and then from right to left.

Programming the forward (left-to-right) summation is straightforward. For the backward summation of exactly the same terms, we must take care as to the starting point in case x^{n-1} is zero to machine accuracy. This is the reason for the structure of the first loop in Program 1.5.5 below which determines "kmax," the actual number of terms to be summed. Table 1.5.6 lists the results of running these two programs for 15,000 terms with different common ratios close to unity. The differences in the results are the effect of different rounding errors and the loss of some terms for which term $< \mu * \text{sum}$ in the forward sums.

Program 1.5.5

```
            var
                    divisor,ratio,sum,term,temp:real;
                    nterms,k,kmax:integer;
            begin
                    write('input the common divisor '); readln(divisor);
                    write('input number of terms to be summed ');
                    readln(nterms);
                    ratio:=1.0/divisor;
                    temp:=1.0;
                    kmax:=0;
                    while (abs(temp)>0) and (kmax<nterms) do begin
                        term:=temp; kmax:=kmax+1; temp:=term*ratio;
                    end;
                    sum:=temp;
                    for k:=1 to kmax do begin
                        sum:=sum+term; term:=term*divisor;
                    end;
                    writeln('sum of first ', kmax, 'terms is ', sum);
            end.
```

The common divisor in the program is the reciprocal of the common ratio x above. In the backward summation, multiplication by "divisor" therefore corresponds to *reducing* the power of x.

TABLE 1.5.6
Summation of the first 15,000 terms of geometric series with common divisor d both forward and backward

d	Forward sum	Backward sum	True sum
1.001	1 000.999 691 5	1 000.999 688 5	1 000.999 691 4
1.000 1	7 769.307 869 5	7 769.310 850 7	7 769.310 863 7
1.000 01	13 929.276 915	13 930.137 841	13 930.137 835
1.000 001	14 888.067 929	14 889.052 951	14 889.052 961

EXERCISES 1.5

1. Use the approximation (1.5.4) to estimate the absolute error in computing $\sqrt{4.00}$, assuming that this is the correctly rounded representation to the number of significant figures shown. Verify that this is indeed a good estimate of the error.

2. For Exercise 1, the derivative becomes unbounded as $x \to 0$. Show that, for the same number of significant figures of accuracy in the original representation, (1.5.4) still gives an accurate estimate of the absolute error in $\sqrt{0.0025}$.

3. Estimate the error in calculating the period of a simple pendulum of length $l = 2$ meters using the formula $\tau = 2\pi\sqrt{l/g}$. Assume that the value of π is accurate to full machine precision, the acceleration due to gravity is taken to be 9.80 meters per second squared, and the measurement of the length of the pendulum is accurate to the nearest millimeter. What is the corresponding estimate of the relative error?

4. Use the mean value theorem for a function of two

variables to derive the absolute error estimate (1.5.7).

5. Verify the absolute error formulas (1.5.9) and (1.5.10) for multiplication and division. Test the accuracy of these estimates with examples. Be sure to include examples of division by a small number.

6. Derive the relative error estimates (1.5.11) to (1.5.13) and the special case (1.5.14).

7. Verify the relative error estimate obtained in Exercise 3 using the error arithmetic of relative error.

8. Find absolute and relative error bounds for the following functions given that $\bar{x} = 1.234$ and $\bar{y} = 234.5$ are accurate to the number of significant figures shown.
   (a) $(x + y)^2$
   (b) $x\sqrt{y}$
   (c) $\sin(x/y)$
   (d) $\sqrt{x^2 + y^2}$
   (e) $y - x/y$

1.6   SUMMATION OF SERIES

The primary purpose of this section is to introduce some of the ideas of analysis of numerical methods by using simple and familiar examples. In particular, we will be concentrating on the efficient estimation of sums of infinite series, taking due account of truncation error and the accumulation of roundoff. Our study of these topics will be centered on the exponential series, the harmonic series, and some series representations of π. The last of these will involve the summation of alternating series which can be enhanced significantly by the use of Euler's method.

It should be observed that the series we consider are not to be thought of as necessarily the most efficient ways of computing the quantities in question but are simply used to provide examples of the sort of considerations which must enter the planning of numerical procedures.

We begin with the simplest of our examples, the evaluation of the exponential function e^x from the defining power series

$$\exp x = e^x = \sum_{i=0}^{\infty} \frac{x^i}{i!} \tag{1.6.1}$$

Consider first the truncation error $T(x, N)$, say, caused by using the first N terms of the series to estimate its sum. This error is simply the sum of the tail, that is, for positive x

$$T(x, N) = \frac{x^{N+1}}{(N+1)!} + \frac{x^{N+2}}{(N+2)!} + \cdots$$

$$\leq \frac{x^{N+1}\{1 + x/(N+2) + [x/(N+2)]^2 + \cdots\}}{(N+1)!} \tag{1.6.2}$$

and, summing the geometric series, this yields (provided that $|x| < N + 2$)

$$T(x, N) \leq \frac{x^{N+1}}{(N+1)![1 - x/(N+2)]} \tag{1.6.3}$$

For negative x, the truncation error is bounded by the first term omitted provided that $|x| < N + 2$ so that the terms of the tail are decreasing in magnitude.

Example 1.6.1. Find the smallest number of terms for which the truncation error in evaluating e^x by the series is less than 10^{-4} for $x = \pm 1$, 10, and 25.

Solution

(*a*) For $x = 1$, we get, using (1.6.3),

$$T(1, N) \le \frac{1}{(N+1)![1 - 1/(N+2)]} = \frac{N+2}{(N+1)(N+1)!}$$

which for $N = 6$ is 2.268×10^{-4}, while for $N = 7$ the above bound is 2.79×10^{-5}. It follows that $N = 7$ will suffice for the evaluation of e with error bounded by 10^{-4}.

(*b*) For $x = -1$, it is enough to stop at the first term smaller than 10^{-4}, which again implies that $N = 7$ will suffice.

(*c*) With $x = 10$, we clearly require significantly more terms; (1.6.3) yields

$$T(10, N) \le \frac{10^{N+1}}{(N+1)![1 - 10/(N+2)]}$$
$$= \frac{10^{N+1}(N+2)}{(N-8)(N+1)!}$$

which is first less than 10^{-4} for $N = 33$ where the above bound is 4.74×10^{-5}.

It is already clear that the standard series will not be a suitable method for the evaluation of the exponential function for large values of x or for high precisions.

(*d*) We cannot compute the bound $T(25, N)$ using the algebraic form described above without causing overflow even on a hand-held calculator with a maximum decimal exponent of 99. We must resort to the formula (1.6.3) computed recursively as in the following algorithm, from which we find that $N = 74$ terms are needed to compute e^{25} with truncation error less than 10^{-4}.

Algorithm 1.6.2 Exponential series truncation error

Input	x, tol
Initialize	n = 1, term = 1
	error > tol
Loop	term:=term $*$ x/n
	n:=n + 1
	IF |term| < tol THEN error:=term/(1 − x/n)
Until	|error| < tol
Output	n − 1 is the number of terms needed.

To obtain a bound for the overall error, it would be necessary to take account of the accumulation of roundoff error as well as the truncation error committed by the method adopted. Although the designer of elementary function routines for the computer or calculator must analyze these errors fully, typically, in practice, this aspect is accounted for by insisting that the computation is performed to greater precision than the final required accuracy.

In the above example, we saw that the number of terms required to evaluate the exponential function from its series can grow rapidly with the argument (and the number of significant figures of accuracy sought). Any efficient routine can use the series for only a very limited range of values. For other values of the argument, it would be necessary to use some range-reduction scheme to bring it into the range where the series is efficient. (It is not necessarily the case that the series is used at all.)

One simple technique for this range reduction is provided in the case of the exponential function by using the identity

$$e^x = (e^{x/2})^2 \qquad (1.6.4)$$

For example, we could compute e^{10} by first evaluating e^5 and then e^{10} as $(e^5)^2$. This ploy could be adopted as often as we wish—halving the argument several times and then squaring the result the same number of times to obtain the required value. The number of terms needed for the basic evaluation of the exponential function will be significantly reduced and so will the overall work needed to obtain the particular value we seek. To illustrate the extent of the reduction, Table 1.6.3 lists the number of terms of the series which are needed to keep the truncation error below 10^{-4} for various numbers of "halvings" of the arguments 10 and 25.

This gain in efficiency is not all free, however. If we look at the propagation of error through the operation of squaring a number, we see that the relative error (approximately) doubles with each squaring; for, if $\bar{x} = x + \delta(\bar{x})$ and therefore $\rho(\bar{x}) = \delta(\bar{x})/x$, then

$$\bar{x}^2 = x^2 + 2x\delta(\bar{x}) + \delta(\bar{x})^2$$

so that $\delta(\bar{x}^2) \simeq 2x\delta(\bar{x})$ and hence

$$\rho(\bar{x}^2) \simeq \frac{2x\delta(\bar{x})}{x^2} = 2\rho(\bar{x}) \qquad (1.6.5)$$

It is therefore necessary to evaluate, say, $e^{1.25}$ to very much more accuracy than the 10^{-4} quoted above. To see how this affects the computation, we consider this case in more detail.

TABLE 1.6.3
Number of terms of the exponential series required to obtain truncation error bounded by 10^{-4}

x	No. of terms	x	No. of terms
10	33	25	74
5	19	12.5	40
2.5	12	6.25	23
1.25	8	3.125	14
0.625	5	1.5625	9
		0.78125	7

Example 1.6.4. The relative error bound sought in e^{10} is about 4×10^{-9}. How many terms in the series for $e^{1.25}$ are necessary to deliver this precision?

Solution. We will again assume that internal calculations are performed to extra precision so that internal rounding errors may be neglected. If we are to use $e^{1.25}$ as the basic evaluation of the exponential function itself, then the argument is to be halved 3 times, and therefore there are to be three squarings to complete the calculation. Each of these results in doubling the relative error, and so the overall relative error will be multiplied eightfold once the evaluation of $e^{1.25}$ is achieved.

We thus require a relative error bounded by 5×10^{-10} in the summation of the series. Now $e^{1.25}$ is approximately 3.5, and so an absolute truncation error bounded by 10^{-9} will be sufficient. This accuracy can be achieved using 13 terms and yields the approximate value

$$\exp 1.25 \simeq 3.490\,342\,957\,4$$

Squaring this quantity 3 times results in the values

$$12.182\,493\,960 \qquad 148.413\,159\,10 \qquad 22\,026.465\,793$$

the last of which indeed satisfies the required tolerance of 10^{-4}.

This example gives a brief introduction to the considerations which must be taken by the designer of efficient numerical software. The next example demonstrates the need for appropriate software so that the user is not led into serious mistakes.

We turn to the example of the "summation" of the harmonic series. We know as mathematicians that this series diverges, but the computer has not had the benefit of that mathematical education and will be perfectly willing to estimate the sum of this infinite series by taking partial sums of the first N terms—just as it did successfully for the exponential series before. To get an idea of the problem, suppose that we try to sum the harmonic series on a machine with just three (decimal) significant figures. The sum is immediately greater than unity, and so any term which is zero to two decimal places will not affect the sum at all. Thus the sum will certainly have stopped increasing by the time it reaches the term $\frac{1}{201}$. At this point the sum obtained by this machine is 6.16—not a very good estimate of infinity!

Of course, the procedure described above does not give a realistic impression of the genuine computational problem. Let us suppose that we allow the summation to continue through all representable integers, as in Program 1.6.5 below. What will be the outcome of this attempt to sum the series?

Program 1.6.5

```
var i: integer; sum: real;
begin
    sum:=0.0; i:=1;
    while 1.0/i〈〉0 do begin
        sum:=sum + 1.0/i;
```

```
                              i:=i+1;
                              writeln(sum:13:10);
                    end;
          end.
```

Since the termination condition is that the term $1/i$ should be *exactly* zero, it is tempting to believe at first that this program would run indefinitely. It will not.

The explanation is simple; since the loop is controlled by the *integer i*, this will eventually reach its maximum value I_{max}. In a Turbo Pascal environment the next term will be the reciprocal of the largest negative integer I_{min}, and so the sum will start to *decrease*, steadily subtracting off all the terms that were previously added in. Eventually the integer i will itself become zero, and the program will terminate with the floating-point error message "division by zero." At that point, the sum would be just $1/I_{min}$ since all the positive terms would have been canceled by their negatives.

We can improve on this attempt by allowing the loop control variable to be of type real. With just that one change to the program, it would never stop since the largest representable quantity Ω has reciprocal greater than ω. Consequently, the sum would be continually incremented by $1/\Omega$.

It is easy to believe that this would lead to rapid growth of the sum and eventual overflow. However, long before i reaches Ω, the terms are too small to affect the sum at all. This sum will remain constant once $1/i < \mu \times$ sum. We can improve, therefore, on our program to estimate the "sum of the harmonic series" by using a better stopping condition which halts the program as soon as the sum can increase no further.

In the previous discussion we commented that the terms would be too small to affect the sum before Ω is reached. In fact, Ω will never be reached by the repeated addition of unity, since if $x > 1/\mu$, then $x \oplus 1 = x$, and so the fixed term would in fact be $1/(1/\mu) = \mu$, which will not affect the sum that is clearly greater than 1.

We may therefore check how many terms are to be summed by testing the difference between 2^k and $2^k - 1$ for successive values of k. For any floating-point system these two quantities will eventually be indistinguishable.

In the case of the IEEE standard single-precision, Turbo Pascal's type *single*, we find that 2^{24} and $2^{24} - 1$ are distinct but that 2^{25} and $2^{25} - 1$ are not. The best we can do with type single is therefore to sum the first 2^{24} terms. The sum obtained from these is 15.40368.

We observed in Sec. 1.5 that the order of summation of the terms of a series can affect the result. This effect is particularly marked for this example. If the same sum is produced by summing the terms in the reverse order, then the same number of terms as above now yields the sum 17.23271. A change of some 12 percent has resulted just from this reversal of the order.

For our final example of this section, we turn to one of the classical problems of mathematics which has been the inspiration for many people and for much mathematics in a variety of widely differing fields. The computation of π was a major challenge for Archimedes and other philosophers of the ancient civilizations. The approaches used have been geometric, number theoretic, analytic, and computational. We are concerned here with just one very simplistic approach as follows. We have

$$\frac{\pi}{4} = \arctan 1 \tag{1.6.6}$$

$$\arctan x = \int_0^x \frac{1}{1+t^2} \, dt \tag{1.6.7}$$

and, for $|t| < 1$,

$$\frac{1}{1+t^2} = 1 - t^2 + t^4 - t^6 + \cdots \tag{1.6.8}$$

Now we may integrate the power series term by term within its radius of convergence, and so combining (1.6.7) and (1.6.8) we get

$$\arctan x = x - \frac{x^3}{3} + \frac{x^5}{5} - \frac{x^7}{7} + \cdots \tag{1.6.9}$$

for $|x| < 1$. The series (1.6.9) is also convergent for $x = 1$ (as can be easily verified by the alternating series test), and so it follows from (1.6.6) that

$$\pi = 4(1 - \tfrac{1}{3} + \tfrac{1}{5} - \tfrac{1}{7} + \cdots) \tag{1.6.10}$$

Since this is an alternating series in which the terms are decreasing, the truncation error is bounded by the first term omitted.

Example 1.6.6. Estimate π with truncation error bounded by 10^{-5} using the series expansion (1.6.10).

Solution. The truncation error in the partial sum terminating after the term $4/(2n-1)$ is bounded by $4/(2n+1)$. This error is less than 10^{-5} if $n \geq 200{,}000$. Clearly, this is not the way to proceed with the accurate computation of π, and we pursue this particular case no further.

We can improve on this performance quite dramatically without changing the basic approach. If instead of (1.6.6) we use the identity

$$\frac{\pi}{6} = \arctan \frac{1}{\sqrt{3}} \tag{1.6.11}$$

then we obtain the series expansion

$$\pi = 6\left(3^{-1/2} - \frac{3^{-3/2}}{3} + \frac{3^{-5/2}}{5} - \frac{3^{-7/2}}{7} + \cdots\right)$$

$$= 2\sqrt{3}\left(1 - \frac{1}{3 \times 3} + \frac{1}{5 \times 3^2} - \frac{1}{7 \times 3^3} + \cdots\right)$$

$$= 2\sqrt{3} \sum_0^\infty \frac{(-1)^n}{(2n+1)3^n} \tag{1.6.12}$$

Example 1.6.7. Use the series expansion (1.6.12) to evaluate π with a truncation error bounded by 10^{-5}.

Solution. Again, the truncation error is bounded by the first term omitted. Thus we require that

$$\frac{2\sqrt{3}}{(2n+1)3^n} < 10^{-5}$$

which is first satisfied for $n = 9$. The value obtained for π from this partial sum is 3.141 599 8, which is in error by approximately 7×10^{-6}.

Before leaving this topic, it is well worthwhile to mention that the summation of alternating series like these can often be significantly improved by the use of *Euler's method*, or the *Euler transformation*. This technique for accelerating the convergence of such series is particularly useful in the case of slowly convergent series such as (1.6.10).

To describe the method, we need to introduce two of the finite difference operators which will also be used extensively in studying interpolation and the numerical solution of differential equations.

We define the shift operator E and the forward difference operator Δ for a sequence (a_n) by

$$E(a_n) = a_{n+1} \tag{1.6.13}$$

and

$$\Delta(a_n) = a_{n+1} - a_n \tag{1.6.14}$$

Example 1.6.8. Let $a_n = 1/n$. Find $E(a_2)$, $\Delta(a_4)$, $E^2(a_1)$, and $\Delta(E(a_3))$.

Solution. Now by the definition (1.6.13), $E(a_2) = a_3 = \frac{1}{3}$. Similarly, using (1.6.14), $\Delta(a_4) = a_5 - a_4$, and so $\Delta(a_4) = \frac{1}{5} - \frac{1}{4} = -\frac{1}{20}$. For the composite operations, we have

$$E^2(a_1) = E(E(a_1)) = E(a_2) = a_3 = \frac{1}{3}$$

and

$$\Delta(E(a_3)) = \Delta(a_4) = -\frac{1}{20}$$

as before.

Consider now the summation of an alternating series. We have

$$\sum (-1)^n a_n = a_0 - a_1 + a_2 - a_3 + \cdots$$
$$= a_0 - Ea_0 + E^2 a_0 - E^3 a_0 + \cdots$$
$$= (1 - E + E^2 - E^3 + \cdots)a_0 \tag{1.6.15}$$

The expression on the right-hand side of (1.6.15) is just the formal binomial expansion of $(1 + E)^{-1} a_0$. Euler's transformation is based on rewriting this using the observation

$$\Delta(a_n) = a_{n+1} - a_n = E(a_n) - a_n = (E-1)a_n$$

from which we deduce that

$$E = 1 + \Delta \qquad (1.6.16)$$

It follows that

$$\sum (-1)^n a_n = (1+E)^{-1} a_0 = (2+\Delta)^{-1} a_0$$
$$= \frac{(1+\Delta/2)^{-1} a_0}{2} \qquad (1.6.17)$$

The right-hand side of (1.6.17) can now be expanded to yield

$$\sum (-1)^n a_n = \frac{1}{2}\left(1 - \frac{\Delta}{2} + \frac{\Delta^2}{4} - \frac{\Delta^3}{8} + \cdots\right) a_0 \qquad (1.6.18)$$

and it is this expression which is often used to accelerate the convergence of the series.

Example 1.6.9. Use Euler's method to compute π with the series (1.6.10).

Solution. We will multiply the result by the factor 4 at the end of the computation, so the series of interest is

$$1 - \tfrac{1}{3} + \tfrac{1}{5} - \tfrac{1}{7} + \cdots$$

That is, $a_n = 1/(2n+1)$. The table below shows, to four decimal places, the first few terms of the series together with their differences. We use the fact that

$$\Delta^{k+1}(a_n) = \Delta^k(a_{n+1}) - \Delta^k(a_n)$$

Table of differences for $a_n = 1/(2n+1)$

a_n	Δa_n	$\Delta^2 a_n$	$\Delta^3 a_n$	$\Delta^4 a_n$	$\Delta^5 a_n$	$\Delta^6 a_n$	$\Delta^7 a_n$
1.0000							
	−0.6667						
0.3333		0.5334					
	−0.1333		−0.4572				
0.2000		0.0762		0.4063			
	−0.0571		−0.0509		−0.3691		
0.1429		0.0253		0.0372		0.3402	
	−0.0318		−0.0137		−0.0289		−0.3166
0.1111		0.0116		0.0083		0.0236	
	−0.0202		−0.0054		−0.0053		
0.0909		0.0062		0.0030			
	−0.0140		−0.0024				
0.0769		0.0038					
	−0.0102						
0.0667							

It can be shown that the entries in the leading diagonal, in other words, the differences $\Delta^k a_0$, are given for this particular series by $(-1)^k (2^k k!)^2/(2k+1)!$ and therefore that the terms of the expansion (1.6.18) are $2^k (k!)^2/(2k+1)!$. For large k, the ratio of successive terms is approximately $1/2$, so the series expansion in (1.6.18) will converge like a geometric series with common ratio $1/2$. The convergence will indeed be much more rapid than for the original series. The value of π computed from just these terms using Euler's method is 3.1371 to four decimal places, which has an error of about 4.5×10^{-3}. The original series, using these terms, gives 3.0171.

Can we do something similar to accelerate the convergence of the better series approximation (1.6.12)? The difficulty here is that the series is already geometric in nature, and so Euler's transformation will have little effect.

A similar expansion to that used above for the series whose general term is

$$a_n = r^n b_n \qquad (1.6.19)$$

eventually simplifies to

$$\sum (-1)^n a_n = \frac{1}{r+1} (1 - \rho\Delta + \rho^2\Delta^2 - \rho^3\Delta^3 + \cdots)b_0 \qquad (1.6.20)$$

where

$$\rho = \frac{r}{1+r} \qquad (1.6.21)$$

For the evaluation of π using (1.6.12), we have $r = \frac{1}{3}$ and therefore $\rho = \frac{1}{4}$ with now $b_n = 1/(2n+1)$. The table of differences is thus the same as that above, and so from the same information we can obtain the value; to four decimal places, 3.1416. This was achieved using the four-decimal-place data above; with greater accuracy of internal calculation we can achieve slightly more accuracy in our approximation, obtaining 3.14158 to five decimal places. With the same terms as we used in Example 1.6.7, Program 1.6.10 gives the approximate value of π correct to five decimal places. With 10 terms the error is reduced to just 2×10^{-7}, while for 15 terms the value is 3.141 592 653 4, which is in error by only 2×10^{-10}.

Program 1.6.10   Computation of π

```
var sum, ratio, rho, mult: real;
    i, n, nterms: integer;
    term, diff1, diff2: array[0..20] of real;
begin
    writeln('Input number of terms to be used ');
    readln(nterms);
    for i:=0 to nterms do begin
        term[i]:=1.0/(2*i+1); diff1[i]:=term[i];
    end;
    ratio:=1.0/3.0; rho:=ratio/(1+ratio); sum:=0;
```

```
        for n:=1 to nterms do begin
            for i:=n to nterms do
                diff2[i]:=diff1[i] − diff1[i − 1];
            for i:=n to nterms do diff1[i]:=diff2[i];
        end;
        mult:=1;
        for i:=0 to nterms do begin
            sum:=sum + abs(diff1[i]) * mult;
            mult:=mult * rho;
        end;
        sum:=sum * 2 * sqrt(3)/(1 + ratio);
        writeln(sum,'        ', pi);
end.
```

Most of this program is self-explanatory, but a few words on the loops generating the arrays diff1 and diff2 are in order. The array diff1 is initialized to contain the original terms, the b_n of the above description. For each value of n, differences of the terms of diff1 from n onward are formed and assigned initially to diff2. The array diff1 is then updated (again from n onward) so that at each stage diff1[i] contains the values $\Delta^i a_0$ for $i \leq n$ and $\Delta^n a_{i-n}$ for $i > n$. You should also observe that the alternating sign of the differences combines with the alternating signs of the terms in (1.6.20) to make every term positive. Hence the *absolute value* of diff1[i] is used in the main summation loop.

Thus far, we have introduced some of the fundamental ideas of computer arithmetic and the analysis of numerical procedures. The last few examples serve to show how the careful organization of the algorithm can improve its performance and reduce the buildup of rounding error and even reduce the truncation error. This can also have the effect of significantly reducing the overall computational effort.

It has also become apparent that there is much more to the question of efficient evaluation of a mathematical function than you probably first thought. We will discuss some of the successful techniques in detail in the next few chapters, but first we consider the classical approach of approximating a function by its Taylor polynomial.

EXERCISES 1.6

1. Find the number of terms needed to evaluate $e^{0.5}$ with a truncation error bounded by 10^{-8}. Estimate the resulting error in e^8 if this value of $e^{0.5}$ is squared 4 times.

2. Obtain a series expansion for the function

$$f(x) = \int_0^x \exp \frac{-t^2}{2} \, dt$$

How many terms are needed to evaluate $f(2)$ with a truncation error bounded by 10^{-8}? What is the value of $f(2)$ given by these terms?

3. (a) Sum the first five terms of the series
$$\sum_{n=0}^{\infty} \frac{(-1)^n}{(n+1)2^n}.$$

   (b) Form the difference table for the terms $b_n = 1/(n+1)$, and use these differences to estimate the sum of the series using Euler's transformation with the same five terms.

(c) Estimate the error in this sum. How many terms of the original series would be necessary to achieve the same accuracy?

4. Consider the evaluation of $\ln 2$ using the series expansion

$$\ln 2 = 1 - \tfrac{1}{2} + \tfrac{1}{3} - \tfrac{1}{4} + \tfrac{1}{5} - \tfrac{1}{6} + \cdots$$

(a) How many terms are needed to reduce the truncation error to less than 10^{-6}?

(b) Obtain a difference table for the first few differences of the terms of this series.

(c) Find a general expression for the differences $\Delta^k a_0$, where $a_n = 1/(n+1)$.

(d) Compute $\ln 2$ from this series using Euler's method, first by hand for six terms and then by computer program to see what accuracy can be obtained.

1.7 TAYLOR SERIES APPROXIMATION

One of the simplest approximation techniques available to us for a sufficiently differentiable function is to use the Taylor series based on a convenient, nearby point. Thus we use the expansion

$$f(a + h) = f(a) + hf'(a) + \frac{h^2 f''(a)}{2!} + \cdots \tag{1.7.1}$$

to approximate values of f close to a. Clearly this requires knowledge of f and its derivatives at the base point a. The series can be truncated using Taylor's theorem to

$$f(a + h) = f(a) + hf'(a) + \cdots + \frac{h^n f^{(n)}(a)}{n!} + \frac{h^{n+1} f^{(n+1)}(c)}{(n+1)!} \tag{1.7.2}$$

where c lies between a and $a + h$ and this "mean value term" provides a formula for the truncation error.

We see that the truncation error in using Taylor's series terminating with the nth derivative is given by $h^{n+1} f^{(n+1)}(c)/(n+1)!$ which is bounded by

$$\max_{a \le t \le a+h} \frac{|f^{(n+1)}(t)| h^{n+1}}{(n+1)!}$$

Example 1.7.1. Find the Taylor series expansion of $\cos x$ based at (a) 0 and (b) $\pi/4$. Compare the truncation errors for the series terminating with the fourth-order term.

Solution

(a) The standard series for the cosine function is the Taylor series based at 0. That is,

$$\cos x = 1 - \frac{x^2}{2!} + \frac{x^4}{4!} + \cdots \tag{1.7.3}$$

which using Taylor's theorem has truncation error bounded by

$$\max_{0 < t \le h} \frac{|\sin t| h^5}{120} \le \frac{h^6}{120}$$

Since the series (1.7.3) is alternating with decreasing terms, the error is in fact bounded by the first term omitted, namely, $h^6/6!$.

For small values of h this truncation error is also small, but for larger values it is easy to see that the error bound grows rapidly. For example, using the fourth-order expansion for $x < 0.1$, we see that the error is bounded by $10^{-6}/720 \approx 1.4 \times 10^{-9}$, whereas for the interval $[0, \pi/2]$ we can only bound the error by about 0.02, which is clearly inadequate for almost all practical purposes.

(b) For the series based at $\pi/4$, we observe that all derivatives take the values $\pm 1/\sqrt{2}$ with signs in alternating pairs after the initial 1. So the Taylor series is

$$\cos x = \frac{1}{\sqrt{2}} \left[1 - \left(x - \frac{\pi}{4} \right) - \frac{(x - \pi/4)^2}{2!} + \frac{(x - \pi/4)^3}{3!} + \cdots \right] \quad (1.7.4)$$

We see that the truncation error for the fourth-order approximation is $[(x - \pi/4)^5 \sin \theta]/120$, where θ is some point between x and $\pi/4$. For $|x - \pi/4| < 0.1$, this is bounded by 6.5×10^{-8}, and so the approximation will again be good for points close to the base point.

This time, for the interval $[0, \pi/2]$ the error will be bounded by $(\pi/4)^5/120 \approx 2.5 \times 10^{-3}$. We observe that the series based at the midpoint $\pi/4$ is much better for approximation over this interval than the one based at an endpoint. This should be expected.

If we wish to use Taylor series methods for approximating a function over a wide range of values of the argument, it will often be necessary to use a high-order approximation—and therefore the expensive computation of a large number of terms of the series—or to use different series in different parts of the range. This *domain decomposition* approach would normally be more economical.

Example 1.7.2. Find Taylor series approximations to $\cos x$ on $[0, \pi/2]$ which have truncation error bounded by 10^{-5}.

Solution

(a) Using just one series based at 0 for the whole range, the derivative in the error term is bounded by unity, and so the truncation error of the $2n$th-order approximation is bounded by $(\pi/2)^{2n+1}/(2n + 1)!$. This is first less than 10^{-5} for $2n + 1 = 11$. $[(\pi/2)^{11}/11! \approx 3.6 \times 10^{-6}.]$ The series approximation is thus

$$\cos x \simeq 1 - \frac{x^2}{2!} + \frac{x^4}{4!} - \frac{x^6}{6!} + \frac{x^8}{8!} - \frac{x^{10}}{10!} \quad (1.7.5)$$

(b) Again using a single series but this time based at $\pi/4$, we get the bound for the nth-order approximation $(\pi/4)^{n+1}/(n + 1)!$ which is bounded by 10^{-5} for $n \geq 7$. Thus we can use the series approximation

$$\cos x \simeq \frac{1}{\sqrt{2}} \left(1 - \Delta x - \frac{\Delta x^2}{2!} + \frac{\Delta x^3}{3!} + \frac{\Delta x^4}{4!} - \frac{\Delta x^5}{5!} - \frac{\Delta x^6}{6!} + \frac{\Delta x^7}{7!} \right)$$

where $\Delta x = x - \pi/4$.

Note here that there are a greater number of terms in this series than in that in (a). Therefore, even though the order of the approximation has been reduced by taking a different base point, the amount of computational effort in using this approximation has increased. For different precisions, the situation could be reversed.

(c) If we were to utilize two series based at $\pi/8$ and $3\pi/8$, then we find that the truncation errors are bounded now by $(\pi/8)^{n+1}/(n+1)!$, which is within the required tolerance for $n = 5$. For values of $x \in [0, \pi/4]$ the series based on $\pi/8$ would be used, while for $x > \pi/4$ the one based at $3\pi/8$ would give the required approximation.

The amount of work saved here relative to (b) is certainly worthwhile. But again the computational effort is comparable with the original MacLaurin series in (a). By further subdivision of the interval, we can achieve much lower-order approximations which deliver the desired accuracy. For example, using intervals of length $\pi/50$ so that $\Delta x \le \pi/100$ allows just second-order Taylor series to be used within each.

One of the major drawbacks to the Taylor series approach to function evaluation is that we must be careful about the values of the argument for which it is used. For example, the commonly used series for the function $\ln(1 + x)$ has radius of convergence 1 and so is only of any value for computing logarithms of numbers in $(0, 2]$.

However, at the left-hand end of this interval of convergence, the series is very slowly divergent and would appear to converge to "$\ln 0$". (For this particular case the series is just the negative of the harmonic series which we have seen in Sec. 1.6 will appear to converge on the computer.) This suggests the need for caution, but probably more important is the question of how to compute values of logarithms of other numbers.

Example 1.7.3. Use a Taylor expansion of the logarithm function to evaluate $\ln 10.1$.

Solution. The simplest approach is just to expand $\ln x$ about the base point $x = 10$. This leads us to the series

$$\ln 10.1 = \ln 10 + \frac{0.1}{10} - \frac{0.1^2}{200} + \frac{0.1^3}{3000} - \cdots$$

which is a rapidly convergent series whose truncation error is already bounded by 10^{-8} using just the terms shown. However, this begs one important question: How do we compute the value $\ln 10$ on which this expansion is built?

Without an answer to the question, we are no further forward. The most obvious method for this task is to use the well-known properties of logarithms to write

$$\ln 10 = \ln 2 + \ln 5 = 2 \ln 2 + \ln 2.5 = 3 \ln 2 + \ln 1.25$$

and to use our standard series for the evaluation of $\ln 2$ and $\ln 1.25$. Both these are alternating series, and so their summation can be greatly improved by the use of Euler's method. (See Exercise 5 and Projects 5 and 6.)

Table 1.7.4 shows the results of computing $\ln 2$ first without and then with the aid of Euler's transformation for varying tolerances. It can be seen from the table that the greatest number of terms used for accuracies down to 10^{-8} is 23. Each additional decimal place accuracy seems to require three more terms

TABLE 1.7.4
Computation of ln 2 using the standard series and Euler's transformation
True value: ln 2 = 0.693 147 180 6

Precision required	Taylor approximation		Euler's transformation	
	n	ln 2	n	ln 2
10^{-1}	5	0.783 333 333 3	3	0.625 000 000 0
10^{-2}	50	0.683 247 160 6	6	0.688 541 666 7
10^{-3}	500	0.692 148 180 6	8	0.692 261 904 8
10^{-4}	5000	0.693 047 190 5	11	0.693 064 856 1
10^{-5}			14	0.693 138 980 4
10^{-6}			17	0.693 146 328 3
10^{-7}			20	0.693 147 089 4
10^{-8}			23	0.693 147 170 6

in the transformed series. Thus about 50 terms would be sufficient for any precision which can be handled in Turbo Pascal.

The entries in the left-hand column stop because this is as far as we can go with the standard integer variable type. Euler's transformation has certainly proved worthwhile with additional precision being obtained very cheaply. Notice how the number of terms required by the Taylor approximation is about half that which would be predicted by a simple truncation error bound, because the truncation error is certainly bounded by the first term omitted but is in fact much closer to half of that.

EXERCISES 1.7

1. Find a bound for the truncation error if the function $\sin x$ is approximated by its MacLaurin series (the Taylor series based at 0). What is the bound for this approximation over the interval $[0, \pi]$? How many terms are needed to reduce this error to 10^{-6}?

2. Find the L_1 and L_2 measures of the error in the approximation of Exercise 1.

3. Obtain the Taylor series expansion for $\sin x$ around the points $\pi/2$ and $\pi/4$, and find bounds for their truncation errors. How many terms are needed to reduce this bound to 10^{-5} over the interval $[0, \pi/2]$? How does this compare with the MacLaurin series?

4. We wish to approximate the sine function over $[0, \pi/2]$ using Taylor series of order not more than three with an error bounded by 10^{-7}. Use the minimum number of such series based at the points of a uniform mesh. How many different base points are required? (The base points will satisfy $x_{k+1} = x_1 + kh$ for some fixed step size h and $k = 1, 2, \ldots, n$ with $x_1 = h/2$ and $x_n = \pi/2 - h/2$.)

5. Find the truncation error in using the nth-order Taylor series approximation for $\ln(1 + x)$ on the interval $-1 < x < 1$. How many terms are needed for this error to be bounded by 10^{-4}? Find the L_1, L_2, and L_∞ measures of the error for $0 \le x \le 1$.

6. Write a program to evaluate the exponential function by summing the series $1 + x + x^2/2! + x^3/3! + \cdots$ using the first 25 terms. Test your program for $x = \pm 1, \pm 2, \ldots, \pm 10$, and compare the results with the built-in exponential function.

PROJECTS

1. Write a procedure to solve a quadratic equation given in terms of its coefficients. In the case of two real roots, the computation should be as accurate as possible. Your procedure should also return the real and imaginary parts of complex roots. Be careful to take full account of the possibility of zero coefficients, especially a leading zero, so that the equation is really linear.

2. Write and run a timings program to obtain information similar to that in Table 1.4.1 for your computer.

3. Write a program to "sum" the harmonic series as accurately as possible. Only include those terms which contribute to the sum. Take care over the order of summation to ensure that any terms which are included are fully accounted for.

4. Write a program to implement the approximation to the sine function obtained in Exercise 4 of Sec. 1.7. Tabulate the values of the approximation and of $\sin x$ for values of x ranging from 0 to 1.5 in steps of 0.05. Graph the output function, the third-order MacLaurin approximation, and the library sine functions on the same axes.

5. Rewrite the series in Exercise 5 of Sec. 1.7 with nonnegative x as a series for $\ln y$ for $1 \le y < 2$. (Alternatively, obtain the series directly as Taylor's series based at $y = 1$.) Use the properties of the logarithm function (such as $\ln 2y = \ln y + \ln 2$) to obtain an algorithm for computing $\ln y$ for larger values of y. Be careful to check the extra precision needed in order to obtain the same accuracy in the final result.

6. Write a program to compute values of $\ln x$ using the Taylor series approach of Project 5 and Euler's method to improve the convergence rate for the summation of the series. (Note that $\ln 2$ will be used extensively within this routine and so should be computed once and stored for subsequent use.)

CHAPTER 2

ITERATIVE SOLUTION OF NONLINEAR EQUATIONS

2.1 INTRODUCTION AND THE BISECTION METHOD

In this chapter we study one of the fundamental problems of mathematics—the solution of an equation in a single unknown. For some very special cases we can write down formulas for the solutions, but this is far from the typical situation. If we consider just simple algebraic polynomial equations, then we can certainly write down formulas which will yield the solutions if the polynomial is of degree one or two. In the latter case, the well-known quadratic formula also gives us the complex roots of the equation if no real roots exist. It is possible to obtain the roots of a cubic polynomial in a similar sort of way, but the formula is already very much more complicated and virtually never used in practice.

In the case of a fourth-degree equation, the formula is excessively complicated, and for polynomials of degree five or more, it can be shown that there can be no such formula. In such cases it is necessary to resort to numerical techniques. In the case of equations which are not polynomial in nature, this is virtually the only approach available apart from some very special examples where knowledge of the functions involved may enable us to find one or more solutions explicitly. We will see in Chap. 3 that even some of the most elementary of mathematical functions are evaluated on the computer or calculator by an equation-solving approach. Foremost among these are the square root and reciprocation functions which in many computer systems are evaluated by solving the equations

$$x^2 - a = 0 \qquad \text{to evaluate } \sqrt{a}$$

or

$$\frac{1}{x} - c = 0 \qquad \text{to evaluate } \frac{1}{c}$$

The techniques we study in this chapter are all *iterative* in nature. This means that we begin our solution process with one or more guesses at the solution being sought and then refine these guesses according to some specific rules of the method. Typically, this results in the generation of a sequence of estimates of the solution which we hope will converge to the true solution.

It is, of course, necessary to analyze the methods to establish conditions under which they do indeed converge. Simply establishing that such an iterative sequence converges is insufficient for many practical purposes. First, we certainly would like to have this knowledge at the beginning of the process and thereby avoid large amounts of fruitless computational effort. Since it is also desirable to have some bound on the error in our computed solution, we also investigate the *rate of convergence* of the iterative schemes. It will be helpful to put these terms into perspective by considering a simple example.

> **Example 2.1.1.** Consider the plight of Pythagoras' young nephew, banished to the island of Kos after some youthful prank had caused embarrassment to his illustrious family. He was told that he could return if he could build himself a raft which would take him back to Athens.
>
> After scouting around the island for a few days, he found several likely looking trees for the main members of the frame of his raft. He needed to calculate the best dimensions for his escape vessel given the trees available—and his desire not to cut down trees he did not need. (This was not due to some ancient sense of conservation, just his instinct for self-preservation.) Finally he decided on the trees for the two long sides and for the crosspieces and that these sides should be in the ratio of 3/2. His next task was to find the right trees for two stout diagonal members. Fortunately, his uncle had taught him the family theorem, and so the young castaway knew that he simply needed to calculate $\sqrt{13}$ to find the necessary length for these diagonals.
>
> But how could he do this with nothing more than the sticks and sand at his disposal? First, he observed that $3^2 = 9 < 13$ and $4^2 = 16 > 13$ and concluded that the number he wanted was somewhere between 3 and 4. Fairly naturally he tried 3.5^2 and found that this is 12.25 and therefore that his solution (and salvation) lay somewhere between 3.5 and 4. Next $3.75^2 = 14.0625$, and so $\sqrt{13} \in (3.5, 3.75)$.
>
> Continuing in this manner he subsequently deduced that
>
> $$3.625^2 \; > 13 \qquad \text{and so} \qquad \sqrt{13} \in (3.5, 3.625)$$
> $$3.5625^2 \; < 13 \qquad \text{and so} \qquad \sqrt{13} \in (3.5625, 3.625)$$
> $$3.59375^2 < 13 \qquad \text{and so} \qquad \sqrt{13} \in (3.59375, 3.625)$$

At this point, Pythagoras' nephew decided that this was at least as accurate as he could hope to produce with the relatively primitive tools he had with him.

What Pythagoras' nephew had in fact discovered was the *method of bisection*, one of the simplest and most reliable of iterative methods for the solution of a nonlinear equation. It is with this method that we begin our discussion. Before proceeding to the details, we should report that, unfortunately, the young man's boat-building skills were not a match for his mathematical reasoning, and he was lost in a storm before he could enlighten the world with his discovery of the bisection technique.

We will see shortly that the convergence of this algorithm is an easy application of the sandwich rule for the convergence of sequences and the intermediate value theorem (IVT)—one of the twin pillars of the convergence theory of iterative processes. The other of these fundamental results on which so much of our theory is based is the mean value theorem (MVT). These two and Taylor's theorem will be cited repeatedly. In our analysis of the bisection method we will also find it easy to establish the rate of convergence of the various iterative sequences which will tell us that, for repeated use within a more complicated setting, more sophisticated approaches are necessary.

Let us suppose then that we seek the solution of the equation

$$f(x) = 0 \tag{2.1.1}$$

for some continuous function f.

In general, the bisection algorithm proceeds by generating sequences to the left and right of the required solution and another sequence of midpoints which is then used in place of the appropriate endpoint. This process for solving $f(x) = 0$ can be achieved with the following Turbo Pascal code:

```
repeat
    mid:=(left + right)/2;
    if f(mid) * f(left) >0 then left:=mid else right:=mid;
until right − left < eps;
```

It is easy to prove, using the sandwich theorem, that for any continuous function f and initial endpoints satisfying

$$f(\text{left}) * f(\text{right}) < 0$$

the sequence of midpoints generated by the bisection method converges to a solution of (2.1.1). (See Exercise 4.)

First, the algorithm requires that we find two points, one on either side of the solution. That is, we must begin the process with two points L and R, say, for which the function f has opposite signs. In other words, we require that

$$f(L)f(R) < 0 \tag{2.1.2}$$

so that, by the IVT, we know there is a point s, say, between L and R for which

$$f(s) = 0 \tag{2.1.3}$$

There may be more than one such solution, but we know there is at least one. It follows that by testing the sign of the function at a point between L and R, we can deduce which part of the interval $[L, R]$ contains the solution. (Strictly, we should say contains an odd number of, and therefore at least one, solution. There may still be solutions in the other part, but there must be an even number of them.) In the bisection method we choose this intermediate point to be the midpoint M, say, of the interval $[L, R]$ and therefore reduce the length of the bracketing interval containing the solution by a factor of $\frac{1}{2}$.

The conditional statement in the code above replaces either L or R with M depending on which of the half-intervals shows a change of sign and therefore, by the IVT, contains the solution. The process can then be repeated until the two endpoints are sufficiently close together. Denoting the sequence of left endpoints, right endpoints, and midpoints by L_n, R_n, and M_n, the proof of convergence proceeds by establishing the following facts. (The details form Exercise 4.)

The sequences (L_n) and (R_n) are monotone increasing and decreasing, respectively.

$L_n < R_n$ for every n, and so the sequences are bounded—and therefore convergent.

$R_n - L_n \to 0$ as $n \to \infty$, so the limits of the two sequences are the same, X, say. By the sandwich rule $M_n \to X$, too.

$f(L_n)f(R_n) \leq 0$ for every n, and so, using the continuity of f, $f(X)^2 \leq 0$.

Now the fact that each of the intervals $[L_n, R_n]$ contains the solution implies that M_n differs from the solution s by at most $e_n = (R_n - L_n)/2$. This error bound is reduced by a factor of 2 on each iteration.

This is an example of a linear rate of convergence since the relation between e_{n+1} and e_n is a simple linear one. To achieve a high degree of accuracy, a large number of iterations may be needed, and so faster and more efficient techniques are desirable. We will study various approaches to this problem in the next few sections.

Example 2.1.2. Find the smallest positive solution of the equation

$$f(x) = x - \tan x = 0$$

to the precision of Turbo Pascal type real.

Solution. Now, for $x \in (0, \pi/2)$, $x < \tan x$, and so $f(x) < 0$ throughout that range. For $x \in (\pi/2, \pi)$, $\tan x < 0$, and so $f(x) > 0$, while $\tan x \to \infty$ as $x \to 3\pi/2$. It follows that the solution sought must lie in the interval $(\pi, 3\pi/2)$. This initial interval is of length $\pi/2$, or somewhat less than 1.6. To achieve the required accuracy, we seek a *relative error* no greater than 2^{-40}. Since the solution lies between 4 and 8, its binary floating-point representation has exponent 3, and this relative accuracy is therefore equivalent to an absolute error of 2^{-37}. To guarantee this level of precision in the midpoint of the final interval, this interval must have length no greater than 2^{-36} which will be achieved by halving 1.6 a minimum of 38 times. The first few iterations yield:

n	L_n	R_n	M_n	$f(L_n)f(M_n)$	So set
0	π	$3\pi/2$	3.927	>0	$L=M$
1	3.927	$3\pi/2$	4.320	>0	$L=M$
2	4.320	$3\pi/2$	4.516	<0	$R=M$
3	4.320	4.516	4.418	>0	$L=M$

by which stage we have reduced the bracketing interval to a length of approximately 0.1, and so the midpoint of this interval will provide the solution with error no greater than 0.05—or, in other words, to one decimal place. Clearly the process requires automation.

The result of running a program to perform the 38 iterations required with the initial interval $[3.142, 4.712]$ or, approximately, $[\pi, 3\pi/2]$, is that the solution $s = 4.493\,409\,457\,9$.

The bisection method can be seen in the above example to be providing steady if unspectacular progress toward the solution of the equation under consideration. However, the number of iterations needed to obtain machine accuracy in the solution is such that if the function f were of a very complicated form which perhaps required the solution of a system of differential equations to evaluate it—as will turn out to be the case for the so-called shooting methods for the solution of two-point boundary value problems (Chap. 12)—then it will be of great importance to reduce the number of iterations needed. For the "one-off" problem of a not-too-difficult nature, such as that faced by Pythagoras' nephew, the bisection algorithm is perhaps as good a technique as any since it is totally reliable and easy to program and/or work by hand. It should be observed that it is only "totally reliable" in situations where the existence of the solution can be deduced from the IVT, that is, where the graph of the function actually crosses the axis. The bisection method cannot locate a double zero, a point at which the x axis is tangent to the curve.

In summary, the convergence of the bisection method is typically slow, but it is a reliable way of obtaining an idea of the location of the root of the equation. It is commonly used for this purpose, namely, to provide a good initial point for a more efficient iteration. Many iterative methods can be written in the form of a fixed-point iteration. This will be the subject of the next section.

EXERCISES 2.1

1. Use the bisection method to obtain an interval of length less than 0.1 containing the solution of $x = \cos x$ in $[0, \pi/2]$. How many more iterations would be needed to reduce the interval length to 10^{-8}?

2. Write a program to implement the bisection algorithm. Use your program to check the result quoted in Example 2.1.2 to machine accuracy on your computer.

3. Show that the equation

$$e^x - Nx - 1 = 0$$

for $N \geq 2$ has its positive solution in the interval $[1, N]$. Use the bisection method for each of the cases $N = 2, 3, \ldots, 10$ to reduce the bracketing interval to length less than 1. Use your computer procedure to locate each of these solutions with error bounded by 10^{-8}.

4. Let f be a continuous function on the interval $[a, b]$ and suppose that

$$f(a) < 0 < f(b)$$

By following the steps outlined in the text, prove that the sequence of midpoints generated by the bisection algorithm starting with $[a, b]$ converges to a solution of the equation $f(x) = 0$.

5. Show that the equation

$$x^4 - 11.6x^3 + 46.86x^2 - 76.676x + 41.6185 = 0$$

has four real roots. Use the method of bisection to locate each of them to five decimal places.

2.2 FIXED-POINT ITERATION

The basic idea of the *fixed-point iteration*, or *function iteration*, approach to the solution of an equation is that the original equation

$$f(x) = 0 \tag{2.2.1}$$

is rewritten or *rearranged* in the form of an equivalent equation

$$x = g(x) \tag{2.2.2}$$

Provided that the two equations (2.2.1) and (2.2.2) are equivalent, it follows that any solution of the original equation is a *fixed point* of the *iteration function* g; that is, it is a point which is unchanged by application of the function g.

The task of solving (2.2.1) is therefore reduced (or perhaps at this stage we should simply say changed) to that of finding a point s, say, satisfying the fixed-point condition (2.2.2). The technique is conceptually very simple.

An initial guess x_0 generates a sequence of estimates of this fixed point by setting

$$x_{n+1} = g(x_n) \qquad (n = 0, 1, 2, \ldots) \tag{2.2.3}$$

The important questions about any iteration are whether it converges, if so whether it has the desired solution as its limit, and how quickly it will achieve any particular precision in the solution.

By way of illustration, we return to the computation of the square root and see that

$$x^2 = a \leftrightarrow x = \frac{a}{x}$$

$$\leftrightarrow 2x = x + \frac{a}{x}$$

$$\leftrightarrow x = \frac{x + a/x}{2} \tag{2.2.4}$$

and this last equation can be used to provide a (very) good iteration for the solution of the original one.

Example 2.2.1. Use the bisection algorithm to obtain an interval of length $\frac{1}{16}$ containing $\sqrt{13}$, and then use its midpoint to start the iteration

$$x_{n+1} = \frac{x_n + 13/x_n}{2}$$

Solution. In Example 2.1.1, we used the bisection method to obtain the bounds $3.5625 < \sqrt{13} < 3.625$, and we start the iteration with the midpoint of this interval, namely, 3.59375. This generates the following values for $x_1, x_2,$ and x_3:

$$3.605\ 570\ 7 \qquad 3.605\ 551\ 3 \qquad \text{and} \qquad 3.605\ 551\ 3$$

The agreement between these last two suggests that the solution is 3.605 551 3 to this accuracy. *Of course, we have not proved this.* (In fact, it is accurate to all the figures shown; we will discuss the proof of this statement a little later.)

To establish the convergence of this iteration, we begin with a brief analysis of the iteration function

$$g(x) = \frac{x + a/x}{2} \tag{2.2.5}$$

We observe that the derivative $g'(x)$ is increasing for positive x. Also

$$g'(x) < 0 \qquad \text{for } x < \sqrt{a}$$

and

$$g'(x) > 0 \qquad \text{for } x > \sqrt{a}$$

It follows from the MVT that $g(x) > \sqrt{a}$ for all positive x and therefore that, for any positive x_0, $x_n > \sqrt{a}$ for $n \geq 1$. Furthermore, if $x > \sqrt{a}$, then $a/x < \sqrt{a}$. We deduce

$$\sqrt{a} < x_{n+1} < x_n < \cdots < x_1 \tag{2.2.6}$$

Therefore (x_n) is a decreasing bounded sequence; thus it converges. Since $\sqrt{a}$ is the only (positive) fixed point of g, it follows that this is indeed the limit.

We can therefore deduce that the limit of the process used in Example 2.2.1 is $\sqrt{13}$, but we still have not established the validity of the statement that the value found above is the solution to seven decimal places. To study the rate of convergence and the precision of the results, we consider the sequence of errors; that is, we study the sequence (e_n) where

$$e_n = x_n - \sqrt{a}$$

For convenience we also denote $x_n + \sqrt{a}$ by s_n. Now,

$$e_{n+1} = x_{n+1} - \sqrt{a} = \frac{x_n + a/x_n - 2\sqrt{a}}{2}$$

$$= \frac{x_n^2 - 2x_n\sqrt{a} + a}{2x_n} = \frac{(x_n - \sqrt{a})^2}{2x_n}$$

$$= \frac{e_n^2}{2x_n} \tag{2.2.7}$$

Similarly, we find that

$$s_{n+1} = \frac{s_n^2}{2x_n}$$

and, therefore,

$$\frac{e_{n+1}}{s_{n+1}} = \frac{e_n^2}{s_n^2} \tag{2.2.8}$$

We already know that $e_n \to 0$. For $e_n < 1$, we see that the error is being reduced as the square of the error; for the case where $s_n > 1$ (which is always true for $a > \frac{1}{4}$ and $n \ge 1$), (2.2.8) implies $e_{n+1} < e_n^2$. An iteration which has the property that $e_{n+1} \simeq c e_n^2$ is said to have *quadratic*, or *second-order*, convergence. This is roughly equivalent to the statement that the number of correct decimal places doubles with each iteration.

Now, if $x_{n+1} = x_n$ to p decimal places, say, then it follows from the definition of e_n that $e_{n+1} = e_n$ to the same (absolute) accuracy. However, in this particular case, $e_{n+1} < e_n^2$, and so it follows that $e_{n+1} = 0$ to the same p decimal places. In the case of Example 2.2.1, it follows that the true solution is indeed 3.605 551 3 to seven decimal places.

We will discuss the details of how this approach can be incorporated into an efficient computer algorithm for computing square roots of floating-point numbers in the next chapter. The particular iteration discussed here is a special case of Newton's (or the Newton-Raphson) method; we will consider this approach in more detail in the next section.

First we develop some of the general theory of iterative methods of solution of equations. It is desirable to develop a theory that enables us to deduce whether or not a particular iteration will converge to the solution of an equation *in advance*. This is another area for application of the MVT and Taylor's theorem.

Suppose then that we wish to solve Eq. (2.2.1),

$$f(x) = 0$$

using the iteration (2.2.3), that is,

$$x_{n+1} = g(x_n)$$

The first and most obvious condition which the iteration function g must satisfy is that the two equations are equivalent; thus we require that

$$x = g(x) \leftrightarrow f(x) = 0 \tag{2.2.9}$$

The harder conditions to check are that the iteration (2.2.3) does indeed converge and that its limit is the required solution of the original equation.

In the case of the above example, the function f is defined as $f(x) = x^2 - a$, while the iteration function is given by $g(x) = (x + a/x)/2$. The equivalence (2.2.9) for this case was established in (2.2.4).

To fix the context of our discussion, assume that Eq. (2.2.1) has a solution in the interval $[a, b] = I$, say. Typically, this information could be obtained from applying the IVT to the observation that $f(a)f(b) < 0$ for the *continuous* function f. We will also assume that the iteration function g is continuous.

Now, if $g(I) \subseteq I$, it again follows from the IVT that g has a fixed point in I. The next theorem establishes conditions under which the iteration will converge to this fixed point, which, by the equivalence (2.2.9) above, is therefore the required solution of (2.2.1). The conditions of the theorem can be seen to be quite natural from a brief look at the Taylor expansion of g about the solution s, say, which gives some idea of the way in which the error in the iterates is reduced.

We denote the error $x_n - s$ by e_n. Then the Taylor series of g gives, assuming sufficient differentiability and recalling that $s = g(s)$,

$$e_{n+1} = x_{n+1} - s = g(x_n) - g(s)$$

$$= (x_n - s)g'(s) + (x_n - s)^2 \frac{g''(s)}{2!} + (x_n - s)^3 \frac{g^{(3)}(s)}{3!} + \cdots$$

$$= e_n g'(s) + e_n^2 \frac{g''(s)}{2!} + e_n^3 \frac{g^{(3)}(s)}{3!} + \cdots \tag{2.2.10}$$

If the sequence is to converge, we might reasonably require that the errors are being reduced with every iteration. In order that $|e_{n+1}| < |e_n|$, it is necessary (neglecting the second- and higher-order terms) that $|g'(s)| < 1$.

Theorem 2.2.2. Let g be a continuously differentiable function which maps the interval I into itself. Thus

$$x \in I \rightarrow g(x) \in I \tag{2.2.11}$$

Suppose further that

$$|g'(x)| < 1 \qquad (x \in I) \tag{2.2.12}$$

Then

(*a*) g has a unique fixed point in I, s say.
(*b*) For any choice $x_0 \in I$, the sequence defined by (2.2.3) converges to s.

Proof. It follows from (2.2.12) and the MVT that g can have at most one fixed point in I, for, if s and t were both fixed points, we would have

$$s = g(s) \qquad \text{and} \qquad t = g(t)$$

and, by the MVT, there exists θ between s and t such that

$$|t - s| = |g(t) - g(s)| = |t - s||g'(\theta)| < |t - s|$$

since $|g'(\theta)| < 1$. This contradiction establishes the uniqueness of the fixed point which we will denote s.

To establish the convergence we again use the MVT. First, note that since $x_0 \in I$, it follows from (2.2.11) that the sequence (x_n) is entirely contained within this interval. Next, we observe that since g' is continuous on the closed bounded interval I, it attains its maximum and minimum values there. It follows that there is a constant $\lambda < 1$ such that

$$|g'(x)| \leq \lambda \qquad (x \in I) \qquad (2.2.13)$$

Now, by the MVT and the fact that s is a fixed point of g, it follows that, for $n = 0, 1, \ldots,$

$$|x_{n+1} - s| = |g(x_n) - g(s)| = |x_n - s| \, |g'(\theta_n)| \leq \lambda |x_n - s|$$

for some θ_n between x_n and s. Hence,

$$|x_{n+1} - s| \leq \lambda^{n+1} |x_0 - s|$$

and since $\lambda < 1$, it follows that $\lambda^n \to 0$ and therefore that $x_n \to s$ as $n \to \infty$, which completes the proof.

In fact, the conditions of the theorem can be weakened somewhat. The differentiability of the iteration function g is not necessary; it can be weakened to a Lipschitz condition with Lipschitz constant $\lambda < 1$.

Definition 2.2.3. A function f is said to be Lipschitz on the interval $[a, b]$ with *Lipschitz constant* λ if

$$|f(x) - f(y)| \leq \lambda |x - y| \qquad (2.2.14)$$

This is an interesting class of functions which will arise from time to time throughout the book. Condition (2.2.14) certainly implies the continuity of f on the interval $[a, b]$ since the right-hand side approaches 0 as $y \to x$. On the other hand, there are Lipschitz functions which are not differentiable. (The simplest example of such a function is the absolute value function on an interval containing 0.) However, any differentiable function whose derivative is bounded on $[a, b]$ is Lipschitz. This is an easy application of the MVT. It follows that the hierarchy goes

$$\text{Bounded derivative} \to \text{Lipschitz} \to \text{continuous}$$

The condition above is, therefore, that g must satisfy

$$|g(x) - g(y)| \leq \lambda |x - y| \qquad \forall x, y \in I \qquad (2.2.15)$$

where $\lambda < 1$ is constant. The details of the (simpler) proof of the theorem for this situation are left as an exercise.

The following corollary shows that our simplistic conclusion based on the Taylor expansion was accurate; the requirement that $|g'(s)| < 1$ is sufficient for the convergence of the iteration.

Corollary 2.2.4. Suppose the iteration function g has a continuous first derivative in a neighborhood of the solution s and that $|g'(s)| < 1$. Then for x_0 sufficiently close to s, the sequence (x_n) converges to s.

Proof. The details of the argument are left as an exercise for the reader. The technique is to establish the existence of a positive number h such that all the conditions of Theorem 2.2.2 are satisfied for the interval $[s - h, s + h]$.

One of the essential tasks facing the numerical analyst is to be able to estimate the accuracy of the solutions obtained to ensure that prescribed tolerances are indeed satisfied. Arguments similar to those of the proof of the theorem provide us with this information. These arguments are again valid whether we use the derivative condition of the theorem or the more general Lipschitz condition on g.

Now, for $m > n$, we get

$$|x_m - x_n| \le |x_m - x_{m-1}| + |x_{m-1} - x_{m-2}| + \cdots + |x_{n+1} - x_n|$$

$$\text{(2.2.16)}$$

and for each k, we have

$$|x_k - x_{k-1}| \le \lambda |x_{k-1} - x_{k-2}| \le \cdots \le \lambda^{k-1} |x_1 - x_0| \qquad \text{(2.2.17)}$$

Combining these two results we obtain

$$|x_m - x_n| \le |x_1 - x_0|(\lambda^{m-1} + \lambda^{m-2} + \cdots + \lambda^n)$$
$$= \lambda^n |x_1 - x_0|(1 + \lambda + \cdots + \lambda^{m-n-1})$$
$$\le \frac{\lambda^n |x_1 - x_0|}{1 - \lambda} \qquad \text{(2.2.18)}$$

Letting $m \to \infty$, we deduce the following bound for the error in x_n:

$$|e_n| = |x_n - s| \le \frac{\lambda^n |x_1 - x_0|}{(1 - \lambda)} \qquad \text{(2.2.19)}$$

The advantage of this particular estimate is that it is entirely computable for any given iteration. All we need is a good estimate of the Lipschitz constant (or the maximum value of the derivative) in the region of the solution.

Example 2.2.5. Consider the following iterations for the possible solution of

$$f(x) = e^x - 1 - 2x = 0$$

(a) $$x_{n+1} = \frac{e^{x_n} - 1}{2}$$

(b) $$x_{n+1} = \ln(1 + 2x_n)$$

Solution. Firstly, we observe that $f(1) < 0 < f(2)$, and so the solution we seek is in the interval $[1, 2]$.

(a) For $g(x) = (e^x - 1)/2$, we have $g'(x) = e^x/2$, which is greater than unity throughout the interval. This iteration will therefore fail to converge.

(b) With $g(x) = \ln(1 + 2x)$, $g'(x) = 2/(1 + 2x)$, which is bounded by $\frac{2}{5}$ and $\frac{2}{3}$ for x

in the interval of interest. Also g is an increasing function of x, and $g(1) = \ln 3$ and $g(2) = \ln 5$ both lie in $[1, 2]$. Thus $g(I) \subseteq I$, and so we can deduce from the theorem that the iteration will converge. The first few iterates, starting from $x_0 = 1.5$, are

1.386 294 4	1.327 761 4	1.296 239 1	1.278 842 3	1.269 109 9
1.263 623 7	1.260 517 8	1.258 755 2	1.257 753 4	1.257 183 7
1.256 859 5	1.256 675 0	1.256 570 0	1.256 510 2	1.256 476 2
1.256 456 8	1.256 445 8	1.256 439 5	1.256 435 9	1.256 433 9

by which stage we probably feel reasonably confident that the solution is 1.2564 to four decimal places.

What does our error bound tell us?

Using the original interval $[1, 2]$, the best value of λ we can use is $g'(1) = \frac{2}{3}$, from which it follows, using (2.2.19), that

$$e_{20} \le 3(1.5 - 1.386\,294\,4)(\tfrac{2}{3})^{20} \simeq 1.0 \times 10^{-4}$$

This error bound shows that even our apparently conservative claim about the accuracy of the solution was slightly too ambitious using the basic information we started with.

However, in the light of this bound we know that the iteration will never give a value below 1.25. This would yield the bound 2/3.5 for λ, which then brings the error bound itself down to 4.8×10^{-6}, not only confirming our claim that the true solution is indeed 1.2564 to four decimal places but also that 1.256 433 9 is correct to five decimal places.

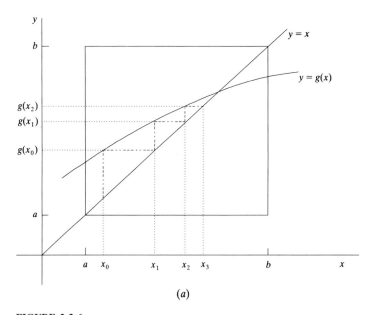

(a)

FIGURE 2.2.6
(a) Monotone convergence for $g'(x) > 0$.

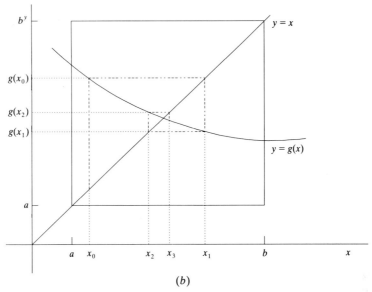

FIGURE 2.2.6
(b) Oscillatory convergence for $g'(x) < 0$.

The progress of fixed-point iterations is illustrated in Fig. 2.2.6. We see that if the g' is positive, the iteration has the "staircase" behavior with monotone convergence to the solution, while if g' is negative, successive iterates bracket the solution. It is also apparent that the closer g' is to zero in the neighborhood of the solution, the more rapid will be the convergence of the iteration.

It is this observation that we capitalize on in the derivation of Newton's method.

EXERCISES 2.2

1. Show that the iteration $x_{n+1} = \cos x_n$ will converge to the solution of $x = \cos x$ for any initial point x_0. Use the IVT to establish that the solution lies in $[0.7, 0.8]$. How many iterations are needed to find this root to six decimal places from $x_0 = 0.75$? (Do *not* perform the iterations.)

2. Find a rearrangement of the equation $e^x - 3x - 1 = 0$ which converges to the solution in $[1, 2]$. Perform the first five iterations starting from 1.5.

3. Show that only one of the iterations

  (i) $x_{n+1} = \tan x_n$

  (ii) $x_{n+1} = \arctan x_n$

  (iii) $x_{n+1} = \pi + \arctan x_n$

will converge to the smallest positive solution of the equation $\tan x = x$.

Use the convergent iteration with an appropriate starting point to find this solution to three decimal places. How many iterations would be necessary to obtain accuracy 10^{-6}? Write a program to implement this, and verify that the desired accuracy is achieved in the predicted number of iterations.

4. Devise suitable fixed-point iterations for the location of the roots of the polynomial equation

$$x^4 - 11.6x^3 + 46.86x^2 - 76.676x + 41.6185 = 0$$

(see Exercise 4, Sec. 2.1), and use them to locate the roots to an accuracy of 10^{-6} using the midpoints of the initial intervals of your bisection solutions as starting points.

5. Show that the iteration $x_{n+1} = (e^{x_n} - 1)/N$ will *not* converge to the positive solution of $e^x - Nx - 1 = 0$ ($N \geq 2$), but that if the iteration $x_{n+1} = \ln(1 + Nx_n)$ is used with a suitable starting point, it will.

6. Write a program to implement the iterations of Exercise 5 with suitable starting points. (See Exercise 3, Sec. 2.1.) Use these to find the solutions to six decimal places.

7. Prove Corollary 2.2.4. That is, show that if $s = g(s)$, g is continuously differentiable, and $|g'(s)| < 1$, then there exists an interval $I = [s - h, s + h]$ for which all the conditions of Theorem 2.2.2 are satisfied and therefore that if $|x_0 - s| \leq h$, the iteration $x_{n+1} = g(x_n)$ converges to s.

8. Show that if in addition to the conditions of Theorem 2.2.2, we also have $g'(x) < 0$ for all $x \in I$ and $x_0 < s$, then the convergence is oscillatory (as in Fig. 2.2.6b) so that

$$x_0 < x_2 < \cdots < x_{2n} < s < x_{2n+1} < \cdots < x_3 < x_1$$

for every $n \geq 2$. (Note that this implies that successive iterates bracket the solution, and so if two iterates agree to, say, p decimal places, then the true solution also has this same value to this accuracy.)

9. Verify that the conditions of Exercise 8 are satisfied for the iteration $x_{n+1} = \cos x_n$. To what accuracy do you *know* the solution after 10 iterations beginning with $x_0 = 1$?

2.3 NEWTON'S METHOD

To obtain an iteration with rapid convergence to the solution of the equation $f(x) = 0$, we seek a rearrangement of that equation satisfying (2.2.9) with the additional property that $g'(s) = 0$. Now

$$f(x) = 0 \leftrightarrow x = x + \lambda(x)f(x)$$

provided that λ has no zeros in the region of the solution. We choose λ so that the iteration function

$$g(x) = x + \lambda(x)f(x) \tag{2.3.1}$$

has zero derivative at the solution s of the equation $f(x) = 0$. Now

$$g'(x) = 1 + \lambda'(x)f(x) + \lambda(x)f'(x)$$

from which we deduce that

$$g'(s) = 1 + \lambda'(s)f(s) + \lambda(s)f'(s) = 1 + \lambda(s)f'(s)$$

since $f(s) = 0$. It follows that $g'(s) = 0$ if $\lambda(s) = -1/f'(s)$, which is certainly satisfied by the choice

$$\lambda(x) = \frac{-1}{f'(x)}$$

provided that f' has no zeros in the neighborhood of the solution.

This yields the iteration function for Newton's method which is given by

$$g(x) = x - \frac{f(x)}{f'(x)} \tag{2.3.2}$$

so the iteration formula itself is

$$x_{n+1} = x_n - \frac{f(x_n)}{f'(x_n)} \tag{2.3.3}$$

Example 2.3.1. For the equation $f(x) = x^2 - a = 0$, we have $f'(x) = 2x$, and so Newton's iteration for the solution of this equation is

$$x_{n+1} = x_n - \frac{x_n^2 - a}{2x_n} = \frac{2x_n^2 - x_n^2 + a}{2x_n} = \frac{x_n + a/x_n}{2}$$

which, of course, is the iteration used before.

We will consider other examples shortly. First we give an alternative derivation and prove some convergence theorems for Newton's method. Consider the first-order Taylor expansion of f about the iterate x_n. We have

$$f(x) \simeq f(x_n) + (x - x_n)f'(x_n)$$

and setting this approximation to zero gives the next iterate x_{n+1} as the solution of $f(x_n) + (x - x_n)f'(x_n) = 0$, which is

$$x_{n+1} = x_n - \frac{f(x_n)}{f'(x_n)}$$

as before.

Graphically, this can be represented as choosing x_{n+1} to be the point at which the tangent to the graph $y = f(x)$ at $x = x_n$ cuts the x axis. This is illustrated in Fig. 2.3.2.

Now the fact that Newton's iteration is chosen so that $g'(s) = 0$ implies that the most significant term in the Taylor expansion (2.2.10) for e_{n+1} in terms of e_n is the second-order term. That is,

$$e_{n+1} = e_n^2 \frac{g''(s)}{2!} + e_n^3 \frac{g'''(s)}{3!} + \cdots$$

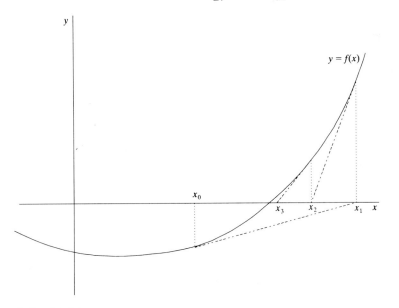

FIGURE 2.3.2
Convergence of Newton's iteration.

and so, if the iteration converges, we see that e_{n+1} is proportional to the square of e_n. That is, the iteration will have second-order convergence.

The question of the convergence of Newton's method can be answered simply—at least locally. Since $g'(s) = 0$, it follows that (provided g' is continuous, which it is if f is twice continuously differentiable) $|g'(x)| < 1$ for all x in some neighborhood of the solution. Corollary 2.2.4 now implies that *provided the initial guess x_0 is good enough*, then Newton's iteration will converge quadratically to the required solution. We state this formally as Theorem 2.3.3.

> **Theorem 2.3.3.** Suppose that f is continuously differentiable, $f(s) = 0$, and $f'(s) \neq 0$. Then there exists an interval $[s - h, s + h]$ within which f' has no zeros and $|g'(x)| < 1$. It follows that Newton's iteration converges quadratically to s.

This local convergence result is not entirely satisfactory in that it demands a starting point which may need to be very close to the solution we seek. Such a starting point can often be found by first using several iterations of the bisection algorithm in the hope that a suitably small interval is obtained. However, we can do better than that in certain cases. This is demonstrated by the next result that establishes conditions under which the convergence can be established less locally.

> **Theorem 2.3.4.** Suppose that f satisfies the following conditions on an interval $[a, b]$:
> (a) $f(a)f(b) < 0$.
> (b) f' has no zeros in $[a, b]$.
> (c) f'' does not change sign in $[a, b]$.
> (d) $|f(a)/f'(a)|, |f(b)/f'(b)| < b - a$.
> Then, for any choice $x_0 \in [a, b]$, Newton's iteration converges to the (unique) solution s of $f(x) = 0$ in $[a, b]$.

Proof. First, from (a), it follows that f has a zero in the interval. Also, since f' is of constant sign in $[a, b]$, f is strictly monotone there, and so this zero is unique.

Without loss of generality we may assume that $f(a) < 0 < f(b)$ and that $f''(x) \geq 0$ for all $x \in [a, b]$. [The other possibilities are all covered by considering $\pm f(\pm x)$.]

It follows that $f'(x) > 0$ throughout the interval and therefore that

$$g'(x) = \frac{f(x)f''(x)}{(f'(x))^2} \quad \begin{cases} \leq 0 & \text{for } x < s \\ \geq 0 & \text{for } x > s \end{cases} \tag{2.3.4}$$

By the MVT, we deduce that if $x < s$, then

$$g(x) - s = g(x) - g(s) = (x - s)g'(\xi) \geq 0$$

since $s > \xi > x$, and so $g'(\xi) \leq 0$. It follows that if $x_0 < s$, then $x_1 \geq s$. Similarly, we find that if $x > s$, then so is $g(x)$. Therefore we have

$$x_n \geq s \quad (n = 1, 2, \ldots) \tag{2.3.5}$$

Next, note that if $x_n > s$, then

$$x_{n+1} = x_n - \frac{f(x_n)}{f'(x_n)} < x_n$$

since $f(x_n), f'(x_n) > 0$.

In summary, we have proved that for $n \geq 1$,

$$x_n > x_{n+1} > s \tag{2.3.6}$$

This bounded monotone sequence must converge, and its limit is a fixed point of g.

It remains to show that it is the required solution s. For this, it is sufficient to establish that the iteration does not leave $[a, b]$ since s is the unique fixed point of g within that interval. The argument used to obtain (2.3.6) yields $g([s, b]) \subseteq [s, b]$. Also, by (d), we have

$$g(a) = a - \frac{f(a)}{f'(a)} < a + (b - a) = b$$

For any $x \in [a, s]$, one last application of the MVT shows that $g(a) - g(x) \geq 0$, and so $g([a, s]) \subseteq [s, b]$, too. Thus for any initial point x_0, the iteration remains in the interval $[a, b]$, and the proof is complete.

It may have been noticed that we did not use the part of condition (d) relating to the other endpoint b. This condition is necessary for some of the other cases mentioned at the beginning of the proof.

At the beginning of this section, we examined the special case of finding square roots by Newton's method. Another important elementary function which can usefully be evaluated in this way is the reciprocal function. To compute the reciprocal of a real number c, we must solve the equation

$$f(x) = \frac{1}{x} - c = 0$$

Now Newton's iteration for this equation is

$$x_{n+1} = x_n(2 - cx_n) \tag{2.3.7}$$

(See Exercise 5.)

Example 2.3.5. Use Newton's iteration to find $\frac{1}{7}$ with the initial guess $x_0 = 0.2$.

Solution. The next few iterates are

0.120 000 0	0.139 200 0	0.142 763 5
0.142 857 0	0.142 857 1	0.142 857 1

so that we see very rapid convergence to the correct result (to the precision shown) after just five iterations.

Now the errors in iteration (2.3.7) satisfy

$$e_{n+1} = x_{n+1} - \frac{1}{c} = -c\left(x_n - \frac{1}{c}\right)^2 = -ce_n^2 \tag{2.3.8}$$

and so if the sequence converges, we see that the convergence is indeed quadratic. The remaining question is to ascertain the range of x_0 for which the iteration converges. It is easy to verify that the conditions of Theorem 2.3.4 are satisfied for the interval $[1/2c, 3/2c]$.

Of course this approach is useful for more than just elementary function evaluations:

Example 2.3.6. Solve the equation $e^x - 2x - 1 = 0$ by Newton's method.

Solution. Here, $f'(x) = e^x - 2$, and so the iteration becomes

$$x_{n+1} = x_n - \frac{e^{x_n} - 2x_n - 1}{e^{x_n} - 2} = \frac{(x_n - 1)e^{x_n} + 1}{e^{x_n} - 2}$$

With the same starting point $x_0 = 1.5$ used in Example 2.2.5, five iterations yield 1.305 902 7, 1.259 058 7, 1.256 439 2, 1.256 431 2, and 1.256 431 2. The greatly improved convergence of Newton's method is readily apparent on comparing these results with those of Example 2.2.5. That the final iterate here is exact to the number of places shown can be deduced in much the same way as for the square root algorithm, since an analysis of this iteration shows that for x_n near the solution, $e_{n+1} < 1.5e_n^2$.

In deriving Newton's method, we made the assumption that $f'(s) \neq 0$. What can we do to recover the quadratic convergence rate in the event that we seek a double zero (or even one of higher multiplicity)?

Recall that a zero, s say, of a function f is said to have *multiplicity* k if

$$f(s) = f'(s) = \cdots f^{(k-1)}(s) = 0 \quad \text{and} \quad f^{(k)}(s) \neq 0$$

Suppose then that we seek the solution s of $f(x) = 0$ and that s is a *double zero*. That is,

$$f(s) = f'(s) = 0 \quad \text{but} \quad f''(s) \neq 0 \tag{2.3.9}$$

and consider the convergence rate of the Newton iteration given by (2.3.2) and (2.3.1). Now

$$g'(x) = \frac{f(x)f''(x)}{(f'(x))^2}$$

and, using L'Hôpital's rule, we find that

$$g'(s) = \lim_{x \to s} \frac{f(x)f''(x)}{(f'(x))^2} = \lim_{x \to s} \frac{f(x)f'''(x) + f'(x)f''(x)}{2f'(x)f''(x)}$$

$$= \frac{1}{2} + \lim_{x \to s} \frac{f(x)f'''(x)}{2f'(x)f''(x)}$$

Since this last limit, by L'Hôpital's rule again, turns out to be zero, we have

$$g'(s) = \frac{1}{2} \tag{2.3.10}$$

It follows that if we define a new iteration function g_2 by

$$g_2(x) = x - \frac{2f(x)}{f'(x)} \tag{2.3.11}$$

then

$$g_2'(x) = 1 - \frac{2f'(x)}{f'(x)} + \frac{2f(x)f''(x)}{(f'(x))^2}$$

and therefore, using (2.3.10), $g_2'(s) = 0$.

A similar argument can be applied to higher multiplicities of zeros. If s is a zero with multiplicity k, then the modified Newton iteration

$$x_{n+1} = g_k(x_n) = x_n - \frac{kf(x_n)}{f'(x_n)} \tag{2.3.12}$$

will exhibit quadratic convergence to s.

Example 2.3.7. Consider the evaluation of arcsin 1 by solution of the equation

$$\sin x - 1 = 0$$

which we know has a double zero at $\pi/2$.

(*a*) The standard Newton iteration generates the sequence by

$$x_{n+1} = x_n - \frac{f(x_n)}{f'(x_n)} = \frac{x_n \cos x_n - \sin x_n + 1}{\cos x_n}$$

With the starting value $x_0 = 1.5$, this yields the following values for the next four iterates:

$$1.535\,413\,0 \qquad 1.553\,106\,5 \qquad 1.561\,951\,6 \qquad 1.566\,374\,0$$

which is converging very slowly to the solution.

(*b*) Using the modified method for the multiplicity 2, we take

$$x_{n+1} = x_n - \frac{2f(x_n)}{f'(x_n)} = \frac{x_n \cos x_n - 2\sin x_n + 2}{\cos x_n}$$

For the same initial point $x_0 = 1.5$, we get for the next three iterates 1.570 825 9, 1.570 796 4, and 1.570 796 4 in which we see that the rapid convergence of Newton's method has indeed been recovered. Note that this "converged" value is out by 1 in the final place because the particular calculator used returns the (exact) value 1 for sin 1.570 796 4.

Of course this recovery of quadratic convergence is dependent on our knowing *in advance* of the computation that we seek a multiple zero. While this may sometimes be true because of some particular knowledge of the equation and its source, it cannot be assumed to be the normal situation. This raises the question of the practical value of this modified Newton algorithm.

Is it possible to ascertain that Newton's method is converging only linearly *from the numerical output* of the algorithm?

The detailed derivation of (2.3.12) shows that for a zero of multiplicity k, the standard Newton iteration function satisfies

$$g'(s) = \frac{k-1}{k} \tag{2.3.13}$$

[Compare (2.3.10) for the case of a double zero, $k = 2$.] The question is how do we determine k from the behavior of the iteration itself? To answer this we must develop a part of the theory we will adopt later for the acceleration of convergence of linearly convergent sequences.

Suppose then that we have an iterative sequence (x_n) which is converging linearly to s. It follows that as the limit is approached, the errors are related (for n sufficiently large) by $e_{n+1} \simeq ce_n$ for some constant c. Hence the difference between two successive iterates satisfies

$$\Delta x_n = x_{n+1} - x_n = (x_{n+1} - s) - (x_n - s)$$

$$= e_{n+1} - e_n \simeq e_n(c - 1) \tag{2.3.14}$$

from which we deduce in turn that

$$\frac{\Delta x_{n+1}}{\Delta x_n} \simeq \frac{e_{n+1}}{e_n} \simeq c \tag{2.3.15}$$

It follows that the ratio of successive differences will be approximately constant and that this constant is the linear convergence rate.

In the case of a root of multiplicity k, (2.3.13) and (2.3.15) show that for Newton's method this ratio will be approximately $(k-1)/k$. If the observed ratio of these differences is indeed a constant, $\bar{c}$ say, then we can recover k as the nearest integer to $1/(1-\bar{c})$.

For the above example the first few iterates of the standard Newton algorithm yield

$$\Delta x_0 = 1.535\,413\,0 - 1.5 = 0.035\,413\,0$$

$$\Delta x_1 = 1.553\,106\,5 - 1.535\,413\,0 = 0.017\,693\,5$$

$$\Delta x_2 = 1.561\,951\,6 - 1.553\,106\,5 = 0.008\,845\,1$$

$$\Delta x_3 = 1.566\,374\,0 - 1.561\,951\,6 = 0.004\,422\,4$$

which have the successive ratios 0.4996, 0.4999, and 0.5000 to four decimal places. This suggests quite clearly that the sequence is converging with a linear rate of $\frac{1}{2}$ and therefore that the root we seek is of multiplicity 2.

It is worth noting here that the above approach could be misleading if the initial estimate of the solution is very inaccurate. In such circumstances it is fairly common for Newton's method still to converge but with a very slow rate over the first few iterations. This could lead the unwary into the false conclusion that the required solution is a multiple one. Nonetheless, the inclusion of the computation of the ratios of successive differences is a worthwhile *and cheap* addition to a program which may be used to recover rapid convergence when a multiple zero is identified. It must also be observed that the derivative of f will be very small in the region of the solution, and so there is a serious risk of numerical instability or even overflow from the division.

Example 2.3.8. The equation $x^4 - 1.3x^3 - 2.97x^2 + 5.929x - 2.662 = 0$ has a solution between 1 and 2. We use Newton's method to locate this root.

Taking $x_0 = 1.5$, the next four iterations, their differences, and their ratios are given by:

n	x_n	Δx_n	$\Delta x_n / \Delta x_{n-1}$
0	1.500 000 000		
		−0.128 440 367	
1	1.371 559 633		0.68634
		−0.088 153 136	
2	1.283 406 497		0.68084
		−0.060 017 992	
3	1.223 388 505		0.67665
		−0.040 611 313	
4	1.182 777 192		

The ratios are all in the region of 0.68, and taking k to be the nearest integer to $1/(1 - 0.68)$ we get $k = 3$. That is, the results above imply that the solution sought is a triple root of the equation. The modified Newton algorithm for a triple root starting with x_4 above yields the next few iterates as

$$1.100\,711\,451 \qquad 1.100\,000\,054 \qquad 1.100\,000\,000 \qquad 1.100\,000\,000$$

The convergence of the modified method has indeed been very fast and the true triple root of the equation, which is $(x - 1.1)^3(x + 2) = 0$, has been found to full accuracy.

EXERCISES 2.3

1. Perform the first five iterations of Newton's method for the solution of $x = \cos x$ from the initial point x_0.

2. Prove the local convergence Theorem 2.3.3 for Newton's method.

3. Find e to five decimal places by solving $\ln x = 1$ using Newton's method starting from 2.5.

4. Write a program to use Newton's method to solve $e^x - Nx - 1 = 0$ $(N = 2, 3, \ldots, 10)$ with appropriate starting values in $[1, N]$.

5. Show that Newton's iteration for finding the reciprocal of a real number c, that is, solving the equation $f(x) = 1/x - c = 0$, is $x_{n+1} = x_n(2 - cx_n)$. Show that the iteration converges for $x_0 \in [1/2c, 3/2c]$.

6. The equation $9x^3 - 2.1x^2 - 9.2x + 4.4 = 0$ has a so-lution in $[0, 1]$. By computing the ratios of differences of Newton iterates starting at 0.5, find the multiplicity of this solution, and use the appropriately modified Newton's method to locate it accurately.

7. Suppose that the equation $f(x) = 0$ has a solution of multiplicity k at $x = s$. By first rewriting $f(x)$ as $(x - s)^k f_k(x)$, show that the standard Newton iteration will have linear convergence rate $(k - 1)/k$. Deduce that the modified Newton algorithm exhibits a quadratic rate of convergence.

8. Write a Turbo Pascal program to implement the modified Newton method for a double zero. Verify that it has quadratic convergence for the evaluation of arcsin 1.

2.4 THE SECANT METHOD

One of the major drawbacks of Newton's method for general equation solving is the need for the derivative of the function whose zero is sought. For the particular "one-off" examples we have dealt with so far this has caused no difficulty, but this would not be the case if the equation were very complicated or was itself the result of a part of a larger computational process. For these situations it is still necessary to have efficient techniques at our disposal.

One of the best approaches is to use simple approximations to the derivative values in Newton's method. If x_n and x_{n-1} are two (good) approximations to the solution s of the equation $f(x) = 0$, then

$$f'(x_n) \simeq \frac{f(x_n) - f(x_{n-1})}{x_n - x_{n-1}} \tag{2.4.1}$$

and so we may replace the Newton iteration formula (2.3.3) with

$$x_{n+1} = x_n - \frac{f(x_n)(x_n - x_{n-1})}{f(x_n) - f(x_{n-1})} = \frac{x_{n-1}f(x_n) - x_n f(x_{n-1})}{f(x_n) - f(x_{n-1})} \tag{2.4.2}$$

The use of this iterative formula for the solution of equations is known as the *secant method*. One drawback of this approach is the need for the previous two iterates in generating the new one. It is usual, though not essential, to start the method with two initial points which bracket the solution of the equation. It can be shown that the eventual rate of convergence of the secant method is *superlinear*; that is,

$$e_{n+1} \simeq c e_n^\alpha \qquad (2.4.3)$$

where $\alpha > 1$. But the convergence rate is no longer quadratic, so $\alpha < 2$.

We will indicate shortly that the actual value of α for the secant method is (approximately) 1.62.

The geometric interpretation of the secant method is similarly straight-forward to that of Newton's method. The latter uses the tangent to the graph of f at x_n to generate x_{n+1} as the point at which this tangent line cuts the x axis. For the secant method, we use the two most recent iterates x_{n-1} and x_n to generate x_{n+1}. This new iterate is at the point where the chord joining the points $(x_{n-1}, f(x_{n-1}))$ and $(x_n, f(x_n))$ cuts the axis. In the limit we anticipate that this chord will be very similar to the tangent since x_{n-1} and x_n will be close together. This proximity of the points and the similarity between the chord and the tangent explain the rapid convergence of the secant algorithm.

Example 2.4.1. Use the secant method to find $\sqrt{7}$ given that $2 < \sqrt{7} < 3$.

Solution. Program 2.4.2 below implements the secant method for this case and yields the following results for the next six iterations starting with $x_0 = 2$ and $x_1 = 3$:

2.600 000 000 0	2.642 857 142 9	2.645 776 566 8
2.645 751 297 2	2.645 751 311 1	2.645 751 311 1

If the starting values are reversed (that is, $x_0 = 3$ and $x_1 = 2$), then we get similar but not identical results:

2.600 000 000 0	2.652 173 913 0	2.645 695 364 2
2.645 751 243 2	2.645 751 311 1	2.645 751 311 1

In either case we see rapid convergence to the true solution without the need for any differentiation. The slight variation can be accounted for by the fact that in the first case the iterates x_1 and x_2 bracket the true solution, while for the second case they do not. In the case where the two iterates bracket the solution, the next iterate is obtained by linear interpolation, while *extrapola-tion* is used for the other case. We will see later that interpolation is a much more reliable process, and so it is desirable for the iterates to retain the bracketing property until the error has been reduced further. Since the ordering of the two initial points is arbitrary, it is sometimes thought desirable to order them so that the bracketing property is retained by x_1 and x_2. In fact, for this simple example the secant method has required just two iterations more than Newton's method to obtain full machine accuracy.

The extrapolation feature of the secant method can on occasion cause difficulties with its convergence—either slowing it down or even preventing it.

There are modifications to the algorithm, most notably one due to Brent (1973) which uses a combination of secant and bisection steps to preserve the interpolation property as much as possible without slowing the convergence down to that of the bisection technique. We do not dwell on the details of this method here.

Program 2.4.2

```
program secant_method;
var iter0, iter1, iter2: real;
    i, niter: integer;
function f(x: real): real; begin f:=x*x − 7; end;
procedure secant (var a, b, c: real);
begin
    c:=(a*f(b) − b*f(a))/(f(b) − f(a));
    a:=b; b:=c;
end;
begin
    write ('Input the number of iterations wanted ');
    readln(niter);
    write('Input the two starting points ');
    readln(iter0, iter1);
    for i:=1 to niter do begin
        secant(iter0, iter1, iter2);
        writeln(i:5, iter2:15:10);
    end;
end.
```

Remarks

1. It should be observed that Program 2.4.2 is not at all robust. For a more robust version we need to replace the fixed For loop with a While or Repeat loop testing for (numerical) convergence to within a given tolerance. This is essential for the secant method because of the division by $f(x_n) - f(x_{n-1})$ which, of course, may become very small, or even zero.

2. The convergence criterion used in the secant method must also be carefully thought out. The nature of the method makes it quite possible to get two successive iterates which are very close together *but on the same side* of the solution and removed from it. It is often desirable, therefore, to incorporate into the convergence test that both $|x_{n+1} - x_n|$ and $|x_{n+1} - x_{n-1}|$ are within the tolerance. While this does still not *guarantee* that the true solution has been found, it is a very strong indicator.

We turn now to the analysis of the convergence rate for this algorithm. Our chief objective is to indicate why the appropriate value of α in (2.4.3) is 1.62. As usual, we denote by e_n the error in the estimate x_n of the solution s. That is,

$$e_n = x_n - s \qquad (2.4.4)$$

Now, from (2.4.2) we obtain

$$x_{n+1} - s = \frac{x_{n-1} f(x_n) - x_n f(x_{n-1})}{f(x_n) - f(x_{n-1})} - s$$

which yields, after simplification of the right-hand side,

$$e_{n+1} = \frac{e_{n-1} f(x_n) - e_n f(x_{n-1})}{f(x_n) - f(x_{n-1})} \tag{2.4.5}$$

Now, from the MVT we have, for each n, since $f(s) = 0$, that

$$f(x_n) = f(x_n) - f(s) = (x_n - s) f'(\xi_n) = e_n f'(\xi_n) \tag{2.4.6}$$

where ξ_n is some point between x_n and s. Substituting (2.4.6) for both n and $n-1$ into the numerator of (2.4.5), we now have

$$e_{n+1} = \frac{e_n e_{n-1} f'(\xi_n) - e_{n-1} e_n f'(\xi_{n-1})}{f(x_n) - f(x_{n-1})} \tag{2.4.7}$$

which reduces to

$$e_{n+1} = e_n e_{n-1} M_n \tag{2.4.8}$$

where

$$M_n = \frac{f'(\xi_n) - f'(\xi_{n-1})}{f(x_n) - f(x_{n-1})}$$

It can be shown that this quantity is bounded—provided we are considering a convergent iteration. It follows that there exists a constant C, say, for which (2.4.8) implies

$$|e_{n+1}| \le C |e_n| |e_{n-1}|$$

which on multiplication by C becomes

$$|C e_{n+1}| \le |C e_n| |C e_{n-1}| \tag{2.4.9}$$

Let us suppose that the initial points of our iteration are chosen sufficiently close to the solution s so that

$$|C e_0|, |C e_1| \le \delta < 1 \tag{2.4.10}$$

Repeated substitution into (2.4.9) using (2.4.10) establishes that

$$|C e_2| \le |C e_1| |C e_0| \le \delta^2$$
$$|C e_3| \le |C e_2| |C e_1| \le \delta^2 \delta = \delta^3$$
$$|C e_4| \le |C e_3| |C e_2| \le \delta^3 \delta^2 = \delta^5$$

In general, we have

$$|C e_{n+1}| \le \delta^{p_n} \delta^{p_{n-1}} = \delta^{p_{n+1}} \tag{2.4.11}$$

It follows that the sequence of powers (p_n) satisfies

$$p_{n+1} = p_n + p_{n-1} \tag{2.4.12}$$

which is to say they are members of a Fibonacci sequence—in fact, the classical one with $p_0 = p_1 = 1$.

Next, if the errors are to satisfy (2.4.3), then it must follow that the ratio of successive members of this sequence must approach the limit α. Therefore, we have

$$\lim \frac{p_{n+1}}{p_n} = \lim \frac{p_n}{p_{n-1}} = \alpha \qquad (2.4.13)$$

Dividing (2.4.12) by p_n and letting $n \to \infty$ then yields the equation

$$\alpha = 1 + \frac{1}{\alpha} \qquad \text{or} \qquad \alpha^2 - \alpha - 1 = 0 \qquad (2.4.14)$$

which has the solutions

$$\alpha = \frac{1 \pm \sqrt{5}}{2}$$

It is clear that the p_n remain positive and so therefore do their ratios; it follows that

$$\alpha = \frac{1 + \sqrt{5}}{2} = 1.6180 \qquad \text{to four decimal places} \qquad (2.4.15)$$

We remark that this last stage of the analysis was based on the fact that if the errors satisfy (2.4.3), then the ratios of the terms of the Fibonacci sequence converge to α. That they do can be deduced directly from the solution of the recurrence relation (2.4.12) which is of the form

$$p_n = A\left(\frac{1 + \sqrt{5}}{2}\right)^n + B\left(\frac{1 - \sqrt{5}}{2}\right)^n$$

Example 2.4.3. The following table illustrates the fact that the secant algorithm has a superlinear rate of convergence with $\alpha \approx 1.62$. The particular equation solved here is $e^x - 10x - 1 = 0$ with two initial points $x_0 = 3$ and $x_1 = 6$. The iterates themselves are tabulated along with their errors and the computed value of e_n / e_{n-1}^α.

n	x_n	e_n	e_n / e_{n-1}^α
2	3.092 667 367 0	0.522 283 060 1	0.12775
3	3.174 286 779 7	0.440 663 647 4	1.26212
4	3.855 852 395 1	0.240 901 968 0	0.90863
5	3.538 197 151 4	0.076 753 275 7	0.77004
6	3.603 027 822 6	0.011 922 604 5	0.76298
7	3.615 596 552 4	0.000 646 125 3	0.84449
8	3.614 945 135 7	0.000 005 291 4	0.77778
9	3.614 950 424 7	0.000 000 002 4	0.82968

The next iterate is exact to the number of figures shown. The entries in the right-hand column are approximately constant after a small number of iterations. This suggests that the constant c of (2.4.3) is somewhere around 0.8 for this particular example. Note that this table requires the use of the solution itself and therefore could not be a part of the solution process. Its inclusion here is purely illustrative.

Before leaving the secant method, we should comment on the formula (2.4.2) used for the approximation of the derivative. This is the simplest of all numerical approximations to the first derivative of a function and can be seen to be completely satisfactory for the present purpose. However, the task of numerical differentiation is generally a very difficult one which is fraught with the dangers of numerical instability. Fortunately, it turns out that many of the important uses of numerical differentiation—within the numerical solution of differential equations and optimization—are not too sensitive to the precision of the approximate differentiation routine used. We will consider these topics in greater detail later.

EXERCISES 2.4

1. Perform the first five iterations of the secant method for the solution of $x^2 - 13 = 0$ with the initial points 3 and 4.
2. Show that the secant method can be applied to the operation of reciprocation without involving any division. Compare its performance for this task with that of Newton's method.
3. Estimate the constants c of Eq. (2.4.3) when the secant method is used for computing the reciprocals of $2, 3, \ldots, 10$.

4. Use the secant method to find the positive solutions of the equations $e^x - Nx - 1 = 0$ for $N = 2, 3, 4, \ldots, 10$. Compare this with the Newton iteration in each case.
5. Derive the iteration formula for the secant method by writing down the equation for the straight line joining the points $(x_{n-1}, f(x_{n-1}))$ and $(x_n, f(x_n))$ and finding the point at which this line crosses the x axis.

2.5   ACCELERATION OF CONVERGENCE

In this section we discuss the possibility of improving the rate of convergence for a slowly convergent sequence of iterates. The approach we adopt is due to Aitken and is an application of a broader technique known as Richardson extrapolation.

In essence the idea is to use the terms of the sequence and their rate of change to predict the future behavior of the iteration. The most striking, though not the most useful, application of this process is that if it is applied to the sequence of partial sums of a convergent geometric series, then the exact sum of the infinite series is obtained immediately. There is a drawback to this spectacular acceleration—it also delivers the "exact sum" of a divergent geometric series immediately! So care is certainly needed.

The process is applicable to any linearly convergent sequence. Although we are only interested here in iterative sequences, since the method has wider applicability, we will derive the acceleration formula and analyze it for a general sequence.

Suppose then that the sequence (x_n) is linearly convergent to the limit s. It follows that there exists a constant c, with $|c| < 1$, such that for sufficiently large values of n,

$$e_{n+1} \simeq c e_n \qquad (2.5.1)$$

where, as usual, $e_n = x_n - s$. From (2.5.1) we can deduce that

$$e_{n+1}^2 \simeq e_{n+2} e_n \qquad (2.5.2)$$

or, equivalently, that

$$(x_{n+1} - s)^2 \simeq (x_{n+2} - s)(x_n - s) \tag{2.5.3}$$

Expanding the two sides of (2.5.3), we get

$$x_{n+1}^2 - 2x_{n+1}s + s^2 \simeq x_{n+2}x_n - (x_{n+2} + x_n)s + s^2$$

from which we find that

$$s \simeq \frac{x_{n+2}x_n - x_{n+1}^2}{x_{n+2} - 2x_{n+1} + x_n} \tag{2.5.4}$$

Note that the denominator of this expression is just the second difference $\Delta^2 x_n$. The approximation to s given by the right-hand side of (2.5.4) will be denoted x_{n+2}^A and the errors $x_n^A - s$ by e_n^A. We will establish that the sequence (x_n^A) is also convergent with the same limit s—but with a significantly faster rate of convergence. The precise result is given as Theorem 2.5.1.

Theorem 2.5.1. Let (x_n) be a linearly convergent sequence with the limit s. Then the sequence (x_n^A) converges to the same limit and

$$\frac{e_n^A}{e_n} \to 0 \qquad \text{as } n \to \infty \tag{2.5.5}$$

Proof. Firstly, we establish that the relationship between e_n^A and e_n is precisely the same as that for x_n^A and x_n; that is,

$$e_{n+2}^A = \frac{e_{n+2}e_n - e_{n+1}^2}{e_{n+2} - 2e_{n+1} + e_n} \tag{2.5.6}$$

Now, $\Delta^2 e_n = e_{n+2} - 2e_{n+1} + e_n = x_{n+2} - 2x_{n+1} + x_n = \Delta^2 x_n$ and

$$\begin{aligned}
e_{n+2}e_n - e_{n+1}^2 &= (x_{n+2} - s)(x_n - s) - (x_{n+1} - s)^2 \\
&= x_{n+2}x_n - x_{n+1}^2 - s\Delta^2 x_n \\
&= x_{n+2}^A \Delta^2 x_n - s\Delta^2 x_n \\
&= e_{n+2}^A \Delta^2 e_n
\end{aligned}$$

The identity (2.5.6) follows.

Next, since (x_n) is linearly convergent, it follows that there exists a constant c, with $|c| < 1$, and a sequence (δ_n) convergent to 0 such that $e_{n+1} = (c + \delta_n)e_n$. From (2.5.6) we now have

$$\begin{aligned}
\frac{e_{n+2}^A}{e_{n+2}} &= \frac{e_n - e_{n+1}^2/e_{n+2}}{\Delta^2 e_n} = \frac{1 - (c + \delta_n)/(c + \delta_{n+1})}{(c + \delta_{n+1})(c + \delta_n) - 2(c + \delta_n) + 1} \\
&\to \frac{0}{(c - 1)^2}
\end{aligned}$$

and, since $c \neq 1$, this completes the proof.

For computational purposes it is desirable to rewrite formula (2.5.4) in the form

$$x_{n+2}^A = x_{n+2} - \frac{(\Delta x_{n+1})^2}{\Delta^2 x_n} \tag{2.5.7}$$

(See Exercise 3.)

Example 2.5.2. Apply Aitken's Δ^2 process to the iteration $x_{n+1} = \cos x_n$ for the solution of the equation $x - \cos x = 0$ with $x_0 = 0$.

Solution. The results of the first few iterations together with their differences and the values x_n^A are:

n	x_n	Δx_n	$\Delta^2 x_{n-1}$	x_n^A
0	0.000 000			
		1.000 000		
1	1.000 000		$-1.459\,698$	
		$-0.459\,698$		
2	0.540 302		0.776 949	0.685 073
		0.317 251		
3	0.857 553		$-0.520\,514$	0.728 010
		$-0.203\,263$		
4	0.654 290		0.342 454	0.733 665
		0.139 191		
5	0.793 480		$-0.231\,302$	0.736 906
		$-0.092\,112$		
6	0.701 369			0.738 050

It is immediately clear that the right-hand column of values of the accelerated sequence are indeed settling much more quickly to a value close to the true solution, which is $0.739\,085$ to six decimal places.

The approach used in the above example is a very inefficient use of the acceleration available through this extrapolation process. Once the entry x_2^A has been evaluated (and is believed to be a better estimate of the true solution), the obvious practical approach is to use this value in place of x_3 to generate the next few iterates.

For the above example, this would lead to the following computation:

$$x_0 = 0.000\,000 \qquad x_1 = 1.000\,000 \qquad x_2 = 0.540\,302$$

as before, from which we deduce

$$x_3 = x_2^A = 0.685\,073 \qquad x_4 = 0.774\,373 \qquad x_5 = 0.714\,860$$

The acceleration is then applied to these three iterates to yield

$$x_6 = x_5^A = 0.738\,660 \qquad x_7 = 0.739\,371 \qquad x_8 = 0.738\,892$$

The next two iterations now yield the values $0.739\,085$ and $0.739\,085$, so the exact solution (to this precision) has been found with very much less effort than the original function iteration would require.

The above scheme is implemented in the following program.

Program 2.5.3

```
Program Aitken;
var a0, a1, eps: real;
function g(x: real): real;
       begin g:=cos(x); end;
procedure Aitken_Iteration(xin: real; var xout: real);
var b, c: real;
begin
     b:=g(xin); c:=g(b);
     xout:=c − sqr(c − b)/(xin + c − 2 ∗ b);
end;

begin
     eps:=1e − 6;
     a1:=0;
     repeat
          a0:=a1;
          Aitken_iteration(a0, a1);
          writeln(a0:15:10, a1:15:10);
     until abs(a1 − a0) < eps;
end.
```

Note that the definition of the function *g* is not essential for this example but is illustrative of the more general situation. This program yields results agreeing with those quoted above and stops after four iterations with the final value 0.739 085 133 2. This final value is in fact accurate to about 10 decimal places, but the next few iterations (if a tighter convergence criterion is adopted) oscillate around this value due to the effect of cancellation in the computation of the second difference in the denominator of (2.5.7), which leads to potential growth in the rounding error.

EXERCISES 2.5

1. Perform the first four iterations of Aitken's method for the solution of $e^x − 2x − 1 = 0$ with $x_0 = 1.5$.

2. Show that $\Delta^2 e_n = \Delta^2 x_n$.

3. Derive the difference form of Aitken's formula, (2.5.7).

4. Apply Aitken's algorithm to the iterations $x_{n+1} = \ln(1 + Nx_n)$ for the solution of $e^x − Nx − 1 = 0$ for $N = 2, 3, \ldots, 10$. Compare the performance with the standard iterations and with the results obtained by the secant method.

5. Modify Program 2.5.3 to compute the differences in Aitken's algorithm to a higher precision than the rest of the computation. Test this and the original program for different values of the tolerance eps.

6. Prove that if (x_n) is the sequence of partial sums of the geometric series $a + ax + ax^2 + \cdots$, then $x_2^A = a/(1 − x)$. Give examples to show both the usefulness and the potential dangers of this result.

2.6 NEWTON'S METHOD FOR TWO NONLINEAR EQUATIONS

Before discussing Newton's method itself, we consider, briefly, the general question of the solution of a system of two *nonlinear equations* in two unknowns. More general systems will be discussed separately later; our main reason for introducing this topic now is to develop techniques for finding quadratic factors (and, therefore, roots) of polynomials, which we do in the next section.

The situation, then, is that we seek values x and y which satisfy the two equations

$$f_1(x, y) = 0 \qquad f_2(x, y) = 0 \tag{2.6.1}$$

simultaneously. One way to visualize this problem is to regard each of Eq. (2.6.1) as defining (implicitly) a curve in the xy plane, in which case our task is to find the crossing points of these curves. (Strictly, we should say meeting points, as the two may be mutually tangential and never in fact cross.)

A simple example is the system of equations

$$x^2 - y^2 = 1 \qquad x^3 y^2 = 1 \tag{2.6.2a, b}$$

which are graphed in Fig. 2.6.1. Let us suppose we seek the solution to these equations in the first quadrant which by inspection of the graphs can be seen to be somewhere close to $(1.2, 0.7)$.

One intuitively simple approach to the solution of this system would be to solve, say, $(2.6.2a)$ for x in terms of y and $(2.6.2b)$ for y in terms of x and use repeated substitution in one and then the other to (we hope) find the solution. This is the natural generalization of the "staircase" and "spider's web" pictures of Fig. 2.2.6 in which the line $y = x$ would be replaced by the second equation. There are significant problems with this apparently easy technique. Not least among these is the fact that the ordering of the two equations is arbitrary, but the decision as to which one to use in which role is certainly not. Indeed it

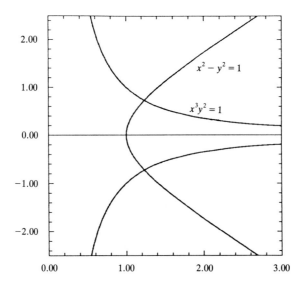

FIGURE 2.6.1
Graphs of the implicit functions defined by Eqs. (2.6.2).

turns out that if using, say, f_1 to find x in terms of y and f_2 for y in terms of x works, then the other choice will diverge. Example 2.6.2 demonstrates this.

Example 2.6.2. Solve the system (2.6.2) using the "alternating variable" iteration described above.

Solution. Solving (2.6.2*a*) for x yields $x = \sqrt{1 + y^2}$, while (2.6.2*b*) gives us $y = \sqrt{1/x^3}$. Beginning with the guess $y_0 = 0.7$, the next few iterations produced by setting

$$x_{n+1} = \sqrt{1 + y_n^2} \qquad y_{n+1} = \sqrt{\frac{1}{x_{n+1}^3}}$$

are $(1.2207, 0.7415)$, $(1.2449, 0.7199)$, and $(1.2322, 0.7311)$ which are converging in a slow and oscillatory manner to the true solution.

However, using the equations in the other order, we get

$$y_{n+1} = \sqrt{x_n^2 - 1} \qquad x_{n+1} = \left(\frac{1}{y_{n+1}^2}\right)^{1/3}$$

which with the more accurate starting value $x_0 = 1.23$ yields the iterates $(1.2493, 0.7162)$, $(1.2127, 0.7488)$, $(1.2855, 0.6861)$, $(1.1529, 0.8078)$, $(1.4483, 0.5737)$, and $(0.9694, 1.0477)$, at which point the next iteration tries to set y to be the square root of a negative quantity.

Clearly, the choice of which equation to use for which variable is critical. We do not pursue this obviously suspect approach any further. There is a more fruitful, and less arbitrary, approach to fixed-point iteration for a system of equations which is, in some ways, a more natural generalization of the single-variable situation. This is derived from considering the original system of equations as a vector equation

$$\mathbf{f(x)} = 0 \tag{2.6.3}$$

where $f: \mathbb{R}^2 \to \mathbb{R}^2$ whose component functions are the scalar functions f_1 and f_2 of (2.6.1).

Such a system of equations can be solved by a fixed-point iteration using a rearrangement of (2.6.3) in the form

$$\mathbf{x = g(x)} \tag{2.6.4}$$

to generate a sequence of points $(x_n, y_n) \in \mathbb{R}^2$ which we hope will converge to the required solution.

Example 2.6.3. Consider the solution of Eqs. (2.6.2) by fixed-point iteration.
The two equations may be rearranged as follows:

$$x^2 - y^2 = 1 \quad \leftrightarrow \quad x = y + \frac{1}{x + y} = g_1(x, y)$$

$$x^3 y^2 = 1 \quad \leftrightarrow \quad y = \sqrt{1/x^3} = g_2(x, y)$$

(Clearly the second of these iteration functions is, in fact, independent of y.) This iteration is very simply implemented as in Algorithm 2.6.4. The results of the first 10 iterations for this example form Table 2.6.5.

Algorithm 2.6.4

Initialize	Define iteration functions g_1 and g_2
	Set tolerance ϵ, starting point (x_1, y_1)
Loop	Repeat
	$(x_0, y_0):=(x_1, y_1)$
	$x_1:=g_1(x_0, y_0)$
	$y_1:=g_2(x_0, y_0)$
	until $\|(x_1, y_1) - (x_0, y_0)\| < \epsilon$
Output	Approximate solution is (x_1, y_1).

This is, of course, only an outline of a simple algorithm. Any realistic implementation must include a limit on the number of iterations and appropriate output in case of failure to converge.

TABLE 2.6.5

n	x_n	y_n
1	1.226 315 789 5	0.760 725 774 3
2	1.263 986 510 4	0.736 370 734 2
3	1.236 281 439 0	0.703 698 040 0
4	1.219 167 410 1	0.727 484 888 6
5	1.241 187 310 3	0.742 856 593 8
6	1.246 877 698 4	0.723 176 008 4
7	1.230 776 383 2	0.718 231 112 7
8	1.231 312 772 0	0.732 371 218 7
9	1.241 618 126 4	0.731 892 713 7
10	1.238 603 890 4	0.722 799 650 1

The values of the iterates here can be seen to be settling down slowly to the solution of the system, but we would still have great difficulty in determining the actual coordinates of this solution to great accuracy.

A similar analysis to that of Sec. 2.2 can be carried out for this situation to determine conditions under which such an iteration will converge. The extra dimension makes this analysis both more difficult and less useful, and we simply state (without proof) one convergence theorem analogous to Theorem 2.2.2. For this we need the following definition of a Lipschitz function of two variables.

Definition 2.6.6. A function $\mathbf{f}: \mathbb{R}^2 \rightarrow \mathbb{R}^2$ is said to satisfy a *Lipschitz condition* with *Lipschitz constant* λ if

$$\|\mathbf{f}(x_0, y_0) - \mathbf{f}(x_1, y_1)\| \leq \|(x_0, y_0) - (x_1, y_1)\|$$

for every $\mathbf{x}_0$ and $\mathbf{x}_1$.

Theorem 2.6.7. Let $\mathbf{I} = I_1 \times I_2 \subseteq \mathbb{R}^2$ be a closed bounded rectangle and suppose that $\mathbf{g}$ is a Lipschitz function on $\mathbf{I}$ with Lipschitz constant $\lambda < 1$ satisfying $\mathbf{g}(\mathbf{I}) \subseteq \mathbf{I}$. Then $\mathbf{g}$ has a unique fixed point $\mathbf{s}$ in $\mathbf{I}$, and the sequence $(\mathbf{x}_n)$ defined for $\mathbf{x}_0 \in \mathbf{I}$ by

$$\mathbf{x}_{n+1} = \mathbf{g}(\mathbf{x}_n) \qquad n = 0, 1, \ldots \tag{2.6.5}$$

converges to $\mathbf{s}$.

The proof of this theorem is almost identical to that of Theorem 2.2.2 except for the use of vector quantities throughout and the use of the Lipschitz condition rather than the MVT. (See Exercise 4.) There are similar results using the two-variable MVT which require that the differential of the iteration function $\mathbf{g}$ must have norm less than unity throughout the rectangle, or two-dimensional interval, $\mathbf{I}$.

For these results, we may use any of the possible norms for $\mathbb{R}^2$. In particular, therefore, we may use the "infinity" or maximum norm,

$$\|(x, y)\|_\infty = \max(|x|, |y|) \tag{2.6.6}$$

in which case the requirement on the differential reduces to the condition

$$\max_i \max_{(x,y) \in \mathbf{I}} |G_{i1}(x, y)| + |G_{i2}(x, y)| \le \lambda < 1 \tag{2.6.7}$$

where $G_{i1} = \partial g_i / \partial x$, $G_{i2} = \partial g_i / \partial y$.

The matrix function G with elements G_{ij} is called the Jacobian matrix for the function $\mathbf{g}$. As with the single-variable case, there are weaker versions of the above theorem which establish the local convergence of the iteration under the condition that the norm of this matrix is less than unity at the solution $\mathbf{s}$.

In the example above, we have

$$g_1(x, y) = y + \frac{1}{x + y} \qquad g_2(x, y) = x^{-3/2}$$

and therefore

$$G_{11}(x, y) = \frac{-1}{(x + y)^2} \qquad G_{12}(x, y) = 1 - \frac{1}{(x + y)^2}$$

$$G_{21}(x, y) = \frac{-3x^{-5/2}}{2} \qquad G_{22}(x, y) = 0$$

It is easy to see that all these are bounded by 1 in the region of the solution near $(1.24, 0.73)$, although $|G_{11}| + |G_{12}| = 1$, and therefore, at best, we may hope this iteration will converge slowly. (Using the 1-norm, we find that $\|G\|_1 < 1$, which does establish the slow convergence found in Example 2.6.3.)

The major difficulty in the fixed-point iteration approach to systems is that it is not usually easy to find an appropriate iteration function. Fortunately, it is relatively straightforward to obtain a generalization of Newton's method which will converge very rapidly—when it does converge. Other methods for the solution of systems of nonlinear equations will be studied later, but for now we content ourselves with a simple derivation of Newton's method and some numerical experiments with it.

Now, the first-order Taylor expansion of the function $\mathbf{f}$ of two variables can be written in component form as

$$f_1(x + h, y + k) \approx f_1(x, y) + hF_{11}(x, y) + kF_{12}(x, y)$$

$$f_2(x + h, y + k) \approx f_2(x, y) + hF_{21}(x, y) + kF_{22}(x, y)$$

(2.6.8)

In matrix form this reduces to

$$\mathbf{f}(\mathbf{x} + \mathbf{h}) \approx \mathbf{f}(\mathbf{x}) + F(\mathbf{x})\mathbf{h}$$

(2.6.9)

The single-variable Newton iteration is derived by setting this approximation to zero. That is precisely the approach here. That is, we set the correction $\mathbf{h}_n$ to $\mathbf{x}_n$ to be the solution of the system of linear equations

$$F(\mathbf{x}_n)\mathbf{h}_n = -\mathbf{f}(\mathbf{x}_n)$$

(2.6.10)

or, equivalently, we put

$$\mathbf{x}_{n+1} = \mathbf{x}_n - F^{-1}(\mathbf{x}_n)\mathbf{f}(\mathbf{x}_n)$$

(2.6.11)

The form of the iteration given by (2.6.11) is precisely the general form which may be used for a system of any dimension, and the derivation of the general formula follows just the same reasoning as we have used here. It is important to note that generally we do not compute the inverse Jacobian F^{-1} but instead solve the linear system (2.6.10) since this is a *much* simpler computational task.

For the case of the 2×2 system under present consideration, it is easy to obtain the solution in the form

$$h_1 = \frac{-D_1}{D} \qquad h_2 = \frac{-D_2}{D}$$

(2.6.12)

where the quantities D, D_1, and D_2 are the determinants given by

$$D = \det F(x_n, y_n) = F_{11}F_{22} - F_{12}F_{21}$$

$$D_1 = f_1 F_{22} - f_2 F_{12} \qquad D_2 = f_2 F_{11} - f_1 F_{21}$$

(2.6.13)

all evaluated at $\mathbf{x}_n$.

Example 2.6.8. Solve the system (2.6.2) using Newton's iteration.

Solution. The system is

$$f_1(x, y) = x^2 - y^2 - 1 = 0$$

$$f_2(x, y) = x^3 y^2 - 1 = 0$$

and the various partial derivatives are

$$F_{11}(x, y) = 2x \qquad F_{12}(x, y) = -2y$$

$$F_{21}(x, y) = 3x^2 y^2 \qquad F_{22}(x, y) = 2x^3 y$$

The first few iterations from the initial point $x_0 = (1.2, 0.7)$ are given in Table 2.6.10(a), which shows rapid convergence to the solution. The second set of results show the very different behavior from the poor starting point $(0.12, 0.01)$. The results were produced by an implementation of the following algorithm.

Algorithm 2.6.9. Newton's method for two equations

Initialize	Define $\mathbf{f}$ and its Jacobian F
	Set tolerance ϵ,
	maximum iteration count maxits,
	starting point (x_1, y_1)
	it:$=0$.

Loop	Repeat
	$(x_0, y_0):=(x_1, y_1)$
	it: $=$ it $+ 1$
	Evaluate f_1, f_2, F_{11}, F_{12}, F_{21}, F_{22} at (x_0, y_0)
	$D:=F_{11}F_{22} - F_{21}F_{12}$
	$D_1:=f_1F_{22} - f_2F_{12}$
	$D_2:=f_2F_{11} - f_1F_{21}$
	$x_1: = x_0 - D_1/D$
	$y_1: = y_0 - D_2/D$
	until $\|(x_1, y_1) - (x_0, y_0)\| < \epsilon$ or it $=$ maxits.

Output	If it $<$ maxits then "solution is (x_1, y_1)"
	else "failed to converge".

TABLE 2.6.10
Newton iteration for solution of (2.6.2)

n	x_n	y_n
	(a) $x_0 = (1.2, 0.7)$	
1	1.238 263 090 7	0.729 879 584 0
2	1.236 511 641 0	0.727 299 544 5
3	1.236 505 703 5	0.727 286 982 5
4	1.236 505 703 4	0.727 286 982 2
	(b) $x_0 = (0.12, 0.01)$	
1	4 931.085 221 7	49 260.357 217
2	4 914.723 856 0	24 875.347 779
3	4 852.396 258 7	12 910.870 565
4	4 643.585 990 9	7 288.813 946 5
5	4 149.136 293 4	4 808.577 893 4
6	3 460.984 266 4	3 600.573 808 4
7	2 801.356 466 7	2 829.635 087 8
8	2 247.824 322 3	2 253.497 303 2
9	1 799.618.658 7	1 800.753 895 8
10	1 439.967 374 6	1 440.194 187 3

This poor choice of initial point is evidently causing more difficulty, although in this case the desired result would eventually be obtained. The near singularity of the Jacobian at $\mathbf{x}_0$ has resulted in a very large correction. For more complicated systems of equations this can be sufficient to prevent convergence altogether.

Algorithm 2.6.9 can easily be modified for larger systems of equations than the 2×2 systems discussed here. However, it should be pointed out that in the higher-dimensional situation, we would *never* obtain the Newton correction by inverting the Jacobian matrix but would instead solve the system of Eqs. (2.6.10). We discuss the solution of such linear systems in Chap. 6.

A more general treatment of Newton's method and the solution of nonlinear systems of equations appears in Chap. 8. The inclusion of this elementary introduction to the subject at this stage is so that we can make use of the ideas in the numerical solution of polynomial equations.

EXERCISES 2.6

1. From the graph below, it is evident that the system of equations

$$x^4 + xy^3 + y^4 = 1 \qquad x^2 + xy - y^2 = 1$$

has a solution close to $(0.9, 0.6)$. Derive the rearrangement

$$x = (1 - xy^3 - y^4)^{1/4} \qquad y = \frac{x + \sqrt{5x^2 - 4}}{2}$$

and perform the first five iterations using this rearrangement. Is this process converging? Try to justify your answer.

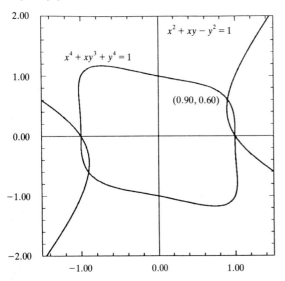

2. Show that the iteration

$$x_{n+1} = \frac{1 - x_n y_n^3 - y_n^4}{x_n^3} \qquad y_{n+1} = \frac{x_n^2 + x_n y_n - 1}{y_n}$$

fails to converge to the solution of the system of Exercise 1 from the very good initial point $(0.9, 0.6)$.

3. Try to find a rearrangement of the equations of Exercise 1 which will converge to the solution near $(0.9, 0.6)$. Try to prove that the iteration you develop does indeed converge, and use it to locate the solution to high accuracy.

4. Prove Theorem 2.6.7. (Use the proof of Theorem 2.2.2 as a model, and modify it for the two-dimensional setting and the Lipschitz condition.)

5. Write a program to apply Newton's method to the solution of the equations in Exercise 1. Obtain the solutions with an error bounded by 10^{-6}.

6. Use Newton's method to solve the system of equations

$$x^2 + y^2 - 2x = 0 \qquad x^2 - y^2 - y = 0$$

Your program should terminate with the appropriate output after 20 iterations or when two successive iterates satisfy $|x_{n+1} - x_n| < \epsilon$, $|y_{n+1} - y_n| < \epsilon$. Use $\epsilon = 10^{-5}$ and three starting points $(1, 1)$, $(1, -1)$, and $(0.6, 1)$. Explain your results.

2.7   POLYNOMIAL EQUATION METHODS

In this section we return to the problem of the solution of a single nonlinear equation in one unknown, but in the very special case of a polynomial equation. Thus we seek solutions to equations of the form

$$a_N x^N + a_{N-1} x^{N-1} + \cdots + a_1 x + a_0 = 0 \qquad (2.7.1)$$

where the coefficients $a_0, a_1, \ldots, a_N$ are real numbers. (This stipulation of real coefficients is more for convenience than out of necessity as much of what is said would apply to complex polynomials too.)

There are many situations in which such problems arise, and the information required may vary greatly. On some occasions, one particular root of the equation may be all that is needed, and we may already have a good idea as to where that is. On other occasions, the largest or smallest root may be the target, while in yet others all the real roots may be required, or even, all the roots—both real and complex. Methods have been developed for all these tasks—with varying degrees of efficiency! We consider just a few of them here.

It should be observed that, in the first of the cases mentioned above, namely, the search for a particular root of (2.7.1) where we know a reasonable approximation to this root, almost certainly the most efficient technique would be a straightforward application of Newton's method to this equation. The only real difficulty that is likely to be encountered is if the desired root turns out to be a multiple one. In this case the modifications described in Sec. 2.3 could be incorporated into the routine.

We will concentrate on the other situations, and we begin with the case where we seek the largest root (in absolute value) of (2.7.1). We study Bernoulli's method which like the QD algorithm that we discuss briefly later is based on the equivalence between polynomial equations and linear constant-coefficient recurrence relations.

Specifically, if the polynomial equation (2.7.1) has roots $\alpha_1, \alpha_2, \ldots, \alpha_N$ which are *all distinct*, then the general solution of the recurrence relation

$$a_N u_{m+N} + a_{N-1} u_{m+N-1} + \cdots + a_1 u_{m+1} + a_0 u_m = 0 \qquad (2.7.2)$$

is given by

$$u_n = A_1 \alpha_1^n + A_2 \alpha_2^n + \cdots + A_N \alpha_N^n \qquad (2.7.3)$$

where the arbitrary constants $A_1, A_2, \ldots, A_N$ may be determined from a set of initial values for (say) $u_0, u_1, \ldots, u_{N-1}$. (In practical application we will often specify values for $u_{-N+1}, u_{-N+2}, \ldots, u_0$.)

We should observe here that the requirement that all the roots be distinct is much stronger than is strictly necessary for the remainder of the argument. We will discuss that point a little later but first develop the theory in the simplest case.

Suppose that the roots are ordered such that

$$|\alpha_1| > |\alpha_k| \qquad (k = 2, 3, \ldots, N) \qquad (2.7.4)$$

and that the initial values are chosen so that the coefficient A_1 in (2.7.3) is nonzero. We see in Exercise 3 that this can be achieved with the initial values

$$u_{-N+1} = u_{-N+2} = \cdots = u_{-1} = 0 \qquad u_0 = 1 \qquad (2.7.5)$$

From (2.7.3), it is apparent that under these conditions the first term $A_1\alpha_1^n$ will dominate as n gets large. It is this domination which allows us to identify α_1 from the sequence of quotients (q_n) defined by

$$q_n = \frac{u_{n+1}}{u_n} \qquad (2.7.6)$$

To see this, we write q_n in full as

$$q_n = \frac{A_1\alpha_1^{n+1} + A_2\alpha_2^{n+1} + \cdots + A_N\alpha_N^{n+1}}{A_1\alpha_1^n + A_2\alpha_2^n + \cdots + A_N\alpha_N^n}$$

which, on dividing numerator and denominator by $A_1\alpha_1^n$ and using the notations

$$R_k = \frac{A_k}{A_1} \qquad \rho_k = \frac{\alpha_k}{\alpha_1} \qquad (2.7.7)$$

reduces to

$$q_n = \frac{\alpha_1 + R_2\rho_2^n\alpha_2 + \cdots + R_N\rho_N^n\alpha_N}{1 + R_2\rho_2^n + \cdots + R_N\rho_N^n} \qquad (2.7.8)$$

Since, by (2.7.4), $|\rho_k| < 1$, it follows that the numerator of (2.7.8) converges to α_1 while the denominator tends to 1 as $n \to \infty$; that is,

$$q_n \to \alpha_1 \qquad \text{as } n \to \infty \qquad (2.7.9)$$

How do we turn this into a method for the approximate location of this largest root α_1? Once a set of starting values has been chosen, say $u_0, u_1, \ldots, u_{N-1}$, then subsequent members of the sequence (u_n) can be generated using a simple rearrangement of (2.7.2):

$$u_n = -\frac{a_{N-1}u_{n-1} + \cdots + a_1 u_{n-N+1} + a_0 u_{n-N}}{a_N} \qquad (2.7.10)$$

The quotients of successive terms can then be computed; we have established that this sequence will converge to α_1 as required.

This method is known as *Bernoulli's method*, and the above analysis thus establishes the convergence of this method to the dominant root of a polynomial equation, provided that the initial values are chosen to ensure that A_1 is nonzero, in the special case where all the roots are distinct. We state a somewhat more general theorem on the convergence of Bernoulli's method a little later.

Example 2.7.1. Consider the polynomial equation $2x^3 - 3x + 2 = 0$.
The corresponding recurrence relation is

$$u_{n+3} = \frac{3u_{n+1} - 2u_n}{2}$$

and we take the initial values, as in (2.7.5), to be $u_{-2} = u_{-1} = 0$, $u_0 = 1$. The next few terms of the sequence (u_n) are then given by

$$u_1 = \frac{3u_{-1} - 2u_{-2}}{2} = 0 \qquad u_2 = \frac{3u_0 - 2u_{-1}}{2} = 1.5$$

$$u_3 = \frac{3u_1 - 2u_0}{2} = -1.0 \qquad u_4 = \frac{3u_2 - 2u_1}{2} = 2.25$$

$$u_5 = \frac{3u_3 - 2u_2}{2} = -3.0$$

from which we get the first few terms of the sequence of quotients which are defined from q_2 onward:

$$q_2 = \frac{u_3}{u_2} = -0.66667 \qquad q_3 = \frac{u_4}{u_3} = -2.25 \qquad q_4 = \frac{u_5}{u_4} = -1.33333$$

Bernoulli's method was used to generate terms up to q_{20} together with the results in Table 2.7.3. Program 2.7.2 implements Bernoulli's method for polynomials of degree at most 20.

Program 2.7.2

```
Program Bernoulli;
type vector = array[0..20] of real;
var
      coeffs, u: vector
      i, j, degree: integer;
Procedure GenerateNextU(n: integer; c: vector; var v: vector);
var k: integer;
begin
      v[n]:=0;
      for k:=0 to n - 1 do v[n]:=v[n] - c[k]*v[k];
      v[n]:=v[n]/c[n];
end;

begin {Main}
      degree:=3;
      u[0]:=0; u[1]:=0; u[2]:=1;
      coeffs[3]:=2; coeffs[2]:=0;
      coeffs[1]:=-3; coeffs[0]:=2;
      for i:=1 to 20 do begin
            GenerateNextU(degree,coeffs,u);
            write(i:5,u[degree]:25);
            if u[degree - 1] = 0 then writeln
                 else writeln(u[degree]/u[degree - 1]:20:6);
            for j:=0 to degree - 1 do v[j]:=v[j + 1];
      end;
end.
```

Remarks

1. The vector u holds just the most recent members of the sequence (u_n), that is, those which are needed for the computation of the next term in that sequence.

2. It is clearly easy to adapt this program to polynomials with other coefficients and different degrees. Note that it is necessary to test for the possibility of individual members of the sequence (q_n) being undefined, although it can be proved that the sequence is well-defined for sufficiently large values of n.

3. The results in Table 2.7.3 show the typical slow convergence of Bernoulli's method to the dominant root, which in this case is in the region of -1.476. That all the entries in the q column are negative is caused by the fact that the signs of the u_n continue to alternate as they did for the first few above. The information gained from the first few iterations of this method could be used to provide a good starting point for Newton's method to locate the root to any required accuracy.

There remain significant difficulties with Bernoulli's method: the rate of convergence is usually slow, we can see even from this very small example that the numerical values of the terms u_n can grow very rapidly (raising the possibility of overflow failure in the program), and the analysis thus far only applies in the very special situation of all roots being distinct and the dominant one being isolated *in absolute value*. This last restriction is one which would be difficult to check in advance of the computation.

The rate of convergence depends critically on the relative magnitudes of the two largest roots. Supposing that the roots have been ordered so that

$$|\alpha_1| > |\alpha_2| > |\alpha_k| \qquad (k = 3, 4, \ldots, N) \qquad (2.7.11)$$

then it is apparent from (2.7.8) that the smaller the value of $|\rho_2|$, the faster will be the convergence.

TABLE 2.7.3
Output from Program 2.7.2; even n

n	u_n	q_{n-1}
2	1.500 000 E + 00	
4	2.250 000 E + 00	−2.250 000
6	4.375 000 E + 00	−1.458 333
8	9.562 500 E + 00	−1.416 667
10	2.109 375 E + 01	−1.454 741
12	4.614 063 E + 01	−1.473 553
14	1.005 234 E + 02	−1.476 928
16	2.188 477 E + 02	−1.476 362
18	4.765 059 E + 02	−1.475 821
20	1.037 634 E + 03	−1.475 672

For the particular example above the relative minimum of the polynomial is positive, and hence there is only one real root. It follows that the other two are complex conjugate and therefore have the same modulus. The product of the absolute values of the roots is 1, and hence we may deduce that the modulus of the complex roots is approximately $\sqrt{1/1.47} \approx 0.825$. Therefore, in this case, we have $|\rho_2| = |\rho_3| \approx 0.825/1.47 \approx 0.56$, which allows for a rate of convergence better than would often be encountered in less well-conditioned problems—nonetheless, the convergence rate is clearly unsatisfactory for accurate location of the root. The rate of convergence can be improved by the use of Aitken's Δ^2 process, and it is reasonably straightforward, though tedious in detail, to establish that if the roots satisfy (2.7.11) so that the two dominant roots are both isolated, then this does indeed lead to improved performance.

Example 2.7.4. Apply Aitken's acceleration to the sequence (q_n) of Example 2.7.1, and compare this with the effect of applying Newton's method to locate the dominant root.

Solution. Now, $q_2 = -0.66667$, $q_3 = -2.25$, $q_4 = -1.33333$, and $q_5 = -1.45833$, and so we get from Aitken's process the approximations

$$q_4^A = -1.66944 \qquad q_5^A = -1.44333.$$

We see that the acceleration is of marginal benefit at best, because in this case, as was pointed out above, the second and third roots are a conjugate pair, and so (2.7.11) is not satisfied. In practice, we would have no way of knowing this in advance.

By contrast, consider the effect of applying Newton's method starting with q_4, that is, at the same point as it would first become possible to apply Aitken's process. For this equation, Newton's method uses the iteration formula

$$x_{n+1} = x_n - \frac{2x_n^3 - 3x_n + 2}{6x_n^2 - 3}$$

which with $x_0 = -1.33333$ yields, to five decimal places, $x_1 = -1.49759$, $x_2 = -1.47610$, $x_3 = -1.47569$, and $x_4 = -1.47569$, which is clearly settling very quickly to the root as we would expect.

We have seen here that Bernoulli's method is likely to be of value in obtaining good initial values for the accurate determination of the roots by other methods; hence, we do not pursue its analysis in detail. However, it is worth stating the following result which establishes that the method can be used to find the modulus of the dominant zero, even when that zero is not isolated, by careful choice of the initial values for the sequence (u_n). For a proof of this result see Henrici (1974), for example.

Theorem 2.7.5. Suppose that the polynomial equation (2.7.1)

$$a_N x^N + a_{N-1} x^{N-1} + \cdots + a_1 x + a_0 = 0$$

has M distinct roots $\alpha_1, \alpha_2, \ldots, \alpha_M$ with respective multiplicities $\nu_1, \nu_2, \ldots, \nu_M$.

If the starting values for Bernoulli's method are

$$u_0 = -\frac{a_{N-1}}{a_N} \qquad u_k = -\frac{(k+1)a_{N-k-1} + a_{N-k}u_0 + a_{N-k+1}u_1 + \cdots + a_{N-1}u_{k-1}}{a_N}$$

for $k = 1, 2, \ldots, N-1$, then the solution of the recurrence relation (2.7.2) is

$$u_n = A_1\alpha_1^n + A_2\alpha_2^n + \cdots + A_M\alpha_M^n$$

and, under the assumption that $|\alpha_1| > |\alpha_k|$ $(k = 2, 3, \ldots, M)$, the sequence (q_n) converges to the dominant root α_1.

The greatest significance of this result lies in the fact that even if the dominant root is a multiple one, Bernoulli's method will still converge to it, provided only that there are no other roots with the same modulus. Also, the fact that this choice of starting values eliminates any polynomial parts in the general solution of (2.7.2) will typically result in (slightly) improved convergence.

The most important case that is still not covered by this result is the possibility of two distinct dominant roots of the same modulus—such as a complex conjugate pair. Although this situation can also be handled by modifications of Bernoulli's method, it is probably better attacked by methods such as Bairstow's which is used to obtain quadratic factors of a polynomial. We study this method shortly.

Before turning to the details of Bairstow's method, we give brief consideration to techniques for the simultaneous location of all the roots of the polynomial equation (2.7.1). Among these, perhaps the simplest is the root-squaring method of Graeffe in which a sequence of polynomials is generated, each of which has as its roots the squares of the roots of the preceding polynomial in the sequence. This is fairly simple to achieve since, if

$$p(x) = C(x - \alpha_1)(x - \alpha_2) \cdots (x - \alpha_N) \tag{2.7.12}$$

then

$$p(x)p(-x) = (-1)^N C^2 (x^2 - \alpha_1^2)(x^2 - \alpha_2^2) \cdots (x^2 - \alpha_N^2)$$

and therefore the polynomial

$$p_1(x) = (-1)^N C^2 (x - \alpha_1^2)(x - \alpha_2^2) \cdots (x - \alpha_N^2) \tag{2.7.13}$$

has zeros at the squares of those of p. By considering p in its standard form

$$p(x) = a_N x^N + a_{N-1} x^{N-1} + \cdots + a_1 x + a_0$$

it is straightforward to obtain the relations for the coefficients $a_0^{(1)}$, $a_1^{(1)}, \ldots, a_N^{(1)}$ of the polynomial p_1; namely,

$$a_j^{(1)} = (-1)^j \left[a_j^2 + 2 \sum_{k=1}^{m} (-1)^k a_{j-k} a_{j+k} \right] \tag{2.7.14}$$

where $m = \min(j, N - j)$. This process can be iterated with the effect that the (relative) separation of the roots is increased. It follows (though we make no attempt to prove it here) that ratios of successive coefficients of the polynomial

p_n, say, will (for sufficiently large n) be close to the roots of that polynomial. Taking the square roots of these n times will yield estimates of the (moduli of the) roots of the original equation.

Graeffe's method has the great virtue of conceptual simplicity—but the great disadvantage that, for all but the simplest of low-degree polynomials, it will fail due to overflow well before useful information is obtained. With a sufficiently robust arithmetic package it may provide a very practical approach to the problem. The interested reader is referred to Henrici (1986) for a much more detailed discussion of this method.

The QD, or quotient-difference, algorithm for the simultaneous location of all roots of the polynomial equation (2.7.1) was originally developed by Rutishauser in 1954. In its simplest form, it too is subject to practical computational difficulties, this time of numerical instability. However, the method can be adapted into a stable form. We content ourselves here with a brief outline of the basic approach which can be viewed as an extension of the ideas of Bernoulli's method.

Denote the sequence (q_n) of Bernoulli's method by $(q_n^{(1)})$, and define the sequence $(e_n^{(1)})$ to be the forward differences of $(q_n^{(1)})$ so that

$$e_n^{(1)} = \Delta q_n^{(1)} = q_{n+1}^{(1)} - q_n^{(1)} \tag{2.7.15}$$

Then for $k = 1, 2, \ldots, N-1$, we define further sequences by

$$q_n^{(k+1)} = \frac{q_{n+1}^{(k)} e_{n+1}^{(k)}}{e_n^{(k)}} \tag{2.7.16}$$

$$e_n^{(k+1)} = q_{n+1}^{(k+1)} - q_n^{(k+1)} + e_{n+1}^{(k)} \tag{2.7.17}$$

[Note that in practice we would not compute the final sequence $(e_n^{(N)})$, which is not relevant to the solution of the original equation.] The process can be summarized as generating an array such as that in Fig. 2.7.6.

$q_0^{(1)}$

$\quad e_0^{(1)}$

$q_1^{(1)} \qquad q_0^{(2)}$

$\quad e_1^{(1)} \qquad e_0^{(2)}$

$q_2^{(1)} \qquad q_1^{(2)} \qquad q_0^{(3)}$

$\quad e_2^{(1)} \qquad e_1^{(2)} \qquad e_0^{(3)}$

$q_3^{(1)} \qquad q_2^{(2)} \qquad q_1^{(3)} \qquad q_0^{(4)}$

$\quad e_3^{(1)} \qquad e_2^{(2)} \qquad e_1^{(3)}$

$q_4^{(1)} \qquad q_3^{(2)} \qquad q_2^{(3)} \qquad q_1^{(4)}$

$\quad e_4^{(1)} \qquad e_3^{(2)} \qquad e_2^{(3)}$

. . . .

FIGURE 2.7.6
The QD array for a polynomial of degree four.

Formulas (2.7.16) and (2.7.17) imply that each q-column entry after the first is obtained as the right-hand point of a rhombus using information from the rest of the vertices of that rhombus while, similarly, the e's are obtained as the lowest vertex of a rhombus. It is evident from the nature of these formulas that there is numerical instability inherent in this process, since (2.7.16) involves the ratio of successive members of an e column and it is expected that those entries should converge to 0.

However, it can be proved that under the assumption that all the roots have distinct moduli, then each q column converges to one of the roots. Specifically, if the roots are ordered so that

$$|\alpha_1| > |\alpha_2| > \cdots > |\alpha_N| \qquad (2.7.18)$$

then, for each $k = 1, 2, \ldots, N$,

$$q_n^{(k)} \to \alpha_k \qquad e_n^{(k)} \to 0 \qquad \text{as } n \to \infty$$

The numerical instability of the method makes it necessary to work to much higher precision than is required in the final results, and even this is often insufficient.

There is an alternative form of the algorithm which proceeds in a row-by-row fashion and which has the property that the divisions involve the entries in the q columns. This removes the worst of the numerical stability difficulty, but it is more difficult to obtain satisfactory starting values. [See Henrici (1974) for a detailed discussion of this method.]

Example 2.7.7. For the equation $x^4 + 2x^3 - 13x^2 - 14x + 24 = 0$ with the initial values $u_0 = u_1 = u_2 = 0$, $u_3 = 1$, the first few rows of the QD array are as follows. In this table, each row represents the entries which can be computed at any one stage of the overall computation. The rows therefore correspond to upward diagonals of Fig. 2.7.6.

$q^{(1)}$	$e^{(1)}$	$q^{(2)}$	$e^{(2)}$	$q^{(3)}$	$e^{(3)}$	$q^{(4)}$
-2.0000						
-8.5000	-6.500					
-2.7059	5.794	2.4120				
-5.6739	-2.968	2.9065	-5.300			
-3.1954	2.478	2.6684	2.730	-1.3745		
-4.8069	-1.612	3.1255	-2.021	-2.3143	-3.670	
-3.5176	1.289	2.8144	1.300	-1.8105	2.525	1.2458
-4.4236	-0.906	3.1082	-0.996	-2.3795	-1.869	1.7616
-3.7182	0.705	2.8948	0.693	-2.0139	1.361	1.4664
-4.2303	-0.512	3.0712	-0.529	-2.3455	-1.024	1.7648
-3.8381	0.392	2.9398	0.380	-2.1155	0.759	1.5676
-4.1273	-0.289	3.0434	-0.289	-2.3073	-0.572	1.7405
-3.9078	0.219	2.9654	0.211	-2.1700	0.426	1.6144
-4.0709	-0.163	3.0257	-0.159	-2.2810	-0.322	1.7255

The true solutions of this equation are $-4, 3, -2$, and 1. The first two q columns are settling down in the vicinity of their respective roots, while the third q column would probably lead to the conclusion that the root is somewhere in the vicinity of -2.2. The final q column which should be converging to 1 starts off tolerably close to that limit but is evidently being contaminated by rounding error before there is any improvement at all.

If the QD process were halted as soon as the first full row had been produced, we would probably guess starting values for Newton's method for locating each of these roots somewhere around $-4.1, 2.95, -2.1$, and 1.25, which would be good enough to allow very rapid convergence to each of the true solutions. The subsequent information would not improve the first two estimates significantly (though it does, of course, strengthen our belief in them) and would probably lead to significantly less good initial guesses at the two smaller roots.

One of the principal difficulties arising in the solution of polynomial equations is the occurrence of complex roots or other repetitions of absolute value such as may arise from repeated or equal and opposite real roots. One of the better approaches to isolating roots of these types is by finding the appropriate quadratic factors of the original polynomial, and this can be achieved through the use of Bairstow's method. Even for isolated real roots, the determination of quadratic factors of the polynomial

$$p(x) = a_N x^N + a_{N-1} x^{N-1} + \cdots + a_1 x + a_0 \qquad (2.7.19)$$

will render the location of the roots themselves very simple.

Bairstow's method proceeds iteratively from an initial guess or estimate at a factor in the form $(x^2 - ux - v)$. The intention is to find u, v so that the remainder on division of p by this factor becomes zero. We therefore write p in the form

$$p(x) = (x^2 - ux - v)q(x) + b_1(x - u) + b_0 \qquad (2.7.20)$$

where this particular form for the remainder is chosen for convenience in the subsequent algebra. The polynomial q is of degree $N-2$ and can be written as

$$q(x) = b_N x^{N-2} + b_{N-1} x^{N-3} + \cdots + b_3 x + b_2 \qquad (2.7.21)$$

Substituting this into (2.7.20) and comparing coefficients, we obtain the identities

$$b_N = a_N \qquad b_{N-1} = a_{N-1} + u b_N$$

$$b_k = a_k + u b_{k+1} + v b_{k+2} \qquad (k = N-2, N-3, \ldots, 1, 0)$$

Setting the artificial coefficients $b_{N+1} = b_{N+2} = 0$, this becomes

$$b_k = a_k + u b_{k+1} + v b_{k+2} \qquad (k = N, N-1, \ldots, 1, 0) \qquad (2.7.22)$$

Note here that all these coefficients b_k are functions of the unknown coefficients u and v.

From (2.7.20), it follows immediately that $x^2 - ux - v$ is a factor of p if, and only if, the coefficients b_1 and b_0 are both zero. The problem is therefore reduced to the solution of the system of equations

$$b_1(u, v) = b_0(u, v) = 0 \qquad (2.7.23)$$

Bairstow's method is the solution of these equations by Newton's method. The one immediately apparent difficulty arises in the need for the elements of the Jacobian matrix for this system, that is, the derivatives $\partial b_1/\partial u$, $\partial b_0/\partial u$, $\partial b_1/\partial v$, $\partial b_0/\partial v$.

This difficulty is more apparent than real since we can obtain simple recurrence relations which yield the desired derivatives by differentiating the relations (2.7.22).

It is convenient to introduce the notation

$$c_k = \frac{\partial b_{k-1}}{\partial u} \qquad d_k = \frac{\partial b_{k-2}}{\partial v} \qquad (2.7.24)$$

so that we seek c_2, c_1, d_3, and d_2. Now differentiating (2.7.22) with respect to u, we get

$$c_{k+1} = \frac{\partial b_k}{\partial u} = b_{k+1} + u \frac{\partial b_{k+1}}{\partial u} + v \frac{\partial b_{k+2}}{\partial u} = b_{k+1} + uc_{k+2} + vc_{k+3} \qquad (2.7.25)$$

Similarly, differentiation with respect to v yields

$$d_{k+1} = b_{k+1} + ud_{k+2} + vd_{k+3} \qquad (2.7.26)$$

Also, we have $b_{N+2} = b_{N+1} = 0$ and $b_N = a_N$ which are all constant, and b_{N-1} is independent of v, so the appropriate starting values for these relations are

$$c_{N+2} = c_{N+1} = 0 \qquad d_{N+2} = d_{N+1} = 0 \qquad (2.7.27)$$

Comparing (2.7.25) and (2.7.26) with these starting conditions clearly implies that, for each k,

$$c_k = d_k \qquad (2.7.28)$$

and so only one of these sequences need be generated.

The Jacobian matrix of the system (2.7.23) is thus

$$J(u, v) = \begin{bmatrix} \dfrac{\partial b_1}{\partial u} & \dfrac{\partial b_1}{\partial v} \\ \dfrac{\partial b_0}{\partial u} & \dfrac{\partial b_0}{\partial v} \end{bmatrix} = \begin{bmatrix} c_2 & c_3 \\ c_1 & c_2 \end{bmatrix}$$

which has the determinant $D = c_2^2 - c_1 c_3$. It follows that

$$J^{-1} = \frac{1}{D} \begin{bmatrix} c_2 & -c_3 \\ -c_1 & c_2 \end{bmatrix}$$

Given an approximate quadratic factor $x^2 - ux - v$ and the corresponding b's and c's, we get the improved estimated coefficients u^+, v^+ from Newton's formula as

$$u^+ = u - \frac{c_2 b_1 - c_3 b_0}{D}$$

$$v^+ = v - \frac{c_2 b_0 - c_1 b_1}{D}.$$

(2.7.29)

Since Bairstow's method is just an application of Newton's method, its convergence follows from that of Newton's iteration.

Example 2.7.8. Apply Bairstow's method to find the complex roots of the equation

$$2x^3 - 3x + 2 = 0$$

which was used in Example 2.7.1.

Solution. In this case we know the real root is near $x = -1.5$ from which we deduce that the quadratic factor relating to the complex roots is approximately $x^2 - 1.5x + \frac{2}{3}$. We thus choose $u = 1.5$ and $v = -0.7$ as starting values for Bairstow's method.

The coefficients of the polynomial are given by

$$a_3 = 2 \qquad a_2 = 0 \qquad a_1 = -3 \qquad a_0 = 2$$

from which we deduce that

$$b_3 = 2 \qquad b_2 = 3 \qquad b_1 = 0.1 \qquad b_0 = 0.05$$

$$c_3 = 2 \qquad c_2 = 6 \qquad c_1 = 7.7$$

Hence $D = 20.6$, and therefore

$$u^+ = 1.5 - \frac{0.6 - 0.1}{20.6} = 1.4757$$

$$v^+ = -0.7 - \frac{0.3 - 0.77}{20.6} = -0.6772$$

for which the corresponding values of b_1 and b_0 are 0.00098 and 0.0028, respectively.

The method is clearly not well-suited to extensive hand calculation but can be seen, even on this little evidence, to be converging very rapidly to the desired quadratic factor.

Another point worthy of mention about the results of this example is that the next values of b_3 and b_2 are, respectively, 2 and 2.9514, so the quotient polynomial is $q(x) = 2x + 2.9514$, which has the root -1.4757 that is, to this accuracy, the real root of the original equation found in Example 2.7.1. (The fact that this root is just $-u^+$ in this case is not typical; it is the result of the fact that $a_2 = 0$ in the original equation.) The point is that we have been able to locate the other root by solving the equation $q(x) = 0$.

This leads naturally to the question of polynomial *deflation*, which is the process whereby having located one or more of the roots or factors of p, we find others by looking only at the quotient polynomial formed by dividing out these factors.

The procedure is very similar to that adopted above. It is also prone to serious buildup of rounding error, since the factors by which we divide are themselves approximations and it is well-known that the roots of polynomials can be extremely sensitive to variations in the coefficients. We will return to this briefly later. [For a full treatment of this topic, the reader is referred to the monumental work of J.H. Wilkinson (1963).]

In the case of deflation by a quadratic factor, the process would simply reduce to finding roots or factors of the polynomial q defined in Eq. (2.7.21) and (2.7.22) for Bairstow's method. We will concentrate on deflation by a linear factor. Suppose then that we have found the (approximate) root α_1 of the polynomial p given by (2.7.19) and that we seek the deflated polynomial q defined by

$$p(x) = (x - \alpha_1)q(x) + b_0 \qquad (2.7.30)$$

Writing

$$q(x) = b_N x^{N-1} + b_{N-1} x^{N-2} + \cdots + b_2 x + b_1 \qquad (2.7.31)$$

setting $b_{N+1} = 0$ as above, and comparing coefficients, we obtain

$$b_k = a_k + \alpha_1 b_{k+1} \qquad (k = N, N-1, \ldots, 1, 0) \qquad (2.7.32)$$

A measure of the accuracy of the root α_1 is provided by the magnitude of b_0. We are now in a position to find further roots or factors of p by working with the deflated polynomial q given by (2.7.31) and (2.7.32). All the same methods which have been proposed for the original polynomial are available for this purpose.

The principal questions which remain concern the overall strategy to be used for polynomial equation solving. We have seen that Bernoulli's method can be used to provide an estimate of the dominant zero, and its extension to the QD algorithm will yield approximations to all the roots. However, neither is likely to be the method of choice for the accurate location of these roots—even in the idealized circumstances under which their convergence is assured. The accurate determination of the roots and factors is probably best achieved by Newton's method and its extension to Bairstow's method for quadratic factors together with polynomial deflation.

However, this deflation must be carefully performed. It is usually advisable to locate the smallest roots first and then use deflation each time to obtain estimates of the larger ones. Note that the smallest root of the polynomial (2.7.19) is the reciprocal of the largest root of

$$a_N x^{-N} + a_{N-1} x^{-N+1} + \cdots + a_1 x^{-1} + a_0 = 0$$

or, equivalently, of

$$a_N + a_{N-1} x + \cdots + a_1 x^{N-1} + a_0 x^N = 0$$

After all roots or factors have been approximated using this deflation, it is advisable to use these approximations as the starting points for Newton's and

Bairstow's method applied to the *original* polynomial to eliminate the effects of the accumulated roundoff error in the repeated deflation.

EXERCISES 2.7

1. Use Bernoulli's method to estimate the dominant root of $x^4 - 4x^3 + 2 = 0$ to two decimal places.

2. Write a program to locate the dominant root of a polynomial using Bernoulli's method to get a one-decimal-place approximation to the root and using this as the starting point of Newton's method to locate the root to eight decimal places. Test it on the example of Exercise 1.

3. Show that the choice of starting values $u_0 = u_1 = \cdots = u_{N-2} = 0$, $u_{N-1} = 1$ leads to solutions of the recurrence relation (2.7.2) in which $A_1 \neq 0$ and hence that under the stated conditions Bernoulli's method will converge to the dominant root of a polynomial equation. (*Hint*: Write down the system

of linear equations for the unknown coefficients $A_1, A_2, \ldots, A_N$ to satisfy $u_0 = u_1 = \cdots = u_{N-2} = 0$. Show that there is no nontrivial solution with $A_1 = 0$.)

4. Find the second real root of $x^4 - 4x^3 + 2 = 0$.

5. Write a program to find quadratic factors of a polynomial by Bairstow's method. Use your program together with polynomial deflation to find all the roots—real and complex—of the polynomial equation $x^6 - 2x^5 + 3x^4 - 7x^3 + 2x^2 - 3x + 6 = 0$ to five-decimal-place accuracy in both the real and imaginary parts. Compare the results obtained by isolating the factors in different orders.

PROJECTS

1. Derive Newton's iteration for evaluating the arc-sine function by solving the equation $\sin x - c = 0$. Write a program to implement this technique, and test it for a range of values of $c \in [-1, 1]$ beginning, in each case, with $x_0 = c$.

2. Show that Newton's method for computing arcsine has linear convergence rate for $c = 1$ but is quadratically convergent for all other nonnegative c. Determine the maximum number of iterations necessary to compute the arcsine function this way to machine precision.

3. Try to improve the arcsine routine of Exercises 4 and 5 by making use of the double-angle formulas

to restrict the range over which Newton's method must be used.

4. Write a program to locate all the roots of a polynomial equation. Initially assume that all roots are real. Use the QD algorithm until one complete row of the array is generated, and use the results to provide starting values for Newton's method. By combining the information from pairs of q columns, obtain estimates for quadratic factors and use Bairstow's method to find these accurately. You can now make your program find complex or repeated roots, too.

CHAPTER
3

APPROXIMATE
EVALUATION
OF ELEMENTARY
FUNCTIONS

3.1 INTRODUCTION

In this chapter we discuss the problem of the evaluation of the elementary functions such as the exponential, logarithmic, and trigonometric functions. Within any computer or calculator system algorithms for these functions must be built-in either in the hardware or by readily accessed software. The techniques used are dependent on the arithmetic system being used as well as the machinery available. In particular we will see that the algorithms employed by most computers are completely different from those used within hand-held calculators. Before we get too deeply involved in the details of any of the particular algorithms, let us first try to answer the simple question, "For what functions and what values of the argument can a computer evaluate $f(x)$ exactly?"

The list is certainly very short. Of course, there are special values of the argument such as $\pi/2$ for which the sine and cosine functions can be computed exactly—but even then not by the computer since it will not have an exact representation of π. In general, all the transcendental functions are excluded from our list of candidates. Because of rounding errors, we eventually conclude that the only functions and arguments in our list are polynomials with integer coefficients and integer arguments. Within this restrictive class, we may still run into problems of integer overflow in some of the intermediate calculations even if the final value is representable.

In view of the above remarks, we will spend a little time on evaluation of this simple class of elementary functions. Consider the evaluation of the function

$$p(x) = a_n x^n + a_{n-1} x^{n-1} + \cdots + a_1 x + a_0 \qquad (3.1.1)$$

Now to evaluate the term $a_k x^k$ (where, of course, coefficients a_k may now be real numbers) we could perform the repeated multiplication as follows:

$$\text{term} := a_k$$
$$\text{for } i = 1 \text{ to } k \text{ do term} := \text{term} * x$$

which entails k multiplications. In total therefore, there are $n + (n - 1) + \cdots + 1$ or $n(n + 1)/2$ multiplications that must be followed by the summation of all the terms, which entails n additions. The complete evaluation uses $n(n + 3)/2$ floating-point operations. (We will return to the efficient computation of integer powers of real numbers in Sec. 3.3.)

This process can be reduced significantly in its computational complexity by the use of *Horner's rule* in which we rewrite the polynomial (3.1.1) as

$$p(x) = \{ \cdots [(a_n x + a_{n-1})x + a_{n-2}]x \cdots + a_1 \}x + a_0 \qquad (3.1.2)$$

which now requires just n multiplications and n additions. This simplification which can easily be programmed therefore reduces the computational effort and, by virtue of the fact that fewer operations are required, should also result in some reduction in the accumulated rounding error.

With the coefficients stored in an array a[0], a[1], . . . , a[n], the computation is achieved by the following simple loop in which the value of the polynomial is assigned to the variable poly:

$$\text{poly} := a[n];$$
$$\text{for } i = n - 1 \text{ downto } 0 \text{ do}$$
$$\text{poly} := \text{poly} * x + a[i];$$

Example 3.1.1. Evaluate the polynomial $p(x) = 5x^5 + 12x^4 - 3x^3 + x^2 + 3x - 16$ by Horner's rule for $x = 2$.

Solution. We write

$$
\begin{aligned}
p(x) &= (\{[(5x + 12)x - 3]x + 1\}x + 3)x - 16 \\
&= (\{[(10 + 12)2 - 3]2 + 1\}2 + 3)2 - 16 \\
&= \{[(44 - 3)2 + 1]2 + 3\}2 - 16 \\
&= [(82 + 1)2 + 3]2 - 16 \\
&= (166 + 3)2 - 16 = 338 - 16 = 322
\end{aligned}
$$

Even when spelled out in this detail the small operation count for the evaluation of the polynomial is evident. Whenever a polynomial is expressed in its standard algebraic form, this is the preferred method of evaluation for the floating-point arithmetic system. (For some highly parallel machines with other arithmetic systems, it may well turn out that this is not the ideal way to proceed, but we do not pursue these ideas here.)

EXERCISES 3.1

1. Use Horner's rule to evaluate the polynomial

$$5x^5 - 4x^4 + 3x^3 - 2x^2 + 6x - 7$$

for $x = 1.2$.

2. Write a procedure for the evaluation of a polynomial by Horner's rule. Test it for the polynomial in Exercise 1.

3.2   ITERATIVE METHODS

Many elementary-function evaluation problems can be reconstrued as equation-solving problems. Two important examples of this are the square root and reciprocal functions: the computation of $\sqrt{a}$ and of $1/c$ can be thought of as finding the solutions x of

$$x^2 = a \qquad (3.2.1)$$

and

$$\frac{1}{x} = c \qquad (3.2.2)$$

respectively. Such equations are natural candidates for numerical solution using an iterative scheme such as Newton's method.

How would we incorporate this into a computer routine for the square root function? We have already seen in Chap. 2 that Newton's iteration for finding the square root of a is

$$x_{n+1} = \frac{x_n + a/x_n}{2}$$

(see Example 2.3.1) and that this iteration is (quadratically) convergent to $\sqrt{a}$ for any positive starting point x_0.

For a normalized binary floating-point representation, we have $x = f \times 2^E$ with $f \in [\frac{1}{2}, 1)$. Now

$$\sqrt{x} = \begin{cases} \sqrt{f} \times 2^{E/2} & \text{for } E \text{ even} \\ \sqrt{\dfrac{f}{2}} \times 2^{(E+1)/2} & \text{for } E \text{ odd} \end{cases}$$

Since $\frac{1}{2} \leq f < 1$, it follows that $\sqrt{f} > 0.7$, and so the initial estimate 0.875 would serve as a good starting point for the iteration in the case of E being even for all possible values of f. (Its maximum possible error is bounded by 0.175.) For the other case, using 0.625 would be similarly efficient. The particular choices here have the property of exact binary representations. A mere four iterations are sufficient to obtain at least 11 decimal places of accuracy—that is, full machine accuracy for real variables in Turbo Pascal.

Example 3.2.1. In the case of $\sqrt{5}$, the binary representation of $a = 5$ is in the form $(\frac{5}{8})(2)^3$.

Here, since 3 is odd, we rewrite the representation as $(\frac{5}{16})(2)^4$, and therefore the square root is obtained as $\sqrt{\frac{5}{16}} \times 2^2$ using the initial guess 0.625 for $\sqrt{\frac{5}{16}}$.

The iterations are (in decimal form) 0.5625, 0.559 027 777 78, 0.559 016 994 48, 0.559 016 994 37, and 0.559 016 994 37.

Indeed, we see that the required accuracy has been achieved in the stated number of iterations. (The extra iteration is included here purely for illustrative confirmation of this statement.)

The process described above is implemented in Program 3.2.2 below. The implementation there is clearly not the most efficient one available. In particular, within a hardware (or built-in software) routine, direct access to the exponent and mantissa of the argument x would be relatively straightforward. Our Turbo Pascal implementation takes no such advantage to illustrate the process of isolating the exponent and mantissa of a floating-point representation. In Table 3.2.3 we show the results of the four iterations and the value of the built-in square root function for each $n = 2, 3, \ldots, 10, 20, \ldots, 100$.

Program 3.2.2

```
program Square_Root;
var
    x, f, sqrtf, sqrtx: real;
    i, n, expt, halfe: integer;
procedure GetEvenExptAndF (x:real;
                var e:integer; var f:real);
begin
    e:=0; f:=x;
    while f>=1 do begin
        f:=f/4; e:=e + 2;
    end;
end;

begin
    for n:=2 to 19 do begin
        if n<=10 then x:=n else x:=10 + 10*(n – 10);
        GetEvenExptAndF(x, expt, f);
        halfe:=expt div 2;
        if f<1/2 then sqrtf:=5/8 else sqrtf:=7/8;
        for i:=1 to 4 do sqrtf:=(sqrtf + f/sqrtf)/2;
        sqrtx:=sqrtf;
        for i:=1 to halfe do sqrtx:=sqrtx*2;
        writeln(x:4:0,sqrtx:20:10,sqrt(x):20:10);
    end;
end.
```

It is immediately apparent that the predicted full accuracy has been achieved with just four iterations from the stated starting points.

Note the use of repeated division by 4 which guarantees that we find the appropriate even exponent and corresponding fraction. Note, too, that this

TABLE 3.2.3

x	sqrt	sqrt(x)
2	1.414 213 562 4	1.414 213 562 4
3	1.732 050 807 6	1.732 050 807 6
4	2.000 000 000 0	2.000 000 000 0
5	2.236 067 977 5	2.236 067 977 5
6	2.449 489 742 8	2.449 489 742 8
7	2.645 751 311 1	2.645 751 311 1
8	2.828 427 124 7	2.828 427 124 7
9	3.000 000 000 0	3.000 000 000 0
10	3.162 277 660 2	3.162 277 660 2
20	4.472 135 955 0	4.472 135 955 0
30	5.477 225 575 1	5.477 225 575 1
40	6.324 555 320 3	6.324 555 320 3
50	7.071 067 811 9	7.071 067 811 9
60	7.745 966 692 4	7.745 966 692 4
70	8.366 600 265 3	8.366 600 265 3
80	8.944 271 910 0	8.944 271 910 0
90	9.486 832 980 5	9.486 832 980 5
100	10.000 000 000 0	10.000 000 000 0

program was written for this specific task and so would not be appropriate for arguments less than unity.

Many advanced computer systems do not have a hardware division operation and must perform division via software, much as they do for the other elementary functions. Newton's method again provides a good basis for such a routine for the normalized binary floating-point system.

In the case of reciprocation, we saw in Eq. (2.3.7) that the Newton iteration formula for finding $1/c$ is $x_{n+1} = x_n(2 - cx_n)$ and that the errors satisfy (2.3.8), namely,

$$e_{n+1} = -ce_n^2 \qquad (3.2.3)$$

It was also established in Exercises 2.3 that this iteration does indeed have quadratic convergence to the required limit $1/c$ for any initial point $x_0 \in [1/2c, 3/2c]$.

We consider next how this can be efficiently implemented for the special case of the reciprocation of a normalized binary floating-point number.

Suppose then that we seek the reciprocal of $C = c \times 2^E$, where $c \in [\frac{1}{2}, 1)$. Then, $1/C = (1/c) \times 2^{-E}$ and $1 < 1/c \le 2$. It follows that $1/2c \in [\frac{1}{2}, 1)$ and $3/2c \in [\frac{3}{2}, 3)$. Thus $[1, \frac{3}{2}]$ is included in the acceptable range of starting values for all possible values of c. It follows that $x_0 = 1.5$ will be a suitable fixed starting point. (We discuss other possible choices shortly.)

Example 3.2.4. Find the reciprocal of the floating-point representation of $C = 7 = (\frac{7}{8})(2)^3 = (0.11100\cdots0)_2(2)^3$. (Compare Example 2.3.5.)

Solution. The reciprocal will be $[1/(\frac{7}{8})](2)^{-3}$ which will then need to be normalized. To find $1/\frac{7}{8}$ with $x_0 = 1.5$, we get

$$1.031\ 200\ 5 \qquad 1.131\ 958\ 0 \qquad 1.142\ 753\ 2 \qquad 1.142\ 857\ 1 \qquad 1.142\ 857\ 1$$

so that seven-decimal-place accuracy has been obtained in just four iterations. The final value would be stored not as $1.142\ 857\ 1 \times 2^{-3}$ but as $0.571\ 427\ 85 \times 2^{-2}$.

From the error equation (3.2.3), we see that, since $c < 1$, $|e_{n+1}| < |e_n|^2$. Now, $|e_0| < \frac{1}{2}$, and so it follows that $|e_n| < 2^{2^{-n}}$. Hence for *single*-precision reals, full accuracy is obtained in at most five iterations, and at most one further iteration is needed for the Turbo Pascal types *real*, *double*, and *extended*. With just a small list of different starting values for different ranges, these iteration counts could be reduced still further.

A natural question to ask at this point is whether a better choice of x_0 can be obtained in a simple manner. To see that the answer is "yes," recall that for floating-point computations we are interested in achieving sufficiently small relative errors. For reciprocation, the relative error ρ_n, say, in the iterate x_n is given by

$$\rho_n = \frac{e_n}{1/c} = ce_n \tag{3.2.4}$$

and using (3.2.3) we obtain

$$|\rho_{n+1}| = c|e_{n+1}| = c|ce_n^2| = (ce_n)^2 = \rho_n^2 \tag{3.2.5}$$

Thus the relative error is going down precisely as the square of the previous relative error.

Write, temporarily, $\delta = 1 - c$ so that $0 < \delta \leq \frac{1}{2}$. Then

$$\frac{1}{c} = \frac{1}{1 - \delta} = 1 + \delta + \delta^2 + \cdots \tag{3.2.6}$$

and so it follows that if we take $x_0 = 1 + \delta$, then the initial absolute error is $e_0 = \delta^2 + \delta^3 + \cdots$ and the initial relative error is

$$\rho_0 = ce_0 = (1 - \delta)(\delta^2 + \delta^3 + \cdots) = \delta^2 \tag{3.2.7}$$

Since $\delta \leq \frac{1}{2}$, it follows that $\rho_0 \leq \frac{1}{4}$ and then from (3.2.5) that subsequent relative errors are bounded by

$$\rho_n \leq (\tfrac{1}{4})^{2^n} = 2^{-2^{n+1}} \tag{3.2.8}$$

It follows that with this choice of initial value one fewer iteration is needed in all the cases mentioned above. That is, four iterations will suffice for single precision, and five are enough for types real, double, or extended. Finally, it should be observed that the definition of x_0 is electronically very simple, since forming δ is simply the operation of taking the two's complement of c, while adding it to unity necessitates nothing more than the insertion of a 1 before the binary point. (Even as a program statement it amounts simply to the assignment $x_0 := 2 - c$.)

In Example 3.2.4, we have $c = \frac{7}{8}$, and so we would set $x_0 = 2 - \frac{7}{8} = 1.125$. The next few iterates are $1.142\,578\,125$, $1.142\,857\,075$, $1.142\,857\,143$, and $1.142\,857\,143$. We see the predicted very rapid convergence to the correct result.

EXERCISES 3.2

1. Write a program to implement Newton's iteration for finding the reciprocal of a floating-point number C, that is, solving the equation $f(x) = 1/x - C = 0$ to the accuracy of type double using the procedures described in Sec. 3.2. In other words, use a routine to find the exponent and mantissa of C and then obtain the reciprocal of this mantissa. Tabulate the results for both the starting values $x_0 = 1.5$ and $x_0 = 1 + \delta$ and compare the results with the value of $1/C$ as given by your machine for $C = 1.1$, $1.2, \ldots, 1, 2, \ldots, 10, 20, \ldots, 100$.

2. Show that the secant method can be applied to the operation of reciprocation without involving any division. Compare its performance for this task with that of Newton's method.

3.3 DESIGN OF COMPUTER ROUTINES

In this section, we examine some of the considerations which must be taken into account in the design of algorithms for the evaluation of the elementary functions in an electronic computer.

We have already made some reference to this problem for the square root and reciprocal functions for each of which Newton's method provides a basis for efficient algorithms within a binary floating-point system. Of course, many computer systems will not need the reciprocal function if they have hardware division. However, even the design of the hardware algorithms is based first on finding a suitable numerical algorithm. It is not necessarily the case that the same algorithms are appropriate for all arithmetic systems. For example, we will see in the next section that the decimal system used in hand-held calculators demands different approaches even for these functions.

For several of the other elementary functions, we saw that summation of series expansions can provide efficient algorithms but that their range of application is often severely limited. Some *range-reduction* procedure is needed to render such approaches useful in practice. It is also not necessarily clear which series expansion (if any) should be used for any particular function once the appropriate range has been determined.

We consider here the logarithmic and exponential functions, sine, cosine, and arctangent, and the formation of integer powers. It is not the intention of this section to provide fully operational software routines for all these but merely to indicate the sorts of approach which may be adopted.

For the case of the function x^n, we can take advantage of the binary representation of the integer n. This is a library function for many computer systems, but not for Turbo Pascal. However, the technique used by most library implementations is similar to that which is adopted by an efficient Turbo Pascal procedure for performing this task. We may suppose, without loss of generality, that the power n is positive. (Negative powers can then be dealt with, for positive x, by reciprocation of the corresponding positive power result.) Now n is expressed in its binary representation as

$$n = \sum_{i=0}^{N} n_i 2^i \qquad (3.3.1)$$

where each n_i is either 0 or 1 and $(N+2)$ is the integer wordlength in bits. (One bit is used for the sign.) It follows from (3.3.1) that

$$x^n = \prod_{i=0}^{N} x^{n_i 2^i} \qquad (3.3.2)$$

Thus x^n is formed as the product of x^{2^i} for each nonzero bit in the binary representation of n. This is easily achieved since $x^{2^{i+1}} = (x^{2^i})^2$, and so the successive factors are simply the squares of the previous ones.

The following Turbo Pascal function, Procedure 3.3.1, includes an efficient implementation of this function based on the binary representation of n. It is essentially a coding of the type of routine which is often built into a computer function library. Of course, the testing of the individual bits could be more directly achieved by the arithmetic unit of the machine, but Turbo Pascal does allow tolerably easy access to the bit representation of integers by way of the integer shift operations such as the *shr* used here. Each subsequent shift to the right puts the bit representing the coefficient of the next power of 2 in the rightmost position. It is sufficient, therefore, to test whether this shifted value is odd to ascertain whether that bit is 0 or 1.

Procedure 3.3.1 Integer power function

```
function xn(x: real; n:integer): real;
var factor, product: real; i: integer;
begin
    i:=n; factor:=x; product:=1;
    repeat
        if odd(i) then product:=factor * product;
        factor:=sqr(factor);
        i:=i shr 1;
    until i = 0;
    xn:=product;
end;
```

The ability to form integer powers efficiently is an important component of any function routine which is based on the evaluation of a power series, although for many such series there is a simple recursive relation between the terms. If the primary purpose of a computer routine was to compute the first 20 powers of x, say, then it would certainly be more efficient to use a simple loop based on the recursion $x^{n+1} = x(x^n)$. Similarly, for example, successive terms of the exponential series would be best obtained by use of the relation

$$\frac{x^{n+1}}{(n+1)!} = \frac{x^n}{n!} \frac{x}{n+1}$$

For this particular case, it is doubly important to use the recurrence since both the numerator and the denominator of the left-hand side would be liable to overflow for very moderate values of n. (In the case of Turbo Pascal, we would also need to exercise care to avoid the integer wraparound effect in the computation of $n!$ since for type integer it turns out that 8! is negative!)

Consider now the efficient evaluation of the exponential function. We have already observed that some range reduction is necessary before the power series can be used to advantage. The use of the identity (1.5.4), namely, $e^x = (e^{x/2})^2$, can reduce the number of terms needed significantly. However, for the binary floating-point system, even greater gains can be made by using the formation of integer powers of the base 2 to as great an extent as possible. This part of the operation is achieved by simply writing that integer into the exponent part of the result. How can we make good use of this observation?

Suppose then that we wish to compute e^x, where x is stored in (normalized) binary floating-point form. Now, if we write

$$y = x \log_2 e \tag{3.3.3}$$

then

$$x = \frac{y}{\log_2 e} = y \ln 2 \tag{3.3.4}$$

and so

$$e^x = e^{y \ln 2} = 2^y \tag{3.3.5}$$

An efficient algorithm could therefore begin by setting

$$y := x \log_2 e$$

$$Y := [y]$$

$$U := y - Y$$

where $[y]$ is the usual integer-part function. If follows from (3.3.5) that

$$e^x = 2^U 2^Y \tag{3.3.6}$$

where, since Y is an integer, we simpy add Y to the exponent of 2^U. Now to compute 2^U, we could set, for some suitable integer n,

$$V := \frac{[2^n U]}{2^n} \qquad W := U - V = \frac{\text{FRAC}(2^n U)}{2^n}$$

where FRAC is the fraction-part function. Then $2^U = 2^V 2^W$ and 2^V can be obtained by *table lookup*. That is, a table of the values of 2^V would be stored in the machine for all the possible values of $V = 0, 1/2^n, 2/2^n, \dots, (2^n - 1)/2^n$. The value of W is then at most 2^{-n}, and this can be computed quickly from an appropriate series.

What value should be chosen for n? Clearly, n must be large enough that the range of values of W is small (so that a short series is sufficient), but yet it must be small enough that the table does not require a great deal of storage. Typically, we might choose $n = 4$ or 5. For $n = 4$, the table would contain values of $2^{k/16}$ for $k = 0, 1, \ldots, 15$, while for $n = 5$, it would have $2^{k/32}$ for $k = 0, 1, \ldots, 31$.

Note that the fraction U would be stored in fixed-point form, and so finding V and W is simple: the first n bits represent V, while replacing these first n bits with zeros in the representation of U yields W. The electronic manipulation of the bits is therefore easy.

How many terms of the series for 2^W are needed to attain the accuracy required? Now the final value of e^x is

$$e^x = 2^V 2^W 2^Y \tag{3.3.7}$$

and, as we have already observed, the multiplication by 2^Y is achieved by adding Y to the exponent of the product of the other two factors. Since 2^W is not much greater than unity, it follows that the appropriate relative precision in this quantity can certainly be achieved by obtaining that same absolute precision. Finally, the power series expansion of 2^W about zero is

$$2^W = e^{W \ln 2} = 1 + W \ln 2 + \frac{(W \ln 2)^2}{2!} + \cdots \tag{3.3.8}$$

For $n = 4$, $W < \frac{1}{16}$, and so $W \ln 2 \le (0.0625)(0.69315) < 0.044$. From our earlier analysis of the truncation error in the exponential series, we see that for this range of values, the truncation error is less than 2^{-24} for $N \ge 4$, where the final term included is $(W \ln 2)^N / N!$. Thus IEEE standard single precision can be obtained using a series of degree just four. For Turbo Pascal type real, $N = 6$ is sufficient. With $n = 5$, so that the table has 32 entries and $W < \frac{1}{32}$, the required values of N are reduced to 3 and 5 for single and real, respectively.

Routines very much like these are, in fact, used by several computer systems.

Example 3.3.2. Compute e^3 using the above algorithm with $n = 4$ with relative error bounded by 10^{-7}. (We use a decimal system bound for simplicity of the representations but still follow the binary routine outlined above.)

Solution. Here $x = 3$, and so $y = 3 \log_2 e = 4.328\,085\,1$. This has integer and fraction parts $Y = 4$ and $U = 0.328\,085\,1$ from which, in turn, we obtain

$$V = \frac{[16U]}{16} = \frac{5}{16} \qquad W = U - V = 0.015\,585\,1$$

Next, 2^V is obtained by table lookup; it is $1.241\,857\,8$. To obtain 2^W to the required accuracy from the power series, we need its absolute error to be bounded by 10^{-7}. For this we need terms up to the fourth order; thus,

$$2^W \simeq 1 + W \ln 2 + \frac{(W \ln 2)^2}{2!} + \frac{(W \ln 2)^3}{3!} + \frac{(W \ln 2)^4}{4!}$$

$$= 1 + 0.010\ 802\ 77 + 0.000\ 058\ 35 + 0.000\ 000\ 21$$

$$+ 0.000\ 000\ 00$$

$$= 1.010\ 861\ 3$$

Thus the final result is

$$e^x = 2^4 2^{5/16}(1.010\ 861\ 3) = (1.255\ 346\ 0)(2)^4 = 20.085\ 536$$

The true value is 20.085 537, and so the relative error achieved is approximately $10^{-6}/20 = 5 \times 10^{-8}$, which is within the tolerance we set.

We turn next to the natural logarithm function ln which we have already considered in some detail in Secs. 1.6 and 1.7. The task then is to compute $\ln x$, where x is a normalized binary floating-point number; that is, $x = f2^E$ where $f \in [\frac{1}{2}, 1)$, and so

$$\ln x = E \ln 2 + \ln f \qquad (3.3.9)$$

Now for $E \geq 1$, $\ln x \geq 0$, while for $E \leq 0$, $\ln x < 0$ and, of course, $\ln f < 0$. In order that the addition in (3.3.9) has two terms of the same sign, we use

$$\ln x = (E - 1)\ln 2 + \ln (2f)$$

for $E > 0$. Thus we define two new quantities F and N by

$$F = \begin{cases} 2f & \text{if } E \geq 1 \\ f & \text{if } E \leq 0 \end{cases} \qquad N = \begin{cases} E - 1 & \text{if } E \geq 1 \\ E & \text{if } E \leq 0 \end{cases} \qquad (3.3.10)$$

and compute

$$\ln x = N \ln 2 + \ln F \qquad (3.3.11)$$

Now the value of $\ln 2$ would be a stored constant, and so the problem is reduced to the computation of $\ln F$ with $F \in [\frac{1}{2}, 2)$. The computation of this quantity can be based upon the Taylor series for $\ln (1 + G)$, where

$$G = F - 1 \qquad (3.3.12)$$

is in the interval $[-\frac{1}{2}, 1)$. Now we have already seen that this Taylor expansion is very slowly convergent, and even the Euler transformation (which is only applicable to the alternating series for $G > 0$) does not accelerate it enough for the whole range. From Table 1.7.4, we see that for IEEE single precision this approach would require the summation of about 23 terms.

We can make significant gains by setting

$$H = \frac{G}{2 + G} \qquad (3.3.13)$$

so that

$$1 + G = \frac{1 + H}{1 - H} \qquad (3.3.14)$$

Now, it follows that

$$\ln F = \ln (1 + G) = \ln \frac{1 + H}{1 - H} = \ln (1 + H) - \ln (1 - H)$$

$$= 2\left(H + \frac{H^3}{3} + \frac{H^5}{5} + \cdots \right)$$

$$= 2H\left(1 + \frac{H^2}{3} + \frac{H^4}{5} + \cdots \right) \tag{3.3.15}$$

Since $G \in [-\frac{1}{2}, 1)$, it follows that $|H| \leq \frac{1}{3}$ and therefore that the series in (3.3.15) converges faster than the geometric series with common ratio $\frac{1}{9}$.

Consider the sum of the tail of this series, T, say. Then

$$|T| = 2|H|\left(\frac{H^{2N}}{2N + 1} + \frac{H^{2N+2}}{2N + 3} + \cdots \right)$$

$$< \frac{2(1 + \frac{1}{9} + \frac{1}{81} + \cdots)}{(2N + 1)3^{2N+1}}$$

$$= \frac{1}{4(2N + 1)3^{2N-1}} \tag{3.3.16}$$

Since $F \leq 2$, it follows from (3.3.11) that to achieve IEEE single precision (that is, a relative error bounded by 2^{-24}) in $\ln x$, it is sufficient for $\ln F$ to be computed to an absolute accuracy no worse than $2^{-24} \ln 2 = \epsilon$, say. From (3.3.16), we see that $|T| < \epsilon$ for $N \geq 7$. That is, the use of the first six terms of the series (3.3.15) is sufficient to deliver the required accuracy. For Turbo Pascal type real, the accuracy requirement is $|T| < 2^{-40} \ln 2$, which is satisfied for $N \geq 12$.

By subdividing the range of values of G still further, this approach can be made yet more efficient and this is indeed done, usually with the help of table lookup, in some systems. The procedure is essentially the same as that described here. The details of this would add nothing to the understanding of the approach.

Example 3.3.3. Use the above algorithm to compute $\ln 6.4$ with the accuracy of Turbo Pascal type real.

Solution. Now $x = 6.4 = (6.4/8)(2)^3 = (0.8)(2)^3$ where, of course, 0.8 would be stored as a binary fraction in the floating-point representation. Since $E = 3$, it follows that we set

$$F = 2f = 1.6 \qquad N = E - 1 = 2$$

so that $x = (1.6)(2)^2$ and we compute $\ln x = 2 \ln 2 + \ln 1.6$. For the latter term, we set $G = 0.6$ and then $H = 0.6/2.6$. (In this case the common ratio H^2 used in the estimate of the truncation error would be approximately 0.054, and so fewer terms than estimated above are in fact needed.)

This process is summarized below in Algorithm 3.3.4. For $x = 6.4$ in type real, for which "Nterms" = 11, we obtain the result ln $6.4 = 1.856\,297\,990\,4$, which agrees with the built-in natural logarithm function, indicating that full accuracy has indeed been achieved.

Algorithm 3.3.4   Natural logarithm function

Input   x, Nterms, ln 2

Compute   f, E where $x = f\,2^E$
                  if E > 0 then F:=2f, N:=E − 1
                              else F:=f, N:=E
                  G:=F − 1
                  H:=G/(2 + G)

Sum   sum:=1, top:=1, fact:=H^2
                  for i = 1 to Nterms
                      top:=top ∗ fact
                      sum:=sum + top/(2i + 1)
                  log F:=2 ∗ sum ∗ H

Output   log x:=log F + N ∗ ln 2

Note here that we have assumed a value of ln 2 is available for use within this routine. Such a value could be obtained by this same algorithm (since in that case $N = 0$) or by the Euler transformation described in Sec. 1.7. It would, of course, be computed just once and stored as a constant.

We turn next to the trigonometric functions sine and cosine and describe just one essentially simple approach to their evaluation. Variations on this theme are used in many computer systems. It is convenient to describe the evaluation of both functions simultaneously; indeed, for machines with even a minimal degree of parallelism, it would be likely for them to be computed simultaneously, too.

Suppose then that we wish to evaluate sin x and/or cos x. The first stage of the computation is the range reduction which is used to produce an argument in a suitably small interval for the series approximations to be acceptable. We saw in Examples 1.7.1 and 1.7.2 that it will not be enough to restrict the range of the series to $[0, \pi/2)$ but that the interval $[0, \pi/4]$ may be acceptable. Thus we set

$$N = \left[\frac{x}{\pi/4}\right] \qquad M = N \bmod 8 \qquad A = x - \frac{N\pi}{4} \qquad (3.3.17)$$

so that we will seek either the sine or the cosine of $A \in [0, \pi/4)$. From consideration of the properties of the trigonometric functions, we see that for

$M = 0$: $x = 2k\pi + A$ $\cos x = \cos A$ $\sin x = \sin A$

$M = 1$: $x = 2k\pi + A + \dfrac{\pi}{4}$ $\cos x = \cos\left(A + \dfrac{\pi}{4}\right)$ $\sin x = \sin\left(A + \dfrac{\pi}{4}\right)$

$M = 2$: $\quad x = (2k + \tfrac{1}{2})\pi + A$ $\qquad\qquad \cos x = -\sin A$ $\qquad\qquad \sin x = \cos A$

$M = 3$: $\quad x = (2k + \tfrac{1}{2})\pi + A + \dfrac{\pi}{4}$ $\qquad \cos x = -\sin\left(A + \dfrac{\pi}{4}\right)$ $\quad \sin x = \cos\left(A + \dfrac{\pi}{4}\right)$

$M = 4$: $\quad x = (2k + 1)\pi + A$ $\qquad\qquad \cos x = -\cos A$ $\qquad\qquad \sin x = -\sin A$

$M = 5$: $\quad x = (2k + 1)\pi + A + \dfrac{\pi}{4}$ $\qquad \cos x = -\cos\left(A + \dfrac{\pi}{4}\right)$ $\quad \sin x = -\sin\left(A + \dfrac{\pi}{4}\right)$

$M = 6$: $\quad x = (2k + \tfrac{3}{2})\pi + A$ $\qquad\qquad \cos x = \sin A$ $\qquad\qquad\quad \sin x = -\cos A$

$M = 7$: $\quad x = (2k + \tfrac{3}{2})\pi + A + \dfrac{\pi}{4}$ $\qquad \cos x = \sin\left(A + \dfrac{\pi}{4}\right)$ $\qquad \sin x = -\cos\left(A + \dfrac{\pi}{4}\right)$

It is therefore sufficient to be able to compute $\sin A$, $\cos A$, $\sin (A + \pi/4)$, and $\cos (A + \pi/4)$. Since $\sin (\pi/2 - x) = \cos x$ and $\cos (\pi/2 - x) = \sin x$, it follows that if we simply replace the arguments for odd values of M with $[\pi/2 - (A + \pi/4)] = \pi/4 - A$ and interchange sine and cosine for those values, we have reduced the original problem to that of computing the sine and cosine for arguments in the interval $[0, \pi/4]$.

The resulting algorithm therefore uses the following modification of (3.3.17):

$$N = \left[\frac{x}{\pi/4}\right] \qquad M = N \bmod 8$$

$$A = \begin{cases} x - \dfrac{N\pi}{4} & M \text{ even} \\[2ex] \dfrac{(N+1)\pi}{4} - x & M \text{ odd} \end{cases} \qquad (3.3.18)$$

This yields the values shown in Table 3.3.5.

This range reduction is incorporated in Algorithm 3.3.6 below. Before describing the agorithm in detail, we must check the accuracy needed in the two series to achieve the desired relative precisions in the floating-point results and then the number of terms necessary to obtain this accuracy.

TABLE 3.3.5

M	$\cos x$	$\sin x$
0	$\cos A$	$\sin A$
1	$\sin A$	$\cos A$
2	$-\sin A$	$\cos A$
3	$-\cos A$	$\sin A$
4	$-\cos A$	$-\sin A$
5	$-\sin A$	$-\cos A$
6	$\sin A$	$-\cos A$
7	$\cos A$	$-\sin A$

By way of illustration, we consider these questions for the case of IEEE single precision. Thus we need a relative precision in the output from the algorithms of 2^{-24}.

For the cases where we compute $\cos A$, we have $\cos A \geq 1/\sqrt{2}$, and so the required relative accuracy will be achieved by insisting on an absolute error bound of $2^{-25} \approx 3 \times 10^{-8}$. Since the series alternates, the error is bounded by the first term omitted, and so we seek K such that $A^{2K}/(2K)! < 2^{-25}$ for all values $A \leq \pi/4$. This is first satisfied for $K = 5$ since $(\pi/4)^{10}/10! \approx 2.5 \times 10^{-8}$. Since this is very close to the required accuracy bound, we must be very careful about the accumulation of rounding error. It would be desirable to take one extra term. That is, we use

$$\cos A \simeq 1 - \frac{A^2}{2!} + \frac{A^4}{4!} - \frac{A^6}{6!} + \frac{A^8}{8!} - \frac{A^{10}}{10!} \qquad (3.3.19)$$

For the $\sin A$ cases, the situation is less straightforward, since $\sin A$ is close to zero when A is and so the relative error can grow rapidly. Now, if $A < 2^{-12}$, it follows that $A - A^3/3! < \sin A < A$, and so the relative error will be bounded by $A^2/6$ which is certainly less than 2^{-24}. Thus for $0 \leq A \leq 2^{-12}$, we can simply use the approximation $\sin A \approx A$. For larger values, we observe that (again using ρ for the relative and δ for the absolute error)

$$\rho(\sin A) = \frac{\delta(\sin A)}{\sin A}$$

which (since $A > 2^{-12}$) is bounded by $2^{12}\delta(\sin A)$. It follows that it will be sufficient to compute $\sin A$ with an absolute error bounded by 2^{-36}. Again we use an alternating series; it is sufficient to use the terms up to $A^{11}/11!$ since $(\pi/4)^{13}/13! < 2^{-36}$. We therefore use the approximations

$$\sin A \simeq \begin{cases} A & A \leq 2^{-12} \\ A - \dfrac{A^3}{3!} + \dfrac{A^5}{5!} - \cdots \dfrac{A^{11}}{11!} & A > 2^{-12} \end{cases} \qquad (3.3.20)$$

Algorithm 3.3.6   Trigonometric functions

Input   x

Compute   N$:=[4x/\pi]$, M$:=$N mod 8
                A$:=$x $-$ N$\pi/4$
                if M odd then A$:=\pi/4 - $A

Initialize   sA$:=$A, sterm$:=$A
                cA$:=$1, cterm$:=$1
                A2$:=A^2$

Loop   for i $= 1$ to 5 do
                k$:=$2i
                cterm$:=-$cterm $*$ A2$/$k(k $- 1)$

$$\text{sterm} := -\text{sterm} * A2/k(k+1)$$
$$\text{cA} := \text{cA} + \text{cterm}, \quad \text{sA} := \text{sA} + \text{sterm}$$

Finalize   if $M \in \{1, 2, 5, 6\}$ then
                    temp$:=$sA, sA$:=$cA, cA$:=$temp
              if $M \geq 4$ then sA$:=-$sA
              if $2 \leq M \leq 5$ then cA$:=-$cA

Output   sinx$:=$sA, cos x$:=$cA

The number of terms used in these series is appropriate for single-precision computation. Table 3.3.7 below shows the results obtained by implementing this algorithm. The computation was performed in type real to control rounding error. In fact, as we can see from the error columns in Table 3.3.7, this program delivered at least 10-significant-figure accuracy for $x = i\pi/21$ ($i = 0, 1, 2, \ldots, 21$). Note that for simplicity in the description six terms of the sine approximation are used for small values of A.

This is by no means the only practical technique for the computation of these functions, and other approaches are adopted by some computer systems. For example, the implementation used on the Intel 8087 mathematical co-processor chip is based on a similar range reduction to that used in Algorithm

TABLE 3.3.7
Computed values of sin x and cos x and their errors using Algorithm 3.3.6

x	sin x	Error	cos x	Error
0.000 000	0.000 000 000 0	0.000E + 00	1.000 000 000 0	5.457E − 12
0.149 600	0.149 042 266 2	0.000E + 00	0.988 830 826 2	0.000E + 00
0.299 199	0.294 755 174 4	4.547E − 13	0.955 572 805 8	1.819E − 12
0.448 799	0.433 883 739 1	4.547E − 13	0.900 968 867 9	0.000E − 00
0.598 399	0.563 320 058 1	−9.095E − 13	0.826 238 774 3	−2.728E − 12
0.747 998	0.680 172 737 8	0.000E + 00	0.733 051 871 8	−6.276E − 11
0.897 598	0.781 831 482 4	−1.728E − 11	0.623 489 801 9	−9.095E − 13
1.047 198	0.866 025 403 8	−9.095E − 13	0.500 000 000 0	−9.095E − 13
1.196 797	0.930 873 748 6	9.095E − 13	0.365 341 024 4	−9.095E − 13
1.346 397	0.974 927 912 2	1.819E − 12	0.222 520 934 0	−1.137E − 12
1.495 997	0.997 203 797 2	1.819E − 12	0.074 730 093 6	−7.958E − 13
1.645 596	0.997 203 797 2	1.819E − 12	−0.074 730 093 6	−5.571E − 12
1.795 196	0.974 927 912 2	1.819E − 12	−0.222 520 934 0	−7.049E − 12
1.944 795	0.930 873 748 6	0.000E + 00	−0.365 341 024 4	−8.640E − 12
2.094 395	0.866 025 403 8	−9.095E − 13	−0.500 000 000 0	−2.728E − 12
2.243 995	0.781 831 482 5	−1.910E − 11	−0.623 489 801 9	−5.457E − 12
2.393 594	0.680 172 737 8	0.000E + 00	−0.733 051 871 8	6.094E − 11
2.543 194	0.563 320 058 1	−3.638E − 12	−0.826 238 774 3	9.095E − 13
2.692 794	0.433 883 739 1	−2.274E − 12	−0.900 968 867 9	−3.638E − 12
2.842 393	0.294 755 174 4	−2.274E − 12	−0.955 572 805 8	−4.547E − 12
2.991 993	0.149 042 266 2	−2.728E − 12	−0.988 830 826 2	0.000E + 00
3.141 593	0.000 000 000 0	−2.728E − 12	−1.000 000 000 0	−5.457E − 12

3.3.6, but this is followed by the computation of tan A, and the values of the sine and cosine functions are obtained from this. The particular algorithm used begins with a stage which is similar to the CORDIC algorithms discussed in the next section. This is followed by a rational function approximation to deliver the final accuracy very efficiently.

Further consideration of the algorithm raises another important question on the evaluation of the trigonometric functions, namely, *"When is it sensible to try to evaluate these functions?"*

Before attempting a full answer to this question, let us consider the breakdown of Algorithm 3.3.6 in a Turbo Pascal environment. (Similar difficulties arise in other systems.) The first stage of the computation is to obtain

$$N := [4x/\pi] \qquad (3.3.21)$$

Now this assignment will fail if $4x/\pi > \mathrm{MAXINT} + 1$. (Recall that MAXINT is the greatest integer representable within type integer.) Thus the routine will fail in these circumstances; this may be easily circumvented by the declaration of N as being of type longint which extends the range of values of N up to about $2^{31} \approx 2 \times 10^9$. That is, the algorithm will still break down for any $x > 2 \times 10^9$. This is only a small part of the possible range of variables of type real, or even type single. A similar *fix-up* could be arranged to extend the range by using an even longer integer type for N, but the question raised above really amounts to asking whether this would be a sensible thing to do.

In the case of single-precision computation, the relative error in x is at best of the order of the single-precision machine unit 2^{-24}, and so if the binary exponent of x is greater than 24 (that is, if x is greater than about 8.4×10^6), then the absolute error in x can be expected to exceed unity. That is, we anticipate

$$\delta(x) > 1 > \frac{\pi}{4}$$

and so it is not even necessarily true that the value of N is well-defined. Certainly Algorithm 3.3.6 loses its validity since the same representation as that of x will be used for all numbers in an interval of length greater than $\pi/4$.

For type real, we can make similar, though less extreme, observations. The relative precision of the floating-point representation of x is now 2^{-40}. For those values of N for which the assignment (3.3.21) fails with N a longint, it follows that the binary exponent of x is at least 30, and therefore we have at most (10-binary or) three-decimal-place absolute accuracy in x and, therefore, in the angle A computed by our algorithm. It follows that this is as much accuracy as we can realistically anticipate in $\sin x$ or $\cos x$. It is probably not sensible to attempt the computation of these quantities for any argument very much larger than this natural threshold; that is, floating-point error halting the program is perhaps the most appropriate outcome here.

Note that within the hardware implementation of algorithms such as these, modified integer formats could be used to allow the computation to proceed for any argument x. The difficulty is that for sufficiently large values of x almost any result in the interval $[-1, 1]$ is equally valid. This is indeed the interval that would be returned by the various interval arithmetic packages for both $\sin x$ and $\cos x$ for very large x.

It should be apparent that all the other trigonometric functions can be easily obtained, using their definitions, from the values of the sine and cosine functions. The last of the elementary functions which we discuss in any detail is the arctangent. Again, the other inverse trigonometric functions can readily be evaluated from this one.

Now Eq. (1.5.9) presented the power series expansion of the arctangent function:

$$\arctan x = x - \frac{x^3}{3} + \frac{x^5}{5} - \frac{x^7}{7} + \cdots \tag{3.3.22}$$

which with its alternating "faster-than-geometric" nature is rapidly convergent for small values of x. This series will therefore provide a good basis for the evaluation of this function when used in conjunction with the following fundamental properties:

$$\arctan (-x) = -\arctan x \tag{3.3.23a}$$

$$\arctan \frac{1}{x} = \frac{\pi}{2} - \arctan x \tag{3.3.23b}$$

$$\arctan x = \arctan a + \arctan \frac{x - a}{1 + ax} \tag{3.3.23c}$$

The last of these is readily derived from the addition formula for $\tan (A - B)$.

From (3.3.23a) and (3.3.23b) it follows that we need only consider arguments $x \in [0, 1]$. Also the series (3.3.22) has truncation error bounded by the first term omitted. For small values of x, $\arctan x \simeq x$, and so the relative truncation error in terminating the series with the term $x^{2n-1}/(2n - 1)$ is (approximately) bounded by $x^{2n}/(2n + 1)$. For IEEE single precision, we require that this quantity be bounded by 2^{-24}, which will be true if $x \le 2^{-5}$ and $n \ge 3$. It follows that the polynomial approximation

$$\arctan x \simeq x - \frac{x^3}{3} + \frac{x^5}{5} \tag{3.3.24}$$

will suffice for this range of x. (Only one further term is required for Turbo Pascal real variables.)

The only remaining problem is to reduce the argument to this interval $[0, \frac{1}{32}]$. This is where (3.3.23c) is important. Recall that we already have $x \in [0, 1]$.

Define

$$a = \frac{[32x]}{32} \qquad b = \frac{x - a}{1 + ax} \tag{3.3.25}$$

It follows that, since $x - a \le \frac{1}{32}$, then $b < 2^{-5}$, too. The value of arctan a can be obtained by table lookup and that of arctan b by the approximation (3.3.24). By (3.3.23c) we have

$$\text{arctan } x = \text{arctan } a + \text{arctan } b$$

This algorithm is summarized as Algorithm 3.3.8, and the results of its implementation together with values of the built-in arctangent function (truncated to single-precision format) are listed in Table 3.3.9 for a range of values of x.

Algorithm 3.3.8   Arctangent function

Input   x

Range reduction

tempx:=x, neg:=false, big:=false
if x < 0 then neg:=true, tempx:=−x
if tempx > 1 then big:=true, tempx:=1/tempx
a:=[32 * tempx]/32, b:=(tempx − a)/(1 + a * tempx)

Compute   atanb:=b$(1 - b^2/3 + b^4/5)$
atanx:=arctan(a) + atanb
if big then atanx:=$\pi/2$ − atanx
if neg then atanx:=−atanx

Output   arctan x:=atanx

Remark. The use of the built-in function arctan(a) is the equivalent of the table lookup stage of the algorithm discussed above. The results in Table 3.3.9 show that the claimed accuracy has indeed been achieved.

TABLE 3.3.9

x	atanx	arctan x	x	atanx	arctan x
0.0	0.000 000 00	0.000 000 00	1.0	0.785 398 16	0.785 398 16
0.1	0.099 668 65	0.099 668 65	2.0	1.107 148 72	1.107 148 72
0.2	0.197 395 56	0.197 395 56	3.0	1.249 045 77	1.249 045 77
0.3	0.291 456 79	0.291 456 79	4.0	1.325 817 66	1.325 817 66
0.4	0.380 506 38	0.380 506 38	5.0	1.373 400 77	1.373 400 77
0.5	0.463 647 61	0.463 647 61	6.0	1.405 647 65	1.405 647 65
0.6	0.540 419 50	0.540 419 50	7.0	1.428 899 27	1.428 899 27
0.7	0.610 725 96	0.610 725 96	8.0	1.446 441 33	1.446 441 33
0.8	0.674 740 94	0.674 740 94	9.0	1.460 139 11	1.460 139 11
0.9	0.732 815 10	0.732 815 10	10.0	1.471 127 67	1.471 127 67

EXERCISES 3.3

1. Find $e^{2.5}$ using the method of Example 3.3.2 with $n = 5$ and relative error bounded by 10^{-6}.
2. Find ln 2.4 with the accuracy of IEEE single precision using the method of Example 3.3.3.
3. In the algorithm for computing e^x, we used the series expansion of 2^W for $W < 2^{-n}$. Show that for Turbo Pascal type real it is sufficient to use a series of degree six for $n = 4$.
4. Write a program to implement the algorithm described for evaluation of the exponential function using $n = 4$. Tabulate the values obtained from this procedure and from the built-in exponential function for $x = 0.1, 0.2, \ldots, 1, 2, \ldots, 10$.
5. Verify that 11 terms of the series (3.3.15) yield a truncation error less than $2^{-40} \ln 2$.
6. Use Algorithm 3.3.6 to compute sin 2.5.
7. Find the absolute accuracy required in the computation of sin A to deliver the full relative accuracy for Turbo Pascal type real values of sin x and cos x. How many terms in the series are needed to achieve this accuracy? Modify and program Algorithm 3.3.6 to compute sin x and cos x to this accuracy. Check your computations against the built-in functions.
8. Compute arctan 30 using Algorithm 3.3.8.

3.4 CORDIC ALGORITHMS

In this section, we discuss the so-called CORDIC algorithms for the evaluation of the elementary functions—and, indeed, for the operations of multiplication and division. These methods are very widely used within electronic calculators. Modifications of these same ideas are also increasingly common on the arithmetic processors of computers, including many of the personal computer architectures and some of the most advanced supercomputer systems.

We begin with a brief discussion on the different requirements of computers and hand-held calculators.

The most important single difference is that a computer will almost invariably operate in the binary system, whereas a hand-held calculator must use the decimal system for all its calculations. The principal reason for this is that a calculator will display the result of every intermediate calculation, which would render the overhead in converting back and forth between the decimal and binary representations very time consuming. This consideration alone is sufficient to insist on decimal arithmetic throughout. (Even the programmable calculators typically use a decimal floating-point system for the sake of speed when used in calculator mode.)

This requirement for decimal arithmetic carries with it some perhaps unexpected complications. For instance the operation of "long multiplication" is a straightforward process for binary numbers but a complicated one for the decimal system. In the former, such multiplication requires just shifts and additions; for a decimal long multiplication algorithm, we require a one-digit multiplication table together with the handling of very frequent carries. The shifts and additions are, of course, still needed, too. By way of example, consider the decimal multiplication 61×46. This uses four single-digit multiplications involving two carries and then the addition of the two terms which entails a further carry. As a binary operation, this same multiplication is

$$61 \times 46 =$$

$$
\begin{array}{r}
111101 \\
\times \quad 101110 \\
\hline
111101 \\
111101 \\
111101 \\
111101 \\
\hline
101011110110
\end{array}
$$

Each of these is just 111101 shifted the appropriate number of binary places. Zero terms are omitted.

This simplicity is the aim of the CORDIC algorithms that are used for the evaluation of the elementary functions in almost every hand-held calculator. It is a matter of some note that the same algorithm can be used for all these functions together with the arithmetic operations of multiplication and division. The mere fact that these operations are more complicated for the decimal system means that any power-series-based algorithm is impractical.

The CORDIC algorithms which we describe here were originally developed by Volder (1959) for the rapid solution of trigonometric problems in the early in-flight navigation computers of the late 1950s. They are based on a decomposition of either the required answer or the argument in terms of simple stored constants. These are combined with addition rules for the elementary functions to provide the desired values. The fact that these algorithms bear the name CORDIC (*co*ordinate *r*otation *di*gital *c*omputer) rather than that of Volder testifies to their having been developed by an engineer.

All the CORDIC algorithms are based on the following decomposition theorem.

Theorem 3.4.1. Suppose the sequence of numbers (ϵ_n) satisfies the following conditions:

$$\epsilon_0 \geq \epsilon_1 \geq \cdots \geq \epsilon_n > 0 \qquad \text{for every } n \tag{3.4.1}$$

$$e_k \leq \sum_{j=k+1}^{n} \epsilon_j + \epsilon_n \qquad \text{for every } k < n \tag{3.4.2}$$

and that

$$|r| \leq \sum_{j=0}^{n} \epsilon_j + \epsilon_n \qquad \text{for every } n \tag{3.4.3}$$

Let $s_0 = 0$, and define

$$s_{k+1} = s_k + \delta_k \epsilon_k \qquad (k = 0, 1, \ldots) \tag{3.4.4}$$

where $\delta_k = \mathrm{sgn}(r - s_k)$. [By convention, we take $\mathrm{sgn}(0) = +1$.] Then, for each $k < n$,

$$|r - s_k| \leq \sum_{j=k}^{n} \epsilon_j + \epsilon_n \tag{3.4.5}$$

for every n, and so, in particular,

$$|r - s_{n+1}| \leq \epsilon_n \tag{3.4.6}$$

Remark. Before proceeding to the proof of this theorem, it may be worthwhile to see exactly what it is saying for one important special case. It should be quite clear that the sequence $\epsilon_n = 2^{-n}$ satisfies (3.4.1) and (3.4.2) and that, for this special case, (3.4.3) reduces to the requirement that $|r| \leq 2$. The conclusion of the theorem is that any such r can be written in signed binary form as

$$r \simeq \pm 1 \pm \tfrac{1}{2} \pm \tfrac{1}{4} \pm \cdots \pm 2^{-n}$$

with an error no greater than 2^{-n}. The theorem for this special case is therefore not especially noteworthy, but its generality is of fundamental importance to the CORDIC algorithms.

> ***Proof.*** We fix n and use a simple induction argument on k. Since $s_0 = 0$, we have, for $k = 0$, by (3.4.3)
>
> $$|r - s_0| = |r| \leq \sum_{j=0}^{n} \epsilon_j + \epsilon_n$$
>
> Next, suppose the result holds for some value k. Since $\delta_k = \mathrm{sgn}(r - s_k)$, it follows that
>
> $$|r - s_{k+1}| = |r - s_k - \delta_k \epsilon_k| = \big||r - s_k| - \epsilon_k\big| \tag{3.4.7}$$
>
> Now using (3.4.2) and the induction hypothesis, we get
>
> $$-\sum_{j=k+1}^{n} \epsilon_j - \epsilon_n \leq -\epsilon_k \leq |r - s_k| - \epsilon_k \leq \sum_{j=k}^{n} \epsilon_j + \epsilon_n - \epsilon_k = \sum_{j=k+1}^{n} \epsilon_j + \epsilon_n$$
>
> which establishes (3.4.5). The error bound (3.4.6) is just the special case of this for $k = n + 1$, in which case the sum on the right-hand side of (3.4.5) is empty and therefore zero.

Remarks

1. Note that allowing $n \to \infty$ in (3.4.2) establishes that each term of the sequence (ϵ_k) is bounded by the tail of the series $\Sigma \, \epsilon_n$, and for all practical purposes, we require this series to be convergent. This condition has the effect of preventing the *sequence* from tending toward zero too rapidly. Indeed, $\epsilon_n = 2^{-n}$ is the fastest converging geometric series which satisfies (3.4.2).

2. Another important sequence with these properties is $\epsilon_n = \arctan 2^{-n}$. This particular choice will be used for all the CORDIC trigonometric function algorithms.

3. For the situation where the series $\Sigma \, \epsilon_n$ is convergent—the only case of practical interest—condition (3.4.3) amounts to the statement that any quantity satisfying

$$|r| \leq \sum \epsilon_n \tag{3.4.8}$$

can be approximated arbitrarily closely by partial sums of the form $\pm \epsilon_0 \pm \epsilon_1 \pm \cdots \pm \epsilon_n$, each error no worse than ϵ_n.

Example 3.4.2. For $\epsilon_n = 2^{-n}$, we can represent $r = 1.5432$ as follows. The decomposition of r begins with $s_0 = 0$, and we then obtain

$\delta_0 = +1$:	$s_1 = 0 + 1 = 1.0$	$r - s_1 = 0.5432$
$\delta_1 = +1$:	$s_2 = 1.0 + \frac{1}{2} = 1.5$	$r - s_2 = 0.0432$
$\delta_2 = +1$:	$s_3 = 1.5 + \frac{1}{4} = 1.75$	$r - s_3 = -0.2068$
$\delta_3 = -1$:	$s_4 = 1.75 - 2^{-3} = 1.625$	$r - s_4 = -0.0818$
$\delta_4 = -1$:	$s_5 = 1.625 - 2^{-4} = 1.5625$	$r - s_5 = -0.0193$
$\delta_5 = -1$:	$s_6 = 1.5625 - 2^{-5} = 1.53125$	$r - s_6 = 0.01195$

It can easily be verified that the errors in the right-hand column each satisfy $|r - s_k| < 2^{-k+1}$. Clearly this can be continued to any required accuracy.

In the introduction to this section we observed that our calculator algorithms need to operate in the decimal system. The sequence $\epsilon_n = 10^{-n}$ does not satisfy the condition (3.4.2). However, simply repeating each of these terms nine $(= 10 - 1)$ times so that $\epsilon_{9k+j} = 10^{-k}$ $(j = 0, 1, \ldots, 8; k = 0, 1, \ldots)$ does satisfy all the conditions of the theorem.

Example 3.4.3. The representation of $r = 1.5432$ using this decimal CORDIC decomposition would be obtained using

$$\delta_0 = \delta_1 = \delta_3 = \delta_5 = \delta_7 = +1 \qquad \delta_2 = \delta_4 = \delta_6 = \delta_8 = -1$$

so that $s_9 = 1.0$ and $r - s_9 = 0.5432$;

$$\delta_9 = \delta_{10} = \delta_{11} = \delta_{12} = \delta_{13} = \delta_{14} = \delta_{16} = +1 \qquad \delta_{15} = \delta_{17} = -1$$

so that $s_{18} = 1.5$ and $r - s_{18} = 0.0432$;

$$\delta_{18} = \delta_{19} = \delta_{20} = \delta_{21} = \delta_{22} = \delta_{24} = \delta_{26} = +1 \qquad \delta_{23} = \delta_{25} = -1$$

so that $s_{27} = 1.55$ and $r - s_{27} = -0.0068$;

$$\delta_{27} = \delta_{28} = \delta_{29} = \delta_{30} = \delta_{31} = \delta_{32} = \delta_{33} = \delta_{35} = -1 \qquad \delta_{34} = +1$$

so that $s_{36} = 1.543$ and $r - s_{36} = +0.0002$; and so on. Again it is easy to check that the claimed accuracy is achieved at every stage of the decomposition.

Despite the early emphasis on the need for decimal computation, throughout the rest of the development of the CORDIC algorithms, we will, for the sake of simplicity, describe the binary equivalents of these algorithms. The corresponding decimal forms are obtained in precisely the way outlined in the above example. The general form of the binary algorithm is described in Algorithm 3.4.4 below. Simple changes in the choices of the parameters ϵ_k and m yield methods for evaluation of almost all the elementary functions and their inverses.

Algorithm 3.4.4   General binary CORDIC algorithm

Set x_0, y_0, z_0.

Loop   For $k = 0$ to N do

$$x_{k+1} = x_k - m\delta_k y_k 2^{-k} \tag{3.4.9a}$$

$$y_{k+1} = y_k + \delta_k x_k 2^{-k} \tag{3.4.9b}$$

$$z_{k+1} = z_k - \delta_k \epsilon_k \tag{3.4.9c}$$

Here m takes one of the values 0 or ± 1, and so these operations consist of just (binary) exponent shifts and additions and/or subtractions.

The choice of the signs is made in one of two ways:

1. The *rotation* mode: $\qquad \delta_k = \text{sgn}(z_k)$
2. The *vectoring* mode: $\qquad \delta_k = -\text{sgn}(y_k)$

The names of these two modes are a further consequence of the history of the methods being developed for navigational purposes. There is a description of the algorithm in which the geometric significance of these terms is explained, but it does not add anything to the mathematical understanding of the method. The choices of the parameters are

Arithmetic operations: $\quad m = 0 \qquad \epsilon_k = 2^{-k}$
Trigonometric functions: $m = 1 \qquad \epsilon_k = \arctan 2^{-k}$
Hyperbolic functions: $\quad m = -1 \quad \epsilon_k = \tanh^{-1} 2^{-k} \qquad (k \geq 1)$

We have already seen that the first of these sequences satisfies (3.4.1) and (3.4.2), and it is easy to establish that the others are both decreasing sequences of positive terms.

To see that the sequence $\epsilon_k = \arctan 2^{-k}$ satisfies (3.4.2), it is sufficient to demonstrate that the terms decrease more slowly than 2^{-k}. That is, we must show that $2\epsilon_{k+1} > \epsilon_k$ for each k. The tangent function is increasing on $[0, 1]$, and so it will be sufficient to establish that $\tan 2\epsilon_{k+1} > \tan \epsilon_k$. Now

$$\tan 2\epsilon_{k+1} = \frac{2\tan \epsilon_{k+1}}{1 - \tan^2 \epsilon_{k+1}} > 2\tan \epsilon_{k+1} = 2(2^{-k-1}) = 2^{-k} = \tan \epsilon_k$$

as desired. We also have $\epsilon_k = \arctan 2^{-k} < 2^{-k}$, and so the series $\Sigma \epsilon_n$ is convergent.

For the final choice, $\epsilon_k = \tanh^{-1} 2^{-k}$, (3.4.2) is *not* satisfied. The sequence decreases just a little faster than the geometric sequence 2^{-k}, and this is sufficient to spoil its behavior. However, it can be proved that if we simply insert a repetition of the terms $\epsilon_4, \epsilon_{13}, \epsilon_{40}, \ldots$, that is, ϵ_{j_n} for $j_n > 1$ satisfying

$$j_n = 3j_{n-1} + 1 \qquad j_0 = 1$$

which are given by

$$j_n = 1 + 3 + 3^2 + \cdots + 3^n = \frac{3^{n+1} - 1}{2}$$

then the resulting sequence does satisfy all the required conditions.

We begin with the simplest cases: the arithmetic operations of multiplication and division performed by the CORDIC algorithms. Since for these operations $m = 0$, it follows from (3.4.9a) that $x_k = x_0$ throughout the calculation which therefore reduces to the evaluation of the two sequences (y_k) and (z_k) given by

$$y_{k+1} = y_k + \delta_k x_0 2^{-k} \tag{3.4.10b}$$

$$z_{k+1} = z_k - \delta_k 2^{-k} \tag{3.4.10c}$$

In the vectoring mode, we choose $z_0 = 0$ and $\delta_k = -\text{sgn } y_k$, and so it follows that $|y_k|$ is being reduced at every iteration. Applying (3.4.10b) repeatedly, we see that

$$y_{k+1} = y_0 + \sum_{n=0}^{k} \delta_n x_0 2^{-n} \tag{3.4.11}$$

from which it follows (provided $|y_0/x_0| < \Sigma 2^{-n} = 2$) that $y_k \to 0$, and by Theorem 3.4.1 for each k we have

$$\left| y_0 + \sum_{n=0}^{k} \delta_n x_0 2^{-n} \right| \leq x_0 2^{-k} \tag{3.4.12}$$

Now

$$z_{k+1} = -\sum_{n=0}^{k} \delta_n 2^{-n} \tag{3.4.13}$$

from which we may deduce, using (3.4.12), that

$$\left| z_{k+1} - \frac{y_0}{x_0} \right| \leq 2^{-k} \tag{3.4.14}$$

Thus we have proved that the vectoring mode CORDIC algorithm with $m = 0$ using $z_0 = 0$ generates a sequence of approximations convergent to y_0/x_0.

In much the same way, we can see that using the *rotation* mode with $y_0 = 0$ yields a sequence (y_k) such that

$$|y_{k+1} - x_0 z_0| \leq x_0 2^{-k} \tag{3.4.15}$$

(See Exercise 4.)

It should be observed here that the range restrictions are not restrictions at all for floating-point arithmetic. In the case of division of two normalized binary floating-point numbers, the ratio of their mantissas can never exceed 2 since each lies in $[\frac{1}{2}, 1)$. Similarly, for multiplication, the error bound (3.4.15) is itself bounded by 2^{-k}. The important implication of this is that the number of iterations required to achieve a specified accuracy can easily be determined in advance.

For example, Turbo Pascal type real requires a relative error no greater than 2^{-40}, and since the ratio of the two mantissas lies in $(\frac{1}{2}, 2)$ it is sufficient to compute this ratio with an absolute error bounded by 2^{-41} which can be achieved in 42 steps of the CORDIC algorithm. (While this may seem a large number of steps compared with the effort expended in reducing the number of terms of various power series, it should be recalled that each of these steps involves nothing more complicated than shifts and additions.)

Example 3.4.5. Consider the CORDIC multiplication of the floating-point binary representations of 5.625 and 3.0625.

Take $X = 5.625 = (0.101101)_2 (2)^3$, $Z = 3.0625 = (0.110001)_2 (2)^2$. Then the product is

$$XZ = (0.101101)_2 (0.110001)_2 (2)^5$$

For the CORDIC algorithm we choose $x_0 = (0.101101)_2$, $z_0 = (0.110001)_2$ together with $y_0 = 0$. To achieve the same relative accuracy as in the original representation, we seek an error no greater than 2^{-9} in the product $x_0 z_0$, for which 10 steps of the algorithm will suffice. This allows for the correct rounding of the result and the possible renormalization shift in the final result. We use the relations (3.4.10b) and (3.4.10c) with $\delta_k = \operatorname{sgn} z_k$. We obtain

$$\delta_0 = +1: \quad y_1 = (0.101101)_2$$

$$z_1 = (0.110001)_2 - 1 = (-0.001111)_2$$

$$\delta_1 = -1: \quad y_2 = (0.101101)_2 - \frac{(0.101101)_2}{2} = (0.0101101)_2$$

$$z_2 = (-0.001111)_2 + \tfrac{1}{2} = (0.010001)_2$$

$$\delta_2 = +1: \quad y_3 = y_2 + \frac{(0.101101)_2}{4} = (0.10000111)_2$$

$$z_3 = z_2 - \tfrac{1}{4} = (0.000001)_2$$

$$\delta_3 = +1: \quad y_4 = y_3 + \frac{(0.101101)_2}{8} = (0.100111011)_2$$

$$z_4 = z_3 - \tfrac{1}{8} = (-0.000111)_2$$

$$\delta_4 = -1: \quad y_5 = y_4 - \frac{(0.101101)_2}{16} = (0.1001001001)_2$$

$$z_5 = z_4 + \tfrac{1}{16} = (-0.000011)_2$$

$$\delta_5 = -1: \quad y_6 = y_5 - \frac{(0.101101)_2}{32} = (0.10001100101)_2$$

$$z_6 = z_5 + \tfrac{1}{32} = (-0.000001)_2$$

$$\delta_6 = -1: \quad y_7 = y_6 - \frac{(0.101101)_2}{64} = (0.100010010101)_2$$

$$z_7 = z_6 + \tfrac{1}{64} = 0$$

At this point, we have in fact found the exact product for this example, but the algorithm would continue to produce $\delta_7 = -1$ together with the corresponding changes to y and z. This error would be gradually corrected by the subsequent choices $\delta_8, \delta_9 = +1$. Rounded to the required 6 bits, the answer above is unchanged; we obtain $x_0 z_0 \simeq (0.100010)_2$, and therefore the final product $XZ \simeq (10001.0)_2 = 17.0$. The true product is $17.226\,562\,5$, so the error is indeed less than $\frac{1}{2}$ in the final place, or $\frac{1}{4}$.

This phenomenon of reaching the exact decomposition of z_0 will always occur for the multiplication case, and so that algorithm would be terminated after the same number of steps as there are bits in the representation. For division, however, this will not generally be true, since the division algorithm decomposes y_0 as $-\Sigma\, \delta_k x_0 2^{-k}$, and this decomposition will not typically be exact for any partial sum.

Consider now the evaluation of the trigonometric functions. We set $m = 1$, and $\epsilon_k = \arctan 2^{-k}$. The rotation mode allows the computation of $\sin\theta$ and $\cos\theta$ by setting $z_0 = \theta$ and decomposing this as $\Sigma\, \delta_k \epsilon_k$. The vectoring mode is used for the arctangent function, square roots, and conversion to polar coordinates.

To justify these statements, we use the standard trigonometric identities for $\sin(a + b)$ and $\cos(a + b)$ to obtain

$$\cos s_{k+1} = \cos(s_k + \delta_k \epsilon_k) = \cos s_k \cos \delta_k \epsilon_k - \sin s_k \sin \delta_k \epsilon_k$$

$$= \cos \epsilon_k (\cos s_k - \sin s_k \tan \delta_k \epsilon_k)$$

$$= \cos \epsilon_k (\cos s_k - \delta_k 2^{-k} \sin s_k) \qquad (3.4.16)$$

and similarly

$$\sin s_{k+1} = \cos \epsilon_k (\sin s_k + \delta_k 2^{-k} \cos s_k) \qquad (3.4.17)$$

where $s_k = \theta - z_k$.

Temporarily, we denote $\cos \epsilon_k$ by c_k, $\cos s_k$ by ξ_k, and $\sin s_k$ by η_k. Equations (3.4.16) and (3.4.17) together with the relation for the z_k's are now

$$\xi_{k+1} = c_k(\xi_k - \delta_k \eta_k 2^{-k}) \qquad (3.4.18a)$$

$$\eta_{k+1} = c_k(\eta_k + \delta_k \xi_k 2^{-k}) \qquad (3.4.18b)$$

$$z_{k+1} = z_k - \delta_k \epsilon_k . \qquad (3.4.18c)$$

Now $|z_{k+1}| \le \epsilon_k$, and so it follows that s_{k+1} approximates θ to this same accuracy and hence that ξ_{k+1} and η_{k+1} approximate $\cos\theta$ and $\sin\theta$ (respectively) with error no greater than 2^{-k}.

As before, the number of terms of these sequences that are required for any specified accuracy can easily be determined. Suppose then that $N + 1$ such steps are required.

Now, $s_0 = 0$, and so $\xi_0 = 1$, $\eta_0 = 0$. In Eq. (3.4.18), we see that the same multiplier c_k is used for both sequences. It follows that we can determine an overall multiplier

$$K_T = \prod_{k=0}^{N} c_k \qquad (3.4.19)$$

which can also be stored. The initial values ξ_0, η_0 are then replaced by $x_0 = K_T$, $y_0 = 0$ and the recurrences (3.4.18a) and (3.4.18b) by

$$x_{k+1} = x_k - \delta_k y_k 2^{-k} \tag{3.4.20a}$$

$$y_{k+1} = y_k + \delta_k x_k 2^{-k} \tag{3.4.20b}$$

which are (3.4.9a), (3.4.9b) with $m = 1$. From our analysis, it follows that x_{N+1}, y_{N+1} approximate $\cos\theta$, $\sin\theta$ with the desired accuracy. Note that we do *not* have $|x_{k+1} - \cos\theta| < 2^{-k}$ or $|y_{k+1} - \sin\theta| < 2^{-k}$ for $k < N$ because of the premultiplication by K_T.

In the vectoring mode, with $z_0 = 0$, we obtain

$$z_{N+1} \simeq \arctan \frac{y_0}{x_0} \tag{3.4.21}$$

$$x_{N+1} \simeq \frac{\sqrt{x_0^2 + y_0^2}}{K_T} \tag{3.4.22}$$

Example 3.4.6. Use the CORDIC algorithm just defined to find $\arctan 1.6875$ with error less than 2^{-6}.

Solution. For simplicity of notation, we will use the decimal representation for all quantities in this example. To achieve the specified accuracy we must compute z_7 in the vectoring mode. The constants used will therefore be

$$\epsilon_0 = \arctan 1 = 0.7854$$
$$\epsilon_1 = \arctan \tfrac{1}{2} = 0.4636$$
$$\epsilon_2 = \arctan \tfrac{1}{4} = 0.2450$$
$$\epsilon_3 = 0.1244 \qquad \epsilon_4 = 0.0624$$
$$\epsilon_5 = 0.0312 \qquad \epsilon_6 = 0.0156$$

to four decimal places which is, of course, much greater accuracy than is required for the output.

From (3.4.21), we see that there is a choice of starting values open to us. We will adopt the simplest option, namely, $x_0 = 1$, $y_0 = 1.6875$. The choice of δ_k is $-\text{sgn } y_k$, and therefore we get

$$\delta_0 = -1: \quad x_1 = 1 + 1.6875 = 2.6875$$
$$y_1 = 1.6875 - 1 = 0.6875$$
$$z_1 = 0 + \epsilon_0 = 0.7854$$

$$\delta_1 = -1: \quad x_2 = 2.6875 + \frac{0.6875}{2} = 3.0313$$
$$y_2 = 0.6875 - \frac{2.6875}{2} = -0.6563$$
$$z_2 = 0.7854 + \epsilon_1 = 1.2490$$

$$\delta_2 = +1: \quad x_3 = 3.0313 + \frac{0.6563}{4} = 3.6876$$

$$y_3 = -0.6563 + \frac{3.0313}{4} = 0.1056$$

$$z_3 = 1.2490 + \epsilon_2 = 1.0040$$

$\delta_3 = -1$: $x_4 = 3.2085$ $y_4 = -0.2954$ $z_4 = 1.1284$

$\delta_4 = +1$: $x_5 = 3.2270$ $y_5 = -0.0937$ $z_5 = 1.0660$

$\delta_5 = +1$: $x_6 = 3.2299$ $y_6 = +0.0071$ $z_6 = 1.0348$

$\delta_6 = -1$: $x_7 = 3.2300$ $y_7 = -0.0434$ $z_7 = 1.0504$

To the accuracy shown, the true value is 1.0358, so our answer has indeed achieved an error less than $\frac{1}{64} = 0.015\,625$. Note here one of the characteristic traits of the CORDIC algorithms: the convergence is not such that the error is being reduced at every iteration. The value of z_6 in this example is very much closer to the true value than is z_7. The important fact is that a guaranteed accuracy can be achieved in a fixed and predetermined number of steps.

We should also observe that the sequence (x_n) generated here is converging to the predicted value $\sqrt{x_0^2 + y_0^2}/K_T$, which for this example is $\sqrt{1 + 1.6875^2}/0.6073 = 3.2299$ to four decimal places.

As a final remark on the CORDIC algorithms for the trigonometric functions, we should pay some attention to the range of validity of the scheme. Now for the trigonometric functions themselves, the decomposition of z_0 as $\Sigma \delta_k \epsilon_k$ will be successful provided $|z_0| \le \Sigma \epsilon_k = E$, say. Now

$$E > \arctan 1 + \arctan 0.5 + \arctan 0.25 + \arctan 0.125 > \frac{\pi}{2}$$

and so a range reduction along the lines of, but simpler than, that used for the computer trigonometric function routines in Sec. 3.3 will suffice to extend the validity of the CORDIC algorithm to all values of the argument. In the case of the arctangent function itself, the fact that $E > 1$ combined with the two fundamental identities (3.3.23a) and (3.3.23b) is similarly sufficient to cover all possible arguments.

The final group of functions for which the CORDIC scheme is well-suited consists of the hyperbolic functions, including the exponential and logarithmic functions. An analysis almost identical to that for the trigonometric functions shows that using $m = -1$ together with $\epsilon_k = \tanh^{-1} 2^{-k}$ $(k = 1, 2, \ldots)$ and the repetitions for $k = 4, 13, \ldots$ then the sequences defined by (3.4.9) can be used to generate the functions cosh, sinh, exp, $\tanh^{-1}$, ln, and the square root.

In this case we define the constant K_H by

$$K_H = \prod \cosh \epsilon_k \qquad (3.4.23)$$

where the product includes the repetitions just mentioned. Then the rotation mode with $x_1 = K_H$, $y_1 = 0$ generates

$$x_{N+1} \approx \cosh z_1 \qquad y_{N+1} \approx \sinh z_1 \qquad (3.4.24)$$

while the vectoring mode with $z_1 = 0$ yields

$$z_{N+1} \simeq \tanh^{-1} \frac{y_1}{x_1} = \frac{\ln w}{2} \tag{3.4.25}$$

where $x_1 = w + 1$, $y_1 = w - 1$ and

$$x_{N+1} \simeq \frac{\sqrt{x_1^2 - y_1^2}}{K_H} = \frac{\sqrt{v}}{K_H} \tag{3.4.26}$$

where $x_1 = v + \frac{1}{4}$, $y_1 = v - \frac{1}{4}$.

This version of the CORDIC algorithm is implemented in Program 3.4.7 below.

The recurrence relations (3.4.9a) and (3.4.9b) for the case $m = -1$ become $x_{k+1} = x_k + \delta_k y_k 2^{-k}$ and $y_{k+1} = y_k + \delta_k x_k 2^{-k}$, and denoting the sum $x_k + y_k$ by u_k, these yield

$$u_{k+1} = u_k + \delta_k u_k 2^{-k} \tag{3.4.27}$$

From Eqs. (3.4.24) we may deduce that

$$u_{N+1} \simeq \cosh z_1 + \sinh z_1 = e^{z_1} \tag{3.4.28}$$

Similarly, the sequence $v_k = x_k - y_k$ satisfies the recurrence relation

$$v_{k+1} = v_k - \delta_k v_k 2^{-k}$$

and generates an approximation to e^{-z_1}. It is thus possible to modify the hyperbolic version of the CORDIC algorithm to work directly with the exponential function and obtain the hyperbolic functions from this. This would perhaps be a preferred option for a computer function library in which the exponential would be the usual standard function.

These algorithms will converge for any argument $|z_1| < \Sigma \, \epsilon_k$, which for this choice of (ϵ_k) is approximately 1.13. One of the easiest ways to reduce the argument z, say, of the exponential function into this range is to set

$$M = \left[\frac{z}{\ln 2} \right] \qquad z_1 = z - M \ln 2$$

so that $e^z = e^{z_1} 2^M$. This is well-suited to the binary floating-point system since $|z_1| < \ln 2$, and so $e^{z_1} \in (\frac{1}{2}, 2)$, which implies that the result will need at most a single binary-place normalization shift.

The number of steps of the CORDIC algorithm implemented in this program is intended to produce single-precision accuracy in the results. However, it will be observed that there are some trailing-figure discrepancies between the computed values and the built-in functions. For the exponential function, these are simply the result of the fact that the exponential function values are obtained by adding those of the sinh and cosh functions, which are themselves computed only to single precision.

Program 3.4.7   CORDIC algorithm; hyperbolic functions

```
program cordic_h;
var
    eps:array[1..45] of real;
    rep:set of byte; (* used for repetitions *)
    KH, x, y, z:real;
    n:integer;
function cosh(x:real):real;
begin cosh:=(exp(x) + exp(-x))/2; end;
    (* Used only for generating constants *)

procedure make_table;
var u:real;
    i:integer;
begin
    u:=1; KH:=1;
    for i:=1 to 28 do begin
        u:=u/2; eps[i]:=ln((1 + u)/(1 - u))/2;
        KH:=KH * cosh(eps[i]);
        if i in rep then KH:=KH * cosh(eps[i]);
    end;
end;

function sgn(x:real):shortint;
begin if x > =0 then sgn:=1 else sgn:=-1; end;

procedure rotation (z:real; var x, y:real);
    (* Computes CORDIC cosh, sinh and exp *)
var i:integer
    d:shortint;
    u, xx:real;
begin
    u:=1; x:=KH; y:=0;
    for i:=1 to 28 do begin
        u:=u/2; d:=sgn(z);
        xx:=x + d*y*u;
        y:=y + d*x*u; x:=xx;
        z:=z - d*eps[i];
        if i in rep then begin
            d:=sgn(z);
            xx:=y + d*y*u;
            y:=y + d*x*u; x:=xx;
            z:=z - d*eps[i];
        end;
    end;
end;

procedure vectoring(y:real; var x, z:real);
    (* Computes CORDIC ln and square root *)
```

```
            var i: integer;
                d: shortint;
                u, xx: real;
        begin
            u:=1; z:=0;
            for i:=1 to 28 do begin
                u:=u/2;
                d:=-sgn(y);
                xx:=x + d * y * u;
                y:=y + d * x * u; x:=xx;
                z:=z - d * eps[i];
                if i in rep then begin
                    d:=-sgn(y);
                    xx:=x + d * y * u;
                    y:=y + d * x * u; x:=xx;
                    z:=z - d * eps[i];
                end;
            end;
        end;

    begin
        rep:=[4, 13, 40];
        make_table;
        for n:=0 to 10 do begin
            rotation (n/10, x, y);
            writeln(x:15:8, y:15:8, x + y:20:8, exp(n/10):18:8);
        end;
        for n:=1 to 10 do begin
            x:=1 + n/10;
            vectoring(n/10 - 1, x, z);
            writeln(2 * z:15:8, ln(n/10):20:8);
        end;
        for n:=1 to 10 do begin
            x:=n/10 + 1/4;
            vectoring(n/10 - 1/4, x, z);
            writeln(x * KH:15:8, sqrt(n/10):20:8);
        end;
    end.
```

Remarks

1. The most marked difference between the computed values and the true ones occurs for ln 0.1. It will be observed that this lies outside the range of convergence of the algorithm. The value obtained is (approximately) twice the negative sum of the ϵ_k.

2. The inclusion of 40 in the set of steps to be repeated is so that the program could be easily amended for a higher precision.

TABLE 3.4.8
Output from the CORDIC algorithm of Program 3.4.7

x	CORDIC cosh x	CORDIC sinh x	CORDIC exp x	Built-in exp x
0.0	1.000 000 00	0.000 000 00	1.000 000 00	1.000 000 00
0.1	1.005 004 17	0.100 166 75	1.105 170 92	1.105 170 92
0.2	1.020 066 76	0.201 336 00	1.221 402 76	1.221 402 76
0.3	1.045 338 51	0.304 520 30	1.349 858 81	1.349 858 81
0.4	1.081 072 37	0.410 752 33	1.491 824 70	1.491 824 70
0.5	1.127 625 96	0.521 095 30	1.648 721 27	1.648 721 27
0.6	1.185 465 22	0.636 653 58	1.822 118 80	1.822 118 80
0.7	1.255 169 01	0.758 583 71	2.013 752 71	2.013 752 71
0.8	1.337 434 95	0.888 105 98	2.225 540 93	2.225 540 93
0.9	1.433 086 39	1.026 516 73	2.459 603 12	2.459 603 11
1.0	1.543 080 63	1.175 201 19	2.718 281 82	2.718 281 83

x	CORDIC ln x	Built-in ln x	CORDIC sqrt x	Built-in sqrt x
0.1	−2.236 346 02	−2.302 585 09	0.316 227 77	0.316 227 77
0.2	−1.609 437 90	−1.609 437 91	0.447 213 60	0.447 213 60
0.3	−1.203 972 81	−1.203 972 80	0.547 722 56	0.547 722 56
0.4	−0.916 290 72	−0.916 290 73	0.632 455 53	0.632 455 53
0.5	−0.693 147 17	−0.693 147 18	0.707 106 78	0.707 106 78
0.6	−0.510 825 63	−0.510 825 62	0.774 596 67	0.774 596 67
0.7	−0.356 674 95	−0.356 674 94	0.836 660 03	0.836 660 03
0.8	−0.223 143 55	−0.223 143 55	0.894 427 19	0.894 427 19
0.9	−0.105 360 51	−0.105 360 52	0.948 683 30	0.948 683 30
1.0	−0.000 000 01	0.000 000 00	1.000 000 00	1.000 000 00

EXERCISES 3.4

1. Find the binary CORDIC decomposition of 1.2345 using $\epsilon_k = 2^{-k}$. Verify that at each stage the error is bounded by 2^{-n}.

2. Show that if (ϵ_k) is any geometric progression with common ratio less than $\frac{1}{2}$ then the condition (3.4.2) is not satisfied and so such a sequence cannot be the basis of a CORDIC algorithm.

3. Use the binary CORDIC algorithm with 10 steps to find $5.6875/3.3125$.

4. Show that the sequence (y_k) generated by the rotation mode with $m = 0$ converges to $x_0 z_0$ and that the error in y_{k+1} is bounded by $x_0 2^{-k}$. How many steps of this algorithm are necessary to yield IEEE single precision in the product of two normalized binary floating-point numbers? Write a program to implement the algorithm; use it to check that this number of steps is sufficient.

5. Write a Turbo Pascal program to implement the CORDIC trigonometric function algorithm. Determine the number of steps necessary to yield the correct precision for variables of type real. Test your program for sin x and arctan x with $x = 0, 0.1, 0.2, \ldots, 1$.

PROJECTS

1. Program Algorithm 3.3.4 for the natural logarithm to deliver IEEE single-precision accuracy. Tabulate the values given by this routine and the built-in ln function to eight significant figures to check the accuracy. Use $x = 0.1, 0.2, \ldots, 1, 2, \ldots, 10$.

2. Write the most efficient routine you can for evaluation of the natural logarithm function. Your program should include a range-reduction scheme to cope with all valid representable arguments and deliver an appropriate error message for nonpositive arguments.

3. Produce a table similar to Table 3.3.5 which gives the value of $\sin x$ and $\cos x$ in terms of $t = \tan A/2$, where A is the result of a range reduction into $[0, \pi/4]$. Write a program to test the validity of your table.

4. Program the arctangent routine of Algorithm 3.3.8 for the evaluation of the arcsine and arccosine functions. Are the same numbers of terms sufficient for these functions? Compare the results with those of the iterative algorithm for arcsine developed in the Chap. 2 projects.

5. Find the number of terms of the arctangent series which would be necessary to attain the precision of Turbo Pascal real variables if the algorithm of Sec. 3.3 is used with $a = [16x]/16$. Modify the algorithm to implement this, and verify that your procedure delivers the required accuracy.

CHAPTER
4

POLYNOMIAL
INTERPOLATION

4.1   INTRODUCTION

In this chapter we begin our study of methods for the approximation of functions other than the elementary functions. Initially, we concentrate on techniques which may be adopted if, for example, we have a table of values of the function which may have been obtained either by some other direct computation or by physical or other experiments.

In a table of values of a Bessel function, for example, we have as data an array of values at points of a uniform mesh, but typically we may need to be able to infer from this information the value of the function at some nontabular point or points. The method of approximation adopted may well vary according to whether we require just one evaluation or a whole set of such function values. The technique adopted may also be dependent on the accuracy of the table entries and on the precision required in the final answer.

Initially, we will concentrate on polynomial interpolation. That is, we will use approximations based on the idea of finding a polynomial which agrees with, or *interpolates*, the data and using this polynomial in place of the original function to estimate its values at other points. This fundamental idea is based on an application of Weierstrass's theorem which, you will recall, states that if f is a continuous function on the interval $[a, b]$, then for any $\epsilon > 0$, there exists a polynomial p such that

$$\|f - p\|_\infty = \max_{a \le x \le b} |f(x) - p(x)| < \epsilon \qquad (4.1.1)$$

That is, any continuous function may be approximated, in the uniform or supremum norm, as closely as we wish by a polynomial.

To put the idea of polynomial interpolation into a familiar framework, it is well-known, and perhaps obvious, that there is a unique straight line passing through any two points in the plane. This straight line is (the graph of) the linear, or first-degree, interpolating polynomial for this data, the coordinates of the two points. In precisely the same way it is easy to show that there is a unique parabola which passes through any three points: this is the graph of the quadratic, or second-degree, interpolating polynomial which agrees with the given data. In general, we will find that there is a unique polynomial of the minimum possible degree n which agrees with a given function at $n + 1$ data points, or *nodes*.

The drawbacks of this polynomial interpolation approach include the following. If the number of data points is large, then the degree of the interpolating polynomial will be high and therefore its evaluation will itself be computationally expensive. Second, few physical laws are of high-degree polynomial nature, and so the approximating function is likely to exhibit less stable behavior than the function or phenomenon being approximated. (Note that high-degree polynomials will often have many turning points, which is again untypical of physical phenomena.) The use of a single interpolation polynomial over a wide range gives equal weight to function values distant from the point of interest as it does to those at nearby points.

It is likely to be more fruitful to use several lower-degree polynomials which utilize data at points close to the required one. This is achieved by the various difference schemes which we will discuss shortly or by the use of spline interpolation. The latter is based on the idea of a piecewise polynomial which interpolates the complete set of data but which is of low degree and therefore simple to evaluate in any subrange.

All these approaches and their associated errors are to be discussed in the next two chapters. We will see that the problem is often reduced to the necessity of solving a (sometimes large) system of linear equations. This important topic is treated separately in Chap. 6.

Before proceeding to the details of any one interpolation scheme, it is necessary to point out one further difficulty which may arise in practice. If the data is itself the result of practical experiment and therefore subject to its own errors, then it is probably unwise to insist that any approximation scheme agrees exactly with this erroneous data. This is one of the primary motivations for the use of approximation algorithms rather than interpolation. The idea here is to find a function whose graph passes close to all the data points without necessarily fitting any of the data *exactly*. This will be the subject of Chap. 7.

4.2 THE LAGRANGE INTERPOLATION POLYNOMIAL

In this section, we derive a formula for the polynomial of minimum degree which takes specified values at a given set of points. We have already observed that there is a unique polynomial of degree at most one whose graph goes through two specified points, and this will turn out to be part of the general pattern—there is a unique polynomial of degree at most N which interpolates $N+1$ given values. There are many different ways of establishing this fact.

Suppose that the values $\{f(x_i): i=0,1,\ldots,N\}$ are given and let p be the minimal-degree interpolation polynomial. Suppose that p is of degree M; in terms of the standard basis, p can be written in the form

$$p(x) = a_M x^M + a_{M-1} x^{M-1} + \cdots + a_1 x + a_0 \qquad (4.2.1)$$

We require that p satisfy the interpolation conditions

$$p(x_i) = f(x_i) \qquad (i=0,1,\ldots,N) \qquad (4.2.2)$$

Substituting expression (4.2.1) into (4.2.2) for each x_i in turn, we obtain the system of equations

$$a_M x_0^M + a_{M-1} x_0^{M-1} + \cdots + a_1 x_0 + a_0 = f(x_0)$$
$$a_M x_1^M + a_{M-1} x_1^{M-1} + \cdots + a_1 x_1 + a_0 = f(x_1) \qquad (4.2.3)$$
$$\cdots\cdots\cdots\cdots\cdots\cdots\cdots\cdots$$
$$a_M x_N^M + a_{M-1} x_N^{M-1} + \cdots + a_1 x_N + a_0 = f(x_N)$$

which can be written in matrix form, reversing the order of the terms on the left-hand sides of (4.2.3), as

$$\begin{bmatrix} 1 & x_0 & x_0^2 & \cdots & x_0^M \\ 1 & x_1 & x_1^2 & \cdots & x_1^M \\ \cdots\cdots\cdots\cdots\cdots \\ 1 & x_N & x_N^2 & \cdots & x_N^M \end{bmatrix} \begin{bmatrix} a_0 \\ a_1 \\ \cdots \\ a_M \end{bmatrix} = \begin{bmatrix} f(x_0) \\ f(x_1) \\ \cdots \\ f(x_N) \end{bmatrix} \qquad (4.2.4)$$

This is a system of $N+1$ equations in the $M+1$ unknown parameters $a_0, a_1, \ldots, a_M$. In general, we expect such a system to have no solution if $N > M$. If, on the other hand, $M > N$, then we do not expect the solution to be unique. This system will have a unique solution if $M = N$ and if the matrix of coefficients above is nonsingular.

This matrix is called a *Vandermonde* matrix. It is reasonably straightforward to establish that its determinant, when $N = M$, is nonzero provided that the interpolation points or nodes are all distinct. (See Exercise 2.) It follows that there is a unique polynomial with degree no greater than N which satisfies Eqs. (4.2.3). We present in the next theorem an alternative, constructive proof of this result whose outcome is the formula for the Lagrange interpolation polynomial.

This particular formula is the result of expressing the equations above in terms of a different basis of the space Π_N of all polynomials of degree no greater than N. This linear space of dimension $N+1$ is sometimes referred to as the space of polynomials of *order* N. The basis we use here consists of polynomials of degree exactly N, each with N specified zeros at interpolation points and the value 1 at the other of them. Since there are $N+1$ such points, there are also $N+1$ of these basis polynomials as required.

Theorem 4.2.1. Given the $N + 1$ values $f(x_0), f(x), \ldots, f(x_N)$ of the function f at the distinct points $x_0, x_1, \ldots, x_N$, there is a unique polynomial $p_N \in \Pi_N$ satisfying (4.2.2); that is,

$$p_N(x_i) = f(x_i) \qquad (i = 0, 1, \ldots, N)$$

Proof. We give here the constructive proof outlined above. If we can find polynomials $l_i(x)$ $(i = 0, 1, \ldots, N)$ of degree no more than N such that

$$l_i(x_j) = \delta_{ij} = \begin{cases} 1 & i = j \\ 0 & i \neq j \end{cases} \qquad (4.2.5)$$

then it will follow that the polynomial given by

$$p_N(x) = \sum_{i=0}^{N} f(x_i) l_i(x) \qquad (4.2.6)$$

will satisfy the interpolation conditions (4.2.2). Conditions (4.2.5) imply that all the zeros of l_i are known. It follows that

$$l_i(x) = C(x - x_0)(x - x_1) \cdots (x - x_{i-1})(x - x_{i+1}) \cdots (x - x_N) \qquad (4.2.7)$$

for some constant C. The value of this constant is obtained from the remaining condition in (4.2.5), namely, that $l_i(x_i) = 1$. We deduce that these basis polynomials are

$$l_i(x) = \frac{(x - x_0)(x - x_1) \cdots (x - x_{i-1})(x - x_{i+1}) \cdots (x - x_N)}{(x_i - x_0)(x_i - x_1) \cdots (x_i - x_{i-1})(x_i - x_{i+1}) \cdots (x_i - x_N)} \qquad (4.2.8)$$

The uniqueness of the interpolation polynomial is also easy to establish. Suppose that a second polynomial q, say, also of degree no more than N, satisfies the interpolation conditions (4.2.2). It follows that for each of the interpolation points, $p_N(x_i) = q(x_i)$ and therefore that the polynomial $p_N - q$ has zeros at all these $N + 1$ nodes. However, $p_N - q$ is a polynomial of degree at most N. From the fundamental theorem of algebra, we deduce that $p_N - q \equiv 0$, that is, $p_N = q$, and the uniqueness is proven.

Remarks

1. Formula (4.2.6) with the l_i given by (4.2.8) is called the *Lagrange interpolation polynomial*. It is the formula on which much of the theory of polynomial interpolation is founded.

2. The basis polynomials l_i form a linearly independent set in the linear space Π_N, and since there are exactly $N + 1$ of them, they are properly referred to as a *basis*. This particular basis is known as the *Lagrange basis* and its members as the *Lagrange basis polynomials*.

Example 4.2.2. Find the Lagrange interpolation polynomial which agrees with the following data. Use it to estimate the value of $f(2.5)$.

i	0	1	2	3
x_i	0	1	3	4
$f(x_i)$	3	2	1	0

Solution. The Lagrange basis polynomials for this case are

$$l_0(x) = \frac{(x-1)(x-3)(x-4)}{(0-1)(0-3)(0-4)}$$

$$l_1(x) = \frac{(x-0)(x-3)(x-4)}{(1-0)(1-3)(1-4)}$$

$$l_2(x) = \frac{(x-0)(x-1)(x-4)}{(3-0)(3-1)(3-4)}$$

$$l_3(x) = \frac{(x-0)(x-1)(x-3)}{(4-0)(4-1)(4-3)}$$

and therefore the interpolation polynomial is

$$
\begin{aligned}
p(x) &= 3l_0(x) + 2l_1(x) + 1l_2(x) + 0l_3(x) \\
&= \frac{3(x^3 - 8x^2 + 19x - 12)}{-12} + \frac{2(x^3 - 7x^2 + 12x)}{6} + \frac{x^3 - 5x^2 + 4x}{-6} \\
&= \frac{-x^3 + 6x^2 - 17x + 36}{12}
\end{aligned}
$$

The estimated value of $f(2.5)$ is $p(2.5) = 1.28125$.

As with any numerical technique, it is important to obtain bounds for the errors involved. In the case of the Lagrange interpolation formula, this is a relatively simple task. It will be convenient here and elsewhere to use the notation

$$L_N(x) = \prod_{i=0}^{N} (x - x_i) = (x - x_0)(x - x_1) \cdots (x - x_N) \qquad (4.2.9)$$

Theorem 4.2.3. Let p_N be the Lagrange interpolation polynomial agreeing with the function f at the $N+1$ distinct points $x_0, x_1, \ldots, x_N$, and suppose that f is at least $N+1$ times differentiable. Then the error in using $p_N(\bar{x})$ to estimate $f(\bar{x})$ is given by

$$R_N(\bar{x}) = f(\bar{x}) - p_N(\bar{x}) = \frac{L_N(\bar{x})f^{(N+1)}(\xi)}{(N+1)!} \qquad (4.2.10)$$

where ξ is some point in the interval I containing $x_0, x_1, \ldots, x_N, \bar{x}$.

Proof. First, if x is any of the nodes $x_0, x_1, \ldots, x_N$, then there is nothing to prove since p_N and f agree at such a point and L_N vanishes there also. Suppose then that x is not such a point and define the function G by

$$G(x) = f(x) - p_N(x) - CL_N(x) = R_N(x) - CL_N(x) \qquad (4.2.11)$$

where the constant C is to be chosen so as to make $G(\bar{x}) = 0$; that such a value C exists follows from the fact that $L_N(\bar{x}) \neq 0$.

We now have that G vanishes at all the $N+2$ distinct points $x_0, x_1, \ldots, x_N, \bar{x}$. Applying Rolle's theorem, it follows that G' vanishes at $N+1$ distinct points, and applying this same theorem repeatedly, we deduce that

$G^{(N+1)}$ vanishes somewhere in the interval I, at ξ say. Now, p_N is a polynomial of degree at most N, while L_N is of degree $N+1$ and has leading coefficient 1. It follows that $p_N^{(N+1)}(x) \equiv 0$ and $L_N^{(N+1)}(x) = (N+1)!$ and therefore that

$$G^{(N+1)}(x) = f^{(N+1)}(x) - C(N+1)! \tag{4.2.12}$$

Since $G^{(N+1)}(\xi) = 0$, it follows that $C = f^{(N+1)}(\xi)/(N+1)!$. The choice of C was such that $G(\bar{x}) = 0$, and so from (4.2.11) it follows that

$$R_N(\bar{x}) = CL_N(\bar{x}) = \frac{L_N(\bar{x})f^{(N+1)}(\xi)}{(N+1)!}$$

which is (4.2.10), as required.

Remarks

1. The *mean value point* ξ in the Lagrange error, or *remainder*, term is, of course, dependent on the point of interest $\bar{x}$. Its actual value could only be found in artificial cases where the function f could be evaluated directly.
2. It is apparent from formula (4.2.10) that the error in Lagrange interpolation is dependent on the choice of the nodes. If there is freedom to choose these, then it is naturally desirable to make the choice so that $|L_N(x)|$ is kept as small as possible throughout the interval I.
3. This last point also gives some insight into why interpolation is much more reliable than *extrapolation*. In the case of interpolation, $\bar{x}$ is within the interval spanned by $x_0, x_1, \ldots, x_N$, whereas for extrapolation, it is outside this interval. The polynomial $L_N(x)$ will increase in magnitude very rapidly outside the span of its zeros, whereas it will remain relatively small within that interval. We will discuss the question of the optimal choice of interpolation points shortly.

Example 4.2.4. Obtain a bound for the error in estimating the function $f(x) = \sin(\pi x/2)$ by the Lagrange interpolation polynomial agreeing with f at $x = 0, \frac{1}{2}, 1$.

Solution. To six decimal places, the data which must be fitted is

n	0	1	2
x_n	0	0.5	1
$f(x_n)$	0.000 000	0.707 107	1.000 000

and therefore the interpolation polynomial itself is

$$p_2(x) = -4(0.707\,107)x(x-1) + x(2x-1)$$

$$= 1.828\,428x - 0.828\,428x^2$$

Now, $f^{(3)}(x) = -(\pi^3/8)\cos(\pi x/2)$, and so the remainder term is given by

$$R_2(x) = -\frac{\pi^3}{8}\cos\frac{\pi\xi}{2}\,\frac{x(x-\frac{1}{2})(x-1)}{6}$$

which is bounded by $M\pi^3/96$, where M is an upper bound for $x(2x-1)(x-1) = 2l_2(x)$. Since L_2 vanishes at both ends of the interval $[0, 1]$, its maximum absolute value must be achieved at a local extremum. These occur at $(1 \pm 1/\sqrt{3})/2$, from which it follows that $M = 1/(6\sqrt{3})$. The bound on the error is therefore $\pi^3/(576\sqrt{3}) = 0.031$.

To test the validity of this error estimate, consider the errors in using p_2 to estimate $f(\frac{1}{3})$ and $f(\frac{2}{3})$. For $x = \frac{1}{3}$, we have $p_2(x) = 1.828\,428x - 0.828\,428x^2 = 0.517\,428$, while for $x = \frac{2}{3}$, $p_2(x) = 0.850\,762$ to six decimal places. These compare with the true values $\sin(\pi/6) = 0.500\,000$ and $\sin(\pi/3) = 0.866\,025$. Both these errors are around 0.017, which is well within (but of the same order of magnitude as) the bound obtained above.

While the Lagrange interpolation formula is at the heart of polynomial interpolation, it is not, by any stretch of the imagination, the most practical way to go about it. Just consider for a moment what is involved in adding in one further data point to the calculations above and you will be quickly convinced that there must be better techniques available. It is to these that we turn in the next few sections.

EXERCISES 4.2

1. Find the interpolation polynomial which agrees with the data $f(0.0) = 0.0000$, $f(0.1) = 0.1823$, $f(0.2) = 0.3365$, and $f(0.3) = 0.4700$, and use it to estimate the values $f(0.15)$, $f(0.28)$.

2. Prove that the Vandermonde determinant [that is, the determinant of the matrix in (4.2.4)] is nonzero by proving that

$$\begin{vmatrix} 1 & x_0 & x_0^2 & \cdots & x_0^N \\ 1 & x_1 & x_1^2 & \cdots & x_1^N \\ \cdots & \cdots & \cdots & \cdots & \cdots \\ 1 & x_N & x_N^2 & \cdots & x_N^N \end{vmatrix} = \prod_{i>j} (x_i - x_j)$$

Use this result to establish an alternative proof of Theorem 4.2.1.

3. Let $x_0, x_1, \ldots, x_N$ be $N+1$ distinct points. Prove that

$$\sum_{i=0}^{N} l_i(x) \equiv 1$$

[Hint: Consider interpolation of the data $f(x_i) = 1$ for every i.]

4. Suppose that $x_i = i$ ($i = 0, 1, \ldots, N$). Show that for every x

$$\sum_{i=0}^{N} il_i(x) = x$$

5. Given that the data in Exercise 1 is taken from the function $\ln(1 + 2x)$, obtain a bound for the interpolation errors. Verify these bounds for the particular values obtained in Exercise 1.

4.3 DIVIDED DIFFERENCE INTERPOLATION

In this section, we consider some of the more practical approaches to polynomial interpolation. To motivate the discussion, consider the process of increasing the degree of an interpolation polynomial by one by incorporating another data point into the formula.

Example 4.3.1. Estimate $\cos 1.12$ using polynomial interpolation with first three and then four points from the data

x	1.0	1.1	1.2	1.3
$\cos x$	0.5403	0.4536	0.3624	0.2675

Solution. In the first case, we use the points 1.0, 1.1, and 1.2 to obtain the estimate

$$\cos 1.12 \approx \frac{0.5403(1.12-1.1)(1.12-1.2)}{(1.0-1.1)(1.0-1.2)}$$

$$+ \frac{0.4536(1.12-1.0)(1.12-1.2)}{(1.1-1.0)(1.1-1.2)}$$

$$+ \frac{0.3624(1.12-1.0)(1.12-1.1)}{(1.2-1.0)(1.2-1.1)}$$

$$= 0.4357 \qquad \text{to four decimal places}$$

For the second case, we incorporate the additional data point $x_3 = 1.3$ and obtain

$$\cos 1.12 \approx \frac{0.5403(1.12-1.1)(1.12-1.2)(1.12-1.3)}{(1.0-1.1)(1.0-1.2)(1.0-1.3)}$$

$$+ \frac{0.4536(1.12-1.0)(1.12-1.2)(1.12-1.3)}{(1.1-1.0)(1.1-1.2)(1.1-1.3)}$$

$$+ \frac{0.3624(1.12-1.0)(1.12-1.1)(1.12-1.3)}{(1.2-1.0)(1.2-1.1)(1.2-1.3)}$$

$$+ \frac{0.2675(1.12-1.0)(1.12-1.1)(1.12-1.2)}{(1.3-1.0)(1.3-1.1)(1.3-1.2)}$$

$$= 0.4357 \qquad \text{to four decimal places}$$

The two values agree to the same accuracy as that of the data and are indeed correct to this precision. The point is, however, that the second calculation had to be started from scratch. There was no simple way of using the work that had already been done in the first part to simplify the second calculation. It is to simplify this process that we seek other representations of the interpolation polynomial.

Among the more straightforward approaches to this problem is the use of *divided difference* formulas. We begin with the necessary definitions and notation.

Definitions 4.3.2. Let $x_0, x_1, \ldots, x_N$ be distinct (interpolation) points. We define the *zeroth divided difference* at x_i by $f[x_i] = f(x_i)$. The *first-order*, or just *first*, *divided difference* at x_i, x_j is defined by

$$f[x_i, x_j] = \frac{f(x_i) - f(x_j)}{x_i - x_j} = \frac{f[x_i] - f[x_j]}{x_i - x_j} \tag{4.3.1}$$

In general, the *k*th *divided difference* $f[x_0, x_1, \ldots, x_k]$ is defined by

$$f[x_0, x_1, \ldots, x_k] = \frac{f[x_0, x_1, \ldots, x_{k-1}] - f[x_1, x_2, \ldots, x_k]}{x_0 - x_k} \tag{4.3.2}$$

Although we have defined some of these divided differences for general points, it is usual to consider the differences for sequences of consecutive nodes, as in the final definition above. Note too that (4.3.1) implies that $f[x_i, x_j] = f[x_j, x_i]$ and that (4.3.2) could have been written as

$$f[x_0, x_1, \ldots, x_k] = \frac{f[x_1, x_2, \ldots, x_k] - f[x_0, x_1, \ldots, x_{k-1}]}{x_k - x_0}$$

We have already seen the use of one divided difference formula in the secant method for the solution of a nonlinear equation. There we used the iteration formula

$$x_{n+1} = x_n - f(x_n) \frac{x_n - x_{n-1}}{f(x_n) - f(x_{n-1})} = x_n - \frac{f(x_n)}{f[x_{n-1}, x_n]} \qquad (4.3.3)$$

which we see is just Newton's iteration with the derivative replaced by this first divided difference. The relation between divided differences and derivatives of the same order will be explored a little latter.

Example 4.3.3. Find the divided differences $f[x_0, x_1]$, $f[x_0, x_1, x_2]$, $f[x_0, x_2, x_1]$ for the data

i	0	1	2
x_i	1.0	1.5	2.5
$f(x_i)$	3.2	3.5	4.5

Solution. Here,

$$f[x_0, x_1] = \frac{3.5 - 3.2}{1.5 - 1.0} = 0.6$$

$$f[x_0, x_2] = \frac{4.5 - 3.2}{2.5 - 1.0} = 0.86667$$

$$f[x_1, x_2] = \frac{4.5 - 3.5}{2.5 - 1.5} = 1.0$$

from which we deduce that

$$f[x_0, x_1, x_2] = \frac{f[x_0, x_1] - f[x_1, x_2]}{x_0 - x_2} = \frac{0.6 - 1.0}{1.0 - 2.5} = 0.26667$$

Similarly,

$$f[x_0, x_2, x_1] = \frac{f[x_0, x_2] - f[x_1, x_2]}{x_0 - x_1} = \frac{0.86667 - 1.0}{1.0 - 1.5} = 0.26667$$

We observe in this last example that the order of the arguments of a divided difference appears not to affect the value of that difference. This turns out to be the general case, as we will see shortly. The result may be established either by direct manipulation or by comparison of the interpolation formulas. The former will be found as one of the exercises, while we pursue the latter within the development of the divided difference formula for polynomial interpolation.

Our task then is to simplify the process of polynomial interpolation. By definition (4.3.2), we have

$$f(x) = f[x_0] + (x - x_0)f[x_0, x] \qquad (4.3.4)$$

from which we see that the first-order interpolation polynomial agreeing with f at x_0 and x_1 is given by

$$p_1(x) = f[x_0] + (x - x_0)f[x_0, x_1] \qquad (4.3.5)$$

It is easy to verify that this polynomial satisfies the interpolation conditions $p_1(x_0) = f(x_0)$ and $p_1(x_1) = f(x_1)$, and then the uniqueness of the Lagrange interpolation polynomial establishes that this is indeed that polynomial. (See Exercise 2.)

Temporarily, we denote the function $f[x_0, x]$ by $f_1(x)$. In the same way as in the derivation of (4.3.4), we obtain

$$f_1(x) = f_1[x_1] + (x - x_1)f_1[x_1, x] \qquad (4.3.6)$$

and substituting this into (4.3.4), we have

$$\begin{aligned} f(x) &= f[x_0] + (x - x_0)f_1(x) \\ &= f[x_0] + (x - x_0)f_1[x_1] + (x - x_0)(x - x_1)f_1[x_1, x] \\ &= f[x_0] + (x - x_0)f[x_0, x_1] + (x - x_0)(x - x_1)f_1[x_1, x] \quad (4.3.7) \end{aligned}$$

Now, using the fact that $f[x_0, x_1] = f[x_1, x_0]$, we deduce that

$$\begin{aligned} (x - x_1)f_1[x_1, x] &= f_1(x) - f_1(x_1) = f[x_0, x] - f[x_0, x_1] \\ &= (x - x_1)f[x_1, x_0, x] \end{aligned}$$

and substituting this into (4.3.7), we obtain

$$f(x) = f[x_0] + (x - x_0)f[x_0, x_1] + (x - x_0)(x - x_1)f[x_1, x_0, x] \quad (4.3.8)$$

It follows that

$$p_2(x) = f[x_0] + (x - x_0)f[x_0, x_1] + (x - x_0)(x - x_1)f[x_1, x_0, x_2] \qquad (4.3.9)$$

is the second-degree interpolation polynomial agreeing with f at x_0, x_1, x_2. The agreement at x_0 and x_1 follows from the fact that the polynomial p_1 defined in (4.3.5) interpolates f at those points (and the extra term in p_2 vanishes there), while the condition at x_2 is obtained by comparison of (4.3.9) and (4.3.8) with $x = x_2$.

Note that the coefficient of x^2 in p_2 is $f[x_1, x_0, x_2]$, which by the uniqueness of the interpolation polynomial must be independent of the order of use of the data points. It follows that this divided difference is also independent of the order of its arguments; therefore, for example, $f[x_0, x_1, x_2] = f[x_1, x_0, x_2]$.

Continuing in this way, we obtain by induction the formula for the interpolation polynomial agreeing with f at $x_0, x_1, \ldots, x_N$:

$$p_N(x) = f[x_0] + (x - x_0)f[x_0, x_1] + (x - x_0)(x - x_1)f[x_0, x_1, x_2]$$
$$+ \cdots + (x - x_0)(x - x_1)\cdots(x - x_{N-1})f[x_0, x_1, \ldots, x_N]$$
$$(4.3.10)$$

This formula is known as *Newton's divided differences interpolation polynomial*.

Example 4.3.4. Repeat the computation of Example 4.3.1 using Newton's divided difference interpolation formula.

Solution. The data is

x	1.0	1.1	1.2	1.3
$\cos x$	0.5403	0.4536	0.3624	0.2675

And so we obtain the following table of divided differences:

TABLE 4.3.5

i	x_i	$f[x_i]$	$f[x_i, x_{i+1}]$	$f[x_{i-1}, x_i, x_{i+1}]$	$f[x_0, x_1, x_2, x_3]$
0	1.0	0.5403			
			−0.8670		
1	1.1	0.4536		−0.2250	
			−0.9120		0.1333
2	1.2	0.3624		−0.1850	
			−0.9490		
3	1.3	0.2675			

from which we make the computations to four decimal places:

$$p_1(1.12) = 0.5403 + (1.12 - 1.0)(-0.8670) = 0.4363$$
$$p_2(1.12) = 0.4363 + (1.12 - 1.0)(1.12 - 1.1)(-0.2250) = 0.4357$$
$$p_3(1.12) = 0.4357 + (1.12 - 1.0)(1.12 - 1.1)(1.12 - 1.2)(0.1333)$$
$$= 0.4357$$

It is plain that the computational effort entailed in adding another data point to our estimation process has been very much reduced.

It is worth remarking here that, in this last example, we took the data points in their natural increasing order. However, on further consideration, it seems more sensible to use the points in the order of their proximity to the point of interest, in this case $x = 1.12$. That is, we would probably do better to take the data in the order $x_0 = 1.1$, $x_1 = 1.2$, $x_2 = 1.0$, and $x_3 = 1.3$. Repeating the calculation for this ordering, we obtain the revised difference table:

i	x_i	$f[x_i]$	$f[x_i, x_{i+1}]$	$f[x_{i-1}, x_i, x_{i+1}]$	$f[x_0, x_1, x_2, x_3]$
0	1.1	0.4536			
			−0.9120		
1	1.2	0.3624		−0.2250	
			−0.8895		0.1333
2	1.0	0.5403		−0.1983	
			−0.9093		
3	1.3	0.2675			

from which we now obtain the values

$$p_1(1.12) = 0.4536 + (1.12 - 1.1)(-0.9120) = 0.4354$$

$$p_2(1.12) = 0.4354 + (1.12 - 1.1)(1.12 - 1.2)(-0.2250) = 0.4357$$

$$p_3(1.12) = 0.4357 + (1.12 - 1.1)(1.12 - 1.2)(1.12 - 1.0)(0.1333) = 0.4357$$

Although the differences are not very great in this case, we can see that the first-order approximation is definitely better. (Note that thereafter the approximations use precisely the same information as in the previous case.) It is also apparent that the order of the arguments has not affected the values of the divided differences; for example, comparing the two tables, we see that

$$f[1.0, 1.1, 1.2] = f[1.1, 1.2, 1.0]$$

In the case of the Lagrange interpolation polynomial we derived an expression for the truncation error in the form given by (4.2.10), namely, that

$$R_N(x) = \frac{L_N(x)f^{(N+1)}(\xi)}{(N+1)!}$$

For Newton's divided difference formula, we obtain, following the same reasoning as above,

$$f(x) = f[x_0] + (x - x_0)f[x_0, x_1] + (x - x_0)(x - x_1)f[x_0, x_1, x_2]$$
$$+ \cdots + (x - x_0)(x - x_1) \cdots (x - x_{N-1})f[x_0, x_1, \ldots, x_N]$$
$$+ (x - x_0)(x - x_1) \cdots (x - x_N)f[x_0, x_1, \ldots, x_N, x]$$

$$(4.3.11)$$

from which it follows that the error is given by

$$f(x) - p_N(x) = (x - x_0)(x - x_1) \cdots (x - x_N)f[x_0, x_1, \ldots, x_N, x]$$
$$= L_N(x)f[x_0, x_1, \ldots, x_N, x] \qquad (4.3.12)$$

Since the interpolation polynomial agreeing with f at $x_0, x_1, \ldots, x_N$ is unique, it follows that these two error expressions must be equal. That is, we have proved the following proposition.

Proposition 4.3.6

$$f[x_0, x_1, \ldots, x_N, x] = \frac{f^{(N+1)}(\xi)}{(N+1)!} \qquad (4.3.13)$$

for some point ξ in the interval spanned by $x_0, x_1, \ldots, x_N$ and x.

By way of illustration, we include here an application of this result which fills in the gap in the proof of convergence of the secant method in Sec. 2.4. We found there that the error e_{n+1} in the iterate x_{n+1} satisfies Eq. (2.4.8), namely, that

$$e_{n+1} = e_n e_{n-1} M_n$$

where the coefficient M_n is given by

$$M_n = \frac{f'(\xi_n) - f'(\xi_{n-1})}{f(x_n) - f(x_{n-1})} \qquad (4.3.14)$$

and it was stated earlier that this quantity can be bounded in the neighborhood of the solution.

Suppose then that the interval spanned by the points x_{n-1}, x_n and the solution s is such that f'' is bounded there and f' does not change sign. Now, from the original analysis, we have

$$f'(\xi_n) = f[x_n, s] \qquad f'(\xi_{n-1}) = f[x_{n-1}, s]$$

from which it follows that

$$f'(\xi_n) - f'(\xi_{n-1}) = f[x_n, s] - f[x_{n-1}, s] = (x_n - x_{n-1}) f[x_{n-1}, x_n, s]$$

Similarly, $f(x_n) - f(x_{n-1}) = (x_n - x_{n-1}) f[x_{n-1}, x_n]$, and hence we have, on substituting into (4.3.14),

$$M_n = \frac{f[x_{n-1}, x_n, s]}{f[x_{n-1}, x_n]} = \frac{f''(\xi)}{2f'(\zeta)}$$

by Proposition 4.3.6, where ξ and ζ are points in the interval spanned by x_{n-1}, x_n, and s.

Finally, the hypothesis on the boundedness of f'' in this interval (and the consequent continuity and boundedness of f') establishes that this ratio is bounded on the interval, which was precisely the statement to be proved.

As a further consequence of the uniqueness of the interpolation polynomial, we may deduce the independence of divided differences relative to the ordering of their arguments. This is an immediate consequence of the following result.

Proposition 4.3.7

$$f[x_0, x_1, \ldots, x_N] = \sum_{i=0}^{N} \frac{f(x_i)}{L_N'(x_i)} \qquad (4.3.15)$$

Proof. Simply comparing the coefficients of x^N in the two formulas (4.2.6) and (4.3.10) and observing that

$$L_N'(x_i) = (x_i - x_0) \cdots (x_i - x_{i-1})(x_i - x_{i+1}) \cdots (x_i - x_N) \qquad (4.3.16)$$

yields the result.

This result illustrates the independence of ordering within divided difference since the right-hand side of (4.3.15) is clearly invariant under permutations of the nodes.

There are several schemes for the efficient implementation of divided difference interpolation, such as that due to Aitken, which are designed for the easy evaluation of the polynomial, taking the points closest to the one of interest first and computing only those divided differences which are actually necessary for the computation. That is, the implementation is iterative in nature; additional data points are included one at a time until successive estimates $p_k(x)$, $p_{k+1}(x)$ of $f(x)$ agree to some specified accuracy—or until all data has been used. We leave the details of Aitken's iterative interpolation to the next section.

Algorithm 4.3.8 describes the use of divided difference interpolation. This is a very simplistic implementation which makes no attempt to use the points in any efficient order. It also uses all interpolation data for every computation so that the differences themselves are computed only once. This is certainly not the most effective technique available, but it is illustrative of the merits and drawbacks of polynomial interpolation.

Algorithm 4.3.8   Divided difference polynomial interpolation

Input:	$N, x_0, x_1, \ldots, x_N, f_0, f_1, \ldots, f_N, \bar{x}$
Generate differences:	$f[x_0], f[x_0, x_1], \ldots, f[x_0, x_1, \ldots, x_N]$
Compute:	$\bar{f} := f[x_0]$; prod$:= \bar{x} - x_0$
	for i = 1 to N do
	$\quad \bar{f} := \bar{f} + \text{prod} * f[x_0, x_1, \ldots, x_i]$
	$\quad \text{prod} := \text{prod} * (\bar{x} - x_i)$
Output:	$f(\bar{x}) \simeq \bar{f}$

Before proceeding any further, it is worth commenting on the procedure used here to generate the divided differences for this algorithm. Only the leading diagonal of the divided difference table is required. This can be achieved by overwriting entries which are no longer needed, as in the following procedure in which the array x holds the node values and d begins holding the data and ends up holding the wanted differences:

```
Procedure GenerateDivDiffs;
    {m is order of difference, n is number of data points}
    var k: integer;
    begin
        for m:=1 to n do
            for k:=n downto m do
                d[k]:=(d[k] − d[k − 1])/(x[k] − x[k − m]);
    end;
```

At each stage the procedure produces the next column of the divided difference table, leaving the entries d[0], . . . , d[m − 1] unchanged and writing mth-order differences into d[m], . . . , d[n].

This is certainly not the most efficient way to proceed, in general, as we would only want to compute those divided differences which are actually to be used in the implementation. We discuss ways to make this process less computationally expensive shortly.

Bear in mind that such schemes will almost inevitably entail repetition of certain calculations if they are used to produce a table of values of the interpolation polynomial, as in Table 4.3.9 below, since computation of the divided differences would then need to be included within the main loop.

Algorithm 4.3.8 produced the results in Table 4.3.9 in which we see that despite the total agreement between the interpolant and the data [taken from the function $1/(1 + x^2)$] at the nodes (in this case the integers), there can be marked disagreement elsewhere.

We turn next to the question of how to make the implementation of the divided difference interpolation formula more efficient. One of the principal advantages of this approach over the Lagrange formulation lies in the ease with which additional data points can be incorporated into the computation. Algorithm 4.3.8 makes no attempt to utilize this feature.

This can be achieved by only using those terms in (4.3.12) which are necessary to achieve agreement of two successive approximations to within the specified tolerance. To maximize the effect of this change, it is desirable to reorder the points in such a way that those closest to the point of interest, $\bar{x}$ in the program, are used first. The third requirement imposed by the goal of maximum efficiency is that only those divided differences which are actually to be used should be computed.

TABLE 4.3.9
Degree-10 polynomial interpolation
Nodes $-5, -4, \ldots, 0, 1, \ldots, 5^*$
Data from $f(x) = 1/(1 + x^2)$

x	$p_{10}(x)$	$f(x)$
−5.00	0.038 461 538 5	0.038 461 538 5
−4.50	1.578 720 990 3	0.047 058 823 5
−4.00	0.058 823 529 4	0.058 823 529 4
−3.50	−0.226 196 289 1	0.075 471 698 1
−3.00	0.100 000 000 0	0.100 000 000 0
−2.50	0.253 755 457 3	0.137 931 034 5
−2.00	0.200 000 000 0	0.200 000 000 0
−1.50	0.235 346 591 3	0.307 692 307 7
−1.00	0.500 000 000 0	0.500 000 000 0
−0.50	0.843 407 429 8	0.800 000 000 0
0.00	1.000 000 000 0	1.000 000 000 0

* The results for positive x were a perfect reflection of those shown.

The first of these objectives amounts only to the replacement of the fixed For loop in Algorithm 4.3.8 by a Repeat loop with a convergence test. The second requires an efficient sorting procedure—a standard exercise in almost all introductory programming courses for which there are several acceptable solutions. The third of these requirements is somewhat more difficult to achieve, although the final coding of it is very brief as we can see in the procedure DivDiffs in Program 4.3.10.

The array diff is initialized to hold the function values at the data points. On each successive call of the procedure the first m entries in the array are modified so that on completion of the loop, diff[0] holds the divided difference $f[x_0, x_1, \ldots, x_m]$, while diff[j] for $j < m$ contains $f[x_j, x_{j+1}, \ldots, x_m]$. On the next call of the procedure with m replaced by m + 1, the first m + 1 entries are overwritten starting with diff[m], which previously held just $f(x_m)$, and is replaced with

$$(\text{diff}[m+1] - \text{diff}[m])/(x[m+1] - x[m]) = f[x_m, x_{m+1}]$$

as required. At this point diff[m] and diff[m − 1] contain the two first-order divided differences $f[x_m, x_{m+1}]$ and $f[x_{m-1}, x_m]$, respectively. This is precisely the information needed for the computation of $f[x_{m-1}, x_m, x_{m+1}]$, which is then written into diff[m − 1], and so on.

Program 4.3.10   Divided difference interpolation using nearest points first

```
program DivDiff;
type vector = array[0..10] of real;
var x, dx, fx, diff:vector;
     xbar, fbar, product, oldf, eps:real;
     i, j, N:integer;

function f(t:real):real;
begin f:=1/(1 + t * t); end;

procedure DivDiffs(m:integer; t:vector; var d:vector);
                   {m is order of difference}
var k:integer;
begin
     for k:=m downto 1 do
          d[k − 1]:=(d[k] − d[k − 1])/(t[m] − t[k − 1]);
end;
     {At finish of loop, d[k] = f[k] = f[k, k + 1, . . . , m] for each k < m
     {where j stands for x[j].
     {Rest of array d[k] contains original values.}

Procedure CycleR(var v:vector; start, finish:integer);
var i:integer; temp:real;
begin
     temp:=v[finish];
     for i:=finish downto start + 1 do v[i]:=v[i − 1];
     v[start]:=temp;
end;
```

```
Procedure OrderData(n : integer; point : real;
                    var x, dx, fx : vector);
var i, j : integer;
begin
    for i := 0 to n do dx[i] := xbar − x[i];
    for i := 1 to n do begin
        j := 0;
        while (abs(dx[j]) < abs(dx[i])) and (j < i) do j := j + 1;
        if j < i then begin
            CycleR(x, j, i); CycleR(dx, j, i); CycleR(fx, j, i);
        end;
    end;
end;

begin {Start main program}
    N := 10; eps := le − 5;
    for i := 0 to N do begin x[i] := i * 5/N; fx[i] := f(x[i]); end;
    writeln('xbar' : 7, 'divdiff(x)' : 18, '1/(1 + x^2)' : 14,'
                    #points' : 12);
    for i := 0 to 30 do begin
        xbar := i * 0.165;
        OrderData(N, xbar, x, dx, fx);
        for j := 0 to N do diff[j] := fx[j];
        fbar := diff[0]; product := dx[0]; j := 0;
        repeat
            j := j + 1; oldf := fbar;
            DivDiffs(j, x, diff);
            fbar := fbar + product * diff[0];
            product := product * dx[j];
        until (abs(fbar − oldf) < eps) or (j = n);
        writeln(xbar : 10 : 7, fbar : 15 : 7, f(xbar) : 15 : 7, j : 8);
    end;
end.
```

When used to tabulate the interpolation polynomial at the same points and with the same data as were used for Table 4.3.9, this program produces identical results using almost all data at all nondata points. For the points used in this program, the results are shown in Table 4.3.11.

It should be observed here that the density of the data points is twice that used for the earlier example, and so the better agreement between the interpolation polynomial and the original function is not too surprising. (Although, as we will see later, it is certainly not a universal cure for inaccuracies in the use of polynomial interpolation.) The other factor which is apparent here is that the interpolation polynomial had significantly greater difficulty in fitting the curve in the region near zero and toward the end of the range. (At zero itself there is no difficulty, of course, because zero is one of the data points.) The difficulty persists at the left-hand end for somewhat longer because some of the derivatives of—and therefore some of the divided

TABLE 4.3.11

xbar	divdiff(x)	$1/(1+x^2)$	No. of points
0.000	1.000 000 0	1.000 000 0	1
0.165	0.983 404 5	0.973 496 6	10
0.330	0.906 109 9	0.901 794 6	10
0.495	0.803 264 4	0.803 196 7	9
0.660	0.695 663 5	0.696 572 9	10
0.825	0.594 529 2	0.595 016 7	10
0.990	0.505 007 9	0.505 025 0	9
1.155	0.428 582 3	0.428 444 4	10
1.320	0.364 734 4	0.364 644 1	10
1.485	0.311 995 7	0.311 990 6	9
1.650	0.268 602 8	0.268 636 7	10
1.815	0.232 844 0	0.232 870 9	10
1.980	0.203 233 6	0.203 235 5	9
2.145	0.178 551 8	0.178 538 7	10
2.310	0.157 810 5	0.157 825 8	6
2.475	0.140 336 4	0.140 338 6	6
2.640	0.125 486 4	0.125 476 8	6
2.805	0.112 772 0	0.112 764 7	7
2.970	0.101 824 4	0.101 823 7	6
3.135	0.092 350 8	0.092 351 2	7
3.300	0.084 106 9	0.084 104 3	6
3.465	0.076 888 4	0.076 886 3	5
3.630	0.070 540 7	0.070 537 3	5
3.795	0.064 924 0	0.064 926 5	6
3.960	0.059 944 1	0.059 946 3	4
4.125	0.055 503 9	0.055 507 4	5
4.290	0.051 533 4	0.051 535 5	7
4.455	0.047 971 5	0.047 968 5	4
4.620	0.044 695 4	0.044 753 9	10
4.785	0.041 695 9	0.041 847 6	10
4.950	0.039 113 3	0.039 211 8	10

differences being used—take large values for small x, making the convergence criterion difficult to satisfy there.

In the next section we discuss an alternative approach to the efficient implementation of polynomial interpolation.

EXERCISES 4.3

1. Find the divided differences $f[x_0, x_1]$, $f[x_0, x_2]$, $f[x_0, x_1, x_2]$, and $f[x_0, x_2, x_1]$ for the data

i	0	1	2
x_i	−1	0	2
$f(x_i)$	1	0	4

2. Verify that p_1 defined by (4.3.5) interpolates f at x_0, x_1.

3. Find the divided difference interpolation polynomial for the data

x	−1	0	2	3	4
f	2	1	5	10	17

4. Prove that any kth-order divided difference of a polynomial of degree less than k is zero.

5. Obtain a direct proof of the fact that the ordering of the arguments of a divided difference does not affect its value. [Use induction and compare the coefficients of $f(x_i)$.]

6. By setting $f_2(x) = f[x_0, x_1, x]$ and using (4.3.4), show that

$$f(x) = f[x_0] + (x - x_0)f[x_0, x_1]$$
$$+ (x - x_0)(x - x_1)f[x_0, x_1, x_2]$$
$$+ (x - x_0)(x - x_1)(x - x_2)$$
$$\times f[x_0, x_1, x_2, x]$$

Hence prove that the cubic interpolation polynomial is

$$p_3(x) = f[x_0] + (x - x_0)f[x_0, x_1]$$
$$+ (x - x_0)(x - x_1)f[x_0, x_1, x_2]$$
$$+ (x - x_0)(x - x_1)(x - x_2)$$
$$\times f[x_0, x_1, x_2, x_3]$$

7. Use induction on N to establish the general form of Newton's divided difference interpolation formula.

8. By considering $f[x, x + h]$ as $h \to 0$, show that $f[x, x] = f'(x)$.

4.4 AITKEN'S ALGORITHM

In this section we discuss one of the particularly efficient techniques for implementing the ideas of divided difference interpolation due to Aitken. Before electronic computers were able to take the drudgery out of extensive calculations, it was especially important to find (often highly ingenious) algorithms for the implementation of the various formulas that had been discovered. The need to avoid unnecessary calculation or wasteful repetition was even greater then than it is today. This necessity was the mother of much inventive algorithmic design; Aitken's algorithm is a good example.

Not uncommonly for numerical methods, the algorithm is based on the formation of a triangular array in which each new row is generated only when it is needed. Specifically then, suppose that we wish to evaluate (or approximate) $f(\bar{x})$ from the data $f(x_0), f(x_1), \ldots, f(x_N)$, where the interpolation points are assumed to be in the correct order of use.

Denote by $A_{k,r}$ $(r \geq k)$ the value at $\bar{x}$ of the interpolation polynomial which agrees with f at the points $x_0, x_1, \ldots, x_{k-1}, x_r$. From (4.3.10) it follows that

$$A_{k,r} = f[x_0] + (\bar{x} - x_0)f[x_0, x_1] + (\bar{x} - x_0)(x - x_1)f[x_0, x_1, x_2]$$
$$+ \cdots + (\bar{x} - x_0)(\bar{x} - x_1) \cdots (\bar{x} - x_{k-1})f[x_0, x_1, \ldots, x_{k-1}, x_r]$$
$$\tag{4.4.1}$$

Thus, for example,

$$A_{0,0} = f(x_0)$$
$$A_{0,1} = f(x_1)$$
$$A_{0,2} = f(x_2)$$
$$A_{1,1} = f(x_0) + (\bar{x} - x_0)f[x_0, x_1]$$
$$A_{1,2} = f(x_0) + (\bar{x} - x_0)f[x_0, x_2]$$
$$A_{2,2} = f(x_0) + (x - x_0)f[x_0, x_1] + (\bar{x} - x_0)(\bar{x} - x_1)f[x_0, x_1, x_2]$$

Aitken's algorithm provides a technique for generating the leading diagonal of this array, namely, the entries $A_{k,k}$, one at a time until convergence. The algorithm is based on the following result, known as *Aitken's lemma.*

Proposition 4.4.1. Let p, q, r be interpolation polynomials satisfying the following interpolation conditions:

$$p(x_i) = q(x_i) = r(x_i) = f(x_i) \qquad (i = 0, 1, \ldots, m) \qquad (4.4.2a)$$

$$r(\xi) = p(\xi) = f(\xi) \qquad r(\eta) = q(\eta) = f(\eta) \qquad (4.4.2b)$$

(Thus p, q each agree with f at $m + 2$ points, while r and f agree at all $m + 3$ of them.) Then r is given by

$$r(x) = \frac{(\xi - x)q(x) - (\eta - x)p(x)}{\xi - \eta} \qquad (4.4.3)$$

Proof. Since the interpolation polynomial agreeing with f at a specified set of points is unique, it suffices to show that the function given by (4.4.3) satisfies the interpolation conditions (4.4.2). For $0 \le i \le m$, we have, since $p(x_i) = q(x_i) = f(x_i)$,

$$r(x_i) = \frac{(\xi - x_i)f(x_i) - (\eta - x_i)f(x_i)}{\xi - \eta} = \frac{(\xi - \eta)f(x_i)}{\xi - \eta} = f(x_i)$$

For $x = \xi$, we have

$$r(\xi) = \frac{0 - (\eta - \xi)p(\xi)}{\xi - \eta} = p(\xi) = f(\xi)$$

and, similarly,

$$r(\eta) = \frac{(\xi - \eta)q(\eta)}{\xi - \eta} = q(\eta) = f(\eta)$$

as required.

The significance of this lemma for our purposes lies in the following corollary.

Corollary 4.4.2. The elements of the Aitken array $A_{k,r}$ satisfy

$$A_{k,r} = \frac{(x_{k-1} - \bar{x})A_{k-1,r} - (x_r - \bar{x})A_{k-1,k-1}}{x_{k-1} - x_r} \qquad (4.4.4)$$

Proof. To see this, we simply set $m = k - 2$, $\xi = x_{k-1}$, $\eta = x_r$, and $x = \bar{x}$ so that $p(\bar{x}) = A_{k-1,k-1}$, $q(\bar{x}) = A_{k-1,r}$, and $r(\bar{x}) = A_{k,r}$ in the lemma.

To see how this result eases the evaluation of the interpolation polynomial, consider Fig. 4.4.3, the diagrammatic representation of the process. To the array described earlier has been added an extra column of information

$A_{0,0}$						$x_0 - \bar{x}$
$A_{0,1}$	$\boxed{A_{1,1}}$					$\boxed{x_1 - \bar{x}}$
$A_{0,2}$	$A_{1,2}$	$A_{2,2}$				$x_2 - \bar{x}$
$A_{0,3}$	$\boxed{A_{1,3}}$	$A_{2,3}$	$A_{3,3}$			$\boxed{x_3 - \bar{x}}$
$A_{0,4}$	$A_{1,4}$	$A_{2,4}$	$A_{3,4}$	$A_{4,4}$		$x_4 - \bar{x}$

FIGURE 4.4.3
The Aitken array for polynomial interpolation.

down the right-hand side containing the array of differences between the interpolation points and the point of interest.

Example 4.4.4. The calculation of $A_{2,3}$ by Corollary 4.4.2 is given by

$$A_{2,3} = \frac{(x_1 - \bar{x})A_{1,3} - (x_3 - \bar{x})A_{1,1}}{x_1 - x_3}$$

which is obtained by the cross-multiplication of the boxed terms divided by the difference of the relevant entries in the right-hand column. For any particular entry, the two previously computed A's which are used are the entry immediately to the left and the main-diagonal entry above that one. For example, in the computation of $A_{2,4}$ the entries $A_{1,4}$ and $A_{1,1}$ would be needed.

At any given stage in the computation, it follows that only the current row and the leading diagonal of the above array are needed. In fact, at the point where we compute a new member of the current row, all that is needed is its immediate predecessor in that row together with the leading diagonal. It is apparent that we should be able to economize on the storage requirements of the resulting algorithm. Before proceeding to these questions of detailed implementation, it is instructive to consider an example worked in detail.

Example 4.4.5. Apply Aitken's algorithm to the approximate evaluation of $f(1.21)$ from the data

x	1.0	1.1	1.2	1.3	1.4	1.5
$f(x)$	0.8415	0.8912	0.9320	0.9636	0.9854	0.9975

Solution. The nodes would be used in the order 1.2, 1.3, 1.1, 1.4, 1.0, and 1.5 from which we compute the array

x_k	$A_{0,k}$	$A_{1,k}$	$A_{2,k}$	$A_{3,k}$	$A_{4,k}$	$x_k - 1.21$
1.2	0.9320					-0.01
1.3	0.9636	0.93516				0.09
1.1	0.8912	0.93608	0.93557			-0.11
1.4	0.9854	0.93467	0.93560	0.93558		0.19
1.0	0.8415	0.93653	0.93557	0.93557	0.93558	-0.21

The entry $A_{1,3} = 0.93467$, for example, resulted from the calculation

$$A_{1,3} = \frac{(0.9854)(-0.01) - (0.9320)(0.19)}{-0.01 - 0.19} = \frac{-0.186934}{-0.2} = 0.93467$$

The above array leads us to the fairly confident conclusion that the true value should be 0.9356 to the same accuracy as the data. Note that although the intermediate computation here was performed to greater accuracy to minimize the effect of accumulated rounding errors, the final result should still be given to no greater precision than that of the original data.

Consider now the implementation of the algorithm just described. There are several points to be considered. First, as with the divided difference scheme discussed in Sec. 4.3, we wish to order the points so that those closest to the point of interest are used first, and the points should be introduced one at a time. In Program 4.3.10 all the data points were sorted into the appropriate order at the outset, but, ideally, all we should do is to find the next closest one each time a new point is needed. This modification is easily accomplished. (See Project 2.)

The other principal efficiency to be sought is in the computation and storage of the elements of the Aitken array. Suppose that the kth approximation $A_{k,k}$ has been computed and that we now wish to compute $A_{k+1,k+1}$.

Within the computer it is sufficient to store the arrays

$$A[k]:=A_{k,k} \qquad x[k]:=x_k \qquad y[k]:=x_k - \bar{x} \qquad f[k]:=f(x_k)$$

Our assumption then is that $A[0], A[1], \ldots, A[k]$ are already known and that the appropriate next data point $x[k+1]$ has been identified. The following few lines of Turbo Pascal code will then be sufficient to generate $A[k+1]$.

```
A[k + 1]:=f[k + 1];
for i:=0 to k do
    A[k + 1]:=(y[i] * A[k + 1] − y[k + 1] * A[i])/(y[i] − y[k + 1]);
```

At the beginning of the ith step of this loop, $A[k+1]$ holds the value of $A_{i,k+1}$, and this value together with $A_{i,i}$ (which is stored as $A[i]$) is used to compute the new value $A[k+1]$ which then holds $A_{i+1,k+1}$ ready for the next step.

The important aspect of this implementation of Aitken's *iterative interpolation* algorithm as it is often called lies in the fact that the divided differences themselves are not computed but just their effect on the interpolation polynomial. The computation is therefore organized in a way which minimizes both computational effort and storage requirements.

In many of the examples considered so far, the data points have been uniformly spaced. This is totally unneccssary for the approaches discussed thus far—and indeed is often not advantageous. However, there is a substantial theory of polynomial interpolation using equally spaced nodes which we will consider in the next section. The use of such regular meshes here is for ease of comparison of the divided difference and *finite difference* approaches.

EXERCISES 4.4

1. Verify directly from formula (4.4.4) that the element $A_{2,2}$ of the Aitken array agrees with the divided difference interpolation formula for the points x_0, x_1, x_2.

2. Show that it is sufficient to store a one-dimensional array to hold the current row of the Aitken array.

3. Write and test a program to implement Aitken's algorithm.

4.5   FINITE DIFFERENCE FORMULAS

In this section we specialize polynomial interpolation to the situation in which the data points are equally spaced. This is precisely the situation of looking up values of a function in a table and estimating the value at some other, nontabular, point. This was once a very common exercise in the use of logarithmic and other tables for the elementary functions. As would be the obvious intuitive approach within that setting, we adopt the same basic principle as for divided difference interpolation in which we want to use information from close to the point of interest first. Clearly, when using four-figure logarithm tables, if we wish to find $\log_{10} 1.2345$, we would begin by estimating that it must lie between $\log_{10} 1.234$ and $\log_{10} 1.235$—and probably about halfway between them. The particular form of the interpolation polynomial will therefore depend on whether the point of interest lies near the beginning, in the middle, or near the end of the table.

We assume throughout this section that

$$x_k = x_0 + kh \qquad (4.5.1)$$

where x_0 is a fixed reference point and h is the constant steplength. The integer values of k may be positive or negative depending on which point in the table we adopt as the fixed reference.

For interpolation near the beginning of the table of values, we use *forward differences* in the definition of the polynomial. These differences have the effect of introducing the points in the order $x_0, x_1, x_2, \ldots$. Similarly, for interpolation in the middle or near the end of the table, we use *central* or *backward differences*, respectively. Before we can study these in detail, it is necessary to define the terms.

Definitions 4.5.1. The first *forward difference* $\Delta f(x_0)$ is defined by

$$\Delta f(x_0) = f(x_0 + h) - f(x_0) = f(x_1) - f(x_0)$$

or, in general,

$$\Delta f(x_r) = f(x_r + h) - f(x_r) = f(x_{r+1}) - f(x_r) \qquad (4.5.2)$$

Higher-order forward differences are then defined recursively by

$$\Delta^k f(x_r) = \Delta(\Delta^{k-1} f(x_r)) = \Delta^{k-1} f(x_{r+1}) - \Delta^{k-1} f(x_r) \qquad (4.5.3)$$

The first *backward difference* $\nabla f(x_n)$ is similarly simply defined by

$$\nabla f(x_n) = f(x_n) - f(x_{n-1}) \qquad (4.5.4)$$

and higher-order differences by

$$\nabla^k f(x_n) = \nabla^{k-1} f(x_n) - \nabla^{k-1} f(x_{n-1}) \qquad (4.5.5)$$

The *central differences* are defined by

$$\delta f(x) = f\left(x + \frac{h}{2}\right) - f\left(x - \frac{h}{2}\right) \qquad (4.5.6)$$

where we note that the point x has not been specified since if x is a tabular point, then $x \pm h/2$ are not. Thus $\delta f(x_{k+1/2})$ is defined and so is

$$\delta^2 f(x_k) = \delta f(x_{k+1/2}) - \delta f(x_{k-1/2}) = f(x_{k+1}) - 2f(x_k) + f(x_{k+1}) \quad (4.5.7)$$

Similarly, we see that all central differences of odd order are defined for $x_{k+1/2}$, while the even-order ones are defined at tabular points. The relationships between these differences are illustrated in Table 4.5.2.

Frequently, we will abbreviate the notation for these finite differences by writing

$$\Delta f(x_r) = \Delta f_r \qquad \delta^2 f(x_r) = \delta^2 f_r \qquad \nabla^3 f(x_r) = \nabla^3 f_r$$

and so on. Thus we have, for example, in Table 4.5.2 that

$$\Delta f_1 = \delta f_{3/2} = \nabla f_2$$

There are several simple, but important, relations which are easy to prove by induction and whose proofs are, for the most part, left as exercises for the reader. It is easy to see some of these for low-order differences. For instance,

$$\Delta f_0 = f_1 - f_0$$

from which we get

$$\Delta^2 f_0 = \Delta f_1 - \Delta f_0 = f_2 - 2f_1 + f_0$$
$$\Delta^3 f_0 = \Delta^2 f_1 - \Delta^2 f_0 = f_3 - 3f_2 + 3f_1 - f_0$$

and, in general,

$$\Delta^k f_r = \nabla^k f_{r+k} = \delta^k f_{r+k/2} = \sum_{i=0}^{k} (-1)^{k-i} \binom{k}{i} f_{r+i} \quad (4.5.8)$$

Similarly, from the definition of the forward differences, we find that

$$\Delta f_0 = f_1 - f_0 = hf[x_0, x_1]$$
$$\Delta^2 f_0 = hf[x_1, x_2] - hf[x_0, x_1] = 2h^2 f[x_0, x_1, x_2]$$

TABLE 4.5.2
The relationship between finite differences

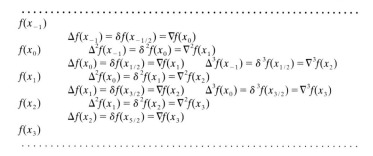

and, in general,

$$\Delta^k f_r = k! h^k f[x_r, x_{r+1}, \ldots, x_{r+k}] \qquad (4.5.9)$$

We content ourselves with the proof of this last result. The proof is by induction, and the truth of the statement for $k = 1$ is trivial. Assuming, then, that (4.5.9) holds for some value of k, it follows from the induction hypothesis that

$$\begin{aligned}
\Delta^{k+1} f_r &= \Delta^k f_{r+1} - \Delta^k f_r \\
&= k! h^k (f[x_{r+1}, x_{r+2}, \ldots, x_{r+k+1}] - f[x_r, x_{r+1}, \ldots, x_{r+k}]) \\
&= k! h^k (x_{r+k+1} - x_r) f[x_r, x_{r+1}, \ldots, x_{r+k+1}] \\
&= k! h^k (k+1) h f[x_r, x_{r+1}, \ldots, x_{r+k+1}] \\
&= (k+1)! h^{k+1} f[x_r, x_{r+1}, \ldots, x_{r+k+1}]
\end{aligned}$$

which is the right-hand side of (4.5.9) for $k + 1$, so that the result follows by induction.

We turn now to the use of these *finite differences*, as the forward, central, and backward differences are collectively known. There are many finite difference interpolation formulas; they are almost all derived from Newton's divided difference formula (4.3.10) and (4.3.11). We recall the formula for the divided difference interpolation polynomial given by (4.3.10) is

$$\begin{aligned}
p_N(x) &= f[x_0] + (x - x_0) f[x_0, x_1] + (x - x_0)(x - x_1) f[x_0, x_1, x_2] \\
&\quad + \cdots + (x - x_0)(x - x_1) \cdots (x - x_{N-1}) f[x_0, x_1, \ldots, x_N]
\end{aligned}$$

For interpolation near the beginning of the table of data, we use forward differences and substitute into the above polynomial using (4.5.9). It follows that

$$\begin{aligned}
p_N(x) &= f[x_0] + (x - x_0) f[x_0, x_1] + (x - x_0)(x - x_1) f[x_0, x_1, x_2] \\
&\quad + \cdots + (x - x_0)(x - x_1) \cdots (x - x_{N-1}) f[x_0, x_1, \ldots, x_N] \\
&= f_0 + \frac{x - x_0}{h} \Delta f_0 + \frac{(x - x_0)(x - x_1)}{2h^2} \Delta^2 f_0 \\
&\quad + \cdots + \frac{(x - x_0)(x - x_1) \cdots (x - x_{N-1})}{N! h^N} \Delta^N f_0 \qquad (4.5.10)
\end{aligned}$$

It is common to rewrite this formula using the so-called dimensionless notation in which we set

$$x = x_0 + sh \qquad P(s) = p_N(x) \qquad (4.5.11)$$

With this notation, (4.5.10) reduces to

$$\begin{aligned}
P(s) &= f_0 + s \Delta f_0 + \frac{s(s-1)}{2} \Delta^2 f_0 + \cdots + \frac{s(s-1) \cdots (s - N + 1)}{N!} \Delta^N f_0 \\
&= \sum_{i=0}^{N} \binom{s}{i} \Delta^i f_0 \qquad (4.5.12)
\end{aligned}$$

where the generalized binomial coefficients are defined by

$$\binom{s}{i} = \frac{s(s-1)\cdots(s-i+1)}{i!} \tag{4.5.13}$$

Formulas (4.5.10) and (4.5.12) are known as *Newton's forward difference formula* (which is believed to be due to Gregory).

Example 4.5.3. Use Newton's forward difference formula to estimate the value of sin 1.14 from the following values:

x	1.0	1.1	1.2	1.3	1.4	
sin x	0.8415	0.8912	0.9320	0.9636	0.9854	

x	1.5	1.6	1.7	1.8	1.9	2.0
sin x	0.9975	0.9996	0.9917	0.9738	0.9463	0.9093

Solution. Although we do not use all the information for this example, we produce the full table of forward differences up to fifth order in Table 4.5.4. Further reference to and use of this table will be made in subsequent examples on other finite difference schemes.

At this stage we must decide on which point in the table to use as the base point for the calculation. Assuming that differences of order at least three—and therefore at least four interpolation points—are to be used, the overall spread of

TABLE 4.5.4
Forward differences for the data of Example 4.5.3

x	f	Δf	$\Delta^2 f$	$\Delta^3 f$	$\Delta^4 f$	$\Delta^5 f$
1.0	0.8415					
		0.0497				
1.1	0.8912		−0.0089			
		0.0408		−0.0003		
1.2	0.9320		−0.0092		−0.0003	
		0.0316		−0.0006		0.0010
1.3	0.9636		−0.0098		0.0007	
		0.0218		0.0001		−0.0011
1.4	0.9854		−0.0097		−0.0004	
		0.0121		−0.0003		0.0007
1.5	0.9975*		−0.0100*		0.0003*	
		0.0021*		0.0000*		−0.0003*
1.6	**0.9996***		**−0.0100***		**0.0000***	
		−0.0079		0.0000		0.0004
1.7	**0.9917**		**−0.0100**		0.0004	
		−0.0179		0.0004		−0.0007
1.8	0.9738		−0.0096		−0.0003	
		−0.0275		0.0001		
1.9	0.9463		−0.0095			
		−0.0370				
2.0	0.9093					

data will be best if $x_0 = 1.0$, even though this is not the closest tabular point to the desired function value.

It follows that we will make use here of the <u>underlined</u> entries in Table 4.5.4. (The other specially marked entries will be used in subsequent examples.)

Of course, there would be a reasonable argument in favor of using $x_0 = 1.1$ since that is the closest tabular point to (the left of) 1.14, but where possible it is the spread of the final set of points to be used that should have the point of interest in the middle of its span, and this is best achieved, for this example, with $x_0 = 1.0$. Thus, in (4.5.12) we have $x_0 = 1.0$, $h = 0.1$, and therefore, $s = (1.14 - 1.0)/0.1 = 1.4$, from which we obtain the following approximations using first, second, third, and then fourth differences:

$$\sin 1.14 \approx 0.8415 + (1.4)(0.0497) = 0.91108$$

$$\sin 1.14 \approx 0.91108 + \frac{(1.4)(0.4)(-0.0089)}{2} = 0.908\,588$$

$$\sin 1.14 \approx 0.908588 + \frac{(1.4)(0.4)(-0.6)(-0.0003)}{6} = 0.908\,605$$

$$\sin 1.14 \approx 0.908605 + \frac{(1.4)(0.4)(-0.6)(-1.6)(-0.0003)}{24} = 0.908\,598$$

from which we can reasonably conclude that $\sin 1.14 = 0.9086$ to four decimal places, which is indeed correct.

For interpolation at points close to the end of the table of data, we use Newton's backward difference formula—which again is believed to be due to Gregory. This formula is based on the point x_N and using the dimensionless notation obtained by setting $x = x_N + hs$ can be written as

$$P(s) = f_N + s\nabla f_N + \frac{s(s+1)}{2}\nabla^2 f_N + \cdots + \frac{s(s+1)\cdots(s+k-1)}{k!}\nabla^k f_N$$

$$(4.5.14)$$

(The detailed derivation of this formula is very similar to that used for the forward difference formula and is left as an exercise for the reader.)

Note that for this formula to be an *interpolation* formula, it is necessary that the value of s in (4.5.14) be negative. Positive values of s would correspond to the process of extrapolation for the estimation of values of the function f at points beyond the end of the table.

Example 4.5.5. Use Newton's backward difference interpolation formula to estimate the value of $\sin 1.87$ from the same data as in Example 4.5.3.

Solution. This time we set $x_N = 1.9$, from which we have $s = -0.3$, and using the boxed differences in Table 4.5.4 yields the approximation

$$\sin 1.87 \approx 0.9463 + (-0.3)(-0.0275) + \frac{(-0.3)(0.7)(-0.0096)}{2}$$

$$+ \frac{(-0.3)(0.7)(1.7)(0.0004)}{6} + \frac{(-0.3)(0.7)(1.7)(2.7)(0.0004)}{24}$$

This expression gives the sequence of estimates 0.9546, 0.95556, 0.95553, and 0.95552 from which we would deduce that the true value should be 0.9555 to four decimal places. To this precision, this is in error by 1 in the final place, almost certainly due to rounding error in the original data. Indeed, studying the difference table itself suggests that there is somewhat greater roundoff error near the two ends of the table, since the final two columns of differences have less pattern to them and their largest entries appear near the two ends.

Careful analysis of tables of finite differences can be used to detect errors in the data because these errors tend to grow as they are propagated further into the table. Furthermore, the relative magnitudes of the propagated errors are in the same proportions as the binomial coefficients, which makes it possible to locate the original error with near certainty. This used to be a very important aspect of numerical analysis when many of the elementary functions and the special functions which arise so often in physical applications were available *only* in the form of tables of their values or by lengthy direct hand calculation. Analysis of tables of differences was one of the few ways in which the accuracy of the tabulated values could be checked. (Since there was at least one major book of mathematical tables which included the statement that there was a deliberate error somewhere as a safeguard against plagiarism, the possibility of performing such a check was vital!)

The above remarks emphasize the point that we have made previously: results should never be quoted to greater accuracy than is present in the data.

To see whether the approximation obtained above has the precision we should expect of it, we must obtain bounds for the errors in the Newton forward and backward difference formulas. This task is made easy since they are both just special cases of the Newton divided difference interpolation formula for which we already have an error bound derived from the original Lagrange remainder formula (4.2.10).

It is not difficult to establish that, in our dimensionless notations, the error bounds for the Newton forward and backward difference formulas each using differences of order up to k are, respectively,

$$E_s = \begin{cases} h^{k+1} \binom{s}{k+1} f^{(k+1)}(\xi) & \text{where } x_0 < x, \; \xi < x_k \quad (4.5.15) \\[2mm] h^{k+1} \binom{s+k}{k+1} f^{(k+1)}(\zeta) & \text{where } x_{N-k} < x, \; \zeta < x_N \quad (4.5.16) \end{cases}$$

In Examples 4.5.3 and 4.5.5, the function f is, of course, just the sine function, and so its derivatives are all bounded by unity and in each case we used approximations with differences of order four. Thus we have $k = 4$ in the above error bounds. For the first example, $s = 1.4$, and so

$$\binom{s}{k+1} = \frac{(1.4)(0.4)(-0.6)(-1.6)(-2.6)}{5!} \approx -0.01165$$

from which we derive the error bound of approximately 1.1×10^{-7} for the forward difference estimate. Thus, with *exact data* and *exact arithmetic* we should anticipate an error of only about 1 in the seventh decimal place.

In the same way for the second example, we get a bound of about 3×10^{-7}. It is evident, therefore, that the error in the final result above is due almost entirely to rounding error, and the major contribution to that certainly comes from the error in the original data; in fact, it can be traced to the tabulated value of sin 1.8 which has almost the maximum possible roundoff error for the precision quoted, its true value being 0.973 847 63 to eight decimal places. This is entirely consistent with the computed error which working to five decimal places was almost exactly 5×10^{-5}.

The most important formulas for interpolation in the middle part of the table are those of Bessel (apparently due to Newton) and Everett (probably discovered by Laplace). A convenient starting point for the derivation of both of these is the Gauss forward formula—believed to be due to Newton and using central differences! (The need for a taxonomist of interpolation formulas is evident. However, it is almost certain that Newton knew all these formulas, so the eventual classification would be much less intriguing than trying to explain the present one.)

For the derivation of the Gauss forward formula, we denote the base point by x_0. Since we will be interested in points both to the left and right of this, we will, of course, be dealing with points such as x_{-1} and x_{-2} as well as x_1 and x_2. The idea is to introduce points in the order $x_0, x_1, x_{-1}, x_2, x_{-2}, \ldots$. This is called the Gauss *forward* formula because we go forward in the table by first using x_k before x_{-k}.

Proceeding in the same way as before, with $x = x_0 + sh$, and substituting in the divided difference formula, we obtain the following:

$$P(s) = f_0 + s\delta f_{1/2} + \frac{s(s-1)}{2}\,\delta^2 f_0 + \frac{s(s-1)(s+1)}{3!}\,\delta^3 f_{1/2}$$

$$+ \frac{s(s-1)(s+1)(s-2)}{4!}\,\delta^4 f_0 + \cdots$$

$$+ \begin{cases} \dfrac{s(s^2-1)(s^2-2^2)\cdots(s^2-(m-1)^2)(s-m)}{(2m)!}\,\delta^{2m} f_0 \\[2mm] \dfrac{s(s^2-1)(s^2-2^2)\cdots(s^2-m^2)}{(2m+1)!}\,\delta^{2m+1} f_{1/2} \end{cases} \qquad (4.5.17)$$

depending on whether we finish with an even- or odd-order difference.

The error terms for these two cases are given by

$$E_s = \frac{h^{2m+1}s(s^2-1)(s^2-2^2)\cdots(s^2-m^2)}{(2m+1)}\,f^{(2m+1)}(\xi)$$

$$E_s = \frac{h^{2m+2}s(s^2-1)\cdots(s^2-m^2)(s-m-1)}{(2m+2)!}\,f^{(2m+2)}(\xi)$$

respectively.

The corresponding Gauss backward formula uses the data points in the order $x_0, x_{-1}, x_1, x_{-2}, x_2, \ldots$, and in the same way as for the forward formula this yields

$$P(s) = f_0 + s\delta f_{-1/2} + \frac{s(s+1)}{2} \delta^2 f_0 + \frac{s(s+1)(s-1)}{3!} \delta^3 f_{-1/2}$$

$$+ \frac{s(s+1)(s-1)(s+2)}{4!} \delta^4 f_0 + \cdots \qquad (4.5.18)$$

The formulas of Bessel and Everett are simple variations on these. Bessel's formula is used in an attempt to cover the interval between x_0 and x_1 uniformly well by taking the average of the Gauss forward formula (4.5.17) and the backward formula (4.5.18), with this latter one based not at x_0 but at x_1. Thus in (4.5.18) we must replace s by $s-1$ everywhere with all differences being of the form $\delta^{2k} f_1$, $\delta^{2k+1} f_{1/2}$. With these substitutions, we get the following formula for the average:

$$P(s) = \mu f_{1/2} + (s - 1/2)\delta f_{1/2} + \frac{s(s-1)}{2} \mu\delta^2 f_{1/2}$$

$$+ \frac{s(s-1)(s-1/2)}{3!} \delta^3 f_{1/2}$$

$$+ \cdots + \begin{cases} \dfrac{s(s^2-1)\cdots[s^2-(m-1)^2](s-m)}{(2m)!} \mu\delta^{2m} f_{1/2} \\[4mm] \dfrac{s(s^2-1)\cdots[s^2-(m-1)^2](s-m)(s-1/2)}{(2m+1)!} \delta^{2m+1} f_{1/2} \end{cases}$$

$$(4.5.19)$$

in which we have used the mean operator μ defined by

$$\mu f_{1/2} = \frac{f_1 + f_0}{2} \qquad (4.5.20)$$

so that, for example, $\mu\delta^2 f_{1/2} = (\delta^2 f_1 + \delta^2 f_0)/2$.

Example 4.5.6. Use the data of Example 4.5.3 to estimate the value of sin 1.52 using Bessel's formula.

Solution. Again we will refer to the difference table, Table 4.5.4. Since 1.52 lies between the tabular points 1.5 and 1.6, we use $x_0 = 1.5$, and therefore $s = 0.2$. The differences used are then those marked with an asterisk (*), and the first few approximations obtained are

$$\sin 1.52 \approx \mu f_{1/2} = \frac{0.9975 + 0.9996}{2} = 0.99855$$

$$\sin 1.52 \approx 0.99855 + (-0.3)(0.0021) = 0.99792$$

$$\sin 1.52 \approx 0.99792 + \frac{(0.2)(-0.8)(-0.0100)}{2} = 0.99872$$

$$\sin 1.52 \approx 0.99872 + \frac{(0.2)(-0.8)(-0.3)(0.0000)}{3} = 0.99872$$

which is correct to the four decimal places of the original data.

The last of the finite difference interpolation schemes we discuss here is that of Everett. This is really just a rearrangement of the Gauss forward formula (4.5.17) terminating with an odd-order difference. The extent of the rearrangement is the straightforward replacement of odd-order central differences by their definition in terms of two central differences of order one lower; that is, we write $\delta^{2k+1}f_{1/2} = \delta^{2k}f_1 - \delta^{2k}f_0$ ($k = 0, 1, \ldots$). Collecting terms based at x_0 and those centered on x_1 separately, we thus obtain from (4.5.17), terminating with the $(2m + 1)$th difference,

$$
\begin{aligned}
P(s) = (1 - s)f_0 &- \frac{s(s-1)(s-2)}{3!}\delta^2 f_0 - \frac{(s+1)s(s-1)(s-2)(s-3)}{5!}\delta^4 f_0 \\
&- \cdots - \frac{(s+m-1)(s+m-2)\cdots(s-m-1)}{(2m+1)!}\delta^{2m}f_0 \\
&+ sf_1 + \frac{(s+1)s(s-1)}{3!}\delta^2 f_1 + \frac{(s+2)(s+1)s(s-1)(s-2)}{5!}\delta^4 f_1 \\
&+ \cdots + \frac{(s+m)(s+m-1)\cdots(s-m)}{(2m+1)!}\delta^{2m}f_1
\end{aligned}
\qquad (4.5.21)
$$

Example 4.5.7. Use the data of Example 4.5.3 and the difference Table 4.5.4 to approximate the value of $\sin 1.63$ with Everett's formula.

Solution. With $x_0 = 1.6$, $x_1 = 1.7$, and $s = 0.3$, we use the differences shown in **bold** print in the table. From just the differences of up to second order we obtain the estimate

$$
\sin 1.63 \approx (0.7)(0.9996) - \frac{(0.3)(-0.7)(-1.7)(-0.0100)}{6}
$$
$$
+ (0.3)(0.9917) + \frac{(1.3)(0.3)(-0.7)(-0.0100)}{6}
$$
$$
= 0.99828
$$

which has an error of about 3×10^{-5}. The rounded value 0.9983 shows an error of 1 in the final position, and, as with the backward difference approximation in Example 4.5.5, this error is almost entirely attributable to the rounding errors in the original data.

In summary, we see in this section that there are several reasonably straightforward formulas which can be used for the task of polynomial interpolation in an equally spaced table of data. While this facility is of much reduced importance relative to its position at the dawning of the computer age—still only some 30-odd years ago—it should not be ignored as a practical computational tool. There are yet many functions in common use whose values are easily available only from tables, and some of these special functions are of great importance to many applications. There will also be a need in the foreseeable future for interpolation techniques in cases where the data is taken from experimental evidence for which there is often no possibility of any direct algorithm for the evaluation of the function in question.

For this last situation, it may be possible to select the data points in advance of the experimentation to facilitate the interpolation process by reducing the error bounds in a particular region of interest. We turn to this question of the choice of the nodes or data points in the next section, but before doing so we should observe that in the case of experimental data, it may be desirable to find an approximating function rather than an interpolating one since the latter preserves any errors in the data. We will discuss the approximation of functions in Chap. 7.

EXERCISES 4.5

1. Using the data of Table 4.5.4, estimate sin 1.23 and sin 1.96 using, respectively, the forward and backward difference formulas.
2. Prove identities (4.5.8).
3. Derive Newton's backward difference interpolation formula (4.5.14).
4. Establish error formulas (4.5.15) and (4.5.16) for the forward and backward interpolation formulas.
5. Write a program to implement Newton's forward and backward difference interpolation formulas. Use it to verify (or improve) the results obtained from the data of Table 4.5.4 for the sine function. Check that your results agree with the error bounds. Tabulate or graph the results of using the appropriate interpolation formula, and compare them with the true values.
6. Obtain error formulas for Bessel's and Everett's interpolation formulas.

4.6 CHOICE OF NODES AND NONCONVERGENCE OF POLYNOMIAL INTERPOLATION

In the previous sections, we developed a little of the theory of polynomial interpolation and saw that it provided tolerably efficient techniques for estimating the value of a function from tabulated data. However, there are serious drawbacks to this approach.

Intuitively, we feel that as the number of nodes is increased, the accuracy of the interpolation should improve, and Weierstrass' theorem appears to support this contention. Unfortunately, this is very far from the case. What Weierstrass' theorem states is that, for any tolerance, there is a polynomial which approximates a continuous function f uniformly on a closed interval within this tolerance. What it does not say is that this approximating polynomial is an interpolation polynomial. In particular, if the interpolation points are uniformly spaced on the interval, then it can happen that there are points in the interval for which the difference between the function and the interpolation polynomial will grow without bound as the number of nodes increases. However, we will also see an example in which *best approximation* is achieved by interpolation at a carefully chosen set of nodes.

The most famous example of the nonconvergence mentioned above is due to Runge. Let f be defined on the interval $[-5, 5]$ by

$$f(x) = \frac{1}{1 + x^2} \tag{4.6.1}$$

For each N let p_N denote the polynomial interpolating f at the equally spaced nodes $x_0, x_1, \ldots, x_N$ given by $x_0 = -5$, $x_i = x_0 + ih$, where $h = 10/N$. It can be shown that $p_N(x) \nrightarrow f(x)$ as $N \to \infty$ for any $|x| > 3.65$. A detailed treatment of this result can be found in Isaacson and Keller (1966), but we content ourselves here with a practical demonstration of the difficulties posed by this function. (This is also the example used in Sec. 4.3 in discussing the use of divided difference interpolation formulas.)

In Fig. 4.6.1a to d we compare the graph of the function f given by (4.6.1) over $[-5, 5]$ with those of the various polynomials using 6, 11, 21, and 41 equally spaced points. It is apparent that the successive approximations fail to provide any acceptable accuracy near the ends of the range. Using 81 points resulted in floating-point overflow with type real and large accumulated roundoff errors near the right-hand end of the graph for type double.

The evaluation of the divided difference polynomial is based on just one computation of the difference table, as in Algorithm 4.3.8.

The obvious nonconvergence is a direct consequence of the choice of equally spaced nodes—and the use of all of them for every point.

What can we do to get around this problem and retain a simple procedure for the estimation of function values?

It is the case that there are selections of interpolation points for which the corresponding polynomials do converge to the true value at every point of the interval. This is achieved by choosing the nodes so as to keep the values of $L_N(x)$ under control throughout the interval of interest. We return to this question shortly.

One of the most attractive answers to this question is to use interpolation but now with different polynomials on different parts of the range.

This can be achieved in many different ways. One of the simplest is that we just use as many terms of Newton's divided difference formula as are necessary to achieve desired accuracy at any given point of interest rather than using all the data for all points of interest. It is this approach which we illustrate in Fig. 4.6.2 below.

An alternative would be to use local low-degree interpolation formulas by taking just the nearby points in the table of data and computing, say, the first three terms of the divided difference interpolation formula. One potential disadvantage of this approach is that the interpolating function itself may not be very smooth. As the point of interest x crosses a data point x_i, say, one of the data points used to the left would be dropped from the interpolant while a new one would be introduced to the right. The mere fact that x_i is a data point implies that continuity will be maintained, but it is unlikely that any higher degree of continuity will be achieved. Results using this approach are illustrated in Fig. 4.6.3. In the next chapter we will study ways of improving the smoothness while retaining the simplicity of this approach.

We examine both these approaches to the problem, again referring to Runge's example. The data is still taken from the same sets of equally spaced

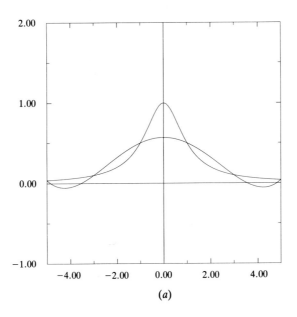

(a)

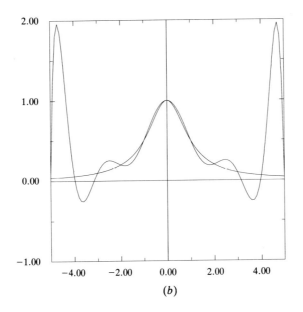

(b)

FIGURE 4.6.1
$f(x) = 1/(1 + x^2)$ and its interpolants using N equally spaced points in $[-5, 5]$. (a) $N = 6$; (b) $N = 11$; (c) $N = 21$; (d) $N = 41$.

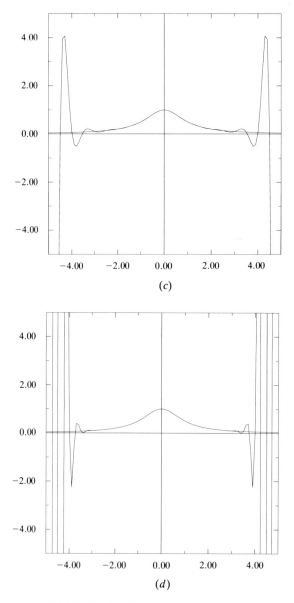

FIGURE 4.6.1 (*Continued*)

nodes as before but for just 21 points. Two parameters were varied, namely, *Nmax*, the maximum number of terms of the divided difference interpolation polynomial to be used, and *eps*, the requested tolerance in the result. The computation of the interpolation polynomial is stopped if either the last term added is smaller than *eps* or if *Nmax* terms have already been used.

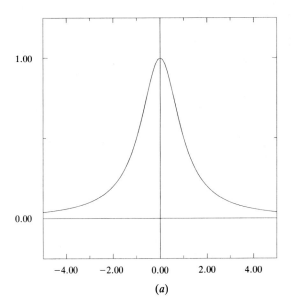

(*a*)

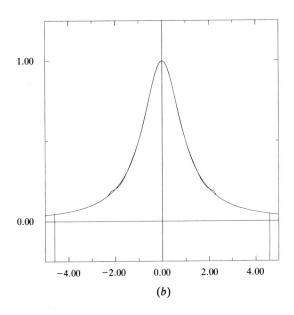

(*b*)

FIGURE 4.6.2

$f(x) = 1/(1 + x^2)$ and its local interpolants using data from 21 equally spaced points in $[-5, 5]$. (*a*) $eps = 10^{-3}$; (*b*) $eps = 10^{-5}$; (*c*) $eps = 10^{-7}$.

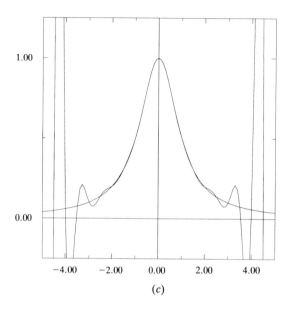

(c)

FIGURE 4.6.2 (*Continued*)

In Fig. 4.6.2, we see the graphs of f and the result of interpolation using just the points nearby the current point of interest. The parameter *Nmax* is thus set to 20, and the tolerance *eps* is varied.

It is apparent that for fairly modest accuracy requirements the computation stops after only a few terms and produces a near-perfect (at that tolerance) fit to the function itself. As the tolerance is tightened, more and more terms are needed at the ends of the range, and eventually the behavior is essentially the same as that observed for the full Lagrange polynomial in Fig. 4.6.1. Notice how the region in which loss of accuracy occurs steadily encroaches from the ends as the tolerance is sharpened.

It should also be observed that the final graph, Fig. 4.6.2c, is generated using the full polynomial for most $|x| > 3.5$. These calculations are essentially equivalent, therefore, to those shown in Fig. 4.6.1c. The graphs are quite different. This is due to the very different arrangement of the computation here. The effect of rearranging the order of the "same" computation is quite obvious.

The alternative suggested above was to use only a low-degree polynomial as a local approximant. This is achieved by varying the parameter *Nmax*. (Varying *eps* has virtually no effect, in this situation, for any value of *eps* smaller than about 10^{-3}.) In Fig. 4.6.3 we see the effect of using *Nmax* = 1 which amounts to using straight-line segments to join neighboring data points and then *Nmax* = 3 so that local cubic interpolants are used. It is clear that the latter copes much better with the high curvature near $x = 0$. It is also apparent

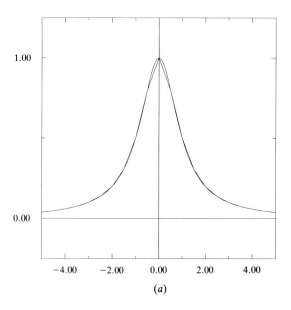

(a)

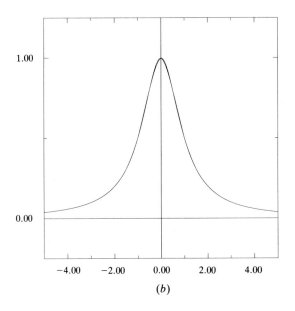

(b)

FIGURE 4.6.3

$f(x) = 1/(1 + x^2)$ and its local interpolants using data from 21 equally spaced points in $[-5, 5]$. (a) $Nmax = 1$; (b) $Nmax = 3$.

that for the relatively modest accuracy requirement of computer graphics, local cubic interpolation is adequate to yield a high level of agreement between the curves.

Of course, the improvements are bought at a considerable computational price. The data must be rearranged for each new point to use only the most relevant information, and then the relevant divided differences must be computed afresh. For the results used in producing Fig. 4.6.1, it is only necessary to compute the divided differences once.

We consider now the question of the choice of nodes in order that the full interpolation polynomial may be used throughout the range but without the nonconvergence exhibited in Fig. 4.6.1 using equally spaced interpolation points. Specifically then, we seek the choice of nodes $x_0, x_1, \ldots, x_N \in [a, b]$ such that the interpolation polynomial p_N is the best approximating polynomial in Π_N for a function f over $[a, b]$ in the uniform norm. Thus we seek nodes such that

$$\|f - p_N\|_\infty = \min\left\{\|f - p\|_\infty : p \in \Pi_N\right\} \tag{4.6.2}$$

where we recall that the ∞ norm (or supremum, or uniform norm) is defined by

$$\|f - p\|_\infty = \sup_{a \le x \le b} |f(x) - p(x)| = \max_{a \le x \le b} |f(x) - p(x)| \tag{4.6.3}$$

since any continuous function achieves its bounds on a closed bounded interval. This problem can be alternatively described as seeking the nodes such that

$$\|f - p_N\|_\infty = \min_p \max_{a \le x \le b} |f(x) - p(x)| \tag{4.6.4}$$

Such an approximation, if it exists, is sometimes referred to as a *minimax* approximation. Of course, we have not established that such a choice of nodes exists in general.

Recall from Theorem 4.2.3 that the error in Lagrange interpolation is given by

$$R_N(x) = \frac{L_N(x)f^{(N+1)}(\xi)}{(N + 1)!} \tag{4.6.5}$$

where ξ is a mean value point in the interval $[a, b]$ and

$$L_N(x) = (x - x_0)(x - x_1) \cdots (x - x_N) \tag{4.6.6}$$

It is probably apparent that the specific choice of interpolation points, if it exists at all, will depend on the function f and that there is little hope of finding a simple general solution to our problem.

At this stage we follow that well-worn path of mathematicians and scientists for centuries past and simplify the problem to one that we can solve. As is commonly the case with such simplifications, the specific problem we will consider is in itself artificial, but, as is also often the case, it does provide us with a good general rule of thumb for the selection of nodes for polynomial interpolation.

Our simplifying assumption is based on the optimistic belief that $f^{(N+1)}$ will vary only slowly over the interval of interest. We thus take the idealized situation that $f^{(N+1)}$ is constant, that is, of course, that f itself is a polynomial of degree $N + 1$. Specifically we consider the situation in which f is a monic polynomial of degree $N + 1$ so that it is given by

$$f(x) = x^{N+1} + a_N x^N + \cdots + a_1 x + a_0 \tag{4.6.7}$$

Our unrealistic problem is thus the search for the best approximation to a polynomial of degree $N + 1$ by a polynomial of degree N. We specialize still further by fixing the interval to be $[-1, 1]$. (This is not a real restriction since any closed bounded interval can easily be mapped into $[-1, 1]$ by a simple linear change of variables.)

Now $f^{(N+1)}(x) \equiv (N + 1)!$, and so (4.6.5) reduces to just

$$R_N(x) = L_N(x) = (x - x_0)(x - x_1) \cdots (x - x_N) \tag{4.6.8}$$

and thus the problem of locating the best choice of nodes is similarly simplified to finding $x_0, x_1, \ldots, x_N$ such that $\max |L(x)|$ is minimized.

This problem has an explicit solution in terms of Chebyshev polynomials. This is an important system of orthogonal polynomials which is widely used in approximation and numerical integration.

Definition 4.6.4. We define the *Chebyshev polynomial* of degree n on $[-1, 1]$ by

$$T_n(x) = \cos (n \cos^{-1} x) \tag{4.6.9}$$

It is probably not immediately apparent that this formula defines a polynomial at all. We establish this fact in the following proposition and go on, in Theorem 4.6.6, to establish that these polynomials have exactly the minimax property we desire.

Proposition 4.6.5
(*a*) $T_0(x) \equiv 1$, $T_1(x) = x$, and

$$T_{n+1}(x) = 2x T_n(x) - T_{n-1}(x) \qquad (n = 1, 2, \ldots) \tag{4.6.10}$$

(*b*) T_n is a polynomial of degree n with leading coefficient 2^{n-1} for $n \geq 1$.

Proof
(*a*) That $T_0(x) = 1$ and $T_1(x) = x$ are immediate from the definition. Denote $\cos^{-1} x$ by θ. Then, for $n \geq 1$, we have

$$T_{n+1}(x) + T_{n-1}(x) = \cos (n + 1)\theta + \cos (n - 1)\theta$$

$$= 2 \cos \frac{(n + 1)\theta + (n - 1)\theta}{2} \cos \frac{(n + 1)\theta - (n - 1)\theta}{2}$$

$$= 2 \cos n\theta \cos \theta$$

$$= 2x T_n(x)$$

as required.

(b) Statement (b) is a simple consequence of (a), and its proof is left as an exercise.

In view of (4.6.8), we seek $x_0, x_1, \ldots, x_N$ such that

$$\max_{-1 \leq x \leq 1} |L_N(x)| \leq \max_{1 \leq x \leq 1} |p(x)|$$

for any monic polynomial p of degree $N + 1$. The next result establishes that the appropriately scaled Chebyshev polynomials have precisely this property.

Theorem 4.6.6

$$\max_{-1 \leq x \leq 1} |2^{-N} T_{N+1}(x)| = 2^{-N} \leq \max_{-1 \leq x \leq 1} |p(x)| \qquad (4.6.11)$$

where p is any monic polynomial of degree $N + 1$.

Proof. From Proposition 4.6.5, we see that $T^*_{N+1} = 2^{-N} T_{N+1}$ is a monic polynomial. Also, by definition, T_{N+1} has maximum absolute value 1 where $\cos [(N + 1) \cos^{-1} x] = \pm 1$, that is, at the points

$$y_k = \cos \frac{k\pi}{N + 1} \qquad (k = 0, 1, \ldots, N + 1) \qquad (4.6.12)$$

It follows that

$$T^*_{N+1}(y_k) = (-1)^k 2^{-N} \qquad (4.6.13)$$

and therefore that T^*_{N+1} has $N + 2$ extreme values in $[-1, 1]$ *which alternate in sign*.

Suppose now that p is a monic polynomial of degree $N + 1$ such that $|p(x)| \leq M < 2^{-N}$ for all $-1 \leq x \leq 1$, and denote by $\delta(x)$ the polynomial defined by $T^*_{N+1}(x) - p(x)$. Now $\delta(y_k)$ must have the same sign as $T^*_{N+1}(y_k)$ for each k since otherwise $|p(y_k)| > 2^{-N}$.

We deduce that δ changes sign at least $N + 1$ times in $[-1, 1]$ (since we have a sequence of $N + 2$ points at which its sign alternates). But T^*_{N+1} and p are both monic of degree $N + 1$. It follows that δ has degree at most N. By the fundamental theorem of algebra, the only polynomial of degree at most N which vanishes at least $N + 1$ times is identically zero. This contradiction completes the proof.

From the remarks preceding the theorem, we may now deduce that T^*_{N+1} is the polynomial we seek and that the required choice of interpolation points is that $x_0, x_1, \ldots, x_N$ are the zeros of T^*_{N+1} or, equivalently, of T_{N+1}. These points satisfy

$$(N + 1) \cos^{-1} x = \frac{(2k + 1)\pi}{2}$$

for some integer k. Thus the desired nodes are at

$$x_k = \cos \frac{(2k + 1)\pi}{2(N + 1)} \qquad (k = 0, 1, \ldots, N) \qquad (4.6.14)$$

Example 4.6.7. Find the best quadratic approximation to $x^3 + x + 1$ over $[-1, 1]$.

Solution. Now according to Theorem 4.6.6 we require the Lagrange interpolation polynomial agreeing with $x^3 + x + 1$ at the zeros of T_3. From the recurrence relation (4.6.10), we obtain

$$T_2(x) = 2xT_1(x) - T_0(x) = 2x^2 - 1$$

$$T_3(x) = 2xT_2(x) - T_1(x) = 2x(2x^2 - 1) - x = 4x^3 - 3x$$

The zeros of T_3 are at $x = 0$, $\pm\sqrt{3}/2$ (that is, at $\cos\pi/6$, $\cos 3\pi/6$, and $\cos 5\pi/6$). The function values to be interpolated are $f(\pm\sqrt{3}/2) = 1 \pm 7\sqrt{3}/8$ and $f(0) = 1$. From either the Lagrange form or the divided difference formula, we deduce that the required interpolation polynomial is

$$p_2(x) = 1 + \frac{7x}{4}$$

We see here that the best quadratic approximation is in fact a straight-line function. It is straightforward to check that the maximum error in this approximation over $[-1, 1]$ is $\frac{1}{4}$ and that this is achieved at the points $\cos(k\pi/3)$, which are -1, $-\frac{1}{2}$, $\frac{1}{2}$, and 1.

We finish this section with a look at how well interpolation using the Chebyshev points works for a situation other than this artificial one of approximating a polynomial of just one higher degree. In Fig. 4.6.8 we show

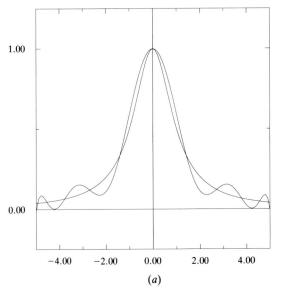

(a)

FIGURE 4.6.8
$f(x) = 1/(1 + x^2)$ and its Chebyshev interpolants using data from N points in $[-5, 5]$. (a) $N = 11$; (b) $N = 21$.

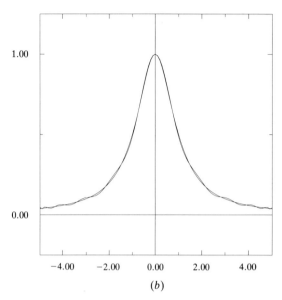

(b)

FIGURE 4.6.8 (*Continued*)

the result of applying interpolation at the Chebyshev points to Runge's example. This is achieved by Algorithm 4.3.8 with the data points simply redefined to be $5\cos\{[(2k+1)\pi]/[2(N+1)]\}$, these being the equivalently placed nodes for Chebyshev interpolation on the interval $[-5, 5]$.

It is apparent from the graphs that this choice of nodes keeps the polynomial behavior of the factor $L_N(x)$ in the Lagrange error term under much tighter control. Very good agreement between the function and the interpolant is achieved throughout the range with a modest number of points. We should stress here that this computation uses the full interpolation polynomial everywhere, and so the graphs are directly comparable with those of Fig. 4.6.1.

The additional control over the errors is achieved by placing more of the nodes toward the ends of the range—though without using the endpoints themselves. This distribution can be seen clearly in Fig. 4.6.8a, where 6 of the 11 nodes are in the previously troublesome region with $|x| > 3.5$.

We will return to the discussion of Chebyshev polynomials and other systems of orthogonal polynomials in Chap. 7.

EXERCISES 4.6

1. Write a program to implement Newton's divided difference interpolation formula with parameters *eps* and *Nmax* such that the computation ceases if the next term is smaller than *eps* or if the polynomial used is of degree *Nmax*. Test your program by reproducing the graphs of this section.

2. Prove that the Chebyshev polynomial T_N is indeed a polynomial of degree N with leading coefficient 2^{N-1}.

3. Write down formulas for the first six Chebyshev

polynomials $T_0, T_1, \ldots, T_5$, and find all their zeros. Use GraphLib or your own program to graph these functions on the same axes. Note the "equal but oscillating" nature of their local extrema.

4. Repeat the experiments of Sec. 4.6 for the function $1/(1 + 10x^2)$. Describe your results, and indicate what you think is the best approach to polynomial interpolation. Take into account the overall computational effort needed.

PROJECTS

1. Use Newton's divided difference interpolation formula to estimate the values of $\ln(1 + i/20)$ ($i = 0, 1, \ldots, 20$) from the data $\ln(1 + i/10)$ ($i = 0, 1, \ldots, 10$). Make the computation as efficient as you can, and produce a graph showing both the interpolation polynomial and the original function.

2. Write a program to implement Aitken's algorithm. (Begin by modifying the data-ordering algorithm to find just the next closest point when it is needed. Make your routine as efficient as possible.)

   Test your program on the function $1/(1 + x^2)$ over $[-5, 5]$. Experiment with the number and positioning of the interpolation points. Does the use of the Chebyshev points help? Can you find the best data set?

3. Write a program to implement Newton's forward and backward difference interpolation formulas and Bessel's formula in appropriate parts of a table of data. Use it to verify (or improve) the results obtained from the data of Table 4.5.4 for the sine

function. Check that your results agree with the error bounds.

4. Find the maximum uniform steplength that can be used in a table of values of $\sin x$ for $0 \leq x \leq \pi/2$ so that using cubic finite difference interpolation will yield an error bounded by 10^{-4}. Write a program to implement this, and check that the desired accuracy is achieved.

5. Examine the effect of roundoff error on the procedure in Project 4 by rounding your data to four- and then five-decimal-place accuracy. Try to analyze this effect.

6. Write a program to implement divided difference interpolation at the Chebyshev nodes for an arbitrary interval $[a, b]$ by transforming this interval into $[-1, 1]$ and then transforming back. The number of nodes should be an input parameter. Test your program for $1/(1 + x^2)$ over $[0, 5]$ with varying numbers of nodes.

CHAPTER
5

OTHER
INTERPOLATION
FUNCTIONS

5.1 INTRODUCTION

In the previous chapter we studied the use of interpolation polynomials for approximating the values of functions. The examples of Sec. 4.6 demonstrate clearly the shortcomings of that approach. The use of local polynomials clearly improved the performance but at the expense of smoothness in the approximating function. Again, careful choice of node placement was a considerable benefit, but that is not always available or convenient. In this chapter we study alternatives to simple polynomial interpolation.

Additional smoothness can be achieved with local polynomial interpolation by using *spline* functions in which different low-degree polynomials are used on each interval $[x_i, x_{i+1}]$ together with the imposition of smoothness conditions to ensure that the overall interpolating function has as high a degree of continuity as possible at each of the nodes, or *knots*, x_i.

> **Definition 5.1.1.** Let $x_0 < x_1 < \cdots < x_N$ be an increasing sequence of knots. The function s is a *spline of degree k* if:
> (*a*) s is a polynomial of degree no more than k on each of the subintervals $[x_i, x_{i+1}]$.
> (*b*) $s, s', \ldots, s^{(k-1)}$ are all continuous on the interval $[x_0, x_N]$.

The familiar connection of points on a graph by straight lines is an example of a linear (or first-degree) spline, that is, a piecewise linear function which is continuous at the breakpoints. By far the most commonly used splines for interpolation purposes are the cubic (or degree three) splines, and it is on these that we will concentrate our attention. First, though, we have some examples of spline functions.

Examples 5.1.2
(a) The function defined by

$$s(x) = \begin{cases} x & 0 \le x \le 1 \\ x^2 - x + 1 & 1 \le x \le 3 \\ 5x - 8 & 3 \le x \le 4 \end{cases}$$

is a second-degree spline. Note that the degree of the pieces is at most two rather than exactly two everywhere. It is not a cubic (third-degree) spline since the second derivative is not continuous; it is 0, 2, and 0 in the three intervals, respectively.

(b) The function

$$s(x) = \begin{cases} 2x - 1 & 0 \le x \le 1 \\ x & 1 \le x \le 2 \end{cases}$$

is a linear spline.

(c) The function

$$s(x) = \begin{cases} x^3 - x^2 + 2x - 1 & -2 \le x \le 0 \\ -x^2 + 2x - 1 & 0 \le x \le 1 \\ x^3 - 4x^2 + 5x - 2 & 1 \le x \le 3 \\ x^3 + x^2 + 25x + 43 & 3 \le x \le 4 \end{cases}$$

is not a cubic spline since s'' is not continuous at $x = 3$. Setting $s(x) = 5x^2 - 22x + 25$ for $3 \le x \le 4$, it becomes a cubic spline.

In the next few sections we concentrate on the use of interpolating splines, specifically cubic spline interpolation. In later sections we turn briefly to other types of functions as a basis for interpolation. Rational functions—that is, quotients of polynomials—have been found to provide very good approximations in many situations. We study their use and their representation as continued fractions in Sec. 5.5.

Trigonometric polynomial interpolation is considered in Sec. 5.6. This is another widely applicable technique which is closely related to the Fast Fourier Transform (FFT), which in turn is a very powerful function approximation method used in many engineering applications.

We conclude the chapter with a very brief introduction to the ideas of bivariate interpolation. The problem of approximating functions of more than one variable is a very difficult one to which there are very few definitive answers available. Indeed, it is not even clearly understood as to what the two-dimensional equivalents of some of the techniques we discuss here are.

EXERCISES 5.1

1. Which of the following functions are splines and of what degree?

(a) $f(x) = \begin{cases} x & x \in [0, 1] \\ \dfrac{x^2 + 1}{2} & x \in [1, 3] \end{cases}$

(b) $f(x) = \begin{cases} x & x \in [0, 1] \\ 2 - x & x \in [1, 2] \end{cases}$

(c) $f(x) = \begin{cases} x^3 + 2x & x \in [0, 1] \\ 3 + 5(x - 1) + 3(x - 1)^2 & x \in [1, 4] \end{cases}$

(d) $f(x) = \begin{cases} x^3 + 2x & x \in [0, 1] \\ 3 + 5(x-1) + 3(x-1)^2 & x \in [1, 4] \\ 45 + 23(x-4) + (x-4)^3 & x \in [4, 5] \end{cases}$

2. Find the linear spline which interpolates the values $\ln(1 + 2x)$ at $x = 0, 0.1, 0.2, 0.3, 0.4,$ and 0.5. Use this linear spline to estimate the values of $\ln 1.1$, $\ln 1.3$, $\ln 1.5$, $\ln 1.7$, and $\ln 1.9$.

5.2   NATURAL CUBIC SPLINE INTERPOLATION

We turn now to the specific problem of obtaining a cubic spline function which interpolates the function f at $x_0, x_1, \ldots, x_N$. It will be convenient to introduce the following notation. In each of the subintervals $I_i = [x_i, x_{i+1}]$ of the interpolation range, s is a polynomial of degree at most three; denote this polynomial by s_i. Then we have

$$s(x) = s_i(x) \qquad x \in I_i, \, i = 0, 1, \ldots, N-1 \tag{5.2.1}$$

A convenient formulation of s_i will be in terms of the distance of x from the two ends of the interval I_i, and so we define new variables u_i by

$$u_i = x - x_i \qquad (i = 0, 1, \ldots, N) \tag{5.2.2}$$

Observe that $du_i/dx = 1$ for every i, and so differentiation or integration with respect to x and with respect to u_i will be equivalent. We denote the steplengths between the knots by

$$h_i = x_{i+1} - x_i = u_i - u_{i+1} \tag{5.2.3}$$

The conditions which must be satisfied are that s must interpolate f at $x_0, x_1, \ldots, x_N$ and s', s'' must be continuous at the interior knots $x_1, x_2, \ldots, x_{N-1}$. We will begin with the last of these, the continuity of s''.

On each of the intervals I_i, s is a cubic, and so s'' is the first-degree polynomial s_i''. Let us denote its (as yet unknown) values at the knots by

$$s''(x_i) = A_i \qquad i = 0, 1, \ldots, N \tag{5.2.4}$$

It follows that $s_i''(x_i) = A_i$ and $s_i''(x_{i+1}) = A_{i+1}$, and since s_i'' is a linear function, we have, for each i,

$$s_i''(x) = \frac{A_{i+1}(x - x_i) - A_i(x - x_{i+1})}{h_i} = \frac{A_{i+1}u_i - A_i u_{i+1}}{h_i} \tag{5.2.5}$$

We may integrate (5.2.5) twice to get

$$s_i(x) = \frac{A_{i+1}u_i^3 - A_i u_{i+1}^3}{6h_i} + cx + d$$

where c and d are constants of integration. This can be conveniently written in the form

$$s_i(x) = \frac{A_{i+1}u_i^3 - A_i u_{i+1}^3}{6h_i} - B_i u_{i+1} + C_i u_i \tag{5.2.6}$$

Thus far, we have enforced the second-derivative continuity of our spline function. We must choose the coefficients in Eq. (5.2.6) for $i = 0, 1, \ldots, N-1$ so that both the interpolation conditions and the first-derivative continuity are satisfied. We thus have $N + 1$ of the A_i's and N each of the B_i and C_i to choose. There are $2N$ interpolation conditions and another $N-1$ equations to be derived from the continuity of s' at all the interior knots. That is, there will be

$3N - 1$ equations in $3N + 1$ unknowns. We will discuss the use of these two extra degrees of freedom later.

Consider first the interpolation conditions. At the point x_i, we have $u_i = 0$ and $u_{i+1} = -h_i$. Denoting $f(x_i)$ by f_i and substituting these values into (5.2.6), we get

$$f_i = \frac{A_i h_i^2}{6} + B_i h_i \qquad (i = 0, 1, \ldots, N - 1)$$

Similarly at x_{i+1}, we have

$$f_{i+1} = \frac{A_{i+1} h_i^2}{6} + C_i h_i \qquad (i = 0, 1, \ldots, N - 1)$$

Solving these two for B_i and C_i yields

$$B_i = \frac{f_i}{h_i} - \frac{A_i h_i}{6}$$

$$\tag{5.2.7}$$

$$C_i = \frac{f_{i+1}}{h_i} - \frac{A_{i+1} h_i}{6}$$

The final system of equations is derived from the first-derivative continuity condition. These equations are obtained by differentiating (5.2.6) with respect to x (remembering that differentiation with respect to x and with respect to u_i or u_{i+1} are the same operation). We obtain

$$s_i'(x) = \frac{A_{i+1} u_i^2 - A_i u_{i+1}^2}{2h_i} - B_i + C_i \tag{5.2.8}$$

from which we may deduce that

$$s_i'(x_i) = C_i - B_i - \frac{A_i h_i}{2} \tag{5.2.9}$$

and, similarly,

$$s_i'(x_{i+1}) = C_i - B_i + \frac{A_{i+1} h_i}{2} \tag{5.2.10}$$

The continuity of s' will be guaranteed if, for every interior knot x_i, we have $s_i'(x_i) = s_{i-1}'(x_i)$ which, on comparing (5.2.9) with (5.2.10) for $i - 1$, yields the equation

$$\frac{(h_{i-1} + h_i) A_i}{2} + B_i - C_i - (B_{i-1} - C_{i-1}) = 0 \tag{5.2.11}$$

for $i = 1, 2, \ldots, N - 1$.

We can subtract the two equations (5.2.7) to obtain

$$B_i - C_i = \frac{(A_{i+1} - A_i) h_i}{6} - \frac{f_{i+1} - f_i}{h_i} \qquad (i = 0, 1, \ldots, N - 1)$$

$$\tag{5.2.12}$$

The final term here is just the divided difference $f[x_i, x_{i+1}]$ which we will denote by d_i. With this notation and substituting (5.2.12) for both i and $i-1$ into (5.2.11), we get

$$\frac{h_{i-1}A_{i-1}}{6} + \frac{(h_{i-1}+h_i)A_i}{3} + \frac{h_iA_{i+1}}{6} = d_i - d_{i-1} \qquad (i = 1, 2, \ldots, N-1)$$
$$(5.2.13)$$

This is a system of $N-1$ equations in the $N+1$ unknown A_i's. As was commented above, there are many ways of using these two extra degrees of freedom. One of the simplest is to simply set

$$A_0 = A_N = 0 \qquad (5.2.14)$$

which gives rise to the so-called natural cubic splines.

The name is derived from the old drafting tool, the *spline*, which was a flexible thin piece of wood that was used to generate a smooth curve passing through specified points. It was therefore an interpolation device. The shape it assumes is the shape with minimum stored energy, and it turns out to be (almost exactly) a natural cubic spline. That it satisfies these end conditions is a natural feature since this implies that the shape is continued in a straight line beyond the extreme knots.

Multiplying Eq. (5.2.13) by 6 and denoting $d_i - d_{i-1}$ by Δd_{i-1}, the natural cubic spline interpolating f at $x_0, x_1, \ldots, x_N$ is obtained with the coefficients A_i satisfying the *tridiagonal* system

$$\begin{bmatrix} 2(h_0+h_1) & h_1 & 0 & \cdots & & 0 \\ h_1 & 2(h_1+h_2) & h_2 & \cdots & & 0 \\ \cdots & & & & & \vdots \\ 0 & \cdots & h_{N-3} & 2(h_{N-3}+h_{N-2}) & h_{N-2} & \vdots \\ 0 & \cdots & 0 & h_{N-2} & 2(h_{N-2}+h_{N-1)}) \end{bmatrix} \begin{bmatrix} A_1 \\ A_2 \\ \vdots \\ A_{N-2} \\ A_{N-1} \end{bmatrix}$$

$$= \begin{bmatrix} 6\Delta d_0 \\ 6\Delta d_1 \\ \vdots \\ 6\Delta d_{N-3} \\ 6\Delta d_{N-2} \end{bmatrix} \qquad (5.2.15)$$

There are other important ways of using the degrees of freedom in system (5.2.13). Foremost among these are the *complete* cubic splines for which the additional conditions are that the derivatives of f and s agree at the two endpoints; that is,

$$s'(x_0) = f'(x_0) \qquad \text{and} \qquad s'(x_N) = f'(x_N)$$

This also leads to a positive definite symmetric tridiagonal system of equations for the $N+1$ unknown coefficients $A_0, A_1, \ldots, A_N$.

We will consider this further in Sec. 5.4, where we study the error analysis of cubic spline interpolation and also discuss other representations for the spline functions in terms of the cubic B-splines (short for basic splines) which are based on the introduction of additional knots at each end of the range. But first we consider some examples and the techniques available for solving these tridiagonal systems.

Example 5.2.1. Find the natural cubic interpolating the data

x	0.0	0.1	0.3	0.6
$f(x)$	0.0000	0.2624	0.6419	1.0296

Solution. Here the steplengths are $h_0 = 0.1$, $h_1 = 0.2$, $h_2 = 0.3$, while the divided differences are $f[x_0, x_1] = 2.6240$, $f[x_1, x_2] = 1.8975$, $f[x_2, x_3] = 1.2923$. System (5.2.13) for this example reduces to the 2×2 system

$$0.6A_1 + 0.2A_2 = 6(1.8975 - 2.6240) = -4.3590$$

$$0.2A_1 + 1.0A_2 = 6(1.2923 - 1.8975) = -3.6312$$

which has the solutions $A_1 = -6.4871$ and $A_2 = -2.3338$.

To complete the solution, we must find the coefficients B_i, C_i $(i = 0, 1, 2)$ which are given by (5.2.7). We get

$$B_0 = 0.0000 \qquad B_1 = 1.5282 \qquad B_2 = 2.2564$$

$$C_0 = 2.7321 \qquad C_1 = 3.2873 \qquad C_2 = 3.4320$$

Finally, we substitute these into (5.2.6), with $u_0 = x - x_0 = x$, $u_1 = x - 0.1$, $u_2 = x - 0.3$, and $u_3 = x - 0.6$, to obtain the formula for the spline interpolant. For this case we get

$$s(x) = \begin{cases} -10.8118x^3 + 2.7321x & 0 \le x \le 0.1 \\ -1.9448(x - 0.1)^3 + 5.4059(x - 0.3)^3 & \\ \qquad - 1.5282(x - 0.3) + 3.2873(x - 0.1) & 0.1 \le x \le 0.3 \\ 1.2966(x - 0.6)^3 - 2.2564(x - 0.6) + 3.4320(x - 0.3) & 0.3 \le x \le 0.6 \end{cases}$$

It is clear from even this very small example that spline interpolation is not well-suited to manual computation. We will discuss the computer implementation of the ideas shortly. Now we consider the important special case where the knots are equally spaced and then give some thought to the solution of tridiagonal systems of linear equations.

If the knots are equally spaced so that

$$x_k = x_0 + kh \qquad (k = 1, 2, \ldots, N) \tag{5.2.16}$$

then Eqs. (5.2.15) can be simplified substantially. First, we can divide throughout by the fixed steplength h. On the right-hand side this has the effect that

$$\frac{\Delta d_i}{h} = \frac{f[x_i, x_{i+1}] - f[x_{i-1}, x_i]}{h} = 2f[x_{i-1}, x_i, x_{i+1}]$$

and denoting this second-order divided difference by D_i, the system becomes

$$\begin{bmatrix} 4 & 1 & 0 & \cdots & \cdots & 0 \\ 1 & 4 & 1 & 0 & \cdots & 0 \\ & \cdots & \cdots & \cdots & \cdots & \\ 0 & \cdots & 0 & 1 & 4 & 1 \\ 0 & \cdots & \cdots & 0 & 1 & 4 \end{bmatrix} \begin{bmatrix} A_1 \\ A_2 \\ \vdots \\ A_{N-1} \end{bmatrix} = \begin{bmatrix} 12D_1 \\ 12D_2 \\ \vdots \\ 12D_{N-1} \end{bmatrix} \qquad (5.2.17)$$

Example 5.2.2. Find the natural cubic spline interpolant to the data

x	0.0	0.1	0.2	0.3	0.4
$f(x)$	0.0000	0.3785	0.6781	0.9260	1.1375

Solution. We obtain the second-order divided differences

$$f[x_0, x_1, x_2] = -3.945 \qquad f[x_1, x_2, x_3] = -2.585 \qquad f[x_2, x_3, x_4] = -1.820$$

and therefore the system of equations

$$4A_1 + A_2 \qquad\quad = -47.34$$
$$A_1 + 4A_2 + A_3 = -31.02$$
$$A_2 + 4A_3 = -21.84$$

This is easily solved: $A_1 = -10.855$, $A_2 = -3.921$, $A_3 = -4.480$. Consideration of Eqs. (5.2.7) for the situation of equally spaced knots immediately implies that $B_1 = C_0$, $B_2 = C_1$, and $B_3 = C_2$. Also, for the present example we have $B_0 = 0$ since both $A_0 = 0$ and $f_0 = 0$. For the coefficients C_i we get the values $C_0 = 3.966$, $C_1 = 6.846$, $C_2 = 9.335$, and $C_3 = 11.375$. We thus have for the interval $[0.0, 0.1]$

$$s_0(x) = -\frac{10.855x^3}{0.6} + 3.966x = -18.092x^3 + 3.966x$$

and, similarly, for the other intervals we use

$$s_1(x) = -6.535(x - 0.1)^3 + 18.092(x - 0.2)^3 - 3.966(x - 0.2) + 6.846(x - 0.1)$$
$$s_2(x) = -7.467(x - 0.2)^3 + 6.535(x - 0.3)^3 - 6.846(x - 0.3)$$
$$\qquad\qquad + 9.335(x - 0.2)$$
$$s_3(x) = 7.467(x - 0.4)^3 - 9.335(x - 0.4) + 11.375(x - 0.3)$$

EXERCISES 5.2

1. Find the cubic spline which interpolates the values $\ln(1 + 2x)$ at $x = 0$, 0.1, 0.2, 0.3, 0.4, and 0.5. Use this spline to estimate the values of $\ln 1.1$, $\ln 1.3$, $\ln 1.5$, $\ln 1.7$, and $\ln 1.9$. Compare the results with those using the linear spline of Exercise 1 in Sec. 5.1.

2. Derive a system of equations for quadratic spline interpolation. (Use a similar process to that used for the cubic spline. You should end up with a bilinear system of equations, and there will be one additional degree of freedom.) Write a program to solve such a bilinear system, and use it to find a quadratic spline interpolating $1/(1 + x^2)$ at -5, $-4, \ldots, 0, 1, \ldots, 5$.

5.3 TRIDIAGONAL SYSTEMS OF EQUATIONS

Our main motivation here is to be able to produce a tolerably efficient code for first obtaining the natural cubic spline interpolant and then evaluating this function. The fact that we seek the *natural* cubic spline is of little importance in itself, and just the same approaches would be valid for any of the other cubic spline interpolants which give rise to a tridiagonal system of equations for their coefficients. Just what do we mean by a *tridiagonal* system of linear equations?

A matrix A is said to be *tridiagonal* if its only nonzero elements are on the main diagonal or immediately above or below that diagonal. That is, the elements a_{ij} of A satisfy

$$a_{ij} = 0 \quad \text{whenever } |i - j| > 1 \tag{5.3.1}$$

It is convenient to use the notation a_i for the elements of the diagonal, b_i for the superdiagonal, and c_i for the subdiagonal. In detail we set

$$a_i = a_{ii} \qquad b_i = a_{i,i+1} \qquad c_i = a_{i+1,i} \tag{5.3.2}$$

Any system of linear equations whose matrix of coefficients is tridiagonal is called a tridiagonal system. Typically, with the notation just introduced, such a system will be of the form

$$
\begin{aligned}
a_1 x_1 + b_1 x_2 &= d_1 \\
c_1 x_1 + a_2 x_2 + b_2 x_3 &= d_2 \\
c_2 x_2 + a_3 x_3 + b_3 x_4 &= d_3 \\
&\qquad\qquad\qquad \cdots \\
c_{N-2} x_{N-2} + a_{N-1} x_{N-1} + b_{N-1} x_N &= d_{N-1} \\
c_{N-1} x_{N-1} + a_N x_N &= d_N
\end{aligned}
\tag{5.3.3}
$$

Such a system lends itself to much more efficient treatment than can be given to a general *full* system of equations. If, as is usually the case for spline interpolation, the system is diagonally dominant—that is to say $|a_i| > |b_i| + |c_{i-1}|$ for each i—then the system is amenable to efficient iterative solution. However, we can usually do even better than this by modifying the standard gaussian elimination *without pivoting* to this special situation.

It is necessary to forego any pivoting strategy since otherwise there is the likelihood of getting fill-in in the matrix during the forward elimination process. This problem of fill-in (that is, zero elements of the matrix being replaced by nonzeros during the solution process) is one of the most important considerations in the design of efficient routines for the solution of large sparse systems on the modern parallel supercomputers. *Sparsity* is the term used to describe systems in which there are few nonzero coefficients in any one of the equations; a tridiagonal system is just one such example. Many others arise in practice in, for example, the numerical solution of partial differential equations whether by finite difference or finite element techniques. We discuss solution of linear systems in general in the next chapter.

Consider now the forward elimination stage of the Gauss elimination process for a tridiagonal system such as that in (5.3.3). The first step would be the subtraction of c_1/a_1 times the first row from the second, after which the second row would have entries in only the diagonal and superdiagonal positions. What is more, the superdiagonal entry b_2 would be unaltered. The second and third equations would now be of the form

$$a_2^{(1)}x_2 + b_2x_3 \qquad = d_2^{(1)}$$
$$c_2x_2 + a_3x_3 + b_3x_4 = d_3$$

which is just the same situation as for the first two rows of the original system. We can continue in this way until the final system is just a bidiagonal system for which back substitution is similarly straightforward.

The forward elimination phase is simply described by the following loop:

for $i = 1$ to $N - 1$

$\qquad m := c_i/a_i; c_i = 0;$

$\qquad a_{i+1} := a_{i+1} - mb_i;$

$\qquad d_{i+1} := d_{i+1} - md_i;$

By computing the multiplier m just once, the total number of arithmetic operations in this stage is reduced to just $5(N - 1)$.

Note that within this loop we are overwriting the entries of the original matrix with the elements of the row-reduced matrix. If within a particular computation it is desired to retain the original matrix, it would, of course, be necessary to keep a separate copy. This overwriting is a common feature of numerical linear algebra routines. For a large tridiagonal system the advantage of storing just three vectors totaling $3N - 2$ entries rather than a full matrix with N^2 entries represents a major advantage.

At the conclusion of this forward elimination, we have a bidiagonal system of equations of the form

$$a_ix_i + b_ix_{i+1} = d_i \qquad (i = 1, 2, \ldots, N - 1)$$
$$a_Nx_N = d_N$$

which is easily solved by the *back substitution* loop:

$\qquad x_N := d_N/a_N$

$\qquad$ for $i = N - 1$ downto 1

$\qquad\qquad x_i := (d_i - b_i * x_{i+1})/a_i$

This back substitution loop entails a total of a further $3N - 2$ arithmetic operations, bringing the total for the complete solution of the system to under $8N$. This makes a very favorable comparison with the gaussian elimination procedure for a full matrix which (as we see later) is of the order of $2N^3/3$.

Indeed, this operation count compares favorably with just two iterations of either the Jacobi or Gauss-Seidel iterative schemes. (See the exercises.)

Our primary purpose in developing this procedure for solving tridiagonal systems here is to use it for spline interpolation. It is to this task that we now address ourselves.

The process consists of the following major steps, which are described in Algorithm 5.3.1 below:

1. Obtain the coefficients and right-hand side of the tridiagonal system from the data. This will entail the formation of divided differences and the computation of the steplengths.
2. Obtain the solution to the tridiagonal system and therefore the coefficients A_i of the natural cubic spline interpolant.
3. Use these coefficients to obtain the remaining coefficients B_i, C_i from Eqs. (5.2.7).
4. Evaluate the spline function at the point (or points) of interest.

Algorithm 5.3.1 Computation of natural cubic spline coefficients

Input: Data $x_0, x_1, \ldots, x_N$ and $f_0, f_1, \ldots, f_N$

Form differences:

$$\text{for } i = 0 \text{ to } N - 1$$
$$h_i := x_{i+1} - x_i; \quad dd_i := (f_{i+1} - f_i)/h_i$$
$$\text{for } i = 1 \text{ to } N - 1$$
$$d_i := 6(dd_i - dd_{i-1})$$

Obtain coefficient matrix:

$$\text{for } i = 1 \text{ to } N - 1$$
$$b_i := c_i = h_i$$
$$a_i := 2(h_{i-1} + h_i)$$

Solve tridiagonal system:

$$\text{for } i = 1 \text{ to } N - 2$$
$$m := c_i/a_i; \quad c_i = 0$$
$$a_{i+1} := a_{i+1} - mb_i$$
$$d_{i+1} := d_{i+1} - md_i$$

$$A_{N-1} := d_{N-1}/a_{N-1}$$
$$\text{for } i = N - 2 \text{ downto } 1$$
$$A_i := (d_i - b_i A_{i+1})/a_i$$

$$A_0 := A_N = 0$$

Compute spline coefficients:

$$\text{for } i = 0 \text{ to } N - 1$$
$$B_i := f_i/h_i - A_i * h_i/6$$
$$C_i := f_{i+1}/h_i - A_{i+1} * h_i/6$$

Output: coefficients $A_0, \ldots, A_N, B_0, \ldots, B_{N-1}, C_0, \ldots, C_{N-1}$

The final stage is the evaluation of this natural cubic spline at a point x, say. This is achieved by identifying the interval $[x_i, x_{i+1}]$ within which the point lies and then using Eq. (5.2.6). Algorithm 5.3.2 describes this.

Algorithm 5.3.2 Evaluation of cubic spline

Input: $x_0, x_1, \ldots, x_N, h_0, h_1, \ldots, h_{N-1}$
              $A_0, \ldots, A_N, B_0, \ldots, B_{N-1}, C_0, \ldots, C_{N-1}$
              x

Compute: Find k such that $x_k \le x \le x_{k+1}$
              $u_0 := x - x_k$
              $u_1 := x - x_{k+1}$
              $s(x) := (A_{k+1}u_0^3 - A_k u_1^3)/6h_k - B_k u_1 + C_k u_0$

Output: $s(x)$

We illustrate this process with finding the natural cubic spline interpolant for $f(x) = 1/(1 + x^2)$ over the interval $[-5, 5]$ with 11 equally spaced knots. The coefficients are listed in Table 5.3.3(a), from which we can get an idea of the changes in the polynomials used in the different subintervals. The spline and the original function are tabulated in Table 5.3.3(b), from which we can see that the level of agreement is vastly superior to that of the divided difference formula in Table 4.3.9. The results are symmetric about $x = 0$, and so only the positive half is included in the table.

In the case of the spline interpolant, we do find that increasing the number of equally spaced knots will result in steady convergence to the original function. However, for spline interpolation, as with polynomial interpolation, the use of equally spaced data points is far from the best possible. For the case of splines there is no simple answer as to what is the best choice of knots. This is still an active area of experiment and research. The essential principle is that more knots should be placed in regions where the nature of the function is changing most rapidly, such as close to turning points of the curve, or where the function changes from, say, rapidly to slowly decreasing.

To see the effect of increasing the number of knots, we consider the natural cubic spline interpolant for the function

$$f(x) = \frac{100}{1 + 10x^2}$$

over the same interval $[-5, 5]$. This function has a sharp spike at $x = 0$ and decreases steadily as $|x|$ increases.

Table 5.3.5 shows the results of this interpolation using 10, 20, and 40 equal subdivisions of the domain of interest. Again the results were symmetric about zero, and so only the results for positive x are tabulated. From the graphs in Figure 5.3.4 below, we can see that the precision of our approximation is significantly improved with the increasing number of knots. The initial spline fit using just 10 equal intervals is clearly incapable of tracking the

TABLE 5.3.3
Natural cubic spline interpolation for $f(x) = 1/(1 + x_2)$ over $[-5, 5]$ using the integers as knots

(a) Coefficients		
i	A_i	B_i	C_i
0	0.000 000 000 0	0.038 461 538 5	0.056 089 847 8
1	0.016 402 089 9	0.056 089 847 8	0.090 120 247 0
2	0.059 278 518 0	0.090 120 247 0	0.183 429 164 3
3	0.099 425 014 4	0.183 429 164 3	0.376 163 095 9
4	0.743 021 424 5	0.376 163 095 9	1.311 918 452 0
5	−1.871 510 712 2	1.311 918 452 0	0.376 163 095 9
6	0.743 021 424 5	0.376 163 095 9	0.183 429 164 3
7	0.099 425 014 4	0.183 429 164 3	0.090 120 247 0
8	0.059 278 518 0	0.090 120 247 0	0.056 089 847 8
9	0.016 402 089 9	0.056 089 847 8	0.038 461 538 5

(b) Values of the spline $s(x)$ and the original function $f(x)$		
x	$s(x)$	$f(x)$
0.00	1.000 000 000 0	1.000 000 000 0
0.25	0.948 323 967 7	0.941 176 470 6
0.50	0.820 530 580 5	0.800 000 000 0
0.75	0.657 471 903 0	0.640 000 000 0
1.00	0.500 000 000 0	0.500 000 000 0
1.25	0.380 482 226 2	0.390 243 902 4
1.50	0.297 347 097 6	0.307 692 307 7
1.75	0.240 538 420 1	0.246 153 846 2
2.00	0.200 000 000 0	0.200 000 000 0
2.25	0.167 247 127 4	0.164 948 453 6
2.50	0.140 081 029 2	0.137 931 034 5
2.75	0.117 874 416 4	0.116 788 321 2
3.00	0.100 000 000 0	0.100 000 000 0
3.25	0.085 823 381 8	0.086 486 486 5
3.50	0.074 681 726 7	0.075 471 698 1
3.75	0.065 905 090 7	0.066 390 041 5
4.00	0.058 823 529 4	0.058 823 529 4
4.25	0.052 836 042 4	0.052 459 016 4
4.50	0.047 617 403 3	0.047 058 823 5
4.75	0.042 911 329 6	0.042 440 318 3
5.00	0.038 461 538 5	0.038 461 538 5

very rapid variations in the function f, even to the extent that it yields negative function values over most of the interval $[1, 2]$. In this interval f is going through the change from very rapid to quite a slow decrease. More knots are obviously required in this region and around the spike itself. It can be seen that the spline functions cannot reproduce the sharpness of the original function even for the case of 21 knots. At the resolution of these graphs the curves for f and the spline using 41 knots were indistinguishable.

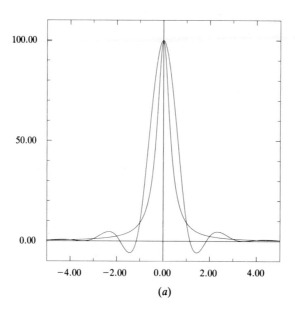

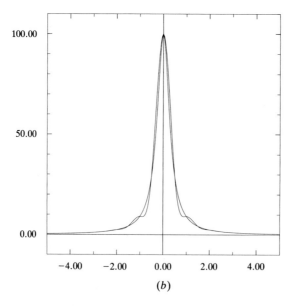

FIGURE 5.3.4
The spike $100/(1 + 10x^2)$ and its spline interpolants using N equally spaced knots. (a) $N = 11$; (b) $N = 21$.

TABLE 5.3.5

x	Spline function values			f(x)
	11 knots	**21 knots**	**41 knots**	
0.000	100.000 000	100.000 000	100.000 000	100.000 000
0.125		92.035 545	86.821 141	86.486 486
0.250	89.395 443	72.809 074	61.538 462	61.538 462
0.375		49.320 922	41.581 106	41.558 442
0.500	64.145 423	28.571 429	28.571 429	28.571 429
0.625		15.937 870	20.307 159	20.382 166
0.750	34.095 419	10.305 281	15.094 340	15.094 340
0.875		8.935 636	11.558 011	11.552 347
1.000	9.090 909	9.090 909	9.090 909	9.090 909
1.125		8.565 178	7.316 694	7.322 654
1.250	−3.428 853	7.280 937	6.015 038	6.015 038
1.375		5.692 785	5.023 516	5.023 548
1.500	−5.649 509	4.255 319	4.255 319	4.255 319
1.625		3.318 994	3.648 137	3.648 803
1.750	−2.162 928	2.817 684	3.162 055	3.162 055
1.875		2.581 117	2.765 644	2.765 774
2.000	2.439 024	2.439 024	2.439 024	2.439 024
2.125		2.254 960	2.166 436	2.166 554
2.250	4.516 601	2.027 780	1.937 046	1.937 046
2.375		1.790 167	1.741 921	1.741 971
2.500	4.238 536	1.574 803	1.574 803	1.574 803
2.625		1.406 705	1.430 455	1.430 487
2.750	2.725 685	1.280 229	1.305 057	1.305 057
2.875		1.182 064	1.195 350	1.195 368
3.000	1.098 901	1.098 901	1.098 901	1.098 901
3.125		1.019 925	1.013 609	1.013 621
3.250	0.240 784	0.944 298	0.937 866	0.937 866
3.375		0.873 676	0.870 267	0.870 275
3.500	0.080 904	0.809 717	0.809 717	0.809 717
3.625		0.753 568	0.755 246	0.755 251
3.750	0.310 578	0.704 348	0.706 090	0.706 090
3.875		0.660 661	0.661 565	0.661 567
4.000	0.621 118	0.621 118	0.621 118	0.621 118
4.125		0.584 559	0.584 254	0.584 261
4.250	0.764 781	0.550 753	0.550 585	0.550 585

TABLE 5.3.5 (*Continued*)

x	Spline function values			f(x)
	11 knots	**21 knots**	**41 knots**	
4.375		0.519 700	0.519 751	0.519 734
4.500	0.737 580	0.491 400	0.491 400	0.491 400
4.625		0.465 741	0.465 246	0.465 319
4.750	0.596 470	0.442 149	0.441 258	0.441 258
4.875		0.419 933	0.419 282	0.419 013
5.000	0.398 406	0.398 406	0.398 406	0.398 406

Table 5.3.6 shows the results of natural cubic spline interpolation using 21 points for the same function as above but using the knots

$$0 \quad \pm 0.25 \quad \pm 0.5 \quad \pm 0.75 \quad \pm 1.0 \quad \pm 1.25 \quad \pm 1.5 \quad \pm 2 \quad \pm 3 \quad \pm 4 \quad \pm 5$$

The level of agreement between the interpolating spline and the original function is now very much better than that achieved using 21 equally spaced points (compare with Table 5.3.5) and is almost as good as that obtained with 41 equally spaced points. Indeed, the agreement between these functions is sufficiently close that their graphs were indistinguishable when plotted using GraphLib.

TABLE 5.3.6

	(a) Coefficients		
i	A_i	B_i	C_i
0	0.000 000 000 0	0.398 406 374 5	0.579 305 550 3
1	0.250 874 772 7	0.579 305 550 3	1.011 079 498 8
2	0.526 929 600 5	1.011 079 498 8	1.969 783 047 8
3	2.815 448 054 6	4.643 428 109 3	7.713 717 699 2
4	9.563 047 183 8	16.622 816 296 4	23.279 248 580 3
5	18.741 643 095 0	23.279 248 580 3	34.621 091 637 9
6	41.821 073 414 5	34.621 091 637 9	56.418 203 049 3
7	95.019 730 589 8	56.418 203 049 3	101.970 247 107 0
8	295.571 212 263 0	101.970 247 107 0	221.415 094 232 0
9	593.730 046 104 0	221.415 094 232 0	489.292 452 883 0
10	−2143.018 869 200 0	489.292 452 883 0	221.415 094 232 0
11	593.730 046 104 0	221.415 094 232 0	101.970 247 107 0
12	295.571 212 264 0	101.970 247 107 0	56.418 203 049 3
13	95.019 730 589 7	56.418 203 049 3	34.621 091 637 9
14	41.821 073 414 6	34.621 091 637 9	23.279 248 580 3
15	18.741 643 095 1	23.279 248 580 3	16.622 816 296 4
16	9.563 047 183 8	7.713 717 699 2	4.643 428 109 3
17	2.815 448 054 7	1.969 783 047 8	1.011 079 498 8
18	0.526 929 600 5	1.011 079 498 8	0.579 305 550 3
19	0.250 874 772 7	0.579 305 550 3	0.398 406 374 5

TABLE 5.3.6 (*Continued*)

(b) Function values

x	s(x)	f(x)
0.000	100.000 000 000 0	100.000 000 000 0
0.125	86.821 140 234 3	86.486 486 486 2
0.250	61.538 461 538 4	61.538 461 538 4
0.375	41.581 112 014 3	41.558 441 558 3
0.500	28.571 428 571 4	28.571 428 571 4
0.625	20.307 138 226 4	20.382 165 605 0
0.750	15.094 339 622 6	15.094 339 622 6
0.875	11.558 089 966 1	11.552 346 570 3
1.000	9.090 909 090 9	9.090 909 090 9
1.125	7.316 400 231 1	7.322 654 462 2
1.250	6.015 037 594 0	6.015 037 594 0
1.375	5.024 613 175 1	5.023 547 880 7
1.500	4.255 319 148 9	4.255 319 148 9
1.625	3.643 006 063 6	3.648 802 736 6
1.750	3.153 757 781 5	3.162 055 336 0
1.875	2.761 216 493 4	2.765 773 552 3
2.000	2.439 024 390 2	2.439 024 390 2
2.125	2.164 471 670 7	2.166 553 825 3
2.250	1.929 440 564 4	1.937 046 004 8
2.375	1.729 461 308 7	1.741 970 604 2
2.500	1.560 064 141 1	1.574 803 149 6
2.625	1.416 779 298 9	1.430 487 259 7
2.750	1.295 137 019 6	1.305 057 096 2
2.875	1.190 667 540 4	1.195 367 949 2
3.000	1.098 901 098 9	1.098 901 098 9
3.125	1.016 023 031 2	1.013 620 525 8
3.250	0.940 839 068 9	0.937 866 354 0
3.375	0.872 810 042 5	0.870 274 680 4
3.500	0.811 396 782 3	0.809 716 599 2
3.625	0.756 060 118 8	0.755 251 357 1
3.750	0.706 260 882 4	0.706 090 026 5
3.875	0.661 459 903 5	0.661 567 087 0
4.000	0.621 118 012 4	0.621 118 012 4
4.125	0.584 704 236 3	0.584 261 457 0
4.250	0.551 720 388 8	0.550 584 996 6
4.375	0.521 676 480 0	0.519 733 636 5
4.500	0.494 082 520 2	0.491 400 491 4
4.625	0.468 448 519 5	0.465 319 179 9
4.750	0.444 284 488 2	0.441 257 584 1
4.875	0.421 100 436 4	0.419 012 701 3
5.000	0.398 406 374 5	0.398 406 374 5

EXERCISES 5.3

1. Write a program to find the natural cubic spline interpolant to a function f. Use your program to graph the cubic spline agreeing with $1/(1 + x^2)$ at $-5, -4, \ldots, 0, 1, \ldots, 5$. Compare this with the quadratic spline found in Exercise 2 of Sec. 5.2.

2. The tridiagonal systems of equations that arise in cubic spline interpolation are diagonally dominant, which makes them amenable to iterative solution. System (5.3.3) can be rewritten as

$$x_i = \frac{d_i - c_{i-1}x_{i-1} - b_i x_{i+1}}{a_i} \qquad (i = 1, 2, \ldots, N)$$

with the convention that $c_0 = x_0 = b_N = x_{N+1} = 0$. Given an estimate $\mathbf{x}^{(k)}$ of the solution, the above system can be used to generate a new estimate by setting

$$x_i^{(k+1)} = \frac{d_i - c_{i-1}x_{i-1}^{(k)} - b_i x_{i+1}^{(k)}}{a_i} \qquad (i = 1, 2, \ldots, N)$$

which gives rise to a simple iterative procedure for the solution of such a system. Write a program to implement this *Jacobi iteration*, and test it for the

cubic spline computations of Tables 5.3.3 and 5.3.6.

3. Repeat Exercise 2 using the *Gauss-Seidel iteration* in which the most recent values for each of the x_i are used throughout; that is, we set

$$x_i^{(k+1)} = \frac{d_i - c_{i-1}x_{i-1}^{(k+1)} - b_i x_{i+1}^{(k)}}{a_i} \qquad (i = 1, 2, \ldots, N)$$

Compare the number of iterations required by the two approaches to achieve particular accuracies, and estimate the floating-point operation counts for these solutions. Compare these with the operation count for the Gauss elimination procedure. (We see later that for such diagonally dominant systems, both these iterations can be proved to converge to the true solution.)

4. Compute the natural cubic spline which interpolates the spike $100/(1 + 10x^2)$ at 21 Chebyshev interpolation points in $[-5, 5]$. Graph this function and the spike on the same axes. Try to find the set of 21 knots which gives the best fit to the original function.

5.4 ANALYSIS OF CUBIC SPLINE INTERPOLATION AND CUBIC B-SPLINES

In this section we return to the development of the theory of spline interpolation. Our principal task is to obtain some error bounds for the process. More precisely, we obtain error estimates as we will establish that the error in cubic spline interpolation behaves like the fourth power of the maximum stepsize used but without deriving an exact formula for the bound. (This distinction will be made clearer later.)

Recall from (5.2.13) that the coefficients A_i of the cubic spline interpolant to the data $f(x_0), f(x_1), \ldots, f(x_N)$, where the knots satisfy $x_0 < x_1 < \cdots < x_N$, are obtained by solving the system

$$h_{i-1}A_{i-1} + 2(h_{i-1} + h_i)A_i + h_i A_{i+1} = 6(d_i - d_{i-1}) \qquad (i = 1, 2, \ldots, N-1)$$

$$(5.4.1)$$

where $h_k = x_{k+1} - x_k$ and $d_k = f[x_k, x_{k+1}]$.

In Sec. 5.2 we discussed one method of using the two degrees of freedom in this system of $N - 1$ equations in the $N + 1$ unknowns, namely, the natural cubic spline given by setting $A_0 = A_N = 0$ or, equivalently,

$$s''(x_0) = s''(x_N) = 0 \qquad (5.4.2)$$

There are two other commonly used cubic splines.

The *complete cubic spline* is obtained by forcing the first derivative of the spline to agree with that of the original function at the two ends of the range. (It is not uncommon in physical applications that values of the first derivative form part of the boundary conditions.) At x_0 this condition is, using (5.2.9) and (5.2.12),

$$f_0' = f'(x_0) = s_0'(x_0) = C_0 - B_0 - \frac{A_0 h_0}{2}$$

$$= d_0 - \frac{(A_1 - A_0)h_0}{6} - \frac{A_0 h_0}{2}$$

from which we deduce that

$$2h_0 A_0 + h_0 A_1 = 6(d_0 - f_0') \tag{5.4.3}$$

Similarly, we have from the condition at x_N the extra equation

$$h_{N-1} A_{N-1} + 2h_{N-1} A_N = 6(f_N' - d_{N-1}) \tag{5.4.4}$$

Augmenting system (5.4.1) with these two equations yields a system of $N + 1$ equations for the unknowns $A_0, A_1, \ldots, A_N$. This system is still a tridiagonal system of equations and so can be solved using the methods of the previous section.

Another method for determining the coefficients is to use the so-called not-a-knot condition. Here, we force x_1, x_{N-1} to be interpolation points but not breakpoints of the spline by insisting on *third-derivative* continuity at these points. Of course, this simply means that the *same* cubic is used in both $[x_0, x_1]$ and $[x_1, x_2]$. Similar comments apply for the other end of the range, but we concentrate on the point x_1 for now.

Now, the third derivative of our cubic spline is given, at points other than knots, by

$$s_i'''(x) = \frac{A_{i+1} - A_i}{h_i}$$

and so the continuity condition at x_1 becomes

$$\frac{A_1 - A_0}{h_0} = \frac{A_2 - A_1}{h_1} \tag{5.4.5}$$

This equation is easily solved for A_0 in terms of A_1 and A_2; substituting into (5.4.1) for $i = 1$ yields

$$(h_0 + 2h_1)A_1 + (h_1 - h_0)A_2 = \frac{6h_1 \Delta d_0}{h_0 + h_1}$$

A similar piece of algebra for the point x_{N-1} reduces the system to a tridiagonal system of $N - 1$ equations for the remaining unknown coefficients.

In summary, we have seen that for all the common choices of additional conditions—the natural or complete splines or the not-a-knot condition—the equations for the coefficients reduce to a tridiagonal system of linear equations.

Furthermore this system is diagonally dominant and therefore nonsingular (see Appendix B). It follows that the system has a unique solution for each of these cases. That is, we have established the following result.

Theorem 5.4.1. For each of the natural, complete, or not-a-knot conditions there is a unique cubic spline that fits the data $f(x_0), f(x_1), \ldots, f(x_N)$ for distinct knots $x_0 < x_1 < \cdots < x_N$.

There are other possible ways of resolving the two extra degrees of freedom, and a similar result applies to them, but we have covered the three principal ones.

At this stage we turn to the error analysis of cubic spline interpolation. Our primary result is that this error is of order h^4, where

$$h = \max_i h_i \tag{5.4.6}$$

Recall that a quantity E is of order h^n, written $E = O(h^n)$, if E/h^n is bounded; that is, there exists a constant M such that

$$|E| \leq M|h|^n \tag{5.4.7}$$

In the case of cubic spline interpolation, it is tolerably easy to establish that provided $f \in C^4[x_0, x_N]$ so that $|f^{(4)}(x)|$ is bounded, then the error satisfies $s(x) - f(x) = O(h^4)$, but it is far less easy to obtain a sharp estimate of the appropriate constant M.

We again use the notation of Sec. 5.2 and set

$$u_i = x - x_i \qquad (i = 0, 1, \ldots, N)$$

from which we also have $u_{i+1} = u_i - h_i$. We will also denote the values of the various derivatives of f at the knots by $f_i' = f'(x_i)$, $f_i'' = f''(x_i)$, and so on.

From Taylor's theorem, we have for $x \in [x_i, x_{i+1}]$

$$f(x) = f_i + u_i f_i' + \frac{u_i^2 f_i''}{2} + \frac{u_i^3 f_i'''}{6} + \frac{u_i^4 f^{(4)}(\xi)}{24}$$

$$= f_i + u_i f_i' + \frac{u_i^2 f_i''}{2} + \frac{u_i^3 f_i'''}{6} + O(h^4) \tag{5.4.8}$$

since $u_i \leq h_i \leq h$ and $f^{(4)}$ is bounded.

Using (5.2.6) and (5.2.7), we can write s_i in the form

$$s_i(x) = \frac{A_{i+1}(u_i^3 - u_i h_i^2) - A_i(u_{i+1}^3 - u_{i+1} h_i^2)}{6h_i} + \frac{f_{i+1} u_i - f_i u_{i+1}}{h_i}$$

which, on factorizing and substituting for u_{i+1} and observing that $(f_{i+1} - f_i)/h_i = d_i$, simplifies to

$$s_i(x) = u_i u_{i+1} \frac{A_{i+1}(u_i + h_i) - A_i(u_{i+1} - h_i)}{6h_i} + u_i d_i + f_i \tag{5.4.9}$$

From (5.4.8) with $x = x_{i+1}$, we may deduce that

$$d_i = f_i' + \frac{h_i f_i''}{2} + \frac{h_i^2 f_i'''}{6} + O(h^3) \tag{5.4.10}$$

Substituting (5.4.10) into (5.4.9) and subtracting (5.4.8), we now obtain

$$s_i(x) - f(x) = u_i u_{i+1} \frac{A_{i+1}(u_i + h_i) - A_i(u_{i+1} - h_i)}{6h_i}$$

$$+ f_i + u_i f_i' + \frac{u_i h_i f_i''}{2} + \frac{u_i h_i^2 f_i'''}{6}$$

$$- f_i - u_i f_i' - \frac{u_i^2 f_i''}{2} - \frac{u_i^3 f_i'''}{6} + O(h^4)$$

since $u_i = O(h)$, $O(h)O(h^3) = O(h^4)$, and the sum of two quantities $O(h^4)$ is still $O(h^4)$. This simplifies to yield

$$s_i(x) - f(x) = \frac{u_i u_{i+1}}{6h_i} [u_i(A_{i+1} - A_i) + h_i(A_{i+1} + 2A_i) - 3h_i f_i'' - h_i(u_i + h_i)f_i''']$$

$$+ O(h^4) \tag{5.4.11}$$

Up to this point, the analysis is valid for any cubic spline interpolant to the data satisfying the underdetermined system (5.4.1).

We now specialize to the case of the complete cubic spline for which we have the additional Eqs. (5.4.3) and (5.4.4) satisfied. From (5.4.3) we have

$$2h_0 A_0 + h_0 A_1 = 6(d_0 - f_0')$$

which can be rewritten using (5.4.10) as

$$h_0(2A_0 + A_1) = 3h_0 f_0'' + h_0^2 f_0''' + O(h^3) \tag{5.4.12}$$

Substituting this into (5.4.11) for $i = 0$ and using the fact that $u_0 u_1/h_0 = O(h)$, we have

$$s_0(x) - f(x) = \frac{u_0^2 u_1}{6} \left(\frac{A_1 - A_0}{h_0} - f_0''' \right) + O(h^4) \tag{5.4.13}$$

Now the complete cubic spline agrees with f at the $N+1$ knots and with its derivative at both x_0 and x_N. It follows from Rolle's theorem that s' and f' must agree at *at least* $N+2$ points in the interval $[x_0, x_N]$. Applying Rolle's theorem once more tells us that s'', f'' must agree at *at least* $N+1$ points. Therefore there exists (at least) one subinterval $[x_i, x_{i+1}]$ in which s'', f'' agree twice. It follows that

$$s''(x) - f''(x) = O(h^2)$$

In particular,

$$s''(x_0) - f''(x_0) = A_0 - f''(x_0) = O(h^2)$$

$$s''(x_1) - f''(x_1) = A_1 - f''(x_1) = O(h^2) \tag{5.4.14}$$

Now, it follows that

$$\frac{f''(x_1) - f''(x_0)}{h_0} - \frac{A_1 - A_0}{h_0} = O(h)$$

and since $(f''(x_1) - f''(x_0))/h_0$ is simply the divided difference approximation to $f'''(x_0)$ which has error $O(h)$, we deduce finally that

$$\frac{A_1 - A_0}{h_0} - f_0''' = O(h) \qquad (5.4.15)$$

Substituting this into (5.4.13) yields the required result

$$s_0(x) - f(x) = O(h^4)$$

for $x_0 \le x \le x_1$. The above argument also implies that the first derivatives of f and s agree up to $O(h^3)$ on this interval. Hence we deduce that

$$s'(x_1) - f'(x_1) = O(h^3)$$

and, using similar reasoning to that employed in deriving (5.4.3), it then follows that

$$2h_1 A_1 + h_1 A_2 = 6(d_1 - f_1') + O(h^3)$$

The remainder of the argument can now proceed exactly as before to extend the result to the next subinterval $[x_1, x_2]$ and then by induction to the whole interval $[x_0, x_N]$.

In summary, we have proved the following theorem for the complete cubic spline.

Theorem 5.4.2. The error in interpolating the data $f(x_0), f(x_1), \ldots, f(x_N)$ at the distinct knots $x_0 < x_1 < \cdots < x_N$ by its complete cubic spline interpolant is $O(h^4)$, where h is an upper bound for the subinterval widths $h_i = x_{i+1} - x_i$. The corresponding error in approximating the first and second derivatives of f by those of the spline are $O(h^3)$ and $O(h^2)$, respectively.

The best value of the constant of proportionality in this error bound is given in de Boor (1978). For the complete cubic spline above, this bound is

$$|s(x) - f(x)| \le \frac{5h^4 K}{384}$$

where K is an upper bound for $f^{(4)}$ on $[x_0, x_N]$.

Similar results can be established, though less easily, for the natural cubic spline interpolant. The constant is, of course, different in this case.

A further attraction of both the natural and complete cubic spline interpolants to f is that they are both minimum energy curves. (Minimum "energy" because they represent solutions to the problem with minimum stored energy which is measured by the size of the second derivative and corresponds to the elastic energy of the drafting spline.) Specifically, for any function σ satisfying the interpolation conditions together with

$$\sigma''(x_0) = \sigma''(x_N) = 0$$

the natural cubic spline s_n satisfies

$$\int_{x_0}^{x_N} |s_n''(x)|^2 \, dx \le \int_{x_0}^{x_N} |\sigma''(x)|^2 \, dx \qquad (5.4.16)$$

Similarly, among all functions satisfying the interpolation conditions and the complete spline conditions

$$\sigma'(x_0) = f'(x_0) \qquad \sigma'(x_N) = f'(x_N) \qquad (5.4.17)$$

the complete cubic spline s_c satisfies

$$\int_{x_0}^{x_N} |s_c''(x)|^2 \, dx \le \int_{x_0}^{x_N} |\sigma''(x)|^2 \, dx \qquad (5.4.18)$$

The arguments needed to establish these results are almost identical. Let s denote either s_n or s_c and consider the function $\delta(x) = s(x) - \sigma(x)$. It follows that $\sigma(x) = s(x) - \delta(x)$ and therefore that (see Exercise 4)

$$\int_{x_0}^{x_N} |\sigma''(x)|^2 \, dx = \int_{x_0}^{x_N} |s''(x) - \delta''(x)|^2 \, dx$$

$$= \int_{x_0}^{x_N} |s''(x)|^2 - 2s''(x)\delta''(x) + |\delta''(x)|^2 \, dx$$

Now the middle term here can be integrated by parts to get

$$\int_{x_0}^{x_N} s''(x)\delta''(x) \, dx = [s''(x)\delta'(x)]_{x_0}^{x_N} - \int_{x_0}^{x_N} s'''(x)\delta'(x) \, dx$$

and the first term vanishes since, for the natural cubic spline, s'' vanishes at both ends of the range, while, for the complete cubic spline, (5.4.17) implies that δ' does.

The final integral can again be integrated by parts—over each subinterval separately—and the interpolation conditions together with the fact that s is a cubic on each interval (so that $s^{(4)} \equiv 0$) establish that this term also vanishes. It follows that

$$\int_{x_0}^{x_N} |\sigma''(x)|^2 \, dx = \int_{x_0}^{x_N} |s''(x)|^2 + |\delta''(x)|^2 \, dx \qquad (5.4.19)$$

from which the two inequalities (5.4.16) and (5.4.18) follow immediately.

Thus far, we have concentrated on the existence, uniqueness, and analysis of spline interpolation without worrying at all about implementational efficiency. The simple approach to finding a cubic spline interpolant that we have described in the last few sections is adequate for the theoretical discussions but is not suitable for extensive calculation. (This is much like the contrast between Lagrange interpolation and the various divided difference schemes in which the former was perfectly suited to the analysis but very inefficient as a computational tool.)

The corresponding improvements in efficiency for spline interpolation are achieved by use of the B-splines (or basis splines) which are usually defined in terms of divided differences, but we do not give a full treatment of them here.

The problem with the original solution is that a completely new definition is necessary for the spline on each of the subintervals. The objective of the B-spline representation is to redefine the spline as a simple linear combination of basis functions, each of which has small *support*, that is, only a small interval in which the function is nonzero.

One of the simplest ways of achieving this is to introduce the truncated power functions defined by

$$x_+^k = \begin{cases} x^k & x \geq 0 \\ 0 & x \leq 0 \end{cases} \tag{5.4.20}$$

and using the truncated cubics $(x - x_i)_+^3$ we can write any of the cubic spline interpolants in the form

$$s(x) = s_0(x) + \sum_{i=1}^{N-1} \alpha_i (x - x_i)_+^3 \tag{5.4.21}$$

where the coefficients α_i are uniquely determined. (See Exercise 6.) Once these coefficients are determined, this would be a somewhat more useful representation to use in extended computation. Especially at the left-hand end of the range, evaluation of s would be easy and fast. However, toward the right-hand end, all the terms in (5.4.21) would be needed and the evaluation of s would be much less efficient.

The support of the cubic B-spline B_i is restricted to the interval $[x_i, x_{i+4}]$. Thus we seek a piecewise cubic function which has second-derivative continuity and which vanishes outside this interval. Let us consider one simple example.

Example 5.4.3. Find a cubic B-spline with knots $-2, -1, 0, 1, 2$.

Solution. Now, the fact that our spline—denote it temporarily by B—vanishes outside $[-2, 2]$ imposes the boundary conditions

$$B(\pm 2) = B'(\pm 2) = B''(\pm 2) = 0$$

while symmetry implies that

$$B(-1) = B(1) = y_1, \text{ say} \quad \text{and} \quad B'(0) = 0$$

Denote $B(0)$ by y_0.

By Taylor's theorem, on the interval $[-2, -1]$ it follows that B has the form $B(x) = y_1(x + 2)^3$ from which we may deduce that

$$B'(-1) = 3y_1 \quad \text{and} \quad B''(-1) = 6y_1$$

Applying Taylor's theorem again, we see that on $[-1, 0]$,

$$B(x) = y_1 + 3y_1(x + 1) + 3y_1(x + 1)^2 + c(x + 1)^3$$

and since $B(0) = y_0$, it follows that $c = y_0 - 7y_1$. Next, we consider $B'(0)$ and find that

$$B'(0) = 3y_1 + 6y_1 + 3(y_0 - 7y_1) = 3y_0 - 12y_1$$

from which we conclude that

$$y_0 = 4y_1 \tag{5.4.22}$$

From the above example, we see that there is still one degree of freedom in the definition of this B-spline. This is available for normalization of the basis. We will see one choice shortly. In Fig. 5.4.4 we illustrate the shape of this curve; clearly the choice of the free parameter y_0 simply affects the vertical scaling. The visually pleasing smoothness of the curve is apparent. These properties are principal reasons for the common choice of cubic splines for fitting physical data.

To determine the right normalization to use, let us consider some of the basic properties of the B-splines. First, the support of B_i is the interval $[x_i, x_{i+4}]$, and so for any knot interval $[x_k, x_{k+1}]$, say, there are exactly four of these B-splines which are nonzero, namely, $B_{k-3}, B_{k-2}, B_{k-1}, B_k$. One of the remarkable properties of these functions is that they sum to a constant on the interval $[x_k, x_{k+1}]$. We state this result more formally shortly but first must consider some implications of the notation we have introduced.

To define the B-splines $B_{N-3}, B_{N-2}, B_{N-1}$, we require additional knots $x_{N+1}, x_{N+2}, x_{N+3}$ which we may place arbitrarily. Similarly, in order that there are still four nonzero B-splines on $[x_0, x_1]$, we require a further set of knots x_{-3}, x_{-2}, x_{-1} which again may be placed arbitrarily. Now since there are precisely four nonzero B-splines on any of the original knot intervals, the sum

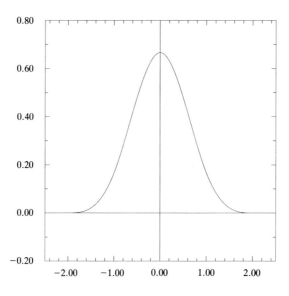

FIGURE 5.4.4
A cubic B-spline for the knots $-2, -1, 0, 1, 2$.

of those four is in fact the sum of the whole set of basis splines. Thus the result that these four sum to a constant can be formalized as follows.

Proposition 5.4.5. Let $y_1 = y_0/4$ be the free parameter of Example 5.4.3; then for any $x \in [x_0, x_N]$,

$$\sum_{i=-3}^{N-1} B_i(x) = 6y_1 \tag{5.4.23}$$

Proof. It is clear that establishing the result for any one knot interval is sufficient, and we content ourselves with the case of equally spaced knots. We illustrate the result for the interval $[0, 1]$ in the case of integer knots.

From Example 5.4.3 we see that the contributions from B_{-3}, B_{-2}, B_{-1}, and B_0 are just translations of the various parts of the spline B of that example. These contributions are

$$y_1(1-x)^3 + y_1 + 3y_1(1-x) + 3y_1(1-x)^2 - 3y_1(1-x)^3 + y_1 + 3y_1x + 3y_1x^2$$
$$- 3y_1x^3 + y_1x^3$$

which on collecting terms reduces to just $6y_1$ as required.

The usual normalization for the cubic B-splines is therefore to take $y_1 = \frac{1}{6}$ and $y_0 = \frac{2}{3}$ so that the constant sum above is 1.

With this normalization, it also turns out that there is a simple general formula for the B-splines, which we do not derive from its usual divided difference definition:

$$B_i(x) = (x_{i+4} - x_i) \sum_{j=i}^{i+4} \frac{(x_j - x)_+^3}{L_i'(x_j)} \tag{5.4.24}$$

where the L_i are local versions of the Lagrange function of polynomial interpolation, namely,

$$L_i(x) = (x - x_i)(x - x_{i+1})(x - x_{i+2})(x - x_{i+3})(x - x_{i+4}) \tag{5.4.25}$$

from which it is easy to deduce that $0 \le B_i(x) \le 1$ everywhere. [This and other properties of the B-splines are treated fully in de Boor (1978).]

The final property that must be included here is that the B-splines do indeed form a basis.

Theorem 5.4.6. If s is a cubic spline with knots $x_0, x_1, \ldots, x_N$, then there exists a unique set of coefficients $\alpha_{-3}, \alpha_{-2}, \alpha_{-1}, \ldots, \alpha_{N-1}$ such that

$$s(x) = \sum_{i=-3}^{N-1} \alpha_i B_i(x) \tag{5.4.26}$$

Again notice that there are $N + 3$ coefficients in (5.4.26), and therefore, just as with the original development of spline interpolation, further conditions need to be added to the $N + 1$ data points $f(x_0), f(x_1), \ldots, f(x_N)$. These degrees of freedom can once more be used to obtain the natural, complete, or any other preferred cubic spline interpolant to the data. Again the reader is referred to de Boor for full details of how to compute the coefficients and their efficient use for interpolation and approximation.

EXERCISES 5.4

1. Obtain the tridiagonal system of equations for complete cubic spline interpolation for equally spaced knots. Find the complete cubic spline interpolant to the spike $100/(1 + 10x^2)$ with 11 and 21 knots. Produce the graphs corresponding to Fig. 5.3.5.

2. Repeat Exercise 1 for the not-a-knot condition.

3. Supply the details of the derivation of Eqs. (5.4.8) through (5.4.11).

4. Let s denote either the natural cubic spline s_n or the complete one s_c and $\delta(x) = s(x) - \sigma(x)$. Show that

$$\int_{x_0}^{x_N} |\sigma''(x)|^2\, dx = \int_{x_0}^{x_N} |s''(x) - \delta''(x)|^2\, dx$$

$$= \int_{x_0}^{x_N} |s''(x)|^2 - 2s''(x)\delta''(x)$$

$$+ |\delta''(x)|^2\, dx$$

Hence derive Eq. (5.4.19):

$$\int_{x_0}^{x_N} |\sigma''(x)|^2\, dx = \int_{x_0}^{x_N} |s''(x)|^2 + |\delta''(x)|^2\, dx$$

5. Show that the truncated power functions x_+^k have $k - 1$ continuous derivatives at $x = 0$.

6. Show that if $s_0(x)$ is a polynomial of degree n which interpolates f at x_0 and x_1, then the nth-degree spline which interpolates f at x_0, x_1, and x_2 can be written in the form

$$s(x) = s_0(x) + \alpha(x - x_1)_+^n$$

and the coefficient α is uniquely determined by the interpolation conditions. This establishes (5.4.21). Once the nature of the interpolating spline has been fixed, the polynomial s_0 is also uniquely determined.

7. Verify that the B-spline definition (5.4.24) agrees with the B-spline found in Example 5.4.3.

8. Find the first derivative of the cubic B-spline (5.4.24) at each of the knots.

5.5 RATIONAL INTERPOLATION AND CONTINUED FRACTIONS

In this section, we consider the use of a more general class of functions for interpolating data. The use of *rational functions*, that is, ratios of polynomials, allows the interpolant to exhibit some very different characteristics such as poles or asymptotic behavior.

As trivial illustrations of this, the behavior of $(x^2 + 1)/x$ near zero or of $1/x$ as x approaches infinity cannot sensibly be imitated by any polynomial. Note that this apparent conflict with Weierstrass' theorem is only an apparent conflict since that result applies to approximating functions uniformly over *closed, bounded* intervals. Thus, far from being in conflict, these examples demonstrate the need for that hypothesis in Weierstrass' theorem.

We need some further notation for the spaces of rational functions we consider here: denote by $\mathscr{P}_{n,m}$ the set of all rational functions ρ of the form

$$\rho(x) = \frac{p(x)}{q(x)} \tag{5.5.1}$$

where $p \in \Pi_n$, $q \in \Pi_m$ so that p and q can be written as

$$p(x) - a_0 + a_1 x + \cdots + a_n x^n$$

$$q(x) = b_0 + b_1 x + \cdots + b_m x^m \tag{5.5.2}$$

Note here that the "space" $\mathscr{P}_{n,m}$ is not a linear space since, for example, it is not closed under addition. (See Exercise 1.)

Clearly, multiplying p and q in (5.5.1) by the same scalar quantity does not change the definition of ρ, and so we may normalize the above equations by setting

$$b_0 = 1 \tag{5.5.3}$$

Note that this normalization is not quite universally acceptable since, as in the examples above, ρ may have singularities at $x = 0$, in which case we would require $b_0 = 0$. We can readily modify our condition to account for this situation by simply insisting that the first nonzero b_i be normalized to unity. However, since a simple translation can be used to move the pole away from the origin, we will simply assume there is no pole there and thus take $b_0 = 1$ throughout.

It follows that there are now $n + m + 1$ parameters in the definition of ρ, and therefore we may suppose that this is the number of data points required for interpolation by a member of the set $\mathscr{P}_{n,m}$. It is to the details of this interpolation problem that we now turn.

Let $x_0, x_1, \ldots, x_N$, where $N = n + m$, be distinct points at which the values $f(x_0), f(x_1), \ldots, f(x_N)$ are known. In order that ρ given by (5.5.1) to (5.5.3) agree with f at these nodes, we have, for each $i = 0, 1, \ldots, N$,

$$f(x_i) = \frac{a_0 + a_1 x_i + \cdots + a_n x_i^n}{1 + b_1 x_i + \cdots + b_m x_i^m} \tag{5.5.4}$$

This system can be rewritten as the linear system

$$a_0 + a_1 x_i + \cdots + a_n x_i^n - f(x_i)(b_1 x_i + b_2 x_i^2 + \cdots + b_m x_i^m) = f(x_i) \tag{5.5.5}$$

Unfortunately, unlike the polynomial and spline interpolation problems this linear system does not necessarily have a solution—and even if it does, the solution need not be unique and may not even satisfy the original interpolation conditions. We illustrate the various possibilities with the following examples.

Examples 5.5.1

(a) Consider the rational interpolation of the data

$$f(-1) = 0 \qquad f(0) = 1 \qquad f(1) = 1 \qquad \text{by } \rho \in \mathscr{P}_{1,1}$$

We seek a rational function of the form

$$\rho(x) = \frac{a_0 + a_1 x}{1 + b_1 x}$$

which takes the prescribed values. System (5.5.5) yields the equations

$$a_0 - a_1 \quad\;\; = 0$$
$$a_0 \qquad\qquad = 1$$
$$a_0 + a_1 - b_1 = 1$$

which clearly has the *unique* solution $a_0 = a_1 = b_1 = 1$. This leads us to the rational function $\rho(x) = (1 + x)/(1 + x)$. Removing the singularity at $x = -1$ by canceling the factor $1 + x$, we have $\rho(x) \equiv 1$, which does not satisfy the first interpolation condition.

We see that even if the linear system has a unique solution, it is not necessarily a solution of the original interpolation problem. If the linear system has a solution which leads to an irreducible rational function—that is, one in which there are no common factors in numerator and denominator—then this will also solve the interpolation problem.

(b) Consider the same problem but with the data

$$f(-1) = 1 \qquad f(0) = 1 \qquad f(1) = 1$$

for which the linear system becomes

$$a_0 - a_1 + b_1 = 1$$
$$a_0 \qquad\qquad = 1$$
$$a_0 + a_1 - b_1 = 1$$

which has infinitely many solutions of the form

$$a_0 = 1 \qquad a_1 = b_1 = a$$

say, for which the corresponding function ρ is given by

$$\rho(x) = \frac{1 + ax}{1 + ax}$$

By taking $a = 0$, this is the irreducible rational function $\rho(x) \equiv 1$ which is, of course, a solution of the original problem.

(c) With a further modification of the data to

$$f(-1) = 2 \qquad f(0) = 1 \qquad f(1) = 2$$

we obtain the linear system

$$a_0 - a_1 + 2b_1 = 2$$
$$a_0 \qquad\qquad = 1$$
$$a_0 + a_1 - 2b_1 = 2$$

which, on eliminating a_0, becomes

$$-a_1 + 2b_1 = 1 \qquad a_1 - 2b_1 = 1$$

which has no solution at all. Since it is apparent that any solution of the interpolation problem (5.5.4) must satisfy the corresponding linear system (5.5.5), it follows that this particular rational interpolation problem has no solution in the set $\mathscr{P}_{1,1}$.

(*d*) As a final case consider the data

$$f(-1) = 2 \qquad f(0) = 1 \qquad f(1) = 3$$

for which the equations are

$$a_0 - a_1 + 2b_1 = 2$$
$$a_0 \qquad\qquad = 1$$
$$a_0 + a_1 - 3b_1 = 3$$

or

$$-a_1 + 2b_1 = 1 \qquad a_1 - 3b_1 = 2$$

which has the unique solution $b_1 = -3$ and $a_1 = -7$. The corresponding rational function

$$p(x) = \frac{1 - 7x}{1 - 3x}$$

interpolates the data as required. Note that p has a pole at $x = \frac{1}{3}$; this singularity is within the span of the data points and indicates that p should not be used as an approximation for f in this region if f is thought to be continuous there. If, however, f is known to have a pole somewhere in the interval $[-1, 1]$, then p may be a good approximation to f.

The most striking feature of these examples is that we really have very little idea in advance as to whether any particular rational interpolation problem will have a solution. This information only becomes available as we solve (or try to solve) the linear system. Even if such a solution is obtained, it must be checked to satisfy the original interpolation conditions. There is one important positive result that is easily established, however.

Proposition 5.5.2. There is at most one solution to the rational interpolation problem (5.5.4).

Proof. See Exercise 4.

At this time you may wonder whether there is any real point to introducing the topic of rational interpolation at all. The answer is a very firm "yes" since there are many examples of excellent rational approximations to elementary functions. Also, there are efficient techniques for the implementation of rational interpolation using *continued fractions*. It is to these that we turn now.

By analogy with polynomial interpolation, what we have so far is a very poor "Lagrangian" theory of rational approximation. In the rest of this section, we develop the equivalent of the newtonian divided difference scheme which is based on *inverse differences*. There is also a more elegant theoretical and algorithmic development comparable to Aitken's lemma and its application. For details of this the reader is referred to Stoer and Bulirsch (1980).

Definition 5.5.3. We define *inverse differences* for the function f at the nodes $x_0, x_1, \ldots, x_N$ recursively as follows:

$$\text{Zeroth inverse difference: } \phi[x_i] = f(x_i)$$

$$\text{First inverse difference: } \phi[x_i, x_j] = \frac{x_i - x_j}{\phi[x_i] - \phi[x_j]}$$

(5.5.6)

In general,

$$\phi[x_i, \ldots, x_p, x_q, x_r] = \frac{x_q - x_r}{\phi[x_i, \ldots, x_p, x_q] - \phi[x_i, \ldots, x_p, x_r]}$$

(5.5.7)

Note that, unlike divided differences, the inverse differences are dependent on the order of the arguments so that, in general,

$$\phi[x_1, x_2, x_3] \neq \phi[x_2, x_3, x_1]$$

for example. It is also the case that some of these inverse differences may become infinite if, for example, two of the function values are equal. (See Exercise 5.)

Example 5.5.4. For the data $x_0 = -1$, $x_1 = 0$, $x_2 = 1$, $x_3 = 2$ with respective function values 2, 1, 3, 2.5, we obtain the following table of inverse differences:

i	x_i	f_i	$\phi[x_0, x_i]$	$\phi[x_0, x_1, x_i]$	$\phi[x_0, x_1, x_2, x_3]$
0	-1	2			
1	0	1	-1		
2	1	3	2	0.3333	
3	2	2.5	6	0.2857	-21.0084

Here each entry is obtained using entries in the previous column which are on the same row and on the main diagonal. Thus, for example, the entry 0.2857 for $\phi[x_0, x_1, x_3]$ is obtained as

$$\frac{x_3 - x_1}{\phi[x_0, x_3] - \phi[x_0, x_1]} = \frac{2 - 0}{6 - (-1)}$$

How do we utilize these inverse differences to obtain useful rational interpolation formulas?

The development of the inverse difference interpolation formula in the form of a *continued-fraction* expression is very similar to the development of Newton's divided difference interpolation formula. By definition, we find that

$$f(x) = f(x_0) + (x - x_0) \frac{f(x) - f(x_0)}{x - x_0}$$

$$= \phi[x_0] + \frac{x - x_0}{\phi[x_0, x]}$$

(5.5.8)

It follows immediately that the function $\rho_{1,0}$ given by

$$\rho_{1,0}(x) = \phi[x_0] + \frac{x - x_0}{\phi[x_0, x_1]} \tag{5.5.9}$$

agrees with f at both x_0 and x_1. [Of course, (5.5.9) is just a more complicated way of writing the first-order divided difference interpolation polynomial.]
Repeating the argument, we find that

$$\phi[x_0, x] = \phi[x_0, x_1] + \frac{x - x_1}{\phi[x_0, x_1, x]}$$

and, substituting this into (5.5.8), we get

$$f(x) = \phi[x_0] + \cfrac{x - x_0}{\phi[x_0, x_1] + \cfrac{x - x_1}{\phi[x_0, x_1, x]}}$$

With x_2 in place of x in this final inverse difference, we obtain the rational interpolant $\rho_{1,1}$ in the form

$$\rho_{1,1}(x) = \phi[x_0] + \cfrac{x - x_0}{\phi[x_0, x_1] + \cfrac{x - x_1}{\phi[x_0, x_1, x_2]}} \tag{5.5.10}$$

which we can see agrees with f at x_0, x_1, x_2.

Example 5.5.5. With the same data as in Example 5.5.4, at this stage we have

$$\rho_{1,1}(x) = 2 + \cfrac{x + 1}{-1 + \cfrac{x}{\frac{1}{3}}} = 2 + \frac{x + 1}{3x - 1} = \frac{7x - 1}{3x - 1}$$

which is the solution obtained in Example 5.5.2(d) for this same data.

From this example we also see readily that expression (5.5.10) is indeed a member of $\mathscr{P}_{1,1}$.
The derivation started above can be continued as long as the inverse differences are defined to obtain a succession of rational interpolants to the data. These functions are taken from the spaces $\mathscr{P}_{1,0}, \mathscr{P}_{1,1}, \mathscr{P}_{2,1}, \dots, \mathscr{P}_{n,n-1}, \mathscr{P}_{n,n}, \dots$ and are given by

$$f(x) = \phi[x_0] + \cfrac{x - x_0}{\phi[x_0, x_1] + \cfrac{x - x_1}{\phi[x_0, x_1, x_2] + \cfrac{x - x_2}{\phi[x_0, x_1, x_2, x_3] + \cdots}}} \tag{5.5.11}$$

Example 5.5.6. Continuing with the continued-fraction representation of rational interpolants to the data in Example 5.5.4, we find that the function $\rho_{2,1}$ is given (with exact arithmetic so that $\phi[x_0, x_1, x_2] = \frac{1}{3}$, $\phi[x_0, x_1, x_3] = \frac{2}{7}$, and $\phi[x_0, x_1, x_2, x_3] = -21$) by

$$\rho_{2,1}(x) = 2 + \cfrac{x+1}{-1 + \cfrac{x}{\frac{1}{3} - \cfrac{x-1}{21}}} = 2 + \cfrac{x+1}{-1 + \cfrac{21x}{8-x}} = 2 + \cfrac{(x+1)(8-x)}{22x-8}$$

$$= \frac{-8 + 51x - x^2}{22x - 8}$$

The simplification of the original formula into an explicit rational form is purely for illustrative purposes, as the original continued fraction is already in a form for efficient computation. Indeed the continued-fraction representation of a rational function is essentially equivalent to using Horner's rule for polynomial evaluation.

The array of inverse differences used here is also well-suited to computer implementation in a similar manner to Newton's divided difference formula for polynomial interpolation, as can be seen from Algorithm 5.5.7 below. However, adding in new data points is much less easy than in that situation. We do not pursue the efficient algorithmic implementation here. As in the case of divided differences we need only store the inverse differences that are required for subsequent computation. This can be achieved by the following procedure, similar to that used for divided differences in Sec. 4.3:

```
Procedure InverseDiffs;
    {m is order of difference, n is number of data points}
var k: integer;
begin
    for m:=1 to n do
        for k:=n downto m do
            d[k]:=(x[k] − x[m − 1])/(d[k] − d[m − 1]);
end;
```

Algorithm 5.5.7   Continued-fraction interpolation

Input: $N, x_0, x_1, \ldots, x_N, f_0, f_1, \ldots, f_N, \bar{x}$

Generate inverse differences:
$$\phi[x_0], \phi[x_0, x_1], \ldots, \phi[x_0, x_1, \ldots, x_N]$$

Compute: fbar$:= \phi[x_0, x_1, \ldots, x_N]$
for $i = N - 1$ downto 0
$$\text{fbar} := \phi[x_0, x_1, \ldots, x_i] + \frac{\bar{x} - x_i}{\text{fbar}}$$

Output: $f(\bar{x}) \simeq \text{fbar}$

In Table 5.5.8 we give the results of using continued-fraction rational interpolation with the nodes $-\pi/2$, $-\pi/4$, 0, $\pi/4$, and $\pi/2$ for the function $\cos x$ over the interval $[-\pi/2, \pi/2]$. The results of using polynomial interpolation with the same nodes are also tabulated along with the true values. The continued-fraction results were obtained using Algorithm 5.5.7.

TABLE 5.5.8
Continued-fraction and polynomial interpolation using five equally spaced nodes in $[-\pi/2, \pi/2]$

x	Continued fraction	Polynomial interpolation	$\cos x$
0.00000	1.00000	1.00000	1.00000
0.15708	0.98760	0.98774	0.98769
0.31416	0.95077	0.95122	0.95106
0.47124	0.89055	0.89126	0.89101
0.62832	0.80861	0.80925	0.80902
0.78540	0.70711	0.70711	0.70711
0.94248	0.58859	0.58729	0.58779
1.09956	0.45581	0.45283	0.45399
1.25664	0.31161	0.30729	0.30902
1.41372	0.15879	0.15479	0.15643
1.57080	0.00000	0.00000	0.00000

For this particularly simple example, there is little difference between the two interpolants, with both showing errors, typically in the third decimal place, although the polynomial tends to show slightly better agreement with the original function. Clearly, Runge's example which proved to be very difficult for polynomial interpolation would be reproduced exactly (apart from rounding errors) by a continued fraction using just five nodes.

EXERCISES 5.5

1. Show that the space of rational functions $\mathcal{P}_{n,m}$ is not a linear space.
2. Find a rational function interpolant $\rho_{2,1} \in \mathcal{P}_{2,1}$ for the data $f(1) = 4$, $f(2) = 3.5$, $f(3) = 4$, and $f(4) = 5$.
3. Modify the data of Exercise 2 so that:
   (a) No solution to the linear system exists.
   (b) The linear system has a solution but the interpolation problem does not.
   (c) The linear system has infinitely many solutions including a solution of interpolation problem.
4. Prove Proposition 5.5.2, namely, that there is at most one solution to the rational interpolation problem (5.5.4).

5. Give an example to show that $\phi[x_0, x_1, x_2] \neq \phi[x_1, x_2, x_0]$ in general.
6. Generate a table of inverse differences for the data of Exercise 2, and use it to obtain the continued-fraction form of that rational interpolant. Verify that you have indeed found the same function.
7. Verify that using data at the nodes $-2, -1, 0, 1, 2$, the continued-fraction interpolant to $1/(1 + x^2)$ in $\mathcal{P}_{2,2}$ reproduces the original function.
8. Write a program for continued-fraction interpolation to the tangent function over the interval $[-1.5, 1.5]$ using nodes $k/10$ for $k = -15, -14, \ldots, 15$.

5.6 TRIGONOMETRIC INTERPOLATION AND FAST FOURIER TRANSFORM

For a large number of practical interpolation and approximation problems, we have advance knowledge of some special properties of the function to be approximated. One common situation is that the function is known to be periodic; that is, there is some constant, τ say—the *period* of the function—such that

$$f(x + \tau) = f(x) \qquad \text{for all } x$$

Most wave theories—electromagnetic, sound, and water waves, for example—exhibit such periodic behavior.

Such functions can often be very well approximated by trigonometric functions. To simplify the theoretical development of trigonometric polynomial interpolation, we will fix the period as being 2π and the interval over which we seek our interpolant to be $[0, 2\pi)$. Other periods or intervals can easily be handled with a simple linear change of the independent variable.

We thus seek a trigonometric polynomial π_n of degree n to interpolate specified function values. Our interpolant has the form

$$\pi_n(x) = a_0 + \sum_{j=1}^{n} a_j \cos jx + b_j \sin jx \tag{5.6.1}$$

which has $2n + 1$ coefficients and should therefore be determined by $2n + 1$ nodes. At first sight you may wonder why such a function is called a trigonometric "polynomial." First, from the basic trigonometric identities, we see that

$$\cos(j + 1)x = \cos jx \cos x - \sin jx \sin x$$
$$\tag{5.6.2}$$
$$\sin(j + 1)x = \sin jx \cos x + \cos jx \sin x$$

from which it follows that (5.6.1) is indeed equivalent to a polynomial of degree n in $\sin x$ and $\cos x$. Alternatively, and essentially equivalently, we know from de Moivre's theorem that

$$e^{ix} = \cos x + i \sin x$$
$$\tag{5.6.3}$$
$$e^{ijx} = (\cos x + i \sin x)^j = \cos jx + i \sin jx$$

from which again we see that (5.6.1) is indeed a trigonometric polynomial. From these identities we also deduce

$$\cos x = \frac{e^{ix} + e^{-ix}}{2} \qquad \sin x = \frac{e^{ix} - e^{-ix}}{2i} \tag{5.6.4}$$

hence it follows that we can write π_n in the form

$$\pi_n(x) = \sum_{j=-n}^{n} \gamma_j e^{ijx} \tag{5.6.5}$$

where the coefficients γ_j are now complex. Write

$$\gamma_j = \alpha_j + i\beta_j \qquad (j = -n, \ldots, n) \tag{5.6.6}$$

Since for x real we require that $\pi_n(x)$ is real also, we may deduce from the conjugacy of e^{ijx} and e^{-ijx} that the coefficients occur as complex conjugate pairs; thus we have

$$\bar{\gamma}_j = \gamma_{-j} \qquad \alpha_j = \alpha_{-j} \qquad \beta_j = -\beta_{-j} \qquad (j = -n, \ldots, n) \qquad (5.6.7)$$

Of course, for our interpolation problem we require the coefficients a_j, b_j of (5.6.1). These are easily obtained from the γ_j's by the relations (see Exercise 2)

$$a_0 = \alpha_0 = \gamma_0 \qquad a_j = \alpha_j + \alpha_{-j} = \gamma_j + \gamma_{-j}$$
$$b_j = -\beta_j + \beta_{-j} = i(\gamma_j - \gamma_{-j}) \qquad (j = 1, 2, \ldots, n) \qquad (5.6.8)$$

Denote the point on the unit circle corresponding to $x \in [0, 2\pi)$ by

$$\xi = e^{ix} \qquad (5.6.9)$$

so that $e^{ijx} = \xi^j$, and therefore the polynomial (5.6.5) may be written as

$$\pi_n(x) = \sum_{j=-n}^{n} \gamma_j \xi^j \qquad (5.6.10)$$

Similarly we denote the points corresponding to the nodes by

$$\xi_k = e^{ix_k} = \cos x_k + i \sin x_k \qquad (5.6.11)$$

Our interpolation problem is thus to find the coefficients in this polynomial such that

$$\pi_n(x_k) = \sum_{j=-n}^{n} \gamma_j \xi_k^j = f_k \qquad (5.6.12)$$

where $f_k = f(x_k)$ $(k = 0, 1, \ldots, 2n)$ are the given function values.

A very slightly generalized version of the Lagrange interpolation theory establishes immediately that this problem has a unique solution provided only that the nodes are all distinct. It is, in general, a difficult task to obtain the coefficients of the polynomial (5.6.10). However, for one very important special case, explicit closed-form expressions for these coefficients can be obtained, and it is on this case of uniformly spaced nodes that we concentrate for the rest of this section.

We develop the theory in the case of an odd number of nodes, although a completely equivalent theory exists for an even number. In the latter case the basic form of interpolant (5.6.1) would have one extra cosine term. The case of an even number of nodes will be used in the subsequent development of the Fast Fourier Transform with which we conclude this section.

Let the nodes x_k satisfy

$$x_k = \frac{2k\pi}{2N+1} \qquad (k = 0, 1, \ldots, 2N) \qquad (5.6.13)$$

so that

$$\xi_k = e^{ix_k} = \cos \frac{2k\pi}{2N+1} + i \sin \frac{2k\pi}{2N+1} \qquad (5.6.14)$$

which are the $(2N + 1)$th roots of unity; that is, they are the complex solutions of

$$z^{2N+1} = 0 \qquad (5.6.15)$$

At this point we need a few elementary properties of these complex roots of unity which we collect into the following result.

Proposition 5.6.1
(a) Let ω be any root of (5.6.15). Then

$$\text{either} \qquad \omega = 1 \qquad \text{or} \qquad 1 + \omega + \cdots + \omega^{2N} = 0$$

(b) $\xi_k = \xi_1^k$ $(k = 0, 1, \ldots, 2N)$, and since $\xi_1^{2N+1} = \xi_0 = 1$, the notation ξ_k can be extended to all integers k by setting $\xi_k = \xi_{k \bmod (2N+1)}$. Then

$$\xi_k^j = \xi_j^k \qquad \text{and} \qquad \overline{\xi_k} = \xi_k^{-1}$$

for all integers j, k.

(c)
$$\sum_{k=0}^{2N} \xi_k^{j-p} = \begin{cases} 2N + 1 & j = p \\ 0 & j \neq p \end{cases}$$

Proof
(a) If ω satisfies (5.6.15) and $\omega \neq 1$, then

$$1 + \omega + \omega^2 + \cdots + \omega^{2N} = \frac{1 - \omega^{2N+1}}{1 - \omega} = 0$$

(b) The first and last parts are immediate from de Moivre's theorem. Using the first part, it follows that $\xi_k^j = \xi^{jk} = \xi_j^k$.
(c) By part (b), $\xi_k^{j-p} = \xi_{j-p}^k$. Now ξ_{j-p} is a root of (5.6.15) and so by (a) is either unity, in which case the sum is $2N + 1$, or else with $\omega = \xi_{j-p}$ in (a), the sum is 0.

We are now in a position to compute the coefficients of the trigonometric interpolation polynomial (5.6.10). The interpolation conditions (5.6.12) are

$$\sum_{j=-N}^{N} \gamma_j \xi_k^j = f_k \qquad (k = 0, 1, \ldots, 2N)$$

and multiplying the kth equation by ξ_k^{-p} and then summing over k we obtain

$$\sum_{k=0}^{2N} f_k \xi_k^{-p} = \sum_{k=0}^{2N} \sum_{j=-N}^{N} \gamma_j \xi_k^j \xi_k^{-p} = \sum_{j=-N}^{N} \gamma_j \sum_{k=0}^{2N} \xi_k^{j-p} = (2N+1)\gamma_p$$

by (c) of Proposition 5.6.1. We thus conclude that

$$\gamma_p = \frac{1}{2N+1} \sum_{k=0}^{2N} f_k \xi_k^{-p} \qquad (p = -N, -N+1, \ldots, N) \qquad (5.6.16)$$

This trigonometric interpolation polynomial is frequently referred to as the *discrete Fourier transform* of the function f. The coefficients used are closely related to those of the standard (or continuous) Fourier transform—indeed, the above computation is equivalent to estimating the normal Fourier coefficients by numerical integration using these nodes. (See Chap. 9.) We will also see in Chap. 7 that this same approximation arises as the linear least squares approximation to f among trigonometric polynomials of degree N.

For many purposes, including the Fast Fourier Transform we are to discuss shortly, it is more convenient to represent the trigonometric polynomial in conventional polynomial form. Since $\xi_k^{2N+1} = 1$, extending the coefficient set by putting $\gamma_{2N+1-j} = \gamma_{-j}$ yields the polynomial

$$p(\xi) = \sum_{j=0}^{2N} \gamma_j \xi^j \qquad (5.6.17)$$

which interpolates the data just as π_N did.

Consider now the task of computing these coefficients using (5.6.16). There are $2N + 1$ coefficients, and each of the equations entails the formation of the sum of $2N + 1$ terms, each of which is a product of one real and one complex factor. There is also the final division by $2N + 1$. A measure of the complexity of this operation can be obtained from the overall operation count measured in terms of real floating-point operations.

In terms of real operations, the multiplications are equivalent to two each, and the summation of $2N + 1$ complex terms requires $4N$ real additions. However, the fact that the coefficients occur as complex conjugate pairs allows us to reduce the number of coefficients to be found to $N + 1$. The total operation count is thus

$$(N + 1)[2(2N + 1) + 4N + 1] = 8N^2 + 11N + 3$$

This is clearly an expensive computation for any moderately large values of N.

For several years this computationally expensive process priced the discrete Fourier transform out of the market for practical function approximation. However, in the mid-1960s methods were developed which make this computation very much more economical for certain choices of the number of nodes. The initial breakthrough in this direction was due to Cooley and Tukey (1965), and it is their approach to the *Fast Fourier Transform* (FFT) which we describe here for the special (most important and most economical) case of 2^N interpolation points.

The idea is that the coefficients are computed in a recursive manner. The recursion has N steps. Step m begins with 2^{N-m+1} trigonometric polynomials, each of which interpolates 2^{m-1} of the data (though not typically at exactly the right points) and produces 2^{N-m} polynomials interpolating 2^m data each. The basic process of the recursion is therefore to obtain the trigonometric polynomial which interpolates a set of data given two lower-degree polynomials which each interpolate half the data. We will see that this reduces the operation count to $O(N \log N)$.

We begin by considering this basic procedure. The task is to find the trigonometric polynomial ρ that interpolates f at the full uniformly spaced data set $x_0, x_1, \ldots, x_{2M-1}$ (so that $x_k = \pi k / M$), assuming that we already know (or have found) two such polynomials σ and τ satisfying the interpolation conditions

$$\sigma(x_{2k}) = f_{2k} \qquad \tau(x_{2k}) = f_{2k+1} \qquad (k = 0, 1, \ldots, M - 1) \qquad (5.6.18)$$

Thus $\sigma(x)$ interpolates half the data—at the even-numbered points—while $\tau(x - \pi/M)$ interpolates the other half since $x_{2k+1} - \pi/M = x_{2k}$.

It follows that

$$\rho(x) = \sigma(x) \frac{1 + e^{iMx}}{2} + \tau\left(x - \frac{\pi}{M}\right) \frac{1 - e^{iMx}}{2} \qquad (5.6.19)$$

satisfies all the interpolation conditions

$$\rho(x_{2k}) = \sigma(x_{2k}) = f_{2k} \qquad \rho(x_{2k+1}) = \tau(x_{2k}) = f_{2k+1}$$

since $Mx_{2k} = 2k\pi$ so that $\exp(iMx_{2k}) = 1$, while $Mx_{2k+1} = (2k + 1)\pi$ and $\exp(iMx_{2k+1}) = -1$.

It remains to find the coefficients of ρ in terms of those of σ and τ. It is convenient here to use the form (5.6.17) for the various polynomials. Denote their respective coefficients for appropriate values of j by $\gamma_j, \delta_j, \varepsilon_j$. From (5.6.19) we have

$$2\rho(x) = 2 \sum_{j=0}^{2M-1} \gamma_j \xi^j = \sum_{j=0}^{M-1} \delta_j \xi^j (1 + e^{iMx}) + \varepsilon_j \xi^j \overline{\xi}_j (1 - e^{iMx}) \qquad (5.6.20)$$

since $\exp[ij(x - \pi/M)] = \exp(ijx) \exp(-ij\pi/M) = \xi^j \overline{\xi}_j$. Comparing the coefficients in the two expansions, we obtain

$$\gamma_j = \frac{\delta_j + \varepsilon_j \overline{\xi}_j}{2}$$

$$\qquad\qquad (j = 0, 1, \ldots, M - 1) \qquad (5.6.21)$$

$$\gamma_{M+j} = \frac{\delta_j - \varepsilon_j \overline{\xi}_j}{2}$$

These relations are at the heart of the recursion used in the FFT. To describe this in detail, we need some further notation. Suppose our goal is the discrete Fourier transform of f of degree $2^N - 1$ so that 2^N interpolation points are to be used in all; these are given by

$$x_k = \frac{2k\pi}{2^N} = \frac{k\pi}{2^{N-1}} \qquad (5.6.22)$$

The recursion begins with 2^N trigonometric polynomials of degree 0, $\tau_k^{(0)}$ $(k = 0, 1, \ldots, 2^N - 1)$ with the interpolation properties

$$\tau_k^{(0)}(x_0) = f_k$$

In general, at step m of the recursion we begin with 2^{N-m+1} polynomials, $\tau_k^{(m-1)}$ with coefficients $\gamma_{k,j}^{(m)}$ $(k = 0, 1, \ldots, 2^{N-m+1} - 1; j = 0, 1, \ldots, 2^{m-1} - 1)$ and satisfying the interpolation conditions

$$\tau_k^{(m-1)}(x_{j2^{N-m+1}}) = f_{2^{m-1}k+j}$$

These are used to generate the 2^{N-m} polynomials, $\tau_k^{(m)}$ with coefficients $\gamma_{k,j}^{(m)}$ satisfying

$$\gamma_k^{(m)}(x_{j2^{N-m}}) = f_{2^m k + j}$$

Relations (5.6.21) yield the recursions

$$\gamma_{k,j}^{(m)} = \frac{\gamma_{k,j}^{(m-1)} + \gamma_{2^{N-m}+k,j}\overline{\xi_{j2^{N-m}}}}{2}$$

$$\gamma_{k,j+2^{m-1}} = \frac{\gamma_{k,j}^{(m-1)} - \gamma_{2^{N-m}+k,j}\overline{\xi_{j2^{N-m}}}}{2}$$

(5.6.23)

for $k = 0, 1, \ldots, 2^{N-m} - 1$; $j = 0, 1, \ldots, 2^{m-1} - 1$. Here we have used the identity

$$e^{-ij\pi/2^{m-1}} = \overline{\xi_{j2^{N-m}}}$$

(5.6.24)

Once the "complex nodes" ξ_k are known (as they must be for either this or the direct computation of the discrete Fourier transform), each step of the recursion requires 2^N complex multiplications and additions. Since a complex multiplication consists of four real multiplications together with two further additions, this amounts to a total of 8×2^N real floating-point operations. There are N such steps. The overall operation count is thus reduced to $8N \times 2^N$, which compares with approximately $8(2^N)^2$ for the direct computation discussed earlier.

For $N = 10$, or 1024 data points, this represents a reduction from about 8,000,000 floating-point operations to around 80,000—a saving by a factor of about 100. The relative saving to be made increases rapidly with the number of data points to be used.

Example 5.6.2. We illustrate the fast Fourier transform with a simple example using just eight data points. Thus $x_0 = 0$, $x_1 = \pi/4$, $x_2 = \pi/2$, $x_3 = 3\pi/4$, $x_4 = \pi$, $x_5 = 5\pi/4$, $x_6 = 3\pi/2$, and $x_7 = 7\pi/4$ so that the ξ_j are given by

$$\xi_0 = 1 \qquad \xi_1 = \frac{1+i}{\sqrt{2}} \qquad \xi_2 = i \qquad \xi_3 = \frac{-1+i}{\sqrt{2}}$$

$$\xi_4 = -1 \qquad \xi_5 = \frac{-1-i}{\sqrt{2}} \qquad \xi_6 = -i \qquad \xi_7 = \frac{1-i}{\sqrt{2}}$$

The data is taken from the 2π – periodic function $\ln[1 + (\sin x)/2]$. So

$$f_0 = 0.00000 \qquad f_1 = 0.30273 \qquad f_2 = 0.40547 \qquad f_3 = 0.30273$$

$$f_4 = 0.00000 \qquad f_5 = -0.43626 \qquad f_6 = -0.69315 \qquad f_7 = -0.43626$$

Initialize. Set $\gamma_{k,0}^{(0)} = f_k$.

Step 1. $j = 0$ throughout this step.

$$k = 0: \qquad \gamma_{0,0}^{(1)} = \frac{f_0 + f_4 \overline{\xi_0}}{2} = 0.00000$$

$$\gamma_{0,1}^{(1)} = \frac{f_0 - f_4 \overline{\xi_0}}{2} = 0.00000$$

$$k = 1: \qquad \gamma_{1,0}^{(1)} = \frac{f_1 + f_5 \overline{\xi_0}}{2} = -0.06677$$

$$\gamma_{1,1}^{(1)} = \frac{f_1 - f_5 \overline{\xi_0}}{2} = 0.36950$$

$$k = 2: \qquad \gamma_{2,0}^{(1)} = \frac{f_2 + f_6 \overline{\xi_0}}{2} = -0.14384$$

$$\gamma_{2,1}^{(1)} = \frac{f_2 - f_6 \overline{\xi_0}}{2} = 0.54931$$

$$k = 3: \qquad \gamma_{3,0}^{(1)} = \frac{f_3 + f_7 \overline{\xi_0}}{2} = -0.06677$$

$$\gamma_{3,1}^{(1)} = \frac{f_3 - f_7 \overline{\xi_0}}{2} = 0.36950$$

Step 2

$$k = 0, \ j = 0: \qquad \gamma_{0,0}^{(2)} = \frac{\gamma_{0,0}^{(1)} + \gamma_{2,0}^{(1)} \overline{\xi_0}}{2} = -0.07192$$

$$\gamma_{0,2}^{(2)} = \frac{\gamma_{0,0}^{(1)} - \gamma_{2,0}^{(1)} \overline{\xi_0}}{2} = 0.07192$$

$$k = 0, \ j = 1: \qquad \gamma_{0,1}^{(2)} = \frac{\gamma_{0,1}^{(1)} + \gamma_{2,1}^{(1)} \overline{\xi_2}}{2} = -0.27466i$$

$$\gamma_{0,3}^{(2)} = \frac{\gamma_{0,1}^{(1)} - \gamma_{2,1}^{(1)} \overline{\xi_2}}{2} = 0.27466i$$

$$k = 1, \ j = 0: \qquad \gamma_{1,0}^{(2)} = \frac{\gamma_{1,0}^{(1)} + \gamma_{3,0}^{(1)} \overline{\xi_0}}{2} = -0.06677$$

$$\gamma_{1,2}^{(2)} = \frac{\gamma_{1,0}^{(1)} - \gamma_{3,0}^{(1)} \overline{\xi_0}}{2} = 0.00000$$

$$k = 1, \ j = 1: \qquad \gamma_{1,1}^{(2)} = \frac{\gamma_{1,1}^{(1)} + \gamma_{3,1}^{(1)} \overline{\xi_2}}{2} = 0.18475 - 0.18475i$$

$$\gamma_{1,3}^{(2)} = \frac{\gamma_{1,1}^{(1)} - \gamma_{3,1}^{(1)} \overline{\xi_2}}{2} = 0.18475 + 0.18475i$$

Step 3. $k = 0$ throughout this step

$$j = 0: \qquad \gamma_{0,0}^{(3)} = \frac{\gamma_{0,0}^{(2)} + \gamma_{1,0}^{(2)} \overline{\xi_0}}{2} = -0.06935$$

$$\gamma_{0,4}^{(3)} = \frac{\gamma_{0,0}^{(2)} - \gamma_{1,0}^{(2)}\bar{\xi}_0}{2} = -0.00258$$

$j = 1$:

$$\gamma_{0,1}^{(3)} = \frac{\gamma_{0,1}^{(2)} + \gamma_{1,1}^{(2)}\bar{\xi}_1}{2} = -0.26797i$$

$$\gamma_{0,5}^{(3)} = \frac{\gamma_{0,1}^{(2)} - \gamma_{1,1}^{(2)}\bar{\xi}_1}{2} = -0.00669i$$

$j = 2$:

$$\gamma_{0,2}^{(3)} = \frac{\gamma_{0,2}^{(2)} + \gamma_{1,2}^{(2)}\bar{\xi}_2}{2} = 0.03596$$

$$\gamma_{0,6}^{(3)} = \frac{\gamma_{0,2}^{(2)} - \gamma_{1,2}^{(2)}\bar{\xi}_2}{2} = 0.03596$$

$j = 3$:

$$\gamma_{0,3}^{(3)} = \frac{\gamma_{0,3}^{(2)} + \gamma_{1,3}^{(2)}\bar{\xi}_3}{2} = 0.00669i$$

$$\gamma_{0,7}^{(3)} = \frac{\gamma_{0,3}^{(2)} + \gamma_{1,3}^{(2)}\bar{\xi}_3}{2} = 0.26797i$$

Equations (5.6.8) with the identification of γ_5, γ_6, γ_7 with γ_{-3}, γ_{-2}, γ_{-1} are now used to obtain the coefficients of the transform itself:

$$a_0 = -0.06935 \qquad a_1 = 0 \qquad a_2 = 0.07192 \qquad a_3 = 0 \qquad a_4 = -0.00258$$

$$b_1 = 0.53594 \qquad b_2 = 0 \qquad b_3 = -0.01338$$

so that the FFT approximation of degree four to f is

$$\ln\left(1 + \frac{\sin x}{2}\right) \simeq -0.06935 + 0.53594 \sin x + 0.07192 \cos 2x$$
$$- 0.01338 \sin 3x - 0.00258 \cos 4x$$

Substituting for $\cos 2x$, $\cos 4x$, and $\sin 3x$ in this expression, we see that the coefficients are close to those of the expansion of $\ln[1 + (\sin x)/2]$ as a power series in $\sin x$:

$$\ln\left(1 + \frac{\sin x}{2}\right) = \frac{\sin x}{2} - \frac{\sin^2 x}{8} + \frac{\sin^3 x}{24} - \cdots$$

See Exercise 3.

It is immediately clear from this example that the FFT is only suitable for computer use—and then only if complex arithmetic is available. For this reason we do not pursue the further implementational details here.

EXERCISES 5.6

1. Express $\sin jx$ and $\cos jx$ as polynomials in $\sin x$ and $\cos x$ for $j = 2, 3, 4$.
2. Show that the coefficients of the trigonometric polynomial interpolant and its complex form are related by Eqs. (5.6.8).
3. Using the results of Exercise 1, rewrite the Fast Fourier Transform of $\ln[1 + (\sin x)/2]$ as a polynomial in $\sin x$, and compare this with the power series expansion.

5.7 BIVARIATE INTERPOLATION

In this final section on interpolation, we provide a brief introduction to the problem of polynomial interpolation in two variables. We content ourselves with the development and analysis of the Lagrangian interpolation theory, leaving the fairly natural extensions to local interpolation with divided differences or finite differences to the reader. At the time of writing there is still no widely accepted standard definition of bivariate cubic B-splines, although much effort has been devoted to obtaining good bivariate spline approximations to functions.

The special case we consider is that of finding a bivariate polynomial which interpolates the function $f(x, y)$ at the regular (though not necessarily uniformly spaced) mesh of data points (x_i, y_j) ($i = 0, 1, \ldots, N$; $j = 0, 1, \ldots, M$) at which the function values are denoted by $f_{i,j}$. Thus it is natural to seek a polynomial $p \in \Pi_{N,M}$ of degree N in x and degree M in y satisfying the conditions

$$p(x_i, y_j) = f_{i,j} \tag{5.7.1}$$

As with the single-variable Lagrangian theory, it is clearly sufficient to find a basis for the linear space $\Pi_{N,M}$, each member of which is unity at one of the mesh points and vanishes at all the others. Such a basis is easily constructed from the usual univariate Lagrange basis.

Denote by $\lambda_i(x)$ the usual Lagrange basis polynomials for the nodes $x_0, x_1, \ldots, x_N$ satisfying

$$\lambda_i(x_k) = \delta_{ik} \tag{5.7.2}$$

and by $\mu_j(y)$ the corresponding basis for the nodes $y_0, y_1, \ldots, y_M$ so that

$$\mu_j(y_k) = \delta_{jk} \tag{5.7.3}$$

According to (4.2.8), we see that these polynomials are given by

$$\lambda_i(x) = \frac{(x - x_0) \cdots (x - x_{i-1})(x - x_{i+1}) \cdots (x - x_N)}{(x_i - x_0) \cdots (x_i - x_{i-1})(x_i - x_{i+1}) \cdots (x_i - x_N)} \tag{5.7.4}$$

and

$$\mu_j(y) = \frac{(y - y_0) \cdots (y - y_{j-1})(y - y_{j+1}) \cdots (y - y_M)}{(y_j - y_0) \cdots (y_j - y_{j-1})(y_j - y_{j+1}) \cdots (y_j - y_M)} \tag{5.7.5}$$

It is now apparent that the polynomials $l_{i,j}(x, y)$ given by

$$l_{i,j}(x, y) = \lambda_i(x)\mu_j(y) \tag{5.7.6}$$

provide a suitable basis and that the required interpolation polynomial p is given by

$$p(x, y) = \sum_{i=0}^{N} \sum_{j=0}^{M} f_{i,j} l_{i,j}(x, y) \tag{5.7.7}$$

In this development we began with the Lagrangian basis for the polynomial spaces in the two variables separately and then combined them to obtain (5.7.7). In practice, we would write this equation in the form

$$p(x, y) = \sum_{i=0}^{N} \sum_{j=0}^{M} f_{i,j} \lambda_i(x) \mu_j(y) \qquad (5.7.8)$$

and use what might be termed *repeated interpolation*. In principle, the order of summation here is immaterial—although there could conceivably be reasons for preferring one order in a particular situation. In the order suggested by (5.7.8) the process can be thought of as interpolation in the y direction first and then interpolation of the results in the x direction. We thus think of

$$p(x, y) = \sum_{i=0}^{N} \lambda_i(x) \sum_{j=0}^{M} f_{i,j} \mu_j(y) \qquad (5.7.9)$$

as first producing a set of interpolants to estimate the values $f(x_i, y)$ and then using these for interpolation in the x direction to estimate $f(x, y)$ itself.

The corresponding formula for the other order of summation produces interpolation estimates of the quantities $f(x, y_j)$ and then uses these for the y interpolation to estimate $f(x, y)$.

This repeated interpolation formulation also enables us to obtain error bounds for $|f(x, y) - p(x, y)|$, again using the single-variable Lagrangian theory provided that f is sufficiently smooth.

We assume hereafter that f has at least $N + 1$ continuous partial derivatives with respect to x and at least $M + 1$ with respect to y. We denote by K_1 and K_2 bounds for the derivatives $\partial^{N+1} f / \partial x^{N+1}$ and $\partial^{M+1} f / \partial y^{M+1}$ over the rectangle $[x_0, x_N] \times [y_0, y_M]$. (Here we have implicitly assumed that x_i, y_j are both increasing sequences. This could easily be replaced with the simple requirement that K_1, K_2 are bounds on some rectangle containing all the mesh points.) We also denote the product functions of the Lagrange theory by

$$\Lambda_1(x) = (x - x_0)(x - x_1) \cdots (x - x_N)$$
$$\Lambda_2(y) = (y - y_0)(y - y_1) \cdots (y - y_M)$$

From the univariate Lagrange interpolation remainder term, Theorem 4.2.3, we deduce that the y interpolation results in an error bounded by

$$|f(x_i, y) - p(x_i, y)| \le \frac{K_2 |\Lambda_2(y)|}{(M + 1)!} \qquad (5.7.10)$$

for each $i = 0, 1, \ldots, N$ since, by definition, we have

$$p(x_i, y) = \sum_{j=0}^{M} f_{i,j} \mu_j(y) \qquad (5.7.11)$$

Furthermore, we have

$$p(x, y) = \sum_{i=0}^{N} \lambda_i(x) p(x_i, y) \qquad (5.7.12)$$

whence we deduce that

$$\begin{aligned}
|f(x, y) - p(x, y)| &= \left| f(x, y) - \sum \lambda_i(x) p(x_i, y) \right| \\
&\leq \left| f(x, y) - \sum \lambda_i(x) f(x_i, y) \right| \\
&\quad + \left| \sum \lambda_i(x) [f(x_i, y) - p(x_i, y)] \right| \\
&\leq \frac{K_1 |\Lambda_1(x)|}{(N+1)!} + \sum \lambda_i(x) \frac{K_2 |\Lambda_2(y)|}{(M+1)!}
\end{aligned}$$

Since, as we saw in Sec. 4.2., $\sum \lambda_i(x) \equiv 1$, it follows, finally, that

$$|f(x, y) - p(x, y)| \leq \frac{K_1 |\Lambda_1(x)|}{(N+1)!} + \frac{K_2 |\Lambda_2(y)|}{(M+1)!} \qquad (5.7.13)$$

It is evident from this analysis that the error bound is independent of the order of summation.

The fact that this bivariate interpolation is achieved as repeated interpolation makes it easy to conceive of similar repeated interpolation formulas based on the divided difference or finite difference schemes discussed in the previous chapter. We do not pursue these details here.

EXERCISES 5.7

1. Find the Lagrange interpolation polynomial $p \in \Pi_{2,3}$ satisfying

$$p(i, j) = i^j \qquad (i = 1, 2; \; j = 1, 2, 3)$$

Use this function to estimate $1.5^{2.5}$.

2. Write a program to find the bivariate Lagrange interpolation polynomial on a regular $N \times M$ mesh of data points. Test it by finding the polynomial in $\Pi_{4,4}$ which agrees with $\sin (x + 2y)$ at 25 equally spaced mesh points in $[-\pi, \pi] \times [-\pi/2, \pi/2]$. Compare the results with the power series expansion of this function using terms of degree up to three.

3. Implement a program to use the nearest 4×4 set of mesh points from an overall $N \times M$ regular mesh of data points for bivariate polynomial interpolation using Newton's divided difference formula in each direction. With an overall mesh of size 21×21, test this for the same function as in Exercise 2, and compare the results with those of that exercise and the true function values.

PROJECTS

1. Experiment with the distribution of first 21 and then 41 knots for natural cubic spline interpolation of the function

$$f(x) = \frac{100}{1 + 100x^2}$$

over the interval $[-10, 10]$. Try to find knot distributions to minimize the discrete least squares error given by

$$\sum_{i=0}^{100} \left| f\left(\frac{i}{10}\right) - s\left(\frac{i}{10}\right) \right|^2$$

2. Set up the equations for complete and natural cubic B-spline interpolation for equally spaced knots as a tridiagonal system. Write a program to implement this interpolation. Verify that this produces the same interpolant as that found in Exercise 1 of Sec. 5.4.

3. Write a program for continued-fraction interpolation to the tangent function over the interval $[-1.5, 1.5]$ using 31 nodes. Compare the results with those of polynomial interpolation and cubic spline interpolation using the same nodes (knots). Try to find the best set of nodes for this problem.

4. Modify your program from Project 3 to use rational interpolation in $\mathscr{P}_{3,2}$ using a continued-fraction expansion based on the nearest six nodes to the point of interest.

5. Find a picture of your favorite car, plane, animal, or person. Take careful measurements of the outline relative to some chosen origin and axes. Use these as data for interpolation. Try to obtain the best reproduction of the outline using the most appropriate data points you can together with the best interpolating function. You may, of course, use different strategies for different parts of the curve.

6. Write a program to obtain contour plots for the original function and the various approximations found in Exercises 2 and 3 of Sec. 5.7. Contour plots provide a very convenient way of visualizing and comparing bivariate functions. Recall that a contour of the function $f(x, y)$ is a curve defined by $f(x, y) = c$ for some constant c. One way of obtaining such a contour plot is to scan in the x direction and solve the resulting equation for y and then to reverse this process. In this case, the equation can be solved using the arcsine function, but usually this step would require a numerical equation solver such as the secant method.

CHAPTER
6

SYSTEMS
OF LINEAR
EQUATIONS

6.1 INTRODUCTION

In the past few chapters we have seen repeatedly the need for the solution of systems of linear equations. Such systems arise within the contexts of polynomial, spline, and trigonometric interpolation. It is also the case that much of the work entailed in the solution of partial differential equations and boundary value problems reduces to the solution of such systems. It is clearly necessary, therefore, to have efficient techniques at our disposal for the solution of linear equations.

The approaches that can be used vary in accordance with the nature of the problem. We saw in the case of cubic spline interpolation that tridiagonal systems were amenable to direct solution by a modification of the standard elimination procedure. For some very large systems, iterative methods are to be preferred. The use of parallel computer architectures modifies the choice further (see Chap. 13).

We begin here with the method of *Gauss* (or *Gaussian*) *elimination* in its simplest form. In later sections we consider modifications and improvements on this idea together with an error analysis. In the last two sections of this chapter we turn to iterative approaches to the solution of linear systems.

The method of Gauss elimination is based on the simple high school algebra approach to the solution of a pair of simultaneous equations in which a multiple of one equation is subtracted from the other to eliminate one of the two unknowns. The resulting equation is then solved for the remaining unknown, and its value is substituted back into one of the original equations to solve for the other. We make these ideas precise in the context of the solution of the equation

$$A\mathbf{x} = \mathbf{b} \qquad (6.1.1)$$

where A is an $m \times n$ matrix, $\mathbf{b}$ is an m vector, and $\mathbf{x}$ is the n vector of unknowns. Thus (6.1.1) represents the system

$$\begin{bmatrix} a_{11} & a_{12} & \cdots & a_{1n} \\ a_{21} & a_{22} & \cdots & a_{2n} \\ \cdots & \cdots & \cdots & \cdots \\ a_{m1} & a_{m2} & \cdots & a_{mn} \end{bmatrix} \begin{bmatrix} x_1 \\ x_2 \\ \cdots \\ x_n \end{bmatrix} = \begin{bmatrix} b_1 \\ b_2 \\ \cdots \\ b_m \end{bmatrix} \qquad (6.1.2)$$

which written in full is

$$a_{11}x_1 + a_{12}x_2 + \cdots + a_{1n}x_n = b_1$$
$$a_{21}x_1 + a_{22}x_2 + \cdots + a_{2n}x_n = b_2 \qquad (6.1.3)$$
$$\cdots\cdots\cdots\cdots\cdots\cdots\cdots\cdots\cdots\cdots\cdots$$
$$a_{m1}x_1 + a_{m2}x_2 + \cdots + a_{mn}x_n = b_m$$

(For most of this chapter we will consider square matrices, but initially we consider the general case.)

The overall procedure is made up of two parts: the *forward elimination* phase and the *back substitution* phase. During the first of these we eliminate x_1 from all the equations except the first, then x_2 from all (the modified) equations after the second one, and so on, until we have reduced the matrix to echelon form (or triangular form for a square system). Once this reduction is complete, we can determine whether the system is over- or underdetermined or has a unique solution, and if it has, use the back substitution procedure to find it.

The forward elimination process begins with the subtraction of a_{i1}/a_{11} times the first row (equation) from the ith row (equation) to leave a system of the form

$$\begin{bmatrix} a_{11} & a_{12} & \cdots & a_{1n} \\ 0 & a'_{22} & \cdots & a'_{2n} \\ \cdots & \cdots & \cdots & \cdots \\ 0 & a'_{m2} & \cdots & a'_{mn} \end{bmatrix} \begin{bmatrix} x_1 \\ x_2 \\ \cdots \\ x_n \end{bmatrix} = \begin{bmatrix} b_1 \\ b'_2 \\ \cdots \\ b'_m \end{bmatrix} \qquad (6.1.4)$$

The coefficients in this reduced system are given by

$$a'_{ij} = a_{ij} - m_{i1}a_{1j} \qquad (1 < i, j)$$
$$b'_i = b_i - m_{i1}b_1 \qquad (1 < i) \qquad (6.1.5)$$

where m_{i1} is the multiplier a_{i1}/a_{11} above. At the next stage, multipliers

$$m_{i2} = \frac{a'_{i2}}{a'_{22}}$$

are used to eliminate the entries in the second column below the leading diagonal. The process can be continued in this way until the final echelon form of the augmented matrix $[A|\mathbf{b}]$ is obtained.

If the final echelon form of this augmented matrix [in which the right-hand side is included as the $(n + 1)$th column] has a row consisting of n zeros and a final nonzero entry, then the system has no solutions. Otherwise solutions to the system do exist. There is a unique solution providing that the first n rows do not contain a complete row of zeros. If either there is a complete row of zeros or there are fewer than n rows and a solution exists, then there are infinitely many solutions.

In the case of a square matrix this alternative reduces to the statement that there is a unique solution if and only if the final $a_{nn} \neq 0$. If $a_{nn} = 0$, then there is no solution if $b_n \neq 0$, while there are infinitely many solutions if $b_n = 0$.

This result takes no account of the possibility that at some stage the diagonal entry which is to be used for the next stage of the elimination is zero, in which case the procedure cannot continue. Nor does it pay any heed to the computer arithmetic and the effect of roundoff errors in the calculation. It is thus a pure theoretical result—but nonetheless a useful one in determining the singularity or otherwise of a matrix on the basis of the computation.

In Algorithm 6.1.2 below, we describe simple Gauss elimination for a square system. The forward elimination procedure there also overlooks the above difficulties in the interests of clarity. We begin with an example worked in detail by hand.

Example 6.1.1. Consider the system of equations given by

$$A = \begin{bmatrix} 1 & 2 & 3 & 4 \\ 2 & 2 & 3 & 4 \\ 3 & 3 & 3 & 4 \\ 4 & 4 & 4 & 4 \end{bmatrix} \qquad \mathbf{b} = \begin{bmatrix} 1.234 \\ 2.234 \\ 3.334 \\ 4.444 \end{bmatrix}$$

The process begins with the augmented matrix tableau

$$\begin{array}{cccccc} 1 & 2 & 3 & 4 & : & 1.234 \\ 2 & 2 & 3 & 4 & : & 2.234 \\ 3 & 3 & 3 & 4 & : & 3.334 \\ 4 & 4 & 4 & 4 & : & 4.444 \end{array}$$

in which we wish to eliminate the entries a_{21}, a_{31}, and a_{41} by subtracting from the second, third, and fourth rows the appropriate multiples of the first row. In this case these multipliers are given by

$$m_{21} = 2 \qquad m_{31} = 3 \qquad m_{41} = 4$$

We obtain the new tableau

$$\begin{array}{cccccc} 1 & 2 & 3 & 4 & : & 1.234 \\ 0 & -2 & -3 & -4 & : & -0.234 \\ 0 & -3 & -6 & -8 & : & -0.368 \\ 0 & -4 & -8 & -12 & : & -0.492 \end{array}$$

Next we eliminate the entries in the a_{32} and a_{42} positions by subtracting the multiples

$$m_{32} = 1.5 \quad \text{and} \quad m_{42} = 2$$

of the second row from the third and fourth rows, respectively, to get the new array

$$
\begin{array}{rrrr r}
1 & 2 & 3 & 4 & : \quad 1.234 \\
0 & -2 & -3 & -4 & : \quad -0.234 \\
0 & 0 & -1.5 & -2 & : \quad -0.017 \\
0 & 0 & -2 & -4 & : \quad -0.024
\end{array}
$$

Finally, subtracting $m_{43} = -2/-1.5 = 1.33333$ times this third row from the fourth one yields the triangular tableau

$$
\begin{array}{rrrr r}
1 & 2 & 3 & 4 & : \quad 1.234 \\
0 & -2 & -3 & -4 & : \quad -0.234 \\
0 & 0 & -1.5 & -2 & : \quad -0.017 \\
0 & 0 & 0 & -1.33333 & : \quad -0.00133
\end{array}
$$

The back substitution begins with this last row (equation) which corresponds to

$$-1.33333x_4 = -0.00133$$

from which we obtain $x_4 = 0.001$. Substituting this into the third equation (row) of the final array,

$$-1.5x_3 - 2x_4 = -0.017$$

yields $x_3 = 0.01$. Now the second equation

$$-2x_2 - 3x_3 - 4x_4 = -0.234$$

gives $x_2 = 0.1$, and substituting all these into the (unchanged) first equation gives the last component of the solution $x_1 = 1$.

It is worthwhile to observe that should the determinant of the original matrix be required, it is easily computed as the product of the diagonal entries in the matrix at the end of the forward substitution. Many commercial Gauss elimination routines overwrite the matrix in the elimination phase as is the case in the algorithm below. It is necessary with such a routine to make a separate copy of the matrix if the original is to be used elsewhere in the program.

The back substitution phase of the solution begins with the last of the equations in the triangular system to solve for x_n. This value is then substituted back into the previous equation to obtain x_{n-1}, and so the process continues until all n unknowns have been found.

Algorithm 6.1.2   Simple Gauss elimination

Input n, an $n \times n$ matrix A, and an n vector **b**

Forward elimination
    for j = 1 to n − 1
        for i = j + 1 to n
            $m_{ij} := a_{ij}/a_{jj}$
            $a_{ij} := 0; \ b_i := b_i - m_{ij} * b_j$

$$\text{for } k = j + 1 \text{ to } n$$
$$a_{ik} := a_{ik} - m_{ij} * a_{jk}$$

Back substitution
$$\text{for } i = n \text{ downto } 1$$
$$x_i := b_i$$
$$\text{for } j = i + 1 \text{ to } n$$
$$x_i := x_i - a_{ij} * x_j$$
$$x_i := x_i / a_{ii}$$

Output   Solution vector **x**.

This algorithm yields the same results as those obtained in Example 6.1.1.

Before we proceed to discuss any of the potential computational difficulties which can arise in Gauss elimination—and possible remedies for these—we consider briefly the computational effort required by this algorithm.

Algorithms for the solution of any particular problem can be compared in terms of their computational efficiency by a simple count of the number of different floating-point operations which must be performed to solve the problem. Comparison of such *operation counts* can be used as an indicator of the relative speeds of algorithms. (We will see similar comparisons for both numerical integration and optimization techniques later—except they will be in terms of the appropriate fundamental operations for those tasks, namely, counting function evaluations.)

For the basic Gauss elimination Algorithm 6.1.2, it is easy to count all the arithmetic operations which are needed. In Table 6.1.3 below we give the numbers of the various operations needed for the forward elimination and back substitution phases together with the totals. Since addition and subtraction are essentially equivalent, these two operations are combined into one count.

By way of justification of the entries in this table, we consider just the multiplications in the forward elimination phase. This procedure has three nested For loops with their indices running through the following ranges:

$$j = 1 \text{ to } n - 1$$
$$i = j + 1 \text{ to } n$$
$$k = j + 1 \text{ to } n$$

TABLE 6.1.3
Operation counts for Gauss elimination on an $n \times n$ system

Operation	Forward elimination	Back substitution	Total
+ or −	$\dfrac{n(n^2 - 1)}{3}$	$\dfrac{n(n - 1)}{2}$	$\dfrac{n(n - 1)(2n + 5)}{6}$
*	$\dfrac{n(n^2 - 1)}{3}$	$\dfrac{n(n - 1)}{2}$	$\dfrac{n(n - 1)(2n + 5)}{6}$
/	$\dfrac{n(n - 1)}{2}$	n	$\dfrac{n(n + 1)}{2}$

There is one multiplication in the innermost loop and another in the middle one. For the innermost loop this corresponds to $n - j$ multiplications, and this loop is performed (for any fixed j) $n - j$ times. With j ranging from 1 to $n - 1$, this amounts to

$$\sum_{j=1}^{n-1} (n - j)^2 = 1^2 + 2^2 + \cdots + (n - 1)^2 = \frac{n(n - 1)(2n - 1)}{6} \qquad (6.1.6)$$

For the multiplication in the middle loop we similarly arrive at the count

$$\sum_{j=1}^{n-1} (n - j) = 1 + 2 + \cdots + (n - 1) = \frac{n(n - 1)}{2}$$

from which we obtain the total count for the forward elimination phase:

$$\frac{n(n - 1)(2n - 1)}{6} + \frac{n(n - 1)}{2} = \frac{n(n^2 - 1)}{3}$$

On most computers division is a significantly slower operation than addition or multiplication. In Table 6.1.3 we see that the number of divisions is only $O(n^2)$, while for both addition and multiplication we have counts of $O(n^3)$. (That is, for example, the number of divisions increases like a constant multiple of n^2 as n gets large.) For any n of even moderate size the relative difference between approximately $2n^3/3$ additions and multiplication and about $n^2/2$ divisions far outweighs the difference in the timings of individual operations. We thus think of the overall procedure of Gauss elimination as entailing approximately $2n^3/3$ floating-point operations, or *flops*.

We will see later on (Chap. 13) that for many highly parallel architectures this operation count can be reduced to $O(n)$ parallel floating-point operations. In this instance, it is also the case that the division count is $O(n)$, and so the relative timings of division and other arithmetic operations become a more important consideration.

From Table 6.1.3 we see further that the forward elimination phase of Gauss elimination is much the more computationally intensive. It is this procedure which accounts for the $O(n^3)$ operation counts, while the back substitution procedure uses only $O(n^2)$ flops. It is not uncommon in various applications that several linear systems need to be solved, each with the same coefficient matrix but with different right-hand sides. Since the major contribution to this operation count (6.1.6) comes from the innermost loop which relates to the updating of the matrix elements, it would clearly be desirable to avoid repeating all that work. This is achieved by the matrix factorization techniques to be discussed in Sec. 6.4.

Prior to that we consider making the Gauss elimination procedure more numerically robust by the use of pivoting, and then, in Sec. 6.3, we study the effect of rounding errors on Gauss elimination. The need for a more robust version of the basic algorithm is easily demonstrated by way of example.

Example 6.1.4. Consider the use of our basic Gauss elimination technique for solving the 3×3 system

$$\begin{bmatrix} 7 & 63 & 0 \\ 2 & 18 & 10 \\ 3 & 30 & 0 \end{bmatrix} \begin{bmatrix} x_1 \\ x_2 \\ x_3 \end{bmatrix} = \begin{bmatrix} 13.3 \\ 3.9 \\ 6.0 \end{bmatrix}$$

After the first stage of the elimination using multipliers

$$m_{21} = \tfrac{2}{7} \qquad \text{and} \qquad m_{31} = \tfrac{3}{7}$$

we obtain the tableau

$$\begin{array}{rrrcl} 7 & 63 & 0 & : & 13.3 \\ 0 & 0 & 10 & : & 0.1 \\ 0 & 3 & 0 & : & 0.3 \end{array}$$

at which point the procedure breaks down since the pivotal element for the next stage is itself zero so that the next multiplier would be given by $\tfrac{3}{0}$.

The obvious remedy is to interchange the second and third rows:

$$\begin{array}{rrrcl} 7 & 63 & 0 & : & 13.3 \\ 0 & 3 & 0 & : & 0.3 \\ 0 & 0 & 10 & : & 0.1 \end{array}$$

in which case the elimination, for this particular example, is complete and the solution is obtained as

$$x_3 = 0.01 \qquad x_2 = 0.1 \qquad x_1 = 1$$

It is tempting to believe that all that is necessary is to incorporate a check for a zero pivot and, should this occur, to interchange the offending row with the next one which does not have a zero entry in the current column of interest. While this may be sufficient for the exact arithmetic performed in the above example, it is certainly not adequate for floating-point arithmetic on the computer.

To see this, consider Example 6.1.4 solved using floating-point arithmetic. Because the multipliers $\tfrac{2}{7}$ and $\tfrac{3}{7}$ are not represented exactly as binary floating-point numbers, the zero entry in the a_{22} position above is computed as a small nonzero number which allows the program to complete its elimination phase, producing the tableau

$$\begin{array}{llll} 7.000E + 00 & 6.300E + 01 & 0.000E + 00 & 1.330E + 01 \\ 0.000E + 00 & 2.910E - 11 & 1.000E + 01 & 1.000E - 01 \\ 0.000E + 00 & 0.000E + 00 & -1.031E + 12 & -1.031E + 10 \end{array}$$

which leads to the solutions, rounded to five decimal places,

$$x_3 = 0.01000 \qquad x_2 = 0.10547 \qquad x_1 = 0.95078$$

The effect of this very small pivot has been an error of about 5 percent in each of x_2 and x_1 despite having still computed the correct value for x_3.

Clearly it is necessary to be able to avoid the use of very small pivots which may well be the result of cancellation in the subtraction of nearly equal quantities. Results of such subtractions are subject to large relative errors, and their use as divisors (in obtaining the various multipliers to be used) will inevitably exaggerate those errors. In the next two sections we turn to the question of the robust implementation of Gauss elimination and its error analysis.

EXERCISES 6.1

1. Solve the following system by Gauss elimination by hand.

$$x_1 + x_2 + x_3 + x_4 + x_5 + x_6 = 11.2345$$
$$x_1 - 2x_2 + 3x_3 - 4x_4 + 5x_5 - 6x_6 = 8.4970$$
$$2x_1 + 2x_2 - 2x_4 + x_6 = 21.9405$$
$$-4x_1 + 10x_3 + 5x_5 - 11x_6 = -37.9855$$
$$x_2 - x_3 + 2x_4 - x_5 = 0.8560$$
$$x_1 - 9x_2 - 20x_6 = 0.9900$$

2. Write a program to implement the Gauss elimination Algorithm 6.1.2. Use it to check your answer to Exercise 1.

3. Determine values of a, b for which the following system has no solution, infinitely many solutions, and a unique solution.

$$x_1 + x_2 + x_3 = 3$$
$$x_1 + 2x_2 + 3x_3 = 6$$
$$2x_1 + 3x_2 + ax_3 = b$$

Use Gauss elimination to find the general solution for the case of infinitely many solutions and for one example of a unique solution.

4. Show that the elementary row operations of Gauss elimination leave the solution of the linear system unchanged:

(a) Show that left multiplication of a matrix A by the matrix M with elements given for fixed i, j by $m_{pp} = 1$ for all p, $m_{ji} = -m$, $m_{pq} = 0$, otherwise, has the effect of subtracting m times the ith row of A from the jth row to form the jth row of the product matrix A'.

(b) Show that the matrix M of (a) is nonsingular and therefore that

$$A\mathbf{x} = \mathbf{b} \leftrightarrow MA\mathbf{x} = M\mathbf{b}$$

(c) Deduce that the solution of the triangular system resulting from the forward elimination phase of Gauss elimination is the solution of the original system.

6.2   PARTIAL PIVOTING

In this section we develop an implementation of Gauss elimination which utilizes the pivoting strategy outlined above. The basic approach is to use the largest (in absolute value) entry on or below the diagonal in the column of current interest as the pivotal element for elimination in the rest of that column.

One immediate effect of this will be to force all the multipliers used to be no greater than 1 in absolute value. This will inhibit the growth of errors in the rest of the elimination phase and in the subsequent back substitution.

At stage j of the forward elimination, it is necessary, therefore, to be able to identify the largest entry from $|a_{jj}|$, $|a_{j+1,j}|$, ..., $|a_{nj}|$, where these a_{kj}'s are the entries in the current partially triangularized coefficient matrix. If this maximum occurs in row p, then the pth and jth rows of the augmented matrix $[A:\mathbf{b}]$ are interchanged and the elimination proceeds as usual.

This maximal element and its position in the column can be found by means of the following simple procedure in which col is the current column number and kpivot is the row number for the pivot element:

Procedure 6.2.1

```
        procedure find_pivot(dim,col: integer; A:matrix;
                        var kpivot: integer);
        var i: integer; pivot: real;
        begin
            pivot := abs(A[col,col]); kpivot := col;
            for i := col + 1 to dim do
                if abs(A[i,col]) > pivot then
                    begin pivot := abs(A[i,col]); kpivot := i; end;
        end;
```

At this stage, a further procedure can be incorporated into a program to interchange rows col and kpivot. This is described in Algorithm 6.2.2 below. The version of Gauss elimination with partial pivoting in this algorithm is certainly not the most efficient available.

Algorithm 6.2.2 Gaussian elimination with partial pivoting

Input n, an $n \times n$ matrix A, and an n vector $\mathbf{b}$

Forward elimination

for $j = 1$ to $n - 1$
  find_pivot(n,j,A,kp)
  exchange rows j and kp of $[A:\mathbf{b}]$
  for $i = j + 1$ to n
    $m := a_{ij}/a_{jj}$; $a_{ij} := 0$
    $b_i := b_i - m * b_j$
    for $k = j + 1$ to n
      $a_{ik} := a_{ik} - m * a_{jk}$

Back substitution

for $i = n$ downto 1
  $x_i := b_i$
  for $j = i + 1$ to n
    $x_i := x_i - a_{ij} * x_j$
  $x_i := x_i/a_{ii}$

Output   Solution vector $\mathbf{x}$.

In Procedure 6.2.1 for finding the next pivot the matrix A is declared as a nonvariable parameter. Since such a parameter can be changed locally within the procedure but must be returned unchanged, it is necessary at execution time for a separate copy of the matrix to be put onto the stack for local use. In Turbo Pascal, the default stack size is of the order of 16K bytes, which corresponds to about 4K single-precision floating-point quantities, and since $4K = 4096 = 64^2$, we should experience stack overflow—a fatal error—for matrices of dimension greater than 64×64. For type real, the corresponding maximum dimension is about 50×50. The simple expedient of declaring A to be variable in this procedure—even though its values are not varied—eliminates the need for this extra copy on the stack and so overcomes this problem. Since in this form only those elements of the matrix which are needed for the procedure (these are just the subdiagonal part of one column) will be fetched from memory, this will almost certainly provide increased speed as well.

The other major saving which can be made derives from the fact that it is unnecessary to perform the actual row interchange. All that is needed is that we know at each stage which rows have already been used as pivotal rows, seek the next pivot from among the remaining rows, and perform the elimination only on those ones. All this is easily achieved by storing a *permutation vector* which simply stores as its jth component the row number which is used for elimination in the jth column.

Both these improvements are incorporated into Algorithm 6.2.5 below from which unnecessary output has also been eliminated. Before that we see the effect of our partial pivoting strategy on the examples considered in the previous section.

Example 6.2.3. For the matrix and vector of Example 6.1.2,

$$A = \begin{bmatrix} 1 & 2 & 3 & 4 \\ 2 & 2 & 3 & 4 \\ 3 & 3 & 3 & 4 \\ 4 & 4 & 4 & 4 \end{bmatrix} \qquad b = \begin{bmatrix} 1.234 \\ 2.234 \\ 3.334 \\ 4.444 \end{bmatrix}$$

Using Algorithm 6.2.2, row 4 is used as the pivotal one on each occasion and gives the following triangular system and solution:

$$
\begin{array}{cccccl}
4.0000 & 4.0000 & 4.0000 & 4.0000 & : & 4.4440 \\
0.0000 & 1.0000 & 2.0000 & 3.0000 & : & 0.1230 \\
0.0000 & 0.0000 & 1.0000 & 2.0000 & : & 0.0120 \\
0.0000 & 0.0000 & 0.0000 & 1.0000 & : & 0.0010
\end{array}
$$

and $x_1 = 1.0000$, $x_2 = 0.1000$, $x_3 = 0.0100$, $x_4 = 0.0010$.

As in Example 6.1.2, we can repeat the calculation in full. From the initial array

$$
\begin{array}{ccccll}
1 & 2 & 3 & 4 & : & 1.234 \\
2 & 2 & 3 & 4 & : & 2.234 \\
3 & 3 & 3 & 4 & : & 3.334 \\
4 & 4 & 4 & 4 & : & 4.444
\end{array}
$$

we select the pivot for the first column as $a_{41} = 4$. After the interchange of this with row 1, we have

$$
\begin{array}{ccccll}
4 & 4 & 4 & 4 & : & 4.444 \\
2 & 2 & 3 & 4 & : & 2.234 \\
3 & 3 & 3 & 4 & : & 3.334 \\
1 & 2 & 3 & 4 & : & 1.234
\end{array}
$$

and multipliers $m_{21} = \frac{1}{2}$, $m_{31} = \frac{3}{4}$, and $m_{41} = \frac{1}{4}$. This leads to the reduced system

$$
\begin{array}{ccccll}
4 & 4 & 4 & 4 & : & 4.444 \\
0 & 0 & 1 & 2 & : & 0.012 \\
0 & 0 & 0 & 1 & : & 0.001 \\
0 & 1 & 2 & 3 & : & 1.123
\end{array}
$$

It is now clear that the largest element in the second column in rows 2, 3, and 4 is in row 4. This row is interchanged with row 2, and then the "elimination" is performed to yield

$$
\begin{array}{cccccl}
4 & 4 & 4 & 4 & : & 4.444 \\
0 & 1 & 2 & 3 & : & 0.123 \\
0 & 0 & 0 & 1 & : & 0.001 \\
0 & 0 & 1 & 2 & : & 0.012
\end{array}
$$

Again row 4 provides the pivot for the third column, and we obtain the final tableau

$$
\begin{array}{cccccl}
4 & 4 & 4 & 4 & : & 4.444 \\
0 & 1 & 2 & 3 & : & 0.123 \\
0 & 0 & 1 & 2 & : & 0.012 \\
0 & 0 & 0 & 1 & : & 0.001
\end{array}
$$

which agrees with the computed array.

Of course, in this case the elimination was in fact complete after the first stage, and we could have completed the hand solution immediately. The final interchanges and the (nonexistent) eliminations using zero multipliers are only needed in preparation for the back substitution.

Example 6.2.4. For the system of Example 6.1.4,

$$
\begin{bmatrix} 7 & 63 & 0 \\ 2 & 18 & 10 \\ 3 & 30 & 0 \end{bmatrix}
\begin{bmatrix} x_1 \\ x_2 \\ x_3 \end{bmatrix}
=
\begin{bmatrix} 13.3 \\ 3.9 \\ 6.0 \end{bmatrix}
$$

Algorithm 6.2.3 produces the final triangular array

$$
\begin{array}{ccccl}
7.0000 & 63.0000 & 0.0000 & : & 13.3000 \\
0.0000 & 3.0000 & 0.0000 & : & 0.3000 \\
0.0000 & 0.0000 & 10.0000 & : & 0.1000
\end{array}
$$

which, of course, yields the true solution (1.0000, 0.1000, 0.0100).

It is apparent that the use of partial pivoting has overcome the initial difficulties with Gauss elimination.

Algorithm 6.2.5 incorporates the more efficient permutation vector approach. The procedure for finding the pivots generates the necessary permutation vector which is used in the slightly modified forward elimination and back substitution procedures. There is, of course, no longer any need for interchanging rows.

For the above examples this algorithm produces precisely the same solutions. The absolute value of the determinant can still be obtained easily—it is just (the absolute value of) the product of all the elements $a_{p_i i}$ of the final matrix.

While it is evident that this algorithm has improved the computational efficiency of the Gauss elimination process significantly by reducing the overhead, there are further improvements to be made. We study methods which provide greater efficiency later.

Algorithm 6.2.5 Gauss elimination with partial pivoting using a permutation vector

Input n, an $n \times n$ matrix A, and an n vector $\mathbf{b}$

Initialize permutation vector

$$p_i := i \ (i = 1, 2, \ldots, n)$$

Forward elimination

for j = 1 to n − 1
  find_pivot(n,j,A,p)
  for i = j + 1 to n
    $m := a_{p_i j}/a_{p_j j}; \ a_{p_i j} := 0$
    $b_{p_i} := b_{p_i} - m * b_{p_j}$
      for k = j + 1 to n
        $a_{p_i k} := a_{p_i k} - m * a_{p_j k}$

Back substitution

for i = n downto 1
  $x_i := b_{p_i}$
  for j = i + 1 to n
    $x_i := x_i - a_{p_i j} * x_j$
  $x_i := x_i/a_{p_i i}$

Output   Solution vector $\mathbf{x}$.

The procedure for finding the pivot must be modified to search through the appropriate rows:

```
procedure find_pivot(dim,col:integer; A:matrix;
                var p: intvec);
var i, piv, temp: integer; pivot: real;
begin
    pivot := abs(A[p[col],col]); piv := col; temp := p[col];
    for i := col + 1 to dim do
        if abs(A[p[i],col]) > pivot then
            begin pivot := abs(A[p[i],col]); piv := i; end;
    p[col] := p[piv]; p[piv] := temp;
end;
```

As a final word on the subject of partial pivoting in Gauss elimination, it is worth mentioning the reason for the name *partial pivoting*. There is an alternative scheme known as *complete pivoting* in which at each stage the largest (subdiagonal) element in any of the remaining rows is used as the pivot. This has the effect—if the interchanges are performed—of permuting the unknowns as well as the equations. Complete pivoting can be implemented in a reasonably efficient manner by storing a permutation *matrix* playing the corresponding role to the vector in the above algorithm. Complete pivoting can yield slightly improved numerical stability but is generally not thought to justify the additional overhead.

EXERCISES 6.2

1. Write a program to implement Gauss elimination with partial pivoting. Test it by reproducing the results of the examples in this section and resolving the system of Exercise 1 in Sec. 6.1.

2. Show that the operation of interchanging two rows of a matrix is equivalent to premultiplying the matrix by a matrix of the form S_{ij} whose elements are given by

$$S_{pp} = 1 \quad (p \neq i, j)$$
$$S_{ij} = S_{ji} = 1$$
$$S_{pq} = 0 \quad \text{otherwise}$$

For example, the 4×4 matrix S_{12} is

$$\begin{bmatrix} 0 & 1 & 0 & 0 \\ 1 & 0 & 0 & 0 \\ 0 & 0 & 1 & 0 \\ 0 & 0 & 0 & 1 \end{bmatrix}$$

Show that premultiplication by such a matrix leaves the solution of a system unchanged.

3. Generate a 5×5 system of equations where the use of partial pivoting is beneficial. Show the solutions obtained both with and without partial pivoting, and estimate their relative errors.

4. Write a program to solve a system of linear equations by Gauss elimination with *complete pivoting*. Repeat the computation of Exercise 1 for this method.

6.3 ERROR ANALYSIS

We do not attempt here a complete error analysis for Gauss elimination since that would be a major undertaking beyond the scope of the present work. The interested reader is referred to the extensive and definitive work of J. H. Wilkinson (1965). What we will do here is to obtain bounds on the error in the solution to the system

$$A\mathbf{x} = \mathbf{b} \tag{6.3.1}$$

where A is assumed to be a nonsingular $n \times n$ matrix. These bounds will be seen to be dependent on the condition number of the matrix A. This quantity will also be seen to be important in estimating the effect of rounding (or experimental) errors in the elements of the matrix and/or the right-hand side $\mathbf{b}$.

In what follows we will denote by $\| \cdot \|$ any of the usual vector norms and its corresponding matrix norm. The three most commonly used are $\| \cdot \|_1$, $\| \cdot \|_2$, and $\| \cdot \|_\infty$ (see Appendix B). Corresponding to these matrix norms are the condition numbers $\kappa_1(A)$, $\kappa_2(A)$, $\kappa_\infty(A)$ given by

$$\kappa_i(A) = \|A\|_i \|A^{-1}\|_i \quad (i = 1, 2, \infty) \tag{6.3.2}$$

[Note that the condition number $\kappa_2(A)$ is often referred to as *the* condition number of the matrix A.] For completeness we recall the definitions of these various norms here:

$$\|\mathbf{x}\|_1 = \sum |x_i|$$

$$\|\mathbf{x}\|_2 = \sqrt{\sum x_i^2}$$

$$\|\mathbf{x}\|_\infty = \max|x_i| \qquad (6.3.3)$$

and in each case $\|A\| = \max\{\|A\mathbf{x}\|/\|\mathbf{x}\|\}$ so that, in particular,

$$\|A\|_\infty = \max_i \sum_j |a_{ij}|$$

$$\|A\|_1 = \max_j \sum_i |a_{ij}| \qquad (6.3.4)$$

which are the maximum row and column sums, respectively.

Note that for any vector norm and its associated matrix norm, it follows immediately from the definitions that

$$\|A\mathbf{x}\| \le \|A\|\,\|\mathbf{x}\| \qquad (6.3.5)$$

Also, if $\lambda_{\max}$ is the largest eigenvalue of A (in absolute value) and has associated eigenvector $\mathbf{x}_{\max}$ (see Appendix B), then

$$A\mathbf{x}_{\max} = \lambda_{\max}\mathbf{x}_{\max}$$

from which it follows that $\|A\mathbf{x}_{\max}\| = |\lambda_{\max}|\,\|\mathbf{x}_{\max}\|$ and therefore that

$$\|A\| \ge |\lambda_{\max}| \qquad (6.3.6)$$

Furthermore, since any eigenvalue of A^{-1} is the reciprocal of an eigenvalue of A, it now follows that

$$\|A^{-1}\| \ge \frac{1}{|\lambda_{\min}|} \qquad (6.3.7)$$

where $\lambda_{\min}$ is the smallest eigenvalue of A in absolute value.

We begin our analysis by obtaining simple bounds for the absolute and relative errors in the solution to the system (6.3.1).

Denote the exact solution by $\mathbf{x}^*$, the computed solution by $\mathbf{x}\hat{}$, and define the *residual vector* $\mathbf{r}$ by

$$\mathbf{r} = \mathbf{b} - A\mathbf{x}\hat{} \qquad (6.3.8)$$

For an estimate of the absolute error, we obtain a bound on $\|\mathbf{x}^* - \mathbf{x}\hat{}\|$. Now, since

$$A\mathbf{x}^* - A\mathbf{x}\hat{} = \mathbf{b} - (\mathbf{b} - \mathbf{r}) = \mathbf{r} \qquad (6.3.9)$$

it follows that

$$\|\mathbf{x}^* - \mathbf{x}\hat{}\| = \|A^{-1}\mathbf{r}\| \le \|A^{-1}\|\,\|\mathbf{r}\| \qquad (6.3.10)$$

The potential difficulty with the solution of linear equations is apparent from (6.3.10) since $\|A^{-1}\|$ can be large, which allows the possibility of large errors in the components of the solution even when the residual vector is small.

A measure of the relative error in this solution can be obtained by dividing the above bound by $\|\mathbf{x}^*\|$ to obtain

$$\frac{\|\mathbf{x}^* - \hat{\mathbf{x}}\|}{\|\mathbf{x}^*\|} \leq \|A^{-1}\| \frac{\|\mathbf{r}\|}{\|\mathbf{x}^*\|} \tag{6.3.11}$$

Now, since $\mathbf{b} = A\mathbf{x}^*$, it follows that $\|\mathbf{b}\| \leq \|A\| \|\mathbf{x}^*\|$ and therefore that $1/\|\mathbf{x}^*\| \leq \|A\|/\|\mathbf{b}\|$. Substituting this last inequality into (6.3.11) gives

$$\frac{\|\mathbf{x}^* - \hat{\mathbf{x}}\|}{\|\mathbf{x}^*\|} \leq \|A\| \|A^{-1}\| \frac{\|\mathbf{r}\|}{\|\mathbf{b}\|} = \kappa(A) \frac{\|\mathbf{r}\|}{\|\mathbf{b}\|} \tag{6.3.12}$$

We can obtain *lower* bounds for these errors too: from (6.3.9) we have $\|\mathbf{x}^* - \hat{\mathbf{x}}\| \geq \|\mathbf{r}\|/\|A\|$, and since $\mathbf{x}^* = A^{-1}\mathbf{b}$, it follows that $1/\|\mathbf{x}^*\| \geq 1/(\|A^{-1}\| \|\mathbf{b}\|)$ and hence that

$$\frac{\|\mathbf{x}^* - \hat{\mathbf{x}}\|}{\|\mathbf{x}^*\|} \geq \frac{\|\mathbf{r}\|}{\kappa(A)\|\mathbf{b}\|} \tag{6.3.13}$$

It is immediately apparent that if these bounds are to be useful, we must obtain estimates of the condition numbers $\kappa(A)$. It is well known that

$$\kappa(A) \geq \left| \frac{\lambda_{max}}{\lambda_{min}} \right| \tag{6.3.14}$$

but this bound is not very useful since we do not wish to obtain a complete eigensolution for our matrix simply to estimate the accuracy of the solution to a system of linear equations. (As we will see later, the matrix eigenvalue problem is significantly harder than the solution of a system of linear equations.)

From the bounds (6.3.12) and (6.3.13), we see immediately that the quantity $\|\mathbf{r}\|/\|\mathbf{b}\|$ will give only a very poor estimate of the relative accuracy of the solution if $\kappa(A)$ is large, whereas for small condition numbers—those near unity—it yields a very good measure of the accuracy of our solution. It is clearly desirable for this purpose to have the condition number as small as possible. It is sometimes possible to reduce the condition number of the matrix by scaling the equations so as to make the elements of A of comparable magnitudes. However, there is no universally accepted approach to this scaling procedure, and we do not discuss it further here.

We address the estimation of condition numbers of matrices shortly. First we investigate the effect of rounding, or other input, errors in the matrix A or the right-hand-side vector $\mathbf{b}$ on the accuracy of the solution.

For the case of errors in $\mathbf{b}$, we suppose that the rounded values are given by $\mathbf{b} + \delta\mathbf{b}$ and estimate the relative error resulting from computing with this right-hand side. So now we let $\hat{\mathbf{x}}$ denote the solution of the system $A\mathbf{x} = \mathbf{b} + \delta\mathbf{b}$, while $\mathbf{x}^*$ denotes the true solution of the "correct" system $A\mathbf{x} = \mathbf{b}$. It follows that

$$A(\hat{\mathbf{x}} - \mathbf{x}^*) = \delta\mathbf{b}$$

Following just the same reasoning as above with $\delta \mathbf{b}$ replacing $\mathbf{r}$, we find that

$$\frac{\|\delta \mathbf{b}\|}{\kappa(A)\|\mathbf{b}\|} \leq \frac{\|\mathbf{x}^* - \hat{\mathbf{x}}\|}{\|\mathbf{x}^*\|} \leq \kappa(A)\frac{\|\delta \mathbf{b}\|}{\|\mathbf{b}\|} \qquad (6.3.15)$$

so that the relative error in the solution is bounded above by the condition number of A times the relative error in $\mathbf{b}$.

In the case where the matrix itself has rounding errors, we in fact solve the system

$$(A + \delta A)\mathbf{x} = \mathbf{b} \qquad (6.3.16)$$

and denoting, now, the exact solution of this system by $\hat{\mathbf{x}}$ we get $A(\mathbf{x}^* - \hat{\mathbf{x}}) = \delta A \hat{\mathbf{x}}$, from which we can deduce that

$$\|\mathbf{x}^* - \hat{\mathbf{x}}\| \leq \|A^{-1}\| \|\delta A\| \|\hat{\mathbf{x}}\|$$

Dividing this last inequality by $\|\hat{\mathbf{x}}\|$ and observing that $\|A^{-1}\| = \kappa(A)/\|A\|$, we arrive at the bound

$$\frac{\|\mathbf{x}^* - \hat{\mathbf{x}}\|}{\|\hat{\mathbf{x}}\|} \leq \kappa(A)\frac{\|\delta A\|}{\|A\|} \qquad (6.3.17)$$

so that the relative error in the solution is bounded by the condition number of A times the relative error in A.

The bounds (6.3.15) and (6.3.17) take no account of further rounding errors incurred in the computation, but we have already seen that these are dependent on the condition number too.

Unfortunately, the only estimate we have for the condition number apart from its definition is given by (6.3.14). This necessitates the computation of either the inverse of A or both the largest and smallest of its eigenvalues—either of which is a more difficult task than the original problem of solving $A\mathbf{x} = \mathbf{b}$. Clearly, we do not want to expend more computational effort on estimating the error in our solution than in finding that solution in the first place.

We can often obtain reasonable estimates fairly easily by using the following result. We will see in the examples that follow that it is sometimes easy to find a singular matrix which is "near" to A and which will provide us with good estimates.

Theorem 6.3.1. Let A be a nonsingular matrix. For any matrix norm associated with a vector norm, and any singular matrix B,

$$\|A^{-1}\| \geq \frac{1}{\|A - B\|} \qquad (6.3.18)$$

and therefore

$$\kappa(A) \geq \frac{\|A\|}{\|A - B\|} \qquad (6.3.19)$$

Proof. For a singular matrix B, there exists a nonzero vector $\mathbf{x}$ such that $B\mathbf{x} = \mathbf{0}$ and therefore

$$\|A\mathbf{x}\| = \|A\mathbf{x} - B\mathbf{x}\| \leq \|A - B\| \|\mathbf{x}\| \qquad (6.3.20)$$

Also $\mathbf{x} = A^{-1}A\mathbf{x}$, and so $\|\mathbf{x}\| \leq \|A^{-1}\| \|A\mathbf{x}\|$. Substituting this into (6.3.20) and dividing by $\|A\mathbf{x}\|$ (which is nonzero since A is nonsingular), we conclude that

$$1 \leq \|A - B\| \|A^{-1}\|$$

from which (6.3.18) follows immediately.

The bound (6.3.19) now follows since $\kappa(A) = \|A\| \|A^{-1}\|$.

Remark. The bound (6.3.18) is a sharp one in the sense that there is a closest singular matrix to A. That there is a closest such matrix follows from the fact that the limit of a sequence of singular matrices is itself singular, and so

$$\sup\left\{ \frac{1}{\|A - B\|} : B \text{ singular}\right\} = \frac{1}{\inf\{\|A - B\|: B \text{ singular}\}}$$

is achieved (and so this supremum is, in fact, a *maximum*). It is less straightforward to establish that this maximum is $\|A^{-1}\|$. The proof of this "nearest singular matrix theorem" is omitted. Explicitly the theorem states [see, for example, Conte and de Boor (1980)] that:

For a nonsingular matrix A, there exists a nearest singular matrix B_0 such that

$$\|A - B_0\| = \min\{\|A - B\|: B \text{ is singular}\}$$

and $\|A^{-1}\| = 1/\|A - B_0\|$.

In the next example we see how to obtain estimates of the condition number of a matrix, and we use these to estimate bounds on the relative errors in the solutions of linear systems.

Example 6.3.2. Consider the system

$$\begin{bmatrix} 1 & 2 & 3 & 5.00001 \\ 2 & 3 & 4 & 7.00002 \\ 4 & 4 & 5 & 9.00003 \\ 5 & 5 & 5 & 10.00004 \end{bmatrix} \begin{bmatrix} u \\ v \\ w \\ x \end{bmatrix} = \begin{bmatrix} 11.00001 \\ 16.00002 \\ 22.00003 \\ 25.00004 \end{bmatrix}$$

which has the exact solution

$$u = v = w = x = 1$$

Using the Gauss elimination with partial pivoting procedures of Algorithm 6.2.5 we obtain the computed solution for type real:

$$u = 1.000\,000\,000 \qquad v = w = 1.000\,005\,457 \qquad x = 0.999\,994\,543$$

which has the following components for the residual vector $\mathbf{r} = \mathbf{b} - A\hat{\mathbf{x}}$:

$$-3.5 \times 10^{-18}, \, 0, \, 1.8 \times 10^{-12}, \, 2.9 \times 10^{-11}$$

We see immediately that the very small residuals do not reflect the much larger errors in the solution.

In this case, using the ∞-norm, we see that $\|A\|_\infty = 25.00004 = \|\mathbf{b}\|_\infty$, while $\|\mathbf{r}\| \approx 3 \times 10^{-11}$. To estimate the condition number $\kappa(A)$ we observe that the matrix B given by

$$\begin{bmatrix} 1 & 2 & 3 & 5 \\ 2 & 3 & 4 & 7 \\ 4 & 4 & 5 & 9 \\ 5 & 5 & 5 & 10 \end{bmatrix}$$

is a singular matrix close to the original matrix A. (To see this, note simply that col. 4 is the sum of cols. 2 and 3.) Clearly $\|B - A\|_\infty = 0.00004$, from which it follows that $\|A^{-1}\|_\infty \geq 1/0.00004 = 2.5 \times 10^4$ and therefore that $\kappa_\infty(A) \geq 6.25 \times 10^5$. Using this value in (6.3.12) suggests the following bound for the relative error in the solution:

$$\frac{6.25 \times 10^5 \times 3 \times 10^{-11}}{25} \approx 7.5 \times 10^{-7}$$

However, comparing the computed solution directly with the given true solution, we see that

$$\frac{\|\mathbf{x}^* - \hat{\mathbf{x}}\|_\infty}{\|\mathbf{x}^*\|_\infty} \approx \frac{4.5 \times 10^{-6}}{1}$$

which is about 6 times larger than the above estimate.

This indicates both that the condition number estimate obtained above is genuinely too small (as the theorem says) and that the actual error can get close to the original bound (6.3.12).

If the fourth column of the matrix is rounded as in matrix B above, it is apparent that the last row is 5 times the difference between the second and first rows. That is also true of the matrix

$$\begin{bmatrix} 1 & 2 & 3 & 5.00001 \\ 2 & 3 & 4 & 7.00002 \\ 4 & 4 & 5 & 9.00003 \\ 5 & 5 & 5 & 10.00005 \end{bmatrix}$$

which differs from the original matrix A by just 0.00001 in the fourth-row–fourth-column element. This new matrix, B_1 say, is, of course, singular and so gives us a new lower bound for $\|A^{-1}\|_\infty$:

$$\|A^{-1}\|_\infty \geq \frac{1}{\|A - B_1\|_\infty} = 10^5$$

from which we obtain $\kappa_\infty(A) \geq 2.5 \times 10^6$. The corresponding relative error estimate for the computed solution is now 3×10^{-6}, a much better estimate of the true error.

Using the 1-norm, $\|\cdot\|_1$, we obtain broadly similar results: the true error is now approximately 3.3×10^{-6}, while the estimate obtained using B as the nearby singular matrix is 1.25×10^{-7}, and the improved estimate using B_1 is 1.25×10^{-6}. (See Exercise 1.)

We next use the bound (6.3.15) to estimate the effect of rounding errors in the right-hand side $\mathbf{b}$. If the data is rounded to the vector $(11, 16, 22, 25)$, then the system has the exact solution $u = 1$, $v = w = 2$, $x = 0$, while the computed solution becomes $u = 1.000\,000\,000$, $v = w = 1.999\,996\,362$, $x = 0.000\,003\,638$. It is immediately obvious that this small change in the data has had a significant effect on the solution; indeed, the relative change in the solution in the ∞-norm is

$$\frac{\|(1, 2, 2, 0) - (1, 1, 1, 1)\|_{\infty}}{\|(1, 1, 1, 1)\|_{\infty}} = 1$$

The bound on this change given by (6.3.15) using B_1 to estimate the condition number of A is $2.5 \times 10^6 \times 4 \times 10^{-5}/25 = 4$, which very clearly indicates that we may expect severe changes in the solution as a result of small changes in the data.

The effect of a similar rounding on the matrix is even more extreme since it becomes a singular matrix. However, the rounding errors in the Gauss elimination with partial pivoting process prevent any division by zero, and so a "solution" is computed: it is approximately,

$$u = 1.000005E + 00 \qquad v = w = 5.497560E + 05 \qquad x = -5.497540E + 05$$

which yields the small residual vector

$$(-3.5E - 18,\ 2.1E - 06,\ 1.8E - 06,\ 1.9E - 06)$$

In many computing environments, the only real data types readily available are IEEE single and double precision (or minor variations thereof), and it is usually the case that execution times for double precision are significantly greater than those for single. Consequently, it is common to use single precision as the default arithmetic. We consider now the effect of solving the same system as above using type *single* throughout.

The solution obtained for the original system is

$$u = 1.000\,002\,5 \qquad v = 1.333\,333\,1 \qquad w = 1.333\,333\,5 \qquad x = 0.666\,666\,69$$

to eight significant figures. The residuals vector is

$$(0,\ 1.9 \times 10^{-6},\ 1.9 \times 10^{-6},\ 1.9 \times 10^{-6})$$

Again we see that a solution very different from the true one yields very small residuals.

The relative error in the solution (in the ∞-norm) is approximately 0.33.

With our best estimate of $\kappa_{\infty}(A)$ of 2.5×10^6 and the above residual vector, we obtain the estimate $2.5 \times 10^6 \times 1.9 \times 10^{-6}/25 \approx 0.19$ for this relative error.

In this instance we can obtain a fuller explanation of this error. Using the fact that, for type single, Turbo Pascal performs its computation to greater accuracy and then rounds the results, we can estimate the effects of the rounding errors in the original matrix and data. (That such rounding errors are significant is a consequence of the fact that the quantities in the fourth column of the matrix and in the vector **b** differ from integers by amounts close to the machine unit of single precision. The program was modified to print out the matrix A and the vector **b** immediately after input. The rounded values of this fourth column and right-hand side were printed to a couple of extra places as

5.000 010 0	7.000 020 0	9.000 029 6	10.000 040 1
11.000 009 5	16.000 019 1	22.000 030 5	25.000 040 1

respectively.)

We can use (6.3.15) and (6.3.17) to estimate the effects of these rounding errors. A reasonable estimate of the overall error in the solution can be obtained by combining these. (This ignores any interaction effects among these errors but should give a decent first-order approximation of the error.)

We have already obtained the contribution 0.19 relative to $\mathbf{x}^*$ from consideration of the residuals above.

For the effect of the rounding on the right-hand side, we have

$$\|\delta \mathbf{b}\|_\infty \approx 5 \times 10^{-7}$$

from which, using (6.3.15), we obtain the relative error contribution of approximately 0.05.

The effect of the rounding error in A is measured by (6.3.17) relative to the true solution of this system which we assume is well-approximated by $\mathbf{x}\hat{}$. Here $\|\delta A\|_\infty \approx 4 \times 10^{-7}$ and $\|\mathbf{x}\hat{}\|_\infty \approx 1.33$, so (6.3.17) gives the estimate 0.03. To obtain the corresponding estimate relative to $\mathbf{x}^*$ this must be multiplied by $\|\mathbf{x}\hat{}\|/\|\mathbf{x}^*\| \approx 1.33$, giving the estimate 0.04.

Summing these three contributions we obtain the overall error estimate 0.28, which is very close to the true relative error calculated earlier.

It is apparent that for this ill-conditioned system, computing in single precision has resulted in substantial errors. It is always advisable, therefore, when working on an IEEE system with possible ill-conditioned matrices to use double-precision arithmetic.

EXERCISES 6.3

1. Repeat the error analysis of the residuals of Example 6.3.2 for the 1-norm. (Recall that $\|\mathbf{x}\|_1 = \Sigma\,|x_i|$ and $\|A\|_1 = \max_j \Sigma_i\,|a_{ij}|$.)

2. Obtain estimates for the errors in your computed solutions to the system in Exercises 1 of Secs. 6.1 and 6.2. Repeat the analysis for computation using other real data types.

3. Develop an example of a 5×5 system of equations with a nearly singular matrix of coefficients. Compute its solution with small perturbations of both the matrix and the right-hand side. Compare the observed errors with those predicted by the error analysis.

6.4 MATRIX FACTORIZATION METHODS

In this section, we consider techniques for improving the efficiency of our Gauss elimination approach to the solution of linear systems. The problem with the algorithms developed thus far is that if the same coefficient matrix was to be used for several different systems of equations, the whole operation would need to be repeated. This is a situation which is quite common in, for example, the fitting of linear least squares approximations to data where the same abscissas may be used for several different sets of data.

The difficulty arises from the fact that we have kept no record of the elimination process. This can be overcome very simply by modifying Gauss elimination to generate a factorization of the original coefficient matrix A as the product of lower and upper triangular matrices. That is, we find L and U such that

L is lower triangular   (so $l_{ij} = 0$ for $j > i$)

U is upper triangular   (so $u_{ij} = 0$ for $i > j$)

and

$$A = LU \tag{6.4.1}$$

It will then follow that any system of equations with coefficient matrix A,

$$A\mathbf{x} = \mathbf{b}$$

say, can be solved in two stages:

1. Find $\mathbf{z}$ such that

$$L\mathbf{z} = \mathbf{b} \tag{6.4.2}$$

2. Find $\mathbf{x}$ such that

$$U\mathbf{x} = \mathbf{z} \tag{6.4.3}$$

since then

$$A\mathbf{x} = L(U\mathbf{x}) = L\mathbf{z} = \mathbf{b}$$

Now Eq. (6.4.2) can be solved by a forward substitution process along the same lines as the back substitution used earlier. This same back substitution is then used for the solution of (6.4.3). The problem thus reduces to the computation of the factors L and U. This is achieved by modifying our existing algorithm.

In the simplest case, the multipliers of the Gauss elimination approach turn out to be precisely the required elements of the lower triangular factor L, while the upper triangular factor consists simply of the triangular system which results from the forward elimination. It follows that the storage of these factors may be achieved by simply overwriting the original matrix. It is this simple case—without pivoting—which we discuss in some detail.

Consider the first stage of the forward elimination phase of the standard Gauss elimination process described by Eqs. (6.1.4) and (6.1.5). (See also Exercise 1 in Sec. 6.1.) Denoting the reduced matrix in (6.1.4) by A', this first stage is summarized by the matrix equation

$$A = M_1 A' \tag{6.4.4}$$

where

$$M_1 = \begin{bmatrix} 1 & 0 & 0 & \cdots & 0 & 0 \\ m_{21} & 1 & 0 & \cdots & 0 & 0 \\ m_{31} & 0 & 1 & 0 & \cdots & 0 \\ \vdots & & & \ddots & & \vdots \\ m_{n-1,1} & 0 & \cdots & 0 & 1 & 0 \\ m_{n1} & 0 & 0 & \cdots & 0 & 1 \end{bmatrix} \tag{6.4.5}$$

In just the same way the next step of the forward elimination produces a matrix A'' from A' satisfying

$$A' = M_2 A''$$

where the matrix M_2 has the multipliers m_{i2} in the subdiagonal entries of its second column while all other elements agree with those of the identity matrix. It follows that

$$A = M_1 M_2 A''$$

and the product $M_1 M_2$ is given by

$$M_1 M_2 = \begin{bmatrix} 1 & 0 & 0 & \cdots & 0 & 0 \\ m_{21} & 1 & 0 & \cdots & 0 & 0 \\ m_{31} & m_{32} & 1 & 0 & \cdots & 0 \\ \vdots & \vdots & & \ddots & & \vdots \\ m_{n-1,1} & m_{n-1,2} & 0 & \cdots & 1 & 0 \\ m_{n1} & m_{n2} & 0 & \cdots & 0 & 1 \end{bmatrix} \qquad (6.4.6)$$

This process can be continued in the obvious manner throughout the forward elimination. Denoting the final reduced matrix (which is upper triangular) by U, it thus follows that

$$A = LU$$

where $L = M_1 M_2 \cdots M_{n-1}$ is given by

$$L = \begin{bmatrix} 1 & 0 & 0 & \cdots & & 0 & 0 \\ m_{21} & 1 & 0 & \cdots & & 0 & 0 \\ m_{31} & m_{32} & 1 & 0 & & \cdots & 0 \\ \vdots & \vdots & & \ddots & \ddots & & \vdots \\ m_{n-1,1} & m_{n-1,2} & \cdots & & m_{n-1,n-2} & 1 & 0 \\ m_{n1} & m_{n2} & \cdots & & m_{nn-2} & m_{nn-1} & 1 \end{bmatrix} \qquad (6.4.7)$$

This particular factorization of the matrix is known as the *Doolittle reduction*. In this case we have a *unit* lower triangular factor L. It is this fact which allows the two factors to be stored in the same location as the original matrix since it is unnecessary to store the diagonal of L explicitly. Implementing the Doolittle reduction, and the solution of a system of linear equations using it, thus entails little more than storing the multipliers m_{ij} as a_{ij} $(i > j)$ at each stage rather than replacing these elements by 0. We will discuss the implementation of these ideas shortly.

It is apparent from Eq. (6.4.1) that the triangular factorization of A is not unique. If D is any nonsingular diagonal matrix (that is, $d_{ij} = 0$ for $i \neq j$, and $d_{ii} \neq 0$) and $A = LU$, then

$$A = (DL)(D^{-1}U) \qquad (6.4.8)$$

where DL is lower triangular and $D^{-1}U$ is upper triangular. (Note that D^{-1} is the diagonal matrix whose diagonal entries are simply the reciprocals of those of D.) Thus corresponding to any such D there is another LU factorization of A. The attraction of the Doolittle form is that the diagonal entries of the lower factor need not be stored.

A similar statement applies to the *Crout reduction* in which the upper triangular factor has a unit diagonal so that its diagonal need not be stored. The Crout reduction can be obtained from the Doolittle one by dividing all superdiagonal elements and multiplying all subdiagonal ones by the diagonal entry on the corresponding row. (It can, of course, be computed directly during the factorization process—as could the factorization corresponding to any particular choice for the diagonal of one of the factors.)

We mention at this point that it is possible to write down explicit formulas for the elements of the Doolittle and Crout factors, but this would add little to the understanding of the algorithm. Nor would it improve the implementation which is our immediate interest.

We will concentrate on the Doolittle factorization.

The triangularization and forward substitution phases are now achieved by the following simple procedures.

Procedure 6.4.1
```
        procedure triangularization(dim: integer; var A: matrix);
        var i,j,k: integer;
        begin
            for j := 1 to dim − 1 do
                for i := j + 1 to dim do begin
                    A[i, j] := A[i, j]/A[j, j];
                    for k := j + 1 to dim do
                        A[i, k] := A[i, k] − A[i, j] * A[j, k];
            end;
        end;
```

Procedure 6.4.2
```
        procedure forward_solve(dim: integer; A: matrix; b: vector;
                    var z: vector);
        var i,j: integer;
        begin
            for i := 1 to dim do begin
                z[i] := b[i];
                for j := 1 to i − 1 do z[i] := z[i] − A[i,j] * z[j];
            end;
        end;
```

The back substitution procedure requires no change from that used in the original Gauss elimination Algorithm 6.1.2. The work of the solution is now achieved with the three procedure calls:
```
        triangularization(n, A);
        forward_solve(n,A,b,z);
        back_substitution(n, A, z, x);
```

Example 6.4.3. Applying this Doolittle reduction solution process to the system of equations used in Example 6.1.1, the stored matrix A at the end of the triangularization procedure consists of (the floating-point form of) the array

$$
\begin{array}{cccc}
1 & 2 & 3 & 4 \\
2 & -2 & -3 & -4 \\
3 & \frac{3}{2} & -\frac{3}{2} & -2 \\
4 & 2 & \frac{4}{3} & -\frac{4}{3}
\end{array}
$$

which represents the lower and upper triangular factors

$$
L = \begin{bmatrix}
1 & 0 & 0 & 0 \\
2 & 1 & 0 & 0 \\
3 & \frac{3}{2} & 1 & 0 \\
4 & 2 & \frac{4}{3} & 1
\end{bmatrix}
\qquad
U = \begin{bmatrix}
1 & 2 & 3 & 4 \\
0 & -2 & -3 & -4 \\
0 & 0 & -\frac{3}{2} & -2 \\
0 & 0 & 0 & -\frac{4}{3}
\end{bmatrix}
$$

It is easy to check that the product LU is indeed the original coefficient matrix

$$
A = \begin{bmatrix}
1 & 2 & 3 & 4 \\
2 & 2 & 3 & 4 \\
3 & 3 & 3 & 4 \\
4 & 4 & 4 & 4
\end{bmatrix}
$$

For the right-hand-side vector $(1.234, 2.234, 3.334, 4.444)$ used earlier, we first must solve the system $L\mathbf{z} = \mathbf{b}$ by forward substitution to obtain

$$
z_1 = 1.234 \qquad z_2 = -0.234 \qquad z_3 = -0.017 \qquad z_4 = -0.001\,333
$$

and then the system $U\mathbf{x} = \mathbf{z}$ is solved by back substitution, yielding the solution $x_4 = 0.001$, $x_3 = 0.01$, $x_2 = 0.1$, $x_1 = 1$.

The system in the above example is, of course, well-conditioned and is therefore amenable to reliable solution without the need for a pivoting procedure. However, we have already seen that this can be very far from the case in general. How, then, do we incorporate pivoting into the Doolittle or Crout reduction schemes?

The answer is fortunately straightforward. As in Algorithm 6.2.5, we can simply store a permutation vector and then perform the factorization of the correspondingly permuted matrix. The procedure find_pivot of that algorithm is thus incorporated into the factorization and then into both the forward solve and back substitution phases of the solution. This is all incorporated into the following algorithm for the Doolittle factorization of the appropriately permuted matrix.

Algorithm 6.4.4 Solution of linear equations by Doolittle factorization with partial pivoting

> *Input* n, an $n \times n$ matrix A, and an n vector $\mathbf{b}$
>
> *Initialize permutation vector*
> $$p_i := i \ (i = 1, 2, \ldots, n)$$
>
> *Factorization*
> for $j = 1$ to $n - 1$
>   find_pivot(n,j,A,p)

$$\text{for } i = j + 1 \text{ to } n$$
$$a_{p_i j} := a_{p_i j} / a_{p_j j}$$
$$\text{for } k = j + 1 \text{ to } n$$
$$a_{p_i k} := a_{p_i k} - a_{p_i j} * a_{p_j k}$$

Forward elimination
$$\text{for } i = 1 \text{ to } n$$
$$z_i := b_{p_i}$$
$$\text{for } j = 1 \text{ to } i - 1$$
$$z_i := z_i - a_{p_i j} * z_j$$

Back substitution
$$\text{for } i = n \text{ downto } 1$$
$$x_i := z_i$$
$$\text{for } j = i + 1 \text{ to } n$$
$$x_i := x_i - a_{p_i j} * x_j$$
$$x_i := x_i / a_{p_i i}$$

Output   Solution vector **x**.

In Table 6.4.5 we show the results of attempting the solution of the systems with the Hilbert coefficient matrix of varying dimension, that is, $a_{ij} = 1/(i + j - 1)$, with the right-hand side chosen so that the exact solution should be $x_i = 1$ for every i.

The results in Table 6.4.5(a) were obtained using type *real* throughout, while those in Table 6.4.5(b) resulted from double precision, or type *double*. Since the exact solution is of the order of unity, type real should represent about 10-decimal-place accuracy; it is reasonable, therefore, to expect solutions which are accurate to at least the five decimal places shown. Clearly, that hope is not realized for even moderate values of *n*. For type double, the corresponding accuracy is of the order of 14 decimal places. We see that for $n = 10$, even this preserves only three- or four-decimal-place accuracy in the computed solution.

We commented earlier—see Eq. (6.4.8)—that the *LU* factorization of a matrix is not unique and that, indeed, we can choose any nonzero diagonal matrix and modify the factorization accordingly.

In the case of a symmetric matrix, it follows that we can choose our factorization in such a way that the upper factor is just the transpose of the lower one. That is, we can factorize A as

$$A = LL^T \tag{6.4.9}$$

This is known as the *Cholesky* factorization of A. An alternative form of this factorization is obtained by setting

$$A = LDL^T \tag{6.4.10}$$

where D is a diagonal matrix and L is *unit* lower triangular. This is also often referred to as the Cholesky factorization of A.

TABLE 6.4.5
Solutions for the Hilbert matrix for increasing dimension

Exact solution: $x_i = 1$

				(a) Type real arithmetic				
n	3	4	5	6	7	8	9	10
x_1	1.00000	1.00000	1.00000	1.00000	1.00000	1.00000	1.00000	1.00001
x_2	1.00000	1.00000	1.00000	1.00000	1.00000	1.00001	1.00003	0.99919
x_3	1.00000	1.00000	1.00000	1.00000	1.00000	0.99986	0.99946	1.01727
x_4		1.00000	1.00000	1.00000	1.00002	1.00075	1.00376	0.84251
x_5			1.00000	1.00000	0.99997	0.99799	0.98659	1.75316
x_6				1.00000	1.00003	1.00283	1.02677	−1.07508
x_7					0.99999	0.99799	0.96985	4.41114
x_8						1.00057	1.01790	−2.30200
x_9							0.99564	2.73607
x_{10}								0.61771

		(b) Type double arithmetic			
n	6	7	8	9	10
x_1	1.000 000 00	1.000 000 00	1.000 000 00	1.000 000 00	1.000 000 00
x_2	1.000 000 00	1.000 000 00	1.000 000 00	1.000 000 02	1.000 000 09
x_3	1.000 000 00	1.000 000 00	0.999 999 97	0.999 999 67	0.999 998 11
x_4	1.000 000 00	1.000 000 01	1.000 000 18	1.000 002 34	1.000 017 14
x_5	1.000 000 00	0.999 999 99	0.999 999 51	0.999 991 44	0.999 918 51
x_6	1.000 000 00	1.000 000 01	1.000 000 69	1.000 017 45	1.000 223 41
x_7		1.000 000 00	0.999 999 51	0.999 979 97	0.999 634 28
x_8			1.000 000 14	1.000 012 09	1.000 352 75
x_9				0.999 997 01	0.999 815 12
x_{10}					1.000 040 60

To obtain these symmetric factorizations of A, partial pivoting must not be used since any row interchanges will destroy the symmetry of the original coefficient matrix A. The simplest way to describe the Cholesky factorization is in terms of the form (6.4.10) as a modification of the Doolittle reduction. The latter produces factors L and U which are, respectively, unit lower and upper triangular. U can be further factorized as

$$U = DU_1$$

where the diagonal matrix D consists of the diagonal entries of U and U_1 is obtained from U by dividing each row by its diagonal entry. U_1 is now a unit upper triangular matrix. It remains to show that $U_1 = L^T$. Now $A = LDU_1$ and $A^T = A$; it follows that

$$A^T = U_1^T D^T L^T = LDU_1$$

and, of course, $D^T = D$, U_1^T is unit lower triangular, and L^T is unit upper triangular. Hence we may deduce (see Exercise 6) that $L = U_1^T$ and $U_1 = L^T$ or $A = LDL^T$ as required for (6.4.10).

To obtain the LL^{T} form of the factorization, we scale the upper triangular factor by dividing by the square roots of its diagonal entries and multiply the columns of L by these same factors.

In either case this scaling is performed as the factorization proceeds rather than as a separate modification of the factors.

Note too that the Cholesky factorization allows the possibility of economizing on storage requirements. A symmetric $n \times n$ matrix requires storage of only $n(n + 1)/2$ entries; either of these symmetric factorizations can be accommodated in the same amount of memory.

Implementation of the Cholesky factorization is left as an exercise for the reader (see Exercise 7).

EXERCISES 6.4

1. Solve the system of Example 6.4.3 using LU factorization with partial pivoting. Verify that the factors do indeed multiply to give the original matrix.

2. Write programs for the Doolittle factorization of a matrix both with and without partial pivoting. Test them on the example above and on the system of Exercises 1 of Secs. 6.1 and 6.2.

3. (a) Verify that the product of the matrices M_1 and M_2 is indeed given by (6.4.6).
   (b) Show that the inverse of M_1 is the matrix with entries $-m_{i1}$ in the subdiagonal part of the first column. (A similar result clearly holds for M_2 and any other such matrix.)

4. (a) Show that the product of lower triangular matrices is lower triangular.
   (b) Show that the inverse of a lower triangular matrix is also lower triangular. [*Hint*: Factorize the matrix as DM, where D is diagonal and M has unit diagonal entries, and use the results of Exercise 3 (b) and part (a).]

5. Write a program to implement the Crout reduction solution of linear equations. Test it on the examples of Exercise 2.

6. Show that if $LU = L_1 U_1$, where L, L_1 are unit lower triangular and U, U_1 are unit upper triangular, then $L = L_1$ and $U = U_1$. Hence complete the derivation of the Cholesky factorization (6.4.10).

7. Write a program to implement the Cholesky factorization of a symmetric matrix. Check your program by solving the system of Example 6.4.3.

8. Use your Cholesky factorization program to solve the Hilbert matrix systems, and compare your results with those of Table 6.4.5. Explain any differences in the results.

6.5 ITERATIVE REFINEMENT

In this section, we take advantage of the factorization methods of the last section to improve our solutions to linear systems of equations in an iterative manner. To justify this as a practical approach it is necessary to consider briefly the operation counts for the factorization methods and compare these with those of Gauss elimination (see Table 6.1.3).

It is sufficient for our purposes to restrict our attention to the Doolittle reduction. The factorization phase of the solution requires

$$\frac{n(n - 1)}{2} \quad \text{divisions}$$

$$\frac{n(n - 1)(2n - 1)}{6} \quad \text{additions or subtractions}$$

$$\frac{n(n-1)(2n-1)}{6} \qquad \text{multiplications}$$

Thus this stage uses approximately $2n^3/3$ floating-point operations. Similarly we find that the forward and back substitution are each of the order of n^2 floating-point operations.

We have already seen that the Gauss elimination process entails $O(2n^3/3)$ floating-point operations. It follows that if we are to solve two systems of equations, each with the same coefficient matrix, then using Gauss elimination twice requires about $4n^3/3$ floating-point operations, while using the Doolittle factorization requires only about $2n^3/3$ together with about $4n^2$ for the two forward and back substitutions. Since $2n^3/3 > 4n^2$ for $n > 6$, it follows that, even for systems of moderate size, the Doolittle factorization will prove more efficient.

How does this affect our objective of improving the accuracy of the computed solutions? Recall the notation of Sec. 6.3 in which $\mathbf{x}^*$ denotes the true solution, $\mathbf{x}\hat{}$ the computed solution of the system $A\mathbf{x} = \mathbf{b}$, and $\mathbf{r}$ denotes the residual vector

$$\mathbf{r} = \mathbf{b} - A\mathbf{x}\hat{} \tag{6.5.1}$$

Let $\delta\mathbf{x}$ denote the solution of the system $A\mathbf{x} = \mathbf{r}$. It follows that

$$\mathbf{x}\hat{} + \delta\mathbf{x} = \mathbf{x}^* \tag{6.5.2}$$

since

$$A(\mathbf{x}\hat{} + \delta\mathbf{x}) = A\mathbf{x}\hat{} + A\delta\mathbf{x} = (\mathbf{b} - \mathbf{r}) + \mathbf{r} = \mathbf{b}$$

This provides us with the basis of the method of *iterative refinement* which consists of the following steps:

1. Obtain the *LU* factorization of the matrix A.
2. Compute the solution $\mathbf{x}\hat{}$ by forward and back substitution.
3. Find the residual vector $\mathbf{r}$.
4. Compute the correction $\delta\mathbf{x}$ by forward and back substitution.
5. Set $\mathbf{x}\hat{} := \mathbf{x}\hat{} + \delta\mathbf{x}$.

How much benefit can be derived from the use of iterative refinement? The final three steps can, of course, be repeated as many times as we wish, but it is often suggested that more than one iteration rarely provides any significant further improvement.

In the case of a well-conditioned system, it is widely thought that little additional accuracy is to be achieved by iterative refinement that is not as easily obtained by the use of pivoting unless the residuals are computed at a higher precision than the rest of the computation. However, there is evidence to suggest that this belief has been overstressed.

For an ill-conditioned system, it is possible to obtain small residuals despite $\hat{x}$ being well-removed from x^*. (See the error analysis of Sec. 6.3.) Here it is highly desirable, it seems, to compute the residuals to greater precision. Otherwise, the analysis of the effect of rounding errors in the right-hand side of (6.3.15) suggests that the relative error in δx may also be large so that the "correction" could be very different from the desired one.

We investigate these statements experimentally. For our experiment, we use a 10×10 Vandermonde matrix with its elements given by

$$a_{ij} = j^{i-1} \tag{6.5.3}$$

and the right-hand-side vector chosen so that the true solution is

$$x_i = \frac{(-1)^{i-1}}{3}$$

The computation was performed with the necessary modifications of the Doolittle reduction program using partial pivoting. This necessitates, of course, the storage of a copy of the original matrix for use in computing the residuals. The main part of the program is then

```
begin
    input;
    A0 := A;
    factorization(n,A,p);
    forward_solve(n,A,b,p,z);
    back_substitution(n,A,z,p,x);
    for k := 1 to maxiter do begin
        mult(n,A0,x,Ax);
        for i := 1 to n do r[i] := b[i] − Ax[i];
        forward_substitution(n,A,r,p,z);
        back_substitution(n,A,z,p,dx);
        for i := 1 to n do x[i] := x[i] + dx[i];
    end;
    output;
end.
```

The main loop (controlled by k) performs *maxiter* iterations of iterative refinement on the original solution generated by the *LU* factorization. The procedure *mult* performs matrix-vector multiplication. We observe that the residuals were computed in type real—the same precision as the rest of the computation.

The results of the original solution and three refinements (maxiter = 3) of it are given in Table 6.5.1.

The matrix given by (6.5.3) is ill-conditioned, as is commonly the case for Vandermonde matrices. In the ∞-norm,

$$\|A\|_\infty = 1 + 2^9 + \cdots + 10^9 \approx 1.57 \times 10^9$$

TABLE 6.5.1
**Results of applying iterative refinement to the *LU*
factorization solution of a Vandermonde system.**

	Initial solution	Refinement iteration		
		1	2	3
x_1	0.333 331 70	0.333 333 35	0.333 333 33	0.333 333 34
x_2	−0.333 322 60	−0.333 333 50	−0.333 333 33	−0.333 333 37
x_3	0.333 274 21	0.333 333 95	0.333 333 31	0.333 333 47
x_4	−0.333 172 94	−0.333 334 68	−0.333 333 25	−0.333 333 61
x_5	0.333 084 55	0.333 335 23	0.333 333 16	0.333 333 69
x_6	−0.333 087 64	−0.333 335 12	−0.333 333 12	−0.333 333 64
x_7	0.333 174 83	0.333 334 46	0.333 333 17	0.333 333 51
x_8	−0.333 268 21	−0.333 333 79	−0.333 333 26	−0.333 333 40
x_9	0.333 317 79	0.333 333 44	0.333 333 31	0.333 333 35
x_{10}	−0.333 331 69	−0.333 333 34	−0.333 333 33	−0.333 333 33

Replacing the second row of A $(1, 2, \ldots, 10)$ with $(5, 5, \ldots, 5)$ yields a singular matrix B. Now

$$\|B - A\| = 4 + 3 + 2 + 1 + 0 + 1 + 2 + 3 + 4 + 5 = 25$$

from which we deduce that $\kappa_\infty(A) \geq 6.28 \times 10^7$.

According to the conventional wisdom outlined above it would seem then that little if any benefit is to be gained from iterative refinement implemented in this precision and used more than once.

The results presented in Table 6.5.1 do *not* support this conventional wisdom.

It is clear that the first iteration has had significant effect despite the residuals being computed in the same precision as the original data and solution. The original solution has a greatest relative error of approximately 7×10^{-4}. This is reduced to less than 6×10^{-6} by the first iteration. There is a further reduction in the maximum relative error to about 6×10^{-7} as a result of the second iterative refinement.

The results show that the rules of thumb outlined above for iterative refinement are no more than that. The first two iterations here appear to contradict them all. However, a third iteration demonstrates clearly the possibility that iterative refinement can result in a less accurate solution in the case of an ill-conditioned matrix.

The experimental investigation of the efficacy of iterative refinement is continued in the exercises.

EXERCISES 6.5

1. Use iterative refinement with double-precision residuals to obtain results corresponding to Table 6.5.1.
2. Repeat the computation of Exercise 1 using other precisions: single precision throughout, single precision with real residuals, and single precision with double-precision residuals.
3. Experiment with iterative refinement on your solutions of the Hilbert matrix systems as in Table 6.4.5.
4. Write a summary of your results from Exercises 1 to 3 and your conclusions on the usefulness of iterative refinement. Add some further examples including some well-conditioned ones to strengthen your conclusions.

6.6 ITERATIVE TECHNIQUES

In this section we turn away from the direct methods of solving a linear system and concentrate on iterative approaches. The most efficient of our direct methods requires about $2n^3/3$ floating-point operations. For a large system, with dimension 1000 say, this cost would be prohibitive on a serial computer, especially if such systems needed to be solved repeatedly. Such large systems arise frequently out of the need to solve partial differential equations.

It is often the case that the large system resulting from the discretization of a partial differential equation will also be a sparse system of equations, that is, one in which many of the matrix elements are zero. As we have already seen for the special case of a tridiagonal matrix in the context of cubic spline interpolation (Secs. 5.2 to 5.4), it may be the case that simplified direct-solution algorithms are available with operation counts which are linear in their dimension n. This depends very much on the structure of the sparsity and whether or not it can be preserved through the solution process.

The simplest alternative to the direct methods are the iterative methods of Jacobi and of Gauss-Seidel.

As usual, we denote the system under consideration by the matrix-vector equation

$$Ax = b \qquad (6.6.1)$$

It will be convenient in the convergence analyses to consider the lower triangular, diagonal, and upper triangular parts of A separately. We denote them by L, D, and U, respectively. L and U each have zero elements on their main diagonal so that

$$A = L + D + U \qquad (6.6.2)$$

We begin with the simplest iterative scheme of all, the *Jacobi iteration*. Here we rewrite the original system of equations as

$$x_1 = \frac{b_1 - (a_{12}x_2 + a_{13}x_3 + \cdots + a_{1n}x_n)}{a_{11}}$$

$$x_2 = \frac{b_2 - (a_{21}x_1 + a_{23}x_3 + \cdots + a_{2n}x_n)}{a_{22}}$$

$$\cdots\cdots\cdots\cdots\cdots\cdots\cdots\cdots\cdots\cdots\cdots\cdots\cdots\cdots \qquad (6.6.3)$$

$$x_n = \frac{b_n - (a_{n1}x_1 + a_{n2}x_2 + \cdots + a_{nn-1}x_{n-1})}{a_{nn}}$$

which, in matrix terms, reduces to

$$\mathbf{x} = D^{-1}(\mathbf{b} - (L + U)\mathbf{x}) \qquad (6.6.4)$$

In the language and notation of Sec. 2.6, extended in the obvious way to higher dimension, this gives rise to an iteration

$$\mathbf{x}_{k+1} = \mathbf{g}(\mathbf{x}_k) \qquad (6.6.5)$$

where the iteration function $\mathbf{g}$ is defined as in (6.6.4) by

$$\mathbf{g}(\mathbf{x}) = D^{-1}(\mathbf{b} - (L + U)\mathbf{x})$$

This general framework provides a simple approach to the convergence analysis of the Jacobi iteration. It was established earlier that such an iteration will converge provided that the iteration function satisfies a Lipschitz condition with Lipschitz constant less than unity. This in turn is assured if, for example, the ∞-norm of the Jacobian matrix of $\mathbf{g}$ is less than unity. (See Theorem 2.6.7.) For the iteration function (6.6.5), this matrix is simply $D^{-1}(L + U)$. Now

$$\|D^{-1}(L + U)\|_\infty = \max_i \sum_{j \neq i} \left| \frac{a_{ij}}{a_{ii}} \right|$$

and so we anticipate convergence provided that this is less than unity. This is the case if A is *diagonally dominant*, which is to say that, for each i,

$$|a_{ii}| > \sum_{j \neq i} |a_{ij}| \qquad (6.6.6)$$

(This condition can be obtained by direct consideration of the errors in a similar way to the analysis of the Gauss-Seidel iteration below. See Exercise 3.)

The condition of diagonal dominance will play an important role in the rest of the discussion of iterative methods.

Consideration of the system of Eqs. (6.6.3) leads to a natural modification of the Jacobi iteration. The Jacobi iteration (6.6.5) written in full, with $x_i^{(k)}$ denoting the ith component of the kth iterate $\mathbf{x}_k$, is

$$x_1^{(k+1)} = \frac{b_1 - (a_{12}x_2^{(k)} + a_{13}x_3^{(k)} + \cdots + a_{1n}x_n^{(k)})}{a_{11}}$$

$$x_2^{(k+1)} = \frac{b_2 - (a_{21}x_1^{(k)} + a_{23}x_3^{(k)} + \cdots + a_{2n}x_n^{(k)})}{a_{22}} \qquad (6.6.7)$$

$$\cdots\cdots\cdots\cdots\cdots\cdots\cdots\cdots\cdots\cdots\cdots\cdots\cdots\cdots$$

$$x_n^{(k+1)} = \frac{b_n - (a_{n1}x_1^{(k)} + a_{n2}x_2^{(k)} + \cdots + a_{nn-1}x_{n-1}^{(k)})}{a_{nn}}$$

It is apparent that by the time even the second of these updates is performed, we are no longer using all the most recent (and, presumably, most accurate) estimates of the solution vector. An obvious modification is to use the latest iterates as soon as they become available. This is the *Gauss-Seidel iteration*, which can be written in matrix terms as

$$\mathbf{x}_{k+1} = D^{-1}(\mathbf{b} - L\mathbf{x}_{k+1} - U\mathbf{x}_k) \qquad (6.6.8)$$

or, in full, as

$$x_1^{(k+1)} = \frac{b_1 - (a_{12}x_2^{(k)} + a_{13}x_3^{(k)} + \cdots + a_{1n}x_n^{(k)})}{a_{11}}$$

$$x_2^{(k+1)} = \frac{b_2 - (a_{21}x_1^{(k+1)} + a_{23}x_3^{(k)} + \cdots + a_{2n}x_n^{(k)})}{a_{22}} \qquad (6.6.9)$$

$$\cdots\cdots\cdots\cdots\cdots\cdots\cdots\cdots\cdots\cdots\cdots\cdots\cdots$$

$$x_n^{(k+1)} = \frac{b_n - (a_{n1}x_1^{(k+1)} + a_{n2}x_2^{(k+1)} + \cdots + a_{nn-1}x_{n-1}^{(k+1)})}{a_{nn}}$$

Before proceeding to the convergence theory for the Gauss-Seidel iteration, we consider an example of its application to a diagonally dominant system. The results were obtained using Procedure 6.6.3 below to perform the iterations.

Example 6.6.1. Solve the following system using the Gauss-Seidel iteration:

$$
\begin{bmatrix}
10 & 2 & -1 & 3 & 1 \\
2 & 10 & 2 & -1 & 3 \\
-1 & 2 & 10 & 2 & -1 \\
3 & -1 & 2 & 10 & 2 \\
1 & 3 & -1 & 2 & 10
\end{bmatrix}
\begin{bmatrix}
x_1 \\ x_2 \\ x_3 \\ x_4 \\ x_5
\end{bmatrix}
=
\begin{bmatrix}
5 \\ -2 \\ 4 \\ -2 \\ 5
\end{bmatrix}
$$

Solution. The results of applying 10 Gauss-Seidel iterations with the starting vector $(0,0,0,0,0)$ are presented in Table 6.6.2 below. We see that there is

TABLE 6.6.2
Gauss-Seidel iterations for the system of Example 6.6.1

Iteration	x_1	x_2	x_3	x_4	x_5
1	0.50000	-0.30000	0.51000	-0.48200	0.68740
2	0.68686	-0.69379	0.77258	-0.76743	0.87020
3	0.85923	-0.86416	0.89926	-0.89808	0.94287
4	0.93790	-0.94010	0.95571	-0.95509	0.97483
5	0.97264	-0.97363	0.98049	-0.98022	0.98892
6	0.98795	-0.98838	0.99141	-0.99129	0.99512
7	0.99469	-0.99488	0.99622	-0.99616	0.99785
8	0.99766	-0.99775	0.99833	-0.99831	0.99905
9	0.99897	-0.99901	0.99927	-0.99926	0.99958
10	0.99955	-0.99956	0.99968	-0.99967	0.99982

steady—if slow—progress toward the true solution $(1, -1, 1, -1, 1)$. Estimates of the rate of convergence will emerge from the analysis to follow, but we observe here that the errors are approximately halved on each iteration.

In this case we simply programmed a fixed number of iterations, whereas a practical program would necessarily be controlled by a While or Repeat loop with a stopping condition based on the change between **x** and **y** measured in some appropriate norm.

Procedure 6.6.3   Gauss-Seidel iteration

```
procedure GSiterate(dim: integer; A: matrix; b,x: vector;
              var y: vector);
(*x is the current estimate, y is the new iterate*)
var i,j: integer;
begin
    for i:= 1 to dim do begin
        y[i] := b[i];
        for j:= 1 to i − 1 do y[i] := y[i] − A[i,j] * y[j];
        for j:= i + 1 to dim do y[i] := y[i] − A[i,j] * x[j];
        y[i] := y[i]/a[i,i];
    end;
end;
```

A simple operation count shows that each iteration of Gauss-Seidel (or Jacobi) requires a total of $n(2n - 1)$ floating-point operations, so any number of iterations greater than about $n/3$ implies a greater computational expense than direct solution by one of the direct methods such as the LU factorization. An iterative approach such as this is, therefore, only likely to be of practical benefit for a large system or one with a large measure of sparsity.

In the above example, we have used 10 (that is $2n$) iterations and still have only about three decimal places correct in our solution. The simplest Gauss elimination program delivers full machine accuracy for this example at a lower cost than the iterations tabulated here. (Pivoting would provide no benefit for this diagonally dominant system.)

In our discussion of the convergence and efficiency of the Gauss-Seidel technique we thus think in terms of its application to large (and probably sparse) systems.

We begin our analysis by establishing the convergence of the Gauss-Seidel iteration in the case of a diagonally dominant matrix A. Recall that the Gauss-Seidel iteration is defined by Eq. (6.6.8) as

$$\mathbf{x}_{k+1} = D^{-1}(\mathbf{b} - L\mathbf{x}_{k+1} - U\mathbf{x}_k)$$

Denoting the exact solution of the system by $\mathbf{x}^*$ and the error vector at iteration k by

$$\mathbf{e}_k = \mathbf{x}_k - \mathbf{x}^* \tag{6.6.10}$$

we obtain, by subtracting $\mathbf{x}^* = D^{-1}(\mathbf{b} - L\mathbf{x}^* - U\mathbf{x}^*)$ from both sides of (6.6.8),

$$\mathbf{e}_{k+1} = -D^{-1}(L\mathbf{e}_{k+1} + U\mathbf{e}_k) \qquad (6.6.11)$$

or, equivalently,

$$(L + D)\mathbf{e}_{k+1} = -U\mathbf{e}_k \qquad (6.6.12)$$

It follows that if the matrix $(L + D)^{-1}U$ has norm less than unity, then $\|\mathbf{e}_k\| \to 0$ as $k \to \infty$. Moreover, $\|(L + D)^{-1}U\|$ provides an estimate of the linear rate of convergence.

In the case of a diagonally dominant matrix, we can consider the elements of $\mathbf{e}_{k+1}$ in (6.6.11) in the ∞-norm. It follows that

$$|e_i^{(k+1)}| \le \lambda_i \|\mathbf{e}_{k+1}\|_\infty + \nu_i \|\mathbf{e}_k\|_\infty \qquad (6.6.13)$$

where

$$\lambda_i = \sum_{j<i} \left| \frac{a_{ij}}{a_{ii}} \right| \qquad \nu_i = \sum_{j>i} \left| \frac{a_{ij}}{a_{ii}} \right| \qquad (6.6.14)$$

Now, if $\|\mathbf{e}_{k+1}\|_\infty = |e_m^{(k+1)}|$, it follows from (6.6.13) that

$$(1 - \lambda_m)\|\mathbf{e}_{k+1}\|_\infty \le \nu_m \|\mathbf{e}_k\|_\infty$$

and therefore that

$$\|\mathbf{e}_{k+1}\|_\infty \le \Lambda \|\mathbf{e}_k\|_\infty \qquad (6.6.15)$$

where

$$\Lambda = \max\left(\frac{\nu_k}{1 - \lambda_k} \right) \qquad (6.6.16)$$

It remains to show that $\Lambda < 1$, for then (6.6.15) applied recursively will imply $\mathbf{e}_k \to 0$, as desired.

First, since A is diagonally dominant, $\lambda_k + \nu_k < 1$ for every k. Since $\lambda_k, \nu_k \ge 0$ it follows that

$$\frac{\nu_k}{1 - \lambda_k} < 1$$

which establishes that $\Lambda < 1$ as required.

In summary, we have proved the following result.

Theorem 6.6.4. If A is a diagonally dominant matrix, the Gauss-Seidel iteration given by Eqs. (6.6.8) or (6.6.9) converges from any initial vector [including $\mathbf{x}_0 = (0, 0, \ldots, 0)$] to the unique solution of $A\mathbf{x} = \mathbf{b}$.

In terms of Example 6.6.1, we find that $\Lambda = 0.75$ for that particular matrix, so the analysis suggests that the error should be reduced by at least 25 percent on each iteration. The evidence of Table 6.6.2 is that the actual performance is somewhat better than this estimate suggests, but it is nonetheless a slow linear convergence rate.

Theorem 6.6.4 covers systems such as that in Example 6.6.1 above and the tridiagonal systems obtained in Chap. 5 for cubic spline interpolation. However, for several discretizations of partial differential equations, the resulting system has tridiagonal form (or some other structured sparsity) but without diagonal dominance. For example, the tridiagonal matrix with rows of the form $(-1, 2, -1)$ arises from using central difference approximations in the discretization; this matrix is not diagonally dominant. It is symmetric and positive definite, however. Recall that a matrix A is positive definite if, for every nonzero real vector $\mathbf{v}$, $\mathbf{v}^T A \mathbf{v} > 0$. (For complex vectors, the corresponding property is $\mathbf{v}' A \mathbf{v} > 0$, where $\mathbf{v}'$ denotes the elementwise complex conjugate transpose of $\mathbf{v}$.)

The following theorem establishes that the Gauss-Seidel method will also converge for such a matrix. The argument is based on Eq. (6.6.12) by establishing that the matrix $(L + D)^{-1} U$ has norm less than unity. Note that since A is symmetric, it follows that $U = L^T$.

> **Theorem 6.6.5.** If A is a symmetric positive definite matrix with positive diagonal entries, then the Gauss-Seidel iteration converges to the unique solution of $A\mathbf{x} = \mathbf{b}$.

> ***Proof.*** We establish that every eigenvalue of $(L + D)^{-1} L^T$ has absolute value less than unity from which the result follows immediately from the above remarks.
>   Let λ be an eigenvalue of $(L + D)^{-1} L^T$ with associated eigenvector $\mathbf{v}$. [Note that $(L + D)^{-1} L^T$ is not symmetric, and so its eigenvalues are in general complex.] It follows that $L^T \mathbf{v} = \lambda (D + L)\mathbf{v}$ and therefore that

$$\mathbf{v}' L^T \mathbf{v} = \lambda \mathbf{v}'(D + L)\mathbf{v} \tag{6.6.17}$$

Now $A = L + D + L^T$, and so

$$\mathbf{v}' A \mathbf{v} = \mathbf{v}'(L + D + L^T)\mathbf{v} = (1 + \lambda)\mathbf{v}'(D + L)\mathbf{v} \tag{6.6.18}$$

Since A is symmetric positive definite, this quantity is positive and, in particular, real so that it is its own conjugate. From (6.6.18) we may also deduce that $\lambda \neq -1$. It follows that

$$(1 + \lambda)\mathbf{v}'(D + L)\mathbf{v} = (1 + \bar{\lambda})\mathbf{v}'(D + L)^T \mathbf{v} \tag{6.6.19}$$

and substituting for $\mathbf{v}' L^T \mathbf{v}$ from (6.6.17) in the right-hand side of (6.6.19) gives

$$(1 + \lambda)\mathbf{v}'(D + L)\mathbf{v} = (1 + \bar{\lambda})\{\mathbf{v}' D \mathbf{v} + \lambda \mathbf{v}'(D + L)\mathbf{v}\}$$

using the fact that $D = D^T$. Simplifying we get

$$(1 - |\lambda|^2)\mathbf{v}'(D + L)\mathbf{v} = (1 + \bar{\lambda})\mathbf{v}' D \mathbf{v} \tag{6.6.20}$$

Finally, multiplying (6.6.20) by $1 + \lambda$ and using (6.6.18), we now have

$$(1 - |\lambda|^2)\mathbf{v}' A \mathbf{v} = |1 + \lambda|^2 \mathbf{v}' D \mathbf{v}$$

Since the elements of D are positive, the right-hand side is clearly positive. Also A is positive definite; it follows that $|\lambda| < 1$, as required.

In Example 6.6.1 we observed a very slow rate of convergence for the Gauss-Seidel iteration, and the analysis establishes that it is always likely to be a slow linear convergence. In the next section we discuss the possibility of accelerating the progress of iterative solution of linear equations.

EXERCISES 6.6

1. Write a program for the Jacobi iteration. Use it to solve a diagonally dominant system such as that of Example 6.6.1. Also find an example for which it fails to converge.

2. Use the Gauss-Seidel iteration to solve the linear system for the coefficients of natural cubic spline interpolation to $1/(1+x^2)$ at integer knots in $[-5, 5]$. (See Table 5.3.3.)

3. Give the direct convergence proof for a diagonally dominant matrix with the Jacobi iteration. Find an estimate of its convergence rate. Compare this theoretical rate with the observed rate in Exercise 1.

4. Experiment with the choice of starting vector for Gauss-Seidel iteration. For example, does choosing

$x_i^{(0)} = b_i/a_{ii}$ improve the performance significantly? Are other simple choices better? (Note that this particular choice is the equivalent of one Jacobi iteration at the outset.)

5. Modify the Gauss-Seidel procedure for the specific case of a tridiagonal system. Repeat the computation of Exercise 3. Make a comparison of the computational efficiency of this solution and the original in terms of operation counts.

6. For cubic spline interpolation at uniformly spaced knots the tridiagonal system has $a_{i-1,i} = a_{i+1,i} = 1$, $a_{ii} = 4$. Estimate the linear rate of convergence of Gauss-Seidel for such a system. How does this compare with the performance found in Exercise 5?

6.7   ACCELERATION AND SUCCESSIVE OVERRELAXATION

In the final section of this chapter we discuss the possibility of improving the performance of the iterative schemes for the solution of linear equations. We saw in the previous section that although the Gauss-Seidel method is likely to converge faster than Jacobi, it is still slow. The 10 iterations used in Example 6.6.1 give only three-decimal-place accuracy, and the error is (approximately) halving on each iteration.

One obvious possibility is to improve on the choice of starting vector. Certainly, we might expect the initial vector **x** given by

$$x_i = \frac{b_i}{a_{ii}} \tag{6.7.1}$$

to yield some improvement at very low cost. This choice is the equivalent of (though cheaper than) performing one Jacobi iteration to begin the process. Since the Jacobi iteration is likely to converge more slowly than Gauss-Seidel, it seems unlikely that this will result in savings of much more than one iteration. What is really needed is a method for accelerating the convergence itself.

The simplest way of achieving this is to use the Aitken Δ^2 process on each of the elements of the vector. This process was proved in Sec. 2.5 to accelerate the convergence of a linearly convergent sequence. We begin by studying the improvement to be gained by applying this process to the Gauss-Seidel iteration applied to the system of Example 6.6.1, which we recall was

$$
\begin{bmatrix}
10 & 2 & -1 & 3 & 1 \\
2 & 10 & 2 & -1 & 3 \\
-1 & 2 & 10 & 2 & -1 \\
3 & -1 & 2 & 10 & 2 \\
1 & 3 & -1 & 2 & 10
\end{bmatrix}
\begin{bmatrix}
x_1 \\ x_2 \\ x_3 \\ x_4 \\ x_5
\end{bmatrix}
=
\begin{bmatrix}
5 \\ -2 \\ 4 \\ -2 \\ 5
\end{bmatrix}
\tag{6.7.2}
$$

The accelerated Gauss-Seidel iteration was implemented using the following Algorithm 6.7.1 with tolerance set at 10^{-4} with two different starting vectors $\mathbf{x}_0 = \mathbf{0}$ and that given by (6.7.1). The results are tabulated in Table 6.7.2 below.

Algorithm 6.7.1   Aitken accelerated Gauss-Seidel iteration

Input n, an $n \times n$ matrix A, an n vector $\mathbf{b}$,
           tolerance ϵ, maximum iteration count *maxit*
           initial vector $\mathbf{x}_1$

Initialize
           itcount $:= 0$

Loop   **Repeat**
              $\mathbf{x} := \mathbf{x}_1$
              Gauss-Seidel iteration using $\mathbf{x}$ to generate $\mathbf{y}$
              Gauss-Seidel iteration using $\mathbf{y}$ to generate $\mathbf{z}$
              Elementwise Aitken Δ^2 acceleration using $\mathbf{x},\mathbf{y},\mathbf{z}$
                  to generate new iterate $\mathbf{x}_1$
              itcount $:=$ itcount $+ 1$
           until $\|\mathbf{x}_1 - \mathbf{x}\| < \epsilon$ **or** itcount $=$ maxit

Output   itcount, $\|\mathbf{x}_1 - \mathbf{x}\|$ and approximate solution $\mathbf{x}_1$.

The two Gauss-Seidel iterations can be achieved just as in the previous section, while the elementwise Aitken acceleration is a simple loop implementing (2.5.4) or (2.5.7) for each component of the vector in turn such as in the following procedure:

Procedure VectorAitken(dim: integer; a,b,c: vector;
              var newa: vector);
  var i: integer;
  begin
      for i := 1 to dim do
          newa[i] := (a[i] * c[i] − b[i] * b[i])/(a[i] + c[i] − 2 * b[i]);
  end;

We note here that the convergence test has been written in terms of a vector norm. Any of them will do, but the easiest to compute are, of course, the 1-norm and the ∞-norm. The reliable computation of the euclidean, or 2-norm, can be problematic for vectors of high dimension or containing very large or very small components.

TABLE 6.7.2
Iterations of the Aitken Gauss-Seidel procedure for (6.7.2)

Iteration	x_1	x_2	x_3	x_4	x_5
		(a) $x_0 = 0$			
1	0.79836	0.95957	1.05127	−1.18191	0.93642
2	0.76006	−0.97359	0.95512	−0.95231	0.98818
3	0.99117	−0.97092	0.99991	−0.99548	0.98447
4	1.00351	−0.99750	0.99698	−0.99493	0.99895
5	0.99886	−0.99630	1.00467	−0.99956	0.99856
6	1.00002	−0.99998	0.99991	−0.99954	0.99998
7	0.99991	−0.99993	0.99990	−0.99997	0.99997
8	0.99999	−1.00007	0.99999	−0.99995	1.00000
9	1.00000	−1.00000	0.99999	−1.00000	1.00000
		(b) $x_i^{(0)} = b_i/a_{ii}$			
1	0.43174	−1.21185	1.37752	−0.97743	0.97155
2	1.02772	−0.96295	1.01400	−1.01873	1.01145
3	1.00236	−1.00380	0.99727	−1.00228	1.00023
4	0.99684	−1.00021	1.00022	−1.00002	1.00008
5	1.00003	−0.99999	1.00002	−1.00004	1.00000
6	1.00000	−1.00001	0.99999	−1.00000	1.00000

We see that the convergence is much better than for the original Gauss-Seidel iteration (see Table 6.6.2), especially in the second case where the ∞-norm of the error is reduced to 10^{-5} in just six Aitken iterations.

The difference in the performance for the two starting vectors is much more marked than we had anticipated. The principal reason for this is that the very first Aitken iteration for $x_0 = 0$ has $x_i = 0$ in the VectorAitken procedure and so minimizes the contribution from the best approximation z. The effect is highlighted by the value for x_2 after the first Aitken iteration being positive. Note that the second iteration does very little more than correct this sign, the other components being adjusted only slightly and not always improved. Progress in the second case is much more uniform.

The amount of work per iteration here is, of course, much greater than for the Gauss-Seidel iteration itself, since each Aitken iteration requires two full Gauss-Seidel cycles together with the acceleration step which entails a further $6n$ floating-point operations per iteration.

On this 5×5 system more work has been used in these six iterations than in a full Gauss elimination or Doolittle reduction with partial pivoting. Such iterative schemes are, therefore, only to be recommended for large systems or those with sufficient structured sparsity (such as tridiagonal or pentadiagonal systems) that efficient special-purpose Gauss-Seidel iterations are reduced to $O(n)$ floating-point operations. However, we recall that for a tridiagonal system, for example, a full LU factorization and solution needs only about $8n$ floating-point operations—the equivalent of about $1\frac{1}{3}$ acceleration steps for the

Aitken algorithm. Memory requirements, however, frequently dictate that large systems must be solved iteratively.

With this, as with many efficiency arguments which are based on operation counts, we may draw significantly different conclusions for implementations on parallel computer architectures (see Chap. 13).

We turn now to an alternative modification of the Gauss-Seidel method which can result in significant improvements in performance. This is the method of *successive overrelaxation*, or SOR method. The idea is that for each component of the updated vector we perform an extrapolation of the Gauss-Seidel iteration by setting

$$x_i^{(k+1)} = \omega y_i^{(k+1)} + (1 - \omega) x_i^{(k)} \tag{6.7.3}$$

where $y_i^{(k+1)}$ denotes the Gauss-Seidel update using $x_j^{(k+1)}$ for $j < i$ and $x_j^{(k)}$ for $j > i$. The parameter ω is the *relaxation parameter*.

Usually this parameter is chosen so that $1 < \omega < 2$, from which it follows that formula (6.7.3) is indeed *extra*polation. It is the fact that $\omega > 1$ which leads to the designation of this method as successive overrelaxation. ("Successive" because the extrapolation is applied to each component successively rather than to the whole vector at the end of the iteration.)

The SOR iteration is easily implemented in Procedure 6.7.3, which is a simple modification of the Gauss-Seidel iteration procedure in which the variable w plays the role of ω.

Procedure 6.7.3   SOR iteration

```
procedure SORiterate(dim: integer; A: matrix; b,x: vector;
                w: real; var y: vector);
var i,j: integer; w1: real;
begin
    w1 := 1 - w;
    for i := 1 to dim do begin
        y[i] := b[i];
        for j := 1 to i - 1 do y[i] := y[i] - A[i,j] * y[j];
        for j := i + 1 to dim do y[i] := y[i] - A[i,j] * x[j];
        y[i] := y[i] * w/A[i,i] + w1 * x[i];
    end;
end;
```

An operation count for this procedure shows that each SOR iteration uses $2n(n + 1)$ floating-point operations. That represents an increase of just $3n$ over straightforward Gauss-Seidel iteration.

We can see from the procedure above that the separate computation of the vector $\mathbf{y}_{k+1}$ implied in (6.7.3) is unnecessary. In the same way as for the Gauss-Seidel and Jacobi iterations, we can write down a matrix-vector equation for this iteration by substituting the Gauss-Seidel iteration formula for $\mathbf{y}_{k+1}$. This yields

$$\mathbf{x}_{k+1} = \omega D^{-1}(\mathbf{b} - L\mathbf{x}_{k+1} - U\mathbf{x}_k) + (1 - \omega)\mathbf{x}_k \tag{6.7.4}$$

or, equivalently,

$$(I + \omega D^{-1}L)\mathbf{x}_{k+1} = \omega D^{-1}\mathbf{b} - \omega D^{-1}U\mathbf{x}_k + (1 - \omega)\mathbf{x}_k \qquad (6.7.5)$$

Since the true solution $\mathbf{x}^*$ is a fixed point of the Gauss-Seidel iteration, it follows that it is also a fixed point for SOR. That is,

$$(I + \omega D^{-1}L)\mathbf{x}^* = \omega D^{-1}\mathbf{b} - \omega D^{-1}U\mathbf{x}^* + (1 - \omega)\mathbf{x}^* \qquad (6.7.6)$$

Subtracting (6.7.6) from (6.7.5), we obtain the following formula for the error vectors $\mathbf{e}_{k+1}$ and $\mathbf{e}_k$:

$$(I + \omega D^{-1}L)\mathbf{e}_{k+1} = -\omega D^{-1}U\mathbf{e}_k + (1 - \omega)\mathbf{e}_k$$

or

$$\mathbf{e}_{k+1} = (I + \omega D^{-1}L)^{-1}((1 - \omega)I - \omega D^{-1}U)\mathbf{e}_k \qquad (6.7.7)$$

To optimize the performance of the SOR method, we need the choice of ω which minimizes the norm of this iteration matrix $(I + \omega D^{-1}L)^{-1}((1 - \omega)I - \omega D^{-1}U)$. Except in very special cases, finding this minimizing value of ω is very difficult. Experiments show that choices around $\omega = 1.2$ or 1.25 seem to work well in many cases.

We restrict our study of this to computational experiment.

Example 6.7.4. We begin by considering the same 5×5 system (6.7.2) that was used in Example 6.6.1 and again for the Aitken acceleration. In Table 6.7.6 we present the results of the first five iterations of the SOR algorithm for a variety of different relaxation parameters ω. In all cases, the iteration was started with $x_i^{(0)} = b_i/a_{ii}$.

The first results are for $\omega = 1$, which is just the Gauss-Seidel method. The next two sets are for $\omega = 1.2$ and $\omega = 1.4$; it is immediately apparent that the former choice has given better approximations than the latter for these early iterations.

For experimental purposes, we use the philosophy of the bisection algorithm and next try $\omega = 1.3$. The results here are certainly an improvement on those for $\omega = 1.4$ but are still not as good as for $\omega = 1.2$. Hence the next experimental value chosen was $\omega = 1.25$, whose results are marginally better than those for $\omega = 1.2$ if the error is measured in the 1-norm while being (even more marginally) worse for the ∞-norm.

For this particular example it indeed appears that the optimal ω is somewhere in the interval $[1.2, 1.25]$.

Example 6.7.5. In this case we choose a larger but sparse system with the matrix A having all elements zero except for

$$a_{ii-1} = a_{ii+1} = 1 \qquad a_{ii} = 4$$

This is the form of the matrix used in natural cubic spline interpolation for equally spaced knots. Since the error is reduced according to Eq. (6.7.7), which is independent of the right-hand-side vector, we choose

$$b_1 = b_n = 5 \qquad b_i = 6 \qquad (i = 2, 3, \ldots, n - 1)$$

so that the exact solution is $\mathbf{x}^* = (1, 1, \ldots, 1)$.

TABLE 6.7.6

Iteration	x_1	x_2	x_3	x_4	x_5
			$\omega = 1.00$ (Gauss-Seidel)		
1	0.59000	−0.56800	0.66260	−0.66632	0.81092
2	0.79866	−0.80216	0.85466	−0.85293	0.91683
3	0.91009	−0.91329	0.93594	−0.93491	0.96355
4	0.96037	−0.96182	0.97174	−0.97135	0.98395
5	0.98255	−0.98318	0.98756	−0.98738	0.99293
			$\omega = 1.20$		
1	0.60800	−0.64592	0.73598	−0.79303	0.93822
2	0.89464	−0.93509	0.96749	−0.97305	0.99126
3	0.99294	−0.99711	0.99744	−0.99979	1.00120
4	1.00019	−1.00042	1.00073	−1.00062	1.00012
5	1.00036	−1.00030	1.00013	−1.00010	1.00008
			$\omega = 1.40$		
1	0.62600	−0.72528	0.81672	−0.93314	1.09260
2	1.00597	−1.08977	1.09353	−1.09394	1.03923
3	1.06980	−1.03945	1.01520	−1.01250	0.99674
4	0.99096	−0.98633	0.99187	−0.98610	0.99180
5	0.99396	−0.99611	0.99628	−0.99914	1.00173
			$\omega = 1.30$		
1	0.61700	−0.68542	0.77542	−0.86134	1.01186
2	0.94830	−1.00914	1.02852	−1.03312	1.01905
3	1.03204	−1.02474	1.01313	−1.01414	1.00515
4	1.00337	−1.00072	1.00103	−0.99877	0.99811
5	0.99908	−0.99892	0.99872	−0.99904	0.99985
			$\omega = 1.25$		
1	0.61250	−0.66562	0.75547	−0.82676	0.97417
2	0.92098	−0.97136	0.99756	−1.00303	1.00605
3	1.01267	−1.01236	1.00680	−1.00875	1.00458
4	1.00348	−1.00229	1.00207	−1.00107	0.99980
5	1.00038	−1.00010	0.99980	−0.99979	0.99996

This system is tridiagonal, positive definite, and diagonally dominant so that we should anticipate the Gauss-Seidel iteration itself performing well and this should be enhanced by the use of the SOR algorithm. The implementation used a modified form of Procedure 6.7.3 which took advantage of the tridiagonal nature of the system. A different data type was defined:

$$\text{tridiag} = \text{array}[1 \ldots 50, -1 \ldots 1];$$

The variable A of type tridiag contained the diagonal elements as A[i, 0], while the subdiagonal and superdiagonal were stored as A[i, −1] and A[i, 1], respectively.

The appropriate SOR iteration procedure is now

```
procedure triSORiterate(dim: integer; A: tridiag;
                b,x: vector; w: real; var y: vector);
var i,j: integer; w1: real;
begin
    w1 := 1 − w;
    y[0] := 0; x[dim + 1] := 0;
    for i := 1 to dim do begin
        y[i] := b[i] − A[i, −1] * y[i − 1] − A[i, 1] * x[i + 1];
        y[i] := y[i] * w/A[i, 0] + w1 * x[i];
    end;
end;
```

The system solved was of dimension 50.

The results are presented in Table 6.7.7 below, again for various values of the relaxation parameter ranging from 0.9 to 1.4 in steps of 0.1. Five iterations were performed for each of these and the ∞-norm and 1-norm of the error vectors computed at each iteration.

We see that this iteration now entails just $8n$ floating-point operations of which $5n$ (the first of the lines generating y[i] together with the division by the diagonal element A[i,0]) correspond to the Gauss-Seidel iteration. We observe that $8n$ is precisely the order of the operation count for Gauss elimination for a tridiagonal system, while $5n$ corresponds to the formation of the Doolittle factorization of the original tridiagonal matrix.

TABLE 6.7.7
Errors in results of five SOR iterations for a tridiagonal system of dimension 50

ω	∞-norm	1-norm	ω	∞-norm	1-norm
0.90	8.75E − 02	2.51E + 00	1.00	1.25E − 01	4.87E + 00
	7.45E − 03	2.56E − 01		2.34E − 02	9.52E − 01
	2.32E − 03	2.95E − 02		4.88E − 03	1.86E − 01
	5.87E − 04	4.17E − 03		9.46E − 04	3.67E − 02
	2.00E − 04	9.13E − 04		1.95E − 04	7.26E − 03
1.10	1.63E − 01	7.15E + 00	1.20	2.00E − 01	9.37E + 00
	5.55E − 02	2.07E + 00		9.70E − 02	3.56E + 00
	1.65E − 02	5.97E − 01		3.94E − 02	1.35E + 00
	4.84E − 03	1.72E − 01		1.52E − 02	5.09E − 01
	1.42E − 03	4.95E − 02		5.83E − 03	1.92E − 01
1.30	2.38E − 01	1.15E + 01	1.40	2.79E − 01	1.36E + 01
	1.48E − 01	5.37E + 00		2.08E − 01	7.46E + 00
	7.66E − 02	2.50E + 00		1.31E − 01	4.09E + 00
	3.67E − 02	1.16E + 00		7.50E − 02	2.24E + 00
	1.72E − 02	5.38E − 01		4.15E − 02	1.23E + 00

We see from these results that for this particular system, none of the choices of ω yields any improvement on the Gauss-Seidel iteration itself ($\omega = 1$). There is slight improvement for the *under*relaxed iteration with $\omega = 0.9$ for the early iterations, but this is not sustained. The ∞-norm of the Gauss-Seidel iterates is being improved by a factor of almost exactly 5 on each iteration, whereas for $\omega = 0.9$ the relative improvement is slowing down with each iteration. It is improving by a factor less than 3 by the fifth iteration.

By way of contrast, it was remarked in Sec. 5.3 that this same system was solved to machine accuracy by the tridiagonal Gauss elimination Algorithm 5.3.1 using fewer floating-point operations than are needed for even two iterations of Gauss-Seidel.

However, iterative schemes such as these are useful particularly in the finite difference or finite element solution of partial differential equations. The structure of the resulting systems is often of a nature which is ideally suited to iterative solution. Their size is often such that storing and manipulating the complete matrix is impossible even if full advantage is taken of their sparsity. Such schemes will be encountered in Chap. 13.

As a final word on this topic, we must point out that there is another, powerful iterative technique—the conjugate gradient method. This method is based on the principle of minimizing the quadratic function $\mathbf{x}^T A \mathbf{x}/2 - \mathbf{b}^T \mathbf{x}$ which has its minimum at $\mathbf{x}^* = A^{-1}\mathbf{b}$. The conjugate gradient and other conjugate direction methods will be discussed in Chap. 8 within the context of optimization.

EXERCISES 6.7

1. Write a program for the Aitken accelerated Gauss-Seidel iteration. Use it to solve the linear system for the coefficients of natural cubic spline interpolation to $1/(1 + x^2)$ at integer knots in $[-5, 5]$. (See Table 5.3.3.)

2. Use the SOR iteration to solve the linear system for the coefficients of natural cubic spline interpolation to $1/(1 + x^2)$ at integer knots in $[-5, 5]$. Vary the relaxation parameter, and plot the resulting splines.

3. Solve a strongly diagonally dominant tridiagonal system in which the diagonal entries are all 10 while the off-diagonal entries are all 1 by Gauss-Seidel and SOR iterations. Experiment with the relaxation parameter ω. Compare your results with those of a direct solution.

4. A well-known example is the following psychological experiment on the intelligence of rats.

   A rat is put into a 5×4 rectangular grid of passages with intersections at the points (i, j) for $i = 1, \ldots, 5$, $j = 1, \ldots, 4$. The rat can only find its food if it emerges at the right-hand side of the grid, that is, if it is at position $(i, 4)$ for some i and chooses to go to the right. Exiting on any other side results in failure to find its food.

   Assuming that the rat makes a random choice at each intersection (each with probability 0.25), what is the probability that the rat finds its food if it starts at position (i, j)?

   Set up a system of linear equations for the probabilities P_{ij} of success as follows:

$$P_{11} = \frac{0 + 0 + P_{12} + P_{21}}{4}$$

$$P_{12} = \frac{P_{11} + 0 + P_{13} + P_{22}}{4}$$

$$P_{13} = \frac{P_{12} + 0 + P_{14} + P_{23}}{4}$$

$$P_{14} = \frac{P_{13} + 0 + 1 + P_{24}}{4}$$

etc., where the terms represent moves to the left, up, right, and down the grid.

Solve the system using Gauss-Seidel and SOR to obtain the probabilities to five decimal places. Estimate the optimal choice of the relaxation parameter.

PROJECTS

1. The $n \times n$ Hilbert matrix has elements

$$a_{ij} = \frac{1}{i+j-1} \qquad (1 \le i, j \le n)$$

Apply the Gauss elimination algorithm to the solution of the system

$$Ax = b$$

for $n = 2, 3, 4, \ldots$ with b chosen so that the exact solution should be $(1, 1, \ldots, 1)$. (That is, choose $b_i = a_{i1} + a_{i2} + \cdots + a_{in}$.) Have your program print out the determinant of the matrix at the end of the forward elimination. Can you explain what is wrong?

2. Repeat the Hilbert matrix experiment of Project 1 using partial pivoting. For what values of n can solutions now be obtained? Compare the relative errors in the solutions obtained by Gauss elimination and Gauss elimination with partial pivoting.

3. Find a singular matrix close to the Hilbert matrix of order n and use it to estimate the condition number of this matrix for $n = 2, 3, 4, \ldots$ for the ∞- and 1-norms.

Use these condition-number estimates to estimate the errors in the solutions obtained in Projects 1 and 2. Compare these results with the observed errors.

4. Write an LU factorization program for solving a tridiagonal system of equations. (Algorithm 5.3.1 provides a simple Gauss elimination procedure for this task.) Test your program by reproducing the graphs of Fig. 5.3.4 for cubic spline interpolation to the spike $100/(1 + 10x^2)$.

5. Modify your program for Project 4 to perform the Cholesky factorization of a symmetric tridiagonal matrix.

6. Show that the operation counts for LU factorization as a method of solving tridiagonal systems are $3(n-1)$ for the factorization, $2(n-1)$ for the forward solve, and $3n - 2$ for the back substitution. Compare the efficiency of the solution to Project 4 with the corresponding solution using iterative methods modified for tridiagonal systems. Graph the output splines. Compare the graphs of the direct solution and the best SOR solution.

CHAPTER
7

APPROXIMATION
OF FUNCTIONS

7.1. INTRODUCTION: WHY APPROXIMATE?

In earlier chapters we have seen various approaches to the problem of estimating functions which are not defined by simple polynomial or rational expressions. The techniques of Chap. 3 were specialized to the situation of special algorithms for the evaluation of elementary functions and are therefore not suited to the general problem.

The interpolation techniques—polynomial, spline, rational, or trigonometric—also have a special purpose of approximating functions for which some values can be obtained to very high accuracy. This is the situation in using, say, a table of values which may be obtained by some expensive but accurate algorithm and employing these values, which can be regarded as exact, to estimate other values.

In such circumstances, the use of interpolation to reproduce the data points *exactly* is justifiable provided the general behavior of the interpolant matches that of the underlying function. We saw that Lagrange interpolation with several data points has the tendency to produce high-degree polynomial behavior which is typically not a good model, especially near the ends of the range. In such circumstances it may well be better to find a good *approximant* rather than an interpolant.

By an *approximant* we mean a function which approximates the true function—well, we hope—but does *not* necessarily *reproduce* the *data* values *exactly*. That is, its graph will *not* in general pass *through* the data points *but* *close* to them. Such a function may be found to stay close to the original function over the full range of interest without the need to use high-degree formulas.

This not only eliminates the high-degree polynomial characteristics of the Lagrange interpolant but may lead to a function which is much more easily computed than say a spline interpolant or a discrete Fourier transform.

There is another, and more important, justification for approximation rather than interpolation in the case of experimental or statistical data. The point here is that data from an experiment is itself subject to errors. An interpolation function will reproduce those errors and pass their effect on to other values obtained from the interpolant.

We illustrate this in the following example where we see that even a very simple relation may be severely corrupted by using interpolation on experimental data.

Example 7.1.1. Consider the approximation of the data:

x	1.38	3.39	4.75	6.56	7.76
y	1.83	2.51	3.65	4.10	5.01

These data points are plotted in Fig. 7.1.2*a* along with the Lagrange interpolant (of degree 4, of course). The other function graphed, in Figure 7.1.2*b*, is a straight-line approximant to this same data. This approximation misses all the data points but still gives a better idea of the general trend of the data. The two curves are compared directly in Fig. 7.1.2*c*.

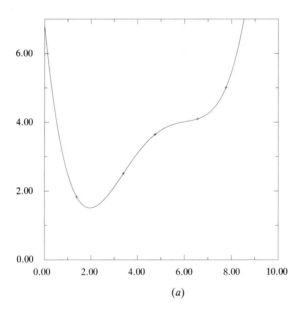

(*a*)

FIGURE 7.1.2
The data of Example 7.1.1, its Lagrange interpolant, and a straight-line approximant.

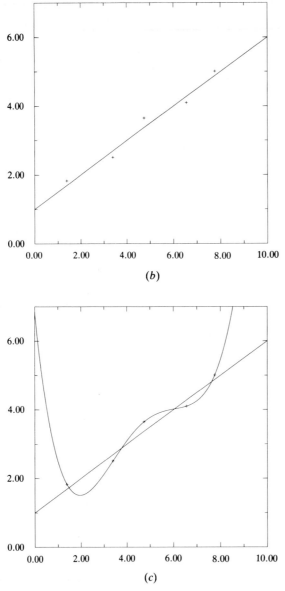

FIGURE 7.1.2 (*Continued*)

If this were indeed experimental data—perhaps measurements of the number of miles (in thousands) traveled by a car after x months—then the straight-line function gives a much better idea of likely subsequent readings than does the polynomial, which would suggest the car would travel its first million miles in just over 6 years!

The straight line graphed here is evidently a *good* approximation to the data—but is probably not the *best* such. However, we have to be careful about such statements since the precise meaning of "best" is dependent on how we measure the quality of the "fit" of the approximation to the data or equivalently measure the error in the approximation.

In this chapter, we will discuss various ways of making these measurements and their implications. For much of the time we will be concerned with polynomial approximations for the same reasons that polynomials dominate the interpolation theory—Weierstrass' theorem and their ease of evaluation.

The idea of best approximation to a function f from functions in a class Π relative to any particular function norm is to find $p \in \Pi$ such that

$$\|f - p\| = \min_{q \in \Pi} \|f - q\| \qquad (7.1.1)$$

Such a best approximation may not exist or, if it does, need not be unique. Such error measures as these are appropriate for the situation where a simple function is wanted to approximate a more complicated one over an interval. In such cases the norm used will usually be one of $\|\cdot\|_1$, $\|\cdot\|_2$, or $\|\cdot\|_\infty$ which give rise to the L_1, L_2, and L_∞ approximations, respectively.

The last of these is often called the *minimax* or best *uniform* approximation since (7.1.1) for the interval $[a, b]$ in this case reduces to

$$\|f - p\|_\infty = \min_{q \in \Pi} \max_{x \in [a,b]} |f(x) - q(x)| \qquad (7.1.2)$$

That is, we seek the function whose maximum deviation from the original function over the interval is minimized.

Finding such minimax approximations is usually difficult. At least in the important case of polynomial approximations over an interval $[a, b]$, however, there is a characterization of best uniform approximation which makes it easy to know when such a solution has been found. We return to this characterization and an algorithm for finding minimax polynomial approximations in the next section.

This algorithm, the *exchange algorithm*, is based on the idea of finding best "discrete L_∞" approximations. That is, we seek functions which minimize

$$\max\{|f(x_i) - p(x_i)|: i = 0, 1, \ldots, n + 1\}$$

where the set of reference points $x_0 < x_1 < \cdots < x_{n+1}$ in $[a, b]$ is adjusted in an iterative manner to reduce the maximum error.

Other approximations also make use of discrete versions of the various norms. These are then also appropriate for the curve-fitting task outlined earlier where we seek an approximant based on a set of data points.

In the case of L_1 approximations from the space Π_n of polynomials of degree at most n, we seek the polynomial $p^*(x)$ such that

$$\|f - p^*\|_1 = \int_a^b |f(x) - p^*(x)| \, dx$$

$$= \min_{a_0, a_1, \ldots, a_n} \int_a^b |f(x) - a_0 - a_1 x - \cdots - a_n x^n| \, dx \qquad (7.1.3)$$

Again there is no easy way of achieving this in its general form. However, if a discrete version of the problem is posed, then we seek instead to minimize

$$\sum_{i=0}^{N} |f(x_i) - a_0 - a_1 x_i - \cdots - a_n x_i^n| \qquad (7.1.4)$$

over all choices of the coefficients $a_0, a_1, \ldots, a_n$. This problem can be rearranged as a linear programming problem and then attacked by methods such as the simplex algorithm or Karmarkar's algorithm. The interested reader is referred to the extensive literature of linear programming for further details.

It is in the case of the L_2 norm that we can most readily establish solutions to the best approximation problem. Such approximations are usually called *least squares* approximations since they minimize the sum (in the discrete case) of the squared errors.

The continuous least squares approximation from Π_n to a function f over the interval $[a, b]$ is the polynomial p^* satisfying

$$\|f - p^*\|_2 = \sqrt{\int_a^b (f(x) - p^*(x))^2 \, dx}$$

$$= \min_{a_0, a_1, \ldots, a_n} \sqrt{\int_a^b (f(x) - a_0 - a_1 x - \cdots - a_n x^n)^2 \, dx} \qquad (7.1.5)$$

over all choices of the coefficients.

Since minimizing the square of this norm is equivalent to minimization of the norm itself, we seek then the set of coefficients $a_0, a_1, \ldots, a_n$ which minimizes

$$\int_a^b (f(x) - a_0 - a_1 x - \cdots - a_n x^n)^2 \, dx$$

This may be regarded as a function of the unknown coefficients, and differentiating with respect to each in turn, we obtain the equations

$$\int_a^b x^i (f(x) - a_0 - a_1 x - \cdots - a_n x^n) \, dx = 0 \qquad (i = 0, 1, \ldots, n) \qquad (7.1.6)$$

which represents a system of $n + 1$ linear equations in the $n + 1$ unknowns.

In the discrete case using a set of data points $x_0, x_1, \ldots, x_N$, where $N > n$, and the discrete version of the L_2 norm we seek p^* such that

$$\sum_{i=0}^{N} (f(x_i) - p^*(x_i))^2 = \min_{a_0, a_1, \ldots, a_n} \sum_{i=0}^{N} (f(x_i) - a_0 - a_i x_i - \cdots - a_n x_i^n)^2 \qquad (7.1.7)$$

and again we can differentiate with respect to the coefficients to obtain the linear system

$$\sum_{i=0}^{N} x_i^j (f(x_i) - a_0 - a_1 x_i - \cdots - a_n x_i^n) = 0 \qquad (j = 0, 1, \ldots, n) \qquad (7.1.8)$$

which can be rearranged as

$$\sum_{i=0}^{N} a_0 x_i^j + a_1 x_i^{j+1} + \cdots + a_n x_i^{n+j} = \sum_{i=0}^{N} x_i^j f(x_i) \qquad (j = 0, 1, \ldots, n)$$

or

$$X\mathbf{a} = \mathbf{f} \tag{7.1.9}$$

where the elements of the matrix X and the right-hand side $\mathbf{f}$ are

$$x_{ij} = \sum_{k=0}^{N} x_k^{i+j-2}$$

$$(i, j = 1, 2, \ldots, n+1) \tag{7.1.10}$$

$$f_i = \sum_{k=0}^{N} x_k^{i-1} f(x_k)$$

These equations (or their equivalent for the continuous case) are called the *normal* equations of the problem.

For n of even moderate size the normal equations are often ill-conditioned. This leads to the desire for alternative bases for our polynomial (or other approximating space) for which these equations take on a simpler form, more amenable to numerical solution.

We consider this in more detail in later sections but finish here with a simple example.

Example 7.1.3. Find the least squares quadratic approximation to $1/x$ on $[1, 4]$.

Solution. The problem is to find the polynomial $a_0 + a_1 x + a_2 x^2$ which minimizes

$$\int_1^4 \left(\frac{1}{x} - a_0 - a_1 x - a_2 x^2 \right)^2 dx$$

The normal equations given by (7.1.6) for this case are

$$\int_1^4 x^i \left(\frac{1}{x} - a_0 - a_1 x - a_2 x^2 \right) dx = 0$$

which we may rearrange to yield the three equations

$$\int_1^4 a_0 + a_1 x + a_2 x^2 \, dx = \int_1^4 \frac{1}{x} \, dx$$

$$\int_1^4 a_0 x + a_1 x^2 + a_2 x^3 \, dx = \int_1^4 x \frac{1}{x} \, dx$$

$$\int_1^4 a_0 x^2 + a_1 x^3 + a_2 x^4 \, dx = \int_1^4 x^2 \frac{1}{x} \, dx$$

These in turn simplify to

$$\begin{bmatrix} 1 & \frac{5}{2} & 7 \\ \frac{5}{2} & 7 & \frac{85}{4} \\ 7 & \frac{85}{4} & \frac{341}{5} \end{bmatrix} \begin{bmatrix} a_0 \\ a_1 \\ a_2 \end{bmatrix} = \begin{bmatrix} \frac{1}{3} \ln 4 \\ 1 \\ \frac{5}{2} \end{bmatrix}$$

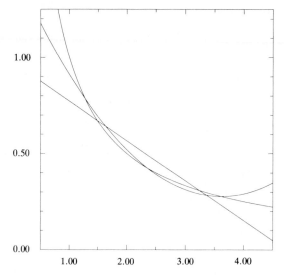

FIGURE 7.1.4
$1/x$ and its linear and quadratic least squares approximations on $[1, 4]$.

which yields the following solution to four decimal places:

$$a_2 = 0.0923 \qquad a_1 = -0.6685 \qquad a_0 = 1.4873$$

The original function and this approximant are plotted in Fig. 7.1.4 below together with the linear least squares approximation which is obtained from the equations

$$\begin{bmatrix} 1 & \frac{5}{2} \\ \frac{5}{2} & 7 \end{bmatrix} \begin{bmatrix} a_0 \\ a_1 \end{bmatrix} = \begin{bmatrix} \frac{1}{3}\ln 4 \\ 1 \end{bmatrix}$$

which have the solution $a_1 = -0.2070$, $a_0 = 0.9796$.

Note how the quadratic approximation has passed through its minimum even before the end of the interval $[1, 4]$. Monotonicity of the original function is not necessarily preserved. You will see in Exercise 2 that solving the same problem over $[1, 2]$, the quadratic approximation, is almost indistinguishable from $1/x$.

EXERCISES 7.1

1. Find the linear, quadratic, and cubic least squares approximations to e^x over $[-1, 1]$.
2. Write a program to find and graph least squares polynomial approximations. Use your best linear equation-solving routine for the solution of the linear system. Test your program by finding approximations to $1/x$ over intervals $[1, b]$ for $b = 2, 3, 4, \ldots, 10$.

7.2 UNIFORM APPROXIMATION

In this section we consider the problem of finding best *uniform* or *minimax* approximations to a continuous function f on the interval $[a, b]$. We restrict our attention to polynomial approximations. Thus we seek $p \in \Pi_n$ such that

$$\| f - p \|_\infty = \min_{q \in \Pi_n} \| f - q \|_\infty = \min_{q \in \Pi_n} \max_{x \in [a,b]} | f(x) - q(x) | \qquad (7.2.1)$$

The approach we use is the *exchange algorithm* which is based on selecting a set of *reference points* $x_0 < x_1 < \cdots < x_{n+1}$ and finding the element of Π_n to minimize $\max_i | f(x_i) - p(x_i) |$. This reference set is adjusted iteratively to improve the overall approximation.

Before considering the details, we must establish a characterization of best uniform approximations which will provide the theoretical basis for the algorithm. As a preliminary step, we establish the following proposition.

Proposition 7.2.1. The polynomial $p_0 \in \Pi_n$ with error function $e_0(x) = f(x) - p_0(x)$ is a best minimax approximation to f on $[a, b]$ if and only if there is no polynomial $q \in \Pi_n$ such that $(f(x) - p_0(x))q(x) > 0$ for every $x \in M$ where M is the set of points at which e_0 attains its norm.

Proof. M is the set of all points in $[a, b]$ at which the error $e_0(x)$ achieves its norm, and we denote this value by ϵ_0; thus

$$\epsilon_0 = \| e_0 \|_\infty = | f - p_0 |_\infty = \max_{x \in [a,b]} | f(x) - p_0(x) | \qquad (7.2.2)$$

and $x \in M \leftrightarrow | e_0(x) | = | f(x) - p_0(x) | = \epsilon_0$.

Suppose first that p_0 is not optimal. It follows that there is an optimal $p^* = p_0 + q$, say. For each $x \in M$, we have $| f(x) - p^*(x) | < | f(x) - p_0(x) |$, or, equivalently,

$$| f(x) - p_0(x) - q(x) | < | f(x) - p_0(x) | \qquad (7.2.3)$$

from which we deduce that $q(x)$ and $e_0(x)$ have the same sign at each $x \in M$; that is,

$$e_0(x)q(x) = (f(x) - p_0(x))q(x) > 0 \qquad (7.2.4)$$

for all $x \in M$.

For the other implication, suppose that (7.2.4) is satisfied by some $q \in \Pi_n$, and we may assume that $\| q \|_\infty = 1$. We will establish that p_0 is not optimal by showing that $p_0 + tq$ has a smaller error bound than ϵ_0 for some small value of t.

Denote by L the set $\{ x \in [a, b] : (f(x) - p_0(x))q(x) < 0 \}$. Note that $L \cap M = \emptyset$. Let $\epsilon_1 = \max_{x \in L} | e_0(x) |$ ($\epsilon_1 = 0$ if $L = \emptyset$). Now $\epsilon_1 < \epsilon_0$, and we let $t = (\epsilon_0 - \epsilon_1)/2$. If $x \in L$, then

$$| f(x) - p^*(x) | = | f(x) - p_0(x) - tq(x) |$$
$$= | f(x) - p_0(x) | + | tq(x) | \leq \epsilon_1 + t < \epsilon_0$$

since $q(x)$ and $f(x) - p_0(x)$ have opposite signs and $\| q \|_\infty = 1$. If, on the other hand, $x \not\in L$, then $q(x)$ and $f(x) - p_0(x)$ have the same sign and we may deduce directly that

$$|f(x) - p^*(x)| = |f(x) - p_0(x) - tq(x)|$$
$$< \max(|e_0(x)|, |tq(x)|) < \epsilon_0$$

since $t < \epsilon_0$ by construction. This completes the proof.

Essentially, this proposition says that if, for a given approximation, we can find another polynomial which has the same sign as the error at all points of maximum error, then this maximum error can be reduced by adding a small multiple of this new polynomial to the approximant (or subtracting it from the error). The intuitive explanation of this statement is that we can reduce these worst errors somewhat without forcing larger errors to occur elsewhere.

With this in mind the main characterization theorem below seems natural. This result states that a best minimax approximation must achieve its maximum error *exactly* $n + 2$ times in the interval $[a, b]$ and that these errors must alternate in sign. The intuitive reasoning behind this is that if there were a uniquely attained maximum error, then we should be able to "push it down" without allowing errors elsewhere to exceed it. The same argument applies for more than one such point until we reach a maximum number of such. A polynomial of degree n can interpolate the original function at $n + 1$ points, and so its error can have n turning points between these. By balancing these errors and those at the ends of the interval we can achieve the $n + 2$ above.

Note that we have already seen one example of this. In Sec. 4.6, we saw that the Chebyshev polynomial of degree n provided the minimax approximations to x^{n+1} by choosing the nodes of Lagrange interpolation so as to minimize the remainder. We observed that the Chebyshev polynomials have precisely the property just described, namely, $n + 2$ equal and alternating extrema. In particular,

$$T_n(x) = \cos(n \cos^{-1} x) \qquad (7.2.5)$$

takes the alternating values 1 and -1 at the points

$$y_k = \cos \frac{k\pi}{n+1} \qquad (k = 0, 1, \ldots, n+1) \qquad (7.2.6)$$

We will make further use of this shortly, but first we prove our characterization of best uniform approximations.

Theorem 7.2.2. The polynomial $p_0 \in \Pi_n$ with error function $e_0(x) = f(x) - p_0(x)$ is a best minimax approximation to f on $[a, b]$ if and only if there exist $n + 2$ points $a \le y_0 < y_1 < \cdots < y_{n+1} \le b$ such that

$$|e_0(y_i)| = \|e_0\|_\infty \qquad (i = 0, 1, \ldots, n+1) \qquad (7.2.7)$$

$$e_0(y_{i+1}) = -e_0(y_i) \qquad (i = 0, 1, \ldots, n) \qquad (7.2.8)$$

Proof. First, suppose that (7.2.7) and (7.2.8) are satisfied. If p_0 is not optimal, then by the proposition there exists $q \in \Pi_n$ such that $q(y_i)$ and $e_0(y_i)$ have the same sign for each i. Now (7.2.8) implies that $q(y_i)$ alternates in sign so that q changes sign at least $n + 1$ times. Since $q \in \Pi_n$, this is a contradiction.

Suppose now that p_0 is optimal. It follows that e_0 must change sign at least $n + 1$ times since otherwise we could find $q \in \Pi_n$ such that $e_0(x)q(x) > 0$ for every x in $[a, b]$ and, in particular, for every $x \in M$—a contradiction of p_0's optimality. Including the endpoints, it follows that there are $n + 2$ local extrema for the error function in $[a, b]$. The intuitive argument above establishes that the magnitudes of these extrema should be equal.

In this particular situation it can be proved that the best uniform approximation is unique, though we omit this proof here.

We turn to the exchange algorithm for computing minimax polynomial approximations. As was outlined above the idea is to use a reference set $a \le x_0 < x_1 < \cdots < x_{n+1} \le b$ and seek the polynomial $p \in \Pi_n$ which minimizes the maximum error on this set. We then adjust the reference set to reduce the maximum error on $[a, b]$. Clearly then there are two major steps to be considered—first how to find the polynomial p and then how to go about improving the reference.

For the first of these we use a corresponding result to Theorem 7.2.2 for discrete minimax approximation which establishes that we require the error function $e(x) = f(x) - p(x)$ to satisfy an equal oscillation condition. That is, the errors at the reference points should be equal in magnitude and alternating in sign:

$$e(x_i) = f(x_i) - p(x_i) = (-1)^i c \qquad (i = 0, 1, \ldots, n + 1) \qquad (7.2.9)$$

for some constant c. This system can be rewritten as a system of linear equations for the coefficients of the polynomial p as follows.

Writing

$$p(x) = a_0 + a_1 x + \cdots + a_n x^n$$

we can rearrange (7.2.9) in the form

$$
\begin{aligned}
a_0 + a_1 x_0 + \cdots + a_n x_0^n + c &= f_0 \\
a_0 + a_1 x_1 + \cdots + a_n x_1^n - c &= f_1 \\
\cdots\cdots\cdots\cdots\cdots\cdots\cdots\cdots\cdots\cdots\cdots\cdots\cdots\cdots & \qquad\qquad (7.2.10)\\
a_0 + a_1 x_n + \cdots + a_n x_n^n + (-1)^n c &= f_n \\
a_0 + a_1 x_{n+1} + \cdots + a_n x_{n+1}^n + (-1)^{n+1} c &= f_{n+1}
\end{aligned}
$$

where as usual f_i represents $f(x_i)$. The equivalent matrix system (7.2.11) below has a Vandermonde-like coefficient matrix which is nonsingular. (See Exercise 2.) It follows that there is a unique solution to the first part of our exchange algorithm iteration.

$$
\begin{bmatrix}
1 & x_0 & x_0^2 & \cdots & x_0^n & +1 \\
1 & x_1 & x_1^2 & \cdots & x_1^n & -1 \\
\hdotsfor{6} \\
1 & x_{n+1} & x_{n+1}^2 & \cdots & x_{n+1}^n & (-1)^{n+1}
\end{bmatrix}
\begin{bmatrix}
a_0 \\
a_1 \\
\cdots \\
a_n \\
c
\end{bmatrix}
=
\begin{bmatrix}
f_0 \\
f_1 \\
\cdots \\
f_{n+1}
\end{bmatrix}
\qquad (7.2.11)
$$

At this point we have a polynomial whose errors at the reference set are equal and alternating in sign. A typical error function for cubic approximation is shown in Fig. 7.2.3 together with the reference points. The reference points x_0, x_1, x_2, x_3, x_4 are the end points and the feet of the various line segments.

Denote by $y_1, y_2, \ldots, y_n$ the turning points of the error function. The "one-point" exchange algorithm uses whichever of these yields the global maximum error as a replacement for one of the current reference set. In the case of Fig. 7.2.3, y_1 would be incorporated into the next reference set in place of one of the existing elements.

The alternating sign property must be preserved, and the "exchange" is made accordingly. For the example above, therefore, y_1 would replace x_1 in the reference for the next iteration. This all assumes that the global maximum error does indeed exceed $|c|$. If not, the polynomial satisfies all the conditions of Theorem 7.2.2, and so it *is* the minimax approximation.

The only remaining question at this stage is how to decide which of the reference points should be omitted to make room for the global maximum point. This is simple.

Let y_i be the point at which the maximum error (in absolute value) is attained. Then y_i lies in one of the intervals $[x_{i-1}, x_i]$ or $[x_i, x_{i+1}]$. The entry to be deleted from the reference is the endpoint of the interval containing y_i at

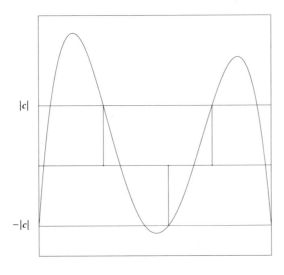

FIGURE 7.2.3
An error function for cubic minimax approximation.

which the error has the same sign as at y_i. The only possible exception to this situation is if the reference does not include the endpoints of the original interval $[a, b]$. If, for example, $a \le y_i < x_0$, then either x_0 or x_{n+1} will be omitted depending on which preserves the alternating sign property. Thus if $e(x_0)$ and $e(y_i)$ have the same sign, then x_0 is deleted. Similar comments apply to the other end of the interval.

There are several questions of detail which must be answered for any implementation, not least of which is how to find the turning points of the error function. This particular point is discussed in detail in Chap. 8 on optimization. For now we will simply suppose that some suitable method is available.

Before presenting the exchange algorithm in detail and considering examples, we should point out that there are other versions of the algorithm available. For example, we could replace more than just one point of the reference on each iteration by using every y_i for which $|e(y_i)| > |c|$.

The iteration will often not be slowed too much by replacing reference points with points where the (absolute value of the) error function exceeds $|c|$ rather than seeking the actual global extremum. It may be that less overall computational effort will be expended by performing more iterations if each of these is significantly cheaper.

This raises the question of the convergence of the exchange algorithm. We do not develop the full theory here but content ourselves with an outline. The first and most important observation is that the value of $|c|$ will *increase* while the maximum error will *decrease* as the iterations progress.

This is sufficient to give us a stopping criterion. At any stage of the iteration, we must have

$$|c| \le \|f - p_0\|_\infty \le \|f - p\|_\infty \tag{7.2.12}$$

where p_0 is the minimax approximation we seek. It follows that if $|c|$ and $|e(y_i)|$ agree to within a specified tolerance, then p and p_0 agree to that same accuracy.

We are now in a position to describe the one-point exchange algorithm for minimax approximation.

Algorithm 7.2.4.   The one-point exchange algorithm

Input   Function to be approximated, f
Degree of approximation, n and tolerance, ϵ
Initial reference $a \le x_0 < x_1 < \cdots < x_{n+1} \le b$

Iterate   Solve (7.2.11) to find $a_0, a_1, \ldots, a_n, c$ such that
$f(x_i) - (a_0 + a_1 x_i + \cdots + a_n x_i^n) = (-1)^i c$ for every i
Find y such that $|e(y)| = \max_{x \in [a,b]} |e(x)|$
If $|e(y)| - |c| > \epsilon$ **then** *update reference* **else** stop

Output   Coefficients of approximation polynomial
$a_0, a_1, \ldots, a_n$.

Update reference

if $y < x_0$ **then**
    if $e(y)e(x_0) > 0$ **then** $x_0 := y$
        else for $i = n + 1$ downto 1 $x_i := x_{i-1}$; $x_0 := y$
if $y \in [x_i, x_{i+1}]$ **then**
    if $e(y)e(x_i) > 0$ **then** $x_i := y$ **else** $x_{i+1} = y$
if $y > x_{n+1}$ **then**
    if $e(y)e(x_{n+1}) > 0$ **then** $x_{n+1} := y$
        else for $i = 0$ to n $x_i := x_{i+1}$; $x_{n+1} := y$

It is apparent from the nature of the algorithm and the input required that the choice of initial reference has a potentially substantial effect on the performance of the exchange algorithm. We consider this question shortly, but first a simple example.

Example 7.2.5. Find the minimax cubic approximation to the exponential function on $[-1, 1]$.

Solution. For our reference, we need five points, and we begin with them uniformly spaced: $x_0 = -1$, $x_1 = -0.5$, $x_2 = 0$, $x_3 = 0.5$, $x_4 = 1$.
The initial system of equations is then

$$\begin{bmatrix} 1 & -1 & 1 & -1 & 1 \\ 1 & -\frac{1}{2} & \frac{1}{4} & -\frac{1}{8} & -1 \\ 1 & 0 & 0 & 0 & 1 \\ 1 & \frac{1}{2} & \frac{1}{4} & \frac{1}{8} & -1 \\ 1 & 1 & 1 & 1 & 1 \end{bmatrix} \begin{bmatrix} a_0 \\ a_1 \\ a_2 \\ a_3 \\ c \end{bmatrix} = \begin{bmatrix} e^{-1} \\ e^{-1/2} \\ 1 \\ e^{1/2} \\ e^1 \end{bmatrix}$$

which yields the solution

$$a_0 = 0.99593 \qquad a_1 = 0.99785 \qquad a_2 = 0.54308 \qquad a_3 = 0.17735$$

and $c = 4.07 \times 10^{-3}$.
    The coefficients of the approximating polynomial are not much different from the first four terms of the power series for the exponential function. The graphs of this approximation and of e^x are almost indistinguishable. The approximation is graphed in Fig. 7.2.6a. The error function corresponding to it is graphed in Fig. 7.2.6b. Its maximum absolute value is 7.68×10^{-3} and occurs at (to three decimal places) 0.725. At this stage we may deduce that the minimax cubic approximation sought will have maximum error at least 4.07×10^{-3} and at most 7.68×10^{-3}. This maximum error occurs between x_3 and x_4 and has the same sign as that at x_3. We therefore replace 0.5 by 0.725 in the reference set for the second iteration. The reference is thus -1, 0.5, 0, 0.725, 1.
    Solving the linear equations (7.2.11) for this reference we get $c = 5.01 \times 10^{-3}$. The new error function is plotted along with the previous one in Fig. 7.2.6c. This has its largest deviation from 0 at $x = -0.675$. The error here is -7.02×10^{-3}. It follows that we replace x_1 with -0.675 for the next iteration.

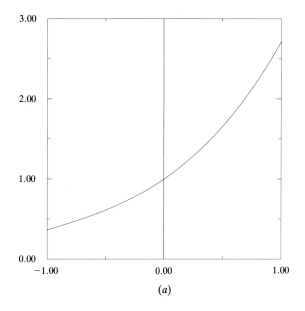

(a)

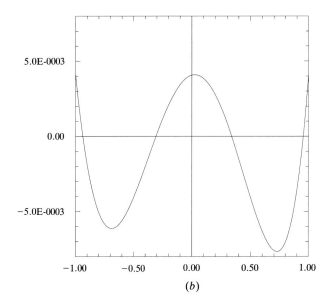

(b)

FIGURE 7.2.6
Minimax cubic approximation to e^x. (a) First iteration; (b) error function, first iteration; (c) iterations 1 and 2; (d) iterations 1, 2, 3.

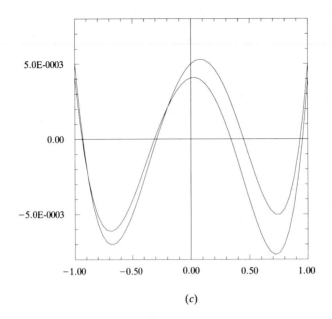

(c)

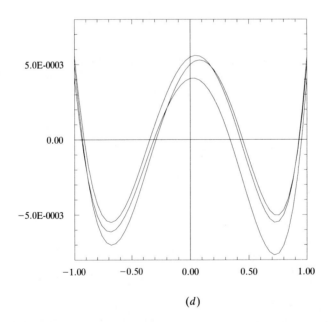

(d)

FIGURE 7.2.6 (*Continued*)

TABLE 7.2.7
Exchange algorithm applied to Example 7.2.5

Reference					$\|c\| \times 10^3$	$\|e(y)\| \times 10^3$	y
−1	−0.5	0	0.5	1	4.07	7.68	0.725
−1	−0.5	0	0.725	1	5.01	7.02	−0.675
−1	−0.675	0	0.725	1	5.50	5.61	0.050
−1	−0.675	0.05	0.725	1	5.53	5.53	0.730

The result of iteration 3 is shown along with the first two in Fig. 7.2.6d, and the results of the first four iterations—sufficient for convergence with $\epsilon = 10^{-5}$—are summarized in Table 7.2.7. We see from the final row of this table that $\|c\|$ and the maximum error agree to the accuracy quoted. This, of course, implies that our approximation agrees with the true minimax approximation to this same accuracy.

Note how the original uniform distribution of the reference set has been modified by the algorithm. The first two iterations have the effect of pulling reference points toward the ends of the interval. Thereafter, only minor modifications of the reference set are made, and in fact we see that the quality of approximation is surprisingly insensitive to changes in this reference.

It is also worth commenting on the "maximization" procedure used here. The program used to generate these results and graphs necessarily needed to evaluate the error at a large number of points of the interval $[-1, 1]$—for these graphs 401 points were used. The interval between these was therefore just 0.005. It was sufficient for our purposes to simply keep track of the largest error recorded and use this as the y of the algorithm. In general, it is not desired to graph all the intermediate approximations, and such a process would be highly inefficient. It is certainly possible to find all the extrema of the error function in many fewer than 400 function evaluations, as we will see in the next chapter.

We have seen that the exchange algorithm can provide rapid convergence to the desired minimax approximation. It remains to discuss the choice of the initial reference. It was remarked earlier that the Chebyshev polynomials have the equioscillation property that we seek for the minimax error function. We also know that interpolation at the zeros of T_{n+1} provides the minimax approximation to a polynomial of degree $n + 1$ from Π_n on $[-1, 1]$. (See Sec. 4.6.)

Now if f is a smooth function, then it may be well-approximated by a Taylor polynomial. If we seek a minimax approximation of degree n to f, then a good starting point would probably be the minimax approximation to its Taylor polynomial of degree $n + 1$. For the interval $[-1, 1]$ this is achieved by interpolation at the zeros of T_{n+1}. For the exchange algorithm, the reference provides a set of points at which the error has the equioscillation property. For

T_{n+1} these points are given by (7.2.6). Rearranging them into increasing order, we have the reference:

$$x_k = \cos \frac{(n+1-k)\pi}{n+1} \qquad k = 0, 1, \ldots, n+1 \qquad (7.2.13)$$

A simple linear transformation provides a good initial reference for other intervals $[a, b]$.

For the example above, we sought a minimax cubic approximation to e^x on $[-1, 1]$, and so the initial reference would be given by (7.2.13) with $n = 3$. That is,

$$x_0 = -1 \qquad x_1 = \frac{-1}{\sqrt{2}} \qquad x_2 = 0 \qquad x_3 = \frac{1}{\sqrt{2}} \qquad x_4 = 1$$

We saw in Table 7.2.7 that the exchange algorithm had the effect of pulling points of the reference toward the ends of the interval. In particular, the first two iterations resulted in deletion of the points at $\pm \frac{1}{2}$ with points close to ± 0.7, which is close to the Chebyshev reference. The results corresponding to Table 7.2.7 for the Chebyshev reference are shown in Table 7.2.8.

We see that the first iteration is indeed significantly improved on that of Table 7.2.7 with agreement between $|c|$ and $|e(y)|$ to nearly five decimal places already. The minor adjustments to the reference make little difference over the next two iterations, but after four iterations we have one more decimal-place accuracy than in the earlier example. We should also remark that the apparent increase in $\|e\|$ between iterations 2 and 3 is the result of the crude maximization discussed above.

The graphs of the various error functions for this case were so similar that no insight is to be gained from them.

As a final comment on minimax approximation, it is worth noting that the maximization process used in the examples has the effect of converting the problem to a discrete minimax approximation in which the functions can only be evaluated at certain tabulated points. In this case these were the points used for graphing the functions. It is apparent, therefore, that the obvious modification of the exchange algorithm for the discrete case provides a highly efficient approach to such problems.

TABLE 7.2.8
Exchange algorithm for Example 7.2.5 using the Chebyshev initial reference

Reference					$\lvert c \rvert \times 10^3$	$\lvert e(y) \rvert \times 10^3$	y
-1	-0.707	0	0.707	1	5.474	5.581	0.050
-1	-0.707	0.050	0.707	1	5.501	5.560	0.730
-1	-0.707	0.050	0.730	1	5.517	5.567	-0.680
-1	-0.680	0.050	0.730	1	5.528	5.528	-1.000

EXERCISES 7.2

1. Show that $\frac{1}{2}$ is the best uniform constant approximation to $\cos x$ on $[-\pi/2, \pi/2]$. Is there a better straight-line approximation? Find the minimax quadratic approximation.

2. Prove that the matrix of Eq. (7.2.11) for the linear system of the exchange algorithm is nonsingular. (*Hint*: One way is to expand the determinant about the right-hand column, noting that each of the subdeterminants is a Vandermonde determinant.)

3. Write a program to implement the exchange algorithm. Check it by reproducing the graphs of Fig. 7.2.6 and the results of Tables 7.2.7 and 7.2.8.

4. Use your program to find minimax approximations to sine and cosine over $[-\pi/2, \pi/2]$ of degrees one through five. Graph the resulting polynomials and the original functions.

5. Show that the "Chebyshev reference" for an interval $[a, b]$ is given by

$$x_k = \frac{a+b}{2} + \frac{b-a}{2} \cos \frac{(n+1-k)\pi}{n+1}$$
$$k = 0, 1, \ldots, n+1$$

Using this initial reference, find minimax approximations of degrees zero through four to $1/(1+x^2)$ on $[-5, 5]$.

6. Find minimax linear, quadratic, and cubic approximations to $\sqrt{x}$ on $[0, 1]$. Graph these approximations and their error functions.

7.3 LEAST SQUARES APPROXIMATION

In this section we turn our attention to what is probably the most widely used type of approximation: *least squares* or best L_2 approximation. There are a great many variations on this theme, and we will explore just some of them. We will consider other aspects of least squares approximation in discussing curve fitting at the end of the chapter.

Initially, we concentrate on continuous least squares approximation in which we seek to approximate $f \in C[a, b]$. In the simplest case we seek polynomial approximations from Π_n. That is, we must find $p \in \Pi_n$ such that

$$\|f - p\|_2^2 = \int_a^b (f(x) - p(x))^2 \, dx \qquad (7.3.1)$$

is minimized. Writing p in its standard form as

$$p(x) = a_0 + a_1 x + a_2 x^2 + \cdots + a_n x^n \qquad (7.3.2)$$

it follows that we seek coefficients $a_0, a_1, \ldots, a_n$ such that

$$F(a_0, a_1, \ldots, a_n) = \int_a^b (f(x) - a_0 - a_1 x - \cdots - a_n x^n)^2 \, dx \qquad (7.3.3)$$

is minimized.

Now,

$$\frac{\partial F}{\partial a_i} = -2 \int_a^b x^i (f(x) - a_0 - a_1 x - \cdots - a_n x^n) \, dx \qquad (7.3.4)$$

for each i. So F will be minimized when $\partial F/\partial a_i = 0$ for every i. Rearranging (7.3.4), we thus obtain the *normal equations*

$$a_0 \int_a^b x^i \, dx + a_1 \int_a^b x^{i+1} \, dx + \cdots + a_n \int_a^b x^{i+n} \, dx = \int_a^b x^i f(x) \, dx \qquad (7.3.5)$$

or

$$Xa = f \qquad (7.3.6)$$

where the matrix X and right-hand-side vector f have elements

$$x_{ij} = \int_a^b x^{i+j-2}\, dx \qquad f_i = \int_a^b x^{i-1} f(x)\, dx \qquad (7.3.7)$$

Again the problem is reduced to the solution of a system of linear equations.

Example 7.3.1. Find the least squares approximations to $\sqrt{x}$ on $[0, 1]$ of degrees one, two, and three.

Solution. Using (7.3.7), we see that for this interval the elements of the matrix X are given by

$$x_{ij} = \frac{1}{i+j-1}$$

while those of f are

$$f_1 = \int_0^1 \sqrt{x}\, dx = \tfrac{2}{3}$$

$$f_2 = \int_0^1 x\sqrt{x}\, dx = \tfrac{2}{5}$$

$$f_3 = \int_0^1 x^2\sqrt{x}\, dx = \tfrac{2}{7}$$

$$f_4 = \int_0^1 x^3\sqrt{x}\, dx = \tfrac{2}{9}$$

For the linear fit, therefore, we must solve

$$\begin{bmatrix} 1 & \tfrac{1}{2} \\ \tfrac{1}{2} & \tfrac{1}{3} \end{bmatrix} \begin{bmatrix} a_0 \\ a_1 \end{bmatrix} = \begin{bmatrix} \tfrac{2}{3} \\ \tfrac{2}{5} \end{bmatrix}$$

which yields $a_0 = 0.267$, $a_1 = 0.800$ to three decimal places. This function and $\sqrt{x}$ are graphed in Fig. 7.3.3a. The remaining two graphs add the quadratic and cubic approximations.

The coefficients of these approximating polynomials are shown in Table 7.3.2 below.

TABLE 7.3.2
Coefficients of least squares approximations to $\sqrt{x}$ on $[0, 1]$

	Linear	Quadratic	Cubic
a_0	0.267	0.171	0.127
a_1	0.800	1.371	1.905
a_2		−0.571	−1.905
a_3			0.889

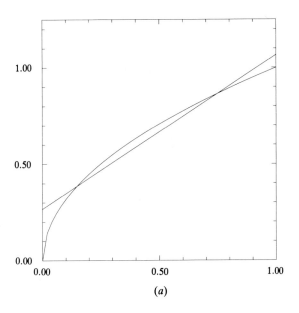

(a)

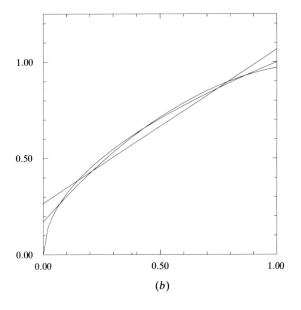

(b)

FIGURE 7.3.3
Least squares approximations to $\sqrt{x}$ on $[0, 1]$. (a) Linear; (b) linear and quadratic; (c) linear, quadratic, and cubic.

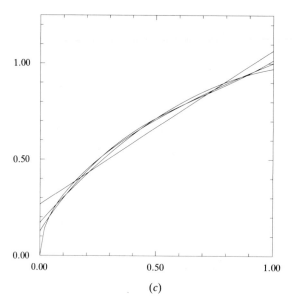

(*c*)

FIGURE 7.3.3 (*Continued*)

There are several comments worth making about this example.

First, we see that the coefficients of the three approximations bear no resemblance to one another. This suggests difficulty in increasing the degree of the approximation. Indeed, just as in the case of Lagrange interpolation, this simple approach necessitates a complete new solution if the degree is increased. For this particular example this is not a major difficulty since the systems are all small and easily solved.

Second, we see that the quality of the fit is steadily improving as the degree increases, but this is a somewhat oversimplistic conclusion. We know from Weierstrass' theorem that we can force the minimax error—and therefore the L_2 error—to be arbitrarily small by allowing the degree of the polynomial to increase. However, if the graphs were extended beyond the specific interval of the approximation, we would see wild fluctuations from the square root function. As with interpolation, it is dangerous to use such approximations beyond the interval on which they are obtained.

There is a further and more troublesome point about increasing the degree of the approximating polynomial. The matrix X of the linear system (7.3.6) for the interval [0, 1] is the Hilbert matrix of appropriate dimension. This matrix is notoriously ill-conditioned—see the projects of Chap. 6, for example. Therefore the solution of the linear system is likely to prove difficult or inaccurate. The theoretical improvement in the approximation may not be realized in practice due to numerical instability.

These computational difficulties can be overcome by using different representations for the approximating polynomials. We use representations in terms of *orthogonal polynomials*. This has the effect of rendering the system of normal equations diagonal so that the coefficients can be computed directly. A further benefit is that increasing the degree of the approximating polynomial reduces to the computation of one more coefficient.

We observe that the quadratic and cubic polynomials both give excellent agreement with the square root over the central part of the interval. But, at the left-hand end especially, there is still significant discrepancy. This is caused in part by the nature of the function being approximated which has a singularity at that endpoint where it has infinite slope. However, the theory presented above only demands continuity of f.

At 0, the three polynomials take the values 0.267, 0.171, and 0.127. It appears that the fit of the polynomial to the original curve is improving only slowly at this end. At 1, the values are 1.067, 0.971, and 1.016, which again are improving—but slowly. In the middle of the interval the agreement is much better, as can be seen from the graphs of the error functions in Fig. 7.3.4. Note how different the behavior of these error functions is from that of minimax approximations. (Compare with Exercise 6 in Sec. 7.2.) There is no equioscillation this time.

In some cases, it may well be beneficial to give greater importance to some parts of the interval. For the above example, it seems that increasing the weighting near the ends would be beneficial. This can be achieved by the introduction of a *weight function*.

We are now in a position to develop the general theory of continuous least squares approximations.

Let $f \in C[a, b]$ and let the *weight function* w be a fixed positive integrable function on $[a, b]$. We seek approximations to f from the finite-dimensional

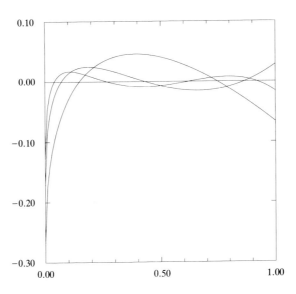

FIGURE 7.3.4
Error functions for the approximations of Example 7.3.1.

subspace $\mathbf{F} \subseteq \mathbf{C}[a, b]$. For much of our discussion $\mathbf{F}$ will be a space of polynomials, but this is not necessary in general.

We define an inner product and therefore a norm on $\mathbf{C}[a, b]$ with respect to w by

$$\langle f, g \rangle = \int_a^b w(x)f(x)g(x) \, dx$$

$$\|f\| = \sqrt{\langle f, f \rangle} = \int_a^b w(x)f^2(x) \, dx \tag{7.3.8}$$

for $f, g \in \mathbf{C}[a, b]$.

Let $\{\phi_0, \phi_1, \ldots, \phi_n\}$ be a basis for the space $\mathbf{F}$. Any element of $\mathbf{F}$ can be written as a linear combination of this basis, and so the inner product and norm for members of $\mathbf{F}$ are also expressible in terms of the inner products of the basis functions.

Let

$$p = \sum_{i=0}^n \alpha_i \phi_i \qquad q = \sum_{i=0}^n \beta_i \phi_i \tag{7.3.9}$$

Then

$$\langle p, q \rangle = \sum_{i=0}^n \sum_{j=0}^n \int_a^b \alpha_i \beta_j w(x) \phi_i(x) \phi_j(x) \, dx = \sum_{i=0}^n \sum_{j=0}^n \alpha_i \phi_{ij} \beta_j \tag{7.3.10}$$

where

$$\phi_{ij} = \langle \phi_i, \phi_j \rangle \tag{7.3.11}$$

Equation (7.3.10) can be rewritten

$$\langle p, q \rangle = \boldsymbol{\alpha}^{\mathrm{T}} \boldsymbol{\Phi} \boldsymbol{\beta} \tag{7.3.12}$$

where the vectors $\boldsymbol{\alpha}$ and $\boldsymbol{\beta}$ consist of the coefficients of the expressions (7.3.9) and $\boldsymbol{\Phi}$ has elements given by (7.3.11).

Our task is to find the least squares approximation to f on $[a, b]$ with respect to w from the space $\mathbf{F}$. That is, we seek $p \in \mathbf{F}$ such that

$$\|f - p\|^2 = \min_{\alpha_0, \alpha_1, \ldots, \alpha_n} \|f - \alpha_0 \phi_0 - \alpha_1 \phi_1 - \cdots - \alpha_n \phi_n\|^2$$

$$= \min_{\alpha_0, \alpha_1, \ldots, \alpha_n} \int_a^b w(x)(f(x) - \alpha_0 \phi_0(x) - \alpha_1 \phi_1(x) - \cdots - \alpha_n \phi_n(x))^2 \, dx \tag{7.3.13}$$

Differentiating with respect to each of the coefficients and setting the derivatives to zero, we arrive at the general form of the normal equations:

$$\boldsymbol{\Phi} \boldsymbol{\alpha} = \mathbf{f} \tag{7.3.14}$$

where the matrix $\boldsymbol{\Phi}$ and vector $\boldsymbol{\alpha}$ are defined by (7.3.9) and (7.3.11) and $\mathbf{f}$ has elements

$$\langle f, \phi_i \rangle = \int_a^b w(x)f(x)\phi_i(x) \, dx \qquad (7.3.15)$$

It is left to the reader to verify that (7.3.14) reduces to the familiar form summarized by (7.3.6) for the case where **F** is the space of polynomials Π_n and the weight function $w(x) \equiv 1$.

Thus far we have included the weight function in our theory. This can be used to force closer agreement between the original function and its approximant in specific parts of the interval.

An important example is a weight function which increases the relative weight of points near the ends of the interval. For the interval $[-1, 1]$ a good choice for such a weight function is

$$w(x) = \frac{1}{\sqrt{1 - x^2}} \qquad (7.3.16)$$

which is clearly positive and integrable over $[-1, 1]$. (Its integral is π.) Also $w(x) \geq 1$ throughout the interval, $w(0) = 1$, and $w(x) \rightarrow \infty$ as $x \rightarrow \pm 1$ so that the weight is increased significantly near the ends of the range.

Example 7.3.5. The function $f(x) = \sqrt{1 - x^2}$ on $[-1, 1]$ represents the upper half of the unit circle centered at the origin. Its approximation presents the same difficulties as in Example 7.3.1—except at both ends of its range. Find the least squares approximations of degrees zero to three to f both with and without the weight function $w(x) = 1/\sqrt{1 - x^2}$.

Solution. First, the function to be approximated is an even function and the interval is symmetric about 0. Also the product of f and any odd function is odd, and the integral of any odd function over $[-1, 1]$ is necessarily 0. It follows that the approximations sought will all be even, too. Therefore the linear and cubic least squares fits will just be the constant and quadratic ones.

Since w is also even, similar comments apply for the second case.

For the constant case with weight function 1, the "system" of equations is simply

$$a_0 \int_{-1}^1 x^0 \, dx = \int_{-1}^1 f(x) \, dx = \frac{\pi}{2}$$

which, of course, yields the approximation $\pi/4$. This has its greatest errors $\pi/4 = 0.7854$ at $x = \pm 1$.

Using the weighted least squares approximation with weight function w, the equation is

$$\alpha_0 \int_{-1}^1 \frac{1}{\sqrt{1 - x^2}} \, dx = \pi\alpha_0 = \int_{-1}^1 w(x)f(x) \, dx = \int_{-1}^1 dx = 2$$

so that the weighted least squares constant (and linear) approximation is $2/\pi = 0.6366$. This represents an improvement of some 19 percent at the ends of the range at a cost of some deterioration in the middle.

For the quadratic (and cubic) approximations, using the observations above, we seek approximations of the form $a_0 + a_2x^2$. In the nonweighted (or weight function 1) case, this results in the equations

$$\begin{bmatrix} 2 & \dfrac{2}{3} \\[2mm] \dfrac{2}{3} & \dfrac{2}{5} \end{bmatrix} \begin{bmatrix} a_0 \\[2mm] a_2 \end{bmatrix} = \begin{bmatrix} \dfrac{\pi}{2} \\[2mm] \dfrac{\pi}{8} \end{bmatrix}$$

which has the solution $a_0 = 21\pi/64$, $a_2 = -15\pi/64$ so that the approximation is

$$p(x) = \frac{3\pi}{64}(7 - 5x^2)$$

This function is graphed with the semicircle itself in Fig. 7.3.6a.

The corresponding graph for the weighted least squares approximation is shown in Fig. 7.3.6b. For this case the equations are

$$\begin{bmatrix} \pi & \dfrac{\pi}{2} \\[2mm] \dfrac{\pi}{2} & \dfrac{3\pi}{8} \end{bmatrix} \begin{bmatrix} \alpha_0 \\[2mm] \alpha_2 \end{bmatrix} = \begin{bmatrix} 2 \\[2mm] \dfrac{2}{3} \end{bmatrix}$$

from which we obtain the approximation

$$p_w(x) = \frac{2}{3\pi}(5 - 4x^2)$$

The errors at the ends of the range are 0.2945 for p and 0.2122 for p_w, which represents a 28 percent improvement. The cost at the midpoint of the range is an increase in the magnitude of the error from 0.0308 for p to 0.0610 for p_w.

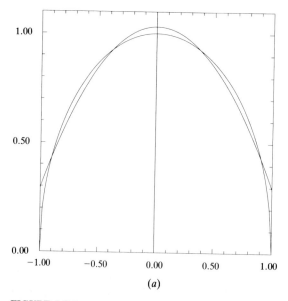

(a)

FIGURE 7.3.6
Least squares approximations to $y = \sqrt{1 - x^2}$, $x \in [-1, 1]$. (a) Standard, $w(x) \equiv 1$; (b) weighted, $w(x) = 1/\sqrt{1 - x^2}$; (c) error functions.

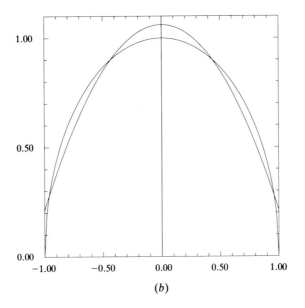

(b)

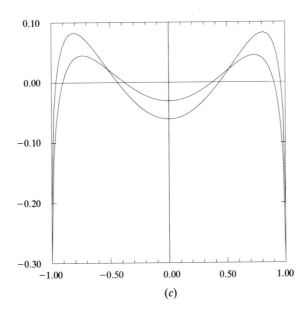

(c)

FIGURE 7.3.6 (*Continued*)

    The two error curves are plotted in Fig. 7.3.6c, which shows clearly the sharper peaks of the error function for the weighted approximation with its greater errors in the middle of the range together with its significantly reduced errors at the ends.

The computation of Example 7.3.5 was straightforward due to the simplification rendered by the even nature of the functions. Even in this case, however, if the degree were allowed to increase substantially, the task would be much less amenable to efficient computation using the standard basis for our polynomial spaces.

For the rest of this section, we are concerned with finding alternative bases for **F** which simplify the computation. This is based on the notion of orthogonality which, in just the same way as for vectors in two- or three-dimensional space, corresponds to the inner product of two elements vanishing.

Definition 7.3.7. We say that two members of $C[a, b]$ are said to be *orthogonal* with respect to the weight function w if their inner product is zero; that is, f and g are orthogonal if

$$\langle f, g \rangle = \int_a^b w(x)f(x)g(x)\, dx = 0$$

Now if we can generate an orthogonal basis for **F**, it will follow that all off-diagonal elements of Φ will be zero, in which case the system (7.3.14) reduces to a diagonal system with the solution

$$\alpha_i = \frac{f_i}{\phi_{ii}} = \frac{\langle f, \phi_i \rangle}{\langle \phi_i, \phi_i \rangle} = \frac{\int_a^b w(x)f(x)\phi_i(x)\, dx}{\int_a^b w(x)\phi_i^2(x)\, dx} \qquad (7.3.17)$$

Such an orthogonal basis can be obtained from any given basis by the Gram-Schmidt process, but in many situations there are easier techniques available. At this stage we present a few simple examples, and in the next section we concentrate briefly on the general theory of orthogonal polynomials.

Example 7.3.8. Find an orthogonal basis for the space Π_3 of polynomials of degree not greater than three with respect to the usual inner product and norm corresponding to $w(x) = 1$ on the interval $[-1, 1]$.

Solution. We will obtain a basis in which each element is of degree one more than its predecessor beginning with a polynomial of degree zero—a constant. The choice of this constant is arbitrary; 1 is convenient and so we set

$$\phi_0(x) = 1$$

Next we seek a linear polynomial ϕ_1 which is orthogonal to ϕ_0. The leading coefficient of this polynomial is again arbitrary and we again choose 1. Thus we may write $\phi_1(x) = x + c$, and we must choose c such that $\langle \phi_0, \phi_1 \rangle = 0$. Thus we require

$$\int_{-1}^1 1(x + c)\, dx = 2c = 0$$

Hence we have

$$\phi_1(x) = x$$

Continuing in like manner, we set $\phi_2(x) = x^2 + c_1x + c_0$ and choose the coefficients so that

$$\int_{-1}^{1} 1(x^2 + c_1x + c_0)\, dx = \int_{-1}^{1} x(x^2 + c_1x + c_0)\, dx = 0$$

This yields $c_1 = 0$ and $c_0 = -\frac{1}{3}$ so that

$$\phi_2(x) = x^2 - \tfrac{1}{3}$$

The same process then gives

$$\phi_3(x) = x^3 - \frac{3x}{5}$$

These are the first few Legendre polynomials.

To see the effect of using orthogonal polynomials in a least squares approximation, we repeat the calculation of Example 7.3.5 using these Legendre polynomials.

We write

$$\sqrt{1 - x^2} \simeq \alpha_0\phi_0(x) + \alpha_1\phi_1(x) + \alpha_2\phi_2(x) + \alpha_3\phi_3(x)$$

and use (7.3.17) to find the coefficients. The same arguments relating to the even nature of f and its approximant imply that $\alpha_1 = \alpha_3 = 0$. For the remaining coefficients, we use

$$\langle \phi_0, \phi_0 \rangle = 2 \qquad \text{and} \qquad \langle \phi_2, \phi_2 \rangle = \tfrac{8}{45}$$

Also,

$$\langle f, \phi_0 \rangle = \int_{-1}^{1} \sqrt{1 - x^2}\, dx = \frac{\pi}{2}$$

$$\langle f, \phi_2 \rangle = \int_{-1}^{1} (x^2 - \tfrac{1}{3})\sqrt{1 - x^2}\, dx = \frac{-\pi}{24}$$

from which we obtain the coefficients $\alpha_0 = \pi/4$, $\alpha_2 = -15\pi/64$ so that the required approximation is

$$\frac{\pi}{4} - \frac{15\pi(x^2 - \tfrac{1}{3})}{64}$$

which can be expanded to yield $21\pi/64 - 15\pi x^2/64$ as before. (Note that this final expansion is included solely to confirm that we have found the same approximation; it would not form part of the usual solution process.)

Using the weight function $w(x) = 1/\sqrt{1 - x^2}$ on the interval $[-1, 1]$ we get the Chebyshev polynomials T_n as the orthogonal family. (See Exercise 5.)

As a final example for this section, we consider an important case of nonpolynomial functions.

Example 7.3.9. The functions $\sin nx$ and $\cos mx$ are all orthogonal on $[-\pi, \pi]$.

Solution. We illustrate this fact for the case of two sine functions:

$$\int_{-\pi}^{\pi} \sin mx \sin nx \, dx = \frac{1}{2}\int_{-\pi}^{\pi} \cos(m-n)x - \cos(m+n)x \, dx$$

and if $m \neq n$ this vanishes. If, however, $m = n \neq 0$, the first term of the integrand is identically 1, and so the integral is just π. The proofs for the other two cases are similar. (See Exercise 7.)

These orthogonality relations form the basis of Fourier approximation for periodic functions.

EXERCISES 7.3

1. Find least squares polynomial approximations of degrees up to three to e^x on $[-1, 1]$. Compare these with the approximations obtained using the weight function $w(x) = 1/\sqrt{1-x^2}$.

2. Obtain least squares approximations to $\ln x$ on $[1, 4]$. Obtain the best fit you can to the original function.

3. Using the weight function $1/\sqrt{25-x^2}$, find polynomial least squares approximations to $1/(1+x^2)$ on $[-5, 5]$.

4. Show that the Legendre polynomial of degree two on $[-1, 1]$ is $x^2 - \frac{1}{3}$ and that the next one is $x^3 - 3x/5$.

5. Prove that the Chebyshev polynomials are orthogonal with respect to $w(x) = 1/\sqrt{1-x^2}$ on $[-1, 1]$. [Recall that the Chebyshev polynomials are given by $T_n(x) = \cos(n \cos^{-1} x)$.]

6. Check the approximations obtained in Example 7.3.5 and Exercise 1 using the appropriate orthogonal polynomials.

7. Prove the remaining orthogonality relations for $\sin nx$ and $\cos mx$. (See Example 7.3.9.) Find the least squares approximation of the form $\sum_{i=1}^{5} a_i \sin ix$ to x on $[-\pi, \pi]$.

7.4 SYSTEMS OF ORTHOGONAL POLYNOMIALS

In this section we develop some of the general theory and properties of orthogonal polynomials on $[a, b]$. We have already seen in examples that systems of such polynomials can be found by the elementary method of undetermined coefficients and commented that the Gram-Schmidt orthogonalization process could be applied to any basis to yield an orthogonal system. Either of these methods would become tedious very rapidly if any reasonable number of members of the system were required.

For the general theory, we fix our notation as follows. We seek and study systems of polynomials $\phi_0, \phi_1, \ldots$ with the properties that:

1. Each ϕ_i is a polynomial of degree i.

2.
$$\langle \phi_i, \phi_j \rangle = \int_a^b w(x)\phi_i(x)\phi_j(x)\, dx = 0 \qquad (i \neq j) \qquad (7.4.1)$$

These properties are not sufficient to define the systems *uniquely* since any constant multiples of the polynomials will also form an orthogonal system. In any particular case, we consider some "normalization" of the system. The most common are:

either each ϕ_i is *monic* (that is has leading coefficient 1)

or
$$\|\phi_i\|^2 = \langle \phi_i, \phi_i \rangle = 1 \qquad (7.4.2)$$

An orthogonal system satisfying (7.4.2) is called *orthonormal*.

There are some elementary properties which should be established before proceeding.

Proposition 7.4.1. Let $\phi_0, \phi_1, \ldots$ be an orthogonal system satisfying conditions 1 and 2 above. Then:

(a) The set $\{\phi_0, \phi_1, \ldots, \phi_n\}$ is linearly independent and is therefore a basis for Π_n.

(b) $\langle \phi_{n+1}, q \rangle = 0$ for every polynomial $q \in \Pi_n$, in particular, for $k \le n$,

$$\langle \phi_{n+1}, x^k \rangle = \int_a^b w(x) x^k \phi_{n+1}(x)\, dx = 0 \qquad (7.4.3)$$

Proof

(a) Suppose that $\{\phi_0, \phi_1, \ldots, \phi_n\}$ is not linearly independent. It follows that the zero polynomial can be expressed as a linear combination of them. Thus, we may write

$$\sum_{i=0}^{n} c_i \phi_i \equiv 0$$

where, of course, not all the coefficients are zero. Forming the inner product with each of the ϕ_i in turn and using the orthogonality condition (7.4.1), we obtain $c_i \langle \phi_i, \phi_i \rangle = c_i \|\phi_i\|^2 = 0$. This contradicts the c_i being nonzero.

(b) The proof of (b) is now straightforward and is left as an exercise.

Part (b) of the proposition establishes that to find a system of orthogonal polynomials by the method of undetermined coefficients, it is sufficient to check that ϕ_{k+1} is orthogonal to $1, x, \ldots, x^k$. However, this is not an efficient way of obtaining more than the first few.

One of the simplest approaches is derived from the following beautiful theorem which establishes that all systems of orthogonal polynomials satisfy a three-term recurrence relation. The theorem represents a simplification of the Gram-Schmidt process for orthogonalizing the standard basis $1, x, x^2, \ldots$.

Theorem 7.4.2. The sequence of orthogonal *monic* polynomials (leading coefficient 1) on $[a, b]$ with respect to the weight function w satisfies

$$\phi_0(x) = 1 \qquad \phi_1(x) = x - a_1$$

$$\phi_n(x) = (x - a_n)\phi_{n-1} - b_n \phi_{n-2} \qquad (n > 1) \qquad (7.4.4)$$

where

$$a_n = \frac{\langle x\phi_{n-1}, \phi_{n-1} \rangle}{\|\phi_{n-1}\|^2} \qquad b_n = \frac{\langle x\phi_{n-1}, \phi_{n-2} \rangle}{\|\phi_{n-2}\|^2} \qquad (7.4.5)$$

Proof. Clearly, each polynomial is monic of the appropriate degree. It follows that the formulas for the coefficients are well-defined. The proof of orthogonality is inductive. First,

$$\langle \phi_1, \phi_0 \rangle = \langle (x - a_1)\phi_0, \phi_0 \rangle = 0$$

by construction.

Suppose now that $\phi_0, \phi_1, \ldots, \phi_{k-1}$ are orthogonal. We must establish that ϕ_k given by (7.4.4) is orthogonal to each of them. Now, for $i < k - 2$, we have

$$\langle \phi_k, \phi_i \rangle = \langle (x - a_k)\phi_{k-1}, \phi_i \rangle - b_k \langle \phi_{k-2}, \phi_i \rangle$$
$$= \langle x\phi_{k-1}, \phi_i \rangle = \langle \phi_{k-1}, x\phi_i \rangle = 0$$

using (b) of Proposition 7.4.1. Also,

$$\langle \phi_k, \phi_{k-2} \rangle = \langle (x - a_k)\phi_{k-1}, \phi_{k-2} \rangle - b_k \|\phi_{k-2}\|^2$$
$$= \langle x\phi_{k-1}, \phi_{k-2} \rangle - b_k \|\phi_{k-2}\|^2$$

using the orthogonality of ϕ_{k-1} and ϕ_{k-2}. The definition of b_k is chosen to make this vanish. Similarly, and finally,

$$\langle \phi_k, \phi_{k-1} \rangle = \langle (x - a_k)\phi_{k-1}, \phi_{k-1} \rangle - b_k \langle \phi_{k-1}, \phi_{k-2} \rangle$$
$$= \langle x\phi_{k-1}, \phi_{k-1} \rangle - a_k \langle \phi_{k-1}, \phi_{k-1} \rangle = 0$$

by the definition of a_k. This completes the induction and the proof.

This result has importance not only in the development of systems of orthogonal polynomials but also in the evaluation of orthogonal polynomial expansions or approximations. First though, we revisit two of the most important examples of orthogonal families.

Example 7.4.3. Find the monic Legendre polynomials $P_0^*, P_1^*, \ldots, P_4^*$ on $[-1, 1]$.

Solution. We recall that the Legendre polynomials are obtained from weight function $w(x) = 1$. Now $P_0^*(x) = 1$, and $a_1 = \int_{-1}^1 x \, dx = 0$ so that $P_1^*(x) = x$. Then

$$a_2 = \frac{\int_{-1}^1 x^3 \, dx}{\int_{-1}^1 x^2 \, dx} = 0 \qquad b_2 = \frac{\int_{-1}^1 x^2 \, dx}{\int_{-1}^1 1 \, dx} = \frac{1}{3}$$

so that

$$P_2^*(x) = x^2 - \tfrac{1}{3}$$

$$a_3 = \frac{\int_{-1}^1 x(x^2 - \tfrac{1}{3})^2 \, dx}{\int_{-1}^1 (x^2 - \tfrac{1}{3})^2 \, dx} = 0 \qquad b_3 = \frac{\int_{-1}^1 x^2(x^2 - \tfrac{1}{3}) \, dx}{\int_{-1}^1 x^2 \, dx} = \frac{4}{15}$$

from which we have

$$P_3^*(x) = xP_2^*(x) - \frac{4P_1^*(x)}{15}$$

(This is, of course, $x^3 - 3x/5$, as before.) For P_4^*, we use

$$a_4 = \frac{\int_{-1}^{1} x(x^3 - 3x/5)^2 \, dx}{\int_{-1}^{1} (x^3 - 3x/5)^2 \, dx} = 0 \qquad b_4 = \frac{\int_{-1}^{1} x(x^3 - 3x/5)(x^2 - \frac{1}{3}) \, dx}{\int_{-1}^{1} (x^2 - \frac{1}{3})^2 \, dx} = \frac{9}{35}$$

to get

$$P_4^*(x) = xP_3^*(x) - \tfrac{9}{35} P_2^*(x) = x^4 - \tfrac{6}{7}x^2 + \tfrac{3}{35}$$

Note that the evaluation of the various integrals here is simplified by the orthogonality since, for example,

$$\int_{-1}^{1} x\left(x^3 - \frac{3x}{5}\right)(x^2 - \tfrac{1}{3}) \, dx = \int_{-1}^{1} x^3\left(x^3 - \frac{3x}{5}\right) dx$$

because P_3^* is orthogonal to all polynomials of degree less than three.

Example 7.4.4. The Chebyshev polynomials are orthogonal over $[-1, 1]$ with respect to the weight function $w(x) = 1/\sqrt{1 - x^2}$. The recurrence relation of Theorem 7.4.2 generates the monic Chebyshev polynomials T_n^*. As in the case of the Legendre polynomials, the inner products $\langle xT_n^*, T_n^* \rangle = 0$ since they are simply integrals of odd functions over symmetric intervals. These polynomials are given by

$$T_0^*(x) = 1 \qquad T_1^*(x) = x$$

and then $T_n^* = xT_{n-1}^* - b_n T_{n-2}^*$. We obtain $T_2^*(x) = x^2 - \frac{1}{2}$, $T_3^*(x) = x^3 - 3x/4$; and, in general, as we discovered in Sec. 4.6, for $n \geq 1$

$$T_n^*(x) = 2^{-n+1}T_n(x) = 2^{-n+1} \cos(n \cos^{-1} x) \qquad (7.4.6)$$

This in turn implies that the recurrence relation of Theorem 7.4.2 for the Chebyshev polynomials becomes

$$T_n^* = xT_{n-1}^* - \frac{T_{n-2}^*}{4} \qquad (n \geq 3) \qquad (7.4.7)$$

The recurrence relations given by (7.4.4) and (7.4.5) make monic orthogonal polynomials particularly suitable for approximation purposes. This advantage is derived from the ease with which we can evaluate an orthogonal polynomial expansion.

The polynomial $p(x) = \sum_{i=0}^{n} \alpha_i \phi_i(x)$ is evaluated according to the following algorithm.

Algorithm 7.4.5   Evaluation of orthogonal expansion

Input	Coefficients $\alpha_0, \alpha_1, \ldots, \alpha_n; a_1, a_2, \ldots, a_n;$ $b_2, b_3, \ldots, b_n; x$
Initialize	$p_{n+2} = p_{n+1} := 0;\ a_{n+1} := 0;\ b_{n+1} = b_{n+2} := 0$
Loop	for $k = n$ downto 0 $p_k := \alpha_k + (x - a_{k+1})p_{k+1} - b_{k+2}p_{k+2}$
Output	$p(x) = p_0$

It is perhaps not immediately obvious what is going on here. The basic principle is that each polynomial ϕ_i in turn is replaced with polynomials of lower degree using (7.4.4). The particular organization of the calculation is such that a mere $2n - 1$ multiplications are required for the complete evaluation. (In the special case of a Chebyshev expansion this can be reduced yet further to only $n + 1$ multiplications.) It is essentially a generalization of Horner's rule for algebraic polynomial evaluation.

To see that we do indeed find the correct sum, we substitute for the coefficients $\alpha_0, \alpha_1, \ldots, \alpha_n$ according to the algorithm, collect terms, and apply the relation (7.4.4) to each term of the resulting sum to obtain

$$p(x) = \sum_{i=0}^{n} \alpha_i \phi_i(x) = \sum_{i=0}^{n} (p_i - (x - a_{i+1})p_{i+1} + b_{i+2}p_{i+2})\phi_i(x)$$

$$= p_0 + p_1(\phi_1(x) - (x - a_1))$$

$$+ \sum_{i=2}^{n} p_i(\phi_i(x) - (x - a_i)\phi_{i-1}(x) + b_i\phi_{i-2}(x))$$

$$= p_0$$

as required.

One of the other areas of computation in which orthogonal polynomials play a central role is in numerical integration where the nodes for gaussian quadrature rules are at the zeros of appropriate orthogonal polynomials. (See Chap. 9.) The power of these rules depends crucially on the fact that these zeros are all simple and lie in the interval $[a, b]$.

Theorem 7.4.6. Let $\phi_0, \phi_1, \ldots$ be a system of orthogonal polynomials on $[a, b]$. Then for each k, ϕ_k has k distinct real zeros in $[a, b]$.

Proof. Suppose that ϕ_k has m zeros in $[a, b]$ at which it changes sign. (Certainly, $0 \leq m \leq k$.) There is a polynomial $p \in \Pi_m$ which changes sign at only those m points. It follows that $p(x)\phi_k(x)$ has no sign changes in $[a, b]$ and therefore that $\langle p, \phi_k \rangle \neq 0$. Since ϕ_k is orthogonal to all polynomials of degree less than k, it follows that $m = k$, and since these zeros are distinct, this completes the proof.

For both the approximation process and for numerical integration, the (weighted) least squares norm of the orthogonal polynomials are used. In the approximation case, the coefficients of the recurrence relations are given by (7.4.5) and the coefficients of the approximation itself by (7.3.17), both of which use the quantities $\|\phi_i\|^2$.

It is clearly often beneficial to use a normalization in which these norms are simple expressions. This means that the monic polynomials generated by (7.4.4) may not be the most convenient for computation. Of course, any other normalization is obtained by a simple scaling of the monic polynomials.

In the case of the Chebyshev polynomials T_n, we have

$$\| T_n(x) \|^2 = \int_{-1}^{1} \frac{\cos^2 (n \cos^{-1} x)}{\sqrt{1 - x^2}} \, dx = \int_0^{\pi} \cos^2 n\theta \, d\theta = \frac{\pi}{2} \tag{7.4.8}$$

for $n \geq 1$ while $\| T_0 \|^2 = \pi$.

For the Legendre polynomials, a convenient renormalization of the P_n^* is less easily obtained. We observed in Example 7.4.3 that the a_n of the recurrence relation (7.4.4) are all zero since P_n^{*2} is even for all n. The convenient normalization turns out to depend on the values of $P_n^*(1)$. We first establish a formula for this quantity.

Proposition 7.4.7

$$P_n^*(1) = \frac{2^n}{\dbinom{2n}{n}} \tag{7.4.9}$$

Proof. The formula is proved by induction. For $n = 0$ or 1, both sides of (7.4.9) are 1. Now suppose the result holds for $n < k$ ($k \geq 2$). Using the fact that $a_k = 0$, we have from (7.4.4) and (7.4.5)

$$P_k^*(1) = P_{k-1}^*(1) - \frac{\langle x P_{k-1}^*, P_{k-2}^* \rangle}{\| P_{k-2}^* \|^2} \, P_{k-2}^*(1) \tag{7.4.10}$$

Now $P_{k-1}^* = x P_{k-2}^* + \text{lower-order terms}$, and so

$$\langle x P_{k-1}^*, P_{k-2}^* \rangle = \langle P_{k-1}^*, x P_{k-2}^* \rangle = \| P_{k-1}^* \|^2$$

since P_{k-1}^* is orthogonal to all polynomials of degree less than $k - 1$. Using integration by parts, it is easy to show (Exercise 5) that

$$\| P_n^* \|^2 = \frac{2}{2n + 1} \, P_n^{*2}(1) \tag{7.4.11}$$

for every n. By the induction hypothesis, (7.4.10) now reduces to

$$P_k^*(1) = \frac{2^{k-1}}{\dbinom{2(k-1)}{k-1}} - \frac{4(2k-3)}{2k-1} \frac{\dbinom{2(k-2)}{k-2}^2}{\dbinom{2(k-1)}{k-1}^2} \frac{2^{k-2}}{\dbinom{2(k-2)}{k-2}}$$

which simplifies to yield

$$P_k^*(1) = \frac{2^{k-1}}{\dbinom{2(k-1)}{k-1}} \frac{k}{2k-1} = \frac{2^k}{\dbinom{2k}{k}}$$

as required.

The standard Legendre polynomials are normalized by setting

$$P_n(1) = 1 \tag{7.4.12}$$

from which it follows that

$$P_n = \binom{2n}{n} \frac{P_n^*}{2^n} \tag{7.4.13}$$

Combining (7.4.9) through (7.4.13), we can now get the corresponding recurrence relation for the P_n:

$$P_n(x) = \binom{2n}{n}\left(\frac{xP_{n-1}^*}{2^n} - \frac{(n-1)^2}{(2n-1)(2n-3)} \frac{P_{n-2}^*}{2^n} \right)$$

$$= \frac{2n-1}{n} xP_{n-1} - \frac{n-1}{n} P_{n-2}$$

or, equivalently,

$$nP_n = (2n-1)xP_{n-1} - (n-1)P_{n-2} \tag{7.4.14}$$

The first few of these Legendre polynomials are therefore given by

$$P_0(x) = 1$$

$$P_1(x) = x$$

$$P_2(x) = \tfrac{3}{2}xP_1(x) - \tfrac{1}{2}P_0(x) = \frac{3x^2}{2} - \frac{1}{2}$$

$$P_3(x) = \tfrac{5}{3}xP_2(x) - \tfrac{2}{3}P_1(x) = \frac{5x^3}{2} - \frac{3x}{2}$$

From (7.4.11), it follows that

$$\|P_n\|^2 = \frac{2}{2n+1} \tag{7.4.15}$$

and therefore that the coefficients in the least squares approximation of f on $[-1, 1]$ in terms of the Legendre polynomials are given by [compare with (7.3.17)]

$$\alpha_n = \frac{2n+1}{2} \int_{-1}^{1} f(x)P_n(x)\, dx \tag{7.4.16}$$

Similarly the coefficients of the Chebyshev expansion, that is, the weighted least squares approximation using $w(x) = 1/\sqrt{1-x^2}$, are given by

$$\alpha_0 = \frac{1}{\pi} \int_{-1}^{1} \frac{f(x)}{\sqrt{1-x^2}}\, dx$$

$$\alpha_n = \frac{2}{\pi} \int_{-1}^{1} \frac{f(x)T_n(x)}{\sqrt{1-x^2}}\, dx \qquad (n \geq 1) \tag{7.4.17}$$

EXERCISES 7.4

1. Prove part (b) of Proposition 7.4.1; that is, show that ϕ_{n+1} is orthogonal to every polynomial in Π_n.
2. Show that the first few monic Chebyshev polynomials are indeed as given in Example 7.4.4 and that they satisfy the recurrence (7.4.7).
3. The Laguerre polynomials are defined to be orthogonal with respect to the weight function $w(x) = e^{-x}$ over $[0, \infty)$. Find the first four monic Laguerre polynomials. (Laguerre polynomials are used in Gaussian quadrature rules for semi-infinite integrals.)
4. By using integration by parts on $\int_{-1}^{1} P_n^{*2}(x)\, dx$, derive Eq. (7.4.11); that is,

$$\|P_n^*\|^2 = \frac{2}{2n+1}\, P_n^{*2}(1)$$

5. Fill in the details of the induction step in the proof of Proposition 7.4.7.
6. Use Eqs. (7.4.16) and (7.4.17) to confirm the approximations obtained in Example 7.3.5 for the unit half circle.
7. Write a program to implement Algorithm 7.4.5 for the evaluation of orthogonal polynomial expansions. Find the coefficients of the Legendre expansion of e^x on $[-1, 1]$ using $P_0, \ldots, P_5$, and use your program for the evaluation of this expansion to obtain a graph of the resulting approximation.

7.5 NONPOLYNOMIAL APPROXIMATION

In this section, we discuss briefly just two approximation schemes which do not use polynomials for the approximating functions. In the first of these we extend the exchange algorithm to rational minimax approximation. The second concerns the least squares approximation of periodic functions by trigonometric polynomials.

We again use the notation of Sec. 5.5 in which $\mathscr{P}_{m,n}$ denotes the set of all rational functions of the form

$$\rho(x) = \frac{p(x)}{q(x)} = \frac{a_0 + a_1 x + \cdots + a_m x^m}{1 + b_1 x + \cdots + b_n x^n} \tag{7.5.1}$$

where we have made the same normalizing assumption as in (5.5.3) that there should be no pole at the origin. In fact, we will make a stronger assumption about our approximating functions: that they should be continuous and so bounded throughout the interval $[a, b]$. For simplicity we standardize this interval to be $[-1, 1]$, in which case the normalization $b_0 = 1$ is certainly admissible.

Approximating functions of the form (7.5.1) have $n + m + 1$ degrees of freedom, and so it is reasonable to expect that the reference set required by the exchange algorithm for rational approximation will have $n + m + 2$ points which we denote by $x_0 < x_1 < \cdots < x_{n+m+1}$.

For such a reference we must find a rational function of the form (7.5.1) for which (as with the polynomial exchange algorithm)

$$\rho(x_i) + (-1)^i h = f(x_i) \qquad (i = 0, 1, \ldots, n + m + 1) \tag{7.5.2}$$

The iteration then proceeds just as in Sec. 7.2 with one point of the reference being replaced by the point at which the maximum absolute error is attained.

The critical step in the iteration is therefore the identification of a rational function which is bounded on $[-1, 1]$ and satisfies (7.5.2). In the same way as for interpolation, we simplify Eqs. (7.5.2) by multiplying each one by $q(x_i)$ to yield

$$p(x_i) = [f(x_i) - (-1)^i h] q(x_i) \qquad (i = 0, 1, \ldots, n + m + 1) \qquad (7.5.3)$$

This may be rewritten more fully as

$$a_0 + a_1 x_i + \cdots + a_m x_i^m = [f(x_i) - (-1)^i h](1 + b_1 x_i + \cdots + b_n x_i^n) \tag{7.5.4}$$

Extending the reasoning used in Exercises 3 and 4 of Sec. 4.2 on Lagrange interpolation, we can readily obtain

$$\sum_{i=0}^{n+m+1} x_i^k l_i(x) \equiv x^k \tag{7.5.5}$$

for each $k \le n + m + 1$, where the functions $l_i(x)$ are the Lagrange basis polynomials for interpolation at $x_0, x_1, \ldots, x_{n+m+1}$. Recall that

$$
\begin{aligned}
l_i(x) &= \frac{(x - x_0) \cdots (x - x_{i-1})(x - x_{i+1}) \cdots (x - x_{n+m+1})}{(x_i - x_0) \cdots (x_i - x_{i-1})(x_i - x_{i+1}) \cdots (x_i - x_{n+m+1})} \\
&= \prod_{\substack{j=0 \\ j \ne i}}^{n+m+1} \frac{x - x_j}{x_i - x_j}
\end{aligned}
\tag{7.5.6}
$$

Now for any $k \le n + m$, the coefficient of x^{n+m+1} in the right-hand side of (7.5.5) is zero, and so it follows that

$$\sum_{i=0}^{n+m+1} \frac{x_i^k}{L'(x_i)} = 0 \tag{7.5.7}$$

where

$$L(x) = \prod_{j=0}^{n+m+1} x - x_j$$

so that

$$L'(x_i) = \prod_{\substack{j=0 \\ j \ne i}}^{n+m+1} x_i - x_j \tag{7.5.8}$$

We can use the identities (7.5.7) to eliminate the coefficients $a_0, a_1, \ldots, a_n$ from Eqs. (7.5.4) and thereby reduce the solution to a linear algebraic problem.

For example, multiplying the ith equation of the system (7.5.4) by $1/L'(x_i)$ and adding we obtain

$$0 = \sum_{i=0}^{n+m+1} \frac{[f(x_i) - (-1)^i h](1 + b_1 x_i + \cdots + b_n x_i^n)}{L'(x_i)}$$

and, in general for $0 \leq k \leq n$, we multiply the ith equation by $x_i^k/L'(x_i)$ to obtain the system of $n+1$ equations for the unknowns h and $b_1, b_2, \ldots, b_n$:

$$0 = \sum_{i=0}^{n+m+1} \frac{[f(x_i) - (-1)^i h](x_i^k + b_1 x_i^{k+1} + \cdots + b_n x_i^{k+n})}{L'(x_i)} \qquad (7.5.9)$$

This is still a nonlinear system of equations, but it is in a convenient form to be rearranged as an eigenvalue problem.

Denote the vector $(1, b_1, \ldots, b_n)$ by $\mathbf{b}$. Then (7.5.9) is of the form

$$F\mathbf{b} = hG\mathbf{b} \qquad (7.5.10)$$

where the matrices F and G have the elements

$$f_{kj} = \sum_{i=0}^{n+m+1} \frac{f(x_i) x_i^{j+k}}{L'(x_i)}$$

$$(j, k = 0, 1, \ldots, n) \qquad (7.5.11)$$

$$g_{kj} = \sum_{i=0}^{n+m+1} \frac{(-1)^i x_i^{j+k}}{L'(x_i)}$$

Equation (7.5.10) is thus in the form of the *generalized eigenvalue problem* where h is the (generalized) eigenvalue and $\mathbf{b}$ the corresponding eigenvector. The eigenvalue problem (Chap. 11) is in general a difficult problem except for small-dimensional matrices where the characteristic equation can be used effectively. For the generalized eigenvalue problem the difficulties are magnified. The equivalent of the characteristic equation for (7.5.10) is to solve the polynomial equation

$$\det F - hG = 0 \qquad (7.5.12)$$

which is not a practical approach for large-dimensional matrices.

Usually it is the smallest eigenvalue h which yields the required value of h, but *not always*; it can happen that this smallest one results in an eigenvector $\mathbf{b}$ such that the denominator polynomial q has a zero in $[-1, 1]$ which must therefore be rejected.

Once the eigenproblem solution has been obtained, the first $m+1$ of Eqs. (7.5.4) form a linear system of equations

$$A\mathbf{a} = \mathbf{r} \qquad (7.5.13)$$

where $\mathbf{a} = (a_0, a_1, \ldots, a_m)$, matrix A has elements x_i^j $(i, j = 0, 1, \ldots, m)$, and $\mathbf{r}$ has components $[f(x_i) - (-1)^i h]q(x_i)$. This is, of course, just the system of equations for Lagrange interpolation of the data r_i at the points $(x_0, x_1, \ldots, x_m)$. The numerator polynomial can be found by solving this system or by any other polynomial interpolation formula such as the divided difference formula (4.3.10).

Example 7.5.1. Find the minimax rational approximation from $\mathcal{P}_{1,1}$ to e^x on $[-1, 1]$.

Solution. We take for our initial reference the equally spaced points $-1, -\frac{1}{3}, \frac{1}{3}, 1$. The values of the $L'(x_i)$ for this case are $-\frac{16}{9}, \frac{16}{27}, -\frac{16}{27}$, and $\frac{16}{9}$, and so the matrices F and G are

$$F = \begin{bmatrix} 0.1762 & 0.5479 \\ 0.5479 & 1.1948 \end{bmatrix} \qquad G = \begin{bmatrix} -4.5 & 0 \\ 0 & -1.5 \end{bmatrix}$$

to four decimal places. Equation (7.5.12) for this iteration is therefore

$$\begin{vmatrix} 0.1762 + 4.5h & 0.5479 \\ 0.5479 & 1.1948 + 1.5h \end{vmatrix} = 6.75h^2 + 5.6409h - 0.0897 = 0$$

which has the solutions $h = 0.0156$ or -0.8513.

Next we require the corresponding eigenvector which we have normalized to be of the form $(1, b_1)$. Now with $h = 0.0156$,

$$F\mathbf{b} = (0.1762 + 0.5479b_1, 0.5479 + 1.1948b_1)$$

$$hG\mathbf{b} = (-0.0702, -0.0234b_1)$$

from which we deduce $b_1 = -0.4497$ so that $q(x) = 1 - 0.4497x$ which has no zeros in the interval $[-1, 1]$. Substituting these values into (7.5.4) for $i = 0, 1$ we get the system

$$\begin{bmatrix} 1 & -1 \\ 1 & -\frac{1}{3} \end{bmatrix} \begin{bmatrix} a_0 \\ a_1 \end{bmatrix} = \begin{bmatrix} 0.5107 \\ 0.8419 \end{bmatrix}$$

which has the solution $a_0 = 1.0075$, $a_1 = 0.4968$ so that the rational function approximation obtained in this first iteration is

$$p(x) = \frac{1.0075 + 0.4968x}{1 - 0.4497x}$$

The error function for this approximation is plotted in Fig. 7.5.3. It attains its (absolute) maximum (among plotted points) at $x = 0.685$ which replaces x_2 in the reference set for the second iteration.

The next two iterations are summarized in Table 7.5.2 below. Their error functions are also plotted in Fig. 7.5.3. There are three degrees of freedom in an approximation from $\mathscr{P}_{1,1}$, and so it is comparable with quadratic polynomial approximation.

TABLE 7.5.2
The reference sets and results of the first three iterations of the rational exchange algorithm for Example 7.5.1

	Iteration 1	Iteration 2	Iteration 3
x_0	-1	-1	-1
x_1	-0.33333	-0.33333	-0.220
x_2	0.33333	0.685	0.685
x_3	1	1	1
h	0.01561	0.02057	0.02095
b_1	-0.44975	-0.43941	-0.43981
a_0	1.00751	1.01764	1.01700
a_1	0.49681	0.51772	0.51749
$|e_{max}|$	0.03582	0.02156	0.02101
x_{max}	0.685	-0.220	0.700

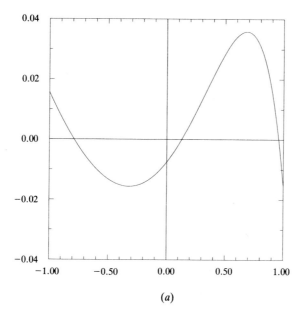

(a)

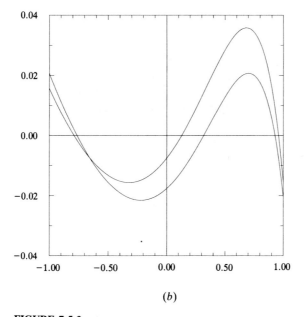

(b)

FIGURE 7.5.3
Error functions of the first three iterations of the exchange algorithm for rational approximation for Example 7.5.1. (a) Iteration 1; (b) iterations 1 and 2; (c) iterations 1 to 3; (d) minimax quadratic error.

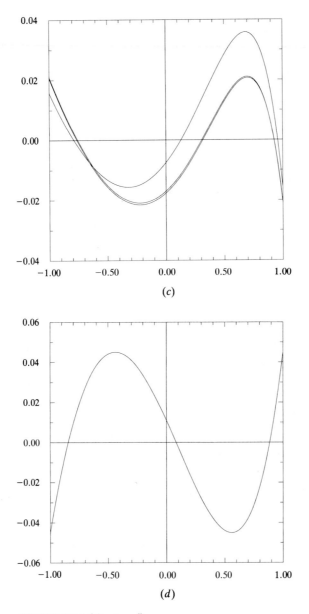

FIGURE 7.5.3 (*Continued*)

Comparing the graphs of the errors for the rational approximants with the minimax quadratic approximation, we see that even the first iteration produces significantly better agreement. The maximum error for the quadratic is 0.0450, so rational approximation has provided an improvement by a factor better than 2. Of course, the rational approximation carries a significantly greater computational overhead.

Comparing h and $|e_{max}|$ for the final iteration, we see that this approximation agrees with the true minimax approximation from the set $\mathscr{P}_{1,1}$ to within 6×10^{-5} so that again the exchange algorithm has shown rapid convergence.

We turn now to the question of least squares approximation to periodic functions. This gives rise to the *continuous Fourier transform*, or *Fourier series* approximation, of such a function.

Suppose then that the function f to be approximated is periodic with period 2π and that we seek the least squares trigonometric polynomial approximation π_N of degree N over the interval $[-\pi, \pi]$. We have already seen in Sec. 7.3 that the basis functions $\sin mx$ and $\cos nx$ are orthogonal over $[-\pi, \pi]$. Also

$$\int_{-\pi}^{\pi} \sin^2 nx \, dx = \int_{-\pi}^{\pi} \cos^2 nx \, dx = \pi \qquad (n \geq 1) \qquad (7.5.14)$$

while, of course,

$$\int_{-\pi}^{\pi} \cos^2 nx \, dx = 2\pi \qquad \text{for } n = 0$$

For convenience, we write the approximation in the form

$$\pi_N(x) = \frac{a_0}{2} + \sum_{k=1}^{N} a_k \cos kx + b_k \sin kx \qquad (7.5.15)$$

It follows, using (7.3.17), that the coefficients of this expansion are given by

$$a_k = \frac{1}{\pi} \int_{-\pi}^{\pi} f(x) \cos kx \, dx \qquad (k = 0, 1, \ldots, N)$$

$$b_k = \frac{1}{\pi} \int_{-\pi}^{\pi} f(x) \sin kx \, dx \qquad (k = 1, 2, \ldots, N)$$

$$(7.5.16)$$

This is just the "Fourier polynomial" approximation to f which is the truncated *Fourier series* whose coefficients are given by allowing $N \to \infty$ in (7.5.16).

Example 7.5.4. Find the coefficients of the Fourier series expansion of the sawtooth wave defined by

$$f(x + 2k\pi) = x \qquad x \in (-\pi, \pi]$$

The graph of this function over $(-2\pi, 2\pi]$ is shown in Fig. 7.5.5 along with the approximations of degree one through four.

Solution. The coefficients a_k of the approximation will all be zero since $x \cos kx$ is an odd function and the integration in (7.5.16) is over a symmetric interval. For the coefficients of the sine terms, we get

$$b_k = \frac{1}{\pi} \int_{-\pi}^{\pi} x \sin kx \, dx = (-1)^{k+1} \frac{2}{k}$$

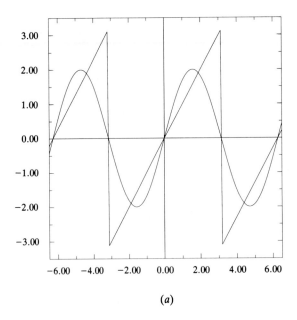

(a)

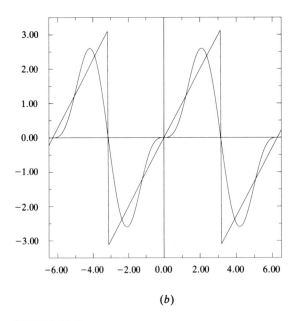

(b)

FIGURE 7.5.5
The sawtooth wave and its Fourier approximations. (a) Degree one; (b) degree two; (c) degree three; (d) degree four.

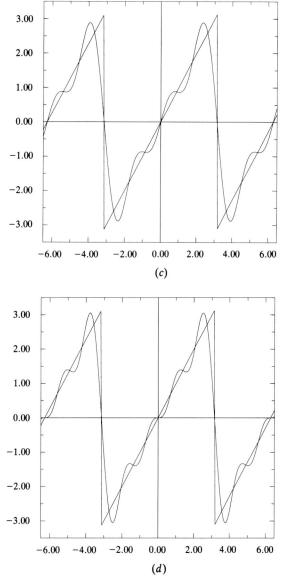

(c)

(d)

FIGURE 7.5.5 (*Continued*)

so that the Fourier series for the function x over $[-\pi, \pi]$ is

$$2 \sum_{k=1}^{\infty} \frac{(-1)^{k+1} \sin kx}{k} = 2 \sin x - \sin 2x + \tfrac{2}{3} \sin 3x - \cdots$$

Note that we have not written that this series equals x since we have said nothing about the convergence of the expansion. We are only interested here in the use of the continuous Fourier transform as an approximation technique.

The important questions of the convergence of such series are left to texts on Fourier analysis.

We see clearly that the successive approximations are getting steadily closer to the original function with the peaks getting steadily larger and moving closer to the required points.

As a final word on Fourier transforms, we return briefly to the discrete Fourier transform which we studied in Sec. 5.6 from the viewpoint of interpolation. This can also be obtained as a least squares approximation in the following manner.

Suppose that we seek such a trigonometric polynomial approximation of degree no more than n to a function f but that we only have the data $f(x_i)$ for the equally spaced points

$$x_i = \frac{2\pi i}{N} \tag{7.5.17}$$

where $N > 2n$. Using arguments similar to those of Proposition 5.6.1, it is easy to see that the following discrete orthogonality properties hold:

$$\sum_{i=0}^{N-1} \cos \frac{2li\pi}{N} \cos \frac{2mi\pi}{N} = \begin{cases} 0 & l \neq m \\ N/2 & l = m \neq 0 \\ N & l = m = 0 \end{cases} \tag{7.5.18}$$

$$\sum_{i=0}^{N-1} \sin \frac{2li\pi}{N} \sin \frac{2mi\pi}{N} = \begin{cases} 0 & l \neq m \\ N/2 & l = m \end{cases} \tag{7.5.19}$$

$$\sum_{i=0}^{N-1} \sin \frac{2li\pi}{N} \cos \frac{2mi\pi}{N} = 0 \tag{7.5.20}$$

That is, the functions $\sin mx$ and $\cos lx$ form an orthogonal family with respect to the inner product

$$\langle f, g \rangle = \sum_{i=0}^{N-1} f(x_i)g(x_i) \tag{7.5.21}$$

Again writing the approximation in the form (7.5.15), it follows from (7.3.17) that the coefficients are given by

$$a_k = \frac{2}{N} \sum_{i=0}^{N-1} f(x_i) \cos \frac{2ki\pi}{N}$$

$$b_k = \frac{2}{N} \sum_{i=0}^{N-1} f(x_i) \sin \frac{2ki\pi}{N} \tag{7.5.22}$$

which corresponds to Eqs. (5.6.16) for the coefficients of the interpolation "polynomial."

This discrete Fourier transform is one example of the use of discrete least squares approximations which form the basis of many curve-fitting techniques for the situation where only certain data points are known. The rest of the chapter is devoted to this topic.

EXERCISES 7.5

1. Find the minimax approximation from $\mathcal{P}_{1,1}$ to $\cos x$ on $[-\pi/2, \pi/2]$.
2. Find the $\mathcal{P}_{2,2}$ minimax approximation to e^x on $[-1, 1]$. Compare it with the polynomial approximation from Π_4.
3. Find the coefficients of the Fourier series for the triangular wave defined by $|x|$ on $[-\pi, \pi]$. Plot the first four approximations.

7.6 DISCRETE LEAST SQUARES APPROXIMATION

In this section we continue with the theme of least squares approximations but in the special context where only data at a discrete set of points is available. The simplest form of this was outlined in the introduction to this chapter where we again found a system of normal equations. We concentrate here on polynomial approximations.

The problem then is to find a polynomial $p \in \Pi_n$ of degree at most n which minimizes the weighted discrete sum of squared errors

$$\sum_{i=0}^{N} w_i(f(x_i) - p(x_i))^2 = \sum_{i=0}^{N} w_i(f(x_i) - a_0 - a_1 x_i - \cdots - a_n x_i^n)^2$$

(7.6.1)

where $x_0, x_1, \ldots, x_N$ are the data points and $w_0, w_1, \ldots, w_N$ the associated positive weights. It is assumed that $N > n$ since otherwise this error measure can be forced to zero by using an interpolation polynomial. In practice it is often the case that the number of data points is significantly greater than the degree of the approximation sought.

As for the continuous case, we can differentiate the sum (7.6.1) with respect to each of the coefficients to obtain the normal equations

$$a_0 \sum_{i=0}^{N} w_i x_i^j + a_1 \sum_{i=0}^{N} w_i x_i^{j+1} + \cdots + a_n \sum_{i=0}^{N} w_i x_i^{j+n}$$

$$= \sum_{i=0}^{N} w_i x_i^j f(x_i) \qquad (j = 0, 1, \ldots, n)$$

(7.6.2)

which we write in matrix form as

$$X\mathbf{a} = \mathbf{f}$$

(7.6.3)

Example 7.6.1. Find the (equally weighted) least squares approximations of degrees one through four for the experimental data:

x	1.47	1.83	3.02	3.56	5.86	8.75	9.45
f	2.09	1.92	2.19	2.64	3.19	3.13	3.61

which comes from a smooth curve but contains experimental errors.

Solution. For the linear approximation, the matrix system is

$$\begin{bmatrix} 7 & 33.94 \\ 33.94 & 227.5084 \end{bmatrix} \begin{bmatrix} a_0 \\ a_1 \end{bmatrix} = \begin{bmatrix} 18.57 \\ 102.0815 \end{bmatrix}$$

which to four decimal places yields the solution

$$a_0 = 1.7252 \qquad a_1 = 0.1913$$

The data and this linear approximation are shown in Fig. 7.6.2a. The successively higher-degree approximations are illustrated in the remaining graphs of Fig. 7.6.2. The coefficients of the approximations are listed in Table 7.6.3.

TABLE 7.6.3
Coefficients of least squares approximations

Coefficient	Linear	Quadratic	Cubic	Quartic
a_0	1.72520	1.46216	1.74035	4.44357
a_1	0.19133	0.32974	0.09383	−2.99877
a_2		−0.01256	0.04061	1.16355
a_3			−0.00336	−0.16135
a_4				0.00749

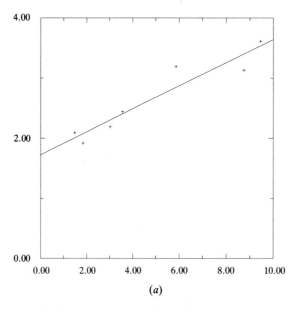

(a)

FIGURE 7.6.2
Discrete least squares approximation. (a) Linear; (b) quadratic; (c) cubic; (d) quartic.

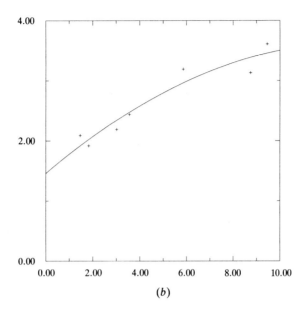

(b)

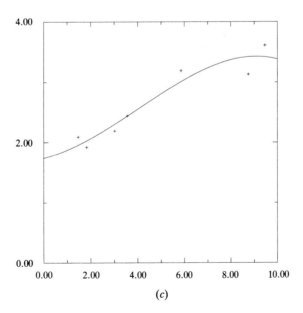

(c)

FIGURE 7.6.2 (*Continued*)

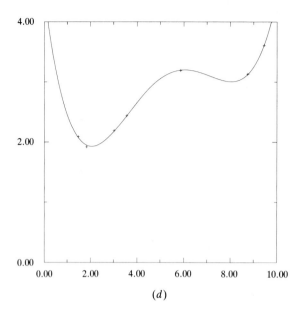

(d)

FIGURE 7.6.2 (*Continued*)

It is immediately apparent that the coefficients of successive approximations bear no resemblance to those of their predecessors. The computation relies on the solution of linear systems which are, typically, ill-conditioned and for which the matrix is full. The only part of the calculation which can be reused if the degree is to be increased is that part of the coefficient matrix and right-hand side which has already been computed.

The problem of increasing the degree of the approximation is not necessarily severe as this may not be a wise policy for improving the fit of the curve anyway. In the above example, for instance, we see that the quartic approximation is very close to interpolating the data but has the least likely overall shape for its graph. The data itself appears to have a slowly increasing trend which is better reflected by the linear and, especially, the quadratic approximations.

It is a common phenomenon for data subject to errors that as the degree of approximation increases close to the number of data points, and so the approximating polynomial gets closer to an interpolation polynomial, so the "shape" of the fit deteriorates.

Fitting a curve to "experimental" data is not the sole purpose of discrete least squares approximations. It may be the case that we seek an approximation to a specific function for which the continuous least squares approach is unsuited. This could be because the integrals required by the normal equations presented special difficulty. One solution to such a problem is to use discrete least squares with a carefully chosen set of data points.

In such situations, we would like to choose the data points so that the computation is stable and straightforward, and we may wish to be able to incorporate a weight function also. There is one especially useful version of this in which all the advantages of an orthogonal system of polynomials are obtained in the discrete case.

For any set of discrete points $x_0, x_1, \ldots, x_N$ and positive weights $w_0, w_1, \ldots, w_N$, we can find a system of polynomials which are orthogonal with respect to the pseudo inner product given by

$$\langle f, g \rangle = \sum_{i=0}^{N} w_i f(x_i) g(x_i) \tag{7.6.4}$$

(This is a *pseudo* inner product since there exist nonzero functions for which $\langle f, f \rangle = 0$.)

One way of generating such a system is easily described in terms of the (weighted) discrete least squares approximation of powers of x by polynomials of degree one less.

Suppose we seek monic polynomials $\phi_0, \phi_1, \ldots$ which are orthogonal with respect to the inner product (7.6.4). Clearly $\phi_0(x) \equiv 1$, and writing $\phi_1(x) = x - a_0$ the orthogonality condition is

$$\langle \phi_1, \phi_0 \rangle = \sum_{i=0}^{N} w_i(x_i - a_0) = 0$$

from which we deduce that

$$a_0 = \frac{\displaystyle\sum_{i=0}^{N} w_i x_i}{\displaystyle\sum_{i=0}^{N} w_i}$$

This is just the least squares constant approximation to x on this data set. In general, $\phi_n(x) = x^n - a_{n-1}x^{n-1} - \cdots - a_1 x - a_0$ is required to be orthogonal to $1, x, x^2, \ldots, x^{n-1}$. The equations corresponding to each of these orthogonality relations are

$$\sum_{i=0}^{N} w_i(x_i^n - a_{n-1}x_i^{n-1} - \cdots - a_1 x_i - a_0) = 0$$

$$\sum_{i=0}^{N} w_i(x_i^{n+1} - a_{n-1}x_i^n - \cdots - a_1 x_i^2 - a_0 x_i) = 0$$

$$\cdots\cdots\cdots\cdots\cdots\cdots\cdots\cdots\cdots\cdots\cdots\cdots\cdots$$

$$\sum_{i=0}^{N} w_i(x_i^{2n-1} - a_{n-1}x_i^{2n-2} - \cdots - a_1 x_i^n - a_0 x_i^{n-1}) = 0$$

which is a system of n linear equations in the n unknown coefficients. This system may be rearranged as

$$a_0 \sum_{i=0}^{N} w_i x_i^j + a_1 \sum_{i=0}^{N} w_i x_i^{j+1} + \cdots + a_{n-1} \sum_{i=0}^{N} w_i x_i^{j+n-1} = \sum_{i=0}^{N} w_i x_i^{j+n}$$

(7.6.5)

for $j = 0, 1, \ldots, n - 1$ which is precisely the system (7.6.2) for the weighted least squares approximation of x^n by a polynomial of degree $n - 1$. The system (7.6.5) remains nonsingular provided $n < N$.

The Chebyshev polynomials have some particularly important discrete orthogonality relations which can be derived from the properties of the appropriate gaussian quadrature rules. (See Chap. 9 for more details.)

From the orthogonality of the Chebyshev polynomials T_n over $[-1, 1]$ with respect to the weight function $1/\sqrt{1 - x^2}$, it follows that the appropriate gaussian integration formula with degree of precision $2N - 1$ uses the zeros of T_N as its nodes. That is, the nodes are

$$x_i = \cos \frac{(2i - 1)\pi}{2N} \qquad (i = 1, 2, \ldots, N)$$

(7.6.6)

for which the corresponding quadrature weights are all π/N. Then if $p \in \Pi_{2N-1}$, we have

$$\int_{-1}^{1} \frac{p(x)}{\sqrt{1 - x^2}} \, dx = \frac{\pi}{N} \sum_{i=1}^{N} p(x_i)$$

It now follows that the Chebyshev polynomials satisfy the orthogonality relations given in Proposition 7.6.4 below.

Proposition 7.6.4. With x_i given by (7.6.6), the Chebyshev polynomials satisfy

$$\sum_{i=1}^{N} T_j(x_i) T_k(x_i) = \begin{cases} 0 & j \neq k \\ N/2 & j = k \neq 0 \\ N & j = k = 0 \end{cases}$$

(7.6.7)

for $j + k < 2N$.

Proof. Since $T_j T_k \in \Pi_{2N-1}$, it follows that

$$\int_{-1}^{1} \frac{T_j(x) T_k(x)}{\sqrt{1 - x^2}} \, dx = \frac{\pi}{N} \sum_{i=1}^{N} T_j(x_i) T_k(x_i)$$

but using the (continuous) orthogonality relations for the Chebyshev polynomials (7.4.8), we see that this integral is 0 for $j \neq k$ while we get $\pi/2$ or π for $j = k \neq 0$ or $j = k = 0$, respectively. The result follows.

The obvious analog of (7.3.17) for the discrete case now allows us to write down the coefficients of discrete Chebyshev least squares approximation to f using the zeros of T_N. Allowing the degree of this approximation to be maximal (that is, $N - 1$), this in turn gives the Chebyshev representation of the Lagrange interpolation polynomial for these same nodes. We summarize these results as follows.

Theorem 7.6.5. With x_i given by (7.6.6), the discrete least squares approximation to f of degree n is

$$\frac{a_0}{2} T_0(x) + \sum_{i=1}^{n} a_i T_i(x) \qquad (7.6.8)$$

where the coefficients are given by

$$a_i = \frac{2}{N} \sum_{j=1}^{N} f(x_j) T_i(x_j) \qquad (7.6.9)$$

In particular, the Lagrange interpolation polynomial agreeing with f at the nodes x_i is

$$\frac{a_0}{2} T_0(x) + \sum_{i=1}^{N-1} a_i T_i(x)$$

Example 7.6.6. Find the least squares approximation to e^x of degrees one and two using the Chebyshev points $\cos[(2i-1)\pi/6]$ $(i = 1, 2, 3)$.

Solution. The nodes are $\pm\sqrt{3}/2, 0$, and $N = 3$. Hence

$$a_0 = \frac{2}{3}(e^{-\sqrt{3}/2} + 1 + e^{\sqrt{3}/2}) = 2.53204$$

$$a_1 = \frac{2}{3}\left(\frac{-\sqrt{3}}{2} e^{-\sqrt{3}/2} + \frac{\sqrt{3}}{2} e^{\sqrt{3}/2}\right) = 1.12977$$

$$a_2 = \frac{2}{3}\left(\frac{1}{2} e^{-\sqrt{3}/2} - 1 + \frac{1}{2} e^{\sqrt{3}/2}\right) = 0.26602$$

so that the linear approximation is

$$1.26602 T_0(x) + 1.12977 T_1(x)$$

and the quadratic is

$$1.26602 T_0(x) + 1.12977 T_1(x) + 0.26602 T_2(x)$$

which also interpolates the exponential function at the three nodes. The exponential function and these two approximations are graphed in Fig. 7.6.7.

Note that the quadratic approximation in fact interpolates e^x at the nodes which are the zeros of the Chebyshev cubic. According to our earlier observations in Sec. 7.2 this should be close to the minimax approximation of the same degree. A brief glance at the error curve for the quadratic approximation in Fig. 7.6.7c shows that this error function is indeed very close to having the "equioscillation" property of minimax approximations.

Of course, in many curve-fitting situations we are not free to choose the data points to be used, and so the excellent fit obtained with just three points here is not typical. We consider this problem in greater detail in the next section.

The great advantages of the Chebyshev polynomials in least squares approximation justify our further study of the efficient computation of the

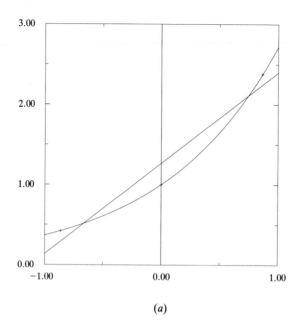

(a)

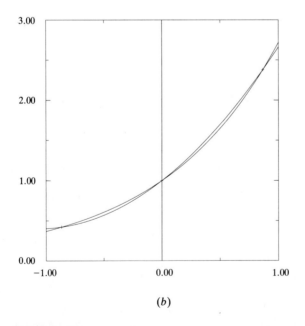

(b)

FIGURE 7.6.7
Chebyshev discrete least squares approximations to e^x using the zeros of T_3 as nodes. (a) Linear; (b) quadratic; (c) error functions.

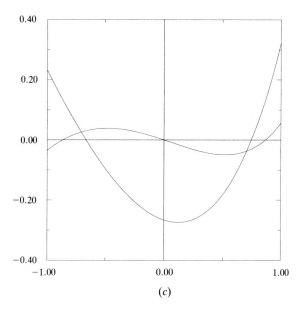

(c)

FIGURE 7.6.7 (*Continued*)

coefficients (7.6.9) and then the evaluation of the Chebyshev expansion given by (7.6.8).

To compute the coefficients, we note that

$$a_0 = \frac{2}{N} \sum_{j=1}^{N} f(x_j)$$

$$a_1 = \frac{2}{N} \sum_{j=1}^{N} x_j f(x_j)$$

(7.6.10)

and for subsequent coefficients we have

$$a_i = \frac{2}{N} \sum_{j=1}^{N} f(x_j) T_i(x_j) = \frac{2}{N} \sum_{j=1}^{N} 2x_j f(x_j) T_{i-1}(x_j) - f(x_j) T_{i-2}(x_j)$$

(7.6.11)

It follows that if two arrays are declared to hold the values of $T_{i-2}(x_j)$ and $T_{i-1}(x_j)$ for $j = 1, 2, \ldots, N$, then the values of $T_i(x_j)$ are obtained using the recurrence relation, and the coefficient a_i is simply the sum of the elements of this array.

The next stage in the approximation procedure is the evaluation of the Chebyshev expansion given by (7.6.8). In Algorithm 7.4.5, we saw that the evaluation of an orthogonal polynomial expansion could be improved by making use of the recurrence relation satisfied by the orthogonal basis under consideration. In the case of the Chebyshev polynomials this task is simplified further by the following algorithm due to Clenshaw for the evaluation of

$$\sum_{i=0}^{N}{}' a_i T_i(x) = \frac{a_0}{2} T_0(x) + \sum_{i=1}^{N} a_i T_i(x) \qquad (7.6.12)$$

Algorithm 7.6.8 Evaluation of Chebyshev expansion

Input	Coefficients $a_0, a_1, \ldots, a_n, x$
Initialize	$\xi := 2x; \; p_{n+2} := p_{n+1} := 0$
Loop	For $k = n$ downto 0
	$\qquad p_k := \xi p_{k+1} - p_{k+2} + a_k$
Output	Value: $= (p_0 - p_2)/2$.

That this algorithm indeed yields the sum (7.6.8) follows from the fact that the p_k satisfy, at each stage,

$$\tfrac{1}{2}(p_k - p_{k+2}) = \sum_{j=0}^{n-k}{}' a_{j+k} T_j(x) \qquad (7.6.13)$$

which is established by a reasonably straightforward induction. (See Exercise 4.)

We can summarize this whole procedure for the calculation and evaluation of least squares approximations at the zeros of T_N as follows.

Algorithm 7.6.9 Discrete Chebyshev least squares approximation

Input	Number of "nodes" N, and maximum degree of approximation n
Initialize	for $i = 1$ to N
	$\qquad x_i := \cos \dfrac{(2i-1)\pi}{2N}, \quad f_i := f(x_i)$
	for $i = 0$ to $n \quad a_i := 0$

Compute coefficients

Initialize	for $i = 1$ to N
	$\qquad \xi_i := 2x_i$
	$\qquad b_i := f_i, \quad a_0 := a_0 + b_i$
	$\qquad c_i := x_i f_i, \quad a_1 := a_1 + c_i$
Loop	for $j = 2$ to n
	$\qquad$ for $i = 1$ to N
	$\qquad\qquad d_i := \xi_i c_i - b_i, \quad a_j := a_j + d_i$
	$\qquad\qquad b_i := c_i, \quad c_i := d_i$
Scale	for $i = 1$ to $n \quad a_i := 2a_i/N$

Evaluation (by Algorithm 7.6.8)

Input	x, coefficients $a_0, a_1, \ldots, a_n$

Initialize	$\xi := 2x, \quad q := r := 0$
Loop	for $k = n$ downto 0
	$\quad p := \xi q - r + a_k$
	$\quad$ if $k > 0$ then $r := q, \ q := p$
Output	$(p - r)/2$

The complete algorithm requires $N(n + 1)$ multiplications for the computation of the coefficients and a further $n + 3$ for evaluation of the expansion at any point. (Discounting multiplications by $0, 2,$ or $\frac{1}{2}$, these operation counts are just Nn and n, respectively.)

EXERCISES 7.6

1. Write a program to obtain discrete least squares polynomial approximations. Use your program to verify the approximations found in Examples 7.6.1 and 7.6.6.

2. Repeat the calculations of Example 7.6.1 but with different weights w_i assigned to the data points. Note how the approximation can be altered quite dramatically by this adjustment.

3. Find the orthogonal polynomials of degrees zero through three for the points $-1, -\frac{1}{2}, 0, \frac{1}{2}, 1$; hence find the least squares quadratic approximation to e^x

using these points. Compare this approximation with Example 7.6.6.

4. Prove (7.6.13) by (reverse) induction—reducing k at each step—and therefore show that Algorithm 7.6.8 yields the sum (7.6.8).

5. Write a program to obtain least squares approximations at the zeros of the Chebyshev polynomial T_N. Use your program to find such approximations of varying degrees to e^x, $\sqrt{1 - x^2}$, and $\tan(\pi x/2)$. Experiment with both N and the degree of the approximation.

7.7 CURVE FITTING: MOVING LEAST SQUARES

This section is concerned with the specific task of fitting a curve to data. The context is that of experimental data or other similar situations in which the analyst has no control over the choice of "nodes" or data points. Thus we are unable to utilize the power of orthogonal polynomial expansions but must use whatever is given. The situation is therefore that of Example 7.6.1 where a number (often a large number) of function values are given which may be subject to error and we must find a curve which passes close to all the points $(x_i, f(x_i))$, $i = 0, 1, \ldots, N$.

Often there will be other requirements on the fitted curve such as particular smoothness properties or monotonicity or some intuitive idea about the "shape" of the curve. There may be particular data points which are to be regarded as especially important for some physical reason or others which appear to be outliers which we may choose to (almost) ignore.

In the last section we considered the standard least squares approximation technique. However, in Example 7.6.1, we used only the uniform weighting $w_i = 1$ for each point. To see the potential effect of weighting different data differently, we return to that example and conduct a few experiments.

Example 7.7.1. Recall the data:

x	1.47	1.83	3.02	3.56	5.86	8.75	9.45
f	2.09	1.92	2.19	2.64	3.19	3.13	3.61

From the distribution of these data points, we observe that the first two and the last three appear to conflict with the general increasing trend of the data.

Our experiment is based on using $w_i = 1$ for most points and varying the weights of one or more to increase or decrease their importance. For each of the graphs in Fig. 7.7.2, the exceptional weights are included in its heading. The resulting quadratic, cubic, and quartic least squares approximations are graphed.

One of the most striking features of these graphs is that the quartic approximation remains almost unchanged for the various different weights. This is due to the fact that, as can be seen from Fig. 7.7.2a which uses the uniform weighting of Example 7.6.1, a quartic polynomial can be found which almost interpolates this data. Inevitably, such a polynomial will retain a small least squares error for any set of weights. In Fig. 7.7.2b, the weights used have the effect of moving the graphs away from the apparently high value at x_0 and the perhaps low one at x_5. This results (approximately) in a slight counterclockwise rotation of the curves which, in turn, gives the cubic approximation a more marked "downturn" at the end of the range and the quadratic a slightly greater initial gradient.

The third of the graphs results from increasing the importance of x_0 while reducing the weight for the two "high" points at x_4 and x_6. The resulting "clockwise rotation" renders the quadratic approximation almost linear—but slightly convex where the previous quadratics had been concave. The cubic appears merely to be flattened out somewhat.

The final set of graphs in Fig. 7.7.2d shows the effect of reducing the weighting of the points in the middle of the range. The resulting quadratic is definitely convex. The cubic approximation is now increasing through the range and indeed would be increasing everywhere since its one inflexion point is in the middle of the graph.

Of course, we have no information as to which of the graphs is the "best" fit to the true function, nor indeed do we know whether any of them is even a good representation of the basic shape. In other words, we have no idea in this case how to adjust the weights to achieve a better graphical representation of the experiment whose results we are studying. What we see though is that some of the basic characteristics of the approximating curve can be altered by simple (and fairly small) changes in the weights.

We have no intrinsic reason to believe that, say, the first data point is more accurate than the second or that any one of the last three is particularly suspect. For this reason it may be desirable to have the weights changing in some smooth manner as the point of interest—the x at which we are currently evaluating the approximation—moves through the region.

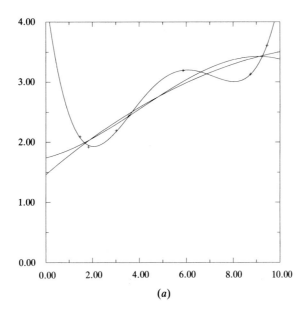

(a)

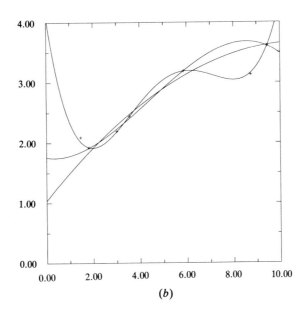

(b)

FIGURE 7.7.2
Weighted least squares approximations to the data of Example 7.7.1. (a) $w_i = 1$; (b) $w_0 = w_5 = 0.01$; (c) $w_0 = 10$, $w_1 = w_4 = w_6 = 0.1$; (d) $w_2 = w_3 = w_4 = 0.1$.

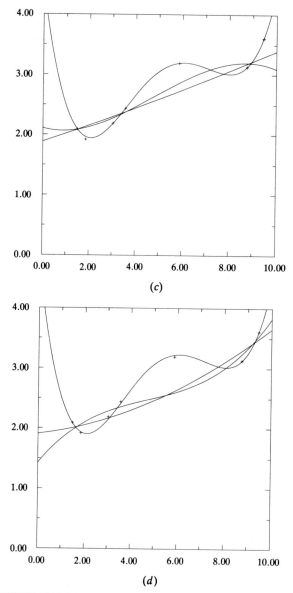

(c)

(d)

FIGURE 7.7.2 (*Continued*)

   This is the basis of the method of *moving least squares* approximation. The principle here is that we assign to each data point not a fixed weight but a *weight function* $w_i(x)$ which determines how much weight is given to the data point x_i when the current point of interest is x. The idea is that the approximation should be more affected by data points close to the current point and less by those that are well removed from it.

The nature of such weight functions is therefore that w_i attains its maximum at x_i and decreases—perhaps quite rapidly—as x moves away from x_i. A commonly chosen form for these functions is

$$w_i(x) = \exp \frac{-(x - x_i)^2}{c} \qquad (7.7.1)$$

for some value of the constant c. This particular form has its basis in polynomial *regression* within statistics where the assumption is that the data points are subject to errors which fit a normal distribution.

The constant c reflects the standard deviation of this error distribution and therefore how narrow (or wide) is the expected range of error: the greater the value of c, the greater the variation which is to be expected. For large values of c, it follows that we have less trust in the specific data values, and therefore we would allow our approximation to be more greatly influenced by distant points than we would for small c.

In Fig. 7.7.3, we show some of these curves for different values of c to give some idea of how quickly the influence of any particular data point attenuates as we move away from it. Obviously the graphs would be symmetric about the y axis.

For our example using the same data as before, we may desire that the attenuation is rapid and use $c = 1$ or 5 or, to include at least some significant influence from the distant points, we could use slow attenuation such as that obtained with $c = 50$.

In Fig. 7.7.4, we illustrate the effects of these weight functions by graphing the "linear" and "quadratic" moving least squares approximations for three values of $c = 1, 10, 50$ in (7.7.1).

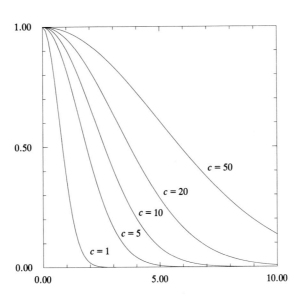

FIGURE 7.7.3
The weight functions (7.7.1) for $x - x_i < 10$.

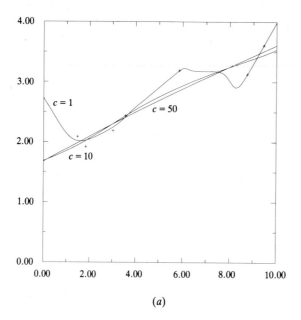

(a)

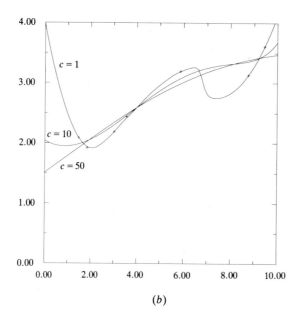

(b)

FIGURE 7.7.4
Moving least squares approximations: data of Example 7.7.1, weight functions (7.7.1). (a) Degree one; (b) degree two.

One immediately striking feature of these approximations is that the graphs of the "linear" approximations are clearly not straight lines. This is because the structure of the approximation is such that the weights used vary with x, and so what is being computed is a *local* straight-line approximation at each point. It is, of course, not the same straight line for each point, and so we get the variation that is evident especially for smaller values of c. Similarly, of course, the "quadratic" approximations do not have parabolic graphs.

It should be immediately apparent that the effect of decreasing the value of c is to "pull" the graph toward each of the data points due to the decreased weight associated with its neighbors and the near-zero weight attached to distant data. One of the effects of this pulling of the graph is a loss of visual smoothness. The more visually pleasing of these approximations are those obtained with $c = 10$ or 50. Indeed the "linear" approximations for both values and the "quadratic" for $c = 50$ are believable fits to this experimental data.

There are several points that need to be considered here on the computational efficiency of such approximations as these.

First, there is the significant advantage that only very low degree approximations are necessary to achieve visually smooth curves which fit the shape of the data very well. Indeed the use of "local constant" approximations corresponding to degree 0 still produces tolerably responsive approximating curves. This is illustrated in Fig. 7.7.5 in which we show the degree-zero approximations for the same three values of c as in Fig. 7.7.4.

These degree-zero approximations are obtained by approximating $f(x)$ by $\sum_{i=0}^{N} w_i(x) f(x_i) / \sum_{i=0}^{N} w_i(x)$ which represents the solution of the "normal equation" for this case. (See Exercise 4.)

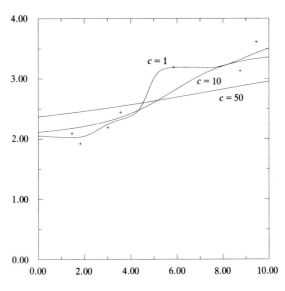

FIGURE 7.7.5
Moving least squares: degree-zero approximations.

In this case, the graph for the case $c = 10$ is a visually smooth curve which fits the data well. Again $c = 1$ results in a curve which is too responsive to the individual data points. We may infer from this that the "experimental errors" in the data are too large to justify this narrow distribution.

A second computational point of importance is that the use of local approximations at each point implies that a linear system of equations must be solved for each point. In the case of the graphs in Figs. 7.7.4 and 7.7.5, each curve was plotted at 400 points. Recomputing the coefficients and then the solution of even a 2×2 system at each of these points is clearly a scarcely acceptable increase in computational effort.

One way of alleviating this would be the use of local approximations in the sense of local interpolation. That is, we could use just the nearest points to the point of interest or just those within a specified distance of it. The drawback of this is that it is difficult to guarantee any degree of smoothness in the resulting approximation—even continuity is not automatic for all choices of weights.

It is often the case in practice that an approximating curve is required which actually interpolates the data. However, as we already know, a full interpolating polynomial is unlikely to be the curve of choice either for its shape or for ease of computation and evaluation. One approach is a modification of the moving least squares method.

It is apparent that moving least squares using $c = 1$ in (7.7.1) produces curves which are close to interpolating the data. This is especially true of the degree-two approximation in Fig. 7.7.4b. In Fig. 7.7.3, we see that the weight function $w_i(x)$ decreases very fast as x moves away from the point x_i. Therefore, at any data point, that data is weighted very heavily relative to any others that are not very nearby.

This idea is at the heart of the interpolatory moving least squares (IMLS) method. Here we choose the weight functions so that

$$w_i(x) \to \infty \text{ as } x \to x_i \tag{7.7.2}$$

which will force the fitted curve to pass through the data points. The penalty for this is that the curve may suffer a loss of smoothness even more severe than that observed in Figs. 7.7.4 and 7.7.5 for $c = 1$.

Of course, the loss of continuity of the weight functions necessitates special care in the implementation of the algorithm at the data points. The IMLS algorithm is described in Algorithm 7.7.6 below and the results presented in graphical form in Fig. 7.7.7 for different choices of k in the weight function

$$w_i(x) = \frac{1}{|x - x_i|^k} \tag{7.7.3}$$

All the additional computation of the moving least squares method is still required in that a new system of equations must be formed and solved for every point.

Algorithm 7.7.6   Interpolatory moving least squares

Input $N, x_0, x_1, \ldots, x_N, f_0, f_1, \ldots, f_N$
weight function $w_i(x)$
degree of approximation n

For each x required:

If $x = x_i$ for some i **then** $p := f_i$

else *Generate matrix system*:

for $i = 0$ to N $w_i := w_i(x)$
for $i = 0$ to n

$$b_i := \sum_{k=0}^{N} w_k f_k x_k^i$$

for $j = 0$ to n $a_{ij} := \sum_{k=0}^{n} w_k x_k^{i+j}$

Solve $A\mathbf{c} = \mathbf{b}$

Evaluate $p := c_n$
for $k = n - 1$ downto 0
$p := p * x + c_k$

Output   Approximation at x is p.

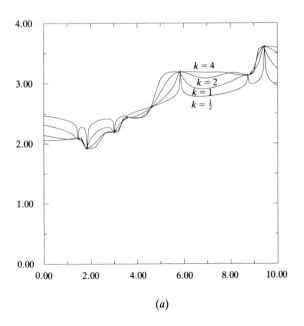

(a)

FIGURE 7.7.7
Interpolatory moving least squares curves: data of Example 7.7.1, weight functions (7.7.3).
(a) Degree zero; (b) degree one; (c) degree two.

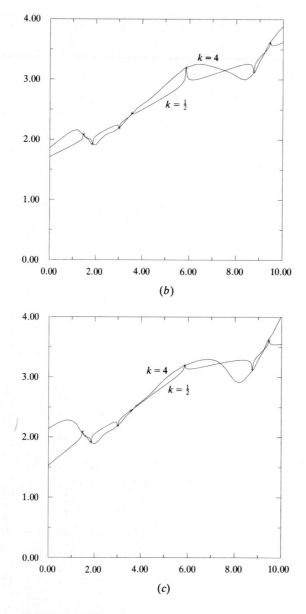

FIGURE 7.7.7 (*Continued*)

One immediately striking feature of these approximations is that for small values of k there are some extremely sharp peaks and troughs to the curves. To achieve any reasonable degree of smoothness it is necessary to use values of at least 2 for k. With such values the resulting curves are smooth interpolants to the data.

The major problem with the particular example used in this section is that the "experimental errors" inserted into the data necessarily result in quite severe variations in any interpolating function. For such situations, in practice, it would be more desirable to use an approximation such as those of Fig. 7.7.2.

However, the moving least squares approach is an illustration of the problem facing the computer-aided designer or the developer of computer graphics or graphical font packages. In these cases a particular shape is required which will pass through (or very close to) specified points where we may assume these points are almost exact.

In the next section we consider alternatives to the moving least squares approach which turn out to be less computationally intensive and avoid the loss of smoothness. The use of Bernstein polynomials and Bezier curves forms a part of several existing interactive computer graphics packages which are used in the desktop publishing industry as well as in computer-aided design (CAD).

EXERCISES 7.7

1. Write a program to implement the moving least squares algorithm. Generate data at a random set of 20 points in $[-5, 5]$ for the function $100/(1 + 10x^2)$. Obtain the "visually best" least squares fit you can using either weighted or moving least squares.

2. Repeat Exercise 1 for the interpolatory moving least squares method.

3. Obtain an operation count for the floating-point multiplications needed for the different algorithms in Exercises 1 and 2 for different choices of weight function and different degrees of approximation. Which gives the best "value for money"?

4. Repeat Exercises 1 to 3 for $\tan x$ on $(-\pi/2, \pi/2)$.

7.8 CURVE FITTING: BERNSTEIN POLYNOMIALS, BEZIER CURVES

In this section we consider another approach to curve fitting which is based on the use of the Bernstein polynomial for a function f on a closed bounded interval $[a, b]$. As a curve-fitting tool this approach has several advantages.

The polynomials are so named because of their central role in Bernstein's constructive proof of the Weierstrass approximation theorem which we have cited repeatedly in this book. The essence of this proof is the construction of a sequence of polynomials which can be shown to converge uniformly to f. It also turns out that the derivatives of these approximations converge uniformly to that of f—assuming the latter exists.

We begin then with the definition of the Bernstein polynomials and of the Bernstein basis polynomials.

Definition 7.8.1. The *Bernstein polynomial* of degree n associated with the function f on $[a, b]$ is defined by

$$B_n(f; x) = \sum_{i=0}^{n} \binom{n}{i}(x-a)^i(b-x)^{n-i}f(x_i) \tag{7.8.1}$$

where the points $x_i = a + ih = a + (i/n)(b-a)$ for $i = 0, 1, \ldots, n$.

In the special case where the interval is $[0, 1]$, Eq. (7.8.1) reduces to

$$B_n(f; x) = \sum_{i=0}^{n} \binom{n}{i}x^i(1-x)^{n-i}f\left(\frac{i}{n}\right) = \sum_{i=0}^{n} f\left(\frac{i}{n}\right)B_{n,i}(x) \tag{7.8.2}$$

where the Bernstein basis polynomials are defined by

$$B_{n,i}(x) = \binom{n}{i}x^i(1-x)^{n-i} \qquad (i = 0, 1, \ldots, n; \, n = 0, 1, \ldots) \tag{7.8.3}$$

We will concentrate throughout the rest of this section on approximation of a function f on $[0, 1]$. The modification to other intervals is achieved by a straightforward transformation of the interval $[a, b]$ into $[0, 1]$.

Readers who have studied any probability theory will immediately recognize the Bernstein basis polynomials as being the probability density functions for a binomial distribution. (Bernstein worked in probability theory as well as in differential equations and approximation theory.) Specifically, $B_{n,i}(x)$ is the probability of achieving exactly i successes in a sequence of n independent trials in which the probability of success on any one trial is x.

In view of this observation, it is clear that we have

$$\sum_{i=0}^{n} B_{n,i}(x) = \sum_{i=0}^{n} \binom{n}{i}x^i(1-x)^{n-i} = (x + (1-x))^n = 1 \tag{7.8.4}$$

from which we may deduce that the Bernstein polynomial for the function $f(x) \equiv 1$ is itself identically 1. That is, this function is reproduced exactly by its Bernstein polynomial.

Now, differentiating (7.8.4) we obtain

$$0 = \sum_{i=0}^{n} \binom{n}{i}x^{i-1}(1-x)^{n-i-1}[i(1-x) - (n-i)x]$$

$$= \sum_{i=0}^{n} \binom{n}{i}x^{i-1}(1-x)^{n-i-1}(i-nx)$$

and multiplying by $x(1-x)$ this yields

$$\sum_{i=0}^{n} \binom{n}{i}x^i(1-x)^{n-i}(i-nx) = 0 \tag{7.8.5}$$

It follows, therefore, that

$$\sum_{i=0}^{n} \binom{n}{i}x^i(1-x)^{n-i}\frac{i}{n} = \sum_{i=0}^{n} \binom{n}{i}x^{i+1}(1-x)^{n-i} = x \tag{7.8.6}$$

using (7.8.4). That is, the Bernstein polynomial for the function $f(x) = x$ reproduces this function exactly.

We may continue in this way: differentiating (7.8.6) we have

$$1 = \sum_{i=0}^{n} \frac{i}{n} \binom{n}{i} x^{i-1}(1-x)^{n-i-1}(i-nx)$$

and on multiplying by $x(1-x)/n$ this yields

$$\frac{x(1-x)}{n} = \sum_{i=0}^{n} \frac{i}{n} \binom{n}{i} x^{i}(1-x)^{n-i}\left(\frac{i}{n}-x\right)$$

$$= \sum_{i=0}^{n} \frac{i^2}{n^2} \binom{n}{i} x^{i}(1-x)^{n-i} - x^2$$

or

$$B_n(x^2; x) = x^2 + \frac{x(1-x)}{n} \tag{7.8.7}$$

Since

$$\max_{0 \le x \le 1} x(1-x) = \tfrac{1}{4}$$

it follows that

$$\|x^2 - B_n(x^2; x)\|_\infty \le \frac{1}{4n} \to 0 \qquad \text{as } n \to \infty$$

The good approximation properties of these Bernstein polynomials give us hope that $B_n(f; x)$ should be a good approximation to $f(x)$ over $[0, 1]$. Indeed, this is at the heart of Bernstein's proof of Weierstrass' theorem in which, specifically, we establish the following.

Theorem 7.8.2 Weierstrass. If f is continuous on $[0, 1]$, then $B_n(f; .)$ converges uniformly to f.

Proof. We must show that for any $\epsilon > 0$,

$$|f(x) - B_n(f; x)| < \epsilon$$

for all $x \in [0, 1]$ and n sufficiently large.

Let $x \in [0, 1]$. Then, using (7.8.4) which implies that

$$f(x) = \sum_{i=0}^{n} \binom{n}{i} x^{i}(1-x)^{n-i} f(x)$$

we get

$$|f(x) - B_n(f; x)| \le \sum_{i=0}^{n} \binom{n}{i} x^{i}(1-x)^{n-i} \left| f(x) - f\left(\frac{i}{n}\right) \right| \tag{7.8.8}$$

Now, since f is continuous on $[0, 1]$, it follows that it is uniformly continuous also. Therefore, there exists δ such that $|f(s) - f(t)| < \epsilon/2$ whenever $|s - t| < \delta$.

Denote by A the set $\{i: |x - i/n| < \delta\}$. Clearly, the sum on the right side of (7.8.8) can be separated into two sums for those i in A and those that are not. For the first of these, we have

$$\sum_{i \in A} \binom{n}{i} x^i (1-x)^{n-i} \left| f(x) - f\left(\frac{i}{n}\right) \right| \leq \sum_{i \in A} \binom{n}{i} x^i (1-x)^{n-i} \frac{\epsilon}{2} \leq \frac{\epsilon}{2}$$

It remains to prove that we can choose n to make the second sum less than $\epsilon/2$ also. Now, f is continuous and so bounded by M, say. Therefore,

$$\sum_{i \notin A} \binom{n}{i} x^i (1-x)^{n-i} \left| f(x) - f\left(\frac{i}{n}\right) \right| \leq 2M \sum_{i \notin A} \binom{n}{i} x^i (1-x)^{n-i} \quad (7.8.9)$$

Multiplying (7.8.5) by x/n, we obtain

$$\sum_{i=0}^{n} \binom{n}{i} x^i (1-x)^{n-i} x \left(\frac{i}{n} - x\right) = 0$$

while, from the derivation of (7.8.7), we have

$$\frac{x(1-x)}{n} = \sum_{i=0}^{n} \frac{i}{n} \binom{n}{i} x^i (1-x)^{n-i} \left(\frac{i}{n} - x\right)$$

Subtracting these two equations, we get

$$\sum_{i=0}^{n} \binom{n}{i} x^i (1-x)^{n-i} \left(\frac{i}{n} - x\right)^2 = \frac{x(1-x)}{n}$$

Since, for $i \notin A$, $\delta \leq |i/n - x|$, it follows that

$$\sum_{i \notin A} \binom{n}{i} x^i (1-x)^{n-i} \delta^2 \leq \sum_{i \notin A} \binom{n}{i} x^i (1-x)^{n-i} \left(\frac{i}{n} - x\right)^2$$

$$\leq \sum_{i=0}^{n} \binom{n}{i} x^i (1-x)^{n-i} \left(\frac{i}{n} - x\right)^2$$

$$= \frac{x(1-x)}{n} \leq \frac{1}{4n}$$

It is therefore sufficient to choose $n > M/\epsilon\delta^2$ to achieve the desired bound since

$$2M \sum_{i \notin A} \binom{n}{i} x^i (1-x)^{n-i} \leq \frac{2M}{4n\delta^2} < \frac{\epsilon}{2}$$

as required.

This proof of Weierstrass' theorem demonstrates that using $B_n(f; x)$ provides a basis for obtaining good approximations to functions. We illustrate this shortly in Example 7.8.4 but first comment on the evaluation of the Bernstein basis polynomials of degree n. As with most computation involving binomial coefficients and factorials, the definition of $B_{n,i}(x)$ is not a numerically efficient method and, in this case, nor is a recursive algorithm which iterates on i since this will often entail division by small quantities. However, it is easy to verify (see Exercise 2) that

$$B_{n,i}(x) = (1-x)B_{n-1,i}(x) + xB_{n-1,i-1}(x) \quad (7.8.10)$$

for $i, n \geq 1$.

This is readily incorporated into an efficient computational algorithm as follows.

Algorithm 7.8.3   Evaluation of the Bernstein basis polynomials

Input	Degree n, and x
Initialize	$B_0 := 1$
Compute	for $i = 1$ to n
	$\quad B_i := x * B_{i-1}$
	$\quad$ for $j = i - 1$ downto 1
	$\qquad B_j := (1 - x) * B_j + x * B_{j-1}$
	$\quad B_0 := (1 - x) * B_0$
Output	for $i = 0$ to n
	$\quad B_{n,i}(x) = B_i$

It should be observed that the operation count for this algorithm is $n(n + 1)$ multiplications so that evaluation of the Bernstein polynomial $B_n(f; x)$ entails $(n + 1)^2$ multiplications *for every point x*. For each degree of polynomial there are also $n + 1$ evaluations of the function f.

Example 7.8.4. Consider the Bernstein polynomial approximation to $\sin 2x$ on $[0, 1]$.

   We begin with finding the first- and second-degree polynomials "by hand" before automating the process. We will also consider the degree of Bernstein polynomial which is necessary to achieve a specific uniform precision over the range.

   For the first-degree approximation, we have

$$B_{1,0}(x) = 1 - x \qquad \text{and} \qquad B_{1,1}(x) = x$$

The "nodes" are 0 and 1 at which the function values are $\sin 0 = 0$ and $\sin 2 = 0.9093$ to four decimal places. Thus we have

$$B_1(\sin 2x; x) = 0.9093x$$

which is, of course, just the straight-line interpolant to the function at these two nodes.

   For degree two, we use

$$B_{2,0}(x) = (1 - x)^2 \qquad B_{2,1}(x) = 2x(1 - x) \qquad B_{2,2}(x) = x^2$$

and the data is, again to four decimals,

$$\sin 0 = 0 \qquad \sin 1 = 0.8415 \qquad \sin 2 = 0.9093$$

so that the approximation is

$$B_2(\sin x; x) = 1.6830x(1 - x) + 0.9093x^2$$

which again interpolates the original function at the two endpoints—*but* its value at 0.5 is 0.6481. Clearly Bernstein polynomials do not in general interpolate the function at more than the endpoints.

   In Fig. 7.8.5, we graph the first few Bernstein polynomials for this function together with $\sin 2x$ itself for comparison.

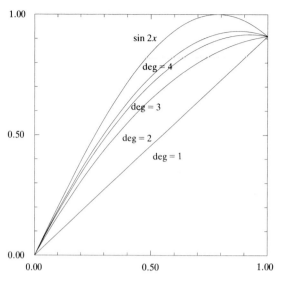

FIGURE 7.8.5
Bernstein polynomial approximation to sin 2x.

It is apparent that the convergence of the Bernstein polynomials is very slow in this case at least. Thus the elegance and power of the Weierstrass approximation theorem and the Bernstein proof thereof do not appear to provide a good practical approach to the approximation of this function.

Let us consider now the question of how the degree of the polynomial required depends on the precision required. First, we need to find δ such that $|\sin 2s - \sin 2t| < \epsilon/2$ for $|s - t| < \delta$. Since, for this case, the derivative is bounded by 2, it follows from the mean value theorem that $|\sin 2s - \sin 2t| < 2|s - t|$. We therefore require $\delta \leq \epsilon/4$. The analysis in the proof of the theorem goes on to establish that we must choose $n > M/\epsilon\delta^2$. With $M = 1$ and $\delta = \epsilon/4$, this reduces to the requirement that $n > 16/\epsilon^3$.

Even for the relatively mild stipulation that our approximation have a uniform accuracy of 0.01, this necessitates some 16 million nodes and therefore some 256 *trillion* multiplications for each required value of x. To graph this function at the 400 points used in Fig. 7.8.5, we would need some 10^{17} multiplications. Even on a very fast—1-gigaflop (that is, 10^9 floating-point operations per second)—computer such a graph would take more than 10^8 seconds, which is something in excess of 3 years!

It appears then that the Bernstein polynomials are far from a practical tool for function approximation. What becomes even more remarkable in the light of all this is just how powerful a tool they can be in CAD. To get a preview of this we consider the problem of obtaining a good smooth approximant to a polygonal, or piecewise linear, function. This is investigated in Example 7.8.6.

The situation is that we have a set of vertices for the polygonal curve in the form (x_i, f_i) for $i = 0, 1, \ldots, N$, where for simplicity we will assume that

$$0 = x_0 < x_1 < \cdots < x_N = 1$$

These points define a piecewise linear interpolant—a linear spline—given, for $x \in [x_i, x_{i+1}]$, by

$$f(x) = \frac{f(x_i)(x_{i+1} - x) + f(x_{i+1})(x - x_i)}{x_{i+1} - x_i} \qquad (7.8.11)$$

It is this function which we approximate with its Bernstein polynomials.

> **Example 7.8.6.** Given the vertices $(0, 0.3)$, $(0.1, 0.9)$, $(0.5, 1.0)$, $(0.8, 0.7)$, and $(1.0, 0.4)$ for the polygonal "curve," we obtain the Bernstein polynomials of degree two through six shown in Fig. 7.8.7 below.

We can see that the curves are being drawn steadily, if slowly, toward the vertices of the polygon. Note, however, that even when the Bernstein polynomial is of degree four or more, it does not interpolate the data. What is achieved is exactly what the theorem demands, namely, that the Bernstein polynomial provides a sufficiently good uniform approximation.

Also, we see that if the polygonal function is concave, then its Bernstein polynomials all lie below the function itself and approach it monotonically. This is a simple consequence of the fact that if f and g are two functions on $[0, 1]$ such that $f(x) \le g(x)$ for every $x \in [0, 1]$, then $B_n(f; x) \le B_n(g; x)$ which in turn follows from the linearity of the mapping $f \mapsto B_n(f; .)$. (See Exercise 5.) A similar observation would apply to a convex path.

However, if the control points do not define a convex polygon when the last point is joined back to the first, it ceases to be the case that the curve will lie "inside" the polygon. For example, if the data used in Example 7.8.6 is

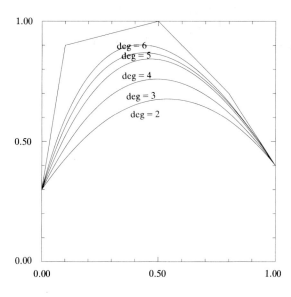

FIGURE 7.8.7
Bernstein polynomials for a polygonal path.

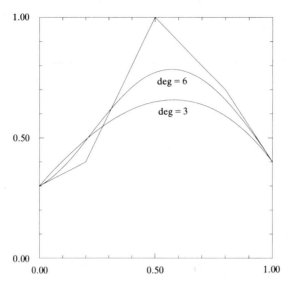

FIGURE 7.8.8
Bernstein polynomials for nonconcave data.

modified by setting $(x_1, f_1) = (0.2, 0.4)$, leaving the remaining data unchanged, the polygonal curve is no longer either concave or convex. The Bernstein polynomials of degree three and six for this data are shown in Fig. 7.8.8. They now intersect the polygonal path itself. The cubic is still concave but the degree-six polynomial has lost this property.

These observations were used by Bezier to develop a powerful tool in computer graphics and computer-aided design. The original development was as a design tool for the Renault car company in the 1960s.

The basic idea is to think of the points (x_i, f_i) not as data points but as "control points" which can be moved to give the resulting Bernstein polynomial a pleasing shape and perhaps to pass through some other specified points, too.

The task then is the choice of the control points (x_i, f_i) which are to be the vertices of a polygonal path. (We will see later that it is desirable to be able to choose not just the f_i but also to adjust the "horizontal" component, but we restrict to the simple case for now.)

Since the Bernstein polynomials always interpolate the two endpoints, we clearly choose $f_0 = y_0$, $f_N = y_N$, where we denote by y_i the (in general, unknown) values of $B_n(f; x_i)$. Also for a concave segment of the curve, we will choose $f_i > y_i$ so that the polygon lies above the desired curve. For the convex segments, it will be necessary to choose $f_i < y_i$.

The process is illustrated in the following example.

Example 7.8.9. We seek a smooth curve satisfying the following requirements. It must pass through the endpoints $(0, 0.1)$ and $(1, 0)$. The curve must be convex and increasing initially and reach its maximum value near $(0.7, 0.8)$. This maximum should be its only stationary point in $[0, 1]$.

The first decisions to be made concern the degree of the Bernstein polynomial (or in this situation, Bezier curve) to be used and the placing of the control points. For the change in curvature it is necessary that the degree be at least three, so we initially try degrees three, four, and five.

A reasonable initial choice for the control points is to take the x_i equally spaced at 0, 0.2, 0.4, 0.6, 0.8, and 1. The interpolation at the endpoints dictates that $f_0 = 0.1$ and $f_5 = 0$. The requirement that the maximum be achieved near $(0.7, 0.8)$ and the fact that the curve is necessarily concave there suggest that we want f_3, f_4 to lie above the curve. We try $f_3 = f_4 = 0.8$. The initial convexity suggests that f_1 should lie below the curve, and to force the initial increase, we try $f_1 = 0.15$. The choice of f_2 appears to be reasonably free as this is likely to be near the inflexion point; we first try $f_2 = 0.45$ which is slightly below the line segment joining (x_1, f_1) and (x_3, f_3).

The results of this first attempt are shown in Fig. 7.8.10a. It is immediately apparent that the peaks of these graphs are located at approximately the desired point but are too low. It is also the case that the cubic appears to offer little hope of achieving sufficient "shape." In the subsequent graphs only the degree-four and degree-five polynomials are plotted.

In Fig. 7.8.10a the maximum value of the degree-five Bezier curve is around 0.6. To raise this we try increasing f_3 and f_4 to 1.0. At the same time we try to put more variation into the curvature by reducing the values of f_1 and f_2. The resulting curves are shown in Fig. 7.8.10b.

The graph in Fig. 7.8.10b comes closer to our requirements, but the maximum is still too low and a little to the left of the desired point. We increase its level by increasing f_3 and f_4, while the shift to the right is achieved by taking $f_4 > f_3$. In order that this increase does not destroy the initial convexity, the values of f_1 and f_2 were decreased further. The result of this is shown in Fig. 7.8.10c: the degree-five curve clearly comes very close to our requirements now. The maximum is very near $(0.7, 0.8)$, the curve is initially convex, but it decreases slightly from $x = 0$. The final graph in Fig. 7.8.10d shows the result of a minor adjustment in which the values of f_1 and f_2 were increased to eliminate this initial downward trend. The final degree-five curve seems to fulfill all our demands.

In this example we used uniformly spaced control points and a fixed number of them. Clearly this is not necessary to define the Bezier curves. For example, it may be desirable to add an extra control point at 0.7 which, of course, becomes x_4 with $f_4 = 1.25$, say, to try to force the maximum to occur very close to 0.7. The result of doing this, with the remaining control points unchanged, is shown in Fig. 7.8.11a. Similarly, we can exaggerate the convexity by reducing the rate of increase near zero. This could be done by inserting a new control point at $(0.3, 0.15)$, for example. Figure 7.8.11b illustrates the effect of making both these changes.

Note that in Fig. 7.8.11b we have allowed the degree of the polynomial to increase further than in the earlier examples. This has the effect of pulling the graph closer to the controls, and so the maximum value is closer to 0.9 for the degree-seven Bezier curve. The point here is simply to see the effect on the overall shape of adjusting the set of control points; the degree-five curve is another satisfactory fit to the requirements that were stated in our problem.

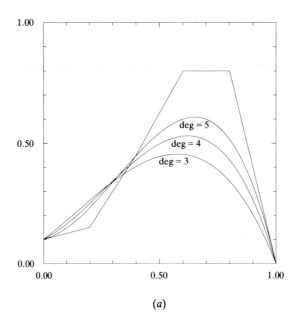

(a)

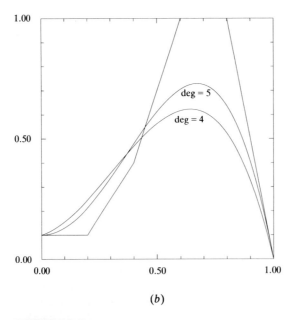

(b)

FIGURE 7.8.10
Bezier curves for Example 7.8.9. (a) $f_1 = 0.15$, $f_2 = 0.45$, $f_3 = f_4 = 0.8$. (b) $f_0 = 0.1$, $f_2 = 0.4$, $f_3 = f_4 = 1.0$. (c) $f_1 = 0$, $f_2 = 0.3$, $f_3 = 1.1$, $f_4 = 1.2$. (d) $f_1 = 0.1$, $f_2 = 0.35$, $f_3 = 1.1$, $f_4 = 1.2$.

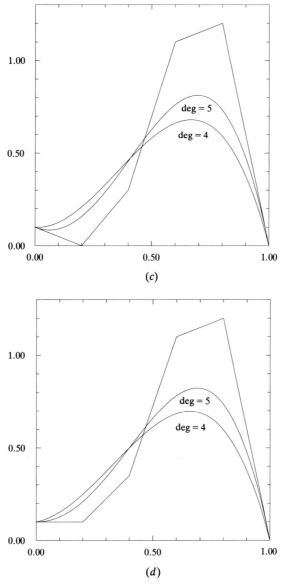

FIGURE 7.8.10 (*Continued*)

The Bezier curves we have considered thus far are just Bernstein polynomials for simple piecewise linear functions. However, in a practical CAD situation we may well have the task of finding a curve which has properties not shared by a function of x. For example a closed curve would not be approximated by such a curve as we have considered so far. There is a simple way of extending the basic idea to this situation.

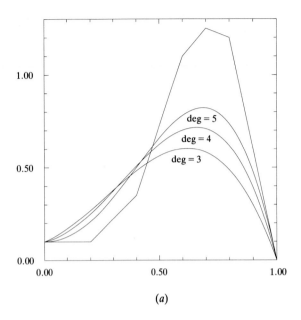

(a)

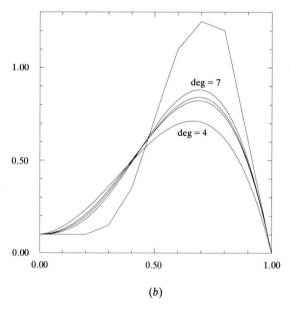

(b)

FIGURE 7.8.11
Bezier curves with additional control points. Control points are those of Fig. 7.8.10d and
(a) $(0.7, 1.25)$, (b) $(0.3, 0.15)$ and $(0.7, 1.25)$.

The idea is to use the Bernstein basis as functions of a parameter t, say, which varies as we move along the curve. The control points are therefore coordinate pairs (x_i, y_i) for $i = 0, 1, \ldots, N$ which are used to define a parametric curve—a Bezier curve—by setting

$$x(t) = \sum_{i=0}^{N} x_i B_{N,i}(t)$$

$$y(t) = \sum_{i=0}^{N} y_i B_{N,i}(t)$$

(7.8.12)

It follows that $(x(0), y(0)) = (x_0, y_0)$ and $(x(1), y(1)) = (x_N, y_N)$ so that the Bezier curve passes through the two endpoints of the curve. It follows that if $(x_0, y_0) = (x_N, y_N)$, then the resulting Bezier curve will be closed. Other than this, the control points behave in just the same way as in the "scalar" case above in pulling the curve toward them. It is no longer the case that the x_i are necessarily increasing, and so it is possible for the curve to intersect itself.

Example 7.8.12. We illustrate the process by attempting to construct an "infinity" symbol using a total of 11 control points. We will try to achieve a width which is half the length and take the central point to be the origin.

If the curve is to start and finish at the left-hand extremity, then fixing a simple scale we take $(x_0, y_0) = (x_{10}, y_{10}) = (-1, 0)$. As a first guess we use

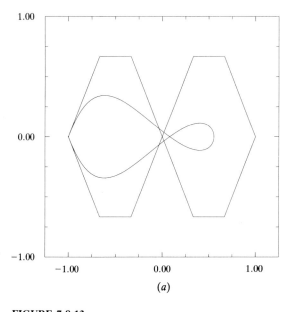

(a)

FIGURE 7.8.13
Bezier curves for approximating an infinity symbol. (a) First attempt; (b) improved; (c) almost acceptable.

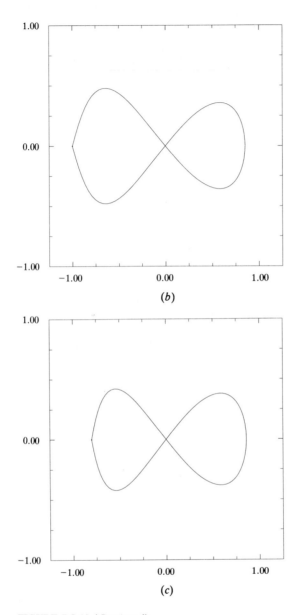

(b)

(c)

FIGURE 7.8.13 (*Continued*)

a symmetric set of control points which in order are $(-1, 0)$, $(-2/3, 2/3)$, $(-1/3, 2/3)$, $(1/3, -2/3)$, $(2/3, -2/3)$, $(1, 0)$, $(2/3, 2/3)$, $(1/3, 2/3)$, $(-1/3, -2/3)$, $(-2/3, -2/3)$, $(-1, 0)$.

The resulting polygonal path and Bezier curve are shown in Fig. 7.8.13a. The right-hand loop is clearly very much smaller than that on the left but has greater smoothness.

In Fig. 7.8.13*b*, we show the result of expanding the right-hand side and increasing the (absolute) *y* values in the left-hand loop. The control points used 1/2, 1, 3/2 in place of 1/3, 2/3, 1 for the *x* coordinates and 1.5 in place of 2/3 in the *y* coordinates on the right. The (absolute) *y* values on the left were set to 1. The curve is clearly a significant improvement. The left loop is still bigger and less smooth but the proportions are about right.

Modifying the *x* coordinate of the left control points to −0.8, −0.6, −0.25 and reducing the (absolute) *y* values for this loop to 0.9 results in the curve in Fig. 7.8.13*c* which is almost acceptable in scale and shape. If it were reduced in scale to that of a single-symbol typographic character, its distortions would be virtually undetectable.

This type of problem is typical of many facing designers of graphical fonts for word processors and laser printers.

EXERCISES 7.8

1. Write down the Bernstein basis polynomials of degrees up to four and therefore the Bernstein polynomial of degrees up to four for the function $\sin \pi x$.

2. Prove the recurrence relation (7.8.10) for the Bernstein basis polynomials, and verify that this yields the same basis polynomials as in Exercise 1.

3. Write a program to implement Bernstein polynomial approximation to the function $\ln(1 + x)$. Find the degree of Bernstein polynomial which is necessary to achieve a uniform precision of 10^{-4}. Plot the original function and its first few Bernstein polynomials.

4. By using a simple transformation of variables, find Bernstein polynomials for the function $\sin x$ on $[-\pi, \pi]$. Graph these approximations.

5. Prove that the Bernstein operator is linear by

showing that

$$B_n(\alpha f + \beta g; x) = \alpha B_n(f; x) + \beta B_n(g; x)$$

Use this result to establish the monotonicity by showing that if $f(x) \leq g(x)$ for every $x \in [0, 1]$, then $B_n(f; x) \leq B_n(g; x)$.

6. Write a program to generate a Bezier curve for a set of control points at the vertices of a simple polygonal path. Use it to generate a satisfactorily smooth curve which initially decreases from $(0, 0.5)$, reaches a minimum near $(0.4, 0)$, and then increases to $(1, 1)$, becoming concave around $x = 0.75$.

7. Generate a computer graphic image of the proportionality symbol $\propto$ by constructing an appropriate Bezier curve.

PROJECTS

1. Consider the function defined by

$$f(x) = \begin{cases} x & 0 \leq x \leq 1 \\ e^{x-1} & 1 \leq x \leq 2 \\ e^{e^{x-2}} & 2 \leq x \leq 3 \end{cases}$$

Try to find the best approximation you can to this function. Supposing that an evaluation of the exponential function is equivalent to 15 floating-point operations, how good an approximation can be achieved more economically than the direct evaluation?

2. Extend the definition of the function in Project 1 by setting

$$f(x) = e^{f(x-1)}$$

for $x > 3$. Repeat the previous exercise for this "generalized exponential function."

3. Find a picture of your favorite shape. Take careful measurements of its outline and try to reproduce it by using discrete least squares, moving least squares, and/or Bezier curves.

CHAPTER
8

OPTIMIZATION

8.1 INTRODUCTION AND CLASSICAL RESULTS

In this chapter we turn to the problem of locating maximum or minimum values of functions. At first sight you may think this problem is already solved by the classical results on critical points for functions of one or several variables. This is far from the case.

A number of complicating factors may be present in any particular optimization problem. Among these is the possibility that the function may be difficult to differentiate, and consequently it may be impractical to consider using the classical results of the calculus. Frequently, there are constraints on the variables of the problem which may be either equations or inequalities that must be satisfied at the solution. The number of variables can be very large, rendering equation-solving techniques impractical. In optimal control problems or variational problems, it is usually the case that the "variables" are themselves functions. At the very least such problems treated classically reduce to the solution of differential equations. In yet other cases, the quantity to be minimized is not given by a closed formula at all. Such is the case with dynamic programming problems including the famous traveling salesman problem.

We illustrate some of the difficulties that may be faced with a few examples, although for the main part of this chapter we will be concerned with the apparently simpler problem of finding extrema (maximum or minimum points) of a function of one or several variables in the absence of constraints.

Example 8.1.1
(a) *Single-variable optimization.* For the minimization of the function te^{-t}, we can simply differentiate and set the derivative to zero to yield $t = 1$ for the minimizer of this function.

By way of contrast, consider the problem of maximizing the lift force on a simple bearing moving with fixed speed U through a fluid with coefficient of viscosity μ, as illustrated in Fig. 8.1.2. The lift force on this bearing is given by

$$F = \frac{6\mu Ul^2}{h_2^2(k-1)^2}\left(\ln k - \frac{2(k-1)}{k+1}\right) \tag{8.1.1}$$

where $k = h_1/h_2 > 1$. The objective is to maximize the lift force F.

The objective function F can easily be seen to be infinitely differentiable, but maximizing this function by solving the equation $F' = 0$ is clearly not straightforward.

(b) *Multivariate unconstrained optimization.* Even for simple functions of just two variables the problem is often not easily solved since the classical necessary conditions for critical points lead to a system of (usually nonlinear) equations to be solved. For greater numbers of variables the problem is more severe.

One practical example which can have many variables is the problem of siting a factory or supply center. Suppose there are suppliers at points with coordinates (x_1, y_1), (x_2, y_2), (x_3, y_3) and outlets at (X_1, Y_1), (X_2, Y_2) with respective total cost per mile transport charges c_1, c_2, c_3, C_1, C_2. The objective is to find the site for the factory which minimizes the total transport cost

$$C = \sum_{i=1}^{3} c_i\sqrt{(x - x_i)^2 + (y - y_i)^2} + \sum_{i=1}^{2} C_i\sqrt{(x - X_i)^2 + (y - Y_i)^2}$$

It is apparent that a more realistic model would need to take account of other factors and that the dimension of the problem would be significantly greater, but even this simplified model would make "analytic" solution impractical.

(c) *Constraints.* Find the maximum and minimum distances from the origin to the intersection of surface

$$4yz + 3zx - 3xy = 2$$

and the plane

$$x + y - z = 1$$

This problem can be solved using Lagrange multipliers, but this requires the solution of a somewhat involved system of nonlinear equations.

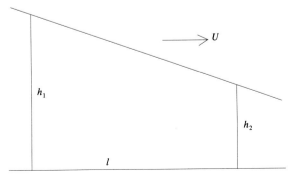

FIGURE 8.1.2
Simple bearing in a viscous fluid.

Other typical constrained optimization problems are on the theme of finding a container of specified shape which minimizes the surface area for a given volume or, equivalently, maximizes the volume for a given surface area. Such problems are significant in the food-canning industry, for example, where the cost of the container should be kept to a minimum.

(d) *Linear programming problems.* This is the name given to the situation of maximizing or minimizing a linear function of several variables subject to linear inequality constraints.

For example, consider the problem

maximize $2x + y$

subject to $x, y \geq 0, y - x \leq 1, 3x + y \leq 2$

It is tolerably easy to see that the solution to such a problem must occur at a vertex of the *feasible region* of points satisfying the constraints. For a problem of this scale it is therefore easy to locate this maximum since there are only four vertices. In a typical realistic problem there may be hundreds or even thousands of variables and constraints. The algorithms used are usually based on the simplex method which for this simple example can be likened to drawing contours of the objective function $2x + y = c$ for various values of c in the feasible region. We can then identify the maximum by finding the corner at which the highest-value contour touches the feasible region. This process is illustrated in Fig. 8.1.3.

It is apparent from the graph that the maximal contour touches the feasible region at the vertex $(\frac{1}{4}, \frac{5}{4})$ at which the objective function takes the value $\frac{7}{4}$.

We do not consider algorithms for linear programming further in this text. Nor do we study problems which may be solved by variational methods such as finding geodesics, shortest paths between two points on a given surface.

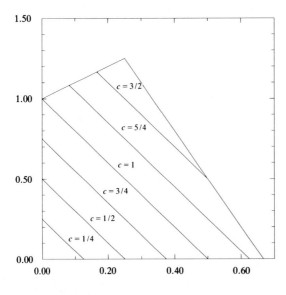

FIGURE 8.1.3
Feasible region and contours for Example 8.1.1(d).

(*e*) *Dynamic programming*. This final example illustrates a large class of problems including some famously difficult problems such as the traveling salesman problem. This is the problem of finding the shortest path among a set of
cities by a path which visits each one of them *exactly* once.

A simple example of a dynamic programming problem is that of finding
the shortest path between two points on a given map. We can illustrate this
with a diagram, as in Fig. 8.1.4. At each junction we must make a decision as
to which direction to follow. For the particular grid shown there are 20
possible paths from *A* to *B* for which a complete enumeration of the options
would entail approximately 1000 additions and then 20 comparisons. The
dynamic programming solution reduces this to just 24 additions and 9
comparisons.

Working backward from the target point *B*, we compute the shortest
route from each vertex and label the vertex accordingly with both the distance
and the direction to be taken *if* the solution passes through that vertex. The
process yields the solution shown by the thick line in Fig. 8.1.4.

For the rest of this chapter we will concentrate on the minimization of a
function $f(x_1, x_2, \ldots, x_n)$ perhaps subject to constraints. An important special
case is single-variable minimization ($n = 1$) which frequently arises as a sub-
problem in multivariable minimization algorithms.

We recall first the classical results of the calculus characterizing critical
points of such functions. In the single-variable case, for a sufficiently differentiable function, we have the standard characterization of stationary points.

Suppose that $f'(x) = f''(x) = \cdots = f^{(2p-1)}(x) = 0$; then:

1. If $f^{(2p)} \begin{cases} >0 \\ <0 \end{cases}$ then x is a $\left\{ \begin{array}{l} \text{local minimum} \\ \text{local maximum} \end{array} \right\}$ of f.

2. If $f^{(2p)}(x) = 0$, $f^{(2p+1)}(x) \neq 0$, then x is a (stationary) point of inflexion of f.

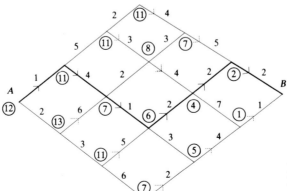

FIGURE 8.1.4
Dynamic programming problem
and solution.

In the case of single-variable minimization over an interval—which is the simplest inequality-constrained optimization problem—then the endpoints of the interval must also be examined as candidates for the extreme values of f.

For functions of more than one variable, the characterization in terms of derivatives of order greater than two is difficult, and so we content ourselves with stating sufficient conditions for a local minimum or maximum. It is helpful to introduce the notation $J(\mathbf{x}_0)$ for the Hessian matrix of f at the point $\mathbf{x}_0$. This is the matrix of second partial derivatives of f whose elements are given by

$$J_{ij}(\mathbf{x}_0) = \frac{\partial^2 f(\mathbf{x}_0)}{\partial x_i \, \partial x_j} \tag{8.1.2}$$

At a stationary point, $\mathbf{x}_0$ say, of the function f it is necessarily the case that the gradient vector vanishes there; that is,

$$\nabla f(\mathbf{x}_0) = \left(\frac{\partial f(\mathbf{x}_0)}{\partial x_1}, \frac{\partial f(\mathbf{x}_0)}{\partial x_2}, \ldots, \frac{\partial f(\mathbf{x}_0)}{\partial x_n} \right)^{\mathrm{T}} = \mathbf{0} \tag{8.1.3}$$

This point is a local minimum of f if $J(\mathbf{x}_0)$ is positive definite. It is a local maximum if the Hessian is negative definite. Recall that a square matrix A is positive definite if

$$\mathbf{x}^{\mathrm{T}} A \mathbf{x} > 0 \tag{8.1.4}$$

for every nonzero vector $\mathbf{x}$.

This is a difficult condition to check for complicated functions of several variables. Even in the two-dimensional case, for which the conditions for a minimum reduce to

$$f_x = f_y = 0 \qquad f_{xx} > 0 \qquad f_{xx} f_{yy} - f_{xy}^2 > 0$$

checking all possible critical points and then comparing their function values can be problematic, even if all such stationary points can be located.

Example 8.1.5. Consider the minimization of the function

$$f(x, y) = x^2 + y^2 - 2 \sin x^2 y$$

which has first partial derivatives

$$f_x = 2x - 4xy \cos x^2 y$$
$$f_y = 2y - 2x^2 \cos x^2 y$$

Clearly there is one stationary point at $(0,0)$ which turns out to be a local minimum. Finding other extrema and classifying them requires a numerical treatment.

In the case of equality-constrained problems we can use the method of Lagrange multipliers which gives us conditions that must be satisfied by a constrained extremum.

Specifically, consider the problem

minimize $f(x_1, x_2, \ldots, x_n)$

subject to $c_j(x_1, x_2, \ldots, x_n) = 0$ $(j = 1, 2, \ldots, m)$

where $m < n$ and the objective function f and the constraint functions c_j are all continuously differentiable.

Suppose $\mathbf{x}_0$ is a local minimum for this problem. Provided $\mathbf{x}_0$ is not a singular point of the constraints, which is to say that the Jacobian matrix with elements $\partial c_j / \partial x_i$ has rank m, then there exist scalars—the *Lagrange multipliers*—$\lambda_1, \lambda_2, \ldots, \lambda_m$—such that the Lagrange equations are satisfied at $\mathbf{x}_0$; that is,

$$\frac{\partial f}{\partial x_i} + \sum_{j=1}^{m} \lambda_j \frac{\partial c_j}{\partial x_i} = 0 \qquad (i = 1, 2, \ldots, n)$$

$$c_j(\mathbf{x}_0) = 0 \qquad (j = 1, 2, \ldots, m)$$

(8.1.5)

which represents $n + m$ equations in the $n + m$ unknowns—the components of $\mathbf{x}_0$ and the Lagrange multipliers.

We finish this introductory section with one simple example of the use of Lagrange multipliers for maximization. (A similar result to that quoted applies to this case.)

Example 8.1.6. Consider the post office problem of maximizing the volume of a parcel that can be fitted into a sack.

The objective function to be maximized is therefore just xyz, and the post office constraints are that no edge should exceed 42 inches and the length plus girth should be no greater than 72 inches; that is, the problem reduces to

maximize xyz

subject to $x, y, z \le 42$ and $x + 2y + 2z \le 72$

It is reasonably easy to see that the last constraint must be satisfied *as an equality* at the solution and that the restrictions on individual edges are not *active* at the solution; that is, $x, y, z < 42$ for the maximum volume.

The problem is thus reduced to maximizing xyz subject to the single constraint $x + 2y + 2z = 72$. The Lagrange equations are

$$yz - \lambda = 0$$
$$xz - 2\lambda = 0$$
$$xy - 2\lambda = 0$$
$$x + 2y + 2z = 72$$

from the first three of which we deduce that $x = 2y = 2z$, and substitution into the constraint equation yields $x = 24$, $y = z = 12$ so that the maximum volume is $24 \times 12 \times 12$ cubic inches, or 2 cubic feet.

EXERCISES 8.1

1. Plot contours of the function $4x^2 - 4xy + 2y^2 + y$ for values $0, 1, 4, 9, 16,$ and 25 in the region $[-3, 3] \times [-2, 2]$. That is, graph the curves $f(x, y) = c$ for the above values of c: this can be done by solving a quadratic equation for one of the variables in terms of the other. Where is the minimum of this function?

2. Find and classify all stationary points of Rosenbrock's function $f(x, y) = 100(x^2 - y)^2 + (1 - x)^2$. Produce a contour plot of this function for $x \in [-1.5, 1.5]$, $y \in [-0.5, 2]$. (We will be using this function throughout the chapter as a test of minimization routines.)

3. Find the maximum and minimum distances from the origin to the intersection of the surface $x^2 + xy - yz = 1$ and the plane $x + y - z = 2$.

4. Find the dimensions of an optimal can of peaches with volume V. That is, minimize the surface area of a closed circular cylinder subject to the constraint $\pi r^2 h = V$, where r is the radius and h is the height of the cylinder. Are you paying more than you need for packaging of canned foods?

5. Find the minimum area of glass sheet needed for an open-topped fish tank of specified volume. (All cross sections are rectangular.)

8.2 SINGLE-VARIABLE MINIMIZATION

In this section we investigate the problem of minimizing a function f of a single real variable. Most numerical methods for solving this problem are based on the idea of reducing an interval which contains the required minimum—a *bracket* for this minimum. It is frequently assumed that the objective function is unimodal inside this bracket, which is to say that it has only one turning point there.

Before considering any method in detail, we observe that here, as in the multivariate situation to be discussed in later sections, we need only concern ourselves with *minimization* since maximizing a function f is equivalent to minimizing $-f$.

One important aspect of the single-variable minimization problem is that several of the multivariate algorithms implement a sequence of single-variable minimizations along carefully chosen lines in the higher-dimensional space. For these *line searches*, it is often the case that we have a good idea as to where the minimum lies, so an initial bracket is fairly easily obtained.

In general this need not be the case. It is worthwhile, therefore, to spend a little time considering how such a bracket for the minimum of the objective function f may be found. The details differ somewhat depending on whether the derivative f' is available within the algorithm. (Indeed, this is an important consideration throughout optimization.)

We can deduce that f has a local minimum in $[a, b]$ if

either $$f'(a) < 0 < f'(b) \tag{8.2.1}$$

or There exists $c \in (a, b)$ such that $f(c) < f(a), f(b)$ $\qquad$ (8.2.2)

The idea behind our bracketing procedure is to choose a starting point and an initial stepsize and then to take repeated steps (increasing their magnitude each time) until either two successive points satisfy (8.2.1) or three satisfy (8.2.2). Specifically, it is common to double the stepsize at each iteration so that with an initial guess x_0 and steplength h, the subsequent points are generated by

$$x_{i+1} = x_i + 2^i h \qquad (8.2.3)$$

The following example illustrates the process both with and without use of the derivative. We will also see the efficiency gained by the doubling of the steplength.

Example 8.2.1. Bracket the minimum of $f(x) = x + 25/x^2$ both with and without using the derivative f' starting from $x_0 = 1$ and using $h = 0.01$.

Solution. According to (8.2.3) the points tried are $1, 1.01, 1.03, 1.07$, etc., until a bracket is obtained. The results obtained are as follows:

i	x_i	$f(x_i)$	$f'(x_i)$
0	1.00	26.0000	-49.0000
1	1.01	25.5174	-47.5295
2	1.03	24.5949	-44.7571
3	1.07	22.9060	-39.8149
4	1.15	20.0536	-31.8758
5	1.31	15.8779	-21.2411
6	1.63	11.0394	-10.5454
7	2.27	7.1216	-3.2746
8	3.55	5.5337	-0.1176
9	6.11	6.7797	0.7808

We see that without the derivative the last three points satisfy (8.2.2), so the minimum lies in the interval $[2.27, 6.11]$. The final two points show a change of sign of the derivative and so imply that the minimum lies in the smaller interval $[3.55, 6.11]$.

The shorter interval obtained using the derivative is typical. The no-derivative bracket uses three points satisfying (8.2.2) with $b - c = 2(c - a)$, whereas the derivative bracket obtained from the same initial point and step-length is either $[a, c]$ or $[c, b]$ which have lengths $(b - a)/3$ and $2(b - a)/3$, respectively. Thus the derivative bracket is either one-third or two-thirds the length of that obtained without the derivative.

It is apparent that the repeated doubling of the steplength in the above bracketing process has resulted in considerably larger bracketing intervals than we would probably require. Of course, a much smaller bracket would be obtained by leaving the steplength unchanged throughout the process. This would result in brackets $[3.67, 3.69]$ without the derivative and $[3.68, 3.69]$ with the derivative. (The true minimum is at $50^{1/3} \approx 3.6840$.) However, the number of function or derivative evaluations needed would be 270 in each case.

The number of such evaluations used above is 10. The algorithms described in the remainder of this section will reduce the larger brackets to acceptable sizes in a small fraction of the number of steps needed for the smaller brackets.

It should be observed that the basic algorithm described here would need several safeguards. It is clearly a possibility for the no-derivative search to go in the wrong direction. If the function is known to be unimodal, the direction of search could be reversed if the first step is uphill. There is also the possibility that the process could miss the minimum altogether. This is illustrated in Fig. 8.2.2 in which the minimum is located in a relatively narrow trough. We do not discuss the details of safeguarding the algorithm here.

We turn now to the question of locating the minimum of f within a known bracket $[a, b]$. There are several approaches to this. Those we discuss fall into two categories: one based on reducing the actual bracketing interval in an efficient manner and one based on iterative estimation of the actual minimum point by minimizing low-degree interpolation polynomials. The interpolation-based methods carry a greater computational overhead but are often better if great accuracy is required.

We begin with the bracket reduction approach. Suppose then that we have a bracket $[a, d]$ for the minimum of the function f. The idea is to place two more points $b < c$ symmetrically in this interval and then reduce the bracket to either $[a, c]$ or $[b, d]$ depending on whether $f(b) < f(c)$. The process is then repeated with this reduced bracket. This step is illustrated in Fig. 8.2.3.

The key to the efficiency of these methods lies in the choice of the additional points within the bracket. At first sight it seems we must make two new function evaluations per iteration, but if the choice is made carefully, then we can use all three of the known points in the new bracket and choose just one more per iteration.

The symmetry requirement is that

$$d - c = b - a \tag{8.2.4}$$

and we will suppose that b, c are placed such that

$$b = a + \alpha(d - a) = (1 - \alpha)a + \alpha d$$

$$c = \alpha a + (1 - \alpha)d \tag{8.2.5}$$

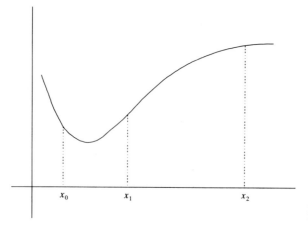

$x_0 \qquad x_1 \qquad x_2$

FIGURE 8.2.2
Failure to bracket the minimum.

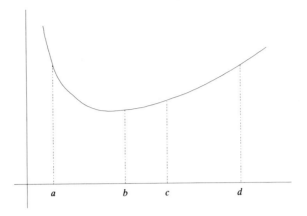

FIGURE 8.2.3
Bracket reduction step. Here $f(b) < f(c)$, so the new bracket is $[a, c]$.

In the situation illustrated in Fig. 8.2.3, the next iteration would begin with the bracket $[a_1, d_1] = [a, c]$ and with the right-hand one of the internal points $c_1 = b$. The new point is

$$b_1 = a_1 + (d_1 - c_1) = (1 - \alpha_1)a_1 + \alpha_1 d_1 \qquad (8.2.6)$$

where

$$\alpha_1 = \frac{d_1 - c_1}{d_1 - a_1} = \frac{c - b}{c - a} = \frac{1 - 2\alpha}{1 - \alpha} \qquad (8.2.7)$$

It is easy to see that the same relation holds for the case where $f(b) > f(c)$ and the new bracket is given by $[b, d]$.

We thus have a simple relation between the α's of (8.2.5) for two successive iterations. Notice that, in practice, we do not have to compute these quantities but define the new point for each iteration according to (8.2.6) and its equivalent form for the other case. However, knowledge of the α's does enable us to analyze the rate of reduction of the bracketing interval.

The simplest choice for the α's is to insist that they are all equal. From (8.2.7) it follows that the required value is $\alpha = (3 - \sqrt{5})/2 \approx 0.382$, which is the golden section. Although this may be the most pleasing aesthetically, it is not the most efficient search of this type. That distinction belongs to the Fibonacci search which is based on the use of the Fibonacci numbers defined by

$$F_{n+1} = F_n + F_{n-1} \qquad F_0 = F_1 = 1 \qquad (8.2.8)$$

Suppose that for some iteration we have

$$\alpha = \frac{F_{n-2}}{F_n} \qquad (8.2.9)$$

then using (8.2.7) and (8.2.8) we have

$$\alpha_1 = \frac{1-2\alpha}{1-\alpha} = \frac{F_n - 2F_{n-2}}{F_n - F_{n-2}} = \frac{F_{n-1} - F_{n-2}}{F_{n-1}} = \frac{F_{n-3}}{F_{n-1}} \qquad (8.2.10)$$

It follows that we can begin a Fibonacci search using (8.2.9) with subsequent α's given by $F_{n-3}/F_{n-1}, \ldots, F_1/F_3, F_0/F_2$. The last of these is, of course, just $\frac{1}{2}$ so that no new points are added at this stage. There are therefore $n-2$ steps in this process.

At each stage, the length of the bracket is reduced by the factor $(1-\alpha)$ for the appropriate α. Now

$$1 - \frac{F_{k-2}}{F_k} = \frac{F_k - F_{k-2}}{F_k} = \frac{F_{k-1}}{F_k}$$

and so the final interval of this Fibonacci search has length

$$\frac{F_{n-1}F_{n-2} \cdots F_2}{F_n F_{n-1} \cdots F_3} L = \frac{2}{F_n} L \qquad (8.2.11)$$

where $L = d - a$ is the length of the original bracket. Deciding which half of this final interval contains the minimum reduces the final length to just L/F_n. It follows that if a final interval of length ϵ is required, then it suffices to choose n so that $F_n > L/\epsilon$.

We have observed that at the conclusion of the $n-2$ steps we have a bracketing interval together with its midpoint so that a further step of the bracket reduction would simply insert a second copy of this midpoint. In practice, however, the "midpoint" generated will be contaminated by rounding error and so will be slightly to one side of the center. We can therefore proceed with one further step of the reduction to determine which half of this bracket contains the minimum. This is one of the very few situations in which roundoff error provides a genuine benefit.

This use of the rounding error is incorporated into the Fibonacci search described below to determine the final bracket.

Algorithm 8.2.4 Fibonacci search

Input	Bracket $[a, d]$ for the minimum of f
	Required tolerance ϵ
Initialize	Find least n such that $F_n > (d-a)/\epsilon$
	$\alpha := F_{n-2}/F_n$
	$b := a + \alpha(d-a); \; c := d - (b-a) = a + d - b$
Loop	for $i = 1$ to $n-2$
	if $f(b) < f(c)$ **then** $d := c; \; c := b; \; b := a + d - c$
	else $a := b; \; b := c; \; c := a + d - b$
Output	**if** $f(b) < f(c)$ **then** final bracket is $[a, c]$
	else final bracket is $[b, d]$.

It should be noted that in the final step it is possible that the roundoff error results in $b > c$, and so care must be exercised over the final decision. (This consideration is excluded from Algorithm 8.2.4 in the interests of simplicity.)

Example 8.2.5. The Fibonacci search does not use derivative information. The appropriate bracket in Example 8.2.1 for the minimum of $x + 25/x^2$ is therefore [2.27, 6.11]. We will use the Fibonacci search to reduce this to a bracket of final length less than 0.1.

The initial bracket is of length 3.84 so that we require $F_n > 3.84/0.1 = 38.4$, which is first achieved for $n = 9$ since $F_8 = 34$ and $F_9 = 55$. It follows that the initial α is given by $\frac{21}{55}$, and so

$$b = 2.27 + 3.84\alpha = 3.7362$$

$$c = 2.27 + 6.11 - 3.7362 = 4.6438$$

to four decimal places. In this case

$$f(b) = 5.5271 < 5.8031 = f(c)$$

and so the bracket is reduced to $[a, c] = [2.27, 4.6438]$.

For the next iteration we use

$$a = 2.27 \qquad c = 3.7362 \qquad d = 4.6438$$

$$b = 2.27 + 4.6438 - 3.7362 = 3.1776$$

This and the subsequent iterations are summarized in Table 8.2.6. These results were computed in single precision, and we see that the final iteration has indeed produced two different "midpoints"—and that their order is reversed. Note that for the quoted b and c of the final iteration, $f(b) < f(c)$ which appears to conflict with the "no" in the final column. The program, of course, interchanges b and c to correct their order before testing for the final bracket.

The total number of function evaluations used in generating these results is 19, which is made up of 10 for the original bracket and a further 9 in this search. Note that if a smaller final bracket had been required, the number of additional steps of the Fibonacci search does not increase very rapidly. For example, since

TABLE 8.2.6
Fibonacci search using bracket obtained in Example 8.2.1

i	a	b	c	d	$f(b) < f(c)$?
1	2.270 000	3.736 182	4.643 818	6.110 000	Yes
2	2.270 000	3.177 637	3.736 182	4.643 818	No
3	3.177 637	3.736 182	4.085 273	4.643 818	Yes
4	3.177 637	3.526 728	3.736 182	4.085 273	No
5	3.526 728	3.736 182	3.875 820	4.085 273	Yes
6	3.526 728	3.666 366	3.736 182	3.875 820	Yes
7	3.526 728	3.596 544	3.666 366	3.736 182	No
8	3.596 544	3.666 366	3.666 359	3.736 182	No

The final bracket is [3.666 359, 3.736 182]

$F_{13} = 377 < 384 < 610 = F_{14}$, only five more iterations would have been sufficient to reduce the interval to less than 0.01 in length. That is, a total of 24 function evaluations would have sufficed—a very significant improvement over the use of very small steps in the bracketing process.

We turn our attention now to the interpolation-based methods which estimate the position of the minimum point itself by minimizing a low-degree interpolation polynomial.

The simplest of these algorithms is the *quadratic search* which uses a quadratic interpolant agreeing with f at three uniformly spaced points.

Suppose then that x_0, x_1, x_2 satisfy

$$x_2 = x_1 + h = x_0 + 2h \qquad f_1 < f_0, f_2 \qquad (8.2.12)$$

where $f_i = f(x_i)$ ($i = 0, 1, 2$). Beginning with either the Gauss central difference formula or Newton's divided difference formula, it is easy to establish (see Exercise 6) that the quadratic polynomial agreeing with f at these three points has its minimum at

$$x^* = x_1 + \frac{h(f_0 - f_2)}{2(f_0 - 2f_1 + f_2)} \qquad (8.2.13)$$

and that this minimum point lies in the interval $[x_1 - h/2, x_1 + h/2]$ which is the middle half of the original bracket.

We can then use either x_1 or x^* as the center of a new interval half the length of the previous one. The choice is simply to take the one which yields the smaller function value. The process can then be iterated until the required accuracy has been obtained. (That the process converges is also easy to prove—see Exercise 7.)

There is just one point at which we must be careful: Eq. (8.2.13) relies on the fact that f_1 is the smallest of the three function values. We must check that this is true for subsequent iterations and, if necessary, shift the interval to either the left or the right by the new h. (Since the original interval is a bracket, this will never be needed more than once for a unimodal function.)

This quadratic search is summarized in Algorithm 8.2.7.

Algorithm 8.2.7 Quadratic search

Input   Bracket $[x_0, x_2]$ for the minimum of f
Required tolerance ϵ
$x_1 := (x_0 + x_2)/2; h := x_1 - x_0$

Loop   **Repeat**
if $f_0 < f_1$ **then** (shift left) $x_2 := x_1; x_1 := x_0; x_0 := x_1 - h$
if $f_2 < f_1$ **then** (shift right) $x_0 := x_1; x_1 := x_2; x_2 := x_1 + h$
$x^* := x_1 + h(f_0 - f_2)/2(f_0 - 2f_1 + f_2)$
if $f(x^*) < f_1$ **then** $x_1 := x^*$
$h := h/2; x_0 := x_1 - h; x_2 := x_1 + h$
until $h < \epsilon$

Output   Minimum is x_1 with error less than h.

The algorithm is illustrated using the same example as for the Fibonacci search, though with a somewhat altered initial bracket to simplify the arithmetic of the first iteration which we cover in detail.

Example 8.2.8. Use the quadratic search to locate the minimum of $x + 25/x^2$ within the initial bracket $[2, 6]$ to an accuracy of 0.01.

Solution. For the first iteration, we have $x_0 = 2$, $x_1 = 4$, $x_2 = 6$ which have the associated function values $f_0 = 8.25$, $f_1 = 5.5625$, $f_2 = 6.6944$. We indeed have a bracket. Using (8.2.13), we have $x^* = 4.4073$ to four decimal places, and $f(x^*) = 5.6943 > f_1$ so that x_1 is retained as the center of the next interval. The three points used for this next iteration are therefore $x_0 = 3$, $x_1 = 4$, $x_2 = 5$ with $f_0 = 5.7778$, $f_1 = 5.5625$, $f_2 = 6.0000$ which again provides a bracket. The subsequent iterations are tabulated below.

x_0	x_1	x_2	f_1	x^*	$f(x^*)$
2.0000	4.0000	6.0000	5.562 500	4.4073	5.694 337
3.0000	4.0000	5.0000	5.562 500	3.8298	5.534 263
3.3298	3.8298	4.3298	5.534 263	3.7200	5.526 568
3.4700	3.7200	3.9700	5.526 568	3.6948	5.526 094
3.5698	3.6948	3.8198	5.526 094	3.6868	5.526 050
3.6243	3.6868	3.7493	5.526 050	3.6847	5.526 047
3.6535	3.6847	3.7160	5.526 047	3.6842	5.526 047
3.6686	3.6842	3.6998	5.526 047	3.6841	5.526 047

The final output from the program reported that the minimum is at 3.684076 with error less than 0.01.

For the quadratic search, each iteration after the first requires three new function evaluations so that the total here is 23 evaluations in addition to those involved in obtaining the original bracket.

Each subsequent iteration reduces the bracket by 50 percent compared with an asymptotic value of about 38 percent for the Fibonacci search. At this accuracy the amount of work required by the two algorithms is comparable, but the quadratic search would clearly be better for refining the solution further.

There are variations on this theme. It is not essential to use uniformly spaced interpolation points, for example. In that case three points which give a bracket are used, and the minimum of the interpolating quadratic is used in place of one of them to reduce the bracket, the process being continued until the required tolerance is achieved. A further modification using the derivative is considered in Exercise 10.

However, if the derivative is to be used, we can make better use of it in the *cubic search*. Here we use a cubic polynomial which agrees with both f and f' at two points x_0 and x_1 which bracket the minimum. Since $[x_0, x_1]$ is a bracket, it follows that

$$f'(x_0) < 0 < f'(x_i) \tag{8.2.14}$$

and therefore that the minimum of the interpolation polynomial must lie in the interval, too. We will again denote this minimum by x^*.

Denote the cubic polynomial by $p(x)$—this is in fact the *Hermite cubic* for the data—and write it in the form

$$p(x) = a(x - x_0)^3 + b(x - x_0)^2 + c(x - x_0) + d \qquad (8.2.15)$$

Now the turning points of p are at

$$x - x_0 = \frac{-b \pm \sqrt{b^2 - 3ac}}{3a} \qquad (8.2.16)$$

Since $p'(x_0) < 0 < p'(x_1)$, exactly one of these lies in the bracket, and it must be the local minimum of p.

It is easy to see that the required turning point always corresponds to choosing the plus sign in (8.2.16): for if $a > 0$, then $p(x) \to \infty$ as $x \to \infty$, and so the local minimum is the right-hand turning point which corresponds to the larger value and thus the plus sign, while if $a < 0$, then the local minimum is the left-hand turning point, and with the denominator in (8.2.16) being negative this too corresponds to the plus sign. Thus we have

$$x^* = x_0 + \frac{\sqrt{b^2 - 3ac} - b}{3a} . \qquad (8.2.17)$$

The algorithm proceeds by simply replacing either x_0 or x_1 by x^* depending on the sign of $f'(x^*)$ to preserve the bracketing property.

It remains then to find the coefficients a, b, c in (8.2.15). (The value of d is easily obtained but does not affect the position of the minimum.) Now,

$$p(x_0) = d = f_0 \quad \text{and} \quad p'(x_0) = c = f_0' \qquad (8.2.18)$$

Setting $h = x_1 - x_0$, the interpolation conditions at x_1 now reduce to

$$ah^3 + bh^2 = f_1 - f_0 - hf_0'$$

$$3ah^2 + 2bh = f_1' - f_0'$$

which have the solutions

$$a = \frac{G - 2H}{h} \qquad b = 3H - G \qquad (8.2.19)$$

where

$$F = \frac{f_1 - f_0}{h} = f[x_0, x_1]$$

$$G = \frac{f_1' - f_0'}{h} = f'[x_0, x_1] \qquad (8.2.20)$$

$$H = \frac{F - f_0'}{h}$$

The process is summarized as Algorithm 8.2.9. Note that in the interests of computational efficiency we compute the quantity $X = 1/h$ to avoid repeated divisions.

Algorithm 8.2.9 Cubic search

Input   Bracket $[x_0, x_1]$ for the minimum of f
Values f_0, f_1, f_0', f_1' and required tolerance ϵ.

Loop   **Repeat**
$X := 1/(x_1 - x_0); F := X(f_1 - f_0);$
$G := X(f_1' - f_0'); H := X(F - f_0')$
$a := X(G - 2H); b := 3H - G; c := f_0'$
$x^* := x_0 + (\sqrt{b^2 - 3ac} - b)/3a$
if $f'(x^*) > 0$ **then** $x_1 := x^*; f_1 := f(x^*); f_1' := f'(x^*)$
else $x_0 := x^*; f_0 := f(x^*); f_0' := f'(x^*)$
until $x_1 - x_0 < \epsilon$

Output   Minimum is x^* with error less than ϵ.

We note immediately that this algorithm is not well-suited to extensive hand calculation but is often highly efficient on the computer, which is why algorithms of this type form the basis of most "line searches" within multivariable algorithms using gradients. Other convergence tests incorporating a requirement on the magnitude of $f'(x^*)$ would also be appropriate for this algorithm, depending on the specific situation.

Example 8.2.10. Minimize $x + 25/x^2$ by the cubic search using the bracket found in Example 8.2.1.

Solution. The bracket obtained using derivatives was $[3.55, 6.11]$. With $x_0 = 3.55$ and $x_1 = 6.11$, we have

$$f_0 = 5.5337 \qquad f_1 = 6.7797 \qquad f_0' = -0.1176 \qquad f_1' = 0.7808$$

The first iteration therefore proceeds by setting

$$X = 0.3906 \qquad F = 0.4867 \qquad G = 0.3509 \qquad H = 0.2361$$

and then

$$a = -0.0473 \qquad b = 0.3572 \qquad c = -0.1176$$

$$x^* = 3.7204$$

Now $f'(x^*) = 0.0290 > 0$, and so for the next iteration we set $x_1 = 3.7204$. Note that the bracket is already reduced to a length less than 0.2. The first three iterations are summarized in the following table. This is sufficient to identify the true minimum to a precision of 0.001.

Iter.	x_0	x_1	x^*	$f'(x^*)$
1	3.550 000	6.110 000	3.720 370	0.029 017 157 1
2	3.550 000	3.720 370	3.683 969	−0.000 051 221 1
3	3.683 969	3.720 370	3.684 032	0.000 000 008 1

The program used then reported

$$\text{Minimum is } 3.684032 \text{ with error less than } 0.0010$$

We note that each iteration of this cubic search requires just two new function (or derivative) evaluations, and so the total cost of obtaining this solution is just 18 evaluations including the 10 derivative evaluations for the initial bracket and then 2 evaluations of f for the required input. It is clear that use of the derivative is worthwhile if it is readily available.

Before leaving the topic of single-variable minimization, we should observe that an alternative approach is to use any of our standard iterative methods for solution of the equation $f'(x) = 0$. One drawback to such an approach is that we make no use of the objective function itself. We must therefore be very careful to ensure that a *minimum* is found. Note too that Newton's method would require evaluation of the second derivative which need not be available even if f' is.

EXERCISES 8.2

1. Find brackets for the minimum of $(1 - \ln x)/(x\sqrt{x})$ beginning with $x = 1$ and using the initial steplength 0.01.

2. Verify that for the golden section search, choosing the value of α to be constant results in $\alpha \approx 0.382$.

3. Use a Fibonacci search to reduce the appropriate bracket from Exercise 1 to a length less than 0.1. Compare this with the golden section search.

4. Write a program to implement the Fibonacci search to reduce an initial bracket of length L to a final bracket of length ϵ. Test your program by obtaining an interval of length less than 0.01 containing the minimum of the function in Exercises 1 and 3.

5. Show that the Fibonacci numbers defined by (8.2.8) are given by

$$F_n = \frac{1}{\sqrt{5}} \left[\left(\frac{1 + \sqrt{5}}{2} \right)^{n+1} - \left(\frac{1 - \sqrt{5}}{2} \right)^{n+1} \right]$$

Use this to show that for a Fibonacci search using n steps, the final interval length is approximately $L\sqrt{5}[2/(1 + \sqrt{5})]^{n+1}$. Show that this is approximately 17 percent better than the golden section.

6. Derive formula (8.2.13) for the minimum of the quadratic interpolation polynomial.

7. Prove that the quadratic search (Algorithm 8.2.7) generates a sequence of points x_1 which converges to the minimum of a unimodal function f.

8. Write a program to implement the quadratic search, and use it to find the minimum of $(1 - \ln x)/(x\sqrt{x})$ with error less than 0.01 beginning with the bracket found in Exercise 1.

9. Modify your quadratic search program to the case where the points are not uniformly spaced. At each iteration, the algorithm should retain the three points which provide the smallest bracket for the minimum. Compare the performance of this algorithm with that of Algorithm 8.2.7.

10. Show that minimizing a function by a quadratic search using an interpolation polynomial which agrees with f_0, f_0', f_1' is equivalent to solving $f'(x) = 0$ by the secant algorithm.

11. Program and test the cubic search algorithm for the function of Exercise 8 using the appropriate bracket from Exercise 1.

8.3 MULTIVARIABLE MINIMIZATION: DIRECT SEARCH METHODS

In this section, we begin our study of the minimization of a function of several variables by considering some of the so-called *direct search* methods which make no use of gradient information. There are many methods available for such problems, but we content ourselves here with describing just one of the more efficient techniques in detail. We will also briefly summarize a few others to give a flavor of the diverse approaches and to introduce some of the ideas to be used in later sections.

Suppose then that we wish to minimize a function $f(x_1, x_2, \ldots, x_n)$.

We begin with the *simplex method* of Nelder and Mead (1965), noting that this has nothing to do with the famous simplex algorithm for linear programming. A simplex in $\mathbb{R}^n$ is a set of points $\mathbf{x}_0, \mathbf{x}_1, \ldots, \mathbf{x}_n$ which form a nondegenerate polyhedron. Equivalently, the vectors $\mathbf{x}_i - \mathbf{x}_0$ $(i = 1, 2, \ldots n)$ are linearly independent. In the case of $\mathbb{R}^2$, which we will use to illustrate the process, a simplex is simply a nondegenerate triangle, while in $\mathbb{R}^3$ it is a tetrahedron.

The simplex method begins with a simplex and adjusts it by replacing one of the vertices by a new one according to a set of fairly intuitive rules. The basic philosophy is that if a move in a particular direction is successful in producing a new best estimate of the minimum, then we try a larger move in that direction, whereas if a move results in a very poor point, then we try reducing that step.

Specifically, let $\mathbf{G}, \mathbf{H}$, and $\mathbf{S}$ denote the vertices of the current simplex with largest, second largest, and smallest function values $f_\mathbf{G}, f_\mathbf{H}$, and $f_\mathbf{S}$, respectively. We will also denote by $\mathbf{X}$ the centroid of the face of the simplex which does not include $\mathbf{G}$. Thus

$$\mathbf{X} = \frac{1}{n} \left(\sum_{i=0}^{n} \mathbf{x}_i - \mathbf{G} \right) \tag{8.3.1}$$

Each iteration begins with a *reflection* step in which $\mathbf{G}$ is reflected in $\mathbf{X}$ to produce a new point $\mathbf{R}$. Denote $f(\mathbf{R})$ by $f_\mathbf{R}$. The rest of the iteration depends on the size of $f_\mathbf{R}$. The reflection step is illustrated in Fig. 8.3.1a. The reflection is achieved by setting

$$\mathbf{R} = \mathbf{X} + a(\mathbf{X} - \mathbf{G}) \tag{8.3.2}$$

where the *reflection coefficient* satisfies $0 < a \leq 1$. The most common choice is $a = 1$.

Now if $f_\mathbf{R} < f_\mathbf{S}$ so that the reflection has resulted in a new best point, then we try a greater move in this same direction, an *expansion* step, as illustrated in Fig. 8.3.1b. This is achieved by setting

$$\mathbf{E} = \mathbf{X} + b(\mathbf{R} - \mathbf{X}) \tag{8.3.3}$$

where the *expansion coefficient* is $b > 1$; usually, $b = 2$. Now if $f_\mathbf{E}$ represents a further improvement, that is, $f_\mathbf{E} < f_\mathbf{R}$, then we replace the vertex $\mathbf{G}$ in our simplex by $\mathbf{E}$ to obtain the simplex for the next iteration. Otherwise $\mathbf{G}$ is replaced by $\mathbf{R}$.

The next situation is that in which $\mathbf{R}$ represents an improvement relative to several vertices of the simplex *but not* relative to $\mathbf{S}$. This is characterized by

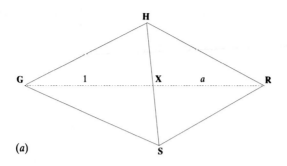

(a)

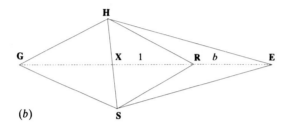

(b)

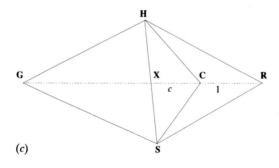

(c)

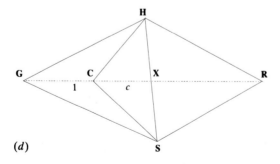

(d)

FIGURE 8.3.1
Nelder and Mead simplex method. (a) Reflection step; (b) expansion step; (c) contraction step $f_R < f_G$; (d) contraction step $f_R \geq f_G$.

the condition $f_{\mathrm{H}} > f_{\mathrm{R}} \geq f_{\mathrm{S}}$, in which case the vertex **G** is again replaced by **R** for the next iteration.

Unfortunately, with any empirical approach such as this, it may well be the case that the reflection results in no real benefit. This may happen in two ways, each of which results in trying a *contraction* step. These two possibilities are illustrated in Fig. 8.3.1*c* and *d*.

In the first case, **R** yields some improvement over **G** and so replaces **G** in the simplex, but the improvement is such that **R** would be the worst point of the next simplex. Thus, we have $f_{\mathrm{G}} > f_{\mathrm{R}} \geq f_{\mathrm{H}}$. This would have the effect of the next iteration simply reflecting this point back toward **G**. (Indeed, for $a = 1$, it would be reflected back to **G** itself.) In this case, and in the worst case where $f_{\mathrm{R}} \geq f_{\mathrm{G}}$, a contraction is made in which the appropriate point is moved in toward **X** by setting

$$\mathbf{C} = \begin{cases} \mathbf{X} + c(\mathbf{R} - \mathbf{X}) & \text{if } f_{\mathrm{G}} > f_{\mathrm{R}} \\ \mathbf{X} + c(\mathbf{G} - \mathbf{X}) & \text{if } f_{\mathrm{R}} \geq f_{\mathrm{G}} \end{cases} \tag{8.3.4}$$

where the *contraction coefficient* satisfies $0 < c < 1$ usually $c = \frac{1}{2}$.

If $f_{\mathrm{C}} < f_{\mathrm{G}}, f_{\mathrm{R}}$ then **C** replaces **G** in the simplex used for the next iteration. Otherwise, none of these moves has resulted in any improvement over the original simplex, and so the whole simplex is reduced by setting

$$\mathbf{x}_i = \frac{\mathbf{x}_i + \mathbf{S}}{2} \qquad (i = 0, 1, \ldots, n) \tag{8.3.5}$$

The whole process is summarized in Algorithm 8.3.2 below. Observe that in all cases where $f_{\mathrm{R}} < f_{\mathrm{G}}$, the vertex **G** is replaced by **R** first irrespective of whether subsequent replacements are to be made. For example, this results in just one case for the contraction step. Note too that the labeling of the vertices **G, H, S** can be made automatic by always storing the vertices in monotone order of their function values—either increasing or decreasing—and placing any newly generated points in the appropriate position in the list. Of course, this requires a complete sorting of the initial vertices.

Algorithm 8.3.2   Nelder and Mead simplex algorithm

Input   Initial simplex with vertices $\mathbf{x}_0, \mathbf{x}_1, \ldots, \mathbf{x}_n$,
     and corresponding function values
    Required tolerance ϵ
    Reflection, expansion, and contraction coefficients a, b, c

Loop   **Repeat**
    Label vertices **G, H, S** for the current simplex with function
      values $f_{\mathrm{G}}, f_{\mathrm{H}}, f_{\mathrm{S}}$ which are greatest, second
      greatest, and smallest.

$$\mathbf{X} := \frac{1}{n} \left(\sum_{i=0}^{n} \mathbf{x}_i - \mathbf{G} \right)$$

    $\mathbf{R} := \mathbf{X} + a(\mathbf{X} - \mathbf{G}); \; f_{\mathrm{R}} := f(\mathbf{R})$
    If $f_{\mathrm{R}} < f_{\mathrm{G}}$ **then G** := **R**; $f_{\mathrm{G}} := f_{\mathrm{R}}$

If $f_R < f_S$ then **E**: = **X** + b(**R** − **X**); f_E: = f(**E**)
    if $f_E < f_R$ then **G**: = **E**; f_G: = f_E
If $f_R \geq f_H$ then **C**: = **X** + c(**G** − **X**); f_C: = f(**C**)
    if $f_C < f_H$ then **G**: = **C**; f_G: = f_C
        else for $i = 0$ to n
            $\mathbf{x}_i$: = $(\mathbf{x}_i + \mathbf{S})/2$; f_i: = $f(\mathbf{x}_i)$
until max $|\mathbf{x}_i - \mathbf{x}_j| < \epsilon$
Find vertex **S** of new simplex with f_S smallest

Output   Minimum value found is f_s at **S**.

Throughout our discussion of methods for the minimization of functions of several variables, we will consider their performance for two different problems. One of these is a simple quadratic function of two variables,

$$f(x, y) = (2x - y)^2 + (y + 1)^2 \tag{8.3.6}$$

A function such as this should be minimized easily and efficiently.

The second of our tests also uses a two-variable function—Rosenbrock's function—

$$f(x, y) = 100(y - x^2)^2 + (1 - x)^2 \tag{8.3.7}$$

which is a very difficult test of a minimization routine. This function has a very steep-sided curved valley following the parabola $y = x^2$.

Before examining the progress of the simplex method on these functions, we illustrate their behavior in Fig. 8.3.3 with contour plots. It is apparent that

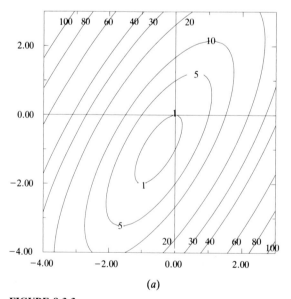

(a)

FIGURE 8.3.3
Contours of the functions (8.3.6) and (8.3.7). (*a*) $(2x - y)^2 + (y + 1)^2$; (*b*) $100(y - x^2)^2 + (1 - x)^2$.

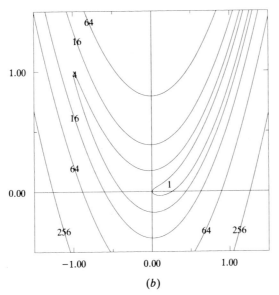

(b)

FIGURE 8.3.3 (*Continued*)

the second of these functions is likely to cause significantly greater difficulty for any minimization routine.

In the case of the quadratic function (8.3.6) the progress of the simplex method is impressive. In Fig. 8.3.4 we show the first few iterations, starting from the simplex with vertices at $(1, 2)$, $(2, 2)$, and $(2, 3)$ on the contour plot of

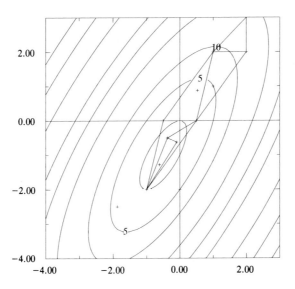

FIGURE 8.3.4
Nelder and Mead simplex method applied to $(2x - y)^2 + (y + 1)^2$.

the function. The extra points plotted (marked with a +) which are not used as vertices of simplexes are the results of reflections before successful expansions or contractions, or expansions which failed to reduce the function value.

The results of these iterations and the subsequent ones which were needed to obtain convergence to a tolerance of ϵ are shown in Table 8.3.5. The specific convergence criterion used was that each side of the simplex (triangle) should be less than the required tolerance in the "taxicab," or L_1, metric

$$\|\mathbf{x} - \mathbf{y}\|_1 = |x_1 - y_1| + |x_2 - y_2|$$

It is apparent that after a few iterations which move the triangle into the vicinity of the minimum, many contraction steps are used to refine the solution.

TABLE 8.3.5
Nelder and Mead simplex method applied to $(2x - y)^2 + (y + 1)^2$

	S	H	G	R	E or C	
x	1.000 000	2.000 000	2.000 000	1.000 000	E	0.500 000
y	2.000 000	2.000 000	3.000 000	1.000 000		0.000 000
f	9.000 000	13.000 000	17.000 000	5.000 000		2.000 000
x	0.500 000	1.000 000	2.000 000	−0.500 000		
y	0.000 000	2.000 000	2.000 000	0.000 000		
f	2.000 000	9.000 000	13.000 000	2.000 000		
x	0.500 000	−0.500 000	1.000 000	−1.000 000	E	−2.000 000
y	0.000 000	0.000 000	2.000 000	−2.000 000		−4.000 000
f	2.000 000	2.000 000	9.000 000	1.000 000		9.000 000
x	−1.000 000	0.500 000	−0.500 000	0.000 000	C	−0.375 000
y	−2.000 000	0.000 000	0.000 000	−2.000 000		−0.500 000
f	1.000 000	2.000 000	2.000 000	5.000 000		0.312 500
x	−0.375 000	−1.000 000	0.500 000	−1.875 000	C	−0.093 750
y	−0.500 000	−2.000 000	0.000 000	−2.500 000		−0.625 000
f	0.312 500	1.000 000	2.000 000	3.812 500		0.332 031
x	−0.375 000	−0.093 750	−1.000 000	0.531 250	C	−0.617 188
y	−0.500 000	−0.625 000	−2.000 000	0.875 000		−1.281 250
f	0.312 500	0.332 031	1.000 000	3.550 781		0.081 299
x	−0.617 188	−0.375 000	−0.093 750	−0.898 438	C	−0.294 922
y	−1.281 250	−0.500 000	−0.625 000	−1.156 250		−0.757 813
f	0.081 299	0.312 500	0.332 031	0.434 814		0.086 868
x	−0.617 188	−0.294 922	−0.375 000	−0.537 109	C	−0.415 527
y	−1.281 250	−0.757 813	−0.500 000	−1.539 063		−0.759 766
f	0.081 299	0.086 868	0.312 500	0.506 668		0.062 795
x	−0.415 527	−0.617 188	−0.294 922	−0.737 793	C	−0.405 640
y	−0.759 766	−1.281 250	−0.757 813	−1.283 203		−0.889 160
f	0.062 795	0.081 299	0.086 868	0.117 215		0.018 351
x	−0.405 640	−0.415 527	−0.617 188	−0.203 979	C	−0.513 885
y	−0.889 160	−0.759 766	−1.281 250	−0.367 676		−1.052 856
f	0.018 351	0.062 795	0.081 299	0.401 457		0.003 423

The algorithm can be seen to be closing in on the minimum in a fairly steady manner. The 10 iterations reported in Table 8.3.5 entail just 22 evaluations of the objective function. The final simplex with its vertices at $(-0.513\,885, -1.052\,856)$, $(-0.405\,640, -0.889\,160)$,   and $(-0.415\,527, -0.759\,766)$ contains the minimum, and subsequent iterations will use contractions and/or reductions to decrease it further.

For a more substantial test, we apply the same algorithm to the minimization of Rosenbrock's function given by (8.3.7). The process was started with the simplex whose vertices are $(0, 0)$, $(0.25, 0)$, and $(0, 0.25)$ which are close enough to the floor of the valley that a good method should be able to locate the minimum without much difficulty.

The results bear this out. They are summarized in Table 8.3.6. We see that the numbers of iterations and function evaluations required increase only slowly as the precision requirement tightens. In the later iterations the algorithm is averaging more than two evaluations per iteration which implies that some reduction steps are needed there.

The experiments with Rosenbrock's function were repeated for an initial simplex around the more difficult starting point $(-1.2, 1)$. Specifically, the vertices used were $(-1.2, 1)$, $(-1.2, 1.2)$, and $(-1, 1.2)$. The most striking feature of these results is the failure of the method for the very modest precision requirement $\epsilon = 0.05$ which is caused by the difficulty the method has in getting the simplex into the valley at the outset. Clearly this results in a very small triangle being used for the next several iterations. Once the algorithm overcomes this initial problem, however, the true solution is found to reasonable accuracy quite economically.

We complete this section with a brief summary of two other methods which can be employed for the minimization of a function of several variables.

The first of these is perhaps the simplest idea of all; the alternating variable method consists of successive *line searches* from the current estimate of the minimum in the coordinate directions. These coordinate directions are

TABLE 8.3.6
Nelder and Mead simplex method; Rosenbrock's function

Tolerance	No. of iters.	No. of evals.	x_{min}	y_{min}	f_{min}
(a) Initial simplex: $(0, 0)$, $(0.25, 0)$, (0.25)					
0.05	41	76	0.98867	0.97785	$1.4E-4$
0.01	48	88	1.00008	1.00039	$5.2E-6$
0.001	57	105	0.99988	0.99976	$1.7E-8$
0.0001	65	123	0.99999	0.99998	$1.1E-10$
(b) Initial simplex: $(-1.2, 1)$, $(-1.2, 1.2)$, $(-1, 1.2)$					
0.05	5	17	-1.06250	1.12500	$4.3E+0$
0.01	91	185	1.00149	1.00296	$2.4E-6$
0.001	98	199	0.99997	0.99994	$4.2E-9$
0.0001	104	211	1.00001	1.00002	$1.4E-10$

taken in turn and the process repeated as often as is necessary for convergence—usually a large number of times!

By a line search we mean locating the minimum of a function of the form

$$\phi(\alpha) = f(\mathbf{x} + \alpha \mathbf{s})$$

Typically, the point $\mathbf{x}$ is the current estimate of the minimum and, for the alternating variable method, $\mathbf{s}$ is one of the unit vectors $\mathbf{e}_i$. On all but the simplest of functions this method will converge *very slowly* to the minimum of f.

The algorithm of Davies, Swann, and Campey (1964) uses the idea of line searches in each of the coordinate directions but just once in each before a new set of search directions is generated.

Each iteration therefore begins with a current estimate of the minimum $\mathbf{x}_0$ and a set of search directions $\mathbf{s}_i$ $(i = 1, 2, \ldots, n)$. Subsequent points $\mathbf{x}_i$ are then generated by setting

$$\mathbf{x}_i = \mathbf{x}_{i-1} + \alpha_i \mathbf{s}_i \qquad (8.3.8)$$

where the steplengths α_i are chosen to minimize $f(\mathbf{x}_{i-1} + \alpha \mathbf{s}_i)$. The results of these line searches are then used to generate a new system of orthonormal search directions which are used for the next iteration. Provided all the steplengths are nonzero, this is done by setting

$$\mathbf{q}_i = \sum_{j=i}^{n} \alpha_j \mathbf{s}_j \qquad (i = 1, 2, \ldots, n) \qquad (8.3.9)$$

and then applying the Gram-Schmidt (or any other) orthonormalization process to this system to generate the new $\mathbf{s}_i$'s. In the event that one or more of the steplengths is zero, the remaining directions are used to generate new search directions, and those corresponding to the zero steps are included as the last members of the new system. (It is a reasonably straightforward exercise to check that such a process yields a new orthonormal system.)

The process can be iterated until the required accuracy is achieved. One useful criterion for this is that $\|\mathbf{q}_1\| < \epsilon$ which, in the L_1 norm, reduces to just $\Sigma |\alpha_i| < \epsilon$.

The method of Rosenbrock is essentially similar to this algorithm but replaces the line searches by a succession of trial steps in the various search directions. The steplengths for these trials are adjusted according to performance. We do not pursue the details of this or other direct search techniques further here.

EXERCISES 8.3

1. Find and classify all stationary points of the function $(x^2 - y)^2 + (y - 2)^2$. Write a program to plot contours of this function in the rectangle $[-2, 2] \times [0, 3]$. Use the simplex method to minimize this function, beginning with the simplex with vertices $(-1, 0)$, $(0, 1)$, and $(0, 0)$.

2. Write a program to implement the simplex algorithm for functions of more than two variables.

(It will be helpful to store the vertices in increasing order of their function values.) Test your program on simple quadratic functions of several variables, on the four-dimensional Rosenbrock function

$$f(x, y, z, w) = 100(x^2 - y)^2 + (1 - x)^2$$
$$+ 100(z^2 - w)^2 + (1 - z)^2$$

and on Wood's four-dimensional modification of this function in which the two-variable components are coupled

$$f(x, y, z, w) = 100(x^2 - y)^2 + (1 - x)^2$$
$$+ 90(z^2 - w)^2 + (1 - z)^2$$

$$+ 10.1((y - 1)^2 + (w - 1)^2)$$
$$+ 19.8(y - 1)(w - 1)$$

Have your program count function evaluations and put a maximum limit on this number.

3. Write a program to implement the alternating variables search for the minimum of a function using a quadratic search for the line searches. How poor is it?

4. Modify your program for Exercise 3 to perform the Davies, Swann, and Campey search.

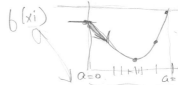

8.4 ELEMENTARY GRADIENT METHODS

In this section we introduce some of the fundamental ideas of multivariate minimization using gradient information. The methods under discussion are not good general-purpose methods but provide a basis for some of the more sophisticated methods to be discussed in the next two sections.

We begin by introducing some notation which will be used throughout the remainder of the chapter and a general description of gradient algorithms. We will denote by $\mathbf{g}(\mathbf{x})$ the gradient vector of the objective function f at the point $\mathbf{x}$. That is, we write

$$\mathbf{g}(\mathbf{x}) = \nabla f(\mathbf{x}) = \left(\frac{\partial f(\mathbf{x})}{\partial x_1}, \frac{\partial f(\mathbf{x})}{\partial x_2}, \dots, \frac{\partial f(\mathbf{x})}{\partial x_n} \right)^{\mathrm{T}} \tag{8.4.1}$$

and we abbreviate this for the points $\mathbf{x}_i$ to just $\mathbf{g}_i$ so that

$$\mathbf{g}_i = \mathbf{g}(\mathbf{x}_i) \tag{8.4.2}$$

The general pattern of gradient (and some nonderivative minimization) algorithms is summarized as Algorithm 8.4.1.

Algorithm 8.4.1   General gradient minimization algorithm

Input	Initial point $\mathbf{x}_0$, $i = 0$
Loop	**Repeat**
	Define a search direction $\mathbf{s}_i$
	Find α_i to minimize (or reduce) $\phi_i(\alpha) = f(\mathbf{x}_i + \alpha \mathbf{s}_i)$
	$\mathbf{x}_{i+1} := \mathbf{x}_i + \alpha_i \mathbf{s}_i$
	$i := i + 1$
	until convergence criterion is satisfied
Output	Approximate minimum point is $\mathbf{x}_i$.

Clearly the principal difference between various methods lies in the manner in which the search direction is chosen. The other important difference lies in how α_i is obtained—whether by a line search for the line minimum or by some "acceptable point" criterion such as we will consider in Sec. 8.7 or by some other means altogether. Since this section is concerned with gradient methods, any line searches should be performed using a derivative method such as the cubic search, Algorithm 8.2.9.

In this section we will consider just two basic approaches to the selection of the search direction: the steepest descent method and the Newton-Raphson (or just Newton's) method.

The steepest descent method is just one example of a descent technique—a class of methods characterized by the simple requirement that the search direction $\mathbf{s}_i$ should be a downhill direction. That is to say that the derivative of the function ϕ_i is negative at $\alpha = 0$ so that the function decreases initially as we move away from $\mathbf{x}_i$ in the direction $\mathbf{s}_i$. This "downhill" condition is therefore

$$\phi_i'(0) = \mathbf{s}_i^T \mathbf{g}_i < 0 \tag{8.4.3}$$

In this case there exists a positive quantity $\bar{\alpha}$ such that

$$f(\mathbf{x}_i + \alpha\mathbf{s}_i) = \phi_i(\alpha) < \phi_i(0) = f(\mathbf{x}_i) \qquad (0 < \alpha < \bar{\alpha}) \tag{8.4.4}$$

The simplest gradient method of all is the *steepest descent* method in which the search direction is simply the negative gradient direction; that is,

$$\mathbf{s}_i = -\mathbf{g}_i$$

This is, of course, the direction which maximizes $|\phi_i'(0)|$ and therefore represents the direction of most rapid reduction in the function value—hence the name.

The problem with the steepest descent method is that it is often very slow in its convergence. Its behavior tends to be very much like that of the alternating variable method—for a function of two variables with $\mathbf{g}_0 = \mathbf{e}_1$ or $\mathbf{e}_2$, it *is* the alternating variable method.

Example 8.4.2. For the quadratic function given by (8.3.6), we have

$$f(x, y) = (2x - y)^2 + (y + 1)^2$$

and so

$$f_x = 4(2x - y) \qquad f_y = 4y - 4x + 2$$

from which we deduce that the minimum in the x direction is given for a fixed y by $x = y/2$ while that in the y direction is at $y = x - \frac{1}{2}$. Taking $\mathbf{x}_0 = (2.5, 2)$, we have $\mathbf{g}_0 = (12, 0)$ so that the initial search direction is just the (negative) x direction. The minimum along this line is at $x = y/2 = 1$ so that $\mathbf{x}_1 = (1, 2)$ and $\mathbf{g}_1 = (0, 6)$. The minimum along the line $(1, 2 - 6\alpha)$ is at $y = \frac{1}{2}$. Subsequent points are $(\frac{1}{4}, \frac{1}{2})$, $(\frac{1}{4}, -\frac{1}{4})$, $(-\frac{1}{8}, -\frac{1}{4})$, $(-\frac{1}{8}, -\frac{5}{8})$, $(-\frac{5}{16}, -\frac{5}{8})$, ... The progress of the algorithm is seen to be very slow even for this very simple function.

For the more demanding Rosenbrock function, we see that the progress is unacceptably slow right from the outset even with $\mathbf{x}_0 = (0, 0)$ which is right in the valley. The progress of the first 10 iterations is plotted in Fig. 8.4.3a and tabulated

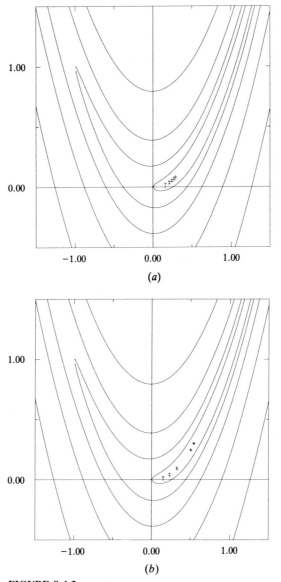

FIGURE 8.4.3
Steepest descent methods applied to Rosenbrock's function. (*a*) Standard line search;
(*b*) modified method.

in Table 8.4.4(a). Again, the initial search direction is the x direction so that the steps are precisely those of the alternating variable method. Even after 10 iterations and some 158 function evaluations, the best point is only $(0.29257, 0.08560)$.

In Fig. 8.4.3*b* we display the result of a simple modification of the steepest descent method in which the exact line search is still performed but then $\mathbf{x}_{i+1}$ is

TABLE 8.4.4
Steepest descent methods; Rosenbrock's function;
$x_0 = (0, 0)$

Iter.	x	y	Evals.
	(a) Standard line searches		
1	0.16126	0.00000	23
2	0.16126	0.02601	37
3	0.21134	0.02601	57
4	0.21134	0.04466	68
5	0.24508	0.04466	88
6	0.24508	0.06007	99
7	0.27112	0.06007	119
8	0.27112	0.07351	130
9	0.29257	0.07351	147
10	0.29257	0.08560	158
	(b) Reduced line searches		
1	0.14514	0.00000	31
2	0.14751	0.02057	48
3	0.22918	0.03244	74
4	0.22791	0.04943	91
5	0.32633	0.08697	122
6	0.32223	0.10032	137
7	0.50555	0.24327	167
8	0.50201	0.24909	182
9	0.53924	0.30247	210
10	0.54650	0.29754	225

defined to be $\mathbf{x}_i + 0.9\alpha_i \mathbf{s}_i$ so that the actual step taken is reduced by 10 percent. The choice of this factor is purely heuristic, although 0.9 has often been found to be successful in improving the performance. It was also necessary here to be more careful about the initial steplength used in the line search algorithm. We see an initial improvement, but it appears that after a few iterations this modified algorithm is becoming "bogged down" just like the original. The results of these iterations are shown in Table 8.4.4(b).

It is also apparent from these results that the somewhat more demanding line search requirement in the reduced algorithm requires a significantly increased number of function evaluations.

Clearly something better than steepest descent is required for an efficient minimization routine. One natural alternative is the Newton-Raphson method which is just the application of Newton's method (see Sec. 2.6) to the system of equations $\mathbf{g}(\mathbf{x}) = 0$.

Thus the "search direction" is defined by the linear system

$$J(\mathbf{x}_i)\mathbf{s}_i = -\mathbf{g}_i \qquad (8.4.5)$$

where J is the Hessian matrix of second derivatives at $\mathbf{x}_i$ defined as in (8.1.2).

The standard Newton method then takes the fixed step $\alpha_i = 1$ so that each step of the algorithm consists of solving the linear system (8.4.5) and setting

$$\mathbf{x}_{i+1} = \mathbf{x}_i + \mathbf{s}_i \qquad (8.4.6)$$

Provided the objective function is convex, the Hessian matrix will be positive definite, and so $\mathbf{s}_i$ will be a downhill direction. This suggests that an alternative to (8.4.6) would be to set

$$\mathbf{x}_{i+1} = \mathbf{x}_i + \alpha_i \mathbf{s}_i \qquad (8.4.7)$$

where the steplength α_i is chosen by a line search. This is called the *modified Newton method*, which is frequently more reliable than Newton's method itself.

We know from the theory of Newton's method that it has quadratic convergence near the solution when it does converge, but unfortunately that is not a reliable condition. We see this lack of reliability by applying Newton's method to Rosenbrock's function.

Example 8.4.5. The gradient vector and Hessian matrix for Rosenbrock's function are given by

$$\mathbf{g}(\mathbf{x}) = \begin{pmatrix} 400x(x^2 - y) + 2(x - 1) \\ -200(x^2 - y) \end{pmatrix}$$

and

$$J(\mathbf{x}) = \begin{bmatrix} 400(3x^2 - y) + 2 & -400x \\ -400x & 200 \end{bmatrix}$$

We begin by applying Newton's method with $\mathbf{x}_0 = (0,0)^{\mathrm{T}}$. It follows that

$$\mathbf{g}_0 = (-2, 0)^{\mathrm{T}} \qquad J = \begin{bmatrix} 2 & 0 \\ 0 & 200 \end{bmatrix}$$

from which we deduce that $\mathbf{s}_0 = (1, 0)^{\mathrm{T}}$ and therefore $\mathbf{x}_1 = (1, 0)^{\mathrm{T}}$.

We observe that $f(\mathbf{x}_1) = 100$ which represents a large *increase* in function value, but for the standard Newton-Raphson method this value need not even be computed.

At $(1, 0)$ we have

$$\mathbf{g}_1 = (400, -200)^{\mathrm{T}} \qquad J = \begin{bmatrix} 1202 & -400 \\ -400 & 200 \end{bmatrix}$$

from which we deduce that $\mathbf{s}_1 = (0, 1)^{\mathrm{T}}$ and $\mathbf{x}_2 = (1, 1)^{\mathrm{T}}$, the true exact minimum of f.

Newton's method is less impressive for the initial point $(-1.2, 1)$ from which the next few iterates are given to four decimal places by

$$(1.1979, -5.1940), (-1.2025, 7.1578), (1.2039, 7.2162), (1.2094, 1.4681),$$

$$(3.8334, 7.8099), \dots$$

which is apparently failing to converge despite having, after four iterations, finally gotten close to the bottom of the valley $y = x^2$ only to move substantially in the wrong direction thereafter.

TABLE 8.4.6
Modified Newton method; Rosenbrock's function

Iter.	x	y	Evals.
(a) $\mathbf{x}_0 = (0, 0)$			
1	0.16132	0.00000	24
2	0.36736	0.10614	57
3	0.63027	0.38027	92
4	0.80761	0.63958	122
5	0.97768	0.95380	157
6	0.99825	0.99670	184
7	1.00002	1.00005	209
8	1.00000	1.00000	234
(b) $\mathbf{x}_0 = (-1.2, 1)$			
1	−1.17518	1.38227	25
2	−0.95950	0.87526	45
3	−0.68974	0.42055	78
4	−0.33388	0.06959	108
5	−0.05079	−0.03609	141
6	0.23893	0.02760	171
7	0.48228	0.20894	201
8	0.70785	0.48560	231
9	0.88438	0.77368	261
10	0.99690	0.99481	290
11	1.00062	1.00127	315
12	1.00000	1.00000	340

Rosenbrock's function is not convex, and so any modified Newton method must be safeguarded to ensure that the search direction is downhill. In the event that it is not, one alternative is to use the steepest descent direction until downhill Newton directions are obtained. This safeguard was incorporated into the program used to produce the results shown in Table 8.4.6(a) for the modified Newton method.

It is apparent that from both starting points, rapid convergence has been achieved—but at the cost of a large number of function evaluations. This certainly still does not provide a satisfactory method for general minimization. In the next few sections we investigate methods which try to combine the reliability of steepest descent methods with the rapid terminal convergence of Newton's method.

EXERCISES 8.4

1. Write a program to implement the steepest descent method and its modification for a reduced step. Test it for the minimization of $(x^2 - y)^2 + (y - 2)^2$.

2. Prove that Newton's method will find the minimum of a positive definite quadratic function

$$f(\mathbf{x}) = c + \mathbf{b}^{\mathrm{T}}\mathbf{x} + \frac{\mathbf{x}^{\mathrm{T}}A\mathbf{x}}{2}$$

where A is a positive definite symmetric matrix.

3. Write programs to implement Newton's method and the modified Newton method. Test it for the function of Exercise 1 and for the four-dimensional Rosenbrock function and Wood's function. (See Exercise 2 in Sec. 8.3.)

4. Use a second-order Taylor series expansion of f about $\mathbf{x}_i$ to obtain the approximation

$$\alpha_i = -\frac{\mathbf{g}_i^T \mathbf{s}_i}{\mathbf{s}_i^T J(\mathbf{x}_i)\mathbf{s}_i}$$

for the minimum of $\phi_i(\alpha) = f(\mathbf{x}_i + \alpha \mathbf{s}_i)$. Derive the approximations

$$\alpha_i = \frac{\mathbf{g}_i^T \mathbf{g}_i}{\mathbf{g}_i^T J(\mathbf{x}_i)\mathbf{g}_i}$$

for the steepest descent method and $\alpha_i \simeq 1$ for Newton's method.

5. Use the approximations obtained in Exercise 4 to carry out four iterations of the steepest descent method for the minimization of $(x^2 - y)^2 + (1 - x)^2$ from the initial point $(0, 0)$. What are the drawbacks of using this approximation in the steepest descent method?

8.5 CONJUGATE DIRECTIONS AND CONJUGATE GRADIENT METHODS

In this section, we begin the development of some of the more powerful methods of numerical minimization of a function of several variables. The theory of these methods is based on the use of conjugate search directions—a term we will define shortly—for the minimization of a positive definite quadratic function. The algorithms that are developed will, of course, be applied to more general functions than just these.

The importance of quadratic functions in this context is that near a local minimum a smooth function can be expected to behave much like a positive definite quadratic. The second-order Taylor expansion of such a function about a local minimum point $\mathbf{x}^*$ can be written as

$$f(\mathbf{x}^* + \mathbf{h}) \simeq f(\mathbf{x}^*) + \frac{\mathbf{h}^T J(\mathbf{x}^*)\mathbf{h}}{2} \tag{8.5.1}$$

since, as in (8.1.3), the gradient vector $\mathbf{g}(\mathbf{x}^*)$ is necessarily zero at $\mathbf{x}^*$. It is a sufficient condition for $\mathbf{x}^*$ to be a local minimum that the Hessian matrix $J(\mathbf{x}^*)$ is positive definite.

The situation is therefore that if a method is to be a good general-purpose minimization technique, then it certainly should perform efficiently on a positive definite quadratic function of the form

$$f(\mathbf{x}) = c + \mathbf{b}^T\mathbf{x} + \frac{\mathbf{x}^T A\mathbf{x}}{2} \tag{8.5.2}$$

where A is a positive definite symmetric matrix, $\mathbf{b}$ a constant vector, and c a constant scalar. Note that for such a function, A is the Hessian matrix and the gradient is given by

$$\mathbf{g}(\mathbf{x}) = A\mathbf{x} + \mathbf{b} \tag{8.5.3}$$

Note that any method for the minimization of a function such as this can also be used as an algorithm for the solution of a system of linear equations, since the minimum of (8.5.2) lies at $\mathbf{x}^*$ which is the solution of the system $\mathbf{g}(\mathbf{x}^*) = 0$; that is,

$$A\mathbf{x}^* = -\mathbf{b} \tag{8.5.4}$$

Indeed one of the first derivations of the conjugate gradient method was as an iterative method for the solution of a system of linear equations.

> **Definition 8.5.1.** Two nonzero vectors $\mathbf{x}, \mathbf{y} \in \mathbb{R}^n$ are said to be *conjugate* with respect to the positive definite $n \times n$ matrix A, or *A-conjugate* if $\mathbf{x}^T A \mathbf{y} = 0$.

This condition simply says that two vectors are A-conjugate if they are orthogonal in the inner product defined by A; that is,

$$(\mathbf{x}, \mathbf{y})_A = \mathbf{x}^T A \mathbf{y}$$

which has an associated metric and norm given by

$$\|\mathbf{x}\|_A^2 = (\mathbf{x}, \mathbf{x})_A = \mathbf{x}^T A \mathbf{x} \tag{8.5.5}$$

There are several basic properties of conjugate directions which will be used repeatedly. Their proofs are left as exercises.

1. *A*-conjugate directions are linearly independent.
2. A set of *A*-conjugate directions has at most n members so that if $\mathbf{s}_0, \mathbf{s}_1, \ldots, \mathbf{s}_{n-1}$ are *A*-conjugate and $\mathbf{x}^T A \mathbf{s}_i = 0$ for each i, then $\mathbf{x} = 0$.
3. If $\mathbf{s}_0, \mathbf{s}_1, \ldots, \mathbf{s}_{n-1}$ are *A*-conjugate, then they form a basis for $\mathbb{R}^n$, and if $\mathbf{x} \in \mathbb{R}^n$, then $\mathbf{x}$ is given in terms of this basis by

$$\mathbf{x} = \sum_{i=0}^{n-1} \frac{\mathbf{x}^T A \mathbf{s}_i}{\mathbf{s}_i^T A \mathbf{s}_i} \mathbf{s}_i \tag{8.5.6}$$

To see the significance of conjugacy to minimization, let us suppose that we have a complete set of n *A*-conjugate directions $\mathbf{s}_0, \mathbf{s}_1, \ldots, \mathbf{s}_{n-1}$. The notation is purposely suggestive of that used for the search directions in the previous section. For a given point $\mathbf{x}_0$ we define a (finite) sequence by

$$\mathbf{x}_{i+1} = \mathbf{x}_i + \alpha_i \mathbf{s}_i \qquad (i = 0, 1, \ldots, n-1) \tag{8.5.7}$$

where the steplengths α_i are chosen to minimize

$$\phi_i(\alpha) = f(\mathbf{x}_i + \alpha \mathbf{s}_i) \tag{8.5.8}$$

We will prove that $\mathbf{x}_n$ is the minimum point $\mathbf{x}^*$ of f. Now,

$$\mathbf{x}_{k+1} = \mathbf{x}_0 + \sum_{i=0}^{k} \alpha_i \mathbf{s}_i \qquad (k = 0, 1, \ldots, n-1) \tag{8.5.9}$$

Since the line searches are exact, it follows that

$$\phi_i'(\alpha_i) = \mathbf{s}_i^T \mathbf{g}(\mathbf{x}_{i+1}) = 0 \tag{8.5.10}$$

for every i. Therefore, using the conjugacy conditions, we have

$$
\begin{aligned}
0 = \mathbf{s}_k^T \mathbf{g}_{k+1} &= \mathbf{s}_k^T(\mathbf{b} + A\mathbf{x}_{k+1}) \\
&= \mathbf{s}_k^T\left(\mathbf{b} + A\mathbf{x}_0 + \sum_{i=0}^{k} \alpha_i A\mathbf{s}_i\right) \\
&= \mathbf{s}_k^T \mathbf{g}_0 + \sum_{i=0}^{k} \alpha_i \mathbf{s}_k^T A\mathbf{s}_i \\
&= \mathbf{s}_k^T \mathbf{g}_0 + \alpha_k \mathbf{s}_k^T A\mathbf{s}_k
\end{aligned}
\tag{8.5.11}
$$

from which we deduce that

$$
\alpha_k = -\frac{\mathbf{g}_0^T \mathbf{s}_k}{\mathbf{s}_k^T A\mathbf{s}_k}
\tag{8.5.12}
$$

Substituting this into (8.5.9), with $k = n - 1$, yields

$$
\mathbf{x}_n = \mathbf{x}_0 - \sum_{i=0}^{n-1} \frac{\mathbf{g}_0^T \mathbf{s}_i}{\mathbf{s}_i^T A\mathbf{s}_i} \mathbf{s}_i
$$

and comparing this with (8.5.6) with $\mathbf{x} = A^{-1}\mathbf{g}_0$, we see that

$$
\mathbf{x}_n = \mathbf{x}_0 - A^{-1}\mathbf{g}_0 = \mathbf{x}_0 - A^{-1}(\mathbf{b} + A\mathbf{x}_0) = -A^{-1}\mathbf{b} = \mathbf{x}^*
$$

as required.

From (8.5.12), we also see that the steplengths used in any one search direction depend only on that direction and the initial point and therefore that the search directions could be used in any order provided that each is used just once. In summary then we have proved the following theorem.

> **Theorem 8.5.2.** Let f be the positive definite quadratic function (8.5.2) and suppose that $\mathbf{s}_0, \mathbf{s}_1, \ldots, \mathbf{s}_{n-1}$ are A-conjugate directions. For any $\mathbf{x}_0 \in \mathbb{R}^n$,
>
> $$
> \mathbf{x}_n = \mathbf{x}_0 - \sum_{i=0}^{n-1} \frac{\mathbf{g}_0^T \mathbf{s}_i}{\mathbf{s}_i^T A\mathbf{s}_i} \mathbf{s}_i
> $$
>
> is the unique minimum point of f, and the search directions can be used in any order.

Thus *if* we can find a system of A-conjugate directions, then a positive definite quadratic can be minimized in at most n iterations. This is called the *finite termination* property of conjugate direction methods.

Clearly the crux of the matter is to be able to generate conjugate directions automatically within a minimization routine. Any method which does generate such directions will have this finite termination property and can also be expected to perform well on more general convex functions.

The first method we consider in detail is Powell's method (1964) which makes no use of the gradient of the objective function. The method does not fit exactly into the general pattern above since the method of generating the search directions is more complicated than in the gradient methods we discuss later.

The structure of the algorithm has something of the flavor of the Davies, Swann, and Campey algorithm. Each iteration entails n line searches and then one more in the direction of overall progress of these. The results of all these line searches are then used to generate the next system of directions consisting of $n - 1$ of the previous ones and this direction of overall progress. We will establish shortly that the new directions generated by Powell's method are A-conjugate, but first we describe the basic algorithm.

Algorithm 8.5.3 Powell's basic method

Input   Initial point $\mathbf{x}_0 \in \mathbb{R}^n$, objective function f
              Tolerance ϵ

Initialize $\mathbf{x}_{n+1} := \mathbf{x}_0$
              for $i = 1$ to n $\mathbf{s}_i := \mathbf{e}_i$

Repeat
    $\mathbf{x}_0 := \mathbf{x}_{n+1}$;
    For $i = 0$ to $n - 1$
        $\mathbf{s}_i := \mathbf{s}_{i+1}$
        Find α_i to minimize $f(\mathbf{x}_i + \alpha\mathbf{s}_i)$
        $\mathbf{x}_{i+1} := \mathbf{x}_i + \alpha_i\mathbf{s}_i$
    $\mathbf{s}_n := \mathbf{x}_n - \mathbf{x}_0$
    Find α_n to minimize $f(\mathbf{x}_n + \alpha\mathbf{s}_n)$
    $\mathbf{x}_{n+1} := \mathbf{x}_n + \alpha_n\mathbf{s}_n$
until $\|\mathbf{x}_{n+1} - \mathbf{x}_0\| < \epsilon$.

Output $\mathbf{x}_{n+1}$ is the approximate minimum of f.

Note that in this basic version of the algorithm, the scheme for updating the system of search directions is simply to delete the first one and incorporate the direction of overall progress as the last one. Thus the first iteration uses the search directions $\mathbf{e}_1, \mathbf{e}_2, \ldots, \mathbf{e}_n$ and then $\mathbf{s}_n$. The second iteration begins with the system $\mathbf{e}_2, \mathbf{e}_3, \ldots, \mathbf{e}_n, \mathbf{s}_n$ and generates a new direction from the overall progress made in these n line searches. We will see that it is these new directions which build up a set of conjugate directions.

We will also see that this simple cyclical replacement strategy is liable to break down in the event that one of the steplengths α_i is zero or very close to zero.

It should also be noted that the convergence test used in this basic algorithm is liable to premature convergence; that is, the stopping criterion can be satisfied at points well-removed from the solution.

Before considering modifications of the basic method, we establish that Powell's method does indeed generate conjugate directions when applied to a positive definite quadratic function. The main result is proved using the following proposition.

$\rightarrow x^T A x > 0$

Proposition 8.5.4. Let f be the positive definite quadratic function defined by (8.5.2). For $\mathbf{x}_0, \mathbf{x}_1, \mathbf{s} \in \mathbb{R}^n$, define two single-variable functions by

$$\psi_i(\alpha) = f(\mathbf{x}_i + \alpha \mathbf{s}) \qquad (i = 0, 1)$$

and denote their minimum points by α_0, α_1, respectively. Let

$$\mathbf{u}_i = \mathbf{x}_i + \alpha_i \mathbf{s} \qquad (i = 0, 1) \qquad \text{and} \qquad \mathbf{w} = \mathbf{u}_1 - \mathbf{u}_0$$

then $\mathbf{w}$ and $\mathbf{s}$ are A-conjugate.

Proof. Since the α_i minimize the functions ψ_i, it follows that

$$\mathbf{s}^T \mathbf{g}(\mathbf{u}_i) = 0 \qquad (i = 0, 1)$$

and therefore that

$$\mathbf{s}^T A \mathbf{w} = \mathbf{s}^T A(\mathbf{u}_1 - \mathbf{u}_0) = \mathbf{s}^T (\mathbf{b} + A\mathbf{u}_1 - (\mathbf{b} + A\mathbf{u}_0)) = \mathbf{s}^T(\mathbf{g}(\mathbf{u}_1) - \mathbf{g}(\mathbf{u}_0)) = 0$$

as required.

We are now in a position to prove the main result on Powell's method.

Theorem 8.5.5. When Algorithm 8.5.3 is applied to the positive definite quadratic function (8.5.2), the new directions generated are A-conjugate.

Proof. The proof is by induction. At each stage, we must consider two successive iterations: we use the superscript $+$ to denote quantities in the later one.

In the first two iterations, two new directions are generated: $\mathbf{s}_n = \mathbf{s}_{n-1}^+$ and $\mathbf{s}_n^+$. Now $\mathbf{s}_n^+ = \mathbf{x}_n^+ - \mathbf{x}_0^+$ and $\mathbf{x}_0^+ = \mathbf{x}_{n+1}$, but $\mathbf{x}_{n+1} = \mathbf{x}_n + \alpha_n \mathbf{s}_n$ and $\mathbf{x}_n^+ = \mathbf{x}_{n-1}^+ + \alpha_{n-1}^+ \mathbf{s}_{n-1}^+$. The search directions $\mathbf{s}_n$ and $\mathbf{s}_{n-1}^+$ are the same, and so, by Proposition 8.5.4, it follows that $\mathbf{s}_n^+ = \mathbf{x}_n^+ - \mathbf{x}_0^+ = \mathbf{x}_n^+ - \mathbf{x}_{n+1}$ and $\mathbf{s}_{n-1}^+$ are A-conjugate.

Suppose now that at some stage in the computation the directions $\mathbf{s}_k, \mathbf{s}_{k+1}, \ldots, \mathbf{s}_n$ are all A-conjugate. The argument used above establishes that $\mathbf{s}_n^+$ and $\mathbf{s}_{n-1}^+ = \mathbf{s}_n$ are conjugate, and we already have that $\mathbf{s}_{k-1}^+, \mathbf{s}_k^+, \ldots, \mathbf{s}_{n-1}^+$ are mutually conjugate. It remains to show that $\mathbf{s}_j^{+T} A \mathbf{s}_n^+ = 0$ $(k-1 \le j < n-1)$. For any such j, $\mathbf{s}_j^{+T} A(\mathbf{x}_{j+1}^+ - \mathbf{x}_{j+2}^+) = 0$ by Proposition 8.5.4 since $\mathbf{s}_{j+1} = \mathbf{s}_j^+$, and the proof is completed by showing that $\mathbf{x}_{j+1}^+ - \mathbf{x}_{j+2}^+$ is a linear combination of $\mathbf{s}_{j+1}^+, \mathbf{s}_{j+2}^+, \ldots, \mathbf{s}_n^+$ and using the known conjugacy conditions. (See Exercise 2.)

We therefore expect that after at most n iterations, Powell's method will produce the exact minimum of a positive definite quadratic. Note that n iterations requires a possible $n(n + 1)$ line searches, so the method may prove to be computationally expensive.

Example 8.5.6. Use Powell's method to minimize the positive definite quadratic function $(2x - y)^2 + (y + 1)^2$ beginning at $\mathbf{x}_0 = (2.5, 2)$, $(2, 2)$, and $(1, 2)$.

Solution

(a) $\mathbf{x}_0 = (2.5, 2)$: With $\mathbf{s}_0 = (1, 0)$, we get $\alpha_0 = -1.5$ and $\mathbf{x}_1 = (1, 2)$ just as in the first iteration of the steepest descent method. Then $\mathbf{s}_1 = (0, 1)$ yields $\alpha_1 = -1.5$ and $\mathbf{x}_2 = (1, 0.5)$. Thus the new search direction is $\mathbf{s}_2 = (-1.5, -1.5)$ which with $\alpha_2 = 1$ gives $\mathbf{x}_3 = (-0.5, -1)$—the true minimum of f.

(*b*) $\mathbf{x}_0 = (2, 2)$: This time $\alpha_0 = -1$ and $\mathbf{x}_1 = (1, 2)$ and then $\alpha_1 = -1.5$, $\mathbf{x}_2 = (1, 0.5)$. This yields $\mathbf{s}_2 = (-1, -1.5)$, and the minimum of f along the line $(1 - \alpha, 0.5 - 1.5\alpha)$ occurs at $\alpha_2 = 1.2$ so that $\mathbf{x}_3 = (-0.2, -1.3)$.

The second iteration begins with the search directions $(0, 1)$ and $(-1, -1.5)$. This yields $\alpha_0 = 0.6$ and $x_1 = (-0.2, -0.7)$ and then $\alpha_1 = 0.24$ and $\mathbf{x}_2 = (-0.44, -1.06)$. The new search direction is then $\mathbf{x}_2 - \mathbf{x}_0 = (-0.24, 0.24)$, and $\alpha_2 = 0.25$ leads to the true minimum $(-0.5, -1)$.

We note that the Hessian matrix for this case is $\begin{bmatrix} 8 & -4 \\ -4 & 4 \end{bmatrix}$ and that the two new search directions introduced, $(-1, -1.5)$ and $(-0.24, 0.24) = 0.24\,(-1, 1)$, are indeed conjugate with respect to this matrix since

$$[-1\ \ 1]\begin{bmatrix} 8 & -4 \\ -4 & 4 \end{bmatrix}\begin{bmatrix} -1 \\ -1.5 \end{bmatrix} = [-12\ \ 8]\begin{bmatrix} -1 \\ -1.5 \end{bmatrix} = 0$$

(*c*) $\mathbf{x}_0 = (1, 2)$: The first iteration uses the initial search directions $(1, 0)$ and $(0, 1)$. This yields $\alpha_0 = 0$ and then $\alpha_1 = -\frac{3}{2}$ so that $\mathbf{x}_2 = (1, 0.5)$. Therefore $\mathbf{s}_2 = (0, -\frac{3}{2})$, and we are already at the line minimum in this direction so that $\alpha_2 = 0$. The next iteration begins with $\mathbf{s}_0 = (0, 1)$ and $\mathbf{s}_1 = (0, -\frac{3}{2})$ which are clearly linearly dependent. Since the current point is the line minimum, no further progress would be made.

Plainly, Powell's method requires modification to cope with this situation. The simplest modification is just to eliminate the first search direction for which the steplength is nonzero. This would be sufficient to ensure that the search directions used for the next iteration are linearly independent, and so the algorithm will still succeed in minimizing the function.

Powell's own suggested modification is based on the fact that if the search directions are normalized so that $\mathbf{s}_i^T A\mathbf{s}_i/2 = 1$, then the determinant of the matrix with columns $\mathbf{s}_0, \mathbf{s}_1, \ldots, \mathbf{s}_{n-1}$ is maximized if these columns are A-conjugate. The replacement policy is therefore to choose the direction to be omitted so as to maximize the resulting determinant.

We next turn to the *conjugate gradient* methods which, as the name suggests, make use of derivative information to generate conjugate search directions.

The argument used in deriving formula (8.5.12) for the steplengths required in a conjugate direction search and in proving the finite termination property can easily be modified to establish that, for our positive definite quadratic function (8.5.2),

$$\mathbf{g}_{k+1}^T\mathbf{s}_i = 0 \qquad (i \le k) \tag{8.5.13}$$

This implies that, for each $k = 1, 2, \ldots, n$, $\mathbf{x}_k$ minimizes f over all points in $\mathbf{x}_0 + \mathbf{L}_k$, where $\mathbf{L}_k$ is the space spanned by $\mathbf{s}_0, \mathbf{s}_1, \ldots, \mathbf{s}_{k-1}$. (See Exercise 4.)

The aim of the conjugate gradient methods is to construct conjugate search directions by taking $\mathbf{s}_0 = -\mathbf{g}_0$ and each subsequent $\mathbf{s}_k$ to a linear combination of the steepest descent direction $-\mathbf{g}_k$ and the previous search direction $\mathbf{s}_{k-1}$. This will eliminate the need to store and update a full system of search directions *and* will allow the minimum of a positive definite quadratic to

conjugate =

be located in just n line searches. This should also provide a good practical method for the accurate minimization of more general functions.

It will be helpful now to introduce two more pieces of notation which will be adopted throughout the discussion of conjugate gradient methods and the next two sections. We use $\boldsymbol{\delta}_i$ and $\boldsymbol{\gamma}_i$ to denote the changes in $\mathbf{x}$ and $\mathbf{g}$ on the ith iteration. Thus

$$\boldsymbol{\delta}_i = \mathbf{x}_{i+1} - \mathbf{x}_i = \alpha_i \mathbf{s}_i \tag{8.5.14}$$

$$\boldsymbol{\gamma}_i = \mathbf{g}_{i+1} - \mathbf{g}_i$$

Note too that for the positive definite quadratic (8.5.2), it follows that

$$\boldsymbol{\gamma}_i = A\boldsymbol{\delta}_i \tag{8.5.15}$$

For the second iteration, we must choose β_0 so that $\mathbf{s}_1 = -\mathbf{g}_1 + \beta_0 \mathbf{s}_0$ and $\mathbf{s}_0$ are A-conjugate. Now, this conjugacy condition

$$\mathbf{s}_1^T A \mathbf{s}_0 = -\mathbf{g}_1^T A \mathbf{s}_0 + \beta_0 \mathbf{s}_0^T A \mathbf{s}_0 = 0$$

is satisfied if $\beta_0 = \mathbf{g}_1^T A \mathbf{s}_0 / \mathbf{s}_0^T A \mathbf{s}_0$. Using (8.5.14) and (8.5.15), we have $A\mathbf{s}_0 = (1/\alpha_0) A\boldsymbol{\delta}_0 = \boldsymbol{\gamma}_0/\alpha_0$ ($\alpha_0 \neq 0$ unless $\mathbf{x}_0$ is the exact minimum of f), and so the conjugacy condition is satisfied by choosing

$$\beta_0 = \frac{\mathbf{g}_1^T \boldsymbol{\gamma}_0}{\mathbf{s}_0^T \boldsymbol{\gamma}_0} \tag{8.5.16}$$

Suppose now that $\mathbf{s}_0, \mathbf{s}_1, \ldots, \mathbf{s}_{k-1}$ are A-conjugate and put

$$\mathbf{s}_k = -\mathbf{g}_k + \sum_{j=0}^{k-1} \beta_j^{(k)} \mathbf{s}_j$$

where the coefficients are to be chosen to achieve the conjugacy of $\mathbf{s}_k$ with all the previous search directions. (Note that to achieve our stated aim we require a solution with $\beta_0^{(k)} = \beta_1^{(k)} = \cdots = \beta_{k-2}^{(k)} = 0$.) From (8.5.13), we have $\mathbf{g}_k^T \mathbf{s}_i = 0$, and therefore $\mathbf{g}_k^T \mathbf{g}_i = 0$ for $i < k$ since $\mathbf{s}_i$ is a linear combination of $\mathbf{g}_i$ and the previous search directions. Hence for any $i < k$, the conjugacy condition is

$$\mathbf{s}_k^T A \mathbf{s}_i = -\mathbf{g}_k^T A \mathbf{s}_i + \sum_{j=0}^{k-1} \beta_j^{(k)} \mathbf{s}_j^T A \mathbf{s}_i$$

$$= -\mathbf{g}_k^T A \mathbf{s}_i + \beta_i^{(k)} \mathbf{s}_i^T A \mathbf{s}_i = 0$$

which is satisfied if $\beta_i^{(k)} = \mathbf{g}_k^T A \mathbf{s}_i / \mathbf{s}_i^T A \mathbf{s}_i$.

As before, $\alpha_i A \mathbf{s}_i = A\boldsymbol{\delta}_i = \boldsymbol{\gamma}_i = \mathbf{g}_{i+1} - \mathbf{g}_i$, and so for $i < k-1$, we have $\mathbf{g}_k^T A \mathbf{s}_i = (1/\alpha_i) \mathbf{g}_k^T (\mathbf{g}_{i+1} - \mathbf{g}_i) = 0$ so that $\beta_i^{(k)} = 0$ ($i = 0, 1, \ldots, k-2$) as required. We thus have, using (8.5.15) as before, that

$$\beta_{k-1} = \beta_{k-1}^{(k)} = \frac{\mathbf{g}_k^T A \mathbf{s}_{k-1}}{\mathbf{s}_{k-1}^T A \mathbf{s}_{k-1}} = \frac{\mathbf{g}_k^T \boldsymbol{\gamma}_{k-1}}{\mathbf{s}_{k-1}^T \boldsymbol{\gamma}_{k-1}} \tag{8.5.17}$$

Finally, using the orthogonality properties $\mathbf{g}_i^T \mathbf{s}_{i-1} = 0$ for $i = k - 1, k$, we see that this formula reduces for $k \geq 1$ to just

$$\beta_{k-1} = \frac{\mathbf{g}_k^T \mathbf{g}_k}{\mathbf{g}_{k-1}^T \mathbf{g}_{k-1}} = \frac{\|\mathbf{g}_k\|^2}{\|\mathbf{g}_{k-1}\|^2} \tag{8.5.18}$$

This formula can, of course, be used for more general minimization than just quadratic functions. Its use in a general minimization algorithm is summarized in Algorithm 8.5.7 below. This particular algorithm, due to Fletcher and Reeves (1964), is just one member of the class of conjugate gradient methods. Probably the best-known of the others is that due to Polak and Ribière (1969) which is derived from (8.5.17) without the assumption that f is quadratic. This results in the final formula $\beta_{k-1} = \mathbf{g}_k^T \gamma_{k-1} / \mathbf{g}_{k-1}^T \mathbf{g}_{k-1}$. The use of this formula is investigated in Exercise 7.

Algorithm 8.5.7 Fletcher-Reeves conjugate gradient method

Input   Initial point $\mathbf{x}$, tolerance ϵ

Initialize $\mathbf{s} := 0$, $\beta := 1$, $\mathbf{g}_0 := \mathbf{g}(\mathbf{x})$

Repeat
  $\mathbf{s} := -\mathbf{g} + \beta \mathbf{s}$
  Find α^* to minimize $f(\mathbf{x} + \alpha \mathbf{s})$
  $\mathbf{x} := \mathbf{x} + \alpha^* \mathbf{s}$; $\mathbf{g}_1 := \mathbf{g}(\mathbf{x})$
  $\beta := \|\mathbf{g}_1\|^2 / \|\mathbf{g}_0\|^2$; $\mathbf{g}_0 := \mathbf{g}_1$
until $\|\mathbf{g}_0\| < \epsilon$

Output   Approximate minimum point is $\mathbf{x}$.

In the results reported below, the line searches were performed using a cubic search with a tolerance of 10^{-3}. The convergence criterion used was $\epsilon = 10^{-4}$.

Example 8.5.8. Use the Fletcher-Reeves conjugate gradient method to minimize Rosenbrock's function.

Solution. The results using $\mathbf{x}_0 = (0, 0)$ and of selected iterations from $\mathbf{x}_0 = (-1.2, 1)$ are shown in Table 8.5.9(a) and (b), respectively, and the overall progress of the algorithm is illustrated in Fig. 8.5.10. In the first case the minimum was found to the required accuracy in 15 iterations, with the best point at this stage being $(0.99998, 0.99996)$. This required just 219 function evaluations. (This count includes one for each evaluation of the objective function itself and two for each evaluation of the gradient vector.)

Note that the second of these tables of output has an "R" against the results of iteration 28, indicating that the algorithm was *restarted* at this stage. A restart is used when the search direction generated by the conjugate gradient method is either uphill or is nearly orthogonal to the steepest descent direction, in which case little improvement in the function value can be expected. The restart consists of using the steepest descent direction and building the subsequent search directions accordingly.

TABLE 8.5.9
Fletcher-Reeves minimization algorithm

Iter.	x	y	Evals.
(a) $x_0 = (0, 0)$			
1	0.16126	0.00000	27
2	0.29306	0.05068	44
3	0.42543	0.14313	59
4	0.58228	0.30574	74
5	0.83823	0.69278	94
6	0.83990	0.70570	102
7	0.84073	0.70581	110
8	0.84312	0.71234	126
9	0.89479	0.79506	145
10	1.00217	1.00478	165
11	1.00242	1.00486	171
12	1.00242	1.00486	179
13	1.00077	1.00144	205
14	0.99998	0.99996	213
15	0.99998	0.99996	219
(b) $x_0 = (-1.2, 1)$			
1	−1.03027	1.06928	7
2	−0.68950	0.43941	37
3	−0.57169	0.27177	45
12	−0.05345	−0.10385	99
13	−0.01545	−0.10918	105
14	0.02146	−0.11145	111
15	0.05763	−0.11092	117
26	0.47970	0.09145	183
27	0.52910	0.13712	189
R 28	0.45778	0.20669	202
29	0.66025	0.42257	232
30	0.92857	0.85914	254
40	0.99065	0.98087	375
47	0.99941	0.99887	453
56	0.99993	0.99986	545

In the iterations immediately preceding the restart very little progress was made, and the algorithm was having difficulty getting into the floor of the valley. This is easily seen in Fig. 8.5.10b, where we see a succession of points outside the $f = 1$ contour followed by a "correction step"—the restart—which takes the algorithm down into the middle of the valley. Thereafter, it progresses well to the minimum which is found, to the required accuracy, in 56 iterations requiring a total of 545 function evaluations. The particular restart criterion used here was to restart if $s^T g / \|g\|^2 > -10^{-4}$. There was only the one restart reported by the program in this case.

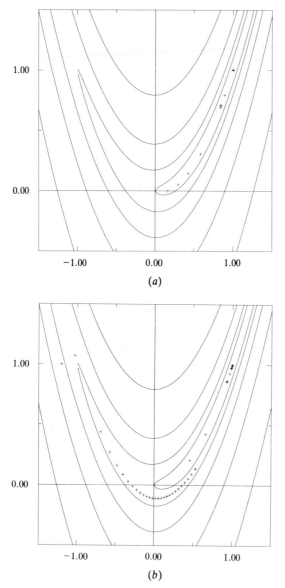

FIGURE 8.5.10
Conjugate gradient method, Rosenbrock's function. (*a*) $\mathbf{x}_0 = (0, 0)$; (*b*) $\mathbf{x}_0 = (-1.2, 1)$.

In both cases, we see satisfactory convergence to the minimum. Note that the exact numbers of iterations and function evaluations used will depend critically on the details of the line search algorithm used including the initial steplength and the convergence criterion. It will also depend on how much information is stored and passed to the line search routine to avoid recomputing any of the evaluations.

(Note: !!)

A final word on the conjugate gradient method is in order. We commented earlier that it was first developed as a technique for the solution of systems of linear equations. Consider the system

$$Ax = b \qquad (8.5.19)$$

where A is a positive definite matrix. The solution of (8.5.19) is equivalent to the minimization of the positive definite quadratic

$$f(\mathbf{x}) = -\mathbf{b}^T\mathbf{x} + \frac{\mathbf{x}^T A \mathbf{x}}{2} \qquad (8.5.20)$$

Now the gradient vector for this function is just $A\mathbf{x} - \mathbf{b} = -\mathbf{r}$, where $\mathbf{r}$ is the usual residual vector for an approximate solution of (8.5.19). We consider the details of the implementation of the conjugate gradient method to this minimization problem.

Now the "search" direction $\mathbf{s}_k$ is defined for this case by

$$\mathbf{s}_k = \mathbf{r}_k + \frac{\|\mathbf{r}_k\|^2}{\|\mathbf{r}_{k-1}\|^2}\mathbf{s}_{k-1} \qquad (8.5.21)$$

and the required steplength is obtained using a similar (but simpler) argument to the derivation of (8.5.12) by

$$\alpha_k = \frac{\mathbf{r}_k^T \mathbf{s}_k}{\mathbf{s}_k^T A \mathbf{s}_k} \qquad (8.5.22)$$

However, the orthogonality of $\mathbf{g}_k$ (and therefore of $\mathbf{r}_k$) and $\mathbf{s}_{k-1}$ allows the further simplification

$$\alpha_k = \frac{\mathbf{r}_k^T \mathbf{r}_k}{\mathbf{s}_k^T A \mathbf{s}_k}$$

The final simplification available for this special case is the updating of the residual vector itself. We have

$$\mathbf{r}_{k+1} = \mathbf{b} - A\mathbf{x}_{k+1} = \mathbf{b} - A(\mathbf{x}_k + \alpha_k\mathbf{s}_k) = \mathbf{r}_k - \alpha_k A \mathbf{s}_k \qquad (8.5.23)$$

and since the vector $A\mathbf{s}_k$ must already be computed to find α_k, this requires no more than a vector addition. In the absence of a good estimate of the solution, it is common to take $\mathbf{x}_0 = 0$ so that $\mathbf{r}_0 = \mathbf{b}$, and the additional matrix-vector multiplication needed to compute $\mathbf{r}_0$ is avoided. This choice is incorporated into the following algorithm.

Algorithm 8.5.11 Conjugate gradient method for linear systems of equations

Input   Positive definite matrix A
Right-hand-side vector $\mathbf{b}$, tolerance ϵ

Initialize $\mathbf{x}: = 0$; $\mathbf{r}_0: = \mathbf{b}$; $\mathbf{s}: = 0$; $\beta: = 0$; $i: = 0$

Repeat
    $i: = i + 1$; $\mathbf{s}: = \mathbf{r}_0 + \beta\mathbf{s}$;

$$\alpha := \frac{\mathbf{r}_0^\mathsf{T} \mathbf{r}_0}{\mathbf{s}^\mathsf{T} A \mathbf{s}}$$

$$\mathbf{x} := \mathbf{x} + \alpha \mathbf{s}; \, \mathbf{r}_1 := \mathbf{r}_0 - \alpha A \mathbf{s}$$

$$\beta := \|\mathbf{r}_1\|^2 / \|\mathbf{r}_0\|^2; \, \mathbf{r}_0 := \mathbf{r}_1$$

until $\|\mathbf{r}_1\| < \epsilon$ or $i = n$

Output   Approximate solution is **x**.

Example 8.5.12. We demonstrate the use of this algorithm with the simple problem of solving the 3×3 Hilbert matrix with right-hand side chosen so that the exact solution is a vector of 1s.

The system of equations is therefore

$$\begin{bmatrix} 1 & \frac{1}{2} & \frac{1}{3} \\ \frac{1}{2} & \frac{1}{3} & \frac{1}{4} \\ \frac{1}{3} & \frac{1}{4} & \frac{1}{5} \end{bmatrix} \begin{bmatrix} x \\ y \\ z \end{bmatrix} = \begin{bmatrix} \frac{11}{6} \\ \frac{13}{12} \\ \frac{47}{60} \end{bmatrix}$$

The results of the three iterations are

Iter.	x	y	z	Residual vector		
1	1.303 161	0.770 049	0.556 805	$-4.0\mathrm{E}-02$	$3.6\mathrm{E}-02$	$4.5\mathrm{E}-02$
2	0.987 212	1.071 478	0.931 043	$3.5\mathrm{E}-05$	$-1.9\mathrm{E}-04$	$1.8\mathrm{E}-04$
3	1.000 000	1.000 000	1.000 000	$7.0\mathrm{E}-10$	$3.9\mathrm{E}-10$	$2.7\mathrm{E}-10$

It is clear that the true solution has been found to very high accuracy with very little effort. From the small residuals in the second iteration it is also easy to believe that this method will frequently deliver the required accuracy for large systems in much less than n iterations.

EXERCISES 8.5

1. Prove the basic properties of conjugate directions:
   (a) A-conjugate directions are linearly independent.
   (b) A set of A-conjugate directions has at most n members so that if $\mathbf{s}_0, \mathbf{s}_1, \ldots, \mathbf{s}_{n-1}$ are A-conjugate and $\mathbf{x}^\mathsf{T} A \mathbf{s}_i = 0$ for each i, then $\mathbf{x} = 0$.
   (c) If $\mathbf{s}_0, \mathbf{s}_1, \ldots, \mathbf{s}_{n-1}$ are A-conjugate, then they form a basis for $\mathbb{R}^n$, and if $\mathbf{x} \in \mathbb{R}^n$, then $\mathbf{x}$ is given in terms of this basis by

$$\mathbf{x} = \sum_{i=0}^{n-1} \frac{\mathbf{x}^\mathsf{T} A \mathbf{s}_i}{\mathbf{s}_i^\mathsf{T} A \mathbf{s}_i} \mathbf{s}_i$$

2. Complete the proof of the induction step of the proof of Theorem 8.5.5. (Specifically, show that $\mathbf{s}_j^+$ and $\mathbf{x}_{j+1}^+ - \mathbf{x}_{j+2}$ are A-conjugate. Write $\mathbf{x}_{j+1}^+ - \mathbf{x}_{j+2}$ as a linear combination of $\mathbf{s}_{j+1}^+, \ldots, \mathbf{s}_n^+$, and deduce the conjugacy of $\mathbf{s}_j^+$ and $\mathbf{s}_n^+$.)

3. Use Powell's basic method to minimize $(x - y)^2 + y^2$ starting at (a) $(4, 2)$, (b) $(3, 1)$, and (c) $(1, 1)$. Modify the method so that the minimum is obtained from the initial point $(1, 1)$.

4. Prove (8.5.13), namely, that using exact line searches in conjugate search directions for the minimization of a positive definite quadratic function yields $\mathbf{g}_{k+1}^\mathsf{T} \mathbf{s}_i = 0$ for every $i \leq k$ and therefore that $\mathbf{x}_{k+1}$ minimizes f over all points of the form $\mathbf{x}_0 + \sum_{i=0}^k a_i \mathbf{s}_i$.

5. Verify that the conjugate gradient algorithm of Fletcher and Reeves locates the minimum of the quadratic function $(2x - y)^2 + (y + 1)^2$ in just two iterations from the starting point $(2.5, 2)$. Check that the two search directions are indeed conjugate with respect to the Hessian matrix of this function.

6. Write a program to implement the Fletcher-Reeves algorithm, and use it to minimize $(x_2 - y)^2 + (1 - x)^2$. Test it further by minimizing Wood's function of four variables. (See Exercise 2 of Sec. 8.3.)

7. Write a program to implement the Polak-Ribiere conjugate gradient method, and compare its performance with that of the Fletcher-Reeves algorithm.

8. Write a program to implement the conjugate gradient method for solving linear systems of equations. Test it on a large-dimensional tridiagonal matrix such as those found in spline interpolation (see Chaps. 5 and 6).

8.6   QUASI-NEWTON METHODS

In Sec. 8.4 we saw that the steepest descent method provided a reliable but very slow technique for minimization, while Newton's method and its modification were unreliable but very fast when they did work. Typically, steepest descent will do a reasonable job of getting into the vicinity of the minimum, and Newton's method will converge very fast once a good starting point has been found. These facts provide the motivation for the quasi-Newton methods which attempt to combine the reliability of steepest descent with the rapid terminal convergence of Newton's method.

The quasi-Newton methods are a subclass of the more general *variable metric* methods, but we content ourselves here with the quasi-Newton approach. The basic form of these algorithms is much the same as the general gradient algorithm (Algorithm 8.4.1) with a special method for the definition of the search directions.

Algorithm 8.6.1   General description of quasi-Newton methods

Input   Initial point $\mathbf{x}_0$, $i = 0$
            Initial positive definite symmetric matrix H_0
            Tolerance ϵ

Repeat
    $\mathbf{s}_i := -H_i \mathbf{g}_i$
    Find α_i to minimize $\phi(\alpha) = f(\mathbf{x}_i + \alpha \mathbf{s}_i)$
    $\mathbf{x}_{i+1} := \mathbf{x}_i + \alpha_i \mathbf{s}_i$
    Define a new "metric matrix" H_{i+1}
    $i := i + 1$
until $\|\mathbf{g}_i\| < \epsilon$

Output   Approximate minimum point is $\mathbf{x}_i$.

Here the notation $\mathbf{g}_i$ is used, as before, for the gradient vector of the objective function at $\mathbf{x}_i$, and we also retain the notation

$$\boldsymbol{\delta}_i = \mathbf{x}_{i+1} - \mathbf{x}_i = \alpha_i \mathbf{s}_i$$

$$\boldsymbol{\gamma}_i = \mathbf{g}_{i+1} - \mathbf{g}_i$$

(8.6.1)

Methods vary according to how the *metric matrix* H_i is updated from iteration to iteration. Once such an *update formula* has been defined, the algorithm is complete. The matrices are called metric matrices since they are to be positive definite and can therefore be used to define a metric on $\mathbb{R}^n$; the search direction defined in the above algorithm is then the steepest descent direction relative to this metric.

The idea is to update H_i in such a way that the search directions used for the first few iterations are close to the (euclidean) steepest descent direction until, as the minimum is approached, they resemble the Newton direction. Thus the matrix H_i can be viewed as an approximation to the inverse Hessian matrix of the objective function f at its minimum $\mathbf{x}^*$. It is usual to take $H_0 = I$ so that the initial search direction *is* the steepest descent direction.

The development of these algorithms began with Davidon (1959), whose Argonne National Laboratory report was improved upon by Fletcher and Powell (1964) to produce the now famous Davidon-Fletcher-Powell, or DFP, method. We begin our study of quasi-Newton update formulas (and the consequent minimization algorithms) with this DFP algorithm. For the development of this theory, we will again restrict our attention to the positive definite quadratic objective function

$$f(\mathbf{x}) = c + \mathbf{b}^T\mathbf{x} + \frac{\mathbf{x}^TA\mathbf{x}}{2} \tag{8.6.2}$$

The algorithms we develop will, of course, have much wider applicability than this.

We recall that for any step $\boldsymbol{\delta}_i = \mathbf{x}_{i+1} - \mathbf{x}_i$, we have

$$\boldsymbol{\gamma}_i = A\boldsymbol{\delta}_i \tag{8.6.3}$$

and so, if H_{i+1} is to be viewed as an approximation to A^{-1}, it is natural to require that

$$H_{i+1}\boldsymbol{\gamma}_i = \boldsymbol{\delta}_i \tag{8.6.4}$$

which is called the *quasi-Newton property*. Since A is positive definite, it is natural to demand that the H_i's should be also. This requirement can also be viewed as the requirement that the search directions are downhill, since the positive definiteness of H_i implies that

$$\mathbf{s}_i^T\mathbf{g}_i = -\mathbf{g}_i^TH_i\mathbf{g}_i < 0 \tag{8.6.5}$$

provided that $\mathbf{g}_i \neq 0$, that is, provided the minimum has not already been found.

The derivation of the particular update formulas is also based on the desire to generate A-conjugate search directions which will guarantee the finite termination property, in which case the final matrix generated would be H_n. Ideally, we would like this to be A^{-1}.

All the requirements are fulfilled by the update formula

$$H_{i+1} = H_i - \frac{H_i\boldsymbol{\gamma}_i(H_i\boldsymbol{\gamma}_i)^T}{\boldsymbol{\gamma}_i^TH_i\boldsymbol{\gamma}_i} + \frac{\boldsymbol{\delta}_i\boldsymbol{\delta}_i^T}{\boldsymbol{\delta}_i^T\boldsymbol{\gamma}_i} \tag{8.6.6}$$

This is just one member of the quasi-Newton family of updates which satisfy these conditions. Clearly this statement requires some proof.

Theorem 8.6.2. Algorithm 8.6.1 using $H_0 = I$ and the update formula (8.6.6) has the following properties when applied to the minimization of the positive definite quadratic function (8.6.2):

(a) H_i is positive definite symmetric for each i for which it is defined.

(b) The search directions generated are A-conjugate.

(c) If $\mathbf{g}_i \neq 0$ $(i = 0, 1, \ldots, n - 1)$, then $\mathbf{x}_n$ is the minimum point $\mathbf{x}^*$ of f and $H_n = A^{-1}$. (If $\mathbf{g}_i = 0$ for some $i < n$, then, of course, $\mathbf{x}_i = \mathbf{x}^*$.)

Proof

(a) We observe first that the update formula has hereditary symmetry—that is, if H_i is symmetric, then so is H_{i+1}—since both the correction terms are of the form $\mathbf{u}\mathbf{u}^\mathrm{T}$ which is symmetric. With $H_0 = I$, this establishes the symmetry of all the H_i's. Similarly, H_0 is positive definite, so it is sufficient to prove that if H_i is positive definite, then so is H_{i+1}.

Suppose then that H_i is positive definite and let $\mathbf{x} \neq 0 \in \mathbb{R}^n$. Then

$$\mathbf{x}^\mathrm{T} H_{i+1} \mathbf{x} = \mathbf{x}^\mathrm{T} H_i \mathbf{x} - \frac{(\mathbf{x}^\mathrm{T} H_i \boldsymbol{\gamma}_i)^2}{\boldsymbol{\gamma}_i^\mathrm{T} H_i \boldsymbol{\gamma}_i} + \frac{(\mathbf{x}^\mathrm{T} \boldsymbol{\delta}_i)^2}{\boldsymbol{\delta}_i^\mathrm{T} \boldsymbol{\gamma}_i} \tag{8.6.7}$$

Now, since H_i is positive definite symmetric, it has a positive definite symmetric square root, B say. Let $\mathbf{p} = B\mathbf{x}$ and $\mathbf{q} = B\boldsymbol{\gamma}_i$. Then (8.6.7) can be rewritten in the form

$$\mathbf{x}^\mathrm{T} H_{i+1} \mathbf{x} = \frac{(\mathbf{p}^\mathrm{T}\mathbf{p})(\mathbf{q}^\mathrm{T}\mathbf{q}) - (\mathbf{p}^\mathrm{T}\mathbf{q})^2}{\mathbf{q}^\mathrm{T}\mathbf{q}} + \frac{(\mathbf{x}^\mathrm{T}\boldsymbol{\delta}_i)^2}{\boldsymbol{\delta}_i^\mathrm{T}\boldsymbol{\gamma}_i} \geq \frac{(\mathbf{x}^\mathrm{T}\boldsymbol{\delta}_i)^2}{\boldsymbol{\delta}_i^\mathrm{T}\boldsymbol{\gamma}_i}$$

by the Cauchy-Schwartz inequality.

Provided $\mathbf{g}_i \neq 0$, it follows that $\mathbf{s}_i$, and so $\boldsymbol{\delta}_i$ are nonzero from which in turn we deduce that $\boldsymbol{\delta}_i^\mathrm{T}\boldsymbol{\gamma}_i = \boldsymbol{\delta}_i^\mathrm{T} A \boldsymbol{\delta}_i > 0$. It follows that $\mathbf{x}^\mathrm{T} H_{i+1}\mathbf{x} \geq 0$, but we must establish the strict inequality $\mathbf{x}^\mathrm{T} H_{i+1}\mathbf{x} > 0$ which we do by contradiction.

If $\mathbf{x}^\mathrm{T} H_{i+1}\mathbf{x} = 0$, then we must have $(\mathbf{p}^\mathrm{T}\mathbf{p})(\mathbf{q}^\mathrm{T}\mathbf{q}) = (\mathbf{p}^\mathrm{T}\mathbf{q})^2$, which implies $\mathbf{p}$ is parallel to $\mathbf{q}$, and $\mathbf{x}^\mathrm{T}\boldsymbol{\delta}_i = 0$. Now $\mathbf{p}$ and $\mathbf{q}$ being parallel is equivalent to $\mathbf{x}$ and $\boldsymbol{\gamma}_i$ being parallel, in which case we have $\boldsymbol{\delta}_i^\mathrm{T}\boldsymbol{\gamma}_i = 0$, which is our contradiction. It follows that H_{i+1} is positive definite and the proof of (a) is complete.

(b) To establish the conjugacy of the search directions we use a coupled induction argument. We show by induction on k that

$$\mathbf{P}(k): \qquad\qquad \boldsymbol{\delta}_i^\mathrm{T} A \boldsymbol{\delta}_j = 0 \qquad (0 \leq i < j < k)$$

$$\tag{8.6.8}$$

$$\mathbf{Q}(k): \qquad\qquad H_k A \boldsymbol{\delta}_i = \boldsymbol{\delta}_i \qquad (0 \leq i < k)$$

for $k = 1, 2, \ldots, n$.

For $k = 1$, the first statement is vacuous and the second is $H_1 A \boldsymbol{\delta}_0 = \boldsymbol{\delta}_0$, which is just a special case, with $i = 0$, of the quasi-Newton property

$$H_{i+1} A \boldsymbol{\delta}_i = H_{i+1} \boldsymbol{\gamma}_i$$

$$= H_i \boldsymbol{\gamma}_i - \frac{H_i \boldsymbol{\gamma}_i (H_i \boldsymbol{\gamma}_i)^\mathrm{T} \boldsymbol{\gamma}_i}{\boldsymbol{\gamma}_i^\mathrm{T} H_i \boldsymbol{\gamma}_i} + \frac{\boldsymbol{\delta}_i (\boldsymbol{\delta}_i^\mathrm{T} \boldsymbol{\gamma}_i)}{\boldsymbol{\delta}_i^\mathrm{T} \boldsymbol{\gamma}_i}$$

$$= \boldsymbol{\delta}_i$$

using the symmetry of H_i.

The induction proceeds by establishing that

$$\mathbf{P}(k), \mathbf{Q}(k) \rightarrow \mathbf{P}(k+1) \qquad \text{and then} \qquad \mathbf{P}(k+1), \mathbf{Q}(k) \rightarrow \mathbf{Q}(k+1)$$

Suppose then that $\mathbf{P}(k)$, $\mathbf{Q}(k)$ hold. As in the last section, the conjugacy of the search directions $\mathbf{s}_i$ for $i < k$ implies that $\mathbf{g}_k^T \boldsymbol{\delta}_i = 0$ as in (8.5.13) from which, using $\mathbf{Q}(k)$, we deduce

$$\boldsymbol{\delta}_i^T A \boldsymbol{\delta}_k = -\alpha_k \boldsymbol{\delta}_i^T A H_k \mathbf{g}_k = -\alpha_k \mathbf{g}_k^T H_k A \boldsymbol{\delta}_i = -\alpha_k \mathbf{g}_k^T \boldsymbol{\delta}_i = 0$$

which establishes $\mathbf{P}(k+1)$ since the other cases are included in $\mathbf{P}(k)$. Combining this with $\mathbf{Q}(k)$, we get, using the symmetry of H_k,

$$(H_k \boldsymbol{\gamma}_k)^T A \boldsymbol{\delta}_i = \boldsymbol{\gamma}_k^T H_k A \boldsymbol{\delta}_i = \boldsymbol{\gamma}_k^T \boldsymbol{\delta}_i = \boldsymbol{\delta}_k^T A \boldsymbol{\delta}_i = 0$$

Hence, for $0 \le i < k$,

$$H_{k+1} A \boldsymbol{\delta}_i = H_k A \boldsymbol{\delta}_i - \frac{H_k \boldsymbol{\gamma}_k (H_k \boldsymbol{\gamma}_k)^T A \boldsymbol{\delta}_i}{\boldsymbol{\gamma}_k^T H_k \boldsymbol{\gamma}_k} + \frac{\boldsymbol{\delta}_k \boldsymbol{\delta}_k^T A \boldsymbol{\delta}_i}{\boldsymbol{\delta}_k^T \boldsymbol{\gamma}_k}$$

$$= H_k A \boldsymbol{\delta}_i = \boldsymbol{\delta}_i$$

by $\mathbf{Q}(k)$.

To complete the proof of $\mathbf{Q}(k+1)$, we need $H_{k+1} A \boldsymbol{\delta}_k = \boldsymbol{\delta}_k$, which is just the quasi-Newton property established above.

(c) That $\mathbf{x}_n = \mathbf{x}^*$ now follows from the conjugacy of the search directions established in (b) and Theorem 8.5.2. Also, (8.6.8) with $k = n$ implies $H_n A \boldsymbol{\delta}_i = \boldsymbol{\delta}_i$ ($0 \le i < n$), and so we have a linearly independent set of n eigenvectors of $H_n A$ each with eigenvalue 1. Since this system is a basis for $\mathbb{R}^n$, it follows that $H_n A = I$ as required.

We observe here that the proof of (a) remains valid for more general functions provided only that $\boldsymbol{\delta}_i^T \boldsymbol{\gamma}_i > 0$. This is an important observation in establishing more general convergence results for the quasi-Newton methods.

This DFP update is just one member of a large class of quasi-Newton update formulas discovered by Broyden in 1967. The general formula, in a slight modification of Fletcher's (1970) parameterization, is given by

$$H_{i+1} = H_i - \frac{H_i \boldsymbol{\gamma}_i (H_i \boldsymbol{\gamma}_i)^T}{\boldsymbol{\gamma}_i^T H_i \boldsymbol{\gamma}_i} + \frac{\boldsymbol{\delta}_i \boldsymbol{\delta}_i^T}{\boldsymbol{\delta}_i^T \boldsymbol{\gamma}_i} + \lambda \mathbf{v}_i \mathbf{v}_i^T \tag{8.6.9}$$

where $\lambda \ge 0$ is the free parameter and

$$\mathbf{v} = \frac{\boldsymbol{\delta}_i}{\boldsymbol{\delta}_i^T \boldsymbol{\gamma}_i} - \frac{H_i \boldsymbol{\gamma}_i}{\boldsymbol{\gamma}_i^T H_i \boldsymbol{\gamma}_i} \tag{8.6.10}$$

Provided that $\boldsymbol{\delta}_i^T \boldsymbol{\gamma}_i > 0$, it is easy to show that if H_i is positive definite, then so is any H_{i+1} given by (8.6.9) and that Theorem 8.6.2 holds for any such formula. (See Exercise 3.) Much effort, both analytic and computational, has been devoted to identifying the "best" quasi-Newton formula or even the best from the much wider class of variable metric methods introduced by Huang (1970). The choice with the widest support is the BFGS algorithm which was derived independently in 1970 in four different ways by Broyden, Fletcher, Goldfarb, and Shanno.

In the parameterization of (8.6.9), this BFGS update formula corresponds to choosing

$$\lambda = \gamma_i^{\mathrm{T}} H_i \gamma_i \tag{8.6.11}$$

Example 8.6.3. Use the DFP and BFGS algorithms to minimize $(2x - y)^2 + (y + 1)^2$ from $\mathbf{x}_0 = (1.5, 2)$.

Solution. As before, the gradient vector for this function is $(4(2x - y), 2(2y - 2x + 1))^{\mathrm{T}}$ so that $\mathbf{g}_0 = (4, 4)^{\mathrm{T}}$. For both algorithms this yields, with the standard choice $H_0 = I$,

$$\mathbf{s}_0 = (-4, -4)^{\mathrm{T}}$$

so that we must minimize f along the line $(1.5 - 4\alpha, 2 - 4\alpha)$. This yields $\alpha = \frac{1}{2}$ so that $\mathbf{x}_1 = (-\frac{1}{2}, 0)$, and then $\mathbf{g}_1 = (-4, 4)^{\mathrm{T}}$. Therefore

$$\boldsymbol{\delta}_0 = (-2, -2)^{\mathrm{T}} \qquad \boldsymbol{\gamma}_0 = H_0 \boldsymbol{\gamma}_0 = (-8, 0)^{\mathrm{T}}$$

$$\boldsymbol{\delta}_0^{\mathrm{T}} \boldsymbol{\gamma}_0 = 16 \qquad \boldsymbol{\gamma}_0^{\mathrm{T}} H_0 \boldsymbol{\gamma}_0 = 64$$

and so the DFP update yields

$$H_1 = \begin{bmatrix} 1 & 0 \\ 0 & 1 \end{bmatrix} + \frac{1}{16} \begin{bmatrix} 4 & 4 \\ 4 & 4 \end{bmatrix} - \frac{1}{64} \begin{bmatrix} 64 & 0 \\ 0 & 0 \end{bmatrix} = \begin{bmatrix} \frac{1}{4} & \frac{1}{4} \\ \frac{1}{4} & \frac{5}{4} \end{bmatrix}$$

Then

$$\mathbf{s}_1 = -H_1 \mathbf{g}_1 = (0, -4)^{\mathrm{T}} \qquad \alpha_1 = \frac{1}{4}$$

$$\mathbf{x}_2 = (-\frac{1}{2}, -1) \qquad \mathbf{g}_2 = (0, 0)^{\mathrm{T}}$$

as expected. We also see that $\boldsymbol{\delta}_1 = (0, -1)^{\mathrm{T}}$, $\boldsymbol{\gamma}_1 = (4, -4)^{\mathrm{T}}$, and $H_1 \boldsymbol{\gamma}_1 = (0, -4)^{\mathrm{T}}$ from which we obtain

$$H_2 = \begin{bmatrix} \frac{1}{4} & \frac{1}{4} \\ \frac{1}{4} & \frac{1}{2} \end{bmatrix}$$

which is indeed the inverse of the Hessian matrix

$$\begin{bmatrix} 8 & -4 \\ -4 & 4 \end{bmatrix}$$

of this function.

For the BFGS method, the first iteration is identical to that above except for the matrix update which also needs the vector $\mathbf{v}$. In this case, we have $\mathbf{v} = (0, -1/8)^{\mathrm{T}}$ so that

$$H_1 = \begin{bmatrix} \frac{1}{4} & \frac{1}{4} \\ \frac{1}{4} & \frac{5}{4} \end{bmatrix} + 64 \begin{bmatrix} 0 & 0 \\ 0 & \frac{1}{64} \end{bmatrix} = \begin{bmatrix} \frac{1}{4} & \frac{1}{4} \\ \frac{1}{4} & \frac{9}{4} \end{bmatrix}$$

which gives $\mathbf{s}_1 = (0, -8)^{\mathrm{T}}$, $\alpha_1 = \frac{1}{8}$, and then $\mathbf{x}_2 = (-\frac{1}{2}, -1)$ as before.

We commented above that the BFGS update is the algorithm of choice among the quasi-Newton family. One of the reasons is that it exhibits greater numerical stability. The DFP update, which corresponds to $\lambda = 0$ in (8.6.9), has a tendency to become singular, or nearly singular, due to accumulation of roundoff error over many iterations on a difficult problem. The BFGS update is somewhat better protected against this by virtue of the fact that at each iteration an additional positive semidefinite matrix is being added to the DFP update. In the next example, we compare these two algorithms' behavior when applied to Rosenbrock's function.

Example 8.6.4. Apply the DFP and BFGS algorithms to Rosenbrock's function.

Solution. The results for $x_0 = (0, 0)$ are shown in Table 8.6.5 and plotted in Fig. 8.6.6.

We see from Table 8.6.5 that the first five iterations are almost identical for the two methods but that eventually the BFGS algorithm performs significantly better. This is probably due to the extra stability of the update formula.

TABLE 8.6.5
Quasi-Newton methods; Rosenbrock's function;
$x_0 = (0, 0)$

Iter.	x	y	Evals.
	(a) DFP		
1	0.16126	0.00000	13
2	0.29284	0.05059	25
3	0.34461	0.12707	37
4	0.46115	0.19625	54
5	0.53789	0.26468	70
6	0.81744	0.66010	95
7	0.81138	0.65650	105
8	0.89434	0.79323	123
9	0.95929	0.91415	141
10	0.96338	0.92834	157
11	0.99219	0.98344	173
12	0.99847	0.99703	189
13	0.99992	0.99983	201
14	1.00000	1.00000	215
	(b) BFGS		
1	0.16126	0.00000	13
2	0.29288	0.05061	28
3	0.34460	0.12704	40
4	0.46094	0.19611	52
5	0.53686	0.26359	66
6	0.85629	0.72796	87
7	0.85251	0.72531	95
8	0.92523	0.85078	113
9	0.99320	0.98750	131
10	0.99862	0.99717	141
11	1.00005	1.00008	157
12	1.00000	1.00000	171

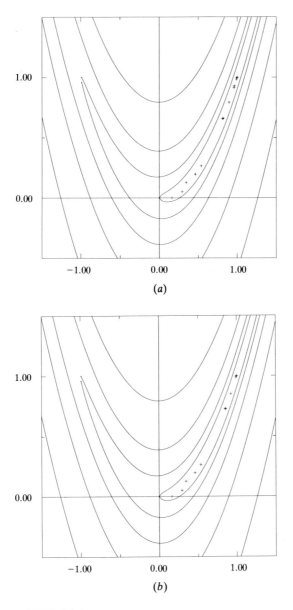

FIGURE 8.6.6
Quasi-Newton methods; Rosenbrock's function. (*a*) DFP: $\mathbf{x}_0 = (0, 0)$; (*b*) BFGS: $\mathbf{x}_0 = (0, 0)$.

Both sets of results show a small "correction" step on iteration 7, but the BFGS seems then to home in on the solution, while the DFP has further difficulty around $x = 0.96$.

Note that both methods have found the minimum to high accuracy in fewer iterations and fewer function evaluations than any of the methods considered up to now. This is certainly not conclusive in the case of DFP for which the results are only marginally better than for either the conjugate gradient or modified Newton methods.

In Fig. 8.6.7 we show the results of applying these same two methods using the starting point $(-1.2, 1)$. The first pair of graphs illustrates the implementation of exactly the same algorithm as in Fig. 8.6.6. In fact, not all iterations are shown because the initial trial step used in the bracketing

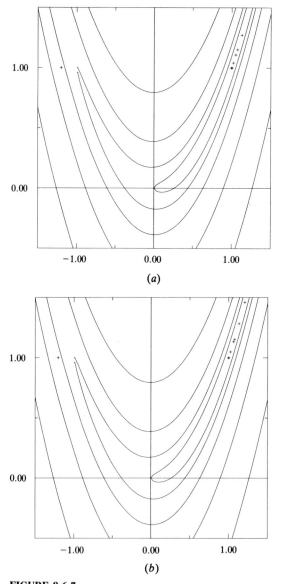

(a)

(b)

FIGURE 8.6.7
Quasi-Newton methods; Rosenbrock's function; $\mathbf{x}_0 = (-1.2, 1)$. (a) DFP; (b) BFGS; modified first step; (c) DFP; (d) BFGS.

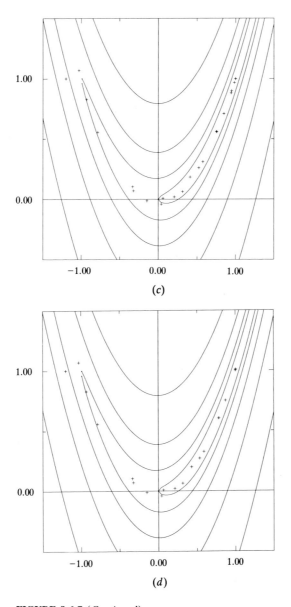

(c)

(d)

FIGURE 8.6.7 (*Continued*)

procedure for the line minimum was sufficiently large that the first iteration jumped right across the "shoulder" of the function into the valley $y = x^2$ close to $x = 2$. The two algorithms then worked their way back down the valley to the minimum with no difficulty.

The DFP algorithm attained the desired accuracy in just 12 iterations and 206 function evaluations, while the BFGS algorithm needed 13 iterations but only 195 evaluations.

In the second pair of graphs in Fig. 8.6.7, we see the results of forcing the algorithms to follow the valley by reducing the initial trial step for the bracketing procedure in iteration 1. This provides us with an estimate of the ability of the two algorithms to follow a steep-sided curved valley. The behavior of both methods is again acceptable:

DFP required 22 iterations and 393 function evaluations.

BFGS needed 20 iterations and only 305 evaluations.

The BFGS method, in particular, compares very favorably with the results of any of the previously discussed methods.

We see in Fig. 8.6.7c and d that the first 13 iterations of the two algorithms are very similar. At this stage the DFP algorithm takes a somewhat smaller step than BFGS, and each follows it with a very small "correction" after which BFGS converges very rapidly while DFP has somewhat more difficulty.

This similarity of performance is typical of the quasi-Newton algorithms. Dixon (1972) established that if the line searches are *exact*, then all the quasi-Newton algorithms will produce the same sequence of iterates when applied to a differentiable convex function. Under these conditions it can also be shown that the quasi-Newton methods will converge superlinearly to the minimum.

EXERCISES 8.6

1. Use the DFP algorithm to minimize the quadratic function $(2x - y)^2 + (y + z)^2 + (x - 2z + 1)^2$. Verify that the search directions generated are indeed conjugate and that H_3 is the inverse Hessian.

2. Repeat Exercise 1 for the BFGS update formula. Verify that exactly the same points are generated by both formulas.

3. Prove Theorem 8.6.2 for the BFGS update. (Most of the proof is unchanged. For the positive definiteness, observe that for the same H_i, the BFGS update is the DFP update plus a term of the form $\mathbf{u}\mathbf{u}^T$.)

4. Write programs to implement the DFP and BFGS algorithms, and test them on Rosenbrock's function of both two and four dimensions and on Wood's function of four dimensions. (See Exercise of Sec. 8.3.)

5. Experiment with the parameter λ of the quasi-Newton family. Do you agree that the BFGS update gives the best algorithm?

6. Prove that all quasi-Newton methods generate the same points when applied to a positive definite quadratic function.

8.7 ELIMINATION OF THE LINE SEARCH

In all the methods we have considered in the last few sections, most of the computational effort has been expended on finding the line minima. In this section, we consider techniques for eliminating these line searches. This, typically, will result in an increase in the number of iterations—but a simultaneous decrease in the number of function evaluations required.

In 1970 Fletcher suggested that this trade-off may be worthwhile and proposed an algorithm for obtaining an "acceptable" point along the line of search. He also established that as the minimum is approached, the choice $\alpha_i = 1$ will be acceptable much of the time. We will discuss Fletcher's algorithm and variations on it shortly.

In 1973, Dixon proposed an algorithm based on the idea of producing the same conjugate directions as would be obtained with exact line searches so that, for a positive definite quadratic, the inverse Hessian has been found in n iterations and one more step will locate the minimum of the function. The algorithm in fact uses a single step of a quadratic search (using one function and two derivative values) to estimate what would have been a better point than the one used. For a positive definite quadratic this is, of course, equivalent to performing the line searches. Dixon's algorithm has not been widely adopted for general-purpose minimization, and we do not discuss its details here.

The idea of these *direct prediction* methods, as they are sometimes known, is to replace the line searches in Algorithm 8.6.1 by a method for choosing an "acceptable point." Thus the algorithms can be summarized as follows.

Algorithm 8.7.1   Quasi-Newton methods without line searches

Input   Initial point $\mathbf{x}_0$, $i = 0$
                Initial positive definite symmetric matrix H_0
                Tolerance ϵ

Repeat
            $\mathbf{s}_i := -H_i\mathbf{g}_i$
            Find α_i to reduce $\phi(\alpha) = f(\mathbf{x}_i + \alpha\mathbf{s}_i)$ an "acceptable" amount
            $\mathbf{x}_{i+1} := \mathbf{x}_i + \alpha_i\mathbf{s}_i$
            Define H_{i+1} by a quasi-Newton formula
            $i := i + 1$
until $\|\mathbf{g}_i\| < \epsilon$

Output   Approximate minimum point is $\mathbf{x}_i$

Again it is common to take $H_0 = I$ so that the initial search direction is the steepest descent direction. (Throughout this section we use the same notation as in the last section.) Clearly the definition of "acceptable" and the means of finding such acceptable points are critical features of these algorithms.

Certainly, we will demand that α_i be such that $f(\mathbf{x}_{i+1}) < f(\mathbf{x}_i)$. This condition is not sufficient on its own as it would potentially allow very small steps which, in turn, might lead to either very slow progress or to a premature conclusion that the algorithm has converged. Most methods use the fact (see Exercise 1) that

$$\frac{f(\mathbf{x}_{i+1}) - f(\mathbf{x}_i)}{\boldsymbol{\delta}_i^{\mathrm{T}}\mathbf{g}_i} \to 1 \qquad \text{as } \alpha_i \to 0 \tag{8.7.1}$$

to ensure that the steplength is not too small.

It is also important that the line minimum should not be overstepped by too much.

Fletcher (1970) advocated that α_i be chosen so that for some small positive quantity μ,

$$\mu \leq \frac{f(\mathbf{x}_{i+1}) - f(\mathbf{x}_i)}{\boldsymbol{\delta}_i^{\mathrm{T}} \mathbf{g}_i} \leq 1 - \mu \tag{8.7.2}$$

The value $\mu = 10^{-4}$ was Fletcher's suggestion and seems to be good in practice. Note that the right-hand inequality here is just the application of (8.7.1).

The significance of the left-hand inequality can most readily be seen by rewriting it in the form

$$\mu \leq \frac{\phi_i(\alpha_i) - \phi_i(0)}{\alpha_i \phi_i'(0)}$$

or, equivalently, since $\phi_i'(0) < 0$,

$$\phi_i(\alpha_i) \leq \phi_i(0) + \mu\alpha_i\phi_i'(0) \tag{8.7.3}$$

The situation is illustrated for a convex function in Fig. 8.7.2. We see that the curve lies above the tangent line. Condition (8.7.3) requires that any acceptable point lie between the two lines $\phi(0) + (1 - \mu)\alpha\phi'(0)$ and $\phi(0) + \mu\alpha\phi'(0)$.

Any acceptable point must lie between $\alpha_{\min}$ and $\alpha_{\max}$. The right-hand condition in (8.7.2) ensures that the steplength cannot be too small, while the left-hand condition forces the acceptable point to lie beneath the upper straight line so that the line minimum cannot be overshot by too much. Note that, in

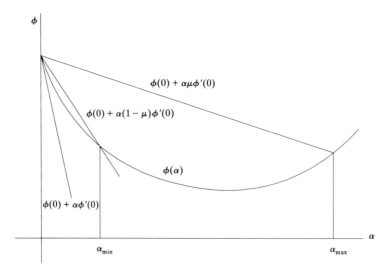

FIGURE 8.7.2
Fletcher's acceptability criterion.

practice, with $\mu = 10^{-4}$, the upper line would be almost horizontal and the lower one would be almost indistinguishable from the tangent.

We have assumed here that the search direction is downhill. It is natural, therefore, to insist that $\boldsymbol{\delta}_i^T \boldsymbol{\gamma}_i > 0$ for every iteration so that the matrices H_i remain positive definite throughout the computation. This condition will necessarily be satisfied if f is convex. (See Exercise 3.)

There remains the question of how to find such an acceptable point. Fletcher's suggestion was simply to begin with $\alpha_i = 1$ and then continually to reduce it by dividing by 10 until an acceptable point is achieved. The upper bound of 1 for the steplength is satisfactory for a positive definite quadratic function but may need to be more flexible for more general functions.

The following algorithm describes one acceptable-point routine which can be used in conjunction with either Fletcher's criterion or that of Powell, which we discuss shortly. The basic principle of this algorithm is that if the initial point is unacceptable, then the range of the search is adjusted accordingly until an acceptable point is located. It is anticipated that in the later stages of the overall minimization, the initial steplength $\alpha_i = 1$ will usually prove acceptable.

Algorithm 8.7.3 Acceptable-point algorithm

> *Input*   Current point **x** and search direction **s**
> Directional derivative $\mathbf{s}^T\mathbf{g}$
> Small positive μ
>
> *Initialize* $\alpha := 1$; mina: $= 0$; maxa: $= 1$
>
> **Repeat**
> $$D := \frac{f(\mathbf{x} + \alpha \mathbf{s}) - f(\mathbf{x})}{\alpha \mathbf{s}^T \mathbf{g}}$$
> **If** $D > 1 - \mu$ (α too small) **then**
>   **if** $\alpha = $ maxa **then** mina: $= \alpha$; maxa: $= 2 * \alpha$; $\alpha :=$ maxa
>   **else** mina: $= \alpha$; $\alpha := (\alpha + \text{maxa})/2$
>   **If** $D < \mu$ (α too large) **then** maxa: $= \alpha$; $\alpha := (\alpha + 3 * \text{mina})/4$
> **until** $\mu \leq D \leq 1 - \mu$
>
> *Output* α is an acceptable steplength.

What is happening here is that if a particular α is found to be too large (small), then it is set to the upper (lower) bound on the steplength and the search continues between the new bounds. The choice of the "weights" in using the midpoint of the new interval when increasing α and the "quarter-point" when decreasing is clearly somewhat arbitrary. However, this choice seems to work fairly well.

Powell's criterion uses the same condition $\mu \leq D$ to prevent overstepping the line minimum by too much. The condition at the other end is replaced by the requirement that

$$\boldsymbol{\delta}_i^T \mathbf{g}_{i+1} \geq \nu \boldsymbol{\delta}_i^T \mathbf{g}_i \tag{8.7.4}$$

where $0 < \mu < \nu < 1$. Powell suggested $\mu = 10^{-4}$ and $\nu = \frac{1}{2}$.

Note that (8.7.4) guarantees retention of positive definiteness for the matrices H_i in Algorithm 8.7.1 since

$$\delta_i^T \gamma_i = \delta_i^T g_{i+1} - \delta_i^T g_i \geq (\nu - 1)\delta_i^T g_i > 0$$

and $\delta_i^T g_i < 0$, $\nu < 1$. A similar algorithm to that above can be used to obtain an acceptable point according to this criterion.

Example 8.7.4. Minimize Rosenbrock's function using quasi-Newton algorithms without line searches.

Solution. For illustrative purposes, we begin by working the first iteration using the DFP update and the Fletcher criterion with $x_0 = (0, 0)$ in detail.

We thus have $g_0 = (-2, 0)^T$; $s_0 = (2, 0)^T$ and $g_0^T s_0 = -4$. Also $f(x_0) = 1$. The "search" begins with $\alpha = 1$, mina $= 0$, and maxa $= 1$. This yields

$$D = \frac{f(2, 0) - f(0, 0)}{-4} = -400$$

which clearly violates the condition since $D < \mu$. We thus set maxa $= 1$ and $\alpha = \frac{1}{4}$. Now, $D = -5.5$ and we set maxa $= \frac{1}{4}$, $\alpha = \frac{1}{16}$ and obtain $D = 0.84$ which satisfies our criterion.

Thus $x_1 = (0.125, 0)$, $\delta_0 = (0.125, 0)^T$, $g_1 = (-0.96875, -3.125)^T$, $\gamma = (1.03125, -3.125)^T$, and we then obtain

$$H_1 = \begin{bmatrix} 1.023 & 0.298 \\ 0.298 & 0.098 \end{bmatrix}$$

The results of the subsequent iterations are illustrated in Fig. 8.7.5a, while those for the BFGS update are in Fig. 8.7.5b.

Both methods converged rapidly:

The DFP update needed 20 iterations and 64 function evaluations.
The BFGS update took 18 iterations and 61 function evaluations.

Both algorithms were able to accept the steplength $\alpha = 1$ on most iterations including the final 16 for the DFP update and the final 8 for BFGS.

Using $x_0 = (-1.2, 1)$, the differences between the two updates are considerably more marked. In Fig. 8.7.6a we see that the DFP update had severe problems generating good search directions in the valley.

The graph shows the first 50 iterations of the algorithm. In fact it had still failed to converge in 250 iterations, having eventually made a few good moves after each of which it had great difficulty getting going again. The situation could be improved slightly by incorporating a test to ensure that $\delta_i^T \gamma_i > \epsilon$ to protect the positive definiteness of the matrix. However, such a strategy was very sensitive to the choice of ϵ and still failed to yield satisfactory performance.

Figure 8.7.6b shows the performance of the BFGS update from the same starting point. It converged in 34 iterations, using just 115 function evaluations.

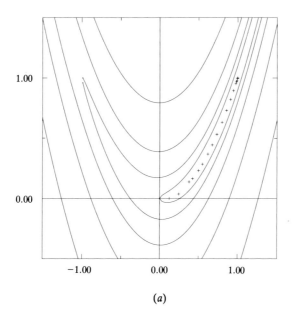

(a)

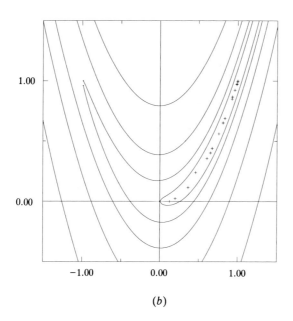

(b)

FIGURE 8.7.5
Quasi-Newton methods, Fletcher's acceptable-point criterion; Rosenbrock's function, $\mathbf{x}_0 = (0, 0)$. (a) DFP update; (b) BFGS update.

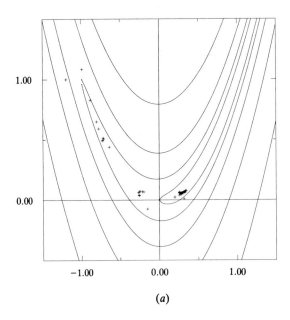

(a)

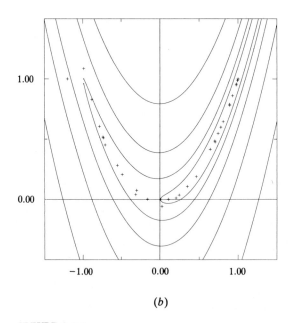

(b)

FIGURE 8.7.6
Quasi-Newton methods, Fletcher's acceptable-point criterion; Rosenbrock's function, $\mathbf{x}_0 = (-1.2, 1)$. (a) DFP update; (b) BFGS update.

The BFGS algorithm had a little difficulty as it "turned the corner" around the origin but still made good and steady progress toward the minimum.

How do we explain such a marked discrepancy between the two updates here? The difficulty lies in the fact that the DFP update is likely to become (nearly) singular due to the buildup of roundoff error. The extra robustness provided by the inclusion of the extra term in the BFGS update has a very marked effect. It is instructive, in this respect, to investigate what happens as the parameter λ in the update formula (8.6.9) is reduced from the BFGS value $\gamma_i^T H_i \gamma_i$ toward 0, the DFP value. This was achieved by setting $\lambda = c\gamma_i^T H_i \gamma_i$ and decreasing c from 1 to 0. A few values of $c > 1$ were also tried, and the results are summarized in Table 8.7.7(a) below. The essential finding here is that performance is changed less than 25 percent by varying c until $c < 0.1$, at which stage the algorithm deteriorates rapidly.

The fact (or, at least, suspicion) that it is the tendency of the DFP update to lose positive definiteness that is at the heart of this deterioration, leads us to speculate that the Powell acceptable-point criterion will prove more robust. This speculation is indeed borne out by the results in Table 8.7.7(b).

For the easier starting point $(0, 0)$, both updates again performed satisfactorily:

DFP needed 21 iterations and 109 evaluations.
BFGS needed 19 iterations and 103 evaluations.

For the Powell criterion, we indeed see that the DFP algorithm has remained stable and converged. However, it used significantly more function evaluations than the other updates needed, and these already represented significant increases over the numbers required using Fletcher's criterion.

The reason that the DFP required so many more evaluations was that, even near the end of the process, the initial $\alpha = 1$ was rarely acceptable. For most iterations $\alpha \geq 2$ was used.

The Fletcher criterion is economical in terms of function evaluations because only the objective function itself is evaluated at unacceptable points;

TABLE 8.7.7

(a) Fletcher criterion; $x_0 = (-1.2, 1)$								
c	5	1(BFGS)	0.5	0.3	0.1	0.05	0.03	0(DFP)
Iters.	31	34	42	42	42	78	78	278
Evals.	114	115	141	138	141	274	259	892

(b) Powell criterion; $x_0 = (-1.2, 1)$					
c	BFGS	0.5	0.2	0.1	DFP
Iters.	31	35	28	30	38
Evals.	166	184	178	184	343

for Powell's criterion the gradient vector must also be evaluated at such points. If $\alpha = 1$ proves to be acceptable, this represents no additional cost–otherwise the number of evaluations is increased by a factor of n (the dimension of the problem) at each unacceptable point tried.

This accounts for the increase in evaluations between Fletcher and Powell criteria and for the increase in the DFP numbers for the Powell criterion. The conclusion of this little experiment is that the Fletcher criterion is to be preferred if good performance is expected, but the slightly more expensive Powell criterion for acceptability may add stability in more difficult cases—especially for the DFP update.

We return now to the theory of these methods to establish convergence at least for positive definite quadratic functions. Stronger results than the one we present have been established, but this will serve to illustrate their nature.

Suppose then that the objective function is given, as in (8.6.2), by

$$f(\mathbf{x}) = c + \mathbf{b}^T\mathbf{x} + \frac{\mathbf{x}^T A \mathbf{x}}{2} \tag{8.7.5}$$

Since $\mathbf{A}$ is positive definite, there exist constants $0 < m \le M$ such that

$$m\|\mathbf{x}\|^2 \le \mathbf{x}^T A \mathbf{x} \le M\|\mathbf{x}\|^2 \qquad (\mathbf{x} \in \mathbb{R}^n) \tag{8.7.6}$$

and for any step $\boldsymbol{\delta}_i$,

$$\boldsymbol{\delta}_i^T \boldsymbol{\gamma}_i = \boldsymbol{\delta}_i^T A \boldsymbol{\delta}_i > 0 \tag{8.7.7}$$

so that the matrices H_i remain positive definite.

We will suppose for simplicity that there exist constants $0 < k \le K$ such that

$$k\|\mathbf{x}\|^2 \le \mathbf{x}^T H_i \mathbf{x} \le K\|\mathbf{x}\|^2 \qquad (\mathbf{x} \in \mathbb{R}^n; i = 0, 1, \ldots) \tag{8.7.8}$$

It is somewhat artificial to use a hypothesis which depends on the performance of the algorithm being analyzed. This is done here only to simplify the argument. Stronger justification than just the hereditary positive definiteness of the matrices—and in certain circumstances a proof—of this hypothesis can be found in the literature.

The critical step in establishing the convergence is to show that the steplength parameters can be bounded away from 0. This is proved for the Fletcher acceptable-point criterion in the following result. A similar result can be established for Powell's criterion.

Proposition 8.7.8. Under the assumptions (8.7.6) and (8.7.8), the numbers

$$\mu_i = \frac{(H_i \mathbf{g}_i)^T A H_i \mathbf{g}_i}{2\mathbf{g}_i^T H_i \mathbf{g}_i} \qquad (i = 0, 1, \ldots)$$

are bounded above and below by positive numbers. Hence, there exist positive constants ϵ, μ and steplengths α_i such that $0 < \epsilon \le \alpha_i \le 1$ and for which the Fletcher criterion (8.7.2) is satisfied.

Proof. By the hypotheses on f, A, and the H_i,

$$\frac{(H_i \mathbf{g}_i)^{\mathrm{T}} A H_i \mathbf{g}_i}{\mathbf{g}_i^{\mathrm{T}} H_i \mathbf{g}_i} \le \frac{M \| H_i \mathbf{g}_i \|^2}{k \| \mathbf{g}_i \|^2} \le \frac{MK^2}{k}$$

and similarly $\mu_i \ge mk^2 / 2K > 0$ which establishes the boundedness.
Now, $f(\mathbf{x}_i) - f(\mathbf{x}_{i+1}) = -\mathbf{g}_i^{\mathrm{T}} \boldsymbol{\delta}_i - \boldsymbol{\delta}_i^{\mathrm{T}} A \boldsymbol{\delta}_i / 2$, and hence

$$\frac{f(\mathbf{x}_{i+1}) - f(\mathbf{x}_i)}{\boldsymbol{\delta}_i^{\mathrm{T}} \mathbf{g}_i} = 1 + \frac{\boldsymbol{\delta}_i^{\mathrm{T}} A \boldsymbol{\delta}_i}{2 \boldsymbol{\delta}_i^{\mathrm{T}} \mathbf{g}_i} = 1 - \alpha_i \mu_i$$

Setting $\mu = \min(10^{-4}, mk^2 / 2K)$, we can find α_i such that $\mu \le \alpha_i \mu_i \le 1 - \mu$ which implies (8.7.2).
It only remains to show that we can choose α_i so that $0 < \epsilon \le \alpha_i \le 1$. Now, by construction, $\mu \le \mu_i$, and so we can choose $\alpha_i \le 1$. Also, putting $\epsilon = 2\mu k / MK^2$, we have $\epsilon \le \mu / \mu_i \le 1$ so that $\alpha_i \ge \epsilon$.

We are now able to establish the main convergence theorem.

Theorem 8.7.9. Under the same hypotheses of Proposition 8.7.8 with the steplengths chosen to satisfy (8.7.2), the sequence $(\mathbf{x}_i)$ generated by any of the quasi-Newton algorithms converges to the minimum point of f.

Proof. Since the search directions are downhill, $\boldsymbol{\delta}_i^{\mathrm{T}} \mathbf{g}_i < 0$, and so (8.7.2) implies that $(f(\mathbf{x}_i))$ is a decreasing sequence which is bounded below by $f(\mathbf{x}^*)$. Hence $(f(\mathbf{x}_i))$ is convergent, and so

$$f(\mathbf{x}_{i+1}) - f(\mathbf{x}_i) \to 0 \qquad \text{as } i \to \infty$$

Since, according to (8.7.2),

$$\frac{f(\mathbf{x}_{i+1}) - f(\mathbf{x}_i)}{\boldsymbol{\delta}_i^{\mathrm{T}} \mathbf{g}_i} \ge \mu > 0$$

it follows that $\boldsymbol{\delta}_i^{\mathrm{T}} \mathbf{g}_i \to 0$ as well. That is, $\alpha_i \mathbf{g}_i^{\mathrm{T}} H_i \mathbf{g}_i \to 0$ as $i \to \infty$. But, $\epsilon \le \alpha_i$ and $\mathbf{g}_i^{\mathrm{T}} H_i \mathbf{g}_i \ge k \| \mathbf{g}_i \|^2$, from which we deduce that $\mathbf{g}_i \to 0$, as required.

The convergence of these algorithms for positive definite quadratic functions gives us confidence that such methods should perform well on more general functions which are convex in the neighborhood of the minimum. We have already seen satisfactory performance at least for Rosenbrock's function.

EXERCISES 8.7

1. Use the definition of the derivative $\phi_i'(0)$ to show

$$\frac{f(\mathbf{x}_{i+1}) - f(\mathbf{x}_i)}{\boldsymbol{\delta}_i^{\mathrm{T}} \mathbf{g}_i} \to 1 \qquad \text{as } \alpha_i \to 0$$

2. Use the Fletcher acceptable-point criterion with the DFP update formula to perform four iterations of the minimization of the quadratic function $(2x - y)^2 + (y + 1)^2$.

3. Show that $\boldsymbol{\delta}_i^{\mathrm{T}} \boldsymbol{\gamma}_i = \alpha_i (\phi_i'(\alpha_i) - \phi_i'(0))$ and therefore that if f is convex, $\boldsymbol{\delta}_i^{\mathrm{T}} \boldsymbol{\gamma}_i > 0$ so that the matrices H_i remain positive definite.

4. Use the DFP and BFGS updates for the minimization of the four-dimensional Rosenbrock function and Wood's function. Compare the results for the two updates using both the Powell and Fletcher acceptable-point criteria.

5. Show that under the same hypotheses as used in Proposition 8.7.8 and Theorem 8.7.9, the quasi-Newton methods using Powell's acceptable-point criterion will converge to the minimum point of a positive definite quadratic function.

8.8   INTRODUCTION TO CONSTRAINED MINIMIZATION

In this section we consider, briefly, the problem of minimizing a function f of several variables subject to certain constraints which must be satisfied at the solution. These constraints may be either *equality constraints*, in which specific equations must be satisfied by the variables, or *inequality constraints* that specify a particular region of $\mathbb{R}^n$, in which we should search for the minimum—or, of course, some combination of these two.

The classical linear programming problem in which the objective function and the constraints are all linear functions of the variables is an important special case for which special algorithms exist. Within this class there is the yet more special case of the integer programming problem where only solutions in which every variable takes an integer value are to be considered. There is an extensive literature on these subjects to which the interested reader is referred. A good starting point would be the book *Linear Programming* by Vasek Chvatal.

Our focus here will be on the more general nonlinear programming problem in which no (or few) assumptions are made about either the objective or the constraint functions. We will discuss just two methods which have great similarities—and some important differences. These are the *penalty function* (or *exterior-point*) and the *barrier function* (or *interior-point*) methods, both of which reduce the problem to a sequence of unconstrained minimization problems. These subproblems can then be solved by the techniques of the last few sections.

Consider, then, the problem

$$\text{minimize} \quad f(\mathbf{x})$$

$$\text{subject to} \quad c_i(\mathbf{x}) \geq 0 \quad (i = 1, 2, \ldots, m) \tag{8.8.1}$$

Note that any equality constraints can be included in this formulation by replacing $c(\mathbf{x}) = 0$ by the two inequality constraints $c(\mathbf{x}) \geq 0$ and $-c(\mathbf{x}) \geq 0$.

Denote by **S** the *feasible region* defined by

$$\mathbf{S} = \{\mathbf{x} \in \mathbb{R}^n : c_i(\mathbf{x}) \geq 0; i = 1, 2, \ldots, m\}$$

One way of transforming the problem (8.8.1) into an unconstrained minimization problem would be to redefine the objective function to be infinite (or very large) outside **S** while leaving it unchanged within **S**. The problem with this is, of course, that such a transformation would destroy any smoothness in the problem. The idea behind the penalty and barrier function methods is to approximate this situation with smooth functions.

We begin with the penalty function approach. The idea is to add to the objective function a "penalty" for violating any constraint and to adjust this penalty to become increasingly severe as the algorithm progresses. Thus we use a *penalty function* $P(\mathbf{x}, r)$ which must be identically zero inside **S** but must grow (rapidly) outside **S**. As the iterations proceed, the parameter r will be increased, and P increases with r.

It is usual for each constraint to be given equal importance, and so typically P has the form

$$P(\mathbf{x}, r) = r \sum_{i=1}^{m} p(c_i(\mathbf{x}))$$

where, in order that P has the desired properties, we insist that:

p is continuous, convex, and nonincreasing on $(-\infty, 0]$.
$p(t) = 0$ if $t \geq 0$, $p(t) > 0$ if $t < 0$, and $p(t) \to \infty$ as $t \to -\infty$.

If it is desired to use gradient methods for the unconstrained subproblems, then greater smoothness would be required of p. Perhaps the simplest suitable function is $p(t) = [\min(t, 0)]^2$ for which

$$P(\mathbf{x}, r) = r \sum_{i=1}^{m} [\min(c_i(\mathbf{x}), 0)]^2 \tag{8.8.2}$$

Note that for any equality constraints, the combination of the two inequality constraints reduces to simply setting $p(c(\mathbf{x})) = (c(\mathbf{x}))^2$ for that term.

Now as r increases, so the penalty increases without creating any discontinuity. The principle is therefore to perform a sequence of unconstrained minimizations of the augmented objective function

$$F(\mathbf{x}, r_k) = f(\mathbf{x}) + P(\mathbf{x}, r_k) \tag{8.8.3}$$

for an increasing sequence $r_k \to \infty$. A common choice is $r_k = 10^k$.

The overall algorithm then consists of simply repeating this step until satisfactory accuracy has been achieved. For the particular penalty function (8.8.2), the algorithm is due to Zangwill (1967).

Algorithm 8.8.1   Zangwill's penalty function method

Input $\mathbf{x}_0$; r_0; $k := 0$; tolerance ϵ

Repeat
$\quad k := k + 1$
$\quad r_k := 10 * r_{k-1}$
$\quad$ Find $\mathbf{x}_k$ to minimize
$$F(\mathbf{x}, r_k) = f(\mathbf{x}) + r_k \sum_{i=1}^{m} [\min(c_i(\mathbf{x}), 0)]^2$$
$\quad$ using $\mathbf{x}_{k-1}$ as initial point
until $c_i(\mathbf{x}) > -\epsilon$ for every i.

Output   Approximate solution is $\mathbf{x}_k$.

The biggest disadvantage of using penalty functions is that (except when the solution is an interior point of **S** so that the problem is ultimately a local unconstrained minimization) the sequence of points generated will all lie outside **S**. Hence the name exterior-point methods. To ensure that the computed solution is a feasible point, it is possible to modify the constraints to

$$d_i(\mathbf{x}) = c_i(\mathbf{x}) - \epsilon \geq 0$$

so that satisfying the convergence criterion for these constraints will ensure that the solution satisfies the original constraints. (Special care must be taken with equality constraints in this modification.)

To establish the convergence of this algorithm, we will need one additional hypothesis which is intuitively reasonable once we have proved the monotonicity of three sequences of function values related to the algorithm.

Proposition 8.8.2. Algorithm 8.8.1 has the following properties for each k:

$$F(\mathbf{x}_k, r_k) \leq F(\mathbf{x}_{k+1}, r_{k+1}) \tag{8.8.4}$$

$$\sum_{i=1}^{m} p(c_i(\mathbf{x}_k)) \geq \sum_{i=1}^{m} p(c_i(\mathbf{x}_{k+1})) \tag{8.8.5}$$

$$f(\mathbf{x}_k) \leq f(\mathbf{x}_{k+1}) \tag{8.8.6}$$

Proof. First, $r > 0$ and $P(\mathbf{x}, r) \geq 0$ for all $\mathbf{x}$. Since $r_k < r_{k+1}$, we have

$$F(\mathbf{x}_{k+1}, r_k) = f(\mathbf{x}_{k+1}) + P(\mathbf{x}_{k+1}, r_k)$$

$$\leq f(x_{k+1}) + P(\mathbf{x}_{k+1}, r_{k+1}) = F(\mathbf{x}_{k+1}, r_{k+1})$$

Also, $\mathbf{x}_k$ minimizes $F(\mathbf{x}, r_k)$ so that

$$F(\mathbf{x}_k, r_k) \leq F(\mathbf{x}_{k+1}, r_k) \leq F(\mathbf{x}_{k+1}, r_{k+1})$$

which establishes (8.8.4).

Similarly, $\mathbf{x}_{k+1}$ minimizes $F(\mathbf{x}, r_{k+1})$, and so we deduce that

$$F(\mathbf{x}_k, r_k) + F(\mathbf{x}_{k+1}, r_{k+1}) \leq F(\mathbf{x}_k, r_{k+1}) + F(\mathbf{x}_{k+1}, r_k)$$

from which, in turn, it follows that

$$P(\mathbf{x}_k, r_k) + P(\mathbf{x}_{k+1}, r_{k+1}) \leq P(\mathbf{x}_k, r_{k+1}) + P(\mathbf{x}_{k+1}, r_k)$$

Combining terms in this last inequality yields

$$(r_{k+1} - r_k) \sum_{i=1}^{m} [p(c_i(\mathbf{x}_k)) - p(c_i(\mathbf{x}_{k+1}))] \geq 0$$

which with the fact that (r_k) is increasing implies (8.8.5).

Finally, since $F(\mathbf{x}_{k+1}, r_k) - F(\mathbf{x}_k, r_k) \geq 0$, we have

$$f(\mathbf{x}_{k+1}) - f(\mathbf{x}_k) \geq r_k \sum_{i=1}^{m} [p(c_i(\mathbf{x}_k)) - p(c_i(\mathbf{x}_{k+1}))] \geq 0$$

proving (8.8.6).

The second of these inequalities establishes that the successive iterates are getting closer to the feasible region in the sense that the total constraint violation is decreasing. The additional hypothesis that is required for the convergence proof is that all the $\mathbf{x}_k$'s and the solution $\mathbf{x}^*$ all lie in a compact set. Since the total constraint violation is decreasing, it follows that this additional hypothesis will be satisfied if the level set $\mathbf{S}_0$ of

$$P(\mathbf{x}_0, 1) = \sum_{i=1}^{m} p(c_i(\mathbf{x}_0))$$

defined by

$$\mathbf{S}_0 = \{\mathbf{x}: P(\mathbf{x}, 1) \le P(\mathbf{x}_0, 1)\}$$

is compact. This will be satisfied if the set $\mathbf{S}_0$ is bounded.

Theorem 8.8.3. If all the iterates $\mathbf{x}_k$ generated by Algorithm 8.8.1 and $\mathbf{x}^*$ lie in a compact set (which is certainly true if $\mathbf{S}_0$ is bounded), then

$$\mathbf{x}_k \to \mathbf{x}^* \qquad \text{as } k \to \infty$$

Proof. Since $(\mathbf{x}_k)$ lies in a compact set, it follows that it has a convergent subsequence $(\mathbf{x}_{k_j})$ with limit $\mathbf{x}'$, say.

Also, $\mathbf{x}^* \in S$, and so $P(\mathbf{x}^*, r) = 0$ for every r. Hence, for each k,

$$f(\mathbf{x}^*) = F(\mathbf{x}^*, r_k)$$

$$\ge F(\mathbf{x}_k, r_k) = f(\mathbf{x}_k) + r_k \sum_{i=1}^{m} p(c_i(\mathbf{x}_k)) \tag{8.8.7}$$

By (8.8.4), $(F(\mathbf{x}_k, r_k))$ is increasing; it is bounded above by $f(\mathbf{x}^*)$ and so converges to L, say. It follows that

$$\lim_{j \to \infty} F(\mathbf{x}_{k_j}, r_{k_j}) = L$$

and, since f is continuous, $f(\mathbf{x}_{k_j}) \to f(\mathbf{x}')$. Therefore,

$$r_{k_j} \sum_{i=1}^{m} p(c_i(\mathbf{x}_{k_j})) = F(\mathbf{x}_{k_j}, r_{k_j}) - f(\mathbf{x}_{k_j}) \to L - f(\mathbf{x}')$$

However, $r_{k_j} \to \infty$, and so

$$\sum_{i=1}^{m} p(c_i(\mathbf{x}')) = \lim_{j \to \infty} \sum_{i=1}^{m} p(c_i(\mathbf{x}_{k_j})) = 0$$

which implies that $\mathbf{x}' \in S$.

Finally, using (8.8.7), we have $f(\mathbf{x}_k) \le f(\mathbf{x}^*)$, and so, using the continuity of f again, $f(\mathbf{x}') \le f(\mathbf{x}^*)$ *and* $\mathbf{x}' \in S$. But $\mathbf{x}^*$ is the minimum point for f in S so that $\mathbf{x}' = \mathbf{x}^*$. Since this holds for *all* convergent subsequences, it follows that $\mathbf{x}_k \to \mathbf{x}^*$, as required.

There are many variations on Zangwill's original algorithm including more recent work on *augmented penalty functions* in an attempt to improve the efficiency. Since each iteration is a full unconstrained minimization, any improvement in the efficiency of the outer iteration will prove worthwhile.

We will illustrate the performance of both penalty and barrier functions by minimizing a simplification of Rosenbrock's functions in the unit circle.

Example 8.8.4. Use the penalty function method to

$$\text{minimize} \quad 10(x^2 - y)^2 + (1 - x)^2$$

$$\text{subject to} \quad x^2 + y^2 \leq 1$$

Solution. The results of the first six outer iterations are shown in Fig. 8.8.5a and tabulated in Table 8.8.6. For the purposes of illustration, the algorithm was started with a small value for the penalty parameter: $r = 0.1$ for the first iteration. The true solution is found to reasonable accuracy very economically. The internal unconstrained minimizations were performed using the BFGS update quasi-Newton method with Powell's acceptable-point criterion. Some care is needed to ensure that the steplengths in the first few iterations of each outer iteration are not too big. The internal unconstrained minimizations were not performed to very high accuracy as they are only providing starting points for subsequent iterations. We see that these six iterations have produced convergence to four decimal places in the components of $\mathbf{x}$. Note that, as predicted by Proposition 8.8.2, the sequences of values $F(\mathbf{x}_k, r_k)$, $f(\mathbf{x}_k)$ are both increasing. We also see that the last two show no difference between the objective function and the augmented function to five decimal places. The last of these shows that $[c(\mathbf{x}_6)]^2 < 10^{-9}$, which means that the boundary of the feasible region has indeed been approached very closely.

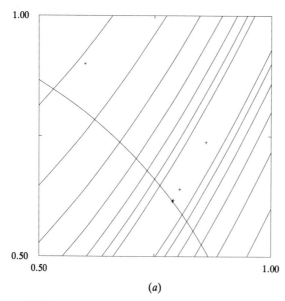

(a)

FIGURE 8.8.5
Penalty and barrier function methods for Example 8.8.4. (*a*) Penalty function; (*b*) barrier function. The contours shown in the two graphs are for $f(x, y) = \frac{1}{2}$, 1, 2, 4, 8, 16, 32; the boundary of the feasible region is also graphed.

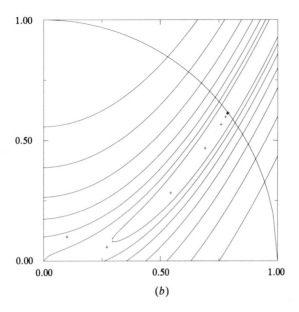

FIGURE 8.8.5 (*Continued*)

We turn now to the interior-point, or barrier function, method which was originally proposed by Fiacco and McCormick (1964). As the name suggests, this method requires an interior point of the feasible region **S** as a starting point. The idea this time is to set up a barrier to prevent the search from leaving the feasible region.

The barrier function $B(\mathbf{x}, r)$ has the property that

$$B(\mathbf{x}, r) \to \infty \qquad \text{as } c_i(\mathbf{x}) \to 0 +$$

so that as **x** approaches the boundary of **S** *from the interior*, the barrier function grows without bound. B also increases with r—or, in practice, decreases as r decreases to 0. The SUMT, or *sequential unconstrained minimization tech-*

TABLE 8.8.6
Penalty function iterations for Example 8.8.4; $\mathbf{x}_0 = (0.6, 0.9)$

r	x	y	$F(\mathbf{x}, r)$	$f(\mathbf{x})$	Total evals.
0.1	0.860 884	0.737 138	0.02761	0.01951	28
1.0	0.803 994	0.639 395	0.04196	0.03891	52
10.0	0.789 986	0.617 405	0.04483	0.04455	70
100.0	0.789 096	0.614 738	0.04514	0.04511	106
1000.0	0.788 693	0.614 847	0.04517	0.04517	124
10000.0	0.788 683	0.614 807	0.04518	0.04518	178

nique, as this method was originally called, proceeds by minimizing

$$G(\mathbf{x}, r_k) = f(\mathbf{x}) + B(\mathbf{x}, r_k) \tag{8.8.8}$$

for a decreasing sequence $r_k \to 0$ which allows the minimizing point to approach the boundary of $\mathbf{S}$.

The standard form of B is

$$B(\mathbf{x}, r) = r \sum_{i=1}^{m} b(c_i(\mathbf{x})) = r \sum_{i=1}^{m} \frac{1}{c_i(\mathbf{x})}$$

although other functions b could be used.

The algorithm is thus very similar to the penalty function algorithm with the essential difference being the way in which the objective function is to be augmented. We observe that the penalty function method is immediately adaptable to equality constraints, while the barrier function is not. The SUMT algorithm handles equality constraints by including penalty terms for these.

Common choices for the sequence (r_k) are 10^{-k} or 10^{-k+1}. All the iterates $\mathbf{x}_k$ generated by this algorithm will lie inside $\mathbf{S}$, but the line searches used in the unconstrained minimization routine must check that no trial point lies outside $\mathbf{S}$. The convergence theory for the barrier function method is very similar to the penalty function method: its details will be left as an exercise.

Thoerem 8.8.7. If the feasible region $\mathbf{S}$ is a compact connected set with nonempty interior, then the barrier function algorithm has the properties:

(a) $G(\mathbf{x}_k, r_k) \ge G(\mathbf{x}_{k+1}, r_{k+1})$

(b) $\sum_{i=1}^{m} b(c_i(\mathbf{x}_k)) \le \sum_{i=1}^{m} b(c_i(\mathbf{x}_{k+1}))$

(c) $f(\mathbf{x}_k) \ge f(\mathbf{x}_{k+1})$

(d) $\mathbf{x}_k \to \mathbf{x}^*$ as $k \to \infty$

In Fig. 8.8.5b we see that the barrier function algorithm is converging steadily toward the minimum point, although its progress is apparently somewhat less than that of the penalty function algorithm. The results of the first 10 iterations of the barrier function algorithm are presented in Table 8.8.8. The value of the parameter r was initialized to 1 and reduced by a factor of 10 on each iteration.

In addition to the greater speed of convergence of the penalty function method, its iterations were much cheaper in terms of function evaluations. Part of the cause of this additional expense is the need for care in the line searches which resulted in using a smaller initial steplength. It also appears that the quadratic growth of the penalty function is sufficient to force that search to converge rapidly to the constraint boundary.

A word of caution is in order: although the difference between these algorithms is quite marked for this particular example, it is often the case that the barrier function algorithm will perform with great efficiency—sometimes more efficiently than the penalty function approach.

TABLE 8.8.8
Barrier function iterations for Example 8.8.4; $x_0 = (0.1, 0.1)$

x	y	$G(x, r)$	$f(x)$	Total evals.
0.270 753	0.057 795	1.61722	0.53421	64
0.545 072	0.282 032	0.36965	0.20933	127
0.691 945	0.467 109	0.12926	0.09626	172
0.759 346	0.566 382	0.06871	0.05896	256
0.778 974	0.597 431	0.05249	0.04973	361
0.785 696	0.610 156	0.04740	0.04644	463
0.787 919	0.613 134	0.04588	0.04557	553
0.788 452	0.614 177	0.04540	0.04531	665
0.788 661	0.614 588	0.04525	0.04521	738
0.788 715	0.614 675	0.04520	0.04519	821

EXERCISES 8.8

1. Carry out the first two iterations of the minimization of $(x + 1)^2$ subject to the constraint $c(x) = x \geq 0$ using the penalty function $P(x, r) = r[\min(c(x), 0)]^2$ and $r_k = 10^{k-1}$. Prove that the resulting sequence (x_k) converges to the required solution.

2. Repeat Exercise 1 for the barrier function method.

3. Prove the convergence theorem, Theorem 8.8.7, for the barrier function method.

4. Write programs to implement the penalty and barrier function algorithms for constrained minimization. Test your programs by minimizing $2x^2 - 2xy + 2y^2 - 6x$ subject to the constraints $x \geq 0$, $y \geq 0$, $3x + 4y \leq 6$, $4y - x \leq 2$.

8.9 SYSTEMS OF NONLINEAR EQUATIONS

We finish this chapter with a very brief look at the problem of solving a system of n nonlinear equations in n unknowns. An introduction to the basic ideas of iterative methods of solving such systems in the two-dimensional case was presented in Sec. 2.6, where we looked briefly at fixed-point iteration and then at Newton's method.

For the example considered there, the function iteration was slowly convergent, while Newton's method was very efficient *when* it worked. Both those approaches generalize naturally to higher-dimensional problems, and the above conclusions are typical.

Specifically, consider the solution of the system of equations

$$\mathbf{f}(\mathbf{x}) = 0 \tag{8.9.1}$$

where $\mathbf{f} : \mathbb{R}^n \to \mathbb{R}^n$ is a differentiable function. It will be convenient at times to write this system in terms of its component functions

$$f_i(x_1, x_2, \ldots, x_n) = 0 \qquad (i = 1, 2, \ldots, n) \tag{8.9.2}$$

There is an obvious generalization of Theorem 2.6.7 for the convergence of a fixed-point iteration $\mathbf{x}_{n+1} = \mathbf{g}(\mathbf{x}_n)$, where

$$\mathbf{x} = \mathbf{g}(\mathbf{x}) \leftrightarrow \mathbf{f}(\mathbf{x}) = 0$$

The higher dimensionality of the problem makes finding a suitable iteration function much harder, however.

Similarly, the Newton iteration generalizes as

$$\mathbf{x}_{n+1} = \mathbf{x}_n - \mathbf{F}^{-1}(\mathbf{x}_n)\mathbf{f}(\mathbf{x}_n) \qquad (8.9.3)$$

where $\mathbf{F}$ is the Jacobian matrix of $\mathbf{f}$. However, this iteration clearly requires the solution of a system of linear equations at each iteration—a similar cost to the Newton method for minimizing a function of several variables.

In Sec. 8.4 we saw that Newton's method for minimization shares the unreliability observed for the equation solver in Sec. 2.6 but that its performance can be safeguarded greatly by including a line search. In the equation-solving situation, that is equivalent to solving the system $\mathbf{F}(\mathbf{x}_n)\mathbf{s} = \mathbf{f}(\mathbf{x}_n)$ and then choosing $\mathbf{x}_{n+1}$ to be the point on the line $\mathbf{x}_n + \alpha\mathbf{s}_n$ which is "closest" to solving the original system (8.9.1).

What do we mean by "closest" here? There is no definitive answer to that question, but one natural measure of closeness would be the use of the least squares metric, in which case α is chosen to minimize $\mathbf{f}(\mathbf{x}_n + \alpha\mathbf{s}_n)^T\mathbf{f}(\mathbf{x}_n + \alpha\mathbf{s}_n)$. It is a natural extension of this to treat the original problem as a minimization problem using the equivalence of (8.9.1) and the minimization of

$$\Phi(\mathbf{x}) = \sum_{i=1}^{n} f_i^2(x_1, x_2, \ldots, x_n) \qquad (8.9.4)$$

Note that this does *not* in general produce the same search directions and iterates as the application of Newton's method to the original system. However, once this principle of *minimizing the sum of squares of residuals* is accepted, then, of course, the minimization can be undertaken by any of our unconstrained minimization techniques—including the highly efficient quasi-Newton methods.

Note that any gradient methods for minimizing $\Phi(\mathbf{x})$ reqire the same derivative information as does Newton's method for the direct solution of (8.9.1) since

$$\frac{\partial \Phi}{\partial x_i} = 2\sum_{k=1}^{n} f_k \frac{\partial f_k}{\partial x_i} = 2(\mathbf{F}^T\mathbf{f})_i \qquad (8.9.5)$$

so that derivatives of this objective function are obtainable directly from the Jacobian matrix of the original system.

We illustrate the performance by applying this technique to the solution of the same system as in Sec. 2.6. See Tables 2.6.5 and 2.6.10 for comparison.

Example 8.9.1. We apply the quasi-Newton algorithm with Powell's acceptable-point criterion to the solution of the system

$$x^2 - y^2 - 1 = 0 \qquad x^3 y^2 - 1 = 0$$

In Sec. 2.6 we saw that the function iteration method was extremely slowly convergent, while Newton's method converged to very high precision in just four

TABLE 8.9.2
Quasi-Newton method for nonlinear equation solution

Iter.	x	y	Φ	Evals.
		(a) $\mathbf{x_0} = (1.2, 0.7)$		
1	1.25556	0.73760	6.953E − 03	10
2	1.22751	0.73373	1.016E − 03	13
3	1.23039	0.72730	4.451E − 04	22
4	1.23643	0.72728	8.280E − 08	25
5	1.23651	0.72729	1.947E − 11	28
		(b) $\mathbf{x_0} = (0.12, 0.01)$		
1	1.06629	−0.06872	1.006E + 00	7
2	0.96182	−0.11020	9.861E − 01	13
3	1.01175	−0.79130	4.866E − 01	22
4	1.07974	−0.92637	4.858E − 01	28
5	1.20047	−0.82963	9.747E − 02	34
6	1.21467	−0.67674	3.242E − 02	37
7	1.23191	−0.71942	1.049E − 03	40
8	1.23787	−0.72905	6.785E − 05	43
9	1.23645	−0.72723	9.711E − 08	46
10	1.23651	−0.72729	9.773E − 12	49

iterations from the good "guess" $\mathbf{x_0} = (1.2, 0.7)$. From the more problematic initial point $(0.12, 0.01)$, Newton's method had produced a "best" point close to $(1440, 1440)$ after 10 iterations. The results of the quasi-Newton algorithm from these same starting points are presented in Table 8.9.2. We observe that in the second case the other solution has been located, but in both cases high satisfactory convergence has been achieved in very few iterations, most of which required only the minimum number of function evaluations.

It is worth observing that the quoted numbers of function evaluations in Table 8.9.2 include each component of the gradient of Φ, which in this situation will frequently be a simple computation once the various components of the original system have been evaluated. The already small numbers of evaluations used here are therefore representative of highly efficient performance.

EXERCISE 8.9

1. Write a program to solve a system of nonlinear equations by minimizing the sum of the squares of residuals using the most efficient optimization routine you can. Test your program by solving

$$x^4 + xy^3 + y^4 = 1 \qquad x^2 + xy - y^2 = 1$$

and

$$x^2 + y^2 - 2x = 0 \qquad x^2 - y^2 - y = 0$$

using suitably chosen starting points.

PROJECTS

1. Experiment with the various coefficients in the Nelder and Mead simplex method for the minimization of Rosenbrock's function and other test functions of varying dimension. Write a summary of your results and conclusions.

2. Write a report to summarize the results of numerical experiments to determine the most efficient minimization algorithm for a collection of functions including Rosenbrock's and Wood's functions extended to dimensions 4, 8, and 20 and the function

$$f = \sum_{i=1}^{10} (e^{ix/10} - e^{iy/10} - e^{-i/10} + e^{-i/2})^2$$

beginning with $\mathbf{x}_0 = (0, 0)$ and other functions.

3. Find the polynomial of degree at most 10 which minimizes

$$\int_0^1 x(p^2(x) + p'^2(x))dx$$

subject to $p(0) = 1$ and $p'(0) = 0$. Repeat this for a general function which satisfies the same initial conditions. Use the function values at n distinct points as the unknowns. (Use divided differences for the derivative and the trapezoidal rule, Chap. 9, for the integrals.) Use the resulting data as input for spline interpolation, and graph the solution function. Use the most efficient minimization routine you can.

CHAPTER

9

NUMERICAL
CALCULUS

9.1 INTRODUCTION

In this chapter we explore techniques for approximating numerically the two fundamental operations of the calculus, differentiation and integration.

If a function has a closed-form representation in terms of the standard functions of the calculus, then its derivatives can be found exactly. However, all that may be known about the function is its values at a discrete set of points obtained by experimental measurement, such as the position of an object measured at regular time intervals. In this case formulas for numerical differentiation can be used to estimate the derivatives of the unknown function. Such formulas also are useful in deriving systems of equations for the unknown solution of a boundary value problem for a differential equation. These problems will be treated in Chap. 12.

The definite integral $\int_a^b f(x)\,dx$ needs to be approximated numerically in situations in which it is impossible or prohibitively difficult to evaluate the integral analytically. Two such examples are discussed below.

Example 9.1.1. Temperature distribution on a thin rod. Suppose that the temperature of a long, thin rod is initially 0 degrees throughout its length and that its left end, $x = 0$, is held at temperature u_0. If the right end is sufficiently far away that its temperature will have negligible effect on the temperature near the left end, then we may take it to be at $x = \infty$ and need specify no particular temperature there. Temperature $u(x, t)$ as a function of position x and time t is then given by (Exercise 1)

$$u(x, t) = u_0 \left[1 - \mathrm{erf}\left(\frac{x}{2\sqrt{\tau}} \right) \right] \tag{9.1.1}$$

419

Here $\tau := kt/c\rho$, where k, c, and ρ are, respectively, the thermal conductivity, specific heat, and density of the rod's material. The gaussian error function erf is defined by

$$\text{erf}(x) = \frac{2}{\sqrt{\pi}} \int_0^x e^{-t^2} \, dt \qquad (9.1.2)$$

Since e^{-t^2} has no closed-form antiderivative, the values of the function erf must be approximated numerically.

Example 9.1.2. Period of an oscillatory motion. A *restoring force* moves a system back toward an equilibrium position. A simple example of this is a block of mass m attached to a spring and subject to negligible damping. If $-\kappa(x)$ is the force exerted by the spring when the displacement from equilibrium is x, then Newton's second law gives the differential equation

$$mx'' = -\kappa(x) \qquad (9.1.3)$$

for the displacement of the block.

A priori all that we know about the force of the spring κ is that $\kappa(0) = 0$ and that $\kappa(x)$ and x have the same sign. Assuming that κ is a smooth function of x, it has the Taylor expansion

$$\kappa(x) = \kappa_1 x + \kappa_2 x^2 + \kappa_3 x^3 + \cdots \qquad (9.1.4)$$

If the maximum displacement x_{max} of the block is small enough, then the first term of (9.1.4) will dominate, giving *Hooke's law*: force $= -\kappa_1 x$. In this case the system is a *harmonic oscillator*. Larger displacements require keeping higher powers of x, in which case the system is sometimes termed an *anharmonic oscillator*.

The differential Equation (9.1.3) has the form

$$x'' + f(x) = 0 \qquad (9.1.5)$$

with $f(x) := \kappa(x)/m$. For f to realistically represent the acceleration due to a restoring force, we must have at minimum $f(0) = 0$ and $xf(x) > 0$ for $0 < |x| \leq x_{max}$, where x_{max} is the amplitude of the oscillation. Another example of such an equation is the differential equation governing the displacement of a simple pendulum:

$$\theta'' + \frac{g}{l} \sin \theta = 0 \qquad (9.1.6)$$

Here θ is the angular displacement from the downward vertical direction, g is the acceleration of gravity, and l is the length of the pendulum.

If f is an odd function, $f(-x) = -f(x)$, then the period of motion is given by

$$T = 2\sqrt{2} \int_0^{x_{max}} \frac{dx}{\sqrt{F(x_{max}) - F(x)}} \qquad (9.1.7)$$

where F is any antiderivative of f (Exercise 2). Only in simple cases such as Hooke's law $[f(x) = \kappa_1 x/m]$ can the integral be evaluated analytically (Exercise 3). In other instances, including the anharmonic oscillator and the pendulum, the integral in (9.1.7) can only be approximated numerically.

EXERCISES 9.1

1. Temperature on a thin rod is governed by the *heat equation*

$$\frac{\partial u}{\partial t} = \frac{k}{\rho c}\frac{\partial^2 u}{\partial x^2} \qquad (9.1.8)$$

The problem described in Example 9.1.1 gives the boundary condition $u(0, t) = u_0$ and the initial condition $u(x, 0) = 0$, $0 \le x \le \infty$. Show that after taking Laplace transforms of the heat equation with respect to the time variable t, the following problem is obtained for the Laplace transform $U(x, s)$ of $u(x, t)$:

$$\frac{\partial^2 U}{\partial x^2} - \frac{\rho cs}{k} U = 0 \qquad U(0, s) = \frac{u_0}{s} \qquad (9.1.9)$$

Solve this differential equation for U and then use the Laplace transform identity

$$\mathcal{L}\left\{1 - \text{erf}\left(\frac{a}{2\sqrt{t}}\right)\right\} = \frac{e^{-a\sqrt{s}}}{s} \qquad a > 0 \qquad (9.1.10)$$

to show that u as given by (9.1.1) is a solution.

2. In this exercise we derive formula (9.1.7) for the period of motion of a system subject to an odd restoring force. Since we are seeking only the period of motion, the starting conditions may be taken to have any convenient form. We choose to start at a point at which the system is farthest from equilibrium, that is, at $x(0) = x_{max}$, $x'(0) = 0$. From the assumption that f is odd, it follows that $x(T/4) = 0$ and $x'(t) < 0$ for $0 < t < T/2$.
    (a) Write Eq. (9.1.5) as $x'' = -f(x)$. Upon multiplying this equation by x', the left-hand side becomes a perfect derivative. Integrate this equation from 0 to t to obtain

$$(x')^2 = 2[F(x_{max}) - F(x)] \qquad (9.1.11)$$

where F is any antiderivative of f.
    (b) Solve (9.1.11) for x' and integrate again from 0 to $T/4$ to obtain (9.1.7).

3. When the restoring term is $f(x) = \kappa_1 x/m$ (Hooke's law), the period of motion can be found analytically. Do this in two ways:
    (a) Use formula (9.1.7).
    (b) Solve the differential Equation (9.1.3) and find the period from the solution.

4. Observe that the integrand in (9.1.7) is singular at $x = x_{max}$. Show that if $f(x) > 0$ in $(0, x_{max}]$, then the integrand is less than $C(x_{max} - x)^{-1/2}$ for some constant C and thus the integral exists. {*Hint:* Show that $(x_{max} - x)/[F(x_{max}) - F(x)]$ is continuous in $[0, x_{max}]$.}

5. While there are numerical integration formulas for singular integrals, it is sometimes preferable to remove the singularity analytically, if possible. The technique outlined in this exercise makes the integrand of (9.1.7) nonsingular when the restoring term f in (9.1.5) is an odd function having an infinite series expansion. Let

$$f(x) = \sum_{n=1}^{\infty} a_{2n-1} x^{2n-1} \qquad (9.1.12)$$

    (a) Show that the change of variables $x = x_{max} \sin \theta$ leads to the formula

$$T = 2\sqrt{2} x_{max} \int_0^{\pi/2} \frac{\cos \theta \, d\theta}{\sqrt{F(x_{max}) - F(x_{max} \sin \theta)}} \qquad (9.1.13)$$

Note that the integrand is now singular at $\theta = \pi/2$.
    (b) Show that if f has the power series expansion (9.1.12) then

$$F(x_{max}) - F(x_{max} \sin \theta) = \sum_{n=1}^{\infty} \frac{a_{2n-1}}{2n}$$
$$\times x_{max}^{2n}(1 - \sin^{2n}\theta) \qquad (9.1.14)$$

    (c) Use the geometric series formula

$$1 - r^n = (1 - r)\sum_{j=0}^{n-1} r^j$$

to show that the integrand of (9.1.13) is

$$\frac{1}{\sqrt{\displaystyle\sum_{n=1}^{\infty} \frac{a_{2n-1}}{2n} x_{max}^{2n} \sum_{m=0}^{n-1} \sin^{2m}\theta}} \qquad (9.1.15)$$

and therefore no longer singular at $\theta = \pi/2$ if $f(x_{max}) \ne 0$.
    (d) Use (9.1.15) to find expressions for the period T for the following two restoring terms:
    i. $f(x) = \kappa_1 x + \kappa_3 x^3$
    ii. $f(\theta) = (g/l)\sin \theta$

6. (*a*) For a simple pendulum the restoring term is $f(\theta) = (g/l)\sin\theta$. Use (9.1.7) to show that the period of a simple pendulum is

$$T = 2\sqrt{\frac{2l}{g}} \int_0^{\theta_{max}} \frac{d\theta}{\sqrt{\cos\theta - \cos\theta_{max}}}$$
(9.1.16)

Note that the requirement $\theta f(\theta) > 0$ for $\theta \in (0, \theta_{max}]$ is violated when $\theta_{max} = \pi$. Interpret this physically, and show that the integral in (9.1.16) does not exist when $\theta_{max} = \pi$.

(*b*) When θ_{max} is small, we have from Taylor's theorem the approximation $\cos\theta \simeq 1 - \theta^2/2$ for $\theta \in [0, \theta_{max}]$. Substitute this approximation into (9.1.16) to find an approximation to the period of a pendulum for small displacements.

(*c*) The integrand of (9.1.16) is singular at $\theta = \theta_{max}$. Make the change of variables $\sin\psi = [\sin(\theta/2)]/[\sin(\theta_{max}/2)]$ to show that

$$T = 4\sqrt{\frac{l}{g}} \int_0^{\pi/2} \frac{1}{\sqrt{1 - \sin^2(\theta_{max}/2)\sin^2\psi}} d\psi$$
(9.1.17)

[*Hint:* $\cos(\theta) = 1 - 2\sin^2(\theta/2)$.] Note that the integrand of (9.1.17) is no longer singular at $\theta = \theta_{max}$. The integral in (9.1.17) is an instance of a *complete elliptic integral of the first kind*.

7. A simple microphone converts the mechanical oscillations of a diaphragm into an oscillatory current. This can be accomplished with an *RL* circuit in which the resistance is made to oscillate with the frequency of the diaphragm. The current I in a circuit with constant inductance L_0, constant voltage E_0, and variable resistance $R = R_0 + R_1\sin\omega t$, where ω is the radian frequency at which the diaphragm vibrates, is described by the differential equation

$$L_0 I' + (R_0 + R_1\sin\omega t)I = E_0 \quad (9.1.18)$$

Solve this first-order linear equation by the use of an integrating factor, assuming the initial current is E_0/R_0. Note that the answer contains an integral which has no closed-form antiderivative.

9.2 NUMERICAL DIFFERENTIATION

In this section we seek ways to approximate numerically the value of the derivative $f'(x)$ of a function $f(x)$.

Forward Difference Quotient

The most obvious way of doing this is to use the definition of derivative:

$$f'(x) = \lim_{h\to 0} \frac{f(x+h) - f(x)}{h} \quad (9.2.1)$$

The approximation

$$f'(x) \simeq \Delta_{forw,h} f(x) := \frac{f(x+h) - f(x)}{h} \quad (9.2.2)$$

is called the *forward difference quotient*. The first question that arises is how does the accuracy of this approximation depend upon h? If f is twice continuously differentiable, then from Taylor's theorem

$$\Delta_{forw,h} f(x) = \frac{f(x) + hf'(x) + h^2 f''(\xi)/2! - f(x)}{h} = f'(x) + \frac{h}{2} f''(\xi) \quad (9.2.3)$$

where $x \le \xi \le x + h$. Thus the *truncation* or *discretization error* made in approximating $f'(x)$ by $\Delta_{\text{forw},h} f(x)$ is

$$e_{\text{trunc}}(h) := \frac{f''(\xi)}{2} h$$

In practical situations neither f'' nor ξ will be known; however, this formula does tell us that as we decrease h, we expect the truncation error made by the forward difference quotient to decrease proportionally.

Definition 9.2.1. Suppose that $A(h)$ is a scheme for numerically approximating some quantity A_0; that is, $\lim_{h \to 0} A(h) = A_0$. The truncation error of the scheme is said to be *order n*, or $O(h^n)$, if there is a constant C, independent of h, for which

$$|e_{\text{trunc}}(h)| := |A(h) - A_0| \le Ch^n \qquad (9.2.4)$$

Thus the forward difference quotient is an order one, or $O(h)$, approximation.

Example 9.2.2. Let us estimate the truncation error made by approximating the derivative of $f(x) = \sin x$ at $x = \pi/3$. If we make the restriction $0 < h \le 0.1$, then we have

$$|e_{\text{trunc}}(h)| = \left| \frac{f''(\xi)}{2} h \right| = \left| \frac{-\sin \xi}{2} h \right| \le \left| \frac{\sin (\pi/3 + 0.1)}{2} \right| h \approx 0.46h$$
$$(9.2.5)$$

This predicts an error of less than 0.046 when $h = 0.1$, 0.0046 when $h = 0.01$, and so forth.

Table 9.2.3 shows the result of approximating $f'(\pi/3) = 0.5$ by $\Delta_{\text{forw},h} f(x)$ using single-precision arithmetic for decreasing values of h. The results are in good agreement with those predicted by the bound (9.2.5) on the truncation error only down to $h = 0.001$. For smaller h the errors are greater than predicted, and we do not see the decrease in error by a factor of 10 that decreasing h by a factor of 10 should produce for an order-one method.

TABLE 9.2.3
Forward difference approximation of the derivative of $f(x) = \sin x$ at $x = \pi/3$

h	$\Delta_{\text{forw},h} f(x)$	Abs. error	Rel. error
$1.000\text{E} - 0001$	$0.455\,901\,62$	$-4.410\text{E} - 0002$	$-8.820\text{E} - 0002$
$1.000\text{E} - 0002$	$0.495\,660\,31$	$-4.340\text{E} - 0003$	$-8.679\text{E} - 0003$
$1.000\text{E} - 0003$	$0.499\,546\,56$	$-4.534\text{E} - 0004$	$-9.068\text{E} - 0004$
$1.000\text{E} - 0004$	$0.499\,486\,98$	$-5.130\text{E} - 0004$	$-1.026\text{E} - 0003$
$1.000\text{E} - 0005$	$0.500\,679\,08$	$6.791\text{E} - 0004$	$1.358\text{E} - 0003$
$1.000\text{E} - 0006$	$0.476\,837\,22$	$-2.316\text{E} - 0002$	$-4.633\text{E} - 0002$
$1.000\text{E} - 0007$	$0.596\,046\,51$	$9.605\text{E} - 0002$	$1.921\text{E} - 0001$
$1.000\text{E} - 0008$	$0.000\,000\,00$	$-5.000\text{E} - 0001$	$-1.000\text{E} + 0000$
$1.000\text{E} - 0009$	$0.000\,000\,00$	$-5.000\text{E} - 0001$	$-1.000\text{E} + 0000$

Since single-precision arithmetic has seven- to eight-digit accuracy, it may seem surprising that the forward difference approximation is at best accurate to about three digits and that the greatest accuracy is achieved at a relatively large $h \sim 10^{-3}$. The reason for this is that as h gets small, the actual values of $f(x + h)$ and $f(x)$ approach each other, but the computed values in general do not because of rounding error in the computation. Indeed, suppose that the value of f in the forward difference formula is subject to an error of magnitude δ. Then in the worst case, where the errors in computing $f(x + h)$ and $f(x)$ have opposite signs, we see that

$$(\Delta_{\text{forw},h} f)_{\text{comp}} = \frac{[f(x + h) + \delta] - [f(x) - \delta]}{h} = \Delta_{\text{forw},h} f + \frac{2\delta}{h} \quad (9.2.6)$$

In Example 9.2.2 the imprecision in the value of f is $\delta \simeq \mu f(x) = 2^{-24} \sin (\pi/3) \simeq 5 \times 10^{-8}$, where $\mu = 2^{-24}$ is the machine unit for single-precision arithmetic. Thus we see that for $h \geq 10^{-4}$, it is $e_{\text{round}}(h) := 2\delta/h \simeq 10^{-7}/h \geq 10^{-3}$, rather than $e_{\text{trunc}}(h) \simeq h/2 \leq 5 \times 10^{-5}$, that is the dominant source of error. In a practical problem where the values of f might be determined by experimental measurement, and thus δ would be substantially larger, the uncertainty in the value of f would place very severe constraints upon how small h should be chosen.

We now have the bound

$$e_{\text{tot}}(h) := e_{\text{trunc}}(h) + e_{\text{round}}(h)$$

$$\leq E(h) := \frac{1}{2} \max_{\xi \in I} |f''(\xi)| h + \frac{2\delta}{h} \quad (9.2.7)$$

for the total error made in approximating the first derivative $f'(x)$ by the forward difference quotient. Here I is some interval containing the points x and $x + h$. Since $\lim_{h \to \infty} E(h) = \infty$ and $\lim_{h \to 0} E(h) = \infty$, it follows that $E(h)$ must have a minimum value for some intermediate h. By finding this h we can get a rough estimate of the h that minimizes e_{tot}. Let $M_2 := \max_{\xi \in I} |f''(\xi)|$. Upon setting

$$E'(h) = \frac{1}{2} M_2 - \frac{2\delta}{h^2} = 0 \quad (9.2.8)$$

and solving for h, we obtain

$$h_{\text{min}} = 2\sqrt{\frac{\delta}{M_2}} \qquad E(h_{\text{min}}) = 2\sqrt{\delta M_2} \quad (9.2.9)$$

For $\delta \simeq 5 \times 10^{-8}$ and $M_2 = \max_{[\pi/3, \pi/3+0.1]} f''(x) \simeq 0.91$, this gives $h_{\text{min}} \simeq 5 \times 10^{-4}$ as an estimate of the optimal choice of h and $E(h_{\text{min}}) \simeq 4 \times 10^{-4}$ as the estimate for the total error in Example 9.2.2. This is in good agreement with what is observed in Table 9.2.3.

Central Difference Quotient

Since any numerical approximation to the derivative will involve a quotient of two small quantities, it seems unlikely that the contribution to the total error due to rounding can be reduced substantially. Thus the only way that we can get more accurate approximations of the first derivative is to find a method that is of higher order and thus which, we hope, will have less truncation error for a given h.

The *central difference quotient* is defined by

$$\Delta_{\text{cent},h} f(x) := \frac{f(x+h) - f(x-h)}{2h} \tag{9.2.10}$$

Note that it is the average of the forward difference quotient and the backward difference quotient

$$\Delta_{\text{back},h} f(x) := \frac{f(x) - f(x-h)}{h}$$

Since when $f''(x) \neq 0$ these two approximations to the slope of f at x bracket the true slope $f'(x)$, there is reason to believe that the average of the two will be a more accurate approximation. See Fig. 9.2.4. Indeed, from Taylor's theorem,

$$\Delta_{\text{cent},h} f(x) = \frac{1}{2h} \left[f(x) + f'(x)h + \frac{f''(x)}{2!} h^2 + \frac{f'''(\xi_1)}{3!} h^3 \right]$$

$$- \frac{1}{2h} \left[f(x) - f'(x)h + \frac{f''(x)}{2!} h^2 - \frac{f'''(\xi_2)}{3!} h^3 \right]$$

$$= f'(x) + \frac{f'''(\xi_1) + f'''(\xi_2)}{12} h^2 \tag{9.2.11}$$

for $x \leq \xi_1 \leq x + h$, $x - h \leq \xi_2 \leq x$, and thus it is easily seen that for the central difference quotient, $e_{\text{trunc}}(h) \leq (M_3/6)h^2$, where $M_3 := \max_{\xi \in I} |f'''(\xi)|$. Hence the central difference quotient is an order-two approximation to the first deriva-

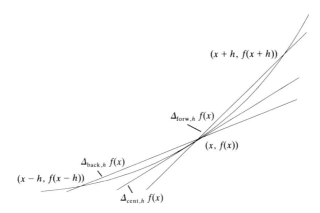

$(x + h, f(x + h))$

$\Delta_{\text{forw},h} f(x)$

$(x, f(x))$

$\Delta_{\text{back},h} f(x)$

$(x - h, f(x - h))$

$\Delta_{\text{cent},h} f(x)$

FIGURE 9.2.4

tive. If δ is the imprecision in the value of f, then $e_{\text{round}} \leq \delta/h$. An argument similar to that leading to (9.2.9) gives

$$h_{\min} = \sqrt[3]{\frac{3\delta}{M_3}} \qquad E(h_{\min}) = \frac{3}{2}\sqrt[3]{\frac{M_3\delta^2}{3}} \qquad (9.2.12)$$

as an estimate of the h that minimizes total error and of the magnitude of the total error.

Example 9.2.5. Table 9.2.6 shows the approximations to $f'(x)$ given by $\Delta_{\text{cent},h}\,f(x)$ for the function of Example 9.2.2, $f(x) = \sin x$, at $x = \pi/3$.

If we restrict $h \leq 0.1$, then $M_3 = \max\limits_{[\pi/3-0.1,\pi/3+0.1]} |-\cos \xi| = 0.58$. Using single-precision arithmetic ($\delta \approx 5 \times 10^{-8}$), Eq. (9.2.12) predicts that $h_{\min} = 6 \times 10^{-3}$ minimizes the total error and that $e_{\text{tot}}(h_{\min}) \approx 1.2 \times 10^{-5}$.

TABLE 9.2.6
Central difference approximations to $f'(\pi/3)$ for $f(x) = \sin x$

h	$\Delta_{\text{cent},h}\,f(x)$	Abs. error	Rel. error
$1.000\text{E} - 0001$	$0.499\ 167\ 14$	$-8.328\text{E} - 0004$	$-1.666\text{E} - 0003$
$1.000\text{E} - 0002$	$0.499\ 993\ 56$	$-6.412\text{E} - 0006$	$-1.282\text{E} - 0005$
$1.000\text{E} - 0003$	$0.500\ 023\ 42$	$2.345\text{E} - 0005$	$4.690\text{E} - 0005$
$1.000\text{E} - 0004$	$0.500\ 083\ 03$	$8.305\text{E} - 0005$	$1.661\text{E} - 0004$
$1.000\text{E} - 0005$	$0.500\ 679\ 08$	$6.791\text{E} - 0004$	$1.358\text{E} - 0003$
$1.000\text{E} - 0006$	$0.476\ 837\ 22$	$-2.316\text{E} - 0002$	$-4.633\text{E} - 0002$
$1.000\text{E} - 0007$	$0.596\ 046\ 51$	$9.605\text{E} - 0002$	$1.921\text{E} - 0001$
$1.000\text{E} - 0008$	$0.000\ 000\ 00$	$-5.000\text{E} - 0001$	$-1.000\text{E} + 0000$
$1.000\text{E} - 0009$	$0.000\ 000\ 00$	$-5.000\text{E} - 0001$	$-1.000\text{E} + 0000$

General Approach to Obtaining Difference Formulas

Suppose that we are given a set of data points (x_i, f_i), $i = 0, \ldots, n$, giving the values of an unknown function f at evenly spaced points $x_0, x_1, \ldots, x_n$ a distance h apart and that we wish to estimate the derivative of f at these same points. The central difference quotient gives us a second-order approximation to the derivative at the points $x_1, \ldots, x_{n-1}$ but cannot be used to estimate the derivatives at the endpoints x_0 and x_n. The forward and backward difference quotients can be used to get approximations at these points, but these are order-one approximations and their accuracy might then be considerably less than those at the interior points.

To find second-order formulas for f' at the endpoints, we start from the three-point Newton divided difference representation of f,

$$f(x) = f[x_0] + (x - x_0)f[x_0, x_1] + (x - x_0)(x - x_1)f[x_0, x_1, x_2]$$
$$+ (x - x_0)(x - x_1)(x - x_2)f[x_0, x_1, x_2, x] \qquad (9.2.13)$$

See Sec. 4.3. The last term in (9.2.13) is the remainder term. It will be used to obtain an estimate of the truncation error. Upon differentiating this formula we obtain

$$f'(x) = f[x_0, x_1] + [(x - x_0) + (x - x_1)]f[x_0, x_1, x_2] + [(x - x_0)(x - x_1)$$
$$+ (x - x_0)(x - x_2) + (x - x_1)(x - x_2)]f[x_0, x_1, x_2, x]$$
$$+ (x - x_0)(x - x_1)(x - x_2) \frac{d}{dx} f[x_0, x_1, x_2, x] \qquad (9.2.14)$$

Let $x_0 = x$, $x_1 = x + h$, and $x_2 = x + 2h$. Then

$$f[x_0, x_1] := \frac{f[x_0] - f[x_1]}{x_0 - x_1} = \frac{f(x + h) - f(x)}{h}$$

$$f[x_1, x_2] := \frac{f[x_1] - f[x_2]}{x_1 - x_2} = \frac{f(x + 2h) - f(x + h)}{h} \qquad (9.2.15)$$

$$f[x_0, x_1, x_2] := \frac{f[x_0, x_1] - f[x_1, x_2]}{x_0 - x_2} = \frac{f(x + 2h) - 2f(x + h) + f(x)}{2h^2}$$

Recall [Eq. (4.3.13)] that $f[x_0, x_1, x_2, x] = f'''(\xi)/3!$ for some ξ contained in the smallest interval containing x_0, x_1, x_2, and x. Substituting all this into (9.2.14) gives

$$f'(x) = \frac{1}{2h}[-3f(x) + 4f(x + h) - f(x + 2h)] + \frac{h^2}{3} f'''(\xi) \qquad 0 \le \xi \le 2h$$
$$(9.2.16)$$

The remainder term indicates that this is indeed an $O(h^2)$ approximation to the derivative.

The formula given by (9.2.16) is suitable for approximation of the derivative at a left endpoint of an interval. A second-order formula for the derivative at the right endpoint can be obtained simply by replacing h by $-h$. Table 9.2.7 summarizes the second-order approximation formulas. Newton divided difference formulas can also be used to obtain the central difference formula (Exercise 7).

TABLE 9.2.7
Second-order difference formulas for f'

	$f'(x)$	Error
Interior	$\frac{1}{2h}(f(x + h) - f(x - h))$	$-\frac{h^2}{6} f^{(3)}(\xi)$
Left end	$\frac{1}{2h}(-3f(x) + 4f(x + h) - f(x + 2h))$	$\frac{h^2}{3} f^{(3)}(\xi)$
Right end	$\frac{1}{2h}(3f(x) - 4f(x - h) + f(x - 2h))$	$\frac{h^2}{3} f^{(3)}(\xi)$

By using divided difference formulas with more points, higher-order approximations can be found. However, these are of little use unless the tabulated values of the function are highly accurate, as the next example indicates.

Example 9.2.8. The second column of Table 9.2.9 shows the values of $f(x) = \sin x$, rounded to two decimal places, at the points $x_i = 0, 0.25, 0.50, 0.75, 1.00$. The remaining columns show the resulting divided differences through order four. The values shown in parentheses are the differences, displayed to four decimal places, that result when values of $\sin x$ which are accurate to 10 digits are used in all computations. As can be seen the effects of the rounding are propagated and compounded as we move right in the table.

To use this table to obtain higher-order approximations, note that, neglecting the remainder term,

$$f(x) = f[x_2] + (x - x_2)f[x_2, x_3] + (x - x_2)(x - x_3)f[x_1, x_2, x_3]$$
$$+ (x - x_1)(x - x_2)(x - x_3)f[x_1, x_2, x_3, x_4]$$
$$+ (x - x_1)(x - x_2)(x - x_3)(x - x_4)f[x_0, x_1, x_2, x_3, x_4]$$
$$\tag{9.2.17}$$

Upon differentiating this equation and setting $x_0 = x - 2h$, $x_1 = x - h$, $x_2 = x$, $x_3 = x + h$, and $x_4 = x + 2h$, we obtain the following approximation:

$$f'(x) = f[x, x + h] - hf[x - h, x, x + h] - h^2 f[x - h, x, x + h, x + 2h]$$
$$+ 2h^3 f[x - 2h, x - h, x, x + h, x + 2h]$$
$$\tag{9.2.18}$$

TABLE 9.2.9
Divided differences for $f(x) = \sin x$

x	$f(x)$			
0.00	0.0000			
	(0.0000)			
		1.0000		
		(0.9896)		
0.25	0.2500		−0.1600	
	(0.2474)		(−0.1231)	
		0.9200		−0.1067
		(0.9281)		(−0.1539)
0.50	0.4800		−0.2400	0.0000
	(0.4794)		(−0.2385)	(0.0198)
		0.8000		−0.1067
		(0.8093)		(−0.1341)
0.75	0.6800		−0.3200	
	(0.6816)		(−0.3390)	
		0.6400		
		(0.6393)		
1.00	0.8400			
	(0.8415)			

The first term of (9.2.18) is the forward difference quotient, an $O(h)$ approximation; the sum of the first two terms is the central difference quotient, an $O(h^2)$ approximation; and the sum of the first three terms and the sum of all four terms are, respectively, $O(h^3)$ and $O(h^4)$ approximations. Setting $x = 0.5$ and $h = 0.25$ gives the following approximations to $f'(0.5) = \cos 0.5 = 0.8776$:

$$O(h): \quad 0.8000$$

$$O(h^2): \quad 0.8000 + 0.25(0.24) = 0.8600$$

$$O(h^3): \quad 0.8600 - 0.25^2(-0.1067) = 0.8667$$

$$O(h^4): \quad 0.8667 + 2(0.25)^3(0.0) = 0.8667 \qquad (9.2.19)$$

Thus it is seen that using the central difference formula results in a fourfold reduction in error, but use of the third-order approximation reduces the error by only another factor of 2, while the fourth-order approximation produces no further improvement. Thus with data of this accuracy there is little reason to use a method of order higher than two.

Formulas for approximating the second and higher derivatives can also be obtained from the Newton divided difference representation. These are given in Table 9.2.10 along with the h^2 term of the error expression.

The interior formula is obtained by differentiating the three-point formula (9.2.14) and setting $x_0 = x - h$, $x_1 = x$, and $x_2 = x + h$. However, the left and right endpoint formulas obtained from (9.2.14) turn out to be only order-one approximations. To get order-two approximations to the second derivative at the endpoints, one must use a four-point formula. See Exercise 7.

TABLE 9.2.10
Second-order difference formulas for f''

	$f''(x)$	Error
Interior	$\dfrac{1}{h^2}\left(f(x + h) - 2f(x) + f(x - h)\right)$	$-\dfrac{h^2}{12} f^{(4)}(\xi)$
Left end	$\dfrac{1}{h^2}\left(2f(x) - 5f(x + h) + 4f(x + 2h) - f(x + 3h)\right)$	$\dfrac{11h^2}{12} f^{(4)}(\xi)$
Right end	$\dfrac{1}{h^2}\left(2f(x) - 5f(x - h) + 4f(x - 2h) - f(x - 3h)\right)$	$\dfrac{11h^2}{12} f^{(4)}(\xi)$

Example 9.2.11. The displacement from equilibrium of a damped spring mass system is governed by the differential equation

$$mx''(t) + bx'(t) + kx(t) = 0 \qquad (9.2.20)$$

Figure 9.2.12 shows the motion of such a system. Let us suppose that the position $x(t)$ can be measured to an accuracy of $\delta = 0.005$. To use (9.2.12) to find a good choice of the stepsize h, we need an estimate of x'''. From Fig. 9.2.12 we see that the quasiperiod of motion is about $T = 3$ time units, the amplitude of the motion

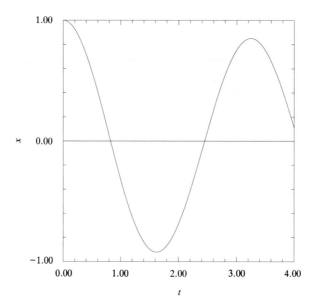

FIGURE 9.2.12

is about 1, and the damping in the system is small. In the absence of damping, solutions to (9.2.20) have the form $x(t) = A\cos(\omega t - \phi)$, where $\omega = 2\pi/T \simeq 2$ and $A \simeq 1$. Thus we make the approximation $|x'''(t)| = A\omega^3 \simeq 8$. This gives $h_{\min} = \sqrt[3]{3\delta/M_3} \simeq 0.1$ and $e_{\text{tot}}(h_{\min}) = \frac{3}{2}\sqrt[3]{M_3\delta^2/3} \simeq 0.06$ for the central difference approximation to the velocity $x'(t)$.

With initial conditions $x(0) = 1$, $x'(0) = 0$, the differential equation (9.2.20), in the case of little damping, has solution $x(t) = e^{-bt/2}[\cos \omega t + (b/2\omega)\sin \omega t]$, where $\omega := \frac{1}{2}\sqrt{4mk - b^2}$. The values of this solution with $m = 1$, $b = 0.1$, $k = 3.75$ were rounded to two decimal places and used to simulate data. Table 9.2.13 shows the accuracy of the approximations obtained with $h = 0.1$ in the interval $0 \le x \le 4$. As can be seen there the accuracy is within the maximum error that was predicted for the central difference formula. At the left and right endpoints the formulas in Table 9.2.7 were used. These formulas are slightly less accurate, both with respect to truncation error and to rounding error.

TABLE 9.2.13
Position and velocity for Example 9.2.11

t	$x(t)$	$v(t)$	Error
0.00	1.00	−0.05	5.0E − 0002
0.50	0.57	−1.55	1.6E − 0002
1.00	−0.32	−1.70	2.9E − 0003
1.50	−0.90	−0.40	1.7E − 0002
2.00	−0.69	1.15	2.6E − 0003
2.50	0.09	1.70	2.8E − 0002
3.00	0.75	0.80	4.7E − 0002
3.50	0.75	−0.75	7.8E − 0003
4.00	0.11	−1.70	1.5E − 0001

EXERCISES 9.2

1. Compute a second-order approximation to f' at each of the three values of x in the table below.

x	0.00	0.25	0.50
$f(x)$	0.00	0.37	0.51

At each value of x estimate the error due to the fact that the values of f are given to only two decimal places.

2. Compute a second-order approximation to f'' at $x = 0$ and $x = 0.5$.

x	0.000	0.500	1.000	1.500
$f(x)$	0.000	0.400	0.462	0.400

At both values of x estimate the error due to the fact that the values of f are given to only three decimal places.

3. Observe that the last two entries in Tables 9.2.3 and 9.2.6 are both zero. Explain in terms of the machine unit μ for single-precision arithmetic why this is.

4. Use formula (9.2.18) to find a third-order approximation to f' at $x = 1.0$, using the data given below.

x	0.800	0.900	1.000	1.100	1.200
$f(x)$	0.577	0.795	1.000	1.195	1.382

Estimate the error due to the fact that the values of f are given to only these decimal places.

5. For each of the functions and points

$$\cos 4x \qquad x = \frac{\pi}{6}$$

$$e^{x/2} \qquad x = 2$$

$$\frac{1}{1+x^2} \qquad x = 1$$

(a) Estimate the size of h that will minimize the total error made by the central difference quotient when four-decimal-place accuracy is used. Also estimate the error at that h.
(b) Approximate f' at the indicated point.
(c) Compare the approximation with the actual value of the derivative.

6. (a) Write a program that will read data on the values of x and $f(x)$, compute approximations to f' and f'', and output x, $f(x)$, $f'(x)$, and $f''(x)$ in four columns.
(b) Test your program by creating a data file that contains the values of the solution to the initial value problem

$$x'' + 0.05x' + 6x = 0 \qquad x(0) = 1 \qquad x'(0) = 0$$
$$(9.2.21)$$

rounded to three decimal places. Choose the spacing between points to be approximately optimal for the central difference formula approximation to velocity.

7. (a) Derive the central difference formula and the truncation error given in Table 9.2.7 from Eq. (9.2.14).
(b) Derive the interior formula and truncation error for f'' given in Table 9.2.10.
(c) Derive the left-hand endpoint formula and truncation error for f'' given in Table 9.2.10.

8. Use the five-point divided difference formula with $x = x_2$ to derive a differentiation rule. What is the order of the truncation error of this rule?

9. (a) Derive the estimates of optimal stepsize and magnitude of error given in (9.2.12) for the central difference formula.
(b) Work out corresponding estimates for:
    i. The left-hand endpoint formula for f' given in Table 9.2.7
    ii. The interior formula for f'' given in Table 9.2.10

10. Rework Example 9.2.8 if the data for $f(x) = \sin x$ is given to three-decimal-place accuracy. Which order approximation gives the most accurate answer?

11. Use the interior approximation formula for f'' to find a formula for $f^{(iv)}$. Use Taylor's theorem to find the order of this approximation.

12. For the boundary value problem

$$x''(t) - 4x(t) = 0 \qquad x(0) = 1 \qquad x(1) = -1$$
$$(9.2.22)$$

replace x'' by its second-order approximation in Table 9.2.10. Find approximations to the unknown solution x at $t = 0.25$, 0.5, and 0.75 by deriving a system of three equations in the three unknowns $x(0.25)$, $x(0.5)$, and $x(0.75)$.

9.3   INTERPOLATORY QUADRATURE

Approximations to the Definite Integral

The most commonly used rules for approximating the definite integral $I = \int_\alpha^\beta f(x)\,dx$ derive from simple geometric reasoning based upon the area interpretation of the integral. A *rectangle rule* (Fig. 9.3.1) simply approximates I, the area beneath f, by a single rectangle of height $f(\bar{x})$, with $\bar{x} \in [\alpha, \beta]$. Thus

$$I \simeq (\beta - \alpha)f(\bar{x}) \qquad (9.3.1)$$

The *trapezoid rule* (Fig. 9.3.2) approximates I as the area beneath a straight-line segment from $(\alpha, f(\alpha))$ to $(\beta, f(\beta))$. Thus

$$I \simeq \frac{\beta - \alpha}{2}\left(f(\alpha) + f(\beta)\right) \qquad (9.3.2)$$

Simpson's rule (Fig. 9.3.3) approximates I as the area beneath a parabola through the three points $(\alpha, f(\alpha))$, $(\nu, f(\nu))$, and $(\beta, f(\beta))$, where

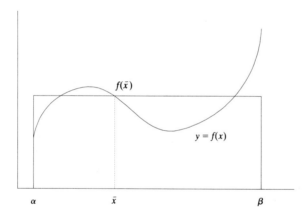

FIGURE 9.3.1

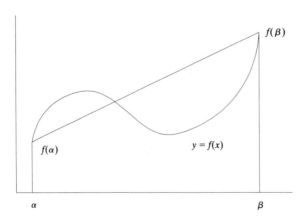

FIGURE 9.3.2

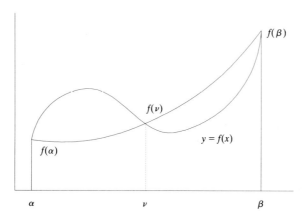

$f(\beta)$

$f(\nu)$

$f(\alpha)$

$y = f(x)$

α ν β **FIGURE 9.3.3**

$\nu := (\alpha + \beta)/2$ is the midpoint of the interval (α, β). This gives the approximation

$$I \simeq \frac{\beta - \alpha}{6} \left[f(\alpha) + 4f(\nu) + f(\beta) \right] \tag{9.3.3}$$

See Exercise 3.

Order of the Methods

It appears geometrically from Figs. 9.3.2 and 9.3.3 that Simpson's rule should give a more accurate approximation than the trapezoid rule in most cases. For rectangle rules the mean value theorem guarantees that there is a choice of the evaluation point $\bar{x}$ for which the approximation will be exact; however, it is clear that there are also choices of $\bar{x}$ that will give answers that are worse than either the trapezoid rule or Simpson's rule.

Example 9.3.4. Table 9.3.5 shows the results of approximating $\int_1^{1.5} (1/x)\, dx = \ln 1.5 \simeq 0.4055$ using a rectangle rule with left endpoint evaluation, a rectangle rule with midpoint evaluation, and the trapezoid and Simpson rules. Note that a rectangle rule with midpoint evaluation turns out to be more accurate than the trapezoid rule for this problem. This is not particularly surprising since the errors made by a rectangular approximation using the height of the function at the midpoint will tend to cancel. See Fig. 9.3.6.

TABLE 9.3.5

Rectangle rule (left endpoint)	0.5000
Rectangle rule (midpoint)	0.4000
Trapezoid rule	0.4167
Simpson's rule	0.4056
Actual	0.4055

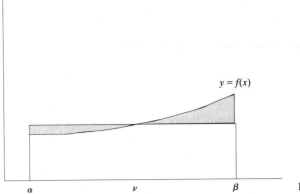

$y = f(x)$

α ν β **FIGURE 9.3.6**

In Definition 9.2.1 the notion of the *order* of a method of approximation was introduced to quantify apparent differences in the accuracy of different methods of approximation. Let us compute the order of the rectangle rule with left endpoint evaluation. Let $I(\theta) = \int_{\alpha}^{\alpha+\theta} f(x)\,dx$ with $\theta := \beta - \alpha$. The error, as a function of the interval width θ, is

$$e(\theta) := I(\theta) - \theta f(\alpha) \qquad (9.3.4)$$

We want to find n and C, independent of θ, for which $|e(\theta)| \leq C\theta^n$. From the fundamental theorem of calculus, $dI/d\theta = f(\alpha + \theta)$. Thus we have

$$e(0) = 0$$
$$e'(\theta) = f(\alpha + \theta) - f(\alpha) \rightarrow e'(0) = 0 \qquad (9.3.5)$$
$$e''(\theta) = f'(\alpha + \theta)$$

From Taylor's theorem

$$e(\theta) = \frac{\theta^2}{2} f'(\alpha + \eta) \qquad (9.3.6)$$

for some $\eta \in (0, \theta)$. If we restrict $\theta \leq \theta_{\max}$, then

$$|e(\theta)| \leq \tfrac{1}{2} \max_{\eta \in [0, \theta_{\max}]} |f'(\alpha + \eta)| \theta^2$$

and thus the rectangle rule with left-endpoint evaluation is an order-two, or $O(\theta^2)$, method, providing that we are integrating a twice continuously differentiable function. Similar arguments based upon the Taylor expansion of the error show that the midpoint and trapezoid rules are $O(\theta^3)$ methods of approximation, while Simpson's rule is an $O(\theta^5)$ method (Exercise 4). The error expressions resulting from the Taylor expansion approach tend to be more complicated than is necessary. We will thus take a different approach and consider the trapezoid and Simpson rules as examples of what are called *Newton-Cotes* methods.

Newton-Cotes Methods

An $(n + 1)$-point Newton-Cotes formula for approximating the definite integral $I = \int_\alpha^\beta f(x)\,dx$ is obtained by replacing the integrand f by the nth-degree Lagrange polynomial that interpolates the values of f at evenly spaced points. If the endpoints α and β of the interval are in the set of interpolating points, then the formula is termed *closed*; otherwise, it is said to be *open*. The trapezoid and Simpson rules are closed formulas, while the rectangle rule with midpoint evaluation is open. In this paragraph we find the general form of a closed method. Open methods are considered in Exercise 7. Let $x_i := \alpha + hi$, $i = 0, \ldots, n$, where $h := (\beta - \alpha)/n$. The Lagrange polynomial (Sec. 4.3)

$$p_n(x) = \sum_{i=0}^n f(x_i) l_i(x)\, dx \qquad \text{where } l_i(x) := \frac{\prod\limits_{j=0, j \neq i}^n (x - x_j)}{\prod\limits_{j=0, j \neq i}^n (x_i - x_j)} \qquad (9.3.7)$$

interpolates the points $(x_i, f(x_i))$, $i = 0, \ldots, n$. The $(n + 1)$-point Newton-Cotes formula is then given by

$$\int_\alpha^\beta f(x)\,dx \simeq \sum_{i=0}^n f(x_i) \int_\alpha^\beta l_i(x)\,dx \qquad (9.3.8)$$

The integrands on the right of (9.3.8) are polynomials and hence explicitly calculable. Indeed, upon making the change of variables $x = \alpha + th$, we obtain

$$\int_\alpha^\beta l_i(x)\,dx = h\lambda_{ni} := h \int_0^n \prod_{j=0, i \neq j}^n \frac{t - j}{i - j}\,dt \qquad (9.3.9)$$

Thus the $(n + 1)$-point Newton-Cotes approximation is

$$\int_\alpha^\beta f(x)\,dx \simeq h \sum_{i=0}^n f(x_i)\lambda_{ni} \qquad (9.3.10)$$

where the coefficients λ_{ni} are independent of α and β.

It is clear on geometric grounds that the two- and three-point Newton-Cotes formulas should yield the trapezoid and Simpson rules. Indeed, we find for $n = 1$ that $\lambda_{10} = \int_0^1 (t - 1)/(0 - 1)\,dt = \frac{1}{2}$ and similarly $\lambda_{11} = \frac{1}{2}$, in which case (9.3.10) gives the trapezoid rule. Likewise, a simple computation yields $\lambda_{20} = \lambda_{22} = \frac{1}{3}$ and $\lambda_{21} = \frac{4}{3}$, and hence (9.3.10) gives Simpson's rule.

Error Estimate for the Trapezoid Rule

We have already mentioned that the trapezoid rule is an $O(h^3)$ approximation to the definite integral; that is, $|E_{\text{trap}}(h)| \leq Ch^3$. It is useful to know more about how the coefficient C depends upon the integrand f.

Theorem 9.3.7. Let $f \in C^2[\alpha, \beta]$. The error that the trapezoid rule makes in estimating $I = \int_\alpha^\beta f(x)\,dx$ is

$$E_{\text{trap}}(f) = -\frac{h^3}{12}\, f''(\xi) \tag{9.3.11}$$

where $h = \beta - \alpha$ and $\xi \in [\alpha, \beta]$.

Proof. From the Lagrange interpolation formula with remainder,

$$f(x) = p_n(x) + \frac{f^{(n+1)}(\xi(x))}{(n+1)!} \prod_{j=0}^{n} (x - x_j) \tag{9.3.12}$$

Upon integrating (9.3.12) we have from (9.3.10) that

$$\int_{\alpha}^{\beta} f(x)\, dx = \sum_{i=0}^{n} f(x_i)\lambda_{ni} + \frac{1}{(n+1)!} \int_{\alpha}^{\beta} f^{(n+1)}(\xi(x)) \prod_{j=0}^{n} (x - x_j)\, dx \tag{9.3.13}$$

For the trapezoid rule ($n = 1$) the remainder term in (9.3.13) is

$$E_{\text{trap}}(f) = \frac{1}{2} \int_{\alpha}^{\beta} f''(\xi(x))(x - \alpha)(x - \beta)\, dx \tag{9.3.14}$$

In the integrand in (9.3.14), $f''(\xi(x))$ is a continuous function of x and $(x - \alpha)(x - \beta)$ is negative on (α, β). The mean value theorem for integrals then can be applied to obtain

$$E_{\text{trap}}(f) = \frac{1}{2} \int_{\alpha}^{\beta} f''(\xi(x))(x - \alpha)(x - \beta)\, dx$$

$$= \frac{1}{2} f''(\xi(\bar{x})) \int_{\alpha}^{\beta} (x - \alpha)(x - \beta)\, dx$$

$$= -\frac{h^3}{12} f''(\xi(\bar{x})) \tag{9.3.15}$$

for some $\bar{x}$, $\xi(\bar{x}) \in [\alpha, \beta]$.

Error Estimate for Simpson's Rule

Theorem 9.3.8. Let $f \in C^4[a, b]$. The error that Simpson's rule makes in estimating $I = \int_{\alpha}^{\beta} f(x)\, dx$ is

$$E_{\text{Simp}}(f) = -\frac{h^5}{90} f^{(4)}(\xi) \tag{9.3.16}$$

where $h = (\beta - \alpha)/2$ and $\xi \in [\alpha, \beta]$.

Proof. From (9.3.13) the error made by approximating the integral over $[\alpha, \beta]$ by Simpson's rule is

$$E_{\text{Simp}}(f) = \frac{1}{6} \int_{\alpha}^{\beta} f^{(3)}(\xi(x))(x - x_0)(x - x_1)(x - x_2)\, dx$$

$$= \int_{\alpha}^{\beta} f[x_0, x_1, x_2, x](x - x_0)(x - x_1)(x - x_2)\, dx \tag{9.3.17}$$

when the divided difference form [Eq. (4.3.13)] of the Lagrange remainder term is used. Let $w(x) := (x - x_0)(x - x_1)(x - x_2)$. In the derivation of the error term

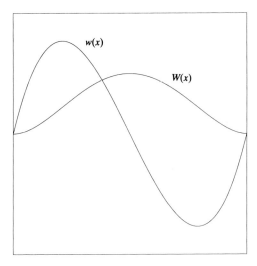

$w(x)$

$W(x)$

FIGURE 9.3.9

for the trapezoid rule we were able to apply the mean value theorem for integrals at this point because the corresponding term was negative. In the present case w changes sign in the interval $[\alpha, \beta]$. However, it is clear from Fig. 9.3.9 that the function $W(x) := \int_\alpha^x w(t)\, dt$ is positive in (α, β). Moreover $W(\alpha) = W(\beta) = 0$. Upon integrating (9.3.17) by parts we obtain

$$E_{\text{Simp}} = -\int_\alpha^\beta W(x) \frac{d}{dx} f[x_0, x_1, x_2, x]\, dx \qquad (9.3.18)$$

The function W does satisfy the hypotheses of the mean value theorem for integrals. Since also $\dfrac{d}{dx} f[x_0, x_1, x_2, x] = f[x_0, x_1, x_2, x, x]$ (see Exercise 8 in Sec. 4.3), we obtain for some $\xi \in (\alpha, \beta)$

$$E_{\text{Simp}}(f) = -f[x_0, x_1, x_2, \xi, \xi] \int_\alpha^\beta W(x)\, dx \qquad (9.3.19)$$

Interchanging the order of integration gives

$$\int_\alpha^\beta W(x)\, dx = \int_\alpha^\beta \int_\alpha^x (t - x_0)(t - x_1)(t - x_2)\, dt\, dx$$

$$= \int_\alpha^\beta \int_t^\beta (t - x_0)(t - x_1)(t - x_2)\, dx\, dt$$

$$= \int_\alpha^\beta (\beta - t)(t - x_0)(t - x_1)(t - x_2)\, dt \qquad (9.3.20)$$

Upon making the change of variables $t = \alpha + \tau h$ in the last integral we obtain

$$\int_\alpha^\beta W(x)\, dx = -h^5 \int_0^2 \tau(\tau - 1)(\tau - 2)^2\, d\tau = \tfrac{4}{15} h^5 \qquad (9.3.21)$$

Since $f[x_0, x_1, x_2, \xi, \xi] = f^{(4)}(\eta)/4!$ for some $\eta \in (\alpha, \beta)$ [see again Eq. (4.3.13)], the error term for Simpson's rule is then seen from (9.3.19) to be (9.3.16).

Higher-Order Newton-Cotes Formulas

An $(n + 1)$-point Newton-Cotes formula is derived by integrating the Lagrange polynomial interpolating $n + 1$ points. Here are the closed Newton-Cotes formulas and error terms for four, five, and nine evenly spaced points.

$n = 3$ *(Newton's $\frac{3}{8}$ rule):*

$$\int_\alpha^\beta f(x) \, dx = \frac{3h}{8} (f_0 + 3f_1 + 3f_2 + f_3) - \frac{3h^5}{90} f^{(4)}(\xi) \qquad (9.3.22)$$

$n = 4$ *(Milne's or Romberg's rule):*

$$\int_\alpha^\beta f(x) \, dx = \frac{2h}{45} (7f_0 + 32f_1 + 12f_2 + 32f_3 + 7f_4) - \frac{8h^7}{945} f^{(6)}(\xi)$$

$$(9.3.23)$$

$n = 8$:

$$\int_\alpha^\beta f(x) \, dx = \frac{4h}{14175} (989f_0 + 5888f_1 - 928f_2 + 10{,}496f_3 - 4540f_4 + 10{,}496f_5$$

$$- 928f_6 + 5888f_7 + 989f_8) - \frac{2368h^{11}}{467{,}775} f^{(10)}(\xi)$$

$$(9.3.24)$$

In the above formulas $f_i = f(a + ih)$.

The approach used to derive the error term for Simpson's rule can be extended to cover the general case of an $(n + 1)$-point Newton-Cotes formula when n is even. The error terms all have the form $E_n = C_n f^{(n+2)}(\xi) h^{n+3}$. A general approach also exists for odd n, but it is more involved and will not be treated here. For odd n the error term has the form $E_n = C_n f^{(n+1)}(\xi) h^{n+2}$. See Issacson and Keller (1966) for a derivation of these results.

Since the order of the error term increases and, for a fixed interval (α, β), the size of h decreases as the number of points in a Newton-Cotes formula increases, higher-order formulas might be expected to produce more accurate approximations. This is not necessarily the case as can be seen in Table 9.3.10, which gives the three-, four-, five- and nine-point approximations to

$$\int_\alpha^\beta \frac{1}{1 + x^2} \, dx = \arctan \beta - \arctan \alpha$$

for $\alpha = 0$, $\beta = 1$ and $\alpha = -5$, $\beta = 5$. Observe that the nine-point approximation for $\alpha = -5$, $\beta = 5$ is worse than the lower-order formulas. The source of the difficulty is that increasing the number of points used in a Lagrange interpolation does not necessarily increase uniformly the accuracy with which the interpolating polynomial approximates the function. This can be seen by examining the graphs of the Lagrange interpolants of the integrand $1/(1 + x^2)$ for different numbers of points in Fig. 4.6.1.

TABLE 9.3.10

Newton-Cotes approximations of $\int_{\alpha}^{\beta} \dfrac{1}{1+x^2}\,dx$

N	Approximation	Error
$\alpha = 0, \beta = 1$		
3	7.833 333 333 4E − 01	−2.064 830 062 4E − 03
4	7.846 153 846 2E − 01	−7.827 787 812 9E − 04
5	7.855 294 117 6E − 01	1.312 483 673 2E − 04
9	7.853 981 684 7E − 01	5.073 161 446 5E − 09
$\alpha = -5, \beta = 5$		
3	6.794 871 794 9E + 00	4.048 070 261 0E + 00
4	2.081 447 963 8E + 00	−6.653 535 700 9E − 01
5	2.374 005 305 0E + 00	−3.727 962 288 5E − 01
9	1.500 488 907 1E + 00	−1.246 312 626 8E + 00

Composite Rules

It is evident that the trapezoid and Simpson rules given above will produce accurate approximations to I only when α and β are close together. As noted in the previous subsection, higher-order Newton-Cotes formulas will not necessarily produce more accurate approximations to an integral.

As a practical matter it is best to use *composite rules* for larger intervals. The interval (a, b) is partitioned into N subintervals (x_{i-1}, x_i), $i = 1, \ldots, N$, of equal width $h := (b - a)/N$, and the rule for a single interval is applied to each subinterval or a grouping of subintervals. For the *composite trapezoid rule*,

$$I \simeq \frac{h}{2} \sum_{i=1}^{N} [f(x_{i-1}) + f(x_i)]$$

$$= \frac{h}{2} \left[f(a) + 2 \sum_{i=1}^{N-1} f(x_i) + f(b) \right] \qquad (9.3.25)$$

To obtain the *composite Simpson rule*, we require the number of intervals N to be an even number and apply (9.3.3) to each of the $N/2$ pairs of subintervals (x_{2i-2}, x_{2i-1}), (x_{2i-1}, x_{2i}). With $h := (b - a)/N$, this gives

$$I \simeq \frac{h}{3} \sum_{i=1}^{N/2} [f(x_{2i-2}) + 4f(x_{2i-1}) + f(x_{2i})]$$

$$= \frac{h}{3} \left[f(a) + 4 \sum_{i=1}^{N/2} f(x_{2i-1}) + 2 \sum_{i=1}^{N/2-1} f(x_{2i}) + f(b) \right] \qquad (9.3.26)$$

To obtain a bound for the error made by the composite trapezoid rule we apply (9.3.11) to each subinterval. Let ξ_j denote the point ξ in (9.3.11) for the jth subinterval. We have

$$\left|\int_a^b f(x)\, dx - \frac{h}{2}\left[f(a) + 2\sum_{i=1}^{N-1} f(x_i) + f(b)\right]\right| \le \frac{h^3}{12}\sum_{j=1}^{N} |f''(\xi_j)|$$

$$\le \frac{h^3}{12} N \max_{a\le x\le b} |f''(x)|$$

$$= \frac{(b-a)h^2}{12} \max_{a\le x\le b} |f''(x)|$$

$$= \frac{(b-a)^3}{12N^2} \max_{a\le x\le b} |f''(x)|$$

$$(9.3.27)$$

We have just shown the following theorem.

Theorem 9.3.11. Let $f \in C^2[a, b]$. The composite trapezoid rule

$$\int_a^b f(x)\, dx \simeq \frac{h}{2}\left[f(a) + 2\sum_{i=1}^{N-1} f(x_i) + f(b)\right] \qquad (9.3.28)$$

makes an error of no more than $(b-a)^3/(12N^2) \max_{a\le x\le b} |f''(x)|$ in approximating the integral. Thus the composite trapezoid rule is order two in the subinterval width h.

A similar analysis can be performed for the composite Simpson rule (Exercise 5).

Theorem 9.3.12. Let $f \in C^4[a, b]$. The composite Simpson rule

$$\int_a^b f(x)\, dx \simeq \frac{h}{3}\left[f(a) + 2\sum_{i=1}^{N/2-1} f(x_{2i}) + 4\sum_{i=1}^{N/2} f(x_{2i-1}) + f(b)\right] \quad (9.3.29)$$

makes an error of no more than $(b-a)^5/(180N^4) \max_{a\le x\le b} |f^{(4)}(x)|$ in approximating the integral. Thus the composite Simpson rule is order four in h.

Example 9.3.13. How many subintervals approximate $\ln 3 = \int_1^3 (1/x)\, dx$ to an accuracy of at least 5×10^{-5} using (a) the composite trapezoid rule and (b) the composite Simpson rule?

Solution
(a) We want

$$E_{C\text{trap}}\left(\frac{1}{x}\right) \le \frac{(3-1)^3}{12N^2} \max_{1\le x\le 3}\left|\frac{2}{x^3}\right| = \frac{4}{3N^2} \le 5 \times 10^{-5} \qquad (9.3.30)$$

which gives $N \ge 163.3$, and thus $N = 164$ is the smallest number of subintervals that ensures the stipulated accuracy.
(b) We want

$$E_{C\text{Simp}}\left(\frac{1}{x}\right) \le \frac{(3-1)^5}{180N^4} \max_{1\le x\le 3}\left|\frac{24}{x^5}\right| \le 5 \times 10^{-5} \qquad (9.3.31)$$

TABLE 9.3.14

Errors made in approximating $\int_1^3 \dfrac{1}{x}\, dx$

N	Trapezoid rule			Simpson's rule		
	Actual	Est.	Last/cur.	Actual	Est.	Last/cur.
10	2.95E − 03	1.33E − 02	· · · · ·	4.83E − 05	7.11E − 05	
20	7.40E − 04	3.33E − 03	3.99E + 00	3.22E − 06	4.44E − 06	1.50E + 01
40	1.85E − 04	8.33E − 04	4.00E + 00	2.05E − 07	2.78E − 07	1.57E + 01
80	4.63E − 05	2.08E − 04	4.00E + 00	1.28E − 08	1.74E − 08	1.59E + 01
160	1.16E − 05	5.21E − 05	4.00E + 00	8.08E − 10	1.09E − 09	1.59E + 01
320	2.89E − 06	1.30E − 05	4.00E + 00	5.09E − 11	6.78E − 11	1.59E + 01
640	7.23E − 07	3.26E − 06	4.00E + 00	0.00E + 00	4.24E − 12	

which gives $N \geq 17.09$. Since $N = 18$ is even, it is the smallest number of subintervals that ensures the stipulated accuracy.

Table 9.3.14 shows the actual errors made by the composite trapezoid and Simpson rules in estimating $\int_1^3 (1/x)\, dx$ along with the error bounds given by Theorems 9.3.11 and 9.3.12. As can be seen the error bound for the trapezoid rule overestimates the true error by about a factor of 4 or 5 for this problem, while the corresponding factor for Simpson's rule is less than 2. In both cases the number of intervals calculated is about twice the number actually needed to attain the specified accuracy.

The error bound in Theorem 9.3.11 predicts that doubling the number of subintervals for the trapezoid rule will decrease the error made by a factor of 4. For Simpson's rule a doubling of N should decrease the error by a factor of 16. The column "*Last/cur.*" in Table 9.3.14 which gives that ratio of the actual errors for $N/2$ and N subintervals indicates that this prediction is borne out very precisely in this example.

Programming the Composite Trapezoid Rule

Algorithm 9.3.15   Composite trapezoid rule

Input   a,b (interval of integration)
         N (number of subintervals)

Define   f(x) (integrand)

h:=(b − a)/N
TrapSum:=(f(a) + f(b))/2
for i = 1 to N − 1
    TrapSum:=TrapSum + f(a + i ∗ h)

Output   h ∗ TrapSum

Algorithm 9.3.15 outlines a program for the composite trapezoid rule. Observe that the summation loop could have been written as follows:

$$x := a$$
$$\text{for } i = 1 \text{ to } N - 1$$
$$\quad x := x + h$$
$$\quad \text{TrapSum} := \text{TrapSum} + f(x)$$

While this approach would be slightly more efficient, saving $N - 1$ multiplications, it might also entail loss of accuracy since the calculation of the mesh points as $x := x + h$ is subject to *cumulative* rounding error, whereas their calculation as $a + ih$ is subject only to the small rounding error entailed in a single addition and multiplication. When the number of subintervals is large and the integrand is changing rapidly near the right endpoint b, inaccuracies can result simply because f is not being evaluated at the correct point. See Exercise 12.

Measuring the Efficiency of a Numerical Integration Algorithm

The best measure of the efficiency of a numerical algorithm is the time it requires to execute on one's particular computer system. Unfortunately, as we saw in Chap. 1 (see especially Table 1.4.1) the relative speeds at which operations such as addition, subtraction, multiplication, division, and function evaluation are performed depend substantively on the hardware and software employed, making general assertions about algorithmic efficiency difficult. However, it is generally true that function evaluation requires at least several times the computational time for any of the four basic arithmetic operations. In a numerical integration routine the number of times the value of the integrand f will need to be computed may be of the same magnitude as the number of other arithmetic operations combined. Thus a simple but effective measure of the efficiency of a numerical integration routine is to count the number of times it computes the integrand f.

Examination of the formulas for the composite trapezoid and Simpson rules with N subintervals indicates that each method requires $N + 1$ evaluations of f. As illustrated in Example 9.3.4 Simpson's rule usually will require substantially fewer subintervals to attain an answer of a given accuracy and consequently will be more efficient. However, since the error made by the trapezoid rule depends upon the magnitude of the second derivative whereas the error made by Simpson's rule depends upon the fourth derivative, this is not certain.

Example 9.3.16. Consider the problem of approximating $\int_0^2 (4t - t^3) e^{t^2} \, dt$ to within 5×10^{-4} using the composite trapezoid and Simpson rules. The anti-derivative

$$\int (4t - t^3) e^{t^2} \, dt = \tfrac{1}{2}(5 - t^2) e^{t^2} + C \tag{9.3.32}$$

can be found analytically, and thus the approximations obtained may be compared with the actual answer.

The second and fourth derivatives of the integrand $f(t):=(4t - t^3)e^{t^2}$ are

$$f''(t) = (18t + 2t^3 - 4t^5)e^{t^2}$$
$$f^{(4)}(t) = (120t + 20t^3 - 80t^5 - 16t^7)e^{t^2}$$

(9.3.33)

Bounds on $0 \leq t \leq 2$ for these two derivatives are not obvious; however, they may be obtained by tabulating the two functions at closely spaced points. Figures 9.3.17 to 9.3.19 show graphs of the two derivatives as well as of f. From these graphs we have

$$\max_{0 \leq t \leq 2} |f''(t)| \leq 4.2 \times 10^3$$
$$\max_{0 \leq t \leq 2} |f^{(4)}(t)| \leq 2.3 \times 10^5$$

(9.3.34)

From Theorems 9.3.11 and 9.3.12 the number of subintervals n that will give approximations of the desired accuracy may be taken to satisfy the inequalities

$$E_{\text{trap}} \leq \frac{2^3 \times 4.2 \times 10^3}{12N^2} \leq 5 \times 10^{-4}$$
$$E_{\text{Simp}} \leq \frac{2^5 \times 2.3 \times 10^5}{180N^4} \leq 5 \times 10^{-4}$$

(9.3.35)

for the trapezoid and Simpson rules, respectively. This gives $N \geq 2367$ for the trapezoid rule and $N \geq 96$ for Simpson's rule. Since as can be seen from Figs. 9.3.18 and 9.3.19 the second and fourth derivatives f increase rapidly near the right endpoint $t = 2$, the bounds used in (9.3.35) are not characteristic of the magnitudes of these derivatives on the whole interval $0 \leq t \leq 2$. Thus it seems possible that fewer subintervals than the number estimated from (9.3.35) are actually required. Indeed, by comparing the trapezoid and Simpson rule approximations for various N with the exact answer computed from (9.3.32), it can be

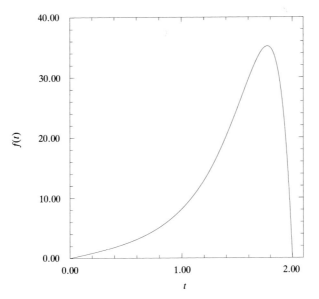

FIGURE 9.3.17

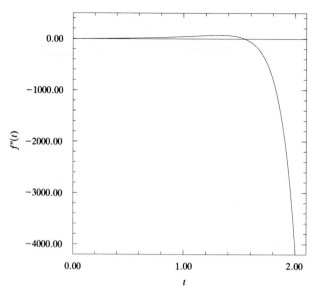

FIGURE 9.3.18

found that $N = 600$ and $N = 50$ will suffice for the trapezoid and Simpson rules, respectively.

Now consider the problem of approximating the function

$$F(x) := \int_0^x (4t - t^3) e^{t^2} \, dt \tag{9.3.36}$$

for $x = 0.02, 0.04, \ldots, 2$. Since we know bounds on f'' and $f^{(iv)}$ that are valid on all of $0 \le x \le 2$, the choices of $N = 2367$ and $N = 96$ for the trapezoid and Simpson

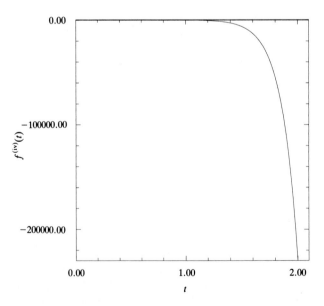

FIGURE 9.3.19

rules, respectively, would suffice for any of the 100 values of x; however, it is clear from Figs. 9.3.18 and 9.3.19 that much smaller choices of N would be sufficient for x in the left part of the interval. A better approach is to note that

$$\max_{0 \leq t \leq x} |f''(t)| \leq (4x^5 + 2x^3 + 18x)e^{x^2}$$

$$\max_{0 \leq t \leq x} |f^{(iv)}(t)| \leq (16x^7 + 80x^5 + 20x^3 + 120x)e^{x^2} \tag{9.3.37}$$

for all $x \in [0, 2]$. These inequalities can be used to make the choice of N appropriate to the value of x and thereby reduce the number of subintervals required for x smaller than 2. See Exercise 15.

Effect of Rounding Error

The error bounds given in Theorems 9.2.11 and 9.2.12 pertain solely to the discretization error, that is, the error resulting from approximating the integrand with an interpolating polynomial. They take no account of rounding error. Rounding error can be shown to be no more than $\frac{2}{3}(b-a)\mu M_0 N$, where μ is the machine unit and $M_0 := \max_{a \leq x \leq b} |f(x)|$ (Exercise 13). Letting $M_2 := \max_{a \leq x \leq b} |f''(x)|$, we have the bound

$$E_{\text{tot}} \leq \frac{(b-a)^3}{12N^2} M_2 + \frac{2}{3}(b-a)\mu M_0 N \tag{9.3.38}$$

Upon differentiating the right-hand side of (9.3.38) with respect to N, setting the result equal to zero, and solving for N we obtain the formula

$$N_{\min} = \sqrt[3]{\frac{(b-a)^2 M_2}{4\mu M_0}} \tag{9.3.39}$$

for the value of N that will minimize the total error. Since both terms in (9.3.38) are bounds for the maximum error and in particular the estimate for the contribution of rounding error likely substantially overestimates the actual rounding error, the value of N given here will only be a rough indication of the point beyond which increasing the number of subintervals may start to produce more rather than less total error. It is important to realize, however, that there is such a point.

> **Example 9.3.20.** Consider the composite trapezoid rule approximation of $\int_1^3 (1/x)\, dx = \ln 3$. For this interval of integration and integrand, $M_0 = 1$ and $M_2 = 2$. At best we can hope to reduce the error in the approximation to about the magnitude of the machine unit μ. For a computation using Turbo Pascal type real without a numeric coprocessor, $\mu = 2^{-39}$. Based on formula (9.3.27) for truncation error alone, this could be accomplished with $N = \sqrt{4/(3\mu)} \approx 800,000$ intervals. On the other hand, formula (9.3.39) predicts that the error will begin to rise after $N_{\min} = \sqrt[3]{2/\mu} \approx 10,000$. Table 9.3.21 shows the actual errors made for various N up to about 150,000 along with the errors estimated by (9.3.27) and (9.3.38) and the ratio of the error for $N/2$ to that of N. It is seen that the bound

TABLE 9.3.21

N	Actual	Est. trunc.	Est. total	Last/cur.
35	2.4179E − 04	1.0884E − 03	1.0884E − 03	
70	6.0463E − 05	2.7211E − 04	2.7211E − 04	3.9989E + 00
140	1.5117E − 05	6.8027E − 05	6.8028E − 05	3.9997E + 00
280	3.7793E − 06	1.7007E − 05	1.7007E − 05	3.9999E + 00
560	9.4482E − 07	4.2517E − 06	4.2531E − 06	4.0000E + 00
1120	2.3620E − 07	1.0629E − 06	1.0656E − 06	4.0000E + 00
2240	5.9057E − 08	2.6573E − 07	2.7116E − 07	3.9996E + 00
4480	1.4763E − 08	6.6433E − 08	7.7298E − 08	4.0004E + 00
8960	3.7289E − 09	1.6608E − 08	3.8339E − 08	3.9590E + 00
17920	9.1131E − 10	4.1521E − 09	4.7614E − 08	4.0918E + 00
35840	2.7649E − 10	1.0380E − 09	8.7961E − 08	3.2961E + 00
71680	1.5643E − 10	2.5950E − 10	1.7411E − 07	1.7674E + 00
143360	−4.5475E − 11	6.4876E − 11	3.4776E − 07	−3.4400E + 00

(9.3.38) substantially overestimates the actual error. However, the departure of the ratio on the last column from its predicted value of 4 indicates that rounding error is of some significance starting around $N = 10,000$, though we do not see an actual increase in the error at this point as predicted by (9.3.39).

For the composite Simpson rule, rounding error will not be expected to pose as much of a problem since the truncation error can be reduced to the error of magnitude of the machine unit with far fewer subintervals. Indeed, in Table 9.3.14 we see that about 600 subintervals sufficed to reduce the actual error and the predicted truncation error to the order of magnitude of the machine unit $\mu \simeq 2 \times 10^{-12}$.

EXERCISES 9.3

1. For each of the following integrands use Theorems 9.3.11 and 9.3.12 to compute the minimum number of subintervals which will ensure that the given accuracy is obtained for the trapezoid and Simpson rules. Take the interval of integration to be $0 \le x \le 3$ in each case.

   (a) $f(x) = \dfrac{1}{1 + 2x}$ tol $= 5 \times 10^{-4}$

   (b) $f(x) = e^{-3x}$ tol $= 5 \times 10^{-5}$

2. Write a program that will find the composite Simpson rule approximation to an integral. Avoid unnecessary loops, unnecessary multiplications where these do not compromise the accuracy of the answer, and any use of an if-then construct.

   Have your program produce a table showing N, the approximation, and the actual error for the integrands and limits of integration given in Exercise 1a and b for $N = 10, 20, \ldots, 200$.

3. Find the parabola $y = ax^2 + bx + c$ that passes through the points $(-h, f(-h))$, $(0, f(0))$, and $(h, f(h))$. Integrate y from $-h$ to h to obtain Simpson's rule (9.3.3).

4. Use Taylor series expansions of the error term as a function of interval width $\theta := \beta - \alpha$ to find the order of (a) the midpoint rule, (b) the trapezoid rule, and (c) Simpson's rule.

5. Derive the error bound for the composite Simpson rule given in Theorem 9.3.12.

6. (a) Verify the coefficients λ_{4i} given in (9.3.23) for the Newton-Cotes formula for $n = 4$.

(b) Follow the derivation of the error term (9.3.16) for Simpson's rule to obtain the error term for $n = 4$ given in (9.3.23).

(c) Use parts (a) and (b) to find a composite Milne's rule and a bound for the error made by this rule.

7. *Open Newton-Cotes formulas.* An n-point open Newton-Cotes formula for approximating $\int_\alpha^\beta f(x)\,dx$ results from integrating the Lagrange interpolating polynomial for the points $(x_i, f(x_i))$, where $x_i = \alpha + ih$, $i = 1, \ldots, n$ with $h = (\beta - \alpha)/(n + 1)$.

(a) Show that the general n-point open Newton-Cotes formula is given by

$$\int_\alpha^\beta f(x)\,dx = h \sum_{i=1}^n f(x_i)\lambda_{ni}$$

$$\text{where } \lambda_{ni} := \frac{\displaystyle\int_0^{n+1} \prod_{j=1, j\neq i}^n (t - j)\,dt}{\displaystyle\prod_{j=1, j\neq i}^n (i - j)}$$

$$(9.3.40)$$

(b) Show that (9.3.40) leads to the approximations

$$n = 2: \int_\alpha^\beta f(x)\,dx \simeq \frac{3h}{2}(f_1 + f_2) \qquad (9.3.41)$$

$$n = 3: \int_\alpha^\beta f(x)\,dx \simeq \frac{4h}{3}(2f_1 - f_2 + 2f_3) \qquad (9.3.42)$$

$$n = 4: \int_\alpha^\beta f(x)\,dx \simeq \frac{5h}{24}(11f_1 + f_2 + f_3 + 11f_4) \qquad (9.3.43)$$

where $f_i = f(x_i)$.

(c) Use Taylor's theorem to show that the lowest-order terms for the error in the approximation schemes given in part (b) are $(3h^3/4)f''(\alpha)$, $(14h^5/45)f^{(4)}(\alpha)$, and $(95h^5/144)f^{(4)}(\alpha)$ for $n = 2$, 3, and 4, respectively.

(d) As with closed Newton-Cotes methods, obtaining simple forms of the error term for open formulas requires different arguments depending upon whether the number of points is even or odd. Use an argument similar to that used to obtain the error term (9.3.16) for Simpson's rule to show that the error made by the three-point approximation

in (9.3.42) is $(14h^5/45)f^{(4)}(\xi)$ for some $\xi \in (\alpha, \beta)$. The error terms for $n = 2$ and 4 turn out to be $(3h^3/4)f''(\xi)$ and $(95h^5/144)f^{(4)}(\xi)$, respectively, but these require different arguments.

(e) Find composite formulas based upon the integration rules given in (9.3.41) to (9.3.43) for $n = 2$ and 3. Use the error terms given in part (d) to find estimates for the error made by these composite rules.

8. The open Newton-Cotes formulas considered in Exercise 7 are useful in approximating integrals with integrands that are undefined at one or both endpoints.

(a) Use each of the three open formulas given in (9.3.41) to (9.3.43) to approximate the integral

$$\int_{-1}^1 \frac{x^6 - 1}{x^2 - 1}\,dx$$

Compare each answer with the actual value of the integral.

(b) The function

$$\text{Si}(x) := \int_0^x \frac{\sin t}{t}\,dt$$

occurs in various applications.

i. Approximate $\text{Si}(0.5)$ using each of the three rules given in (9.3.41) to (9.3.43).

ii. Use one of the composite rules that were derived in Exercise 7e to obtain a graph of $\text{Si}(x)$ for $0 \leq x \leq 2$. Use enough intervals so that no plotted point is in error by more than 0.01.

9. Use the temperature formula (9.1.1) and the composite Simpson rule to generate graphs of temperature u against position x at times $t = 10$ seconds, and $t = 1$ minute for rods composed of each of the following materials:

i. Pure copper: $\rho = 8954\ \text{kg/m}^3$, $c = 0.3831\ \text{kJ/(kg} \cdot {}^\circ\text{C)}$, $k = 386\ \text{W/(m} \cdot {}^\circ\text{C)}$.

ii. Window glass: $\rho = 2700\ \text{kg/m}^3$, $c = 0.84\ \text{kJ/(kg} \cdot {}^\circ\text{C)}$, $k = 0.78\ \text{W/(m} \cdot {}^\circ\text{C)}$.

Take the left-hand side of the rod to be held at 20°C. Your graphs should extend to the right until the temperature at $t = 1$ minute is essentially 0. Take the number of subintervals to be sufficient to ensure an accuracy of at least 5×10^{-3}.

10. Use the formula given in Exercise 7 of Sec. 9.1 and the composite trapezoid rule to calculate the period of an anharmonic oscillator, $f(x) = x - \kappa_3 x^3$. Using a maximum displacement of $x_{max} = 1$, make a table of the periods, accurate to two decimal places, for $\kappa_3 = 0.1, 0.2, \ldots, 0.9$.

11. Use formula (9.1.16) for the period of a pendulum of length $l = 1$ foot to generate a graph of the period against the maximum placement between $\theta_{max} = \pi/10$ and $\theta_{max} = 9\pi/10$. Use the composite trapezoid rule, and compute answers to an accuracy of at least two decimal places.

12. Write a program based upon Algorithm 9.3.15 to compute the trapezoid rule approximation to $\int_0^b 1/(1 - x^2)^2 \, dx$. Use 200,000 subintervals and compute approximations for $b = 0.9, 0.99, 0.999$, and 0.9999. Compare the trapezoid approximations to the actual values of the integral. Modify the program to use the alternative summation loop given just after Algorithm 9.3.15 and again compute the approximations for the same values of b.

13. (a) In this exercise we derive the error bound for the rounding error made in computing the composite trapezoid rule approximation. This estimate was used in (9.3.38). In Algorithm 9.3.15 the trapezoid rule approximation is computed as

$$h\left[\frac{f(a) + f(b)}{2} + \sum_{i=1}^{N-1} f(x_i)\right]$$

Since division by 2 entails no rounding error, it suffices to find a bound for the error made in computing the sum $\Sigma_{i=0}^{N} f_i$ of $N + 1$ numbers f_i. Assume first that $|f_i| \le 1$ for all i and that $N = 2^M$. Show that in the worst case of no guard digits the combined error in adding a number of binary exponent n to a number of binary exponent 0 due to the exponent shift and carry-store operations is $(2^{n+1} - 1)\mu$. Show that the maximum possible error in computing this sum $\Sigma_{i=0}^{N} f_i$ is $\mu \Sigma_{n=0}^{M-1} (2^{n+1} - 1)2^n$, and hence from the formula for summing a geometric series the error is bounded by $(2\mu/3)(4^M - 1) - \mu(2^M - 1)$. Conclude by making the substitution $M = \log_2 N$ that the maximum rounding error

made in calculating $\Sigma_{i=0}^{N} f_i$ is no more than $\frac{2}{3}\mu N^2$. Argue then that the rounding error for the trapezoid rule is bounded by $(2(b-a)N/3)\mu \max|f(x_i)|$. Test the bound $\frac{2}{3}\mu N^2$ by comparing it to the actual errors made in computing the sum $\Sigma_{i=0}^{10000} f$ for each of $f = 0.9, 0.99$, and 0.999.

(b) The bound derived in part (a) can also be used to estimate the rounding error made in Simpson's rule. Assume that the computation is done as indicated in formula (9.3.35). Use the bound in part (a) to find an estimate comparable to (9.3.38) for the number of subintervals that minimizes total error.

14. For $b = 0.5, 1, 1.5, 2, 2.5, 3$ make tables showing the error the trapezoid rule makes in approximating $\int_0^b 1/(1 + x^2) \, dx$ for $N = 1000, 2000, 3000, \ldots, 20,000$ subintervals. In each case note the smallest N for which the error in the approximation increased rather than decreased. In each case compare this value to that predicted by (9.3.39).

15. (a) Write programs that will approximate

$$F(x) = \int_0^x (4t - t^3)e^{t^2} \, dt$$

for $x = 0.02, 0.04, \ldots, 2$ using the trapezoid and Simpson rules. Use the values $N = 2367$ for the trapezoid rule and $N = 96$ for Simpson's rule derived in Example 9.3.16 as the number of subintervals for all values of x. Record the time required for each program to execute.

(b) Modify the programs of part (a) so that the number of subintervals used for a given value of x is determined from the inequalities (9.3.37) and Theorems 9.3.11 and 9.3.12. Have your programs keep track of the total number of function evaluations required to find the approximations at all 100 points. Compare these numbers with the number of evaluations of the integrand required by the approach of part (a). Also record the times required for the programs to execute and compare these to part (a). Do these timings bear out the contention that the number of evaluations of the integrand is a good measure of the efficiency of the program?

9.4   ADAPTIVE QUADRATURE

In the last section we saw that given a desired accuracy to which to approximate a definite integral, we could determine how many subintervals would be required for the composite trapezoid or Simpson rules. The virtue of this approach is that, excluding rounding error, the answer is mathematically certain to be within the specified accuracy. A difficulty with the approach is that a considerable amount of analysis may be required to find the number of subintervals needed. First, the second or fourth derivative must be calculated to use Theorems 9.3.11 or 9.3.12. If a symbolic algebra package is available, this poses no difficulty. Second, a bound on the relevant derivative over the range of integration must be found. As was seen in Sec. 9.1, integrands arising in physical problems frequently depend upon parameters, and the value of the integral may be needed for a number of different sets of the parameters. Finding a bound on the derivative as a function of the parameters in the integrand can be a difficult task.

Because of the difficulties just cited, *adaptive methods* of numerical integration are often preferable. For such methods the user of the program is required only to enter the desired accuracy of the approximation, and no further analysis is required. Here is a simple adaptive algorithm based upon the trapezoid rule.

Algorithm 9.4.1.   Simple adaptive scheme

$\quad$ *Input* $\quad\quad$ a,b (interval of integration)
$\quad\quad\quad\quad\quad\quad$ tol (desired accuracy)

$\quad$ *Define* $\quad\quad$ f(x) (integrand)

$\quad$ N:=1
$\quad$ Current:=(b − a) * (f(a) + f(b))/2 (Trapezoid rule for one interval)

$\quad$ repeat
$\quad\quad\quad$ Last:=Current
$\quad\quad\quad$ N:=2*N
$\quad\quad\quad$ h:=(b − a)/N
$\quad\quad\quad$ TempSum:=(f(a) + f(b))/2
$\quad\quad\quad$ for i = 1 to N − 1
$\quad\quad\quad\quad\quad$ TempSum:=TempSum + f(a + i * h)
$\quad\quad\quad$ Current:=h * TempSum
$\quad$ until |Current − Last| ≤ tol

$\quad$ *Output* $\quad\quad$ Current, the best approximation to the integral.

This scheme, starting with the trapezoid rule for a single subinterval, repeatedly doubles the number of subintervals and compares the trapezoid rule approximation for this number of subintervals with the previous approximation until the two agree to within the specified accuracy tol.

The adaptive method of Algorithm 9.4.1 given above has two defects. First of all, it is not guaranteed to produce an answer that is actually within the given tolerance, as the next example shows.

Example 9.4.2. Let us use Algorithm 9.4.1 to compute $\int_0^{2\pi} \sin^2 x \, dx$. The first approximation found is the trapezoid rule approximation for one subinterval: $\pi(\sin^2 0 + \sin^2 2\pi) = 0$. The second approximation, the trapezoid rule for two subintervals, is $(\pi/2)(\sin^2 0 + 2\sin^2 \pi + \sin^2 2\pi) = 0$, and thus the first two approximations agree to any specified tolerance. The actual value of the integral is π, however.

As we will see, no adaptive method is certain to give an answer to within the given accuracy, and thus this first difficulty is inherent to adaptive methods.

A second criticism of Algorithm 9.4.1 is that it is inefficient. In Sec. 9.3 it was mentioned that much of the computer time consumed in approximating an integral numerically is spent in evaluating the integrand f. For instance, in Example 9.3.13 it was found that 164 subintervals suffice to obtain a trapezoid rule approximation to $\int_1^3 (1/x) \, dx$ that is accurate to within 5×10^{-5}. If Algorithm 9.4.1 is used, however, $2 + 3 + 5 + \cdots + 257 + 513 = 1033$ evaluations of the integrand are necessary, and thus there is a considerable loss of efficiency.

The inefficiency of Algorithm 9.4.1 can be remedied to some extent by making a simple observation. When we compute the trapezoid rule approximation for N and $2N$ subintervals, the computation for $2N$ requires the evaluation of f at all the points used in the approximation for N. If these redundant evaluations can be avoided, a gain in efficiency results. Algorithm 9.4.3 shows how to accomplish this.

Algorithm 9.4.3. Efficient iterative trapezoid scheme

Input   a,b (interval of integration)
            tol (desired accuracy)

Define   f(x) (integrand)

N:=1
Current:=(b − a) * (f(a) + f(b))/2

repeat
      Last:=Current
      N:=2 * N
      h:=(b − a)/N
      OddSum:=0
      for i = 1 to N/2
          OddSum:=OddSum + f(a + (2 * i − 1) * h)
      Current:=Current/2 + h * OddSum
until |Current − Last| ≤ tol

Output   Current as the approximation to the integral.

Algorithm 9.4.3 is easily seen to require only as many function evaluations as are needed for the trapezoid rule approximation for the final value of N used. Thus in the example mentioned above, approximating $\int_1^3 (1/x)\, dx$, only 513 function evaluations are necessary, about one-half as many as required by Algorithm 9.4.1. While this is an improvement, it is possible to do better. In this and succeeding sections we will investigate more sophisticated methods of adaptive quadrature that are substantially more efficient than Algorithm 9.4.3.

An iterative scheme analogous to Algorithm 9.4.3 based upon Simpson's rule can also be devised. See Exercise 3.

Adaptive Simpson Quadrature

Adaptive Simpson quadrature determines the subinterval width according to the local magnitude of the fourth derivative. Different subinterval widths may be used on different parts of the domain of integration, thereby avoiding the use of a small subinterval width where it is unnecessary. Let $S(\alpha, \beta)$ denote the Simpson rule approximation to $I_{\alpha\beta} := \int_\alpha^\beta f(x)\, dx$ for just two subintervals; that is,

$$S(\alpha, \beta) = \frac{\beta - \alpha}{6}\left[f(\alpha) + 4f\left(\frac{\alpha + \beta}{2}\right) + f(\beta)\right] \qquad (9.4.1)$$

With this notation the composite Simpson rule approximation for four subintervals can be written $S(\alpha, \nu) + S(\nu, \beta)$, where $\nu := (\alpha + \beta)/2$ is the midpoint of (α, β).

The building block of adaptive Simpson quadrature is a test that decides upon the basis of the values of $S(\alpha, \beta)$, $S(\alpha, \nu)$, and $S(\nu, \beta)$ whether the four-subinterval approximation $I_{\text{test}} := S(\alpha, \nu) + S(\nu, \beta)$ is accurate to a specified accuracy τ. To derive such a test we note from (9.3.16) that

$$I_{\alpha\beta} = S(\alpha, \beta) - \frac{(2h)^5}{90} f^{(4)}(\xi_{\alpha\beta})$$

$$I_{\alpha\beta} = S(\alpha, \nu) - \frac{h^5}{90} f^{(4)}(\xi_{\alpha\nu}) + S(\nu, \beta) - \frac{h^5}{90} f^{(4)}(\xi_{\nu\beta})$$

$$(9.4.2)$$

where $h = (\beta - \alpha)/4$, $\nu = (\alpha + \beta)/2$, $\xi_{\alpha\beta} \in (\alpha, \beta)$, $\xi_{\alpha\nu} \in (\alpha, \nu)$, and $\xi_{\nu\beta} \in (\nu, \beta)$. Equating the two expressions for $I_{\alpha\beta}$ in (9.4.2) gives

$$S(\alpha, \beta) - S(\alpha, \nu) - S(\nu, \beta) = \frac{(2h)^5}{90} f^{(4)}(\xi_{\alpha\beta}) - \frac{h^5}{90} f^{(4)}(\xi_{\alpha\nu}) - \frac{h^5}{90} f^{(4)}(\xi_{\nu\beta})$$

$$(9.4.3)$$

If α and β are sufficiently close together, then the three fourth-derivative terms in (9.4.3) will have similar magnitudes. Replacing the three terms by a common value f_4 yields

$$S(\alpha, \beta) - S(\alpha, \nu) - S(\nu, \beta) \approx \left[\frac{(2h)^5}{90} - \frac{h^5}{90} - \frac{h^5}{90}\right] f_4 = \frac{h^5}{3} f_4 \qquad (9.4.4)$$

From the second equation in (9.4.2), $I_{\alpha\beta} - S(\alpha, \nu) - S(\nu, \beta) \approx (h^5/45)f_4$. Thus we conclude from (9.4.4) that the approximation $I_{\alpha\beta} \approx S(\alpha, \nu) + S(\nu, \beta)$ will be accurate to within τ when

$$|S(\alpha, \beta) - S(\alpha, \nu) - S(\nu, \beta)| < 15\tau \qquad (9.4.5)$$

Adaptive Simpson quadrature approximates the integral $I_{\alpha\beta} = \int_a^b f(x)\,dx$ to within accuracy tol by a recursive application of the following procedure:

(a) Calculate $S(\alpha, \beta)$, $S(\alpha, (\alpha + \beta)/2)$, and $S((\alpha + \beta)/2, \beta)$.
(b) Calculate $I_{\text{test}} := S(\alpha, (\alpha + \beta)/2) + S((\alpha + \beta)/2, \beta)$.
(c) Accept the approximation $I_{\alpha\beta} \approx I_{\text{test}}$ when $|S(\alpha, \beta) - I_{\text{test}}| < 15\tau$.

At the outset $\alpha = a$, $\beta = b$, and $\tau = $ tol. If the approximation in (b) is not accepted, then the divide-and-compare process is repeated for each of the intervals (α, μ) and (μ, β) separately with $\tau = $ tol/2. This is continued until an acceptable approximation is obtained for each of the subintervals generated by the process. The final approximation I_{ab} is the sum of all the accepted approximations.

Example 9.4.4. Use adaptive Simpson quadrature to approximate $\int_0^1 e^{3x}\,dx$ with an error of no more than tol $= 5 \times 10^{-4}$.

Solution. We will let I denote the running sum of the accepted approximations. Initially $I = 0$.

    Level 1. Interval $= (0, 1)$, $\tau = 5 \times 10^{-4}$.
(a) $S(0, 1) = 6.50205$, $S(0, 0.5) = 1.16247$, $S(0.5, 1) = 5.20985$.
(b) $I_{\text{test}} = 6.37232$.
(c) $|S(0, 1) - I_{\text{test}}| = 1.3 \times 10^{-1} > 15\tau = 7.5 \times 10^{-3}$.
    Reject: repeat process for interval $(0, 0.5)$ and $(0.5, 1)$.
    Level 2. Interval $= (0, 0.5)$, $\tau = 2.5 \times 10^{-4}$.
(a) $S(0, 0.5) = 1.16247$, $S(0, 0.25) = 0.37237$, $S(0.25, 0.5) = 0.78831$.
(b) $I_{\text{test}} = 1.16069$.
(c) $|S(0, 0.5) - I_{\text{test}}| = 1.7 \times 10^{-3} < 15\tau = 3.75 \times 10^{-3}$
    Accept: $I = I + I_{\text{test}} = 1.16069$.
    Level 2. Interval $= (0.5, 1)$, $\tau = 2.5 \times 10^{-4}$.
(a) $S(0.5, 1) = 5.20985$, $S(0.5, 0.75) = 1.66886$, $S(0.75, 1) = 3.53298$.
(b) $I_{\text{test}} = 5.20184$.
(c) $|S(0.5, 1) - I_{\text{test}}| = 8.0 \times 10^{-3} > 15\tau = 3.75 \times 10^{-3}$.
    Reject: repeat process for $(0.5, 0.75)$ and $(0.75, 1)$.
    Level 3. Interval $= (0.5, 0.75)$, $\tau = 1.25 \times 10^{-4}$
(a) $S(0.5, 0.75) = 1.16247$, $S(0.5, 0.625) = 0.37237$, $S(0.625, 0.75) = 0.98898$.
(b) $I_{\text{test}} = 1.66869$.
(c) $|S(0.5, 0.75) - I_{\text{test}}| = 1.68892 \times 10^{-4} < 15\tau = 1.875 \times 10^{-3}$
    Accept: $I = I + I_{\text{test}} = 2.82938$.
    Level 3. Interval $= (0.75, 1)$, $\tau = 1.25 \times 10^{-4}$.
(a) $S(0.75, 1) = 3.53298$, $S(0.75, 0.875) = 1.43896$, $S(0.875, 1) = 2.09367$.

(b) $I_{\text{test}} = 3.53262.$

(c) $|S(0.75, 1) - I_{\text{test}}| = 3.6 \times 10^{-4} < 15\tau = 1.875 \times 10^{-3}.$

Accept: $I = I + I_{\text{test}} = 6.36200.$

The final approximation $I = 6.36200$ differs from the actual value $I = 6.36185$ by about 2×10^{-4}, and thus the specified tolerance was achieved. The number of evaluations of the integrand required in the above process was 13, assuming that care was taken to reuse calculations from the previous level. Use of the inequality in Theorem 9.3.12 shows that 12 subintervals would suffice to attain the given accuracy simply using the composite Simpson rule. This is again 13 function evaluations, and thus adaptive Simpson quadrature is about as efficient but, of course, does not require the preliminary calculation of the number of subintervals by the program user. The iterative Simpson approach of Exercise 3 is less efficient, requiring 33 function evaluations. The actual minimum number of subintervals that are required for Simpson's rule to give an answer of the indicated accuracy is 10, that is, 11 evaluations of the integrand.

Trust, but Verify

Observe that adaptive Simpson quadrature is by no means certain to produce an approximation which meets the specified tolerance. In the derivation three different fourth derivatives were equated. Assuming that the divide-and-compare process has continued to the point at which the interval under consideration is small, such an assumption is reasonable. Unfortunately the test (9.4.5) may lead to acceptance fortuitously when the subinterval width h is far too large to assume that the fourth derivative is approximately constant.

Example 9.4.5. Figure 9.4.6 shows the graph of the function

$$F(x) = \int_0^x \sin^2 t \, dt \tag{9.4.6}$$

on the interval $0 \le x \le 10\pi$. Note that when $x = 4\pi$, the adaptive Simpson algorithm will compare $S(0, 4\pi) = 0$ to $S(0, 2\pi) + S(2\pi, 4\pi) = 0 + 0$ and conclude that the value of the integral is 0. While this answer is obviously wrong, note that $F(x)$ is also seriously in error at values of x near multiples of 2π but that in isolation these answers might not be so conspicuously wrong. As seen in Example 9.4.2, the iterative trapezoid Algorithm 9.4.3 fares no better on this example, nor does Romberg integration, another widely used adaptive method that will be treated in the next section. In fact, any method of numerical integration that estimates the value of the integral on the basis of the values of the integrand at a fixed set of points chosen irrespective of the integrand can easily be deceived. See Exercise 9.

When using methods such as iterative or adaptive Simpson quadrature that evaluate the integrand at evenly spaced points, one must approach periodic and quasiperiodic functions circumspectly. The simplest means of avoiding the problem illustrated in this example is to compute the integral as a sum of separate integrals over each period.

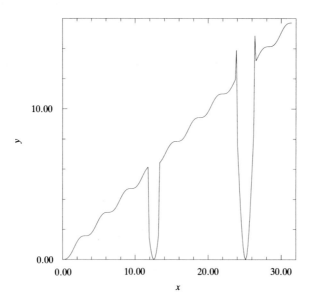

FIGURE 9.4.6

Programming Adaptive Simpson Quadrature

The primary difficulty that a program for adaptive Simpson quadrature must treat is the unpredictable way in which the divide-and-test process will create new subintervals to which the test must be applied recursively. A standard means for dealing with recursive processes is to put the results on a stack and then pull them off later. Program 9.4.7 outlines a Turbo Pascal program for doing this. Completion of two of the procedures in the program is left as Exercise 4. The data for each subinterval that must be saved are the left and right endpoints, the tolerance, and the values of the integrand at the endpoints and midpoint. These are stored in the record variable SimpsonRec. Recent implementations of most programming languages have aggregate data types similar to the Pascal type *record*. If such a type is not available, then several separate stack arrays must be maintained and acted upon by the stacking and unstacking procedures Push and Pop.

Program 9.4.7

```
Program AdaptiveSimpson;

type
      UniVar = function(x: real): real;
      SimpsonRec = record
                       leftend, rightend: real;
                       tau: real;
                       leftf,midf,rightf: real
             end;
```

```
const
    StackSize = 100;
    StackTop: word = 0;

var stack: array[0..StackSize] of SimpsonRec;

Procedure Push(s: SimpsonRec);
begin
    inc(StackTop);
    stack[StackTop]:=s
end;

Procedure Pop(var s: SimpsonRec);
begin
    s:=stack[StackTop];
    dec(StackTop)
end;

Procedure Subdivide (f: UniVar; oldrec: SimpsonRec;
            var leftrec,rightrec: SimpsonRec);
```

Exercise: This procedure takes the information in oldrec and uses it to generate records for the left and right halves of the interval represented by oldrec. Be careful to use the function evaluations already stored in oldrec where possible.

```
Function Accepted(oldrec,leftrec,rightrec: SimpsonRec;
            var itest: real): boolean;
```

Exercise: This function uses the information in the three records oldrec, leftrec, and rightrec to decide whether to accept the four-interval Simpson rule approximation itest.

```
{$F+} Function f(x: real): real; {$F−}
begin
    f:=exp(3*x)
end;

var
    currec,leftrec,rightrec: SimpsonRec;
    a,b,smallint,int, tol: real;

begin
    a:=0; b:=1; tol:=5e−4
    with currec do begin
        leftend:=a; rightend:=b;
        tau:=tol;
        leftf:=f(a); midf:=f((a+b)/2); rightf:=f(b)
    end;
    push(currec);
    int:=0
    repeat
        pop(currec);
        subdivide(f,currec,leftrec,rightrec);
```

```
            if accepted(currec, leftrec, rightrec, smallint) then
                int:=int + smallint
            else begin
                push(leftrec); push(rightrec)
            end;
        until StackTop = 0;
        writeln('Adaptive Simpson approximation is ', int)
end.
```

Example 9.4.8. Table 9.4.9 shows the results of using Program 9.4.7 to tabulate the values of

$$F(x) = \int_0^x (4t - t^3)e^{t^2} \, dt \qquad (9.4.7)$$

to within $tol = 5 \times 10^{-4}$ at 21 evenly spaced points between $x = 0$ and $x = 2$ (see Example 9.3.16). Also shown are the results of applying the iterative Simpson algorithm of Exercise 3. The cumulative function evaluations required for the iterative Simpson approach were 640 compared to 264 for adaptive Simpson. Observe that at $x = 1.40$ the actual error is slightly higher than the specified tolerance. Occasional near-misses are not surprising since the acceptance test (9.4.5) is based upon the approximation $f^{(4)}(x) \equiv$ constant in the subinterval under consideration.

TABLE 9.4.9

	Iterative Simpson		Adaptive Simpson	
x	Error	Evals.	Error	Evals.
0.10	1.3E − 0009	5	1.3E − 0009	5
0.20	8.5E − 0008	5	8.5E − 0008	5
0.30	9.9E − 0007	5	9.9E − 0007	5
0.40	5.8E − 0006	5	5.8E − 0006	5
0.50	2.3E − 0005	5	2.3E − 0005	5
0.60	4.5E − 0006	9	7.2E − 0005	5
0.70	1.2E − 0005	9	1.9E − 0004	5
0.80	2.8E − 0005	9	4.5E − 0004	5
0.90	3.7E − 0006	17	5.9E − 0005	9
1.00	6.8E − 0006	17	1.1E − 0004	9
1.10	1.1E − 0005	17	1.8E − 0004	9
1.20	1.3E − 0005	17	2.4E − 0004	9
1.30	3.8E − 0006	17	1.4E − 0004	9
1.40	−4.9E − 0005	17	−5.1E − 0004	9
1.50	−1.5E − 0005	33	2.0E − 0005	21
1.60	−3.2E − 0006	65	−9.9E − 0006	21
1.70	−9.3E − 0006	65	−1.6E − 0006	25
1.80	−2.5E − 0005	65	−1.2E − 0004	25
1.90	−4.1E − 0006	129	3.2E − 0005	37
2.00	−1.1E − 0005	129	2.3E − 0005	41

Choosing the Size of the Stack

In Program 9.4.7 the variable StackSize was set equal to 100. This is in fact larger than is necessary for any floating-point type in Turbo Pascal. In fact the stack cannot grow much beyond the number of bits in the mantissa of floating-point type used. For by the time the recursion level is equal to the number of bits in the mantissa (or a few more if the arithmetic uses guard digits), the interval under consideration has the form $(c, c(1 + \mu))$, where μ is the machine unit. The calculation of the midpoint of the interval then yields

$$\frac{c \oplus c(1 \oplus \mu)}{2} = \frac{2c}{2} = c \tag{9.4.8}$$

Thus the decision calculation (9.4.5) is

$$S(c, c(1 + \mu)) - S(c, c) - S(c, c(1 + \mu)) = 0$$

and the four-subinterval approximation is accepted. This, of course, by no means implies that the actual error tolerance has been met.

EXERCISES 9.4

1. Do a level-by-level analysis of adaptive Simpson quadrature like that of Example 9.4.4 for

   (a) $\int_0^3 \frac{x}{1 + x^2}\, dx$, tol $= 5 \times 10^{-4}$

   (b) $\int_0^3 xe^{-x^2}\, dx$, tol $= 10^{-4}$

   In each case compare the result obtained with the actual value of the integral.

2. Write a program implementing Algorithm 9.4.1. Test your program on the two examples in Exercise 1.

3. Write a program implementing an iterative Simpson procedure similar to that for the trapezoid rule given by Algorithm 9.4.3. Make sure that the program never evaluates the integrand at the same point twice. Test your program on the two examples in Exercise 1. (*Hint:* It is only necessary to calculate the function at the odd-subscripted nodes at each iteration.)

4. Complete Program 9.4.7 by writing the procedures *Subdivide* and *Accepted*. Test your program on the two examples in Exercise 1.

5. Use the temperature formula (9.1.1) to generate graphs of temperature against position at times $t = 10$ seconds and $t = 1$ minute for each of the sets of the material parameters given in Exercise 9

in Sec. 9.3. Do this using both iterative Simpson quadrature (Exercise 3) and adaptive Simpson quadrature. Install a counter in the call to the function that evaluates the integrand which records the cumulative total function evaluations. Compare the efficiency of the two methods on this basis and on the basis of the total time required to generate the graphs.

6. Use adaptive Simpson quadrature to generate a graph of the solution I of Exercise 7 in Sec. 9.1 from 0 to 4π. Use $\omega = 2\pi$, $E_0 = 100$ volts, $R_0 = 10$ ohms, and $L_0 = 0.5$ henry.

7. Use adaptive Simpson quadrature to graph the period of a pendulum of length 1 foot against its initial displacement from $\theta_{max} = 0$ to $\theta_{max} = 5\pi/6$ using formula (9.1.16).

8. (a) Use the error term for the trapezoid rule given by (9.3.11) to generate an adaptive trapezoid rule quadrature similar to adaptive Simpson quadrature.

   (b) Write a program implementing the adaptive trapezoid rule algorithm that you developed in part (a). Test it (i) on the two examples in Exercise 1 and (ii) by generating the temperature versus position graphs of Exercise 5. Compare the efficiency of your adaptive trapezoid rule program with that of adaptive Simpson quadrature.

9. (a) Explain why adaptive Simpson quadrature obtained a value of 0 for the integral (9.4.6) at $x = 8\pi$.
   (b) Show that if a method of numerical integration approximates the value of the integrand at a set of points $x_0, \ldots, x_n$, then there is a nonnegative function with integral 1 for which the method will return an approximation of 0.
10. Revise the adaptive Simpson quadrature program

9.4.7 so that it takes as input a parameter N which causes the integral to be calculated as

$$\int_a^b f(x)\, dx = \sum_{i=0}^{N-1} \int_{x_i}^{x_{i+1}} f(x)\, dx \qquad (9.4.9)$$

where $\{x_0, \ldots, x_N\}$ is a uniform partition of the interval $[a, b]$ into N subintervals. Use this program to produce a corrected version of Fig. 9.4.6.

9.5 RICHARDSON EXTRAPOLATION AND ROMBERG INTEGRATION

In the last section we studied one effective method of adaptive integration, adaptive Simpson quadrature, which uses a method of fixed order but varies the subinterval width with the magnitude of the fourth derivative of the integrand. An alternative to this is to compute successively higher-order approximations to the integral until two agree to within the specified tolerance. This is the approach of *Romberg integration*. This method is a particularization of a general scheme of variable-order numerical approximation called *Richardson extrapolation* which we discuss first.

Richardson Extrapolation

Let $A(h)$ be a scheme for approximating numerically a quantity $A(0)$ which depends upon a parameter h in such a way that $\lim_{h \to 0} A(h) = A(0)$. Suppose moreover that the error made in approximating $A(0)$ by $A(h)$ has for some $N \geq 1$ a power series expansion in h of the form

$$A(0) - A(h) = \sum_{j=1}^{N} a_{2j} h^{2j} + O(h^{2N+1}) \qquad (9.5.1)$$

Example 9.5.1. The central difference formula. In Sec. 9.2 we considered the central difference formula for approximating the first derivative

$$f'(x) \simeq \nabla_{\text{cent},h} f := \frac{f(x+h) - f(x-h)}{2h} \qquad (9.5.2)$$

Let us show that the expansion of the error made by the central difference formula has the form (9.5.1). For $f \in C^{2N+1}$ we have from Taylor's theorem

$$\nabla_{\text{cent},h} f = \frac{1}{2h}\left[\sum_{j=0}^{2N} \frac{f^{(j)}(x) h^j}{j!} + O(h^{2N+1}) - \sum_{j=0}^{2N} \frac{f^{(j)}(x)(-h)^j}{j!} + O(h^{2N+1}) \right] \qquad (9.5.3)$$

Observe that the even powers of h cancel whereas the odd powers sum together. This gives

$$f'(x) - \nabla_{\text{cent},h} f = -\sum_{k=1}^{N-1} f^{(2k+1)}(x) h^{2k} + O(h^{2N}) \qquad (9.5.4)$$

which is of the form (9.5.1).

The process of Richardson extrapolation consists of successively eliminating terms in the error expansion to produce approximations of higher order. From (9.5.1) we have

$$A(0) = A(h) + \sum_{j=1}^{N} a_{2j}h^{2j} + O(h^{2N+1})$$

$$A(0) = A\left(\frac{h}{2}\right) + \sum_{j=1}^{N} a_{2j}\left(\frac{h}{2}\right)^{2j} + O(h^{2N+1})$$

(9.5.5)

Multiplying the second equation in (9.5.5) by 4 and subtracting the first yields

$$3A(0) = 4A\left(\frac{h}{2}\right) - A(h) + \sum_{j=2}^{N} \left(\frac{1}{2^{2j-2}} - 1\right)a_{2j}h^{2j} + O(h^{2N+1}) \quad (9.5.6)$$

The multiplicative factor 4 was chosen to make the h^2 terms cancel. Equation (9.5.6) shows that

$$A_1(h) := \frac{4A(h/2) - A(h)}{3} \quad (9.5.7)$$

is an $O(h^4)$ approximation to $A(0)$. Observe that we did not actually need to know the value of the coefficient a_2 but only that the error expansion had the form (9.5.1). The process can be continued. From (9.5.6) we know that $A(0) = A_1(h) - \frac{3}{4}a_4h^4 + \cdots$. By eliminating the h^4 term we obtain an order-six approximation (Exercise 9a),

$$A_2(h) = \frac{16A_1(h/2) - A_1(h)}{15} \quad (9.5.8)$$

In general, one obtains recursively the $O(h^{2n+2})$ approximations (Exercise 9b)

$$A_0(h) := A(h)$$

$$A_n(h) := \frac{4^n A_{n-1}(h/2) - A_{n-1}(h)}{4^n - 1} \quad (9.5.9)$$

for the Richardson extrapolation process.

Example 9.5.2. Extrapolation of the central difference formula. If we substitute the central difference approximation (see Example 9.5.1) $A(h) = [f(x+h) - f(x-h)]/2h$ into (9.5.7), we obtain the fourth-order approximation

$$A_1(h) = \frac{f(x-h) - 8f(x-h/2) + 8f(x+h/2) - f(x+h)}{6h} \quad (9.5.10)$$

to the first derivative. In turn, substituting (9.5.10) into (9.5.8) yields the sixth-order approximation

$$A_2(h) = \frac{1}{30h}\left[-f(x-h) + 16f\left(x - \frac{h}{2}\right) - 64f\left(x - \frac{h}{4}\right) + 64f\left(x + \frac{h}{4}\right)\right.$$

$$\left. - 16f\left(x + \frac{h}{2}\right) + f(x+h)\right] \quad (9.5.11)$$

Clearly, using the extrapolation formulas (9.5.9) to generate formulas for higher-order approximations becomes very cumbersome. Fortunately, as we will see, it is not necessary to have explicit formulas available to calculate the numerical values of the higher-order approximations.

The Aitken Δ^2 scheme for accelerating the rate of convergence of a linearly convergent sequence considered in Sec. 2.5 also turns out to be an instance of Richardson extrapolation.

Programming Richardson Extrapolation

Note that to find $A_2(h)$ we must calculate $A_1(h/2)$, which in turn requires the computation of $A(h/4)$. Table 9.5.3 illustrates the dependencies.

TABLE 9.5.3
Richardson extrapolation

Level	$O(h^2)$		$O(h^4)$		$O(h^6)$		$O(h^8)$
0	$A(h)$						
		↘					
1	$A(h/2)$	→	$A_1(h)$				
		↘		↘			
2	$A(h/4)$	→	$A_1(h/2)$	→	$A_2(h)$		
		↘		↘		↘	
3	$A(h/8)$	→	$A_1(h/4)$	→	$A_2(h/2)$	→	$A_3(h)$
...	...		...		...		...

Example 9.5.4. Derivative of $f(x) = \sin x$ at $x = \pi/3$. Let us calculate an $O(h^{10})$ approximation to the derivative $f'(\pi/3) = \cos \pi/3 = \frac{1}{2}$ of $f(x) = \sin x$ at $x = \pi/3$ using extrapolation on the central difference formula $A(h) := [f(x + h) - f(x - h)]/(2h)$. We take $h = \pi$ as the starting stepsize and use seven-digit accuracy for the values of $\sin x$. The first row of Table 9.5.3 is the central difference formula with $h = \pi$:

$$A(\pi) = \frac{\sin(\pi/3 + \pi) - \sin(\pi/3 - \pi)}{2\pi} = 0.000\,000\,0 \qquad (9.5.12)$$

The second line is the central difference formula with $h = \pi/2$ followed by $A_1(\pi)$:

$$A\left(\frac{\pi}{2}\right) = \frac{\sin(\pi/3 + \pi/2) - \sin(\pi/3 - \pi/2)}{\pi} = 0.318\,309\,9$$

$$A_1(\pi) = \frac{4A(\pi/2) - A(\pi)}{3} = 0.424\,413\,2 \qquad (9.5.13)$$

Table 9.5.5 shows the result of continuing this calculation to the fourth level. Observe that whereas the best approximation obtained directly from the central difference formula, $A(\pi/16)$, was in error by about 3×10^{-3}, the final result of the extrapolation process at level 4 is in error only by 10^{-7}.

TABLE 9.5.5

Level	$O(h^2)$	$O(h^4)$	$O(h^6)$	$O(h^8)$	$O(h^{10})$
0	0.000 000 0				
1	0.318 309 9	0.424 413 2			
2	0.450 158 2	0.494 107 6	0.498 753 9		
3	0.487 247 8	0.499 611 0	0.499 977 9	0.499 997 3	
4	0.496 793 4	0.499 975 2	0.499 999 5	0.499 999 9	0.499 999 9

The caveats made in Sec. 9.2 concerning the sensitivity of numerical differentiation to inaccuracy in the values of the function being differentiated remain in effect, however. Table 9.5.6 shows the results of the same extrapolation using data on the sine function that is accurate to only two places. Extrapolation is a very powerful means of refining numerical approximations, but it is not capable of alchemy.

TABLE 9.5.6

Level	$O(h^2)$	$O(h^4)$	$O(h^6)$	$O(h^8)$	$O(h^{10})$
0	0.000 000 0				
1	0.318 309 9	0.424 413 2			
2	0.452 000 0	0.496 563 4	0.501 373 4		
3	0.483 831 0	0.494 441 4	0.494 299 9	0.494 187 6	
4	0.509 295 8	0.517 784 1	0.519 340 3	0.519 737 7	0.519 837 9

An adaptive procedure effecting Richardson extrapolation is programmed by performing the calculations indicated by Table 9.5.3 row by row until the approximations of two successive orders agree to within the specified tolerance. Algorithm 9.5.7 gives the details.

Algorithm 9.5.7. Adaptive procedure for Richardson extrapolation

Define $A_0(h)$ (Numerical scheme to be extrapolated)

Constant   MaxLevel (Maximum permissible table depth)

Declare   Arrays LastRow() and CurrRow() of dimension
             MaxLevel + 1

Input   h (beginning step size)
             tol (desired error tolerance)

Level := 0
CurrRow(0) := $A_0(h)$

Repeat
    h := h/2
    LastRow() := CurrRow()
    CurrRow(0) := $A_0(h)$
    Level := Level + 1
    Coeff := 1 [The coefficient 4^j in (9.5.9)]

```
        for j = 1 to Level
            Coeff:=4 * Coeff
            CurrRow(j):=(Coeff * CurrRow(j − 1) − LastRow(j − 1))/(Coeff − 1)
    until |CurrRow(Level) − LastRow(Level − 1)| < tol or Level > MaxLevel
```

Output   Answer: CurrRow(Level)
              Warning that the tolerance may not have been achieved if
              Level exceeded MaxLevel.

In Algorithm 9.5.7 the use of the array LastRow is for the sake of clarity and is not essential. The algorithm could simply have successively overwritten the elements of CurrRow as the extrapolation proceeded along each row. See Exercise 8.

Euler-Maclaurin Summation Formula

Observe that while in (9.5.1) we assumed that the error expansion had only even powers of h, this is not strictly necessary. The process could be carried out if the error expansion was of the form

$$A(0) − A(h) = \sum_{j=1}^{N} a_j h^j + O(h^{N+1}) \tag{9.5.14}$$

however, at each level of the extrapolation table the order of the approximation on the diagonal would be only one greater than on the succeeding level rather than two greater for the extrapolation table for (9.5.1). See Exercise 3. For this reason extrapolation is more effective when the underlying method of approximation $A(h)$ has an error expansion of the form (9.5.1).

Romberg integration is Richardson extrapolation applied to the trapezoid rule for numerical integration. The reason for the choice of the trapezoid rule as the base method of approximation is that its error expansion is of the form (9.5.1); that is, it has only even powers of h. This is the content of the *Euler-Maclaurin summation formula*, which is used as well in other branches of mathematics such as number theory. The formula involves *Bernoulli numbers* which will be discussed shortly.

Theorem 9.5.8. The Euler-Maclaurin summation formula. For $f \in C^{2n+1}[a, b]$,

$$\int_a^b f(x)\, dx = h \sum_{i=0}^{N-1} f(x_i) + \frac{h}{2}[f(b) − f(a)]$$

$$− \sum_{j=1}^{n} \frac{B_{2j} h^{2j}}{(2j)!} [f^{(2j−1)}(b) − f^{(2j−1)}(a)] + O(h^{2n+1}) \tag{9.5.15}$$

where $x_i = a + ih$, $i = 0, \ldots, N$, with $h = (b − a)/N$, and the B_{2j} are Bernoulli numbers.

The first two terms on the right-hand side of (9.5.15) together constitute the composite trapezoid rule. Hence the Euler-Maclaurin formula states that the error expansion for the trapezoid rule approximation to a definite integral has the form (9.5.1):

$$E_{C\text{trap}}(f) = -\sum_{j=1}^{n} \frac{B_{2j}h^{2j}}{(2j)!} [f^{(2j-1)}(b) - f^{(2j-1)}(a)] + O(h^{2n+1}) \quad (9.5.16)$$

Bernoulli Numbers

The *Bernoulli numbers* are defined by

$$B_j := \lim_{t \to 0} \frac{d^j}{dt^j} \left(\frac{t}{e^t - 1} \right) \quad j = 0, 1, 2, \dots \quad (9.5.17)$$

From this definition they are easily seen to be the numerators of the coefficients of the Maclaurin expansion

$$\frac{t}{e^t - 1} = \sum_{j=0}^{\infty} \frac{B_j}{j!} t^j \quad (9.5.18)$$

The Bernoulli numbers may be calculated from the following theorem.

Theorem 9.5.9. The Bernoulli numbers satisfy the recurrence relation

$$B_0 = 1 \qquad B_j = -\frac{1}{j+1} \sum_{k=0}^{j-1} \binom{j+1}{k} B_k \qquad j = 1, 2, \dots \quad (9.5.19)$$

Proof. From (9.5.18) we have

$$t = \left(\sum_{j=0}^{\infty} \frac{B_j}{j!} t^j \right)(e^t - 1)$$

$$= \left(\sum_{j=0}^{\infty} \frac{B_j}{j!} t^j \right)\left(\sum_{k=1}^{\infty} \frac{t^k}{k!} \right)$$

$$= \sum_{j=0}^{\infty} \sum_{k=0}^{j-1} \frac{B_k t^k}{k!} \frac{t^{j-k}}{(j-k)!}$$

$$= \sum_{j=1}^{\infty} t^j \sum_{k=0}^{j-1} \frac{B_k}{k!(j-k)!} \quad (9.5.20)$$

Equating like powers of t gives $B_0 = 1$ and

$$\sum_{k=0}^{j-1} \frac{B_k}{k!(j-k)!} = 0$$

for $j = 2, 3, \dots$. Solving for B_{j-1} yields

$$B_{j-1} = -(j-1)! \sum_{k=0}^{j-2} \frac{B_j}{k!(j-k)!} = -\frac{1}{j} \sum_{k=0}^{j-2} \binom{j}{k} B_k \quad (9.5.21)$$

which is (9.5.19).

From (9.5.19) it can be verified that $B_0 = 1$, $B_1 = -\frac{1}{2}$, $B_2 = \frac{1}{6}$, $B_3 = 0$, and $B_4 = -\frac{1}{30}$ (Exercise 11a). It turns out that $B_j = 0$ for odd $j \geq 3$ (Exercise 11b). Algorithm 9.5.10 uses the recurrence (9.5.19) to calculate Bernoulli numbers. The binomial coefficients are generated from Pascal's triangle.

Algorithm 9.5.10. Calculation of the Bernoulli numbers

Constant   NMax (subscript of largest Bernoulli number to be calculated)

Declare   bc(,) ((*NMax* + 1) × (*NMax* + 1) array of binomial coefficients)
Ber() (array of Bernoulli numbers)

Generate binomial coefficients from Pascal's triangle:
for i = 1 to NMax + 1
bc(i, 0):=1, bc(i, i):=1
for j = 1 to i − 1
bc(i, j):=bc(i − 1, j − 1) + bc(i − 1, j)

Calculate Bernoulli numbers from (9.5.19):
for i = 1 to NMax
Ber(i):=0
for j = 0 to i − 1
Ber(i):=Ber(i) + bc(i + 1, j) Ber(j)
Ber(i):=−Ber(i)/(i + 1)

Output: array Ber()

Table 9.5.11 shows the first 21 Bernoulli numbers as determined by Algorithm 9.5.10. As is evident from examining the odd-subscripted Bernoulli numbers, there is a loss of accuracy of about 10^{-9} by $n = 19$ due to the fact that errors compound in a recursive calculation. Beyond $n = 20$ the even-subscripted Bernoulli numbers continue to ascend in magnitude and eventually become quite large. For instance, $B_{60} \simeq -2.1 \times 10^{34}$ (see Exercise 12).

TABLE 9.5.11

n	B_n	n	B_n
0	$1.00000E + 00$	11	$6.06330E - 13$
1	$-5.00000E - 01$	12	$-2.53114E - 01$
2	$1.66667E - 01$	13	$-3.53404E - 11$
3	$0.00000E + 00$	14	$1.16667E + 00$
4	$-3.33333E - 02$	15	$4.36557E - 11$
5	$6.06330E - 13$	16	$-7.09216E + 00$
6	$2.38095E - 02$	17	$-5.17401E - 10$
7	$-5.68434E - 13$	18	$5.49712E + 01$
8	$-3.33333E - 02$	19	$-1.49012E - 09$
9	$-5.27507E - 12$	20	$-5.29124E + 02$
10	$7.57576E - 02$		

Note that since $B_2 = \frac{1}{6}$, the first term in the error expansion (9.5.16) is $(-h^2/12)[f'(b) - f'(a)]$. An application of the mean value theorem gives $f'(b) - f'(a) = (b - a)f''(\xi)$ for some $\xi \in (a, b)$. This is in agreement with Theorem 9.3.7 with $h = b - a$.

Proof of the Euler-Maclaurin Summation Formula

Let

$$I(f) := \int_a^b f(x)\, dx \qquad R_N(f) := h \sum_{i=0}^{N-1} f(x_i) \qquad E_N(f) := I(f) - R_N(f)$$

Observe that $R_N(f)$ is simply the Riemann sum approximation to $I(f)$ for N subintervals with left endpoint evaluation and that $E_N(f)$ is the error made by this approximation. We will show that for $f \in C^{m+1}[a, b]$,

$$E_N(f) = -\sum_{j=1}^{m} \frac{B_j h^j}{j!} I(f^{(j)}) + O(h^{m+1}) \tag{9.5.22}$$

Upon setting $m = 2n$ and noting that $B_1 = -\frac{1}{2}$, $B_{2j+1} = 0$ for $j \geq 1$, and $I(f^{(j)}) = f^{(j-1)}(b) - f^{(j-1)}(a)$, this becomes the Euler-Maclaurin formula (9.5.15).

Note that

$$E_N(f) = \sum_{i=0}^{N-1} \left[\int_{x_i}^{x_i+h} f(x)\, dx - hf(x_i) \right] =: \sum_{i=0}^{N-1} e_i(h) \tag{9.5.23}$$

Let us first find the Maclaurin series expansion for the error $e_i(h)$ on the ith subinterval with respect to the subinterval width h. Let $f \in C^{m+1}[a, b]$. Immediately $e_i(0) = 0$, and since $e_i'(h) = f(x_i + h) - f(x_i)$, $e_i'(0) = 0$ also. From then on $e_i^{(j)}(0) = f^{(j-1)}(x_i)$. This gives the Maclaurin expansion

$$e_i(h) = \sum_{j=2}^{m+1} \frac{h^j}{j!} f^{(j-1)}(x_i) + \frac{h^{m+2}}{(m+2)!} f^{(m+1)}(\xi_i) \tag{9.5.24}$$

where $\xi_i \in (x_i, x_{i+1})$. Substituting (9.5.24) into (9.5.23),

$$E_N(f) = \sum_{i=0}^{N-1} \sum_{j=2}^{m+1} \frac{h^j}{j!} f^{(j-1)}(x_i) + \sum_{i=0}^{N-1} \frac{h^{m+2}}{(m+2)!} f^{(m+1)}(\xi_i) \tag{9.5.25}$$

Now

$$\left| \sum_{i=0}^{N-1} \frac{h^{m+2}}{(m+2)!} f^{(m+1)}(\xi_i) \right| \leq \frac{h^{m+1}}{(m+2)!} \frac{(b-a)}{N} \sum_{i=0}^{N-1} |f^{(m+1)}(\xi_i)|$$

$$\leq \frac{h^{m+1}(b-a)}{(m+1)!} \max_{a \leq x \leq b} |f^{(m+1)}(x)| \tag{9.5.26}$$

Thus the remainder term in (9.5.25) is $O(h^{m+1})$. Upon interchanging the order of summation in (9.5.25), we obtain

$$E_N = \sum_{j=2}^{m+1} \frac{h^j}{j!} \sum_{i=0}^{N-1} f^{(j-1)}(x_i) + O(h^{m+1}) = \sum_{j=2}^{m+1} \frac{h^{j-1}}{j!} R_N(f^{(j-1)}) + O(h^{m+1})$$

(9.5.27)

After a shift of the index j this gives

$$I(f) = R_N(f) + \sum_{j=1}^{m} \frac{h^j}{(j+1)!} R_N(f^{(j)}) + O(h^{m+1})$$

(9.5.28)

Substituting $I_N - E_N$ for R_N yields finally

$$E_N(f) = \sum_{j=1}^{m} \frac{h^j}{(j+1)!} [I(f^{(j)}) - E_N(f^{(j)})] + O(h^{m+1})$$

(9.5.29)

Note that a particular consequence of (9.5.29) is that if $f \in C^1[a, b]$, then $E_N(f)$ is $O(h)$. Thus if $f \in C^2[a, b]$, then $f' \in C^1[a, b]$ whence $E_N(f')$ is $O(h)$, and again from (9.5.29) it follows that

$$E_N(f) = \frac{h}{2} I(f') + O(h^2) = -B_1 h + O(h^2)$$

(9.5.30)

which is (9.5.22) for $m = 1$.

We now establish (9.5.22) in general by induction on the degree of differentiability $m + 1$. Assume that for $l < m$, $f \in C^{l+1}[a, b]$ implies that

$$E_N(f) = -\sum_{k=1}^{l} \frac{B_k h^k}{k!} I(f^{(k)}) + O(h^{l+1})$$

(9.5.31)

Let us show that (9.5.31) holds for $l = m$. If $f \in C^{m+1}[a, b]$, then $f^{(j)} \in C^{m-j+1}[a, b]$ whence (9.5.31) holds with $l = m - j + 1$. Substituting this into (9.5.29) gives

$$E_N(f) = \sum_{j=1}^{m} \frac{h^j}{(j+1)!} \left[I(f^{(j)}) + \sum_{k=1}^{m-j} \frac{B_k h^k}{k!} I(f^{(j+k)}) \right] + O(h^{m+1})$$

$$= \sum_{j=1}^{m} \frac{h^j}{(j+1)!} I(f^{(j)}) + \sum_{j=1}^{m} \sum_{k=1}^{m-j} \frac{B_k h^{j+k}}{k!(j+1)!} I(f^{(j+k)}) + O(h^{m+1})$$

(9.5.32)

Rearrangement of the double sum in (9.5.32) yields in turn

$$E_N(f) = \sum_{j=1}^{m} \frac{h^j}{(j+1)!} I(f^{(j)}) + \sum_{j=1}^{m} h^j I(f^{(j)}) \sum_{k=1}^{j-1} \frac{B_k}{k!(j-k+1)!} + O(h^{m+1})$$

$$= \sum_{j=1}^{m} \frac{h^j I(f^{(j)})}{(j+1)!} \left[B_0 + \sum_{k=1}^{j-1} \binom{j+1}{k} B_k \right] + O(h^{m+1})$$

$$= \sum_{j=1}^{m} \frac{h^j I(f^{(j)})}{(j+1)!} (-j! B_j) + O(h^{m+1})$$

(9.5.33)

from Theorem 9.5.9. This demonstrates (9.5.22) and hence the Euler-Maclaurin formula.

Romberg Integration

The Euler-Maclaurin formula establishes that the error expansion of the composite trapezoid rule in powers of the subinterval width h has the form (9.5.1). Thus Richardson extrapolation (Algorithm 9.5.7) can be applied with the trapezoid rule as the basc method of approximation. This is called *Romberg integration* and is outlined in Algorithm 9.5.12. $T(f, a, b, N)$ denotes the composite trapezoid rule approximation with N subintervals to $\int_a^b f(x)\, dx$. For the sake of efficiency the trapezoid rule approximations should be calculated to avoid redundant evaluations of the integrand by using the information contained in the trapezoid rule approximation for half as many intervals which is stored as Last(0):

$$T(f, a, b, N) = \frac{1}{2}\, \text{Last}(0) + h \sum_{i=1}^{N/2} f(a + (2i - 1)h) \qquad (9.5.34)$$

where $h = (b - a)/N$.

Algorithm 9.5.12. Romberg integration

Constant	MaxLevel (Maximum permissible table depth)
Declare	Arrays LastRow( ) and CurrRow( ) of dimension MaxLevel + 1
Input	a, b (endpoints of interval of integration) tol (desired error tolerance)

Level := 0
N := 1 (number of subintervals)
CurrRow(0) := $T(f, a, b, 1)$

Repeat
    N := 2 * N
    LastRow( ) := CurrRow( )
    CurrRow(0) := $T(f, a, b, N)$ [see Eq. (9.5.34)]
    Level := Level + 1
    Coeff := 1 [The coefficient 4^j in (9.5.9)]

    for j = 1 to Level
        Coeff := 4 * Coeff
        CurrRow(j) := (Coeff * CurrRow(j − 1) − LastRow(j − 1))/(Coeff − 1)
until |CurrRow(Level) − LastRow(Level − 1)| < tol or Level > MaxLevel

Output	Answer: CurrRow (Level) Warning that the tolerance may not have been achieved if Level exceeded MaxLevel.

Observe that in Romberg integration all evaluations of the integrand occur in the trapezoid rule approximations in the first column of the extrapolation table. The remaining columns are computed from Eq. (9.5.9) and involve only simple arithmetic computations, which nonetheless greatly increase the order and thus, one hopes, the accuracy of the answer. This makes Romberg integration a very effective means of getting accurate answers efficiently.

It is also of interest to note that the $O(h^4)$ approximation resulting from Romberg integration is none other than the composite Simpson rule approximation (Exercise 10a). However, Romberg integration does not in general give the composite Newton-Cotes formula for a given order.

Example 9.5.13. Table 9.5.14 shows the error made at each stage of the extrapolation process when Romberg integration is employed to approximate $\int_0^1 e^{3t}\,dt$ to within a tolerance of 5×10^{-4}. Note that since the error made by the trapezoid rule falls at best by a factor of 4 when the number of subintervals is doubled, another five to six iterations would be necessary to attain the specified error using the iterative trapezoid rule procedure of Algorithm 9.5.10. Also observe that because of the rapidity with which the error decreased, the final approximation was far more accurate than required. This is a common occurrence with extrapolation methods.

TABLE 9.5.14

4.2E + 0000				
1.2E + 0000	1.4E − 0001			
3.0E − 0001	1.0E − 0002	1.8E − 0003		
7.4E − 0002	6.9E − 0004	3.5E − 0005	6.4E − 0006	
1.9E − 0002	4.4E − 0005	5.7E − 0007	3.1E − 0008	5.7E − 0009

A Comparison of Adaptive Simpson and Romberg Integration

Comparison of two effective methods of approximation is often inconclusive. This is the case with adaptive Simpson quadrature and Romberg integration. Table 9.5.15 shows the result of applying both methods to approximating $\int_0^x (4t - t^2)e^{t^2}\,dt$ to within a tolerance of 5×10^{-4} for $x = 0.1, 0.2, \ldots, 2.0$. Since this is an example of an integrand that has a large variation in the magnitude of the fourth derivative over the interval of integration, adaptive Simpson quadrature might be expected to do better. We see that this is indeed the case, at least insofar as the cumulative number of evaluations of the integrand is concerned. Adaptive Simpson quadrature required about two-thirds of the function evaluation needed by Romberg integration (264 versus 406). Such a difference would be of significance only in long computations. Moreover, Romberg integration requires significantly less use of the four basic arithmetic operations, and thus Romberg integration may take less computational time than the figure of 50 percent more function evaluations indicates. In fact, when the integral in Example 9.5.13 was evaluated for $x = 0.01$, $0.02, \ldots, 2.0$ on an IBM AT class computer using implementations in Turbo Pascal 5.5, adaptive Simpson quadrature took about 86 percent of the time required for Romberg integration if a numeric coprocessor was used and about 94 percent of the amount of time if a coprocessor was not used. Recall from Table 1.4.1 that function evaluation is more time-consuming relative to the basic arithmetic operations when there is no numeric coprocessor.

TABLE 9.5.15

	Romberg		Adaptive Simpson	
x	Error	Evals.	Error	Evals.
0.10	2.1E − 0008	3	1.3E − 0009	5
0.20	5.6E − 0010	5	8.5E − 0008	5
0.30	1.4E − 0008	5	9.9E − 0007	5
0.40	1.4E − 0007	5	5.8E − 0006	5
0.50	8.3E − 0007	5	2.3E − 0005	5
0.60	−5.1E − 0009	9	7.2E − 0005	5
0.70	−3.6E − 0008	9	1.9E − 0004	5
0.80	−2.1E − 0007	9	4.5E − 0004	5
0.90	−9.8E − 0007	9	5.9E − 0005	9
1.00	−4.1E − 0006	9	1.1E − 0004	9
1.10	−1.6E − 0005	9	1.8E − 0004	9
1.20	−3.9E − 0007	17	2.4E − 0004	9
1.30	−1.4E − 0006	17	1.4E − 0004	9
1.40	−1.2E − 0008	33	−5.1E − 0004	9
1.50	−4.5E − 0008	33	2.0E − 0005	21
1.60	−1.6E − 0007	33	−9.9E − 0006	21
1.70	−5.8E − 0007	33	−1.6E − 0006	25
1.80	−2.0E − 0006	33	−1.2E − 0004	25
1.90	−8.1E − 0009	65	3.2E − 0005	37
2.00	−3.0E − 0008	65	2.3E − 0005	41

As shown in Fig. 9.4.6, adaptive Simpson quadrature can produce very inaccurate answers when its test criterion is fulfilled "coincidentally." Romberg integration suffers from the same difficulty. In fact, since Romberg integration can accept an approximation on the basis of as few as three evaluations of the integrand, it would return 0 as the approximation to $\int_0^{2\pi} \sin^2 t \, dt$. Thus one must be careful when applying either adaptive Simpson quadrature or Romberg integration to periodic or quasiperiodic functions.

EXERCISES 9.5

1. Use the central difference formula and extrapolation to find an $O(h^6)$ approximation to the first derivative at $x = 0$ from the following functional data.

x	$f(x)$	x	$f(x)$
−0.6	−0.56464	0.15	0.19867
−0.3	−0.38942	0.3	0.38942
−0.15	−0.19867	0.6	0.56464
0.0	0.00000		

2. Following Algorithm 9.5.7 write a program that will find the first derivative of a function to a given accuracy using the central difference formula and extrapolation. Assume that a formula for the function which is to be differentiated is available to the program. Test your program on the following examples:

(a) $f(x) = e^{-3x}$, at $x = 1$ with tol $= 5 \times 10^{-4}$.
(b) $f(x) = 1/(1 + x^2)$, at $x = 0.5$ with tol $= 5 \times 10^{-3}$.

3. (a) Work out formulas analogous to (9.5.9) when the expansion of the error made by the base method is of the form

$$A(0) - A(h) = \sum_{j=1}^{N} a_j h^j + O(h^{N+1})$$

(9.5.35)

(b) Use the formulas in part (a) and the forward difference approximation to the first derivative to find an $O(h^4)$ approximation to the first derivative at $x = 0$ from the following functional data:

x	0.0	0.1	0.2	0.4	0.8
$f(x)$	1.00000	1.22140	1.49182	2.22554	4.95303

4. Write a program that will find right-hand derivatives to a specified accuracy using the forward difference formula and extrapolation. Test your program by finding the derivative of

$$f(x) = \begin{cases} 0 & x = -1 \\ \exp - \dfrac{1}{1 - x^2} & x > -1 \end{cases}$$

(9.5.36)

5. (a) Show that the error expansion of the approximation

$$f''(x) \simeq \frac{f(x + h) - 2f(x) + f(x - h)}{h^2}$$

(9.5.37)

has the form (9.5.1).

(b) Write a program that uses formula (9.5.37) and extrapolation to approximate the second derivative of a function to a specified tolerance.

6. Use Romberg integration to find an $O(h^6)$ approximation to the definite integral $\int_0^3 f(x)\, dx$ from the following functional data:

x	0.00	0.75	1.50	2.25	3.00
$f(x)$	1.00000	0.47237	0.22313	0.10540	0.04979

7. Write a program efficiently implementing Romberg integration. Test your program on the following integrals.

(a) $\displaystyle\int_0^3 \frac{x}{1 + x^2}\, dx$, tol $= 5 \times 10^{-5}$

(b) $\displaystyle\int_0^3 xe^{-x^2}\, dx$, tol $= 5 \times 10^{-4}$

8. (a) Rewrite Algorithm 9.5.7 so that it is necessary to declare only the single array CurrRow.

(b) Rewrite Algorithm 9.5.12 for Romberg integration to incorporate the modification made in part (a).

9. (a) Derive formula (9.5.8) for the second-level of extrapolation.

(b) Use induction to derive the general nth-level extrapolation formula given by (9.5.9).

10. (a) Show that applying extrapolation to the composite trapezoid rule formula for N and $2N$ subintervals leads to the composite Simpson rule for $2N$ subintervals.

(b) What is the formula for the $O(h^6)$ approximation resulting from extrapolation of the composite trapezoid rule?

11. (a) Use (9.5.19) to verify the values given for the first five Bernoulli numbers.

(b) Show that $B_{2j+2} = 0$ for $j \geq 1$. [*Hint:* Show that $t/(e^t - 1) - B_1 t$ is an even function in t.]

12. Write the recurrence relation (9.5.19) for the Bernoulli numbers in a form that takes into account the fact that $B_{2j-1} = 0$ for $j \geq 2$. Use this recurrence relation to find the Bernoulli numbers B_{2j}, $j = 1, \ldots, 30$.

13. The *second Euler-Maclaurin summation formula* is given by

$$\int_a^b f(x)\, dx = h \sum_{i=1}^{N} f(m_i)$$

$$- \sum_{j=1}^{n} \frac{B_{2j} h^{2j}}{(2j)!} (1 - 2^{-2j+1})[f^{(2j-1)}(b)$$

$$- f^{(2j-1)}(a)] + O(h^{2n+2}) \quad (9.5.38)$$

where $f \in C^{(2N+2)}[a, b]$ and m_i is the midpoint of the ith subinterval. Since this representation is an *open formula*, that is, it does not use the values of f at the endpoints a and b, it may be used to derive approximation methods for improper integrals. In this exercise we derive this formula from first principles, that is, without using the first Euler-Maclaurin summation formula. See the next exercise for a shorter derivation based upon the first formula. To obtain the second formula we start with the Riemann sum approximation with midpoint evaluation

$$R_N(f) = h \sum_{i=1}^{N} f(m_i)$$

(a) Show that this gives the following analog of (9.5.23):

$$E_N(f) = \sum_{i=1}^{N} \left[\int_{m_i - h/2}^{m_i + h/2} f(x)\, dx - hf(m_i) \right]$$

$$=: \sum_{i=1}^{N} e_i(h) \qquad (9.5.39)$$

(b) Show that this leads to the following analog of (9.5.29):

$$E_N(f) = \sum_{j=1}^{N} \frac{1}{(2j+1)!} \left(\frac{h}{2}\right)^{2j} [I(f^{(2j)})$$

$$- E_N(f^{(2j)})] + O(h^{2N+2}) \qquad (9.5.40)$$

(c) Use the result of part (b) to derive the second Euler-Maclaurin formula.

14. Use the first Euler-Maclaurin summation formula to derive the second.

15. Use Romberg integration and the temperature formula (9.1.1) to generate graphs of temperature against position at times $t = 10$ seconds and $t = 1$ minute for each of the sets of material parameters given in Exercise 9 in Sec. 9.3.

16. Use Romberg integration to generate a graph of the solution I of Exercise 7 in Sec. 9.1 from 0 to 4π. Use $\omega = 2\pi$, $E_0 = 100$ volts, $R_0 = 10$ ohms, and $L_0 = 0.5$ henry.

17. Use Romberg integration to graph the period of a pendulum of length 1 foot against its initial displacement from $\theta_{max} = 0$ to $\theta_{max} = 5\pi/6$.

18. Alter the Romberg integration algorithm so that it takes as an input a parameter N which causes the integral to be calculated as

$$\int_a^b f(x)\, dx = \sum_{i=0}^{N-1} \int_{x_i}^{x_{i+1}} f(x)\, dx \quad (9.5.41)$$

where $\{x_0, \ldots, x_N\}$ is a uniform partition of the interval $[a, b]$ into N subintervals. Use this program to produce a corrected version of Fig. 9.4.6.

9.6 GAUSSIAN INTEGRATION

In Sec. 9.3 we derived Newton-Cotes formulas such as the trapezoid rule

$$\int_a^b f(x)\, dx \simeq \frac{b-a}{2} [f(a) + f(b)] \qquad (9.6.1)$$

and Simpson rule

$$\int_a^b f(x)\, dx \simeq \frac{b-a}{6} \left[f(a) + 4f\left(\frac{a+b}{2}\right) + f(b) \right] \qquad (9.6.2)$$

for approximating the definite integral. In general a closed N-point Newton-Cotes formula has the form

$$\int_a^b f(x)\, dx \simeq \sum_{i=1}^{N} w_i f(x_i) \qquad (9.6.3)$$

where $x_i = a + (i-1)h$, $i = 1, \ldots, N$, with $h := (b-a)/(N-1)$. We will call the points x_i the *nodes* and the coefficients w_i the *weights*. Recall that the Newton-Cotes formulas were derived by integrating the Lagrange polynomial which interpolates the points $(x_i, f(x_i))$.

As noted in Sec. 9.3 the error made by an N-point Newton-Cotes formula is proportional to $f^{(N)}$ when N is even and to $f^{(N+1)}$ when N is odd. Consequently, an N-point Newton-Cotes formula will compute definite integrals of polynomials of degree no more than $N-1$ exactly. From (9.6.3) it follows that the weights w_i in the Newton-Cotes formulas are solutions to the linear system of equations

$$\sum_{i=1}^{N} w_i = \int_a^b dx$$

$$\sum_{i=1}^{N} x_i w_i = \int_a^b x \, dx$$

$$\sum_{i=1}^{N} x_i^2 w_i = \int_a^b x^2 \, dx \qquad (9.6.4)$$

$$\cdots\cdots\cdots\cdots\cdots\cdots\cdots$$

$$\sum_{i=1}^{N} x_i^{N-1} w_i = \int_a^b x^{N-1} \, dx$$

The coefficient matrix of this system is the transpose of the Vandermonde matrix (see Sec. 4.2) which is nonsingular, and thus (9.6.4) uniquely determines the coefficients w_i. Hence solving (9.6.4) for the weights w_i is an alternative means of deriving the Newton-Cotes formulas.

Gaussian integration assumes an approximation of the form

$$\int_{-1}^{1} f(x) \, dx = \sum_{i=1}^{N} w_i f(x_i) \qquad (9.6.5)$$

but rather than taking the x_i to be equally spaced, as is the case for the Newton-Cotes formulas, chooses them, as well as the w_i, so as to compute exactly the integral of as high a degree polynomial as possible. Choosing the interval of integration to be $(-1, 1)$ in (9.6.5) entails no restriction, for upon making the change of variables

$$\bar{x} := \frac{2}{b-a} \left[x - \left(\frac{a+b}{2} \right) \right] \qquad (9.6.6)$$

we have

$$\int_a^b f(x) \, dx = \int_{-1}^{1} \frac{b-a}{2} f\left(\frac{b-a}{2} \bar{x} + \frac{a+b}{2} \right) d\bar{x} \qquad (9.6.7)$$

Let us find the gaussian integration formula for $N = 2$. The availability of four unknowns x_1, x_2, w_1, w_2 suggests that we should be able to compute the integrals of cubic polynomials exactly. The requirement that $1, x, x^2$, and x^3 be computed exactly by (9.6.5) gives from (9.6.4) the system of equations

$$w_1 + w_2 = 2$$

$$x_1 w_1 + x_2 w_2 = 0$$

$$x_1^2 w_1 + x_2^2 w_2 = \tfrac{2}{3} \qquad (9.6.8)$$

$$x_1^3 w_1 + x_2^3 w_2 = 0$$

The nodes x_i will be the roots of some polynomial $p(x) = (x - x_1)(x - x_2) = x^2 + a_1 x + a_0$. Adding a_0 times the first equation in (9.6.8) plus a_1 times the second equation to the third equation gives

$$w_1 p(x_1) + w_2 p(x_2) = \tfrac{2}{3} + 2a_0 \qquad (9.6.9)$$

that is, $a_0 = -\tfrac{1}{3}$ since x_1 and x_2 are roots of p. Repeating this procedure for the second through fourth equations in (9.6.8) gives

$$w_1 x_1 p(x_1) + w_2 x_2 p(x_2) = \tfrac{2}{3} a_1 \qquad (9.6.10)$$

and therefore $a_1 = 0$. Thus $p(x) = x^2 - \tfrac{1}{3}$ whence $x_1 = -\sqrt{\tfrac{1}{3}}$, $x_2 = \sqrt{\tfrac{1}{3}}$. Solving the first pair of equations in (9.6.8) gives $w_1 = 2x_2/(x_2 - x_1) = 1$, $w_2 = -2x_1/(x_2 - x_1) = 1$. Consequently, the two-point Gauss formula

$$\int_{-1}^{1} f(x)\, dx = f\left(-\frac{\sqrt{3}}{3}\right) + f\left(\frac{\sqrt{3}}{3}\right) \qquad (9.6.11)$$

will compute the integrals of polynomials of degree three exactly. Observe that the trapezoid rule, which is the corresponding two-point Newton-Cotes formula, computes polynomials of degree no more than one exactly.

Example 9.6.1. Let us compare the accuracy of the trapezoid rule, Simpson's rule, and the two-point Gauss rule (9.6.11) for the integral $I = \int_0^2 e^x\, dx = 6.389$ to three decimal places. We get $I \approx 8.389$ for the trapezoid rule and $I \approx 6.421$ for Simpson's rule. To obtain the Gauss rule approximation we apply (9.6.11) to the function

$$\frac{b-a}{2} f\left(\frac{b-a}{2}\bar{x} + \frac{a+b}{2}\right) = f(\bar{x}+1) = e^{\bar{x}+1}$$

[see (9.6.7)] to find that $I \approx e^{-\sqrt{1/3}+1} + e^{\sqrt{1/3}+1} = 6.368$. Thus for this problem the Gauss rule is slightly more accurate than Simpson's and substantially more accurate than the trapezoid rule. Observe, however, that the Gauss rule requires only two function evaluations rather than the three required by Simpson's rule.

Legendre Polynomials

It is evident that the method employed to obtain the Gauss two-point rule will become increasingly cumbersome for more nodes and weights. Fortunately the Gauss nodes x_i, $i = 1, \ldots, N$, have a simple characterization, they are the zeros of the Nth-degree Legendre polynomial. The *Legendre polynomials* are given by the formula

$$P_n(x) = \frac{1}{2^n n!} \frac{d^n}{dx^n} (x^2 - 1)^n \qquad (9.6.12)$$

Table 9.6.2 lists some important properties of the Legendre polynomials. The first four Legendre polynomials are

$$P_0(x) = 1 \qquad P_1(x) = x \qquad P_2(x) = \tfrac{1}{2}(3x^2 - 1) \qquad P_3(x) = \tfrac{1}{2}(5x^3 - 3x)$$

$$(9.6.13)$$

TABLE 9.6.2
Properties of the Legendre polynomials

1. $P_n(x)$ has n zeros in the interval $(-1, 1)$

2. $(n + 1)P_{n+1}(x) = (2n + 1)xP_n(x) - nP_{n-1}(x)$

3. $\int_{-1}^{1} P_m(x)P_n(x)\, dx = 0 \qquad$ for $m \neq n$

4. $\int_{-1}^{1} [P_n(x)]^2\, dx = \dfrac{2}{2n + 1}$

5. $\int_{-1}^{1} x^k P_n(x)\, dx = 0 \qquad k = 0, \dots, n - 1$

6. $\int_{-1}^{1} x^n P_n(x)\, dx = \dfrac{2^{n+1}(n!)^2}{(2n + 1)!}$

Legendre polynomials are odd or even functions as their degree is odd or even. See Sec. 7.4 for a further discussion of Legendre polynomials. Note that the zeros of $P_2(x)$ are $x = \pm\sqrt{3}/3$. These are, as we found, the Gauss nodes for $N = 2$.

Properties 3 and 4 are important for our purposes. In the language of inner product spaces, they say that the set $\mathcal{P}_m := \{P_0, P_1, \dots, P_m\}$ is *orthogonal with respect to the inner product* $(f, g) := \int_{-1}^{1} f(x)g(x)\, dx$. $\mathcal{P}_m$ is a basis for the space Π_m of polynomials of degree no more than m, in fact the one that results from applying the process of *Gram-Schmidt orthogonalization* to the basis $(1, x, x^2, \dots, x^m)$. Again see Sec. 7.4 for further discussion.

N-Point Gauss Formula

On the basis of what we found for the Gauss two-point formula, we expect that an N-point formula will compute integrals of polynomials of degree no more than $2N - 1$ exactly. If we accept for the moment that zeros of the Legendre polynomial $P_N(x)$ are the appropriate choice for the Gauss nodes x_i for the N-point formula, then the weights w_i can be determined from the nonsingular system

$$\sum_{i=1}^{N} w_i = \int_{-1}^{1} dx = 2$$

$$\sum_{i=1}^{N} x_i w_i = \int_{-1}^{1} x\, dx = 0 \qquad\qquad (9.6.14)$$

$$\dotsb$$

$$\sum_{i=1}^{N} x_i^{N-1} w_i = \int_{-1}^{1} x^{N-1}\, dx = \frac{1}{N}\left(1 - (-1)^N\right)$$

Theorem 9.6.3. In formula (9.6.5) let the nodes x_i be the zeros of the Legendre polynomial P_N and let the weights w_i be determined by (9.6.14). Then (9.6.5) computes $\int_{-1}^{1} p(x)\, dx$ exactly for every $p \in \Pi_{2N-1}$.

Proof. Let $p \in \Pi_{2N-1}$. Then a polynomial division gives

$$p(x) = q(x)P_N(x) + r(x) \tag{9.6.15}$$

where $q, r \in \Pi_{N-1}$. Since $\mathscr{P}_{N-1}$ is a basis for Π_{N-1}, we can write

$$q(x) = \sum_{j=0}^{N-1} q_j P_j(x) \qquad \text{and} \qquad r(x) = \sum_{j=0}^{N-1} r_j P_j(x)$$

Thus the left-hand side of (9.6.5) is

$$
\begin{aligned}
\int_{-1}^{1} p(x)\,dx &= \int_{-1}^{1} (q(x)P_N(x) + r(x))\,dx \\
&= \sum_{j=0}^{N-1} q_j \int_{-1}^{1} P_j(x)P_N(x)\,dx + \sum_{j=0}^{N-1} r_j \int_{-1}^{1} P_j(x)\,dx \\
&= 2r_0
\end{aligned} \tag{9.6.16}
$$

from properties 3, 5, and 6 of Table 9.6.2. On the other hand, the right-hand side of (9.6.5) is

$$
\begin{aligned}
\sum_{i=1}^{N} w_i p(x_i) &= \sum_{i=1}^{N} w_i (q(x_i)P_N(x_i) + r(x_i)) \\
&= \sum_{i=1}^{N} w_i r(x_i) \\
&= \sum_{j=0}^{N-1} r_j \sum_{i=1}^{N} w_i P_j(x_i)
\end{aligned} \tag{9.6.17}
$$

since the x_i are the zeros of P_N. Because the weights w_i were chosen to integrate any polynomial in Π_{N-1} exactly, we have from properties 5 and 6 of Table 9.6.2 that

$$\sum_{i=1}^{N} w_i P_j(x_i) = \int_{-1}^{1} P_j(x)\,dx = \begin{cases} 2 & j = 0 \\ 0 & j = 1, \ldots, N-1 \end{cases} \tag{9.6.18}$$

Hence (9.6.17) implies that the right-hand side of (9.6.5) is also equal to $2r_0$.

A simple consequence of Theorem 9.6.3 is a representation for the weights w_i in terms of the Lagrangian basis functions

$$l_i(x) := \alpha_i \prod_{j=1, j \neq i}^{N} (x - x_j) \qquad \text{with } \alpha_i := \frac{1}{\displaystyle\prod_{j=1, j \neq i}^{N} (x_i - x_j)} \tag{9.6.19}$$

Corollary 9.6.4. The gaussian weights are given by the formula

$$w_i = \int_{-1}^{1} [l_i(x)]^2\,dx \tag{9.6.20}$$

Proof. Since $[l_i(x)]^2 \in \Pi_{2N-2}$, gaussian integration will compute its integral exactly. Thus

$$\int_{-1}^{1} [l_i(x)]^2\,dx = \sum_{j=1}^{N} w_j [l_i(x_j)]^2 = w_i \tag{9.6.21}$$

Identities involving Legendre polynomials permit alternative formulas for the weights. For instance,

$$w_i = \frac{2(1 - x_i^2)}{n^2 [P_{N-1}(x_i)]^2} \qquad (9.6.22)$$

See Exercise 16. Since the zeros of the Legendre polynomials are symmetric about the origin, Eq. (9.6.22) implies that the weights w_i are the same for $\pm x_i$.

Programming Gaussian Integration

Programming formula (9.6.5) is straightforward once the nodes x_i and the weights w_i are known. The zeros of the Legendre polynomials must in general be computed approximately by numerical methods. For known values of the nodes the weights can be calculated from (9.6.22). Table 9.6.5 gives numerical values of the nodes and weights for selected values of N up to 12.

TABLE 9.6.5

N	$\pm x_i$	w_i
2	0.577 350 269 189 626	1.000 000 000 000 000
3	0.000 000 000 000 000	0.888 888 888 888 888
	0.774 596 669 241 483	0.555 555 555 555 556
4	0.339 981 043 584 856	0.652 145 154 862 546
	0.861 136 311 594 053	0.347 854 845 137 454
5	0.000 000 000 000 000	0.568 888 888 888 889
	0.538 469 310 105 683	0.478 628 670 599 366
	0.906 179 845 938 664	0.236 926 885 056 189
6	0.238 619 186 083 197	0.467 913 934 572 691
	0.661 209 386 466 265	0.360 761 573 048 139
	0.932 469 514 203 152	0.171 324 492 379 170
8	0.183 434 642 495 650	0.362 683 783 378 362
	0.525 532 409 916 329	0.313 706 645 877 887
	0.796 666 477 413 627	0.222 381 034 453 374
	0.960 289 856 497 536	0.101 228 536 290 376
10	0.148 874 338 981 631	0.295 524 224 714 753
	0.433 395 394 129 247	0.269 266 719 309 996
	0.679 409 568 299 024	0.219 086 362 515 982
	0.865 063 366 688 985	0.149 451 349 150 581
	0.973 906 528 517 172	0.066 671 344 308 688
12	0.125 233 408 511 469	0.249 147 045 813 403
	0.367 831 498 998 180	0.233 492 536 538 355
	0.587 317 954 286 617	0.203 167 426 723 066
	0.769 902 674 194 305	0.160 078 328 543 346
	0.904 117 256 370 475	0.106 939 325 995 318
	0.981 560 634 246 719	0.047 175 336 386 512

Algorithm 9.6.6 outlines the approximation of an integral using an N-point Gauss rule for $N \le 6$.

Algorithm 9.6.6. Gaussian integration

Input $N \le 6$ (Number of nodes)
            a,b (Interval of integration)

Initialize   Arrays x(,) and w(,) for the Gauss nodes and weights in Table 9.6.5, x(N, i) is the ith nonnegative node for the Gauss N-point formula, and w(N, i) is the corresponding weight.

h:=(b − a)/2
m:=(a + b)/2
x:=h * x(N, 1)
if N is odd then Ans:=h * w(N, 1) * f(x)
else Ans:=h * w(N, 1) * (f(−x + m) + f(x + m))

for i = 2 to $\left[\dfrac{N+1}{2} \right]$

    x:=h * x(N, i)
    Ans:=Ans + h * w(N, i) * (f(−x + m) + f(x + m))

Output   Ans

Example 9.6.7. Table 9.6.8 shows the errors made when Algorithm 9.6.6 is applied to the integral of Example 9.6.1. The errors are quite small even when the integrand is evaluated at only four or five points.

TABLE 9.6.8
Errors made by Gauss
N-point rule in approximating
$\int_0^2 e^x \, dx$

N	Error
2	−2.1E − 0002
3	−1.8E − 0004
4	−8.0E − 0007
5	−1.6E − 0009

Hermite Interpolation

In Chap. 4 we treated Lagrange interpolation, the problem of finding the polynomial of least degree that interpolates a given set of functional values. The *Hermite interpolant* of a function f is the polynomial h of least degree for which

$$h(x_i) = f_i := f(x_i) \qquad h'(x_i) = f'_i := f'(x_i) \qquad (9.6.23)$$

at given nodes x_i, $i = 1, \ldots, N$. It is easily verified that the solution to this problem is

$$h(x) = \sum_{i=1}^{N} [1 - 2l_i'(x_i)(x - x_i)][l_i(x)]^2 f_i + (x - x_i)[l_i(x)]^2 f_i' \qquad (9.6.24)$$

where the l_i are given by (9.6.19). See Exercise 12. Evidently $h \in \Pi_{2N-1}$.

The derivation of an expression for the error made by the Hermite interpolant is similar to that for the Lagrange interpolant (Exercise 13). The result is

$$f(x) - h(x) = \frac{f^{(2N)}(\xi)}{(2N)!} [L_N(x)]^2 \qquad (9.6.25)$$

where $L_N(x) := \Pi_{i=1}^{N} (x - x_i)$ and ξ is in an interval containing x and all the x_i.

Hermite Interpolation and Gaussian Integration

Hermite interpolation provides a framework for gaussian integration analogous to that which Lagrange interpolation provides for the derivation and error analysis of the Newton-Cotes formulas. From (9.6.24)

$$\int_{-1}^{1} f(x)\, dx \simeq \int_{-1}^{1} h(x)\, dx = \sum_{i=1}^{N} w_i f(x_i) + v_i f'(x_i) \qquad (9.6.26)$$

where

$$w_i := \int_{-1}^{1} [1 - 2l_i'(x_i)(x - x_i)][l_i(x)]^2\, dx$$

$$v_i := \int_{-1}^{1} (x - x_i)[l_i(x)]^2\, dx \qquad (9.6.27)$$

By comparison with (9.6.5) we see that gaussian integration may be viewed as the problem of selecting the nodes x_i so that $v_i = 0$, $i = 1, \ldots, N$. Let us verify that this occurs when the x_i are chosen to be the zeros of the Nth Legendre polynomial P_N. Observe that $(x - x_i)l_i(x) = \alpha_i L_N(x)$, where α_i is given in (9.6.19). If the x_i are the zeros of P_N, then L_N and P_N are polynomials of degree N having the same zeros and hence are identical up to a constant factor. Indeed it is easily seen that

$$L_N(x) = \frac{2^N (N!)^2}{(2N)!} P_N(x) \qquad (9.6.28)$$

(Exercise 14). Consequently v_i is proportional to the integral $\int_{-1}^{1} P_N(x)l_i(x)\, dx$, which is zero from property 5 of Table 9.6.2 since $l_i \in \Pi_{N-1}$.

Given that the v_i are zero, (9.6.26) now implies that

$$w_i = \int_{-1}^{1} [l_i(x)]^2\, dx \qquad (9.6.29)$$

which is the same formula for the weights in (9.6.5) as in (9.6.20). Thus the approach of integrating the Hermite interpolant does indeed yield the gaussian integration formula.

Example 9.6.9. Figure 9.6.10 shows the Lagrange interpolant for the function $f(x) = 1/(1 + x^2)$ with nodes $x_i = -1, 0, 1$. The area beneath this curve is the Simpson rule approximation to $\int_{-1}^{1} 1/(1 + x^2)\, dx$. Also shown is the Hermite interpolant of the same function for the Gauss nodes $x_i = -0.775, 0, 0.775$. The area beneath this curve is the Gauss three-point approximation and is evidently much more accurate.

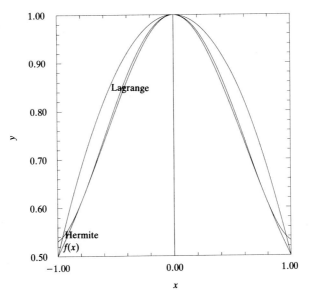

FIGURE 9.6.10

Error Analysis for Gaussian Integration

Theorem 9.6.11. The error made by gaussian integration in approximating the integral $\int_a^b f(x)\, dx$ is

$$E_N(f) = \frac{(b - a)^{2N+1}(N!)^4}{(2N + 1)[(2N)!]^3}\, f^{(2N)}(\xi) \tag{9.6.30}$$

for some $\xi \in [a, b]$.

Proof. We will prove the result for $(a, b) = (-1, 1)$. Formula (9.6.30) follows after a change of variables (Exercise 14). Let $h \in \Pi_{2N-1}$ be the Hermite polynomial that interpolates the values of f and f' at the zeros x_i of the Legendre polynomial P_N. Then from the error formula (9.6.25),

$$E_N(f) = \int_{-1}^{1} (f(x) - h(x))\, dx$$

$$= \int_{-1}^{1} \frac{f^{(2N)}(\xi(x))}{(2N)!} \, [L_N(x)]^2 \, dx$$

$$= \frac{f^{(2N)}(\bar{\xi})}{(2N)!} \int_{-1}^{1} [L_N(x)]^2 \, dx \qquad (9.6.31)$$

for some $\bar{\xi} \in [a, b]$. The last equality follows from the mean value theorem for integrals. The error formula (9.6.30) [with $(a, b) = (-1, 1)$] now follows from (9.6.28) and property 4 of Table 9.6.2.

Table 9.6.12 gives the magnitude of the numerical factor

$$\frac{2^{2N+1}(N!)^4}{(2N+1)[(2N)!]^3}$$

multiplying the derivative in (9.6.30) when $(a, b) = (-1, 1)$. Clearly N-point gaussian integration will yield very accurate answers for relatively small N as long as the width of the interval of integration and the $2N$th derivative are not too large.

TABLE 9.6.12

N	Factor	N	Factor
1	3.33E − 0001	7	2.13E − 0015
2	7.41E − 0003	8	2.22E − 0018
3	6.35E − 0005	9	1.82E − 0021
4	2.88E − 0007	10	1.20E − 0024
5	8.08E − 0010	11	6.52E − 0028
6	1.54E − 0012	12	2.96E − 0031

Adaptive Gaussian Quadrature

On the basis of our examples and Table 9.6.12, we expect gaussian integration to give us very accurate answers with a relatively small number of evaluations of the integrand. On the other hand, the error expression (9.6.30), depending as it does on the $2N$th derivative of the integrand, does not provide a practical means of estimating the error. Hence an adaptive scheme requiring only the input of an error tolerance is desirable. Algorithm 9.6.13 gives such a scheme.

Algorithm 9.6.13. Adaptive gaussian integration

Input   tol (Error tolerance)
            a,b (Limits of integration)

Initialize   Arrays x(,) and w(,) containing the positive Gauss nodes and weights for the even-point formulas up to $N = 12$, x(n, i) is the ith positive node for N = 2n.

Define   Function *Gauss*(n, h, m):

    GAns:=0
    for i = 1 to n
        x:=h*w(n, i)
        GAns:=GAns + h*w(n, i)*(f(−x + m) + f(x + m))
    Return: GAns

    h:=(b − a)/2
    m:=(a + b)/2
    n:=1

    Ans:=Gauss(1, h, m)
    repeat
        Lastans:=Ans
        n:=n + 1
        Ans:=Gauss(n, h, m)
    until |Ans-Lastans| < tol or $n > 6$

Output   Ans.
        If $n > 6$, warning message that tolerance may not have been
        attained.

The reason that the Gauss formulas for even N, rather than all N, were used is as follows. Suppose that gaussian integration first meets the specified tolerance with a $(2m + 1)$-point formula. A scheme that computed approximations for $N = 2, 3, 4, \ldots$ would require

$$2 + 3 + 4 + \cdots + 2m + 1 + 2m + 2 = (m + 1)(2m + 3) - 1 \quad (9.6.32)$$

evaluations of the integrand, while a scheme computing the approximation for $N = 2, 4, 6, \ldots$ would require

$$2 + 4 + 6 + \cdots + 2m + 2 + 2m + 4 = (m + 2)(m + 3) \quad (9.6.33)$$

evaluations. Thus the former scheme will be more efficient (9 versus 12 evaluations) only if $m = 1$. If the tolerance is first met with a $2m$-point formula, then the first scheme requires $(2m + 1)(m + 1) - 1$ versus $(m + 1)(m + 2)$ for the second, and thus the second is again better unless $m = 1$.

Comparison of Adaptive Gaussian, Romberg, and Adaptive Simpson Integration

Table 9.6.14 compares the adaptive Gauss scheme set forth in Algorithm 9.6.13 to Romberg (Sec. 9.5) and adaptive Simpson (Sec. 9.4) integration for the problem of approximating $\int_0^x (4t - t^3)e^{t^2}\, dt$ to within a tolerance of 5×10^{-4} for $x = 0.1, 0.2, \ldots, 2$. The cumulative number of evaluations of the integrand is 286 for the adaptive Gauss scheme as opposed to 264 for adaptive Simpson quadrature and 406 for Romberg integration. Thus despite the fact that, for a given number of evaluations of the integrand, the Gauss formula produces far more accurate answers than the corresponding Newton-Cotes

TABLE 9.6.14

x	Adap. Gauss Error	Evals.	Romberg Error	Evals.	Adap. Simpson Error	Evals.
0.10	$5.7E - 0013$	6	$2.1E - 0008$	3	$1.3E - 0009$	5
0.20	$9.1E - 0013$	6	$5.6E - 0010$	5	$8.5E - 0008$	5
0.30	$2.5E - 0012$	6	$1.4E - 0008$	5	$9.9E - 0007$	5
0.40	$8.7E - 0011$	6	$1.4E - 0007$	5	$5.8E - 0006$	5
0.50	$1.4E - 0009$	6	$8.3E - 0007$	5	$2.3E - 0005$	5
0.60	$2.7E - 0012$	12	$-5.1E - 0009$	9	$7.2E - 0005$	5
0.70	$2.5E - 0011$	12	$-3.6E - 0008$	9	$1.9E - 0004$	5
0.80	$2.2E - 0010$	12	$-2.1E - 0007$	9	$4.5E - 0004$	5
0.90	$1.5E - 0009$	12	$-9.8E - 0007$	9	$5.9E - 0005$	9
1.00	$8.6E - 0009$	12	$-4.1E - 0006$	9	$1.1E - 0004$	9
1.10	$4.4E - 0008$	12	$-1.6E - 0005$	9	$1.8E - 0004$	9
1.20	$2.0E - 0007$	12	$-3.9E - 0007$	17	$2.4E - 0004$	9
1.30	$8.4E - 0007$	12	$-1.4E - 0006$	17	$1.4E - 0004$	9
1.40	$2.6E - 0009$	20	$-1.2E - 0008$	33	$-5.1E - 0004$	9
1.50	$1.3E - 0008$	20	$-4.5E - 0008$	33	$2.0E - 0005$	21
1.60	$6.0E - 0008$	20	$-1.6E - 0007$	33	$-9.9E - 0006$	21
1.70	$2.6E - 0007$	20	$-5.8E - 0007$	33	$-1.6E - 0006$	25
1.80	$1.1E - 0006$	20	$-2.0E - 0006$	33	$-1.2E - 0004$	25
1.90	$6.4E - 0009$	30	$-8.1E - 0009$	65	$3.2E - 0005$	37
2.00	$3.0E - 0008$	30	$-3.0E - 0008$	65	$2.3E - 0005$	41

formula, the scheme of Algorithm 9.6.13 is no more than competitive with the other adaptive schemes insofar as the number of function evaluations is concerned. The reason for this is that in adaptive Simpson and Romberg integration every evaluation of the integrand contributes to the final approximation. In the adaptive gaussian scheme no evaluations prior to the ones used in the accepted approximation are involved in the final answer. They are in some sense wasted. Thus some of the inherent superiority of the Gauss method is lost.

Table 9.6.15 compares the results of approximating six different integrals using adaptive Simpson, Romberg, and adaptive gaussian integration. Integrals of the form $\int_a^x f(t)\, dt$ were computed for N evenly spaced points between limits a and b. The six problems were:

1. $\int_0^x (4t - t^3)e^{t^2}\, dt$, $0 < x \le 2$, $N = 300$, tol $= 5 \times 10^{-4}$

2. $\int_0^x e^{2t}\, dt$, $0 < x \le 2$, $N = 300$, tol $= 5 \times 10^{-4}$

3. $\int_0^x te^{-t^2}\, dt$, $0 < x \le 2$, $N = 200$, tol $= 5 \times 10^{-4}$

4. $\int_0^x \frac{1}{1 + t^2}\, dt$, $0 < x \le 1$, $N = 500$, tol $= 5 \times 10^{-4}$

TABLE 9.6.15
Comparison of adaptive Simpson, Romberg, and adaptive gaussian quadrature on an IBM AT class computer

Times are in seconds and accurate to ± 0.05 second

	Coprocessor			No coprocessor		
Prob.	Adap. Simpson	Romberg	Adap. Gaussian	Adap. Simpson	Romberg	Adap. Gaussian
1	4.41	4.71	3.15	11.05	12.90	8.36
2	3.10	2.59	2.02	7.74	6.93	5.25
3	1.33	1.41	1.39	3.65	4.07	4.04
4	1.53	1.71	1.58	3.21	3.71	2.61
5	2.25	2.69	2.14	9.17	12.15	9.17
6	2.93	3.45	2.90	6.71	8.27	6.43

5. $\int_0^x \dfrac{1}{\sqrt{1-t^2}}\, dt$, $0 < x \le 0.95$, $N = 500$, tol $= 5 \times 10^{-4}$

6. $\int_0^x \sin^2 t\, dt$, $0 < x \le \pi$, $N = 400$, tol $= 5 \times 10^{-4}$

As can be seen from Table 9.6.15 adaptive gaussian quadrature is as or more efficient than the other two methods, though the difference between it and adaptive Simpson quadrature is significant only on the first two problems. Moreover, as is evident from the error formula (9.6.31), gaussian integration is very sensitive to the width of the interval of integration, and it is sometimes necessary to compute an integral as the sum of several integrals over smaller intervals. Many feel that an adaptive scheme such as Algorithm 9.6.13 does not fully exploit the inherent superiority of the Gauss method, and research on better adaptive gaussian schemes continues.

Gaussian Integration for Improper Integrals

An improper integral such as

$$\int_{-1}^{1} \frac{f(x)}{\sqrt{1-x^2}}\, dx \qquad (9.6.34)$$

where $f(x)$ is continuous poses difficulties for any of the methods of integration we have studied thus far. Neither adaptive Simpson quadrature nor Romberg integration can be applied directly since they use the values of the integrand at the endpoints. These methods can only be applied indirectly by computing

$$\int_{-1+\epsilon}^{1-\epsilon} \frac{f(x)}{\sqrt{1-x^2}}\, dx \qquad (9.6.35)$$

for a sequence of smaller and smaller values of ϵ. Moreover, since the errors made by these methods grow with the magnitude of the derivatives of the integrand, which become infinite at the endpoints in (9.6.34), convergence of the method can be expected to be quite slow for small ϵ.

Gaussian integration does not require evaluation of the integrand at the endpoints, and thus the Gauss formula can be applied to (9.6.34) directly. However, the error in the approximation again depends upon the magnitude of the derivatives, which will be large near the endpoints, and thus the accuracy of such an approximation is suspect.

Example 9.6.16. The result of applying gaussian integration to the integral $\int_{-1}^{1} (x^2/\sqrt{1-x^2})\, dx$ is given in Table 9.6.17. The errors are large and improving only very slowly. Hence gaussian integration is an ineffective means of approximating integrals with integrands that are singular at the endpoints.

TABLE 9.6.17

n	Error
2	$-7.54\text{E} - 0001$
4	$-3.96\text{E} - 0001$
6	$-2.71\text{E} - 0001$
8	$-2.06\text{E} - 0001$
10	$-1.67\text{E} - 0001$
12	$-1.40\text{E} - 0001$

Approximating integrals with singular integrands or infinite limits requires an extension of gaussian integration. Recall from Sec. 7.4 that a set of polynomials $\{p_j(x)\}$ is *orthogonal on* (a, b) *with respect to the weight function* $w(x)$ if

$$\int_a^b w(x)p_j(x)p_k(x)\, dx = 0 \tag{9.6.36}$$

for $j \neq k$. If an orthogonal set of polynomials $\{p_j(x)\}$ is known for a given a, b, and w, then the approximation

$$\int_a^b w(x)f(x)\, dx \simeq \sum_{i=1}^{N} w_i f(x_i) \tag{9.6.37}$$

where the x_i are the roots of p_N and

$$w_i = \int_a^b w(x)[l_i(x)]^2\, dx \tag{9.6.38}$$

will compute the integral exactly when $f \in \Pi_{2N-1}$. With $w(x) = 1$, the set of orthogonal polynomials is the Legendre polynomials, and we obtain the formula (9.6.5) with the w_i given by (9.6.20). What has thus far been referred to as gaussian integration is called *Gauss-Legendre integration* in this more general context. The proofs that polynomials of degree $2N - 1$ are integrated exactly by (9.6.37) and that the weights are given by (9.6.38) follow the lines of those given for Gauss-Legendre integration and are left as exercises (Exercise 17).

In Sec. 7.2 it was shown that the *Chebyshev polynomials* $T_N(x):=\cos(N \arccos x)$, which have zeros $x_i = \cos[(2i-1)\pi/2N]$, are orthogonal with respect to the weight function $w(x) = 1/\sqrt{1-x^2}$. The weights have the simple formula $w_i = \pi/N$. This formula may be established by an argument similar to that leading to formula (9.6.52) of Exercise 16 for the weights in Gauss-Legendre integration. Consequently the *Gauss-Chebyshev integration formula*

$$\int_{-1}^1 \frac{1}{\sqrt{1-x^2}} f(x)\,dx \simeq \frac{\pi}{N}\sum_{i=1}^N f(x_i) \qquad (9.6.39)$$

is exact when f is a polynomial of degree at most $2N-1$. Thus if (9.6.39) is applied to the integral $\int_{-1}^1 (x^2/\sqrt{1-x^2})\,dx$, the errors will be of the order of magnitude of the machine unit μ.

Example 9.6.18. In Example 9.1.2 we considered the period of oscillation of a mass subjected to a restoring term $f(x)$. For odd f it was found that the period was

$$T = 2\sqrt{2}\int_0^{x_{max}} \frac{1}{\sqrt{F(x_{max})-F(x)}}\,dx \qquad (9.6.40)$$

where $F(x)$ is any antiderivative of f. Since F is even, we can instead write

$$T = \sqrt{2}\int_{-x_{max}}^{x_{max}} \frac{1}{\sqrt{F(x_{max})-F(x)}}\,dx = x_{max}\sqrt{2}\int_{-1}^1 \frac{1}{\sqrt{F(x_{max})-F(x_{max}\bar{x})}}\,d\bar{x} \qquad (9.6.41)$$

where the last integral is obtained upon making the change of variables $\bar{x} = x/x_{max}$.

For an anharmonic oscillator $f(x):=f_1 x + f_3 x^3$, we have $F(x) = \frac{1}{2}f_1 x^2 + \frac{1}{4}f_3 x^4$. The requirement that the restoring force not vanish for $0<|x|\le x_{max}$ gives $|f_3|x_{max}^2 < f_1$. With this restriction

$$F(x_{max})-F(x_{max}\bar{x}) = (1-\bar{x}^2)[\tfrac{1}{2}f_1 x_{max}^2 + \tfrac{1}{4}f_3 x_{max}^4(1+\bar{x}^2)] \qquad (9.6.42)$$

and the second factor does not vanish in $[-1,1]$. This gives

$$T = 2\int_{-1}^1 \frac{1}{\sqrt{1-\bar{x}^2}}\frac{1}{\sqrt{f_1 + (f_3/2)x_{max}^2(1+\bar{x}^2)}}\,d\bar{x} \qquad (9.6.43)$$

where the second factor in the integrand is continuous in $[-1,1]$. Table 9.6.19

TABLE 9.6.19

N	Approximation	Change
2	5.358 318 433 9E + 0000	
4	5.366 646 717 1E + 0000	8.328E − 0003
6	5.366 659 333 6E + 0000	1.262E − 0005
8	5.366 659 355 2E + 0000	2.160E − 0008
10	5.366 659 355 2E + 0000	3.638E − 0011
12	5.366 659 355 2E + 0000	7.276E − 0012
14	5.366 659 355 2E + 0000	−7.276E − 0012
16	5.366 659 355 2E + 0000	7.276E − 0012

shows the results of applying the Gauss-Chebyshev integration formula (9.6.37) to (9.6.43). Even the two-point approximation is accurate to one digit past the decimal point. Eleven-digit accuracy is attained by $N = 8$. The third column indicates the change from the previous approximation.

In Example 9.6.18 it was important that we were able to explicitly factor out the $(1/\sqrt{1 - \bar{x}^2})$ from $F(x_{max}) - F(x_{max}\bar{x})$. For the problem of calculating the period of a pendulum, this poses a difficulty. From (9.6.41) with $F(\theta) = \cos \theta$,

$$T = \theta_{max}\sqrt{2} \int_{-1}^{1} \frac{1}{\sqrt{\cos \theta_{max} - \cos (\theta_{max}\theta)}} \, d\theta \qquad (9.6.44)$$

We can write this in the form (9.6.39) with

$$f(\theta) = \sqrt{\frac{1 - \theta^2}{\cos \theta_{max} - \cos (\theta_{max}\theta)}} \, d\theta \qquad (9.6.45)$$

However, the function f, while it is infinitely differentiable, is difficult to calculate numerically in this form since near $\theta = \pm 1$ the quotient of two nearly zero quantities must be found. This is the same problem that arises in numerical differentiation and, as we have seen, is inherently difficult. The pendulum problem is considered further in Exercise 18.

While the above discussion has been directed to the application of Gauss-Chebyshev integration to improper integrals, it may be employed in approximating any integral. Indeed,

$$\int_{-1}^{1} f(x) \, dx = \int_{-1}^{1} \frac{1}{\sqrt{1 - x^2}} \sqrt{1 - x^2} f(x) \, dx$$

$$= \frac{\pi}{N} \sum_{i=1}^{N} \sqrt{1 - \cos^2\left[\frac{(2i - 1)\pi}{2N}\right]} f(x_i)$$

$$= \frac{\pi}{N} \sum_{i=1}^{N} \sin\left[\frac{(2i - 1)\pi}{2N}\right] f(x_i) \qquad (9.6.46)$$

The approach of gaussian integration also works for other improper integrals. In particular, integrals of the form

$$\int_{0}^{\infty} e^{-x} f(x) \, dx \qquad (9.6.47)$$

can be approximated by a gaussian formula based upon the zeros of the *Laguerre polynomials* which are orthogonal in $(0, \infty)$ with respect to the weight function e^{-x}, and integrals of the form

$$\int_{-\infty}^{\infty} e^{-x^2} f(x) \, dx \qquad (9.6.48)$$

can be calculated by a gaussian formula using the zeros of the *Hermite polynomials*.

EXERCISES 9.6

1. Apply the two- and three-point Gauss-Legendre integration formulas to approximate the following integrals. Compare the approximations with the exact answers and the approximations given by the trapezoid and Simpson rules.

   (a) $\int_{-1}^{1} \frac{1}{1+x^2} \, dx$

   (b) $\int_{0}^{3} xe^{-x^2} \, dx$

2. Use error formula (9.6.30) to estimate the number of points that would be needed in the Gauss-Legendre formula to obtain an approximation of accuracy 5×10^{-4} for the following integrals.

   (a) $\int_{0}^{\pi} \sin x \, dx$

   (b) $\int_{0}^{2} e^{-x} \, dx$

3. Write a program implementing Algorithm 9.6.6. Print out the approximations and the errors made for $N = 2, \ldots, 6$ for the following problems.

   (a) $\int_{0}^{3} \frac{x}{1+x^2} \, dx$

   (b) $\int_{0}^{3} xe^{-x^2} \, dx$

4. Find the two- and three-point Gauss-Legendre approximations to $\int_{0}^{x} \sin^2 t \, dt$ for $x = \pi/2, \pi, 3\pi/2, 2\pi$.

5. Use two- and three-point Gauss-Chebyshev integration to find approximations to the following.

   (a) $\int_{-1}^{1} \frac{e^{-x}}{\sqrt{1-x^2}} \, dx$

   (b) $\int_{0}^{\pi} \frac{\cos x}{\sqrt{\pi^2 - x^2}} \, dx$

6. Use formula (9.6.46) to find two- and three-point Gauss-Chebyshev approximations to the following.

   (a) $\int_{-1}^{1} \frac{1}{1+x^2} \, dx$

   (b) $\int_{0}^{3} xe^{-x^2} \, dx$

Compare these approximations to the actual answers and the Gauss-Legendre, trapezoid, and Simpson approximations of Exercise 1.

7. Write a program that will calculate the N-point Gauss-Chebyshev approximation to

   $$\int_{-1}^{1} \frac{f(x)}{\sqrt{1-x^2}} \, dx$$

   Use the program to approximate the two integrals in Exercise 5.

8. Write a program implementing Algorithm 9.6.13 for adaptive gaussian integration. Have the program approximate the integrals in Exercise 3 to the tolerances (a) 5×10^{-5} and (b) 5×10^{-4}.

9. Write an adaptive integration program based upon the Gauss-Chebyshev formula (9.6.46). Have the program approximate the integrals in Exercise 3 to the tolerances (a) 5×10^{-5} and (b) 5×10^{-4}.

10. Derive the Gauss three-point formula by solving system (9.6.4) for the weights and nodes when $N = 3$.

11. (a) Derive the trapezoid rule by assuming an approximation of the form (9.6.3) and choosing the weights to integrate exactly as high a degree as possible. What is the highest-degree polynomial that is in fact integrated exactly?

    (b) Derive the Simpson rule by assuming an approximation of the form (9.6.3) and choosing the weights to integrate exactly as high a degree as possible. What is the highest-degree polynomial that is in fact integrated exactly?

12. Verify that the polynomial h given by (9.6.24) satisfies conditions (9.6.23).

13. Following the derivation of the error made by the Lagrange interpolant, derive formula (9.6.25) for the error made by the Hermite interpolant.

14. (a) Derive formula (9.6.28).
    (b) Starting from formula (9.6.33), complete the derivation of the error formula (9.6.30).

15. Compare the efficiency of an adaptive scheme based upon the Gauss-Legendre N-point formulas for $N = 3, 6, 9, \ldots$ with the one used in Algorithm 9.6.13.

16. In this exercise we derive formula (9.6.22) for the weights for Gauss-Legendre integration.

(a) Show $w_i = \int_{-1}^{1} l_i(x) \, dx$. (*Hint:* See the proof of Corollary 9.6.4.)

(b) Derive *Christoffel's* identity

$$(t - x) \sum_{i=0}^{n} (2i + 1) P_i(x) P_i(t)$$

$$= (n + 1)[P_{n+1}(t) P_n(x) - P_n(t) P_{n+1}(x)]$$

(9.6.49)

Hint: Use property 2 of Table 9.6.2 to show that

$$(2i + 1)(t - x) P_i(x) P_i(t) = (i + 1)$$

$$\times [P_{i+1}(t) P_i(x) - P_i(t) P_{i+1}(x)]$$

$$- i[P_i(t) P_{i-1}(x) - P_{i-1}(t) P_i(x)] \quad (9.6.50)$$

(c) Use Christoffel's identity to show that for a zero x_k of P_n,

$$\int_{-1}^{1} \frac{P_n(x)}{x - x_k} \, dx = \frac{-2}{(n + 1) P_{n+1}(x_k)} \quad (9.6.51)$$

(d) Show

$$w_k = \frac{2}{n P_n'(x_k) P_{n-1}(x_k)} \quad (9.6.52)$$

[*Hint:* Note that

$$l_i(x) = \frac{L(x)}{L'(x_i)(x - x_i)}$$

and use the result of parts (a) and (c).]

(e) Show that $(1 - x^2) P_n'(x) + nx P_n(x) = P_{n-1}(x)$. [*Hint:* Argue that

$$(1 - x^2) P_n'(x) + nx P_n(x) = \sum_{j=0}^{n-1} a_j P_j(x)$$

for some a_j. Multiply by $P_k(x)$, $0 \le k \le n - 1$, integrate this equation, and use the properties of Legendre polynomials to show $a_k = 0$, $k < n - 1$, and $a_{n-1} = n$.]

(f) Use (9.6.52) and the result of part (e) to establish (9.6.22).

17. Suppose that $\{p_j\}$ with $p_j \in \Pi_j$ is orthogonal on (a, b) with respect to a weight function $w(x)$. Suppose moreover that $\{p_j\}_{j=0}^{N}$ is a basis for Π_N. Let

$$\langle f, g \rangle := \int_a^b w(x) f(x) g(x) \, dx \quad (9.6.53)$$

(a) Show that $\langle p, p_N \rangle = 0$ for all $p \in \Pi_{N-1}$.

(b) Show that

$$\begin{bmatrix} p_0(x_1) & \cdots & p_0(x_N) \\ \cdots & \cdots & \cdots \\ p_{N-1}(x_1) & \cdots & p_{N-1}(x_N) \end{bmatrix} \quad (9.6.54)$$

is nonsingular for any mutually distinct choice of the points $\{x_i\}$ in (a, b).

(c) Show that if a set of weights $\{w_i\}_{i=1}^{N}$ is chosen to satisfy the equations

$$\sum_{i=1}^{N} x_i^j w_i = \int_a^b w(x) x^j \, dx \qquad j = 0, \ldots, N - 1$$

(9.6.55)

where $\{x_i\}_{i=1}^{N}$ is the set of zeros of p_n, then the approximation formula

$$\int_a^b w(x) f(x) \, dx \simeq \sum_{i=1}^{N} w_i f(x_i) \quad (9.6.56)$$

integrates any $p \in \Pi_{2N-1}$ exactly and the w_i are given by (9.6.38).

18. (a) Use formula (9.1.16) to determine the period of a pendulum with $g/l = 1$ and $\theta_{max} = \pi/2$ to an accuracy of six decimal places. Also approximate the period using:

   i. Gauss-Legendre integration for $N = 2, 4, \ldots, 12$ applied to formula (9.6.44).

   ii. Gauss-Chebyshev integration for $N = 2, 4, \ldots, 12$ applied to the formula

$$T = \theta_{max} \sqrt{2} \int_{-1}^{1} \frac{1}{\sqrt{1 - \theta^2}}$$

$$\times \frac{\sqrt{1 - \theta^2}}{\sqrt{\cos \theta_{max} - \cos (\theta_{max} \theta)}} \, d\theta \quad (9.6.57)$$

(b) As we saw in part (a) straightforward applications of gaussian integration to (9.6.44) yield only moderate accuracy. To obtain further accuracy we need to explicitly factor $1 - \theta^2$ from $\cos (\theta_{max}) - \cos (\theta_{max} \theta)$ and apply Gauss-Chebyshev integration. Show that

$$\cos \theta_{max} - \cos (\theta_{max} \theta)$$

$$= (1 - \theta^2) \sum_{k=0}^{\infty} \theta^{2k} \left[\cos \theta_{max} - \sum_{n=0}^{k} \frac{\theta_{max}^{2n}}{(2n)!} \right]$$

(9.6.58)

Argue that the coefficient of θ^{2k} is bounded by $1/(k + 1)!$. Based upon this observation, write a subroutine that will calculate

$$\sqrt{\frac{1 - \theta^2}{\cos \theta_{max} - \cos (\theta_{max} \theta)}}$$

to a specified accuracy. Use this subroutine and Gauss-Chebyshev integration to calculate the period of a pendulum to six decimal places.

PROJECTS

1. Write a program that will calculate the outward unit-normal vector to a surface $z = f(x, y)$ for input values of (x, y). Also write a program that will estimate the outward normal vector from a file of values of a bivariate function at a rectangular grid of points.

2. Use Laplace transforms to show that the solution to the initial value problem

$$mx'' + bx' + kx = f(t) \qquad x(0) = x_0 \qquad x'(0) = x_0'$$

which describes a forced, damped harmonic oscillator is

$$x(t) = x_0 e^{-bt/2m} \cos \omega t + \frac{bx_0 + mx_0'}{m\omega} e^{-bt/2m} \sin \omega t$$

$$+ \frac{1}{\omega m} \int_0^t e^{-bv/2m} \sin \omega v f(t - v) \, dv$$

when $b^2 < 4mk$ and $\omega := \sqrt{k/m - b^2/(4m^2)}$. Write a program to make graphs of the solution for various initial conditions when the external force f is a periodic function such as a sine wave or a square wave. Use numerical integration to calculate the integral in the formula above. Pay particular attention to the case where the angular frequency of the forcing function is at or near the resonant value of $\sqrt{k/m - b^2/(2m^2)}$.

3. The Fourier series of a function f which is periodic with period $2L$ is given by

$$f(x) \sim \frac{a_0}{2} + \sum_{n-1}^{\infty} a_n \cos \frac{n\pi x}{L} + b_n \sin \frac{n\pi x}{L}$$

where

$$a_n = \frac{1}{L} \int_{-L}^{L} f(x) \cos \frac{n\pi x}{L} \, dx$$

$$b_n = \frac{1}{L} \int_{-L}^{L} f(x) \sin \frac{n\pi x}{L} \, dx$$

Write a program that will draw graphs of the partial sums of the Fourier series of a function using numerical integration to compute the coefficients a_n and b_n. Have the program graph the Fourier series of discontinuous as well as continuous functions. When the function is discontinuous, you will encounter *Gibbs' phenomenon*. Refer to texts on differential equations and Fourier series to find out more about the phenomenon. It can be somewhat ameliorated by the use of σ factors. See Lanczos'

Applied Analysis (1988), for instance. Incorporate σ factors into your program, and produce graphs illustrating their effect.

4. Work out a Romberg integration scheme based upon the second Euler-Maclaurin summation formula (Exercise 13 in Sec. 9.5). Test it upon integrals having singularities at the endpoints such as the function Si defined in Exercise 8 in Sec. 9.3 and the formulas for the period of an anharmonic oscillator and a simple pendulum.

5. Use one of the root-finding schemes considered in Chap. 2 to find all nonnegative roots of each of the Legendre polynomials $P_{14}, P_{16}, \ldots, P_{20}$ to as high an accuracy as possible. Use the values of these roots to calculate the weights for Gauss-Legendre integration and thereby extend Table 9.6.5. Add the nodes and weights to the adaptive gaussian integration scheme of Algorithm 9.6.13.

6. Find out more about Gauss-Hermite or Gauss-Laguerre integration. In particular make a table of properties similar to Table 9.6.2 for the properties of Legendre polynomials, prove analogs of Theorems 9.6.3 and 9.6.11, and find derivations of the formulas for the weights. Write a program implementing the integration scheme you are studying.

7. *Multidimensional integration.* (*a*) Given a region $\Re = \{(x, y): a \le x \le b, g(x) \le y \le f(x)\}$, the two-dimensional integral $\int \int_\Re h(x, y) \, dx \, dy$ can be calculated as the iterated integral $\int_a^b \int_{g(x)}^{f(x)} h(x, y) \, dy \, dx$. Use one of the adaptive methods developed in this chapter to write a program to calculate such an iterated integral to within a specified accuracy. Apply your program to calculating the following integrals to an accuracy of 5×10^{-3}:

i. $\displaystyle\int_{\pi/6}^{\pi/4} \int_0^{\sin x} e^y \cos x \, dy \, dx$

ii. $\displaystyle\int_0^1 \int_x^1 \frac{1}{y} \sin y \cos \frac{x}{y} \, dy \, dx$

In each case compare your approximation with the actual answer. Record the number of times that the integrand needed to be evaluated to attain the specified accuracy.

(*b*) The approach of part (*a*) is limited to the situation in which a two-dimensional integral can be expressed as an iterated integral and is, in any event, rather inefficient. An alternative is to develop composite integration rules analogous to the trapezoid and Simpson rules for one-dimensional integration. A compendium of two- and three-dimensional integration formulas that are accurate over small polygonal regions can be found in chap. 25 of Abramowitz and Stegun (1965). Select one of these formulas and write a composite integration program based upon it. Apply your program to the examples of part (a) and to approximating $\int_{x^2+y^2\leq 1} e^{-(x^2+y^2)/\sigma} \, dx \, dy$ for $\sigma = 1$ and 10.

(c) Use one of the polygonal formulas in Abramowitz and Stegun (1965) to develop an adaptive scheme of two-dimensional integration analogous to adaptive Simpson quadrature. Test your program on the examples of parts (a) and (b).

CHAPTER
10

NUMERICAL SOLUTION OF DIFFERENTIAL EQUATIONS

10.1 INTRODUCTION: EULER'S METHOD AND EXAMPLES

An ordinary differential equation of order n relates an unknown function $x(t)$ to its derivatives $x', x'', \ldots, x^{(n)}$. The importance of differential equations in science stems from the fact that it is often relatively easy to reason about how an unknown function changes relative to its current value.

> **Example 10.1.1 Temperature of building.** Given adequate circulation of air, the temperature of a building may be characterized by a single value $T(t)$. At the surface of the building there will be a loss or gain of heat due to convection. *Newton's law of cooling* is an empirical law stating that the change in the temperature of an object resulting from convection is proportional to the discrepancy between the temperature of the object and the temperature $A(t)$ of its surroundings (often called the *ambient* temperature). This gives the differential equation
>
> $$T'(t) = k[A(t) - T(t)] \qquad (10.1.1)$$
>
> where k is a positive constant. This is a first-order linear differential equation. If the temperature of the building at time $t = 0$ is T_0, then the solution to (10.1.1) is given by
>
> $$T(t) = T_0 e^{-kt} + ke^{-kt} \int_0^t A(\tau)e^{k\tau}\, d\tau \qquad (10.1.2)$$
>
> See Exercise 1. Depending upon the form of the ambient temperature function $A(t)$, the integral in (10.1.2) may or may not have a closed-form representation.

For instance, if the ambient temperature is a constant A_0, then the solution is simply

$$T(t) = (T_0 - A_0)e^{-kt} + A_0 \qquad (10.1.3)$$

and thus the temperature of the building approaches the ambient temperature. If the integral in (10.1.2) cannot be expressed in closed form, then it may be approximated by one of the methods of numerical integration discussed in Chap. 9. However, methods such as Romberg integration, adaptive Simpson quadrature, and Gauss quadrature require at least three to six evaluations of the integrand. Thus use of such methods in the calculation of the solution given by (10.1.2) may be efficient when the temperature is needed at only a few values of t. However, if we wish to construct a graph of the solution and hence need $T(t)$ at many values of t, it turns out to be more efficient to approximate $T(t)$ directly from the differential equation. Various ways to do this are the subject of this chapter.

The differential equation (10.1.1) takes into account only the heating or cooling that results from the difference between the temperature of the building and its surroundings. Other factors such as heat emitted internally by machinery or by people occupying the building or the presence of an air-conditioning or heating system will also affect the temperature. For instance, the differential equation

$$T'(t) = k[A(t) - T(t)] + R(T(t)) \qquad (10.1.4)$$

where

$$R(T) = \begin{cases} -U_0 & T \geq T_{\text{crit}} \\ 0 & T < T_{\text{crit}} \end{cases} \qquad (10.1.5)$$

describes the temperature of a building with an air-conditioning system. The air-conditioning system is off when the temperature of the building is less than some value T_{crit} set by the thermostat and cools the building at a fixed rate of U_0 degrees per unit time when the temperature of the building exceeds T_{crit}. There is no simple formula to represent the solution $T(t)$ to (10.1.4), and the best recourse is to approximate the solution numerically.

Example 10.1.2 Logistic model of population growth. The rate at which a population of organisms grows clearly depends upon the number of organisms $P(t)$ currently alive. Let us take as a prototype for a model of population growth the differential equation

$$P'(t) = r(P(t))P(t) \qquad (10.1.6)$$

The function $r(P)$ is the net birthrate per organism, that is, the birthrate minus the death rate. Because the growth of any population will be constrained by limited supplies of food and space, it is expected that the net birthrate will fall as P increases, eventually becoming zero at some value P_{max} representing the *carrying capacity* of the environment. The exact form of r is not clear and may well depend upon the particular population being modeled. In the absence of any specific knowledge, one often starts with the simplest form of an unknown function that is consistent with expectations. In the case at hand, $r(P) = a - bP$, $a, b > 0$, which predicts a simple linear decrease in the birthrate with increasing population, fits our requirements. This gives the *logistic model* of population growth

$$P'(t) = [a - bP(t)]P(t) \qquad a, b > 0 \tag{10.1.7}$$

To obtain a specific solution to the equation we must know the initial size of the population $P(0) = P_0$. When a and b are constant, the differential equation (10.1.7) is separable and thus may be solved analytically. The solution is

$$P(t) = \frac{P_0 a e^{at}}{a + P_0 b(e^{at} - 1)} \tag{10.1.8}$$

(Exercise 2). Solution curves for the choice $a = 1$, $b = 0.01$, and various initial conditions are shown in Fig. 10.1.3. Observe that all solutions shown approach the carrying capacity $P_{max} = 100$ as time t becomes large. Indeed, it is easily verified from (10.1.8) that $\lim_{t \to \infty} P(t) = a/b$.

A first-order differential equation is *autonomous* if it has the form $x'(t) = f(x)$; that is, it depends upon time only through the unknown function $x(t)$. An *equilibrium point* of an autonomous first-order differential equation is a value of x at which $x' = 0$. Note that $P = a/b$ and $P = 0$ are equilibrium points of (10.1.7). In general we expect equilibrium points to influence the behavior of solutions lying near to them. However, as Fig. 10.1.3 indicates, the nature of that influence may not be the same for all equilibria. The equilibrium point $P = a/b$ is a *stable node*, that is, solutions starting near it tend to approach it, while $P = 0$ is *unstable*, solutions starting near it tend to move away from it.

The logistic model derived in the example above is probably too simple to accurately model an actual population outside of a laboratory. However, the means by which it was obtained represent a reasonable approach to modeling a

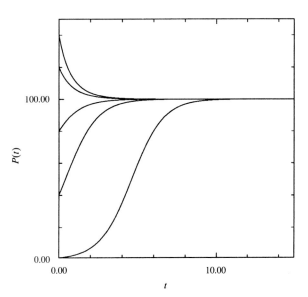

FIGURE 10.1.3

physical process. In general one begins with the simplest differential equation consistent with the fundamental facts, analyzes the behavior of its solutions, and then serially adds to or modifies the equation to ascertain whether such alterations improve the solution's agreement with the observed behavior of the system. Usually, after very few such modifications, the equation can no longer be solved analytically.

For instance, in the logistic model (10.1.7) if a population of herbivores is under consideration, then the food supply, and hence the carrying capacity of the environment, would vary from year to year depending upon the weather. The constraint parameter b is then a function of time, and the solution (10.1.8) is no longer valid. Equation (10.1.7) with time-dependent a and b is a *Bernoulli equation*. Such equations can be solved analytically, but as in the case of the solution (10.1.2) for the temperature of a building, the solutions may contain unevaluable integrals that must be approximated numerically (see Exercise 3). Again, direct numerical approximation of the solution may be the most efficient approach.

Direction Fields and Euler's Method

The central concern of this chapter is approximating the solution to an initial value problem for a differential equation. Let us first consider the initial value problem for a first-order differential equation

$$x' = f(t, x) \qquad x(t_0) = x_0 \qquad\qquad (10.1.9)$$

The models for temperature of a building and logistic population growth discussed above have this form. Note that according to the differential equation in (10.1.9), the slope of a solution passing through a given point (t, x) is $f(t, x)$. A useful aid in visualizing the behavior of solutions to a first-order differential equation is a *direction field*. A lattice of points is chosen in a region of interest in the tx plane, and the slope is calculated from the differential equation at each lattice point. A short line segment having that slope is then drawn at each lattice point.

> **Example 10.1.4.** Figures 10.1.5 and 10.1.6 show direction fields for solutions to a logistic differential equation $x' = [1 - b(t)x]x$ when $b(t) \equiv 0.01$ and when $b(t) = 0.01[(1 + 0.2t)/(1 + 0.1t)]$, respectively. The latter choice of $b(t)$ represents an increase over time of the parameter b from 0.01 at $t = 0$ to 0.02 for large t.

Euler's method is a simple scheme for approximating the solution to the initial value problem (10.1.9). We partition the time interval (t_0, t_f) into N subintervals separated by nodes $t_i = t_0 + ih$ with $h = (t_f - t_0)/N$ and seek approximations ξ_i to $x(t_i)$. The parameter h is called the *stepsize*. Euler's method generates approximations simply by following the direction field. The slope at the point (t_0, x_0) given by the initial condition is $f(t_0, x_0)$. The line passing through (t_0, x_0) having this slope is then $x - x_0 = f(t_0, x_0)(t - t_0)$. The Euler approximation is the ordinate of this line when $t = t_1 = t_0 + h$; that is,

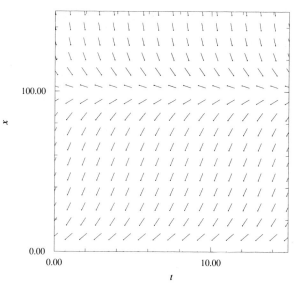

FIGURE 10.1.5

$\xi_1 = x_0 + hf(t_0, x_0)$. Having established an approximation ξ_1 to $x(t_1)$, the process can be repeated to obtain the approximation $\xi_2 = \xi_1 + hf(t_1, \xi_1)$ to $x(t_2)$. *Euler's method* then is the iteration

$$\xi_0 = x_0$$

$$\xi_{i+1} = \xi_i + hf(t_i, \xi_i) \qquad i = 0, \ldots, N-1$$

(10.1.10)

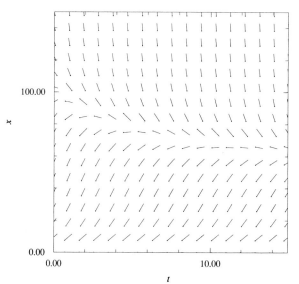

FIGURE 10.1.6

Algorithm 10.1.7   Euler's method for a first-order equation

Input t_0 and x_0 (independent and dependent coordinates specified by the
            initial condition)
            t_f (right endpoint of interval of approximation)
            N (number of subintervals to use)

Define   function $f(t, x)$ (right-hand side of the differential equation)

$t := t_0, \; x := x_0$
$h := (t_f - t_0)/N$
for $i = 1$ to N
        $\{x := x + hf(t, x)$
        $t := t_0 + ih$
        Output: $t, x\}$

Figure 10.1.8 shows the Euler method approximations to the initial value
problem

$$x' = (1 - 0.01x)x \qquad x(0) = 1 \qquad\qquad (10.1.11)$$

for $N = 5, 10, 20$, and 40. The points at which approximations were calculated
are marked by the symbol "+." These curves represent approximations to the
curve through $(0, 1)$ shown in Fig. 10.1.3. As the graph for $N = 5$ indicates, if
the stepsize is chosen too large, then approximations may be based upon
direction field information at points which lie far from the actual solution,
resulting in increasingly inaccurate answers.

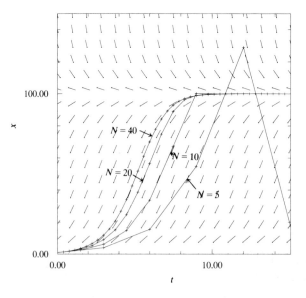

FIGURE 10.1.8

Euler's Method for Systems of Differential Equations

Euler's method extends readily to initial value problems for systems of first-order differential equations

$$
\begin{aligned}
x_1'(t) &= f_1(t, x_1, x_2, \ldots, x_n) & x_1(t_0) &= x_{10} \\
x_2'(t) &= f_2(t, x_1, x_2, \ldots, x_n) & x_2(t_0) &= x_{20} \\
&\cdots\cdots\cdots\cdots\cdots\cdots\cdots & & \\
x_n'(t) &= f_n(t, x_1, x_2, \ldots, x_n) & x_n(t_0) &= x_{n0}
\end{aligned}
\tag{10.1.12}
$$

The function $f_i(t, x_1, \ldots, x_n)$ gives the slope of the solution $x_i(t)$ at the point $(t, x_1, \ldots, x_n)$, and the basic Euler formula $\xi_{i+1} = \xi_i + h\ slope$ is simply applied to each equation separately. This gives the iteration

$$
\begin{aligned}
\xi_{m,0} &= x_{m0} \qquad m = 1, \ldots, n \\[4pt]
\xi_{1,i+1} &= \xi_{1,i} + hf_1(t_i, \xi_{1,i}, \xi_{2,i}, \ldots, \xi_{n,i}) \\
\xi_{2,i+1} &= \xi_{2,i} + hf_2(t_i, \xi_{1,i}, \xi_{2,i}, \ldots, \xi_{n,i}) \\
&\cdots\cdots\cdots\cdots\cdots\cdots\cdots\cdots\cdots\cdots \qquad i = 0, \ldots, N-1 \\
\xi_{n,i+1} &= \xi_{n,i} + hf_n(t_i, \xi_{1,i}, \xi_{2,i}, \ldots, \xi_{n,i})
\end{aligned}
\tag{10.1.13}
$$

where $\xi_{m,i}$ is the approximation to $x_m(t_i)$.

The resemblance of Euler's method for systems of first-order equations to Euler's method for a single first-order equation can be emphasized by the use of vector notation. With $\mathbf{x}(t) := [x_1(t), \ldots, x_n(t)]^{\mathrm{T}}$ and $\mathbf{f}(t, \mathbf{x}) = [f_1(t, \mathbf{x}), \ldots, f_n(t, \mathbf{x})]^{\mathrm{T}}$, the initial value problem (10.1.12) is

$$
\mathbf{x}' = \mathbf{f}(t, \mathbf{x}) \qquad \mathbf{x}(t_0) = \mathbf{x}_0
\tag{10.1.14}
$$

and with $\boldsymbol{\xi}_i := [\xi_{1,i}, \ldots, \xi_{n,i}]^{\mathrm{T}}$, the Euler method iteration (10.1.13) is

$$
\begin{aligned}
\boldsymbol{\xi}_0 &= \mathbf{x}_0 \\
\boldsymbol{\xi}_{i+1} &= \boldsymbol{\xi}_i + h\mathbf{f}(t_i, \boldsymbol{\xi}_i)
\end{aligned}
\tag{10.1.15}
$$

Example 10.1.9 Predator-prey models. The logistic model of population growth of Example 10.1.2 does not explicitly take into account the interaction of the population with other populations of organisms for which it is either predator or prey. To model such an interaction, let $P_1(t)$ denote the number of organisms in a prey population at time t and $P_2(t)$ denote the number of its predators. We use the same growth model for both species:

$$
\begin{aligned}
P_1'(t) &= r_1(P_1, P_2)P_1 \\
P_2'(t) &= r_2(P_1, P_2)P_2
\end{aligned}
\tag{10.1.16}
$$

If, as with the logistic model, we assume the simplest forms for r_1 and r_2 that are consistent with our expectation about such an interaction, then we obtain the *Lotka-Volterra* predator-prey model

$$P_1'(t) = (a - bP_2)P_1$$

$$P_2'(t) = (-c + dP_1)P_2$$

(10.1.17)

where a, b, c, and d are positive constants. This system cannot be solved analytically for x_1 and x_2 as a function of time t. Hence its solutions must be approximated numerically.

Higher-Order Differential Equations

The solution to initial value problems for higher-order differential equations

$$x^{(n)}(t) = f(t, x, x', \ldots, x^{(n-1)})$$

$$x(t_0) = x_0 \qquad x'(t_0) = x_0', \ldots, x^{(n-1)}(t_0) = x_0^{(n-1)}$$

(10.1.18)

is usually approximated numerically by writing the higher-order equation as an equivalent first-order system. Let $x_1(t) := x(t)$, $x_2(t) := x'(t), \ldots, x_n(t) := x^{(n-1)}(t)$. Then (10.1.18) is equivalent to the initial value problem

$$\begin{aligned} x_1' &= x_2 & x_1(t_0) &= x_0 \\ x_2' &= x_3 & x_2(t_0) &= x_0' \\ &\cdots\cdots\cdots\cdots\cdots\cdots\cdots\cdots\cdots\cdots \\ x_{n-1}' &= x_n & x_{n-1}(t_0) &= x_0^{(n-2)} \\ x_n' &= f(t, x_1, x_2, \ldots, x_n) & x_n(t_0) &= x_0^{(n-1)} \end{aligned}$$

(10.1.19)

Example 10.1.10 Duffing's equation. In Example 9.1.2 the motion of an object of mass m subject to a restoring force $\kappa(x)$ such as that exerted by a spring was considered. The motion is governed by the differential equation $mx'' + \kappa(x) = 0$. Other forces such as friction and externally applied forces may be present. The functional form of the frictional force depends upon the nature of the friction. *Viscous* friction arises from the contact of the object's surface with a liquid medium and is approximately proportional to the velocity of the object. If the motion is resisted by viscous damping and also subject to an external force $f(t)$ and the spring force is taken to be $\kappa(x) = kx + \epsilon x^3$, then from Newton's second law, the system is governed by the *forced Duffing equation*

$$mx'' + bx' + kx + \epsilon x^3 = f(t)$$

(10.1.20)

Appropriate initial conditions are to specify the position x_0 and velocity v_0 of the mass at some starting time, which we take to be $t = 0$. Writing this as an initial value problem for a system of first-order differential equations of the form (10.1.19) gives

$$\begin{aligned} x_1' &= x_2 & x_1(0) &= x_0 \\ x_2' &= -bx_2 - kx_1 - \epsilon x_1^3 + f(t) & x_2(0) &= v_0 \end{aligned}$$

(10.1.21)

Except when $\epsilon = 0$ (a forced, damped harmonic oscillator), this system cannot be solved analytically.

Nondimensionalization of Problems

Example 10.1.11 Height of a projectile. The height h of a projectile fired upward at velocity v_0 from the surface of a planet is governed by the initial value problem

$$h'' = -\frac{g}{(1 + h/R)^2} \qquad h(0) = 0 \qquad h'(0) = v_0 \qquad (10.1.22)$$

where R is the radius of the planet and g is the acceleration due to gravity at the surface of the planet (Exercise 13a). Air resistance has been neglected. If we define the dimensionless variables $h^* := h/R$ and $t^* := \sqrt{g/R}\,t$, then (10.1.22), takes the simpler form

$$\frac{d^2 h^*}{dt^{*2}} = -\frac{1}{(1 + h^*)^2} \qquad h^*(0) = 0 \qquad h^{*'}(0) = \frac{v_0}{\sqrt{gR}} \qquad (10.1.23)$$

The dimensionless form (10.1.23) of the initial value problem has the advantage that the three parameters g, R, and v_0 occurring in (10.1.22) have coalesced into the single dimensionless parameter $v_0^* := v_0/\sqrt{gR}$. Changes in the three original parameters can affect $h(t)$ only to the extent they change the value of v_0^*. This is important because one often wishes to assess the effect of various parameters on the solution of a differential equation. If such an assessment cannot be accomplished analytically, then one recourse is to examine numerical solutions for different sets of parameters. Clearly the success of such an approach depends upon having as few parameters as possible.

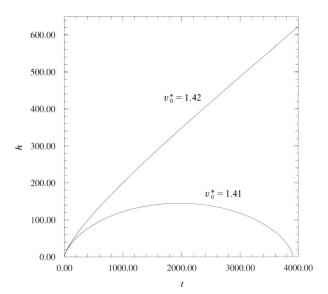

FIGURE 10.1.12

To illustrate this let us attempt to find by numerical experimentation the smallest value of v_0 that will cause the projectile to escape the planet's gravitational field. Figure 10.1.12 shows the numerical solution of (10.1.23) for $v_0^* = 1.41$ and $v_0^* = 1.42$. For $v_0^* = 1.42$ the height becomes a linearly increasing function of time as time becomes large, and thus the projectile is apparently escaping the planet's gravitational field. In terms of the original parameters of the problem this gives the approximate formula $v_0 \simeq 1.42\sqrt{gR}$ for the escape velocity of the projectile. By an analytical argument the actual formula is found to be $v_0 = \sqrt{2gR}$ (Exercise 14). Observe that it would have been very difficult to arrive at the approximate formula by experimenting with numerical solutions to the original initial value problem (10.1.22), which involved the three parameters g, R, and v_0.

That nondimensionalizing the projectile problem (10.1.22) produced a problem (10.1.23) depending upon fewer parameters is not coincidental. The *Buckingham Π theorem* states that a physical problem rendered in nondimensional form contains the fewest parameters possible. See Lin and Segal (1974) for further discussion of this result.

Example 10.1.13 Nondimensional Duffing equation. Let us consider the Duffing equation (10.1.20) with a sinusoidal forcing term, no damping, and zero initial conditions:

$$mx'' + kx + \epsilon x^3 = F_0 \cos \omega t \qquad x(0) = 0 \qquad x'(0) = 0 \qquad (10.1.24)$$

How small does the coefficient ϵ of the nonlinear term in the spring force have to be for it to be neglected? This is a difficult question, as the answer may depend upon the magnitudes of the other four parameters m, k, F_0, and ω in the problem. However, noting that the solution in the linear case, $\epsilon = 0$, has the form

$$x(t) = c_1 \cos\sqrt{\frac{k}{m}}\, t + c_2 \sin\sqrt{\frac{k}{m}}\, t + \frac{F_0}{|k - m\omega^2|} \cos(\omega t - \phi) \quad (10.1.25)$$

suggests that the change to the dimensionless variables $t^* := \omega t$ and $x^* := |k - m\omega^2|/F_0$ should scale the magnitude of x^* to be roughly unity. In the new variables, (10.1.24) becomes

$$\frac{d^2 x^*}{dt^{*2}} + k^* x^* + \epsilon^* x^{*3} = |k^* - 1| \cos t^* \qquad (10.1.26)$$

with

$$k^* = \frac{k}{m\omega^2} \qquad \epsilon^* = \frac{\epsilon F_0^2}{m\omega^2 (k - m\omega^2)^2} \qquad (10.1.27)$$

The solution depends now upon the two dimensionless parameters k^* and ϵ^*. With x^* being of roughly unit magnitude, the condition that the nonlinear term be inconsequential is $\epsilon^* \ll k^*$ or, in terms of the original parameters,

$$\epsilon \ll \frac{k(k - m\omega^2)^2}{F_0^2} \qquad (10.1.28)$$

where $\ll$ is rendered "much smaller than." Such a notion is not mathematically precise, depending at minimum upon the desired accuracy of the approximation; however, it represents a considerable refinement of the intuitive $\epsilon \ll k$. Nor may "$\ll$" be independent of the value of other dimensionless parameters in the problem. Figure 10.1.14 shows that numerical solutions to (10.1.26) for $\epsilon^* = 0$

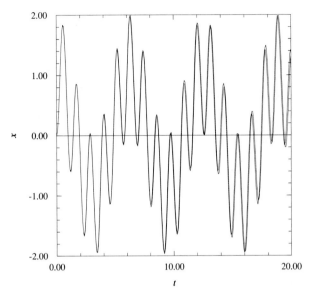

FIGURE 10.1.14

and $\epsilon^* = 0.1$ are nearly identical when $k^* = 30$. However, Fig. 10.1.15 shows that solutions corresponding to the same two values of ϵ^* differ considerably when $k^* = 2$. Thus a relation such as $\epsilon^* \ll |k^* - 1|/k^*$ based upon the relative distance of k^* from its resonant value of $k^* = 1$ might be more appropriate. It is nevertheless clear that the change to dimensionless form has clarified and simplified the problem and indicated the direction that numerical experiments should take.

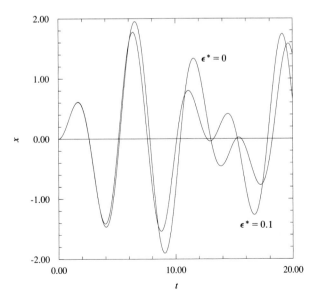

FIGURE 10.1.15

EXERCISES 10.1

1. (a) Derive the solution to the differential equation (10.1.1) given by (10.1.2).
   (b) For $T(0) = 50°, 60°$, and $80°$ find and graph the solutions given by (10.1.2) when $k = 0.5 \text{ hour}^{-1}$ and
   i. $A(t) = 70°$
   ii. $A(t) = 70° + 20° \sin(\pi t / 12)$

2. Show that the solution to the logistic equation (10.1.7) with the initial condition $P(0) = P_0$ is as given by (10.1.8).

3. (a) A *Bernoulli equation* has the form

$$x'(t) = a(t)x(t) + b(t)x(t)^n \quad (10.1.29)$$

   Show that this equation becomes a first-order linear equation upon making the change of variables $u = x^{1-n}$, and then use integrating factors to find a formula for x.

   (b) Suppose that the constraint parameter b in the logistic model (10.1.7) varies sinusoidally with time, $b(t) = B_0 + B_1 \sin \omega t$, where $B_0 > B_1$. Use the formula derived in part (a) to graph the population over two cycles of b when $a = 0.01$, $\omega = 2\pi$, $B_0 = 0.0001$, $B_1 = 0.00001$, and $P(0) = 20$.

4. (a) Find the Euler method approximation to $x(2)$ for the initial value problem $x' = 5t - 1 - 2x$, $x(0) = 1$ using stepsizes of $h = 1$ and $h = 0.5$. Compare your answers with the actual solution.

   (b) Find the Euler method approximation to $x(3)$ for the initial value problem $x' = (x^2 + x)/t$, $x(1) = 1$ using a stepsize of $h = 0.5$. Also find the actual solution analytically and discuss the validity of the approximation.

5. Write a computer program that graphs direction fields for a given first-order differential equation. Use your program to plot a direction field for the equation $x' = \sin tx$ in the square $0 \le t \le 5$, $0 \le x \le 5$.

6. Write a computer program implementing Euler's method for a single first-order differential equation. Apply it to tabulate approximations to the initial value problem $x' = 5t - 2x$, $x(0) = 1$ on $0 \le t \le 2$ for stepsizes of $h = 0.2, 0.1$, and 0.05. For each stepsize also have the program print out the value of the actual solution and the error made by the Euler approximation.

7. Use your Euler method program (Exercise 6) to construct graphs of the solutions to each of the two building temperature problems of Exercise 1b using $N = 20$ and $N = 40$ steps. On each graph also plot the actual solution.

8. (a) Use your Euler method program to find and graph the temperature of an air-conditioned building [see Eq. (10.1.4)] when $k = 0.5 \text{ hour}^{-1}$, $A(t) = 70°$, $U_0 = 4°/\text{hour}$, $T_{crit} = 65°$, and $T(0) = 50°$. Use a stepsize h that is small enough so that the same graph results from using stepsizes h and $h/2$.

   (b) Find and graph the solution of an air-conditioned building when the ambient temperature is given by $A(t) = 70 + 20 \sin(\pi t / 12)$. Use the parameters given in part (a).

9. Use your Euler method program to approximate the solution that you graphed in Exercises 3b. Use $N = 25, 50$, and 100 steps.

10. Calculate the Euler method approximations to $x_1(1)$ and $x_2(1)$ for the initial value problem

$$x_1' + x_2 = t^2 \qquad x_1(0) = 1$$
$$x_2' - x_1 = 1 \qquad x_2(0) = 0 \qquad (10.1.30)$$

   for stepsizes of $h = 1$ and $h = 0.5$. Compare with the values of the actual solution at $t = 1$.

11. Write the initial value problem $x'' + x = e^{-t}$, $x(0) = 1$, $x'(0) = 0$ as a system of two first-order differential equations. Use Euler's method to approximate $x(t)$ at $t = 0.5$ for stepsizes of $h = 0.5$ and $h = 0.25$. Compare these approximations to the actual value of the solution at $t = 0.5$.

12. (a) A spring is suspended from a ceiling. An object of mass m is attached to the spring and then lowered to its equilibrium position. The distance s that the spring is stretched in this process is $s = mg/k$. The spring is then set in motion. Show that if the spring obeys Hooke's law and the position of the object is measured from its equilibrium position, then its motion is still governed by the differential equation $mx'' + kx = 0$.

   (b) If the force exerted by the spring in part (a) instead obeys the nonlinear law $\kappa(x) = kx + \epsilon x^3$, what differential equation describes the motion of the spring if position is again measured from equilibrium?

13. Newton's law of universal gravitation states that the gravitational force exerted by a body of mass m_2 upon a body of mass m_1 is given by

$$\mathbf{F} = -\frac{Gm_1 m_2}{|\mathbf{r}|^3}\mathbf{r} \qquad (10.1.31)$$

where $\mathbf{r}$ is the vector from the position of the body of mass m_2 to the position of the body of mass m_1.

(a) Consider a body of mass m falling toward the surface of a planet of mass M and radius R. Supposing that air resistance is proportional to the square of the body's velocity, show that the height h of the body above the planet's surface satisfies the initial value problem

$$h'' - b(h')^2 + \frac{g}{(1 + h/R)^2} = 0 \qquad (10.1.32)$$

$$h(0) = h_0 \qquad h'(0) = v_0$$

where g is the acceleration due to gravity at the planet's surface.

   i. Letting $h^* := h/R$, $\ t^* := \sqrt{g/R}\,t$, write (10.1.32) in the nondimensional form

$$\frac{d^2 h^*}{dt^{*2}} - b^* \frac{dh^*}{dt^*} + \frac{1}{(1 + h^*)^2} = 0$$

Give a formula for b^*.

   ii. Write this initial value problem for the dimensionless equation as an initial value problem for a system of first-order equations.

(b) Find a system of six first-order differential equations for the position and velocity of a planet subject to the gravitational attraction of another planet.

14. Show that (10.1.22) can be written as

$$v\frac{dv}{dh} = -\frac{g}{(1 + h/R)^2} \qquad v(0) = v_0 \qquad (10.1.33)$$

if velocity $v := h'(t)$ is regarded as a function of height h. Solve this separable first-order differential equation, and use your solution to argue that the escape velocity of the projectile is $v_0 = \sqrt{2gR}$.

15. The damped Duffing equation

$$mx'' + bx' + kx + \epsilon x^3 = F_0 \cos \omega t \qquad (10.1.34)$$

has the steady state solution

$$x_{ss}(t) = \frac{F_0}{\sqrt{(k - m\omega^2)^2 + b^2\omega^2}} \cos(\omega t - \phi) \qquad (10.1.35)$$

when $\epsilon = 0$. Write (10.1.34) in terms of dimensionless variables and parameters in which the amplitude of the steady state would be about unity. Give a condition in terms of the original parameters under which the nonlinear term ϵx^3 would have little effect on the steady state solution.

16. (a) Show that the system of two masses and three springs shown in Fig. 10.1.16 satisfies the system of equations

$$m_1 x_1'' = -k_1 x_1 + k_2(x_2 - x_1)$$
$$m_2 x_2'' = -k_2(x_2 - x_1) - k_3 x_2 \qquad (10.1.36)$$

where the positions x_1 and x_2 are measured from the equilibrium positions of the two masses and the springs are assumed to obey Hooke's law. Write this system as a system of first-order differential equations.

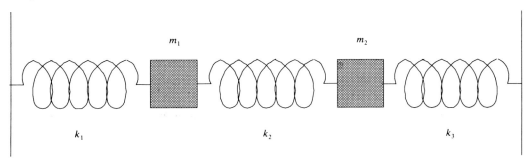

m_1 m_2

k_1 k_2 k_3

FIGURE 10.1.16

(*b*) What system of first-order differential equations would result if we assume instead that each spring exerts a restoring force of the form $\kappa(x) = kx + \epsilon x^3$?

17. Write a computer program implementing Euler's method for a system of two first-order differential equations. Apply it to tabulate approximations to the initial value problem

$$x_1' + x_2 = t^2 \qquad x_1(0) = 1$$
$$x_2' - x_1 = 1 \qquad x_2(0) = 0 \qquad (10.1.37)$$

on $0 \le t \le 1$ for stepsizes of $h = 0.2, 0.1$, and 0.05. For each stepsize also have the program print out the value of the actual solutions and the error made by the Euler approximations.

18. The angular displacement θ of a simple pendulum, measured from the downward vertical direction, is governed by the differential equation $\theta'' + g/l \sin\theta$, where g is the acceleration due to gravity and l is the length of the pendulum. Use Euler's method to graph the angular displacement of a pendulum of length 1 foot over two periods of its motion when $\theta(0) = \pi/2$, $\theta'(0) = 0$. Choose the stepsize h so that stepsizes of h and $h/2$ yield the same graph.

19. Consider the initial value problem

$$\theta'' + \frac{g}{l}\sin\theta \qquad \theta(0) = \theta_0 \qquad \theta'(0) = 0 \qquad (10.1.38)$$

describing a simple pendulum released from rest with an angular displacement θ_0. The Taylor expansion $\sin\theta = \theta - (1/3!)\theta^3 + \cdots$ suggests that for "small" θ_0 the solution to the *linearized pendulum equation*

$$\theta'' + \frac{g}{l}\theta = 0 \qquad \theta(0) = \theta_0 \qquad \theta'(0) = 0 \qquad (10.1.39)$$

which can be solved exactly, should be a good approximation to the solution of (10.1.38). Nondimensionalize the initial value problem (10.1.38) to find a condition in terms of g, l, and θ_0 which better expresses the condition under which solutions to (10.1.39) will be good approximations to those of (10.1.38).

20. Use Euler's method to graph the solutions to the Lotka-Volterra model (10.1.17). Take $a = 2$, $b = 1$, $c = 1$, $d = 1$ and use initial conditions $x_1(0) = 1$,

$x_2(0) = 0.1$. Graph solutions over the time interval $0 \le t \le 10$. Choose the stepsize h so that stepsizes of h and $h/2$ yield the same graph.

21. The initial value problem for the predator-prey model (10.1.17)

$$x_1' = (a - bx_2)x_1 \qquad x_1(0) = x_{10}$$
$$(10.1.40)$$
$$x_2' = (-c + dx_1)x_2 \qquad x_2(0) = x_{20}$$

depends upon six dimensioned parameters. If we take the dimensions of x_1 and x_2 to be organisms, then a and c have the dimensions of $1/\text{time}$, b and d have the dimensions of $1/(\text{organisms} \cdot \text{time})$, and x_{10} and x_{20} have the dimensions of organisms. Note that $\bar{x}_1 = c/d$, $\bar{x}_2 = a/b$ is an equilibrium point of the system (10.1.40). Show that by changing to the dimensionless variables

$$x_1^* = \frac{d}{c}x_1 \qquad x_2^* = \frac{b}{a}x_2 \qquad (10.1.41)$$

and making a suitable choice for a dimensionless time variable, the predator-prey equations can be put in the nondimensional form

$$\frac{dx_1^*}{dt^*} = (1 - x_2^*)x_1^* \qquad x_1^*(0) = x_{10}^*$$
$$(10.1.42)$$
$$\frac{dx_2^*}{dt^*} = \sigma(-1 + x_1^*)x_2^* \qquad x_2^*(0) = x_{20}^*$$

which depends upon only the three dimensionless parameters σ, x_{10}^*, and x_{20}^*.

22. The predator-prey equations (10.1.40) do not take into account resource constraints on the prey population. Adding a logistic term gives the system

$$x_1' = (a - bx_2 - ex_1)x_1$$
$$(10.1.43)$$
$$x_2' = (-c + dx_1)x_2$$

Find a dimensionless form for these equations.

23. (*a*) Derive a model similar to the Lotka-Volterra model (10.1.17) for the interaction of two species that compete for the same resources. *The law of competitive exclusion* states that two species cannot occupy the same competitive niche; one species will drive the other to extinction. Use Euler's method to investigate whether your model is consistent with the law of competitive exclusion.

(b) Following Exercise 21, put the initial value problem for the competing species model in dimensionless form.

24. In deriving the logistic model of population growth we started from the fundamental form $P' = f(P)P$. If we assume that there is a population P_{max} for which $f(P_{max}) = 0$, then from Taylor's theorem

$$f(P) = f'(P_{max})(P - P_{max}) + \frac{f''(P_{max})}{2!} (P - P_{max})^2$$

$$+ \cdots$$

(a) Show that retaining only the first term of this series results in the logistic growth model.

(b) For the model that results from retaining the first two terms of the Taylor expansion:

   i. Find the equilibrium points and classify each one as stable or unstable. What range of values of $f''(P_{max})$ leads to realistic conclusions about the growth of the population?

   ii. Use Euler's method to graph solutions to the model in part i for various initial conditions to illustrate its behavior in relation to the equilibrium points that you found.

25. The *Lorenz equations* [Lorenz (1963)]

$$x_1' = s(x_2 - x_1)$$
$$x_2' = rx_1 - x_2 - x_1x_3 \qquad (10.1.44)$$
$$x_3' = -bx_3 + x_1x_2$$

model the convective motion of a fluid heated from below. The unknown x_1 measures the intensity of the convective motion, x_2 measures the temperature difference between the ascending and descending currents, and x_3 measures the deviation of the vertical temperature profile from a linear one. Find the three equilibrium points of this system.

10.2 ANALYSIS OF EULER'S METHOD

In Sec. 10.1 we introduced Euler's method

$$\xi_0 = x_0$$

$$\xi_{i+1} = \xi_i + hf(t_i, \xi_i) \qquad i = 0, \ldots, N - 1 \qquad (10.2.1)$$

for approximating the solution to the first-order initial value problem

$$x' = f(t, x) \qquad x(t_0) = x_0 \qquad (10.2.2)$$

Here it is assumed that we are seeking N evenly spaced approximations on some interval (t_0, t_f) and that $t_i = t_0 + ih$ with $h := (t_f - t_0)/N$. As is usual in numerical analysis the question is how to measure the accuracy and efficiency of an approximation scheme such as Euler's method. As has been the case before, one very useful characterization of a method is its *order*. In the context of a numerical method of approximating the solution to an initial value problem, we will say that a method is order n if there is a constant C, independent of the stepsize h (or equivalently the number of steps N) such that

$$|x(t_i) - \xi_i| < Ch^n \qquad i = 1, \ldots, N \qquad (10.2.3)$$

> **Example 10.2.1.** Figure 10.2.2 shows Euler method approximations to the initial value problem $x' = 1 - 2x$, $x(0) = -1$ for $N = 5, 10$, and 20 steps, as well as the actual solution $x(t) = \frac{1}{2} - \frac{3}{2}e^{-2t}$. Points at which the approximate solution was calculated are marked with a "+." Observe that at points at which the solution

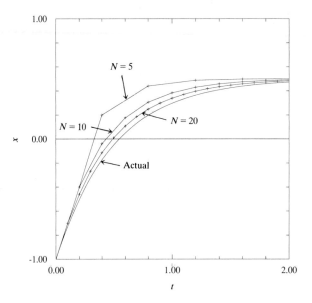

FIGURE 10.2.2

was approximated for all three values of N, a halving of the stepsize resulted in about a halving of the error. Thus it appears the Euler's method is order one.

Heuristic confirmation that Euler's method is order one is provided by Taylor's theorem. We have for any $t_i, i = 0, \ldots, N-1$,

$$x(t_{i+1}) = x(t_i) + hx'(t_i) + \frac{x''(\tau_i)}{2!} h^2 \qquad (10.2.4)$$

where $\tau_i \in (t_i, t_{i+1})$. Since from the differential equation (10.2.2) $x'(t_i) = f(t_i, x(t_i))$, we can write (10.2.4) as

$$x(t_{i+1}) - [x(t_i) + hf(t_i, x(t_i))] = \frac{x''(\tau_i)}{2} h^2 \qquad (10.2.5)$$

Note that the quantity in brackets on the left side of (10.2.5) becomes the Euler approximation to $x(t_{i+1})$ when the actual solution $x(t_i)$ is replaced by its Euler approximation ξ_i. The quantity $x(t_{i+1}) - [x(t_i) + hf(t_i, x(t_i))]$ is called the *local truncation error* of Euler's method. Equation (10.2.5) tells us that at each step Euler's method has a local truncation error proportional to h^2. However, such errors can be expected to compound, since at each step the approximate rather than the exact solution is being used in the calculation. For instance, obtaining the Euler approximation at $t = t_f$ requires $N = (t_f - t_0)/h$ steps, and thus intuitively the error of approximation of $x(t_f)$ will be proportional to $h^2(1/h) = h$; that is, Euler's method will be of order one. In the remainder of this section our principal objective is to establish the order of Euler's method rigorously. A similar approach will be used to establish the order of other methods that will be treated in later sections.

Existence of Solutions to Initial Value Problems

Implicit in the definition (10.2.3) of order for a method of approximating a solution to an initial value problem is the assumption that the actual solution exists and is unique. It is not, however, certain that this is the case.

Example 10.2.3. Figure 10.2.4 shows the (rather implausible) Euler method approximation to the initial value problem

$$x' = -\frac{t}{x} \qquad x(0) = 1 \tag{10.2.6}$$

with $N = 400$ steps taken in both the forward and backward t directions. The difficulty is that the solution to (10.2.6), *as an explicit function of t, is $x(t) = \sqrt{1-t^2}$* and thus exists only in the interval $-1 \le t \le 1$. It is clear that any numerical scheme such as Euler's method which steps rightward, or leftward, at fixed increments h will produce a similarly erroneous result. Thus it is worthwhile to explore under what circumstances we can assert the existence of solutions to an initial value problem.

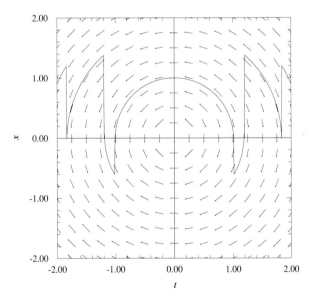

FIGURE 10.2.4

Theorem 10.2.5 Existence theorem. Suppose that the right-hand side $f(t, x)$ in (10.2.2) is continuous in the rectangle $\Re = \{(t, x) : |t - t_0| \le a, |x - x_0| \le b\}$ and moreover there is a constant K such that the function f satisfies $|f(t, x)| \le K$ in $\Re$. Then the initial value problem (10.2.2) has a continuously differentiable solution in the interval $|t - t_0| \le \min(a, b/K)$.

The proof of this theorem may be found in Nemytskii and Stepanov (1989) for instance. Let us see what it does, and does not, tell us in various situations.

Example 10.2.6. Let us consider four initial value problems.

(a) $x' = 1/(1 + x^2)$, $x(0) - x_0$. Since $f(t, x) := 1/(1 + x^2)$ is continuous and bounded by $K = 1$ in any rectangle of the form $\{(t, x): |t| \le a, |x - x_0| \le b\}$, Theorem 10.2.5 asserts the existence of a solution in the interval $|t| \le b$ for arbitrarily large b.

(b) $x' = (1 - x)x$, $x(0) = \frac{1}{2}$. Since $f(t, x) := x(1 - x)$ is continuous in any rectangle of the form $\{(t, x): |t| \le a, |x - 1/2| \le b\}$, Theorem 10.2.5 asserts that a solution exists in any interval of the form $|t| \le b/K$, where $K = \max_{|x - 1/2| \le b} |x(1 - x)|$. Thus for $b = \frac{1}{2}$ it is easily seen that $K = \frac{1}{4}$, and thus a solution exists in the interval $|t| \le 2$. It may be shown (Exercise 1) that the largest interval in which the theorem can be used to assert the existence of a solution is $|t| \le 2\sqrt{2}$. In actuality, however, this problem is an instance of the logistic model of population growth (Example 10.1.2), and the solution exists in the interval $-\infty \le t < \infty$, as indicated by Eq. (10.1.8).

(c) $x' = 1 + x^2$, $x(0) = 0$. The best bound that we can obtain for $1 + x^2$ in $-b \le x \le b$ is $K = 1 + b^2$. Consequently we can assert existence of solutions in any interval of the form $|t| \le b/(1 + b^2)$. The right side of this inequality takes its maximum $\frac{1}{2}$ at $b = 1$. Hence solutions exist in at least the interval $|t| \le \frac{1}{2}$. The solution to this problem is $x(t) = \tan t$, and thus in fact the solution exists only in the interval $|t| < \pi/2$.

(d) $x' = -t/x$, $x(0) = 1$. Since, for $t \ne 0$, $-t/x$ approaches infinity as x approaches 0, we must require $b = 1 - \epsilon$ for some $0 < \epsilon < 1$. We then have the bound $|-t/x| \le a/\epsilon$ in the rectangle $\{(t, x): |t| \le a, |x - 1| \le 1 - \epsilon\}$. Thus solutions will exist on any interval of the form $|t| \le \min(a, \epsilon(1 - \epsilon)/a)$. Since $\epsilon(1 - \epsilon) \le \frac{1}{4}$ for $\epsilon \in (0, 1)$ and the quantity $\min(a, 1/(4a))$ takes its largest value at $a = \frac{1}{2}$, Theorem 10.2.5 asserts the existence of solutions in the interval $|t| \le \frac{1}{2}$. As we saw above, the solution in fact exists in $[-1, 1]$.

Lipschitz Continuity

While the existence of solutions can be asserted when the function $f(t, x)$ in (10.2.2) is simply bounded and continuous, establishing that a solution to an initial value is the unique solution to that problem requires a somewhat stronger condition.

Definition 10.2.7. A function $f(t, x)$ is *Lipschitz continuous* in a domain $\Re \in \mathbb{R}^2$ if there is a constant $L > 0$ for which

$$|f(t, x_1) - f(t, x_2)| < L|x_1 - x_2| \tag{10.2.7}$$

for all (t, x_1), $(t, x_2) \in \Re$.

Example 10.2.8. To show that $f(t, x) := t^3(x^2 - t)$ is Lipschitz continuous in $\Re := \{(t, x): -2 \le t \le 2, 0 \le x \le 2\}$, we note that

$$|t^3(x_1^2 - t) - t^3(x_2^2 - t)| = |t^3| |x_1^2 - x_2^2|$$

$$= |t^3| |x_1 + x_2| |x_1 - x_2| \tag{10.2.8}$$

$$\le 8 \cdot 4 |x_1 - x_2|$$

Thus the Lipschitz condition is satisfied with $L = 32$.

The Lipschitz condition is more stringent than continuity in the variable x since, for instance, $f(t, x) = x^{1/3}$ is continuous in $\Re := \{(t, x): 0 \leq t \leq 1, 0 \leq x \leq 1\}$ but not Lipschitz continuous (Exercise 3a). It is weaker than differentiability since, for instance, $f(t, x) := |x|$ is Lipschitz continuous in $\Re := \{(t, x): 0 \leq t \leq 1, -1 \leq x \leq 1\}$ but not differentiable with respect to x there (Exercise 3b).

The next theorem permits us to assert Lipschitz continuity for a large class of functions.

Theorem 10.2.9. Let $f(t, x)$ be differentiable with respect to x in a rectangle $\Re$. If there is a constant L such that

$$\left| \frac{\partial f}{\partial x}(t, x) \right| \leq L \qquad \text{for all } (t, x) \in \Re \tag{10.2.9}$$

then f is Lipschitz continuous with constant L.

Proof. Exercise 4.

Uniqueness of Solutions

Theorem 10.2.10. If the function $f(t, x)$ in (10.2.2) is continuous in a rectangle $\Re = \{(t, x): |t - t_0| \leq a, |x - x_0| \leq b\}$ and is Lipschitz continuous in the variable x, then there is at most one solution to the initial value problem (10.2.2) lying in $\Re$.

Proof. Again, see Nemytskii and Stepanov (1989).

Example 10.2.11. Figure 10.2.12 shows the Euler method approximations to the initial value problem

$$x' = (t - 1)x^{1/3} \qquad x(0) = x_0 \tag{10.2.10}$$

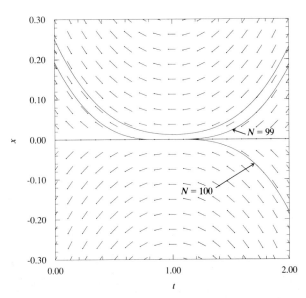

FIGURE 10.2.12

for $x_0 = 0.25$ with $N = 100$ steps and for $x_0 = 3^{-3/2}$ with $N = 99$ and $N = 100$ steps. While the solution for $x_0 = 0.25$ appears to be unique, there is an obvious problem at $x_0 = 3^{-3/2}$. As noted after Example 10.2.8 the right-hand side of (10.2.10) is not Lipschitz continuous in any rectangle containing $x = 0$. Discussion of this example is continued in Exercise 6.

The usefulness of the existence and uniqueness Theorems 10.2.5 and 10.2.10 is that they indicate curves in the tx plane where the solution may cease to exist or be unique. As shown in the examples above, they do not tell us whether the solution will actually intersect such a curve. However, *should* such an intersection occur, the validity of the numerical solution thereafter is questionable.

Order of Euler's Method

In establishing that Euler's method is order one, we will use the following lemma.

Lemma 10.2.13. Let $a > 0$, $b \geq 0$ and let x_n, $n = 0, 1, \dots$, be a sequence of nonnegative numbers satisfying the inequality

$$x_{n+1} \leq (1 + a)x_n + b \tag{10.2.11}$$

Then

$$x_n \leq e^{na}x_0 + b\,\frac{e^{na} - 1}{a} \tag{10.2.12}$$

Proof. We have

$x_1 \leq (1 + a)x_0 + b$

$x_2 \leq (1 + a)x_1 + b \leq (1 + a)[(1 + a)x_0 + b] + b = (1 + a)^2 x_0 + (1 + a)b + b$

$$\tag{10.2.13}$$

$x_3 \leq (1 + a)x_2 + b$

$\qquad \leq (1 + a)[(1 + a)^2 x_0 + (1 + a)b + b] + b = (1 + a)^3 x_0 + (1 + a)^2 b + (1 + a)b + b$

Continuing in this way we see that

$$x_n \leq (1 + a)^n x_0 + b \sum_{j=0}^{n-1} (1 + a)^j$$

$$= (1 + a)^n x_0 + b\,\frac{(1 + a)^n - 1}{a} \tag{10.2.14}$$

from the finite geometric series formula. From the Maclaurin expansion of e^a, we have $1 + a < e^a$ for $a > 0$ and thus that $(1 + a)^n < e^{na}$. Substituting this into (10.2.14) gives the result.

We are now in a position to show that Euler's method is order one.

Theorem 10.2.14. Let x be the solution to (10.2.2) and suppose that $c \leq x(t) \leq d$ for $t \in (t_0, t_f)$. Let ξ_i, $i = 0, \ldots, N$, be the Euler method approximation to $x(t_i)$, where $t_i = t_0 + ih$ with $h = (t_f - t_0)/N$. If $|x''(t)| \leq M$ for some constant M and all $t \in (t_0, t_f)$ and the function f in (10.2.2) is Lipschitz continuous with constant L in the rectangle $\Re = \{(t, x): t_0 \leq t \leq t_f, c \leq x \leq d\}$, then

$$|x(t_i) - \xi_i| \leq \frac{Mh}{2L} \left(e^{L(t_i - t_0)} - 1 \right) \qquad i = 1, \ldots, N \qquad (10.2.15)$$

Remark. Note that in particular (10.2.3) is now verified with $C = (M/2L) \times (e^{L(t_f - t_0)} - 1)$. Thus we have established that the order of Euler's method is one.

Proof of Theorem 10.2.14. Let $E_i := x(t_i) - \xi_i$ be the error made by Euler's method at the ith step. From Taylor's theorem applied to $x(t)$ and (10.2.2) we have

$$E_{i+1} = x(t_{i+1}) - \xi_{i+1}$$

$$= x(t_i) + hx'(t_i) + \tfrac{1}{2}h^2 x''(\tau_i) - [\xi_i + hf(t_i, \xi_i)]$$

$$= E_i + h[x'(t_i) - f(t_i, \xi_i)] + \tfrac{1}{2}h^2 x''(\tau_i)$$

$$= E_i + h[f(t_i, x(t_i)) - f(t_i, \xi_i)] + \tfrac{1}{2}h^2 x''(\tau_i) \qquad (10.2.16)$$

for some $\tau_i \in (t_i, t_{i+1})$. Since x'' is bounded by M and f is Lipschitz continuous,

$$|E_{i+1}| \leq |E_i| + hL|x(t_i) - \xi_i| + \frac{Mh^2}{2} = |E_i|(1 + hL) + \frac{Mh^2}{2} \qquad (10.2.17)$$

Thus from Lemma 10.2.13,

$$|E_i| \leq \frac{Mh^2}{2} \frac{e^{ihL} - 1}{hL} = \frac{Mh}{2L} \left(e^{ihL} - 1 \right) \qquad (10.2.18)$$

since $E_0 = 0$. The result now follows upon observing that $ih = t_i - t_0$.

Example 10.2.15. Note that the inequality (10.2.15) permits exponential growth in the error. Figure 10.2.16 shows the Euler method approximation with $h = 0.1$ to the initial value problem $x' = 1 + 2x$, $x(0) = 0$ on $0 \leq t \leq 1$ as well as the graph of the actual solution $x(t) = -\tfrac{1}{2} + \tfrac{1}{2}e^{2t}$. From the differential equation we see that the Lipschitz constant is $L = 2$, and from the actual solution it is easily seen that $M = \max_{0 \leq t \leq 1} |x''(t)| \leq 2e^2$. Also plotted in Fig. 10.2.16 are the curves $x(t) \pm E(t)$, where $E(t) = (Mh/2L)(e^{L(t - t_0)} - 1)$ with $t_0 = 0$, $L = 2$, and $M = 2e^2$. Thus it is seen that (10.2.15) substantially overestimates the error. As we saw in Example 10.2.1 (see Fig. 10.2.2), the error made by Euler's method will not always grow with t. Further examples are considered in Exercise 8.

The Effect of Rounding Error

The error estimated by Theorem 10.2.14 is the *global truncation* or *discretization error*, that is, the cumulative error due to the fact that Euler's method approximates the solution curve by a series of straight-line segments. Errors

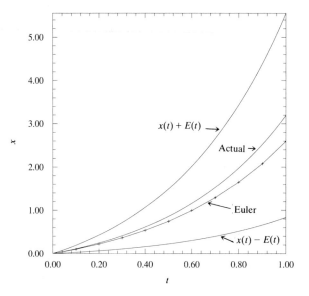

FIGURE 10.2.16

due to rounding are not included. Let δ_i denote the error due to rounding made in the ith step of Euler's method. The computed Euler approximations are

$$\xi_0 = x_0 + \delta_0$$

$$\xi_{i+1} = \xi_i + hf(t_i, \xi_i) + \delta_{i+1}$$

(10.2.19)

A rounding error term is included for the initial condition because the exact value may be a number such as $\frac{1}{3}$ or π that is not exactly representable in computer floating-point arithmetic. An argument similar to that used in proving Theorem 10.2.14 establishes the following theorem.

Theorem 10.2.17. Let $\delta = \max_{0 \leq i \leq N} |\delta_i|$. Then under the same hypotheses as Theorem 10.2.14,

$$|x(t_i) - \xi_i| \leq \left(\frac{hM}{2L} + \frac{\delta}{hL} \right)(e^{L(t_i - t_0)} - 1) + |\delta_0| e^{L(t_i - t_0)}$$

(10.2.20)

Proof. Exercise 9.

Note that the right-hand side of (10.2.20) approaches infinity as h approaches either 0 or ∞. Thus the minimum taken by the right-hand side must occur at some intermediate value of h. An elementary calculus argument shows that the minimum occurs at

$$h_{\min} = \sqrt{\frac{2\delta}{M}}$$

(10.2.21)

We do not expect rounding to be the dominant source of error until $h < h_{min}$. Unless the computation of f is very involved, the rounding errors δ_i at each step of Euler's method will be of the order of the machine unit μ times the magnitude of f. For Turbo Pascal's type real, for which $\mu = 2^{-39}$ or 2^{-40} depending upon whether a numeric coprocessor is used, this gives a value for h_{min} of around $10^{-7}/\sqrt{M}$ when f is of unit magnitude. Thus, so long as the accuracy requirements are modest, it appears that rounding error will not be a significant factor in a Euler method approximation for a machine unit this small. We will see in the next section that when very high accuracies are required, a higher-order method should be used instead of Euler's method.

Example 10.2.18. Another way to see the effect of rounding error is to compare the accuracy of solutions as the number of steps is repeatedly doubled. Since Euler's method is of order one, each such doubling should result in a halving of the global truncation error. To the extent that rounding error corrupts the computation, this will not happen. Table 10.2.19 shows the results of applying Euler's method to find $x(1)$ for the initial value problem $x' = -x$, $x(0) = 1$ for successively larger numbers of steps. Were rounding introducing significant errors, the ratio of the error for N steps to the error for $N/2$ which is shown in the last column would increase above the predicted value of $\frac{1}{2}$. As we see, this does not happen, at least down to $h \approx 10^{-6}$. This is consistent with the estimate given by (10.2.21).

TABLE 10.2.19

N	Error	$\dfrac{\text{Err}(N)}{\text{Err}(N/2)}$
1,000	1.840E-04	
2,000	9.199E-05	4.999E-01
4,000	4.599E-05	4.999E-01
8,000	2.299E-05	5.000E-01
16,000	1.150E-05	5.000E-01
32,000	5.748E-06	5.000E-01
64,000	2.874E-06	5.000E-01
128,000	1.437E-06	4.999E-01
256,000	7.180E-07	4.998E-01
512,000	3.583E-07	4.990E-01
1,024,000	1.778E-07	4.963E-01

First-Order Systems of Differential Equations

The system (10.1.12) of first-order differential equations in vector notation is

$$\mathbf{x}'(t) = \mathbf{f}(t, \mathbf{x}) \qquad \mathbf{x}(t_0) = \mathbf{x}_0 \qquad\qquad (10.2.22)$$

Upon choosing a suitable vector norm, the definition of Lipschitz continuity extends readily to the case of systems of first-order differential equations. We will work with the 1-norm

$$\|\mathbf{x}\|_1 := \sum_{j-1}^{n} |x_j| \tag{10.2.23}$$

though the 2- or ∞-norm would work as well.

Definition 10.2.20. A vector-valued function $\mathbf{f}(t, \mathbf{x})$ is Lipschitz continuous in a region $\Re \subset \mathbb{R}^{n+1}$ if there is constant L for which

$$\|\mathbf{f}(t, \mathbf{x}_1) - \mathbf{f}(t, \mathbf{x}_2)\|_1 \leq L\|\mathbf{x}_1 - \mathbf{x}_2\|_1 \tag{10.2.24}$$

for every $(t, \mathbf{x}_1)$, $(t, \mathbf{x}_2) \in \Re$.

Example 10.2.21. Let us show that the system

$$x_1' = (1 - x_2)x_1$$
$$x_2' = (-1 + x_1)x_2 \tag{10.2.25}$$

has a Lipschitz continuous right-hand side in any rectangle

$$\Re = \{(t, x_1, x_2): 0 \leq t \leq t_f, 0 \leq x_1 \leq c, 0 \leq x_2 \leq d\} \tag{10.2.26}$$

Observe that this is an instance of the predator-prey model discussed in Example 10.1.9. We have with $\mathbf{u} = [u_1, u_2]^T$ and $\mathbf{v} = [v_1, v_2]^T$,

$$\|\mathbf{f}(t, \mathbf{u}) - \mathbf{f}(t, \mathbf{v})\|_1 = |(u_1 - u_1 u_2) - (v_1 - v_1 v_2)| + |(-u_2 + u_1 u_2) - (-v_2 + v_1 v_2)|$$
$$\leq |u_1 - v_1| + 2|u_1 u_2 - v_1 v_2| + |u_2 - v_2|$$
$$\leq |u_1 - v_1| + 2|u_1 u_2 - v_1 u_2| + 2|v_1 u_2 - v_1 v_2| + |u_2 - v_2|$$
$$= (1 + 2|u_2|)|u_1 - v_1| + (1 + 2|v_1|)|u_2 - v_2|$$
$$\leq (1 + 2\max(c, d))\|\mathbf{u} - \mathbf{v}\|_1 \tag{10.2.27}$$

Thus $\mathbf{f}$ is Lipschitz continuous in $\Re$ with $L = 1 + 2\max(c, d)$.

The existence Theorem 10.2.5 holds provided that $\mathbf{f}$ is continuous and bounded in the 1-norm. Likewise the uniqueness Theorem 10.2.10 holds for systems provided that $\mathbf{f}$ is Lipschitz continuous in the sense of Definition 10.2.20.

EXERCISES 10.2

1. Show that the largest interval in which Theorem 10.2.5 can be used to assert the existence of solutions for the initial value problem of Example 10.2.6b is $|t| \leq 2\sqrt{2}$.

2. Find whether each of the following functions is Lipschitz continuous in the rectangle $\Re = \{(t, x): 0 \leq t \leq 1, -\infty < x < \infty\}$.
   (a) $3t^2 x$
   (b) $tx^3 - 1$
   (c) $x - 1/x$

3. (a) Show $f(t, x) = x^{1/3}$ is continuous in $\Re :=$ $\{(t, x): 0 \leq t \leq 1, 0 \leq x \leq 1\}$ but not Lipschitz continuous.
   (b) Show that $f(t, x) := |x|$ is Lipschitz continuous in $\Re := \{(t, x): 0 \leq t \leq 1, -1 \leq x \leq 1\}$.

4. Prove Theorem 10.2.19 by using the mean value theorem.

5. Use the existence Theorem 10.2.5 to find an interval $|t| \leq a$ on which solutions to the following initial value problems exist.

(a) $x' = \dfrac{t}{1 + x^2}, x(0) = 0$

(b) $x' = \dfrac{t}{x}, x(0) = 1$

(c) $x' = 1 - x^2, x(0) = 0$

(d) $x' = \dfrac{1}{1 - x^2}, x(0) = 0$

6. Show that the initial value problem (10.2.10) with $x_0 = 0$ has three continuously differentiable solutions in $0 \le t < \infty$. Explain why this does not contradict Theorem 10.2.10.

7. *Definition.* A domain $D \subset \mathbb{R}^2$ is *convex* if for $0 \le \lambda \le 1$ the point $(\lambda t_1 + (1 - \lambda)t_2, \lambda x_1 + (1 - \lambda)x_2) \in D$ whenever $(t_1, x_1), (t_2, x_2) \in D$.

   Thus a region is convex when any two points contained in it can be connected by a straight line lying entirely within the region. Prove Theorem 10.2.9 when the domain is a convex region rather than simply a rectangle.

8. Make graphs illustrating the error bound (10.2.15) similar to Fig. 10.2.16 for the following initial value problems.
   (a) $x' = x, x(0) = 1, 0 \le t \le 2, h = 0.1$
   (b) $x' = 1 - 2x, x(0) = -1, 0 \le t \le 2, h = 0.1$
   (c) $x' = -x, x(0) = 1, 0 \le t \le 10, h = 2$

9. Prove Theorem 10.2.17.

10. Show that the right-hand sides of each of the following systems of differential equations are Lipschitz continuous in any rectangle $\mathfrak{R} = \{(t, x_1, x_2): 0 \le t \le t_f, |x_1| \le c, |x_2| \le d\}$.
    (a) $x_1' = x_2, x_2' = -x_1 - 2x_2$
    (b) $x_1' = x_1 - x_1 x_2, x_2' = x_2 - x_1 x_2$
    (c) $x_1' = t x_2, x_2' = x_1^2 - x_2$

10.3 RUNGE-KUTTA METHODS

One-Step Methods

In the last section we considered Euler's method

$$\xi_0 = x_0$$
$$\xi_{i+1} = \xi_i + hf(t_i, \xi_i) \tag{10.3.1}$$

for approximating the solution to the initial value problem

$$x' = f(t, x) \qquad x(t_0) = x_0 \tag{10.3.2}$$

As in Secs. 10.1 and 10.2 we assume that an approximation to the initial value problem (10.3.2) is sought at N evenly spaced points on an interval $[t_0, t_f]$ and thus that $h = (t_f - t_0)/N$ and $t_i = t_0 + ih$.

Theorem 10.2.14 indicates that we can make the Euler method approximation to the solution of an initial value problem as accurate as we like by choosing the stepsize h sufficiently small, at least to within the limitations imposed by rounding error. However, as the next example shows, if the stepsize is too large, the Euler approximation can be quite inaccurate.

Example 10.3.1. Figure 10.3.2 shows the graph of the Euler approximation to the initial value problem $x' = 5(t - 1)x, x(0) = 5$ for $h = \frac{1}{4}, \frac{1}{8}$, and $\frac{1}{16}$ along with the direction field and the actual solution $x(t) = 5e^{(5/2)t^2 - 5t}$. The stepsize $h = \frac{1}{4}$ is sufficiently large to carry the Euler approximation below the x axis, where the direction field is very different from what it is near the exact solution. For the two smaller stepsizes, $h = \frac{1}{8}$ and $h = \frac{1}{16}$, the Euler method approximation departs from the trajectory of the solution less dramatically, but, because the global truncation error for Euler's method is order one in the stepsize h, decreasing the stepsize from $h = \frac{1}{8}$ to $h = \frac{1}{16}$ results only in a halving of the error.

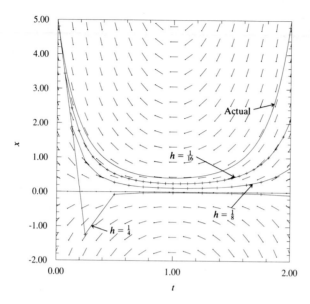

FIGURE 10.3.2

Euler's method obtains its approximation ξ_{i+1} to $x(t_{i+1})$ by following a line of slope $f(t_i, \xi_i)$ from the point (t_i, ξ_i). As the above example indicates, this can lead to substantial errors if the stepsize h is too large relative to the rate at which the direction field is changing. A way to improve upon Euler's method is to find a better way to select the direction in which the approximate solution moves from the point (t_i, ξ_i). A *one-step method* has the form

$$\xi_0 = x_0$$

$$\xi_{i+1} = \xi_i + h\Phi(t_i, \xi_i; h)$$

(10.3.3)

The function $\Phi(t, x; h)$ predicts the direction that the solution will take from the point (t_i, ξ_i). Thus it should meet the requirement that $\lim_{h \to 0} \Phi(t, x; h) = f(t, x)$. If Φ satisfies this condition, then method (10.3.3) is said to be *consistent*. Methods of the form (10.3.3) are called one-step methods because they rely solely upon the approximation ξ_i to $x(t_i)$ to calculate the approximation ξ_{i+1} to $x(t_{i+1})$. *Multistep methods*, which may use several previous approximations $\xi_i, \xi_{i-1}, \ldots$ to calculate ξ_{i+1}, will be treated in the next section.

Modified Euler and Midpoint Methods

A means of improving upon Euler's method is to reduce the dependence of the direction Φ in (10.3.3) upon the value ξ_i of the approximation at t_i by calculating the slopes at other nearby points and computing Φ as the weighted average of these slopes.

For instance, instead of using the Euler approximation $\xi_{\text{Euler}} = \xi_i + hf(t_i, \xi_i)$ as the approximation to $x(t_{i+1})$, we can simply use it to locate another point near the trajectory of $x(t)$ and then take the estimated slope Φ to be the average of the slopes at (t_i, ξ_i) and $(t_{i+1}, \xi_{\text{Euler}})$. This gives the *modified Euler method*

$$\xi_0 = x_0$$

$$m_1 = f(t_i, \xi_i)$$

$$m_2 = f(t_i + h, \xi_i + hm_1) \tag{10.3.4}$$

$$\xi_{i+1} = \xi_i + h\,\frac{m_1 + m_2}{2}$$

As can be seen in Fig. 10.3.3, the modified Euler method gives substantially better approximations for the initial value problem of Example 10.3.1 when the same stepsizes are used.

Another possibility for an improved one-step method is to use Euler's method to approximate the solution at the midpoint $t_i + h/2$ of (t_i, t_{i+1}) and take the estimated slope Φ to be f at this point. This gives the *midpoint method*

$$\xi_0 = x_0$$

$$m_1 = f(t_i, \xi_i)$$

$$m_2 = f\left(t_i + \frac{h}{2}, \xi_i + \frac{hm_1}{2}\right) \tag{10.3.5}$$

$$\xi_{i+1} = \xi_i + hm_2$$

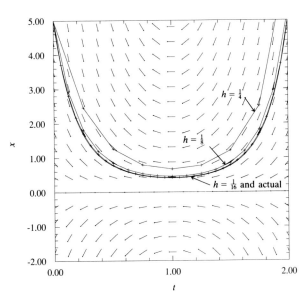

FIGURE 10.3.3

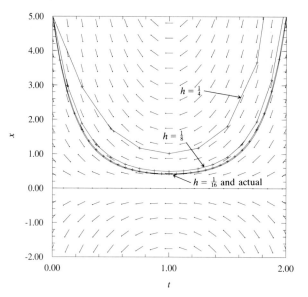

FIGURE 10.3.4

Figure 10.3.4 shows the result of applying the midpoint method to the initial value problem of Example 10.3.1. The results are again noticeably better than those of Euler's method.

In Sec. 10.2 it was shown that the global truncation error made by Euler's method is order one in the stepsize h. On the basis of what we have seen in Figs. 10.3.3 and 10.3.4, the errors made by the modified Euler method and the midpoint method appear to be of higher order. Both these methods are in fact order two in h, as we now show.

Definition 10.3.5. The *local truncation error* of a one-step method is

$$\epsilon_i(h) := x(t_{i+1}) - x(t_i) - h\Phi(t_i, x(t_i); h) \tag{10.3.6}$$

Observe that the right-hand side of (10.3.6) is simply the difference between the exact solution and the approximation that the one-step method would generate if it started from the exact solution point $(t_i, x(t_i))$. Of course, the method will actually start at the point (t_i, ξ_i), and thus the approximation will be subject as well to the accumulation of errors made at previous time steps. Hence the term *local* truncation error.

Let

$$\epsilon(h) := \max_{0 \le i \le N-1} |\epsilon_i(h)| \tag{10.3.7}$$

We will show in Theorem 10.3.7 that the order of $\epsilon(h)$ in h is one greater than the global order of the error made by the one-step method.

Example 10.3.6. Let us find the order of the local truncation error for the modified Euler method under the assumption that f in (10.3.2) has continuous third partial derivatives in a rectangle containing the solution $x(t)$. From Taylor's theorem

$$x(t_{i+1}) - x(t_i) = x'(t_i)h + \frac{1}{2!}x''(t_i)h^2 + \frac{1}{3!}x'''(t_i)h^3 + O(h^4) \quad (10.3.8)$$

Also, upon differentiating the differential equation $x' = f(t, x)$ we obtain

$$x'' = \frac{\partial f}{\partial t}(t, x) + \frac{\partial f}{\partial x}(t, x)x' = \frac{\partial f}{\partial t}(t, x) + \frac{\partial f}{\partial x}(t, x)f(t, x)$$

$$x''' = \frac{\partial^2 f}{\partial t^2}(t, x) + 2\frac{\partial^2 f}{\partial t\partial x}(t, x)f(t, x) + \frac{\partial^2 f}{\partial x^2}(t, x)f(t, x)^2 \quad (10.3.9)$$

$$+ \frac{\partial f}{\partial t}(t, x)\frac{\partial f}{\partial x}(t, x) + \left(\frac{\partial f}{\partial x}(t, x)\right)^2 f(t, x)$$

whence with $x_i := x(t_i)$, $f_i := f(t_i, x_i)$, $f_{i,t} := (\partial f/\partial t)(t_i, x_i)$, and so forth, we have from (10.3.8)

$$x_{i+1} - x_i = hf_i + \frac{h^2}{2!}(f_{i,t} + f_{i,x}f_i)$$

$$+ \frac{h^3}{3!}(f_{i,tt} + 2f_{i,tx} + f_{i,xx}f_i^2 + f_{i,t}f_{i,x} + f_{i,x}^2 f_i) + O(h^4) \quad (10.3.10)$$

Upon applying Taylor's theorem of two variables (Appendix A) to the slope estimate Φ for the modified Euler method, we obtain

$$\Phi(t_i, x_i; h) = \tfrac{1}{2}[f(t_i, x_i) + f(t_i + h, x_i + hf(t_i, x_i))]$$

$$= \tfrac{1}{2}[f_i + f_i + f_{i,t}h + f_{i,x}f_ih + \tfrac{1}{2}(f_{i,tt}h^2 + 2f_{i,tx}f_ih^2 + f_{i,xx}f_i^2h^2)] + O(h^3)$$

$$= f_i + \tfrac{1}{2}f_{i,t}h + \tfrac{1}{2}f_{i,x}f_ih + \tfrac{1}{2}f_{i,tt}h^2 + f_{i,tx}f_ih^2 + \tfrac{1}{2}f_{i,xx}f_i^2h^2 + O(h^3)$$
$$(10.3.11)$$

Subtracting (10.3.11) from (10.3.10) then yields

$$\epsilon_i(h) = (-\tfrac{1}{12}f_{i,tt} - \tfrac{1}{6}f_{i,t}f_i - \tfrac{1}{12}f_{i,xx}f_i^2 + \tfrac{1}{6}f_{i,t}f_{i,x} + \tfrac{1}{6}f_{i,x}^2 f_i)h^3 \quad (10.3.12)$$

It follows that if the second partial derivatives of f are continuous in a rectangle which encloses the solution x, then $|\epsilon_i(h)| \le Mh^3$ for some M that is independent of i and h. Thus the local truncation error of the modified Euler method is of order three in h.

The next theorem relates the maximum local truncation error ϵ to the order of the global truncation error made by the method.

Theorem 10.3.7. Let Φ in (10.3.3) be Lipschitz continuous with constant L in the variable x in a rectangle $\mathfrak{R} = \{(t, x); t_0 \le t \le t_f, c \le x \le d\}$, where $c \le x(t_i)$, $\xi_i \le d$ for $t_i = 0, \ldots, N$. Then

$$|x(t_i) - \xi_i| \le \frac{\epsilon(h)}{hL}(e^{L(t_i - t_0)} - 1) \quad (10.3.13)$$

Proof. As in the case of Euler's method the proof is based upon Lemma 10.2.13. We have from (10.3.3) and (10.3.6) with $x_i:=x(t_i)$ that

$$|x_{i+1} - \xi_{i+1}| = |x_{i+1} - [\xi_i + h\Phi(t_i, \xi_i; h)]|$$
$$= |x_{i+1} - [x_i + h\Phi(t_i, x_i; h)]$$
$$+ [x_i + h\Phi(t_i, x_i; h)] - [\xi_i + h\Phi(t_i, \xi_i; h)]|$$
$$= |\epsilon_i(h) + x_i - \xi_i + h[\Phi(t_i, x_i; h) - \Phi(t_i, \xi_i; h)]|$$
$$\leq \epsilon(h) + (1 + hL)|x_i - \xi_i| \qquad (10.3.14)$$

from the assumption of Lipschitz continuity for Φ. Application of Lemma 10.2.13 now gives the result.

Example 10.3.8. Let us show that the global truncation error made by the modified Euler method is $O(h^2)$. Since we know from Example 10.3.6 that the local truncation error is order three, this follows from Theorem 10.3.7 once we have established that

$$\Phi(t, x; h) := \tfrac{1}{2}[f(t, x) + f(t + h, x + hf(t, x))]$$

is Lipschitz continuous. We show this under the assumption that f itself is Lipschitz continuous with constant L_f. We have

$$|\Phi(t, x_1; h) - \Phi(t, x_2; h)|$$
$$= |\tfrac{1}{2}[f(t, x_1) + f(t + h, x_1 + hf(t, x_1))] - \tfrac{1}{2}[f(t, x_2) + f(t + h, x_2 + hf(t, x_2))]|$$
$$\leq \tfrac{1}{2}|f(t, x_1) - f(t, x_2)| + \tfrac{1}{2}|f(t + h, x_1 + hf(t, x_1)) - f(t + h, x_2 + hf(t, x_2))|$$
$$\leq \tfrac{1}{2}L_f|x_1 - x_2| + \tfrac{1}{2}L_f|[x_1 + hf(t, x_1)] - [x_2 + hf(t, x_2)]|$$
$$\leq \tfrac{1}{2}L_f|x_1 - x_2| + \tfrac{1}{2}(L_f|x_1 - x_2| + h|f(t, x_1) - f(t, x_2)|)$$
$$\leq (L_f + \tfrac{1}{2}hL_f^2)|x_1 - x_2| \qquad (10.3.15)$$

Thus Φ is Lipschitz continuous with constant $L_\Phi := L_f + \tfrac{1}{2}hL_f^2$.

The midpoint method can also be shown to have a global truncation error of order two by a similar sequence of calculations (Exercise 5).

Consistency and Convergence

As mentioned earlier a one-step method is *consistent* if $\lim_{h\to0} \Phi(t, x; h) = f(t, x)$; that is, the slope estimator Φ approaches the actual slope of the solution as the stepsize becomes small. A method is *convergent* if $\lim_{h\to0} \max_{1\leq i\leq N} |x(t_i) - \xi_i| = 0$; that is, the approximations approach the actual solution as h becomes small. The question then arises as to whether the consistency of a method suffices to guarantee its convergence. With certain conditions placed upon the uniformity of the convergence of Φ to f, it can be shown that consistency implies that $\epsilon(h)$ approaches 0 as h approaches 0 and thus from Theorem 10.3.7 that the method converges (Exercise 6a).

Runge-Kutta Methods

The modified Euler and midpoint methods are members of the *Runge-Kutta* family of methods. These one-step methods estimate the slope as a weighted average of the slopes of the direction field at various points in the strip between $t = t_i$ and $t = t_{i+1}$ that are expected to lie near the trajectory of the solution. The most general Runge-Kutta method involving two slope calculations is

$$\xi_0 = x_0$$
$$m_1 = f(t_i, \xi_i)$$
$$m_2 = f(t_i + \alpha h, \xi_i + h\beta m_1) \tag{10.3.16}$$
$$\xi_{i+1} = \xi_i + h(\omega_1 m_1 + \omega_2 m_2)$$

where $\alpha \in (0, 1]$. This gives the modified Euler method when $\alpha = \beta = 1$ and $\omega_1 = \omega_2 = \frac{1}{2}$ and the midpoint method when $\alpha = \beta = \frac{1}{2}$ and $\omega_1 = 0$, $\omega_2 = 1$. Not every choice of α and β will lead to a method that has order-three local truncation error, however. Indeed, with $x_i := x(t_i)$, $f_i := f(t_i, x_i)$, $f_{i,t} := (\partial f/\partial t)(t_i, x_i)$, and so forth, we have from Taylor's theorem by a sequence of computations similar to that used in Example 10.3.6

$$\epsilon_i(h) = x_{i+1} - x_i - h[\omega_1 f_i + \omega_2 f(t_i + \alpha h, x_i + \beta h f_i)]$$
$$= x_i' h + \tfrac{1}{2} x_i'' h^2 - [\omega_1 h f_i + \omega_2 h(f_i + f_{i,t}\alpha h + f_{i,x}\beta h f_i)] + O(h^3)$$
$$= (1 - \omega_1 - \omega_2) h f_i + (\tfrac{1}{2} - \alpha \omega_2) f_{i,t} h^2 + (\tfrac{1}{2} - \beta \omega_2) f_{i,x} f_i h^2 + O(h^3) \tag{10.3.17}$$

Thus, since $f(t, x)$ is arbitrary, for ϵ_i to be $O(h^3)$ we must have

$$\omega_1 + \omega_2 = 1 \qquad \alpha \omega_2 = \tfrac{1}{2} \qquad \beta \omega_2 = \tfrac{1}{2} \tag{10.3.18}$$

This in turn implies that

$$\beta = \alpha \qquad \omega_2 = \frac{1}{2\alpha} \qquad \omega_1 = 1 - \frac{1}{2\alpha} \tag{10.3.19}$$

It can be shown that no choice of α can lead to an order of local truncation error greater than three (Exercise 8).

The general form of a Runge-Kutta method involving n slope calculations is

$$\xi_0 = x_0$$
$$m_1 = f(t_i, \xi_i)$$
$$m_2 = f(t_i + \alpha_2 h, \xi_i + \beta_{21} h m_1)$$
$$\cdots\cdots\cdots\cdots\cdots\cdots\cdots\cdots\cdots \tag{10.3.20}$$
$$m_n = f\left(t_i + \alpha_n h, \xi_i + h \sum_{j=1}^{n-1} \beta_{nj} m_j\right)$$
$$\xi_{i+1} = \xi_i + h \sum_{j=1}^{n} \omega_j m_j$$

Thus a three-slope Runge-Kutta method has the form

$$\xi_0 = x_0$$
$$m_1 = f(t_i, \xi_i)$$
$$m_2 = f(t_i + \alpha_2 h, \xi_i + \beta_{21} m_1 h)$$
$$m_3 = f(t_i + \alpha_3 h, \xi_i + \beta_{31} m_1 h + \beta_{32} m_2 h) \tag{10.3.21}$$
$$\xi_{i+1} = \xi_i + h(\omega_1 m_1 + \omega_2 m_2 + \omega_3 m_3)$$

By expanding the local truncation error in a manner similar to that used to obtain (10.3.18), the following relations among the coefficients are found:

$$\omega_1 + \omega_2 + \omega_3 = 1$$
$$\beta_{21}\omega_2 + (\beta_{31} + \beta_{32})\omega_2 = \tfrac{1}{2}$$
$$\alpha_2\beta_{21}\omega_2 + \alpha_3(\beta_{31} + \beta_{32})\omega_3 = \tfrac{1}{3}$$
$$\alpha_2\beta_{31}\omega_3 = \tfrac{1}{6}$$
$$\alpha_2\omega_2 + \alpha_3\omega_3 = \tfrac{1}{2} \tag{10.3.22}$$
$$\tfrac{1}{2}\alpha_2^2\omega_2 + \tfrac{1}{2}\alpha_3^2\omega_3 = \tfrac{1}{6}$$
$$\tfrac{1}{2}\beta_{21}^2\omega_2 + \tfrac{1}{2}(\beta_{31} + \beta_{32})^2\omega_3 = \tfrac{1}{6}$$
$$\beta_{21}\beta_{31}\omega_3 = \tfrac{1}{6}$$

(Exercise 9a). This in turn implies that

$$\beta_{21} = \alpha_2 \qquad \beta_{31} + \beta_{32} = \alpha_3 \tag{10.3.23}$$

which leads to the simplified relations

$$\omega_1 + \omega_2 + \omega_3 = 1$$
$$\beta_{21}\omega_2 + (\beta_{31} + \beta_{32})\omega_3 = \tfrac{1}{2}$$
$$\beta_{21}^2\omega_2 + (\beta_{31} + \beta_{32})^2\omega_3 = \tfrac{1}{3} \tag{10.3.24}$$
$$\beta_{21}\beta_{31}\omega_3 = \tfrac{1}{6}$$

(Exercise 9b). These relations will be used in Sec. 10.5 to derive a Runge-Kutta-Felberg variable-step method.

A commonly used member of the Runge-Kutta family is *classical Runge-Kutta*:

$$\xi_0 = x_0$$
$$m_1 = f(t_i, \xi_i)$$
$$m_2 = f\left(t_i + \frac{h}{2}, \xi_i + \frac{hm_1}{2}\right)$$

$$m_3 = f\left(t_i + \frac{h}{2}, \; \xi_i + \frac{hm_2}{2}\right)$$

$$m_4 = f(t_i + h, \; \xi_i + hm_3)$$

$$\xi_{i+1} = \xi_i + \frac{h}{6}\,(m_1 + 2m_2 + 2m_3 + m_4) \qquad (10.3.25)$$

This is an order-four method (Exercise 11).

Example 10.3.9. Table 10.3.10 shows the errors made at selected values of t by the Euler, modified Euler, midpoint, and classical Runge-Kutta methods when applied to the initial value problem of Example 10.3.1, $x' = 5(t-1)x$, $x(0) = 5$. Errors are shown for $N = 20$ and $N = 40$ steps. Since the stepsize has been halved, the error for $N = 40$ would be expected to decrease by a factor of 2 for the order-one Euler method, 4 for the order-two modified Euler and midpoint methods, and 16 for the order-four classical Runge-Kutta method. This is about what we see for $t \le 1$; however, for $1 < t \le 2$ the errors for the modified Euler, midpoint, and classical Runge-Kutta methods decrease by a factor somewhat greater than is expected for methods of their order. Indeed, at $t = 2$ the decrease is by a factor that suggests a method one order higher. The reason for this is the symmetry that the solution possesses about the line $t = 1$. This is explored further in Exercise 12.

TABLE 10.3.10

t	Euler	Modified Euler	Midpoint	Classical Runge-Kutta
		$N = 20$		
0.00	$0.000\mathrm{E}+00$	$0.000\mathrm{E}+00$	$0.000\mathrm{E}+00$	$0.000\mathrm{E}+00$
0.20	$-6.578\mathrm{E}-01$	$8.684\mathrm{E}-02$	$1.257\mathrm{E}-01$	$9.114\mathrm{E}-04$
0.40	$-4.732\mathrm{E}-01$	$6.226\mathrm{E}-02$	$9.638\mathrm{E}-02$	$5.689\mathrm{E}-04$
0.60	$-3.308\mathrm{E}-01$	$4.148\mathrm{E}-02$	$6.826\mathrm{E}-02$	$3.561\mathrm{E}-04$
0.80	$-2.621\mathrm{E}-01$	$3.104\mathrm{E}-02$	$5.347\mathrm{E}-02$	$2.640\mathrm{E}-04$
1.00	$-2.467\mathrm{E}-01$	$2.784\mathrm{E}-02$	$4.903\mathrm{E}-02$	$2.388\mathrm{E}-04$
1.20	$-2.817\mathrm{E}-01$	$3.044\mathrm{E}-02$	$5.356\mathrm{E}-02$	$2.639\mathrm{E}-04$
1.40	$-3.949\mathrm{E}-01$	$4.003\mathrm{E}-02$	$6.936\mathrm{E}-02$	$3.547\mathrm{E}-04$
1.60	$-6.834\mathrm{E}-01$	$6.072\mathrm{E}-02$	$1.039\mathrm{E}-01$	$5.676\mathrm{E}-04$
1.80	$-1.461\mathrm{E}+00$	$9.738\mathrm{E}-02$	$1.707\mathrm{E}-01$	$9.949\mathrm{E}-04$
2.00	$-3.838\mathrm{E}+00$	$1.233\mathrm{E}-01$	$2.634\mathrm{E}-01$	$1.327\mathrm{E}-03$
		$N = 40$		
0.00	$0.000\mathrm{E}+00$	$0.000\mathrm{E}+00$	$0.000\mathrm{E}+00$	$0.000\mathrm{E}+00$
0.20	$-2.877\mathrm{E}-01$	$1.840\mathrm{E}-02$	$2.583\mathrm{E}-02$	$4.832\mathrm{E}-05$
0.40	$-2.257\mathrm{E}-01$	$1.329\mathrm{E}-02$	$1.977\mathrm{E}-02$	$3.058\mathrm{E}-05$
0.60	$-1.661\mathrm{E}-01$	$8.932\mathrm{E}-03$	$1.403\mathrm{E}-02$	$1.926\mathrm{E}-05$
0.80	$-1.358\mathrm{E}-01$	$6.730\mathrm{E}-03$	$1.102\mathrm{E}-02$	$1.430\mathrm{E}-05$
1.00	$-1.307\mathrm{E}-01$	$6.065\mathrm{E}-03$	$1.011\mathrm{E}-02$	$1.294\mathrm{E}-05$
1.20	$-1.524\mathrm{E}-01$	$6.659\mathrm{E}-03$	$1.103\mathrm{E}-02$	$1.430\mathrm{E}-05$
1.40	$-2.194\mathrm{E}-01$	$8.762\mathrm{E}-03$	$1.416\mathrm{E}-02$	$1.922\mathrm{E}-05$
1.60	$-3.942\mathrm{E}-01$	$1.312\mathrm{E}-02$	$2.066\mathrm{E}-02$	$3.056\mathrm{E}-05$
1.80	$-8.852\mathrm{E}-01$	$1.973\mathrm{E}-02$	$3.120\mathrm{E}-02$	$5.101\mathrm{E}-05$
2.00	$-2.469\mathrm{E}+00$	$1.556\mathrm{E}-02$	$3.258\mathrm{E}-02$	$4.162\mathrm{E}-05$

Efficiency of Runge-Kutta Methods

The descriptions of the modified Euler, midpoint, and classical Runge-Kutta methods above make it clear that for one-step methods higher order is attained at the expense of increasing the amount of computation required to generate the approximation at the next step. Table 10.3.11 indicates the number of additions, multiplications, and evaluations of the right-hand side (RHS) of (10.3.2) that is required for the three higher-order methods discussed.

Table 10.3.12 shows the result of applying the Euler, modified Euler, midpoint, and classical Runge-Kutta methods to approximate the solution to the initial value problem of Example 10.3.1, $x' = 5(t-1)x$, $x(0) = 5$ at $t = 1$. The second column indicates the total number of evaluations of the right-hand side $5(t-1)x$, the fourth column the time required to execute a Turbo Pascal implementation of the method on a 386 PC system without numeric coprocessor. The machine unit was $\mu = 2 \times 10^{-12}$, and the target accuracy was 5×10^{-8}. As can be seen in Table 10.3.12 the higher-order classical Runge-Kutta method is far more efficient than the second-order methods, despite the greater amount of computation required per step.

Observe in Table 10.3.12 that Euler's method produced an answer accurate to a little better than 10^{-6} after 2 million steps. Since it is an order-one method, achieving an error of 5×10^{-8} would be expected to require around 100 million steps. However, as indicated in Sec. 10.2, rounding error would likely severely corrupt the answer were this many steps used. Thus Euler's method probably cannot attain accuracies on the order of 10^{-8} for this problem with the arithmetic precision employed.

Table 10.3.13 shows the time required by the four methods that we have discussed to approximate the solution to Example 10.3.1 at $t = 1$ using 8000 steps. Note that while the increased computational complexity of the second-

TABLE 10.3.11

Method	Adds.	Mults. and divs.	RHS evals.
Euler	1	1	1
Modified Euler	4	3	2
Midpoint	3	4	2
Classical Runge-Kutta	9	11	4

TABLE 10.3.12

Method	Steps	Error	Time, s
Euler	2,000,000	1.4×10^{-6}	342.7
Modified Euler	6600	4.9×10^{-8}	2.2
Midpoint	8300	5.0×10^{-8}	2.7
Classical Runge-Kutta	80	4.5×10^{-8}	0.1

TABLE 10.3.13

Method	Error	Time, s	Time/step
Euler	3.4×10^{-4}	1.2	1.5×10^{-4}
Modified Euler	3.3×10^{-8}	2.6	3.3×10^{-4}
Midpoint	5.4×10^{-8}	2.6	3.3×10^{-4}
Classical Runge-Kutta	1.4×10^{-11}	5.8	7.3×10^{-4}

TABLE 10.3.14

Order	2	3	4	5	6	$n > 6$
Function evals./step	2	3	4	6	7	$n + 2$

order methods does make the computation time per step more than double that of Euler's method, the error made by the second-order methods is smaller by a factor of 10^4. The fact that gains in accuracy of similar magnitude were not achieved for classical Runge-Kutta over the second-order methods will be discussed further in the next subsection.

For the three higher-order Runge-Kutta methods discussed thus far, the number of evaluations of the right-hand side of the differential equation required by the method coincides with the order of the truncation error. As Table 10.3.14 shows, this is not the case in general. However, the number of evaluations continues to grow only linearly with the order of the method. The number of other additions and multiplications grows, at worst, quadratically with the order of the method (Exercise 13). Since the error made by a method of order n falls as h^n with decreasing stepsize h, an nth order Runge-Kutta method can be expected to be more efficient in attaining answers of a given accuracy than any method of order less than n for sufficiently small h.

Effect of Rounding Error on Higher-Order Methods

As indicated in Sec. 10.2, rounding error is not in most cases a significant consideration in Euler's method unless the desired accuracy requires an extremely small stepsize. For higher-order methods such as order-four classical Runge-Kutta, rounding error would be expected to affect the answer in fewer steps because of the higher computational complexity of the algorithm. However, it is often the case that fourth-order methods are able to achieve their maximum attainable accuracy in sufficiently few steps that rounding error is of little consequence. Table 10.3.15 illustrates this for the initial value problem of Example 10.3.1 at $t = 1$. The machine unit for the arithmetic used was $\mu = 2 \times 10^{-12}$, and thus the classical Runge-Kutta approximation with 640 steps is in error by only about 4μ, which is all the accuracy that one could hope to attain given the precision of the arithmetic employed.

TABLE 10.3.15

N	Error	Prev. error/error
5	4.9E − 03	
10	2.4E − 04	2.1E + 01
20	1.3E − 05	1.8E + 01
40	7.5E − 07	1.7E + 01
80	4.5E − 08	1.7E + 01
160	2.8E − 09	1.6E + 01
320	1.7E − 10	1.7E + 01
640	8.2E − 12	2.0E + 01
1280	3.6E − 12	2.3E + 00
2560	8.2E − 12	4.4E − 01
5120	−2.3E − 11	3.6E − 01

The third column indicates the ratio of the error of approximation using N steps to that obtained with $N/2$ steps. For a fourth-order method one would expect this ratio to be about 16. Obviously once the truncation error has been reduced to near the magnitude of the machine unit, no further improvement by a factor of 16 is possible. The fact that higher-order methods such as classical Runge-Kutta can attain very high accuracies with relatively large stepsizes should be borne in mind in selecting a stepsize, especially when efficiency is a consideration.

Runge-Kutta Methods for Systems of Equations

Runge-Kutta methods can also be applied to initial value problems for systems of first-order differential equations

$$\mathbf{x}' = \mathbf{f}(t, \mathbf{x}) \qquad \mathbf{x}(t_0) = \mathbf{x}_0 \tag{10.3.26}$$

In vector notation the generalization is straightforward; for instance, the modified Euler method is

$$\boldsymbol{\xi}_0 = \mathbf{x}_0$$
$$\mathbf{m}_1 = \mathbf{f}(t_i, \boldsymbol{\xi}_i)$$
$$\mathbf{m}_2 = \mathbf{f}(t_i + h, \boldsymbol{\xi}_i + h\mathbf{m}_1) \tag{10.3.27}$$
$$\boldsymbol{\xi}_{i+1} = \boldsymbol{\xi}_i + \frac{h}{2}(\mathbf{m}_1 + \mathbf{m}_2)$$

Unfortunately, most implementations of most computer languages currently do not support arithmetical operations on vectors. Thus a program for the modified Euler method is more apt to look like the following algorithm.

Algorithm 10.3.16   Modified Euler for a system of two equations

Input t_0, t_f (endpoints of the time interval)
            x_{01}, x_{02} (initial values of the dependent variables)
            N (number of steps)

Define   functions $f_1(t, x_1, x_2)$, $f_2(t, x_1, x_2)$ (right-hand sides of the two equations in the system)

$h := (t_f - t_0)/N$
$t := t_0, x_1 := x_{01}, x_2 := x_{02}$
for $i = 1$ to N
   $\{m_{11} := f_1(t, x_1, x_2), m_{12} = f_2(t, x_1, x_2)$
   $m_{21} := f_1(t + h, x_1 + hm_{11}, x_2 + hm_{12}), m_{22} := f_2(t + h, x_1 + hm_{11}, x_2 + hm_{12})$
   $x_1 := x_1 + 0.5h(m_{11} + m_{21}), x_2 := x_2 + 0.5h(m_{12} + m_{22})$
   $t := t_0 + ih$
   Output: $t, x_1, x_2\}$

The difficulty with Algorithm 10.3.16 is that it is somewhat inconvenient to change the dimension of the system. A program in which the dimension can be changed more readily will need to make use of the specific constructs available in the language employed. Here is a Turbo Pascal program for the modified Euler method in which the dimension of the system can be changed more easily.

Program 10.3.17   Modified Euler method for systems

```
Program ModEulersys;
(**Part that depends upon the particular initial value problem **)

const Dim = 2;

type DepVect = array[1..Dim] of real;

const
     N: LongInt = 200;
     t0: real = 0; tf: real = 1;
     x0: DepVect = (1, 0.2); {Initial conditions}

Function f(i: integer; t: real; x: DepVect): real;
begin
     case i of
          1: f := (x[1] − 0.5 * x[2]) * x[1];
          2: f := (−0.1 * x[2] + x[1]) * x[2];
     end
end;

(* * * * * * * * * * * * * * * * * * * * * * * * * *)
```

```
Procedure OutputAns(t: real; x: DepVect);
var i: integer;
begin
    write (t:0.4, '  ');
    for i := 1 to Dim do write (x[i]:0:4, '  ');
    writeln
end;

var
    i, j: integer;
    t,h; real;
    x, xeuler: DepVect;
    m1, m2: array[1..Dim] of real;

begin
    t:=t0; x: = x0;
    h:=(tf − t0)/N;
    for j:=1 to N do begin
        for i := 1 to Dim do m1[i] := f(i, t, x);
        for i := 1 to Dim do xeuler[i] := x[i] + h * m1[i];
        for i := 1 to Dim do m2[i] := f(i, t + h, xeuler);
        for i := 1 to Dim do x[i] := x[i] + 0.5 * h * (m1[i] + m2[i]);

        t := t0 + j * h;
        OutputAns(t, x)
    end
end.
```

Remark. Note in the above program that divisions by 2 in statement (10.3.27) of the modified Euler method are actually computed as multiplication by 0.5. As seen in Table 1.4.1, it may be the case that multiplications are substantially faster than divisions.

Finally, we note that up through order four, the order of the global truncation error for a Runge-Kutta method is the same when applied to a first-order system as it is when applied to a single differential equation. Beyond order four the order of the truncation error of a system may be the same as for a single equation or it may be lower.

EXERCISES 10.3

1. Use the
  (a) Modified Euler method
  (b) Midpoint method
  (c) Classical Runge-Kutta method
to find approximations to $x(2)$ for the initial value problem $x' = 5t − 2x$, $x(0) = 1$ using stepsizes of $h = 1$ and $h = 0.5$. Compare your answers with the actual solution.

2. Write a computer program implementing the
  (a) Modified Euler method
  (b) Midpoint method
  (c) Classical Runge-Kutta method
for a single first-order differential equation. Apply the method to tabulate approximations to the initial value problem $x' = 5t − 2x$, $x(0) = 1$ on the interval $0 \le t \le 2$ for stepsizes of $h = 0.2, 0.1$, and

0.05. For each stepsize also have the program print out the value of the actual solution and the error made by the approximation.

3. Use the
   (a) Modified Euler method
   (b) Midpoint method
   (c) Classical Runge-Kutta method
   to find and graph the temperature of an air-conditioned building [see Equation (10.1.4)] when $k = 0.5$ hour^{-1}, $A(t) = 70°$, $U_0 = 4°$/hour, $T_{\text{crit}} = 65°$, and $T(0) = 50°$. Use a stepsize h that is small enough so that the same graph results from using stepsizes of h and $h/2$.

4. Repeat Exercise 3 when the ambient temperature is $A(t) = 70° + 20° \sin(\pi t/12)$.

5. Use Theorem 10.3.7 to show that the global error made by the midpoint method is order two.

6. (a) Show that if x is a continuously differentiable solution to (10.3.2) and if $\lim_{h \to 0} \Phi(t, x; h) = f(t, x)$ uniformly in t and x in some rectangle containing the graph of x, then $\lim_{h \to 0} \epsilon(h) = 0$. Hence, under these conditions, consistency implies convergence from Theorem 10.3.7.
   (b) Show that if f and its first partial derivatives are continuous in a rectangle containing the solution, then the modified Euler method satisfies the hypotheses of part (a).

7. The choice of $\alpha = \frac{2}{3}$, $\beta = \frac{2}{3}$, $\omega_1 = \frac{1}{4}$, and $\omega_2 = \frac{3}{4}$ in (10.3.16) gives *Heun's method*. Write a program implementing Heun's method for the initial value problem for a single first-order differential equation. For the initial value problem in Exercise 2, compare the accuracy of Heun's method with that of the modified Euler and midpoint methods for stepsizes of $h = 0.2, 0.1$, and 0.05. Also compare the results obtained for Example 10.3.9 to those for the modified Euler and midpoint methods.

8. In (10.3.17) calculate the coefficient of the h^3 term in terms of $f(t, x)$ and its partial derivatives. Show that no choice of α will cause this coefficient to vanish for arbitrary $f(t, x)$.

9. (a) Derive the formulas given in (10.3.22).
   (b) Show that the formulas (10.3.22) imply the formulas (10.3.24).

10. (a) Show that the Runge-Kutta method

$$\xi_0 = x_0$$

$$m_1 = f(t_i, \xi_i)$$

$$m_2 = f(t_i + h, \xi_i + hm_1)$$

$$m_3 = f\left(t_i + \frac{h}{2}, \xi_i + \frac{h(m_1 + m_2)}{4}\right)$$

$$\xi_{i+1} = \xi_i + \frac{h}{6}(m_1 + m_2 + 4m_3) \quad (10.3.28)$$

has local truncation error of order four.
   (b) Show that the method given by (10.3.28) has global truncation error of order three.

11. Show that the classical Runge-Kutta method has a global truncation error that is of order four.

12. (a) For any one-step method show that

$$x_N - \xi_N = x_N - x_0 - h \sum_{k=0}^{N-1} \Phi(t_k, \xi_k; h)$$

   (b) For the initial value problem (10.3.2) on an interval $[t_0, t_f]$, suppose that the actual solution is symmetric about the midline $t = t_m := (t_0 + t_f)/2$ and that f has the property $f(t_m + t, x) = -f(t_m - t, x)$. Use part (a) with $t_N = t_f$ to show that for Euler's method $x_N - \xi_N = -hf(t_0, x_0) + O(h^2)$, and hence Euler's method remains order one unless $f(t_0, x_0) = 0$. Show, however, that for the modified Euler and midpoint methods, $x_N - \xi_N = O(h^3)$. This explains the observations of Example 10.3.9.

13. Show that the general n-slope Runge-Kutta method given by (10.3.20) requires no more than $\frac{1}{2}n^2 + \frac{3}{2}n - 1$ additions and no more than $\frac{1}{2}n^2 + \frac{5}{2}n - 1$ multiplications beyond those required in the n evaluations of the right-hand-side f.

14. Calculate the
    (a) Modified Euler method
    (b) Midpoint method
    (c) Classical Runge-Kutta method
    approximations to $x_1(1)$ and $x_2(1)$ for the initial value problem

$$x_1' + x_2 = t^2 \qquad x_1(0) = 1$$

$$x_2' - x_1 = 1 \qquad x_2(0) = 0 \qquad (10.3.29)$$

for stepsizes of $h = 1$ and $h = 0.5$. Compare with the values of the actual solution.

15. Write a program implementing the
    (a) Modified Euler method
    (b) Midpoint method

(c) Classical Runge-Kutta method for a system of differential equations. Apply it to tabulate approximations to $x_1(t)$ and $x_2(t)$ for the initial value problem

$$x_1' + x_2 = t^2 \qquad x_1(0) = 1$$
$$x_2' - x_1 = 1 \qquad x_2(0) = 0 \tag{10.3.30}$$

on the interval $0 \le t \le 1$ for stepsizes of $h = 0.2, 0.1$, and 0.05. For each stepsize also have the program print out the value of the actual solutions and the error made by the approximations.

16. Compare the execution times and accuracies given in Tables 10.3.12 and 10.3.13 with those for your computer system. Try to account for any differences that you observe.

17. Use the
(a) Modified Euler method
(b) Midpoint method
(c) Classical Runge-Kutta method
to graph solutions to the Lotka-Volterra model (10.1.17). Take $a = 2$, $b = 1$, $c = 1$, $d = 1$ and use the initial conditions $x_1(0) = 1$, $x_2(0) = 0.1$. Graph solutions over the time interval $0 \le t \le 10$. Choose the stepsize h so that stepsizes of h and $h/2$ yield the same graph.

Use one of the methods studied in this section for Exercises 18 to 23.

18. Write a program to construct graphs of the solutions x_1 and x_2 against time to the mass-spring system modeled by the system of differential equations given by (10.1.36). Use $m_1 = m_2 = 1$ and $k_1 = 1$, $k_2 = 2$, $k_3 = 3$. Also graph x_2 against x_1 with time as a parametric variable. Does the motion of the system appear to be periodic?

19. (a) For the nondimensional differential equation for the height of an object above a planet's surface given in Exercise 13a of Sec. 10.1, find by numerical experimentation an approximate value for the greatest height h^* at which an object can be released from rest without air resistance significantly affecting its time of descent. Make this determination for $b^* = 1, 0.1, 0.01$, and 0.001.
(b) For streamlined objects such as spheres and ellipsoids, the parameter b in the dimensioned equation (10.1.32) is approximately $0.1\rho A$,

where ρ is the density of the air and A is the cross-sectional area of the object perpendicular to the direction of motion. Express the nondimensional results that you found in part (a) in terms of the dimensional parameters g, R, ρ, and A.

20. (a) The angular displacement θ of a simple pendulum, measured from the downward vertical direction, is governed by the differential equation $\theta'' + (g/l) \sin \theta = 0$, where g is the acceleration due to gravity and l is the length of the pendulum. For a pendulum of length 1 foot with the initial conditions $\theta(0) = \pi/2$, $\theta'(0) = 0$, graph the angular displacement of the pendulum over two periods of its motion.
(b) Using the dimensionless form of the pendulum equation (see Exercise 19 in Sec. 10.1) and numerical experimentation, construct a table that can be used to determine the approximate period of a pendulum from its initial displacement θ_0 when $0 < \theta < \pi$.

21. Write a program to plot solutions to a second-order autonomous system

$$x_1' = f_1(x_1, x_2)$$
$$x_2' = f_2(x_1, x_2) \tag{10.3.31}$$

in phase portrait form, that is, x_2 plotted against x_1 with time t as a parametric variable. Use your program to investigate the effect of the parameter σ on the phase trajectories of the dimensionless predator-prey model given in Exercise 21 in Sec. 10.1. Also investigate the effect of adding a resource constraint term to the model. See Exercise 22 in Sec. 10.1.

22. (a) Write a program to plot solutions to the initial value problem for a single second-order differential equation. Use your program to investigate how small ϵ^* must be so that solutions to the dimensionless Duffing equation (10.1.26) do not visibly differ from those for which $\epsilon^* = 0$. Use $k^* = 2$ and plot over the interval $0 \le t \le 20$ (see Fig. 10.1.15). Express the result in terms of the original parameters of the problem.
(b) Attempt to make the conclusion of part (a) independent of the value of k^* by experimenting with relations such as the one $\epsilon \ll |k^* - 1|/k^*$ suggested in Example 10.1.13.

(c) Do a similar analysis for the damped Duffing equation (Exercise 15 in Sec. 10.1).

23. (a) Plot the solution x_1 to the Lorenz equations (Exercise 25 in Sec. 10.1) on the interval $0 \le t \le 25$ using 500 steps of the method you have selected and then 1000 steps. Take the parameters in the equation to be $s = 10$, $r = 28$, and $s = \frac{8}{3}$. Use the initial conditions $x_1(0) = x_2(0) = x_3(0) = 1$. What you observe

is not necessarily a defect of the method that you selected but rather the extreme sensitivity of the Lorenz equations to their initial data.

(b) Plot the solution x_3 against x_1 with time as a parametric variable. Describe the behavior of the system with respect to the equilibrium points that you found in Exercise 25 in Sec. 10.1.

10.4 MULTISTEP METHODS

The Runge-Kutta family of one-step methods discussed in Sec. 10.3 generates the approximation ξ_{i+1} to the solution $x(t_{i+1})$ to the initial value problem

$$x' = f(t, x) \qquad x(t_0) = x_0 \tag{10.4.1}$$

using only the approximation ξ_i to $x(t_i)$. *Multistep methods* may use several previous approximations $\xi_i, \xi_{i-1}, \ldots$ in determining ξ_{i+1}. As before we assume that approximations $\xi_i \simeq x(t_i)$ are sought in an interval (t_0, t_f) at the $N+1$ evenly spaced points $t_0, \ldots, t_N = t_f$ with $h := (t_f - t_0)/N$ being the distance between points.

Two Multistep Methods

One means of deriving a multistep method is to use one of the higher-order difference formulas discussed in Sec. 9.2 to approximate x' in (10.4.1). For instance, if we use the central difference formula $x'(t) \simeq [x(t+h) - x(t-h)]/2h$, then we obtain the multistep method

$$\xi_{i+1} = \xi_{i-1} + 2hf(t, \xi_i) \tag{10.4.2}$$

sometimes referred to as the multistep *midpoint rule*. In a manner similar to that for one-step methods, the *local truncation error* of a multistep method is the error made by one step of the method, when the exact solution is used in lieu of previous approximations. For the midpoint rule (10.4.2) the local truncation error can be calculated from Taylor's theorem. For $x \in C^3[t_0, t_f]$,

$$x(t_{i+1}) - \xi_{i+1} = x(t_{i+1}) - [x(t_{i-1}) + 2hf(t_i, x(t_i))]$$
$$= [x(t_i) + hx'(t_i) + \tfrac{1}{2}h^2x''(t_i) + \tfrac{1}{6}h^3x'''(\tau_i)]$$
$$- [x(t_i) - hx'(t_i) + \tfrac{1}{2}h^2x''(t_i) - \tfrac{1}{6}h^3x'''(\tau_{i-1})]$$
$$- 2hf(t_i, x(t_i))$$
$$= \tfrac{1}{6}h^3(x'''(\tau_i) + x'''(\tau_{i-1})) \tag{10.4.3}$$

where $\tau_{i-1} \in (t_{i-1}, t_i)$ and $\tau_i \in (t_i, t_{i+1})$. Thus the order of the local truncation error is three, the same as that of the modified Euler and midpoint Runge-Kutta methods discussed in Sec. 10.3.

Example 10.4.1. Figure 10.4.2 shows the result of applying method (10.4.2) to the initial value problem $x' = 5(t-1)x$, $x(0) = 5$ of Example 10.3.1 with stepsizes of $h = \frac{1}{4}$ and $h = \frac{1}{8}$. The value of ξ_1 was computed using the modified Euler method of Sec. 10.3. Compared to the approximations generated by the modified Euler and midpoint methods for the same problem (Figs. 10.3.3 and 10.3.4), we see that those of (10.4.2) are not particularly good. The tendency to oscillate which is very evident for $h = \frac{1}{4}$ is characteristic of some multistep methods. This will be discussed more in Sec. 10.7.

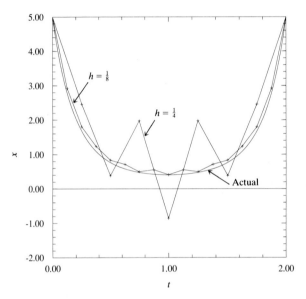

FIGURE 10.4.2

Another means of obtaining multistep methods is to note that for any differentiable function $x(t)$,

$$x(t_{i+1}) = x(t_i) + \int_{t_i}^{t_{i+1}} x'(t)\, dt \qquad (10.4.4)$$

The Lagrange interpolating polynomial through the points $(t_{i-1}, x'(t_{i-1}))$ and $(t_i, x'(t_i))$ is given by

$$p_2(t) := -\frac{1}{h}(t - t_i)x'(t_{i-1}) + \frac{1}{h}(t - t_{i-1})x'(t_i) \qquad (10.4.5)$$

Replacing $x'(t)$ in (10.4.4) by $p_2(t)$ and integrating we obtain

$$x(t_{i+1}) \simeq x(t_i) + h[\tfrac{3}{2}x'(t_i) - \tfrac{1}{2}x'(t_{i-1})] \qquad (10.4.6)$$

For a solution to (10.4.1), $x'(t_i) = f(t_i, x(t_i))$. This gives the *Adams-Bashforth* two-point method

$$\xi_{i+1} = \xi_i + \frac{h}{2}[3f(t_i, \xi_i) - f(t_{i-1}, \xi_{i-1})] \qquad (10.4.7)$$

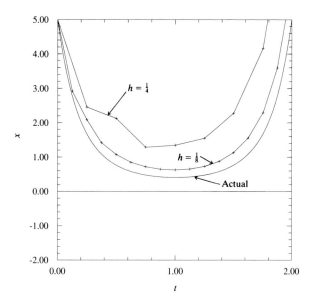

FIGURE 10.4.3

It may be shown by a computation similar to (10.4.3) that the local truncation error of the Adams-Bashforth method (10.4.7) is $O(h^3)$ (Exercise 3). Higher-order Adams-Bashforth methods can be obtained by approximating x' by higher-degree Lagrange interpolating polynomials. We will see later (Theorem 10.4.20) that for multistep methods obtained in this way, the order of the global truncation error in h is one less than that of the local truncation error, just as it was for one-step methods. Figure 10.4.3 shows the results of applying (10.4.7) to the problem of Example 10.4.1. Comparison with Figs. 10.3.3 and 10.3.4 shows that for this problem, the accuracy of the Adams-Bashforth two-point formula is a bit less than that of the Runge-Kutta formulas of comparable order. This is not surprising since second-order Runge-Kutta methods use information on the direction field at t_i and a second point in the interval (t_i, t_{i+1}) to approximate the solution at t_{i+1}, whereas the Adams-Bashforth two-point formula uses the less current information on the slopes at t_{i-1} and t_i. However, multistep formulas have the advantage that in general they can be implemented more efficiently than those of the Runge-Kutta family.

Programming an Adams-Bashforth Method

Observe that (10.4.7) cannot be used to calculate ξ_1, the approximation at $t_1 := t_0 + h$. Typically, a Runge-Kutta method of comparable order is used to generate the approximations necessary to start a multistep method. As we will see later, (10.4.7) has a global truncation error that is order two in the stepsize h. Hence we will use the order-two modified Euler method of Sec. 10.3 to generate ξ_i.

Algorithm 10.4.4 Adams-Bashforth two-point formula

Input t_0, t_f (interval over which approximation is sought)
x_0 (initial value of the dependent variable)
N (number of subintervals)

Define $f(t, x)$ (right-hand side of the differential equation)

$t := t_0, x := x_0$
output: t, x
$h := (t_f - t_0)/N$
$f_0 := f(t, x)$
$t := t_0 + h$
$x := x + 0.5h[f_0 + f(t + h, x + hf_0)]$ (Modifier Euler approximation)
output: t, x

for $i = 2$ to N
$\quad \{f_1 := f(t, x)$ (only new evaluation required)
$\quad t := t_0 + ih$
$\quad x := x + h(1.5f_1 - 0.5f_0)$
$\quad$ output: t, x
$\quad f_0 := f_1$ (save value of f for next iteration)}

The important thing to note about Algorithm 10.4.4 is that once the multistep formula is being used, only one new evaluation of the right-hand-side f of (10.4.1) is required to obtain the next approximation, providing that previous evaluations of f have been saved. An order-two Runge-Kutta method in contrast requires two function evaluations per step. The arithmetic computation other than the evaluation of f is also a bit less than that for Runge-Kutta methods of comparable order. Thus the Adams-Bashforth two-step method is more efficient than the Runge-Kutta methods of comparable order when the same stepsize is used.

Implicit Methods

If in (10.4.4) we interpolate the points $(t_i, x(t_i))$ and $(t_{i+1}, x(t_{i+1}))$ rather than $(t_{i-1}, x(t_{i-1}))$ and $(t_i, x(t_i))$, then the multistep method obtained is the *Adams-Moulton two-point formula*

$$\xi_{i+1} = \xi_i + \frac{h}{2} [f(t_i, \xi_i) + f(t_{i+1}, \xi_{i+1})] \qquad (10.4.8)$$

which is also sometimes called the *implicit trapezoid method*. Derivation of this method, which has $O(h^3)$ local truncation error, is Exercise 4.

A formula such as (10.4.8) is said to be an *implicit* method since the sought-after approximation ξ_{i+1} occurs on both sides of the equation. Methods such as the Adams-Bashforth formula and the one-step Runge-Kutta formulas in which ξ_{i+1} is directly calculable from the right-hand side of the formula are termed *explicit*. Since implicit multistep methods use more current direction field information than explicit multistep methods, they can be expected to be more accurate.

Example 10.4.5. For the initial value problem $x' = 5(t-1)x$, $x(0) = 5$ of Example 10.4.1 the Adams-Moulton formula (10.4.8) is

$$\xi_{i+1} = \xi_i + \frac{h}{2}\left[5(t_i - 1)\xi_i + 5(t_{i+1} - 1)\xi_{i+1}\right] \qquad (10.4.9)$$

For this particular problem we may solve for ξ_{i+1} to obtain the explicit formula

$$\xi_{i+1} = \frac{2 + 5h(t_i - 1)}{2 - 5h(t_{i+1} - 1)}\,\xi_i \qquad (10.4.10)$$

Figure 10.4.6 shows the approximations generated by (10.4.10). As we can see they are better than the Adams-Bashforth approximations for the same problem (Fig. 10.4.3) and comparable to Runge-Kutta methods of the same local truncation error such as the modified Euler and midpoint methods (Figs. 10.3.3 and 10.3.4).

The approach in the above example obviously is limited to situations in which the implicit relation (10.4.8) can be solved explicitly for ξ_{i+1}, and it thus lacks general applicability. For instance, it could not be used on the differential equation $x' = \sin tx$. However, this difficulty can be circumvented by solving (10.4.8) numerically, say by Newton's method. If we let

$$G(\xi_{i+1}) := \xi_{i+1} - \xi_i - \frac{h}{2}\left[f(t_i, \xi_i) + f(t_{i+1}, \xi_{i+1})\right] \qquad (10.4.11)$$

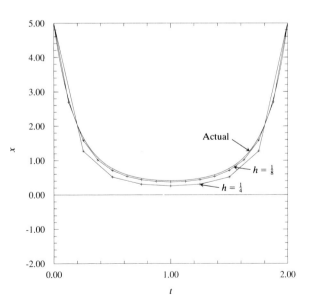

FIGURE 10.4.6

then

$$G'(\xi_{i+1}) = 1 - \frac{h}{2} \frac{\partial}{\partial x} f(t_{i+1}, \xi_{i+1}) \qquad (10.4.12)$$

In Algorithm 10.4.7 below we first use Euler's method to obtain an initial guess $\xi_{i+1}^{(0)}$ for ξ_{i+1} and then refine it with the Newton iteration

$$\xi_{i+1}^{(n+1)} = \xi_{i+1}^{(n)} - \frac{G(\xi_{i+1}^{(n)})}{G'(\xi_{i+1}^{(n)})} \qquad (10.4.13)$$

until two successive values $\xi_{i+1}^{(n)}$, $\xi_{i+1}^{(n+1)}$ differ by less than some specified tolerance.

Algorithm 10.4.7 Implicit trapezoid rule

Input t_0, t_f (endpoints of interval)
            x_0 (initial value of dependent variable)
            N (number of steps)
            tol (error tolerance for Newton's method)

Define   functions $f(t, x)$ (right-hand side of differential equation)
            $G(t, x_0, x_1, h)$ [function given by (10.4.11)]
            $dG(t, x_1, h)$ [function given by (10.4.12)]

$h := (t_f - t_0)/N$
$t := t_0$
output: t_0, x_0
for $i = 1$ to N
$\quad \{\xi_1 := x_0 + hf(t, x_0)$ (Use Euler approximation for initial guess)
$\quad$ repeat (Newton's method)
$\qquad \xi_0 := \xi_1$
$\qquad \xi_1 := \xi_1 - \dfrac{G(t, x_0, \xi_1, h)}{dG(t, \xi_1, h)}$
$\quad$ until $|\xi_1 - \xi_0| <$ tol
$\quad t := t_0 + ih$
$\quad$ output: t, ξ_1
$\quad x_0 := \xi_1\}$

Table 10.4.8 shows the errors made at selected points by the modified Euler, Adams-Bashforth, and implicit trapezoid rule methods for the initial value problem of Example 10.4.1. As can be seen in the table the implicit trapezoid rule generates answers for this problem that are comparable to those of the modified Euler method and noticeably better than those of the explicit Adams-Bashforth method. However, it is clear that because several evaluations of the functions defined by (10.4.11) and (10.4.12) may be needed for each step, the implicit trapezoid rule will not be competitive with the other two methods insofar as efficiency is concerned. The last column of the table shows the results of applying the *Adams predictor-corrector method*, which will be discussed in the next subsection.

TABLE 10.4.8

t	Modified Euler	Adams-Bashforth	Implicit trapezoid	Adams predictor-corrector
			$N = 20$	
0.00	0.000E + 00	0.000E + 00	0.000E + 00	0.000E + 00
0.20	8.684E − 02	2.531E − 01	−5.581E − 02	−1.972E − 02
0.40	6.226E − 02	2.227E − 01	−4.382E − 02	−5.678E − 02
0.60	4.148E − 02	1.746E − 01	−3.191E − 02	−4.572E − 02
0.80	3.104E − 02	1.453E − 01	−2.554E − 02	−3.682E − 02
1.00	2.784E − 02	1.369E − 01	−2.362E − 02	−3.389E − 02
1.20	3.044E − 02	1.499E − 01	−2.554E − 02	−3.699E − 02
1.40	4.003E − 02	1.905E − 01	−3.191E − 02	−4.838E − 02
1.60	6.072E − 02	2.750E − 01	−4.382E − 02	−7.609E − 02
1.80	9.738E − 02	4.256E − 01	−5.581E − 02	−1.446E − 01
2.00	1.233E − 01	5.874E − 01	1.455E − 11	−3.362E − 01
			$N = 40$	
0.00	0.000E + 00	0.000E + 00	0.000E + 00	0.000E + 00
0.20	1.840E − 02	6.379E − 02	−1.363E − 02	−1.193E − 02
0.40	1.329E − 02	5.573E − 02	−1.081E − 02	−1.269E − 02
0.60	8.932E − 03	4.211E − 02	−7.921E − 03	−9.671E − 03
0.80	6.730E − 03	3.427E − 02	−6.359E − 03	−7.782E − 03
1.00	6.065E − 03	3.195E − 02	−5.885E − 03	−7.185E − 03
1.20	6.659E − 03	3.478E − 02	−6.359E − 03	−7.810E − 03
1.40	8.762E − 03	4.397E − 02	−7.921E − 03	−1.002E − 02
1.60	1.312E − 02	6.255E − 02	−1.081E − 02	−1.502E − 02
1.80	1.973E − 02	9.141E − 02	−1.363E − 02	−2.568E − 02
2.00	1.556E − 02	9.086E − 02	7.276E − 12	−4.764E − 02

An alternative to using Newton's method to solve for ξ_{i+1} in (10.4.8) is simply to iterate

$$\xi_{i+1}^{(n+1)} = \xi_i + \frac{h}{2} \left[f(t_i, \xi_i) + f(t_{i+1}, \xi_{i+1}^{(n)}) \right] \tag{10.4.14}$$

until two successive iterates agree to within the specified tolerance. This approximation process will converge more slowly than Newton's method, but it is simpler to implement, especially in the case of systems of differential equations where the use of Newton's method (Sec. 2.6) would require either analytic calculation or numerical approximation of the jacobian matrix. There are, however, classes of differential equations, particularly *stiff* systems of equations, which will be discussed in Sec. 10.7, for which a simple iteration may not converge. In such cases Newton's method must be used.

Predictor-Corrector Methods

The modified Euler method of Sec. 10.3 can be regarded as an application of the implicit trapezoid method (10.4.8) in which the occurrence of the unknown ξ_{i+1} on the right-hand side is replaced by the value $\xi_i + hf(t_i, \xi_i)$ that Euler's method predicts for it. Such a scheme is called a *predictor-corrector method*. In this subsection we will see that the pairing of an explicit Adams-Bashforth predictor with an implicit Adams-Moulton corrector can produce particularly efficient predictor-corrector methods.

Algorithm 10.4.9 gives a predictor-corrector scheme based on the order-two Adams-Bashforth and Adams-Moulton methods. The accuracy achieved by this method on the test problem of Example 10.4.1 is indicated in Table 10.4.8.

Algorithm 10.4.9   Order-two Adams predictor-corrector method

Input t_0, t_f (endpoints of domain interval)
             x_0 (initial value of dependent variable)
             N (number of steps)

Define   function $f(t, x)$ (right-hand side of differential equation)

$t := t_0, x := x_0$
output: t_0, x_0
$h := (t_f - t_0)/N$
$f_0 := f(t, x)$
$x := x + 0.5h[f_0 + f(t + h, x + hf_0)]$ (Use modified Euler for the first step)
$t := t + h$
output: t, x
for $i = 2$ to N
    $\{f_1 := f(t, x)$
    $t := t_0 + ih$
    $x_{pred} := x + h(1.5f_1 - 0.5f_0)$ (Adams-Bashforth predictor)
    $x := x + 0.5h[f_1 + f(t, x_{pred})]$ (Adams-Moulton corrector)
    output: t, x
    $f_0 := f_1$ (Save the value of f for the next iteration)$\}$

While Table 10.4.8 shows that the order-two Adams-Bashforth-Moulton predictor-corrector method has an accuracy comparable to that of the modified Euler for this problem, its efficiency would not be expected to be much greater since it requires the same number of evaluations of the right-hand side $f(t, x)$ of the differential equation. However, if we use the order-four Adams-Bashforth and Adams-Moulton methods, which are obtained by using a four-point interpolation in (10.4.4), in the predictor-corrector scheme

$\xi_0 = x_0$

ξ_1, ξ_2, ξ_3 determined by order-four classical Runge-Kutta

$$\xi_{i+1}^{pred} = \xi_i + \frac{h}{24}[55f(t_i, \xi_i) - 59f(t_{i-1}, \xi_{i-1}) + 37f(t_{i-2}, \xi_{i-2}) - 9f(t_{i-3}, \xi_{i-3})]$$

$$\xi_{i+1} = \xi_i + \frac{h}{24}[9f(t_{i+1}, \xi_{i+1}^{pred}) + 19f(t_i, \xi_i) - 5f(t_{i-1}, \xi_{i-1}) + f(t_{i-2}, \xi_{i-2})]$$

$$(10.4.15)$$

TABLE 10.4.10

t	Adams-Bashforth 4	Adams predictor-corrector 4	Classical Runge-Kutta
		$N = 20$	
0.00	$0.000E + 00$	$0.000E + 00$	$0.000E + 00$
0.20	$9.114E - 04$	$9.114E - 04$	$9.114E - 04$
0.40	$4.676E - 02$	$-6.589E - 03$	$5.689E - 04$
0.60	$6.067E - 02$	$-7.323E - 03$	$3.561E - 04$
0.80	$4.952E - 02$	$-6.158E - 03$	$2.640E - 04$
1.00	$4.279E - 02$	$-5.691E - 03$	$2.388E - 04$
1.20	$4.621E - 02$	$-6.275E - 03$	$2.639E - 04$
1.40	$6.114E - 02$	$-8.403E - 03$	$3.547E - 04$
1.60	$9.553E - 02$	$-1.378E - 02$	$5.676E - 04$
1.80	$1.710E - 01$	$-2.803E - 02$	$9.949E - 04$
2.00	$3.264E - 01$	$-7.268E - 02$	$1.327E - 03$
		$N = 40$	
0.00	$0.000E + 00$	$0.000E + 00$	$0.000E + 00$
0.20	$2.909E - 03$	$-3.341E - 04$	$4.832E - 05$
0.40	$4.693E - 03$	$-5.802E - 04$	$3.058E - 05$
0.60	$3.630E - 03$	$-4.448E - 04$	$1.926E - 05$
0.80	$2.929E - 03$	$-3.518E - 04$	$1.430E - 05$
1.00	$2.722E - 03$	$-3.230E - 04$	$1.294E - 05$
1.20	$3.001E - 03$	$-3.555E - 04$	$1.430E - 05$
1.40	$3.928E - 03$	$-4.721E - 04$	$1.922E - 05$
1.60	$5.980E - 03$	$-7.565E - 04$	$3.056E - 05$
1.80	$9.946E - 03$	$-1.465E - 03$	$5.101E - 05$
2.00	$1.455E - 02$	$-3.482E - 03$	$4.162E - 05$

then the result is an order-four method that still requires only two new evaluations, $f(t_i, \xi_i)$ and $f(t_{i+1}, \xi_{i+1}^{pred})$, of the right-hand side of the differential equation per step. This is one-half the number of evaluations necessary for an order-four Runge-Kutta method. Table 10.4.10 compares the errors made by the order-four Adams-Bashforth, Adams predictor-corrector, and classical Runge-Kutta methods for the initial value problem of Example 10.4.1. As can be seen there the use of the Adams-Moulton corrector substantially improves the accuracy of the Adams-Bashforth predictions, though the errors remain larger than those of classical Runge-Kutta for this problem.

Comparison of Fourth-Order Runge-Kutta and Adams Methods

To further compare the accuracy and efficiency of Runge-Kutta and Adams methods, Table 10.4.12 shows the time required by classical Runge-Kutta and the fourth-order Adams predictor-corrector scheme (10.4.15) to attain a specified accuracy for five different problems. The five problems are given in Table 10.4.11.

TABLE 10.4.11
Five test problems

1. Approximate $x(1.25)$ for

$$x' = 5(t-1)x \qquad x(0) = 5$$

Solution:

$$x(t) = 5e^{(5/2)t^2 - 5t}$$

2. Approximate $x(1.5)$ for

$$x' = 1 + x^2 \qquad x(0) = 0$$

Solution:

$$x(t) = \tan t$$

3. Approximate $x(30)$ for

$$x' = \cos\frac{\pi t}{12} - x \qquad x(0) = 50$$

Solution:

$$x(t) = \frac{\cos(\pi t/12) + (\pi/12)\sin(\pi t/12)}{1 + (\pi/12)^2} + \left[50 - \frac{1}{1 + (\pi/12)^2}\right]e^{-t}$$

4. Approximate $x_1(3)$, $x_2(3)$, and $x_3(3)$ for

$$x_1' = -2x_1 - x_2 + e^{-3t} \qquad x_1(0) = 1$$
$$x_2' = 2x_1 - x_2 + x_3 \qquad x_2(0) = 0$$
$$x_3' = 2x_2 - 2x_3 - 2e^{-3t} \qquad x_3(0) = 0$$

Solution:

$$x_1(t) = -2e^{-t} + (4 + 2t)e^{-2t} - e^{-3t}$$
$$x_2(t) = 2e^{-t} - 2e^{-2t}$$
$$x_3(t) = 4e^{-t} - (6 + 4t)e^{-2t} + 2e^{-3t}$$

5. Approximate $x_1(20)$ and $x_2(20)$ for

$$x_1'' = -2x_1 + \tfrac{1}{2}x_2 \qquad x_1(0) = 0 \qquad x_1'(0) = 0$$
$$x_2'' = 2x_1 - 2x_2 + 10\cos 2t \qquad x_2(0) = 0 \qquad x_2'(0) = 0$$

Solution:

$$x_1(t) = \tfrac{5}{3}\cos 2t + \tfrac{5}{6}\cos t - \tfrac{5}{2}\cos\sqrt{3}t$$
$$x_2(t) = -\tfrac{20}{3}\cos 2t + \tfrac{5}{3}\cos t + 5\cos\sqrt{3}t$$

Graphs of the solutions of problems 3, 4, and 5 are shown in Figs. 10.4.13 to 10.4.15. For problem 5 the given system was converted to an initial value problem for a system of four first-order differential equations

$$
\begin{aligned}
u_1' &= u_2 & u_1(0) &= 0 \\
u_2' &= -2u_1 + \tfrac{1}{2}u_3 & u_2(0) &= 0 \\
u_3' &= u_4 & u_3(0) &= 0 \\
u_4' &= 2u_1 - 2u_3 + 10\cos 2t & u_4(0) &= 0
\end{aligned}
\tag{10.4.16}
$$

TABLE 10.4.12
Steps required by fourth-order Adams predictor-corrector and classical Runge-Kutta methods to find a solution to within a target tolerance

Prob.	Target error	Adams predictor-corrector			Classical Runge-Kutta		
		Steps	Actual error	Time, s	Steps	Actual error	Time, s
1	5E − 09	380	4.6E − 09	0.17	180	5.0E − 09	0.12
2	5E − 07	1510	4.9E − 07	0.63	530	4.8E − 07	0.32
3	5E − 09	290	4.7E − 09	0.31	470	4.7E − 09	0.90
4	5E − 09	120	4.1E − 09	0.37	110	4.7E − 09	0.58
5	5E − 06	1320	5.0E − 06	2.84	970	4.9E − 06	3.16

Algorithm 10.4.9 generalizes easily to the case of the initial value problem $\mathbf{x}' = \mathbf{f}(t, \mathbf{x})$, $\mathbf{x}(0) = \mathbf{x}_0$. Errors reported in Tables 10.4.12 and 10.4.16 for the two system problems, 4 and 5, are the maximum errors of the three and four components, respectively. Times given are for implementations in Turbo Pascal running on a 386 IBM PC class computer without a numeric coprocessor. Because the clock has a very low resolution, ±0.05 seconds, repeated runs were used so that the total time of execution was in excess of 10 seconds. This makes the timings accurate to about two decimal places.

Except for problem 3 the classical Runge-Kutta method was able to attain the target accuracy in fewer steps than the Adams predictor-corrector method. It is expected that this will often be the case since classical Runge-Kutta computes its approximations for $x(t_{i+1})$ based upon four slopes in the interval

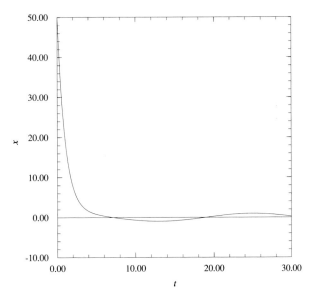

FIGURE 10.4.13

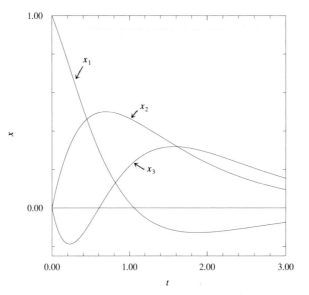

FIGURE 10.4.14

(t_i, t_{i+1}), whereas the Adams method uses less current slope data from the points t_{i-1}, t_{i-2}, and t_{i-3} in its calculations. In problems 1 and 2 the accuracy of classical Runge-Kutta was sufficiently greater to make it more efficient than the Adams method. In problems 3, 4, and 5, where the number of steps of the two methods was more comparable, the fact that the Adams method requires only two evaluations of the right-hand side of the differential equation, rather than the four necessary for classical Runge-Kutta, made it as or more efficient. This

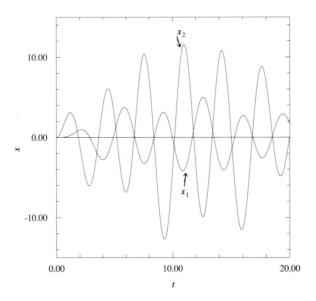

FIGURE 10.4.15

TABLE 10.4.16
Time required for 500 steps of fourth-order Adams predictor-corrector and classical Runge-Kutta methods

Prob.	Adams predictor-corrector 4		Classical Runge-Kutta	
	Error	Time, s	Error	Time, s
1	1.5E − 09	0.22	8.3E − 11	0.34
2	2.9E − 05	0.21	6.0E − 07	0.31
3	4.5E − 10	0.53	3.7E − 09	0.95
4	1.2E − 11	1.51	1.1E − 11	2.63
5	2.3E − 04	1.07	6.9E − 05	1.62

advantage will be especially great in problems in which the right-hand side of the differential equation contains transcendental functions such as exp and cos which are relatively costly to evaluate as was the case in problems 3, 4, and 5.

It should be noted, however, that the number of steps necessary to obtain the specified accuracy was determined beforehand by comparison with the value of the actual solution, which is, of course, not known in a realistic problem. Since the truncation errors made by approximation methods depend upon the derivatives of the unknown solution, there is in general no way to determine the number of steps required a priori. Table 10.4.16 indicates that for a fixed number of steps the Adams predictor-corrector method will almost always require less time to execute, though possibly with some loss in accuracy. Thus as a practical matter it is necessary to iteratively compute approximations with smaller and smaller stepsizes until two successive approximations agree to the desired accuracy. Applying this approach to the five problems in Table 10.4.11 is Exercise 16. More efficient methods of error control will be considered in Secs. 10.5 and 10.6.

In some applications the right-hand side of the differential equation may have discontinuities. Such is the case for the air-conditioning problem of Example 10.1.1. Discontinuities create problems for multistep methods, since if t_{i+1} is the first point past the discontinuity, then the direction field information at the trailing points $t_i, t_{i-1}, \ldots$ may be wholly misleading. Moreover, some of the direction field information will remain erroneous for several steps. For Runge-Kutta methods the effect of the discontinuity in the direction field is confined to a single step, and thus these methods are preferable in such situations.

Example 10.4.17. The differential equation

$$x' = \begin{cases} 1 - x & t < \pi \\ -5x & t > \pi \end{cases} \tag{10.4.17}$$

with the initial condition $x(0) = 0$ has solution

$$x(t) = \begin{cases} 1 - e^{-t} & t < \pi \\ (1 - e^{-\pi})e^{-5(t-\pi)} & t > \pi \end{cases} \tag{10.4.18}$$

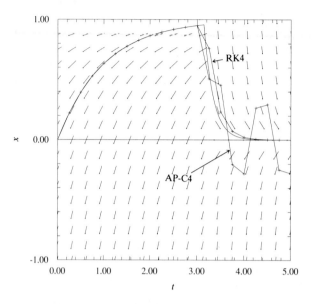

FIGURE 10.4.18

Figure 10.4.18 shows the classical Runge-Kutta approximation with a stepsize of $h = \frac{1}{4}$ and the fourth-order Adams predictor-corrector approximations with $h = \frac{1}{4}$. To attain an accuracy of 5×10^{-5} at $t = 5\pi/4$, the Adams method required 3740 steps while classical Runge-Kutta required 700.

Derivation of the Adams Methods

As mentioned earlier the Adams methods are derived by replacing the integrand $x'(t)$ of (10.4.4) by a Lagrange interpolating polynomial. To obtain an n-point Adams-Bashforth formula we interpolate the points $(t_i, x'(t_i))$, $(t_{i-1}, x'(t_{i-1})), \ldots, (t_{i-n+1}, x'(t_{i-n+1}))$. In divided difference form (Sec. 4.3)

$$
x'(t) = x'[t_i] + x'[t_i, t_{i-1}](t - t_i) + x'[t_i, t_{i-1}, t_{i-2}](t - t_i)(t - t_{i-1})
$$
$$
+ \cdots + x'[t_i, t_{i-1}, \ldots, t_{i-n+1}](t - t_i)(t - t_{i-1}) \cdots (t - t_{i-n+2})
$$
$$
+ \frac{x^{(n+1)}(\tau_i)}{n!} (t - t_i)(t - t_{i-1}) \cdots (t - t_{i-n+1})
$$

where $\tau_i \in (t_{i-n+1}, t_i)$. Substituting this into (10.4.4) gives

$$
x(t_{i+1}) = x(t_i) + x'[t_i] \int_{t_i}^{t_{i+1}} dt + x'[t_i, t_{i-1}] \int_{t_i}^{t_{i+1}} (t - t_i)\,dt
$$
$$
+ x'[t_i, t_{i-1}, t_{i-2}] \int_{t_i}^{t_{i+1}} (t - t_i)(t - t_{i-1})\,dt + \cdots
$$
$$
+ x'[t_i, t_{i-1}, \ldots, t_{i-n+1}] \int_{t_i}^{t_{i+1}} (t - t_i)(t - t_{i-1}) \cdots (t - t_{i-n+2})\,dt
$$
$$
+ \frac{x^{(n+1)}(\tau_i)}{n!} \int_{t_i}^{t_{i+1}} (t - t_i)(t - t_{i-1}) \cdots (t - t_{i-n+1})\,dt \qquad (10.4.19)
$$

For instance, to obtain the four-point Adams-Bashforth formula we note that

$$x'[t_i, t_{i-1}] = \frac{x'(t_i) - x'(t_{i-1})}{h}$$

$$x'[t_i, t_{i-1}, t_{i-2}] = \frac{x'(t_i) - 2x'(t_{i-1}) + x'(t_{i-2})}{2h^2} \qquad (10.4.20)$$

$$x'[t_i, t_{i-1}, t_{i-2}, t_{i-3}] = \frac{x'(t_i) - 3x'(t_{i-1}) + 3x'(t_{i-2}) - x'(t_{i-3})}{6h^3}$$

and

$$\int_{t_i}^{t_{i+1}} (t - t_i)(t - t_{i-1}) dt = \int_0^h \eta(\eta + h) d\eta = \tfrac{5}{6} h^3$$

$$\int_{t_i}^{t_{i+1}} (t - t_i)(t - t_{i-1})(t - t_{i-2}) dt = \tfrac{9}{4} h^4 \qquad (10.4.21)$$

$$\int_{t_i}^{t_{i+1}} (t - t_i)(t - t_{i-1})(t - t_{i-2})(t - t_{i-3}) dt = \tfrac{251}{30} h^5$$

Substituting this into (10.4.19) with $n = 4$ gives

$$x(t_{i+1}) = x(t_i) + \frac{h}{24} [55x'(t_i) - 59x'(t_{i-1}) + 37x'(t_{i-2}) - 9x'(t_{i-3})]$$

$$+ \tfrac{251}{720} x^{(5)}(\tau_i) h^5 \qquad (10.4.22)$$

which is the four-point Adams-Bashforth formula since $x'(t_i) = f(t_i, x(t_i))$ from the differential equation.

Similarly, by writing out the divided difference form of the Lagrange interpolant of the points $(t_{i+1}, x'(t_{i+1}))$, $(t_i, x'(t_i))$, ..., $(t_{i-n+2}, x'(t_{i-n+2}))$ and substituting it into (10.4.4), we obtain a general formula for the implicit Adams-Moulton methods (Exercise 5). The two-, three-, and four-point methods along with their local truncation error terms are given in Table 10.4.19. For brevity the notation $x_i := x(t_i)$ is used.

Observe that the local truncation error terms for the implicit Adams-Moulton methods are in each case substantially smaller than those of the Adams-Bashforth method of the same order. This indicates that Adams predictor-corrector methods may be expected to give more accurate answers than the Adams-Bashforth methods used alone. For instance, let us compute the local error of the two-point predictor-corrector scheme

$$\xi_{i+1}^{pred} = \xi_i + \frac{h}{2} [3f(t_i, \xi_i) - f(t_{i-1}, \xi_{i-1})]$$

$$\xi_{i+1} = \xi_i + \frac{h}{2} [f(t_{i+1}, \xi_{i+1}^{pred}) + f(t_i, \xi_i)] \qquad (10.4.23)$$

Since we are finding the local error, we use the exact values $x_i := x(t_i)$ and $x_{i-1} := x(t_{i-1})$ in lieu of the approximate values ξ_i and ξ_{i-1}. From the

TABLE 10.4.19
Adams-Bashforth and Adams-Moulton formulas

Adams-Bashforth

Two-point:

$$x_{i+1} = x_i + \frac{h}{2} [3f(t_i, x_i) - f(t_{i-1}, x_{i-1})] + \frac{5}{12} x'''(\tau_i) h^3$$

Three-point:

$$x_{i+1} = x_i + \frac{h}{12} [23f(t_i, x_i) - 16f(t_{i-1}, x_{i-1}) + 5f(t_{i-2}, x_{i-2})] + \frac{3}{8} x^{(4)}(\tau_i) h^4$$

Four-point:

$$x_{i+1} = x_i + \frac{h}{24} [55f(t_i, x_i) - 59f(t_{i-1}, x_{i-1}) + 37f(t_{i-2}, x_{i-2}) - 9f(t_{i-3}, x_{i-3})] + \frac{251}{720} x^{(5)}(\tau_i) h^5$$

Adams-Moulton

Two-point:

$$x_{i+1} = x_i + \frac{h}{2} [f(t_{i+1}, x_{i+1}) + f(t_i, x_i)] + \frac{1}{12} x'''(\tau_i) h^3$$

Three-point:

$$x_{i+1} = x_i + \frac{h}{12} [5f(t_{i+1}, x_{i+1}) + 8f(t_i, x_i) - f(t_{i-1}, x_{i-1})] - \frac{1}{24} x^{(4)}(\tau_i) h^4$$

Four-point:

$$x_{i+1} = x_i + \frac{h}{24} [9f(t_{i+1}, x_{i+1}) + 19f(t_i, x_i)$$
$$- 5f(t_{i-1}, x_{i-1}) + f(t_{i-2}, x_{i-2})] - \frac{19}{720} x^{(5)}(\tau_i) h^5$$

---

two-point Adams-Bashforth formula we have $\xi_{i+1}^{\text{pred}} = x_{i+1} - \frac{5}{12} x'''(\tau_i) h^3$. Thus in (10.4.23)

$$\xi_{i+1} = x_i + \frac{h}{2} [f(t_{i+1}, x_{i+1} - \frac{5}{12} x'''(\tau_i) h^3) + f(t_i, x_i)]$$

$$= x_i + \frac{h}{2} [f(t_{i+1}, x_{i+1}) - \frac{5}{12} h^3 x'''(\tau_i) \frac{\partial f}{\partial x}(t_{i+1}, x_{i+1}) + O(h^6) + f(t_i, x_i)]$$

$$= x_i + \frac{h}{2} [f(t_{i+1}, x_{i+1}) + f(t_i, x_i)] + O(h^4)$$

$$= x_{i+1} - \frac{1}{12} x'''(\bar{\tau}_i) h^3 + O(h^4) \tag{10.4.24}$$

This indicates that the local truncation error term in the predictor-corrector scheme is that of the implicit method. Thus for the two-point predictor-corrector scheme (10.4.23) the corrected value ξ_{i+1} would be expected to be about 5 times more accurate than the predicted value ξ_{i+1}^{pred}. Indeed, in Table 10.4.8 we do see improvement by a factor of 4 to 5 between the Adams-Bashforth and Adams predictor-corrector approximations.

Global Order of the Adams Methods

Adams methods have the form

$$\xi_{i+1} = \xi_i + h \sum_{j=0}^{n} \beta_j f(t_{i-j+1}, \xi_{i-j+1}) \tag{10.4.25}$$

The *local truncation error* of an Adams method is

$$\epsilon_i(h) = x(t_{i+1}) - x(t_i) - h \sum_{j=0}^{n} \beta_j f(t_{i-j+1}, x(t_{i-j+1})) \tag{10.4.26}$$

That Adams methods converge to the actual solution of a differential equation is established by the next theorem.

> **Theorem 10.4.20.** Let f in (10.4.1) be Lipschitz continuous with constant L in the variable x in a rectangle $\Re = \{(t, x): t_0 \le t \le t_f, c \le x \le d\}$, where $c \le x(t_i), \xi_i \le d$ for $i = 0, \ldots, N$. Then for $\epsilon(h) := \max_{0 \le i \le N-1} |\epsilon_i(h)|$,
>
> $$|x(t_i) - \xi_i| \le \frac{\epsilon(h)}{h\lambda} \left(e^{\lambda(t_i - t_0)} - 1 \right) \tag{10.4.27}$$
>
> where $\lambda := hL \sum_{j=0}^{n} |\beta_j|$.

This establishes that the order of the global truncation error of the Adams methods is one less than that of the local truncation error. The proof of this theorem, which is similar to that of Theorem 10.3.7, is Exercise 6.

EXERCISES 10.4

1. Use the

 (a) Adams predictor two-point method
 (b) Adams predictor-corrector two-point method
 (c) Implicit trapezoid method

 to find approximations to $x(2)$ for the initial value problem $x' = 5t - 2x, x(0) = 1$ using stepsizes of $h = 1$ and $h = 0.5$. Compare your answers with the actual solution.

2. Write a computer program implementing the

 (a) Adams-Bashforth four-point method
 (b) Adams predictor-corrector four-point method
 (c) Implicit trapezoid method

 for a single first-order differential equation. Apply the method to tabulate approximations to the initial value problem $x' = 5t - 2x, x(0) = 1$ on the interval $0 \le t \le 2$ for stepsizes of $h = 0.2, 0.1$, and 0.05. For each stepsize also have the program print out the value of the actual solution and the error made by the approximation.

3. Show by use of a Taylor series argument that the Adams-Bashforth two-point scheme (10.4.7) has $O(h^3)$ local truncation error.

4. (a) Derive the Adams-Moulton formula (10.4.8) by following the derivation of the Adams-Bashforth formula (10.4.7).

 (b) Use a Taylor series argument to show that the local truncation order of (10.4.8) is $O(h^3)$. [*Hint*: Use Taylor's theorem of two variables to expand the term $f(t_{i+1}, x(t_{i+1}))$ about the point $(t_i, x(t_i))$.]

5. (a) Derive a general formula for Adams-Moulton methods similar to (10.4.19) for Adams-Bashforth formulas.

 (b) Use the formula you found in part (a) to derive the four-point Adams-Moulton formula given in Table 10.4.19.

6. Prove Theorem 10.4.20.

7. From the right-hand endpoint differentiation formula of Table 9.2.7

$$x'(t) = \frac{1}{2h} [3x(t) - 4x(t - h) + x(t - 2h)]$$

$$+ \frac{h^3}{3} x'''(\xi) \qquad (10.4.28)$$

Such a formula is sometimes referred to as a *backward differentiation formula*. Use this formula to obtain the implicit method

$$\xi_{i+1} = \tfrac{4}{3}\xi_i - \tfrac{1}{3}\xi_{i-1} + \tfrac{2}{3}hf(t_{i+1}, \xi_{i+1})$$

$$(10.4.29)$$

Write a computer program similar to Algorithm 10.4.7 which uses the implicit formula given by (10.4.29) and Newton's method. Compare the results of this program with those given for the implicit trapezoid rule in Table 10.4.8.

8. (*a*) Show that the Adams-Moulton two-point formula of Table 10.4.19 can be derived from (10.4.4) using the trapezoid rule with error term.

(*b*) Use the open Newton-Cotes method (9.3.42) and Simpson's rule to derive *Milne's predictor-corrector method*

$$\xi_{i+1}^{\text{pred}} = \xi_{i-3} + \frac{4h}{3} [2f(t_i, \xi_i) - f(t_{i-1}, \xi_{i-1})$$

$$+ 2f(t_{i-2}, \xi_{i-2})]$$

$$(10.4.30)$$

$$\xi_{i+1} = \xi_{i-1} + \frac{h}{3} [f(t_{i+1}, \xi_{i+1}^{\text{pred}}) + 4f(t_i, \xi_i)$$

$$+ f(t_{i-1}, \xi_{i-1})]$$

State the local truncation error for each of the methods separately. Find the local truncation error for the predictor-corrector method.

(*c*) Write a computer program implementing Milne's method. Test it on the problem of finding $x(1)$ with a stepsize of $h = 0.1$ for the initial value problems

$$x' = x \qquad x(0) = 1 \qquad (10.4.31)$$

and

$$x' = -x \qquad x(0) = 1 \qquad (10.4.32)$$

Compare with the results obtained when the predictor alone is used.

9. Compare the results of applying the implicit trapezoid rule and the four-point Adams predictor-corrector method to the problem of approximating $x(2)$ for the initial value problem $x' + 20x = e^{-t}$, $x(0) = 1$ with a stepsize of $h = 0.1$.

10. Note in the fourth-order Adams predictor-corrector scheme (10.4.15) that the corrected value ξ_{i+1} may be further refined by again substituting it into the corrector equation. Write a program that permits more than one iteration of the corrector in the fourth-order Adams predictor-corrector scheme. Apply the program to make a table showing the errors made at the points $t_i = 0.1, 0.2, \ldots, 2$ for one, two, and three iterations of the corrector for the initial value problem $x' = 5(t - 1)x$, $x(0) = 5$. Would more than three iterations be useful? Support your contention by testing the program on the other problems in Table 10.4.11.

11. Use the

(*a*) Adams-Bashforth four-point method
(*b*) Adams predictor-corrector four-point method
(*c*) Implicit trapezoid method

to find and graph the temperature of an air-conditioned building [see Eq. (10.1.4)] when $k = 0.5$ hour^{-1}, $A(t) = 70°$, $U_0 = 4°$/hour, $T_{\text{crit}} = 65°$, and $T(0) = 50°$. Use a stepsize h that is small enough so that the same graph results from using stepsizes of h and $h/2$.

12. Repeat Exercise 11 when the ambient temperature is $A(t) = 70° + 20° \sin(\pi t/12)$.

13. Calculate the

(*a*) Adams-Bashforth two-point method
(*b*) Adams predictor-corrector two-point method
(*c*) Implicit trapezoid method

approximations to $x_1(1)$ and $x_2(1)$ for the initial value problem

$$\begin{array}{ll} x_1' + x_2 = t^2 & x_1(0) = 1 \\ x_2' - x_1 = 1 & x_2(0) = 0 \end{array} \qquad (10.4.33)$$

for stepsizes of $h = 1$ and $h = 0.5$. Compare with the value of the actual solution.

14. Write a program implementing the

(*a*) Adams-Bashforth four-point method
(*b*) Adams predictor-corrector four-point method

for a system of differential equations. Apply it to tabulate approximations to $x_1(t)$ and $x_2(t)$ for the initial value problem

$$\begin{array}{ll} x_1' + x_2 = t^2 & x_1(0) = 1 \\ x_2' - x_1 = 1 & x_2(0) = 0 \end{array} \qquad (10.4.34)$$

on the interval $0 \le t \le 1$ for stepsizes of $h =$ 0.2, 0.1, and 0.05. For each stepsize also have the program print out the value of the actual solutions and the error made by the approximations.

15. Compare the execution times and accuracies given in Tables 10.4.12 and 10.4.16 with those for your computer system. Try to account for any differences that you observe.

16. (a) Write a procedure that will calculate the four-point Adams predictor-corrector approximation to an initial value problem for N steps. Call this function in a program that repeatedly doubles the number of steps until two successive approximations agree to a specified tolerance. Apply this program to each of the five problems in Table 10.4.11, recording the time of execution for each.
    (b) Repeat part (a) using the classical Runge-Kutta method instead of the Adams predictor-corrector method. Which method is more efficient on the whole?

17. Use the
    (a) Adams-Bashforth four-point method
    (b) Adams predictor-corrector four-point method

    to graph solutions to the Lotka-Volterra model (10.1.17). Take $a = 2$, $b = 1$, $c = 1$, $d = 1$ and use the initial conditions $x_1(0) = 1$, $x_2(0) = 0.1$. Graph solutions over the time interval $0 \le t \le 10$. Choose the stepsize h so that stepsizes of h and $h/2$ yield the same graph.

Use one of the methods studied in this section for the exercises given below.

18. Exercise 18 in Sec. 10.3
19. Exercise 19 in Sec. 10.3
20. Exercise 20 in Sec. 10.3
21. Exercise 21 in Sec. 10.3
22. Exercise 22 in Sec. 10.3
23. Exercise 23 in Sec. 10.3

10.5 VARIABLE-STEP METHODS

Sections 10.3 and 10.4 treated two families of *fixed-step* methods, the one-step Runge-Kutta methods and the multistep Adams methods. The interval h between values of t at which approximations were calculated was an input to the method. The local truncation error estimates such as (10.3.12) for the modified Euler method and those given in Table 10.4.19 for the Adams method depend upon the unknown value x of the solution, and thus there is no means a priori to find the stepsize h that will suffice to give a specified accuracy. In this section we examine means of finding an approximation at a given point t_f to the solution $x(t_f)$ of the initial value problem

$$x' = f(t, x) \qquad x(t_0) = x_0 \qquad (10.5.1)$$

that is accurate to within a specified tolerance.

A Variable-Step Runge-Kutta Method

As indicated by Eq. (10.3.12) for the local truncation error of the modified Euler method, error terms for members of the Runge-Kutta family are rather complicated, even for a method that is locally $O(h^3)$. They become more so for higher-order methods such as classical Runge-Kutta which is locally $O(h^5)$. Fortunately it turns out that reasonably effective estimates of stepsize required to attain a specified local truncation error can be found that use only the order of the local truncation error and do not require further knowledge of the form of the error term. The first variable-step method we consider is based upon comparison of the estimates for one and two steps for the value of x at some time obtained by a Runge-Kutta method with local truncation error term that is of the form Ch^n, where C is not known. Suppose that we are already in possession of an estimate ξ_t for $x(t)$ and a candidate stepsize h_0. The Runge-Kutta method is used to calculate $\xi_{t+h_0}^{(1)}$ and $\xi_{t+h_0}^{(2)}$, approximations to $x(t + h_0)$ using stepsizes of h_0 and $h_0/2$, respectively. If $E_{est} := |\xi_{t+h_0}^{(1)} - \xi_{t+h_0}^{(2)}|$ is less than tol, then the more accurate of the two approximations, $\xi_{t+h_0}^{(2)}$, is accepted as the approximation for $x(t + h_0)$. Whether or not the approximation is accepted, we need a new estimate h_{tol} of a stepsize that will produce an approximation within the specified tolerance. If the approximation was accepted, this value will be used as h_0 in the next step; if not, then it will be used as h_0 in repeating the current step. To find h_{tol} we first note that

$$E_{est} = |\xi_{t+h_0}^{(1)} - \xi_{t+h_0}^{(2)}| \approx \left| Ch_0^n - C\left(\frac{h_0}{2}\right)^n \right| = (1 - 2^{-n})Ch_0^n \quad (10.5.2)$$

This gives the estimate

$$C \approx \frac{E_{est}}{(1 - 2^{-n})h_0^n} \quad (10.5.3)$$

Since h_{tol} is to satisfy $\text{tol} \approx Ch_{tol}^n$, we will use the formula

$$h_{tol} = \left[\frac{(1 - 2^{-n})\,\text{tol}}{E_{est}} \right]^{1/n} h_0 \quad (10.5.4)$$

At the start of the application of this variable-step process we have $\xi_{t_0} = x_0$. Lacking any better information, h_0 is taken to be $t_f - t_0$.

Example 10.5.1. Let us illustrate the process described above by finding an approximation to $x(0.3)$ to within an accuracy of $\text{tol} = 0.05$ for the initial value problem $x' = 5(t - 1)x$, $x(0) = 5$. The modified Euler method will be used, and hence $n = 3$ in (10.5.4). The process is started with $\xi_0 = 5$, $h_0 = 0.3$. Thus we have

Step 1. First try:

$$t = 0 \qquad h_0 = 0.3 \qquad t + h_0 = 0.3$$

Applying the modified Euler method with one and then two steps gives

$$\xi_{0.3}^{(1)} = 2.5625 \qquad \xi_{0.3}^{(2)} = 1.5980 \qquad E_{est} = 0.9645$$

Thus E_{est} is substantially larger than tol. From (10.5.4) it is estimated that the accuracy requirement can be achieved with a stepsize of $h_{tol} = 0.1070$. Thus we now have

Second try:

$$t = 0 \qquad h_0 = 0.1070 \qquad t + h_0 = 0.1070$$

The modified Euler method now gives

$$\xi^{(1)}_{0.1070} = 3.1072 \qquad \xi^{(2)}_{0.1070} = 3.0331 \qquad E_{\text{est}} = 0.0741$$

which is still slightly above the specified tolerance. On the basis of the new value of E_{est}, formula (10.5.4) now predicts that $h_{\text{tol}} = 0.0898$ will suffice. This gives
Third try:

$$t = 0 \qquad h_0 = 0.0898 \qquad t + h_0 = 0.0898$$

The modified Euler approximations are

$$\xi^{(1)}_{0.0898} = 3.3151 \qquad \xi^{(2)}_{0.0898} = 3.2694 \qquad E_{\text{est}} = 0.0457$$

and thus the estimated error is now within the given tolerance. Hence we have the approximation $x(0.0898) \approx 3.2694$. In fact, from the actual solution, $x(0.0898) = 3.2563$, and thus the true error, 0.013, is smaller than the estimated error E_{est}. On the basis of the new value of E_{est}, formula (10.5.4) predicts that a stepsize of $h_{\text{tol}} = 0.0885$ is required to attain the error tolerance. We use this as the initial h_0 in the next step.

Step 2. First try:

$$t = 0.0898 \qquad h_0 = 0.0885 \qquad t + h_0 = 0.1783$$

The modified Euler approximations are

$$\xi^{(1)}_{0.1783} = 2.2560 \qquad \xi^{(2)}_{0.1783} = 2.2347 \qquad E_{\text{est}} = 0.0213$$

and thus the tolerance is achieved the first try. Hence $x(0.1783) \approx 2.2347$, and formula (10.5.4) gives the estimate $h_{\text{tol}} = 0.1125$ for use in the next step. The actual value of the solution is $x(0.1783) = 2.2198$, and thus at this point we are in error by 0.015.

Step 3. First try:

$$t = 0.1783 \qquad h_0 = 0.1125 \qquad t + h_0 = 0.2907$$

The modified Euler approximations are

$$\xi^{(1)}_{0.2907} = 1.4787 \qquad \xi^{(2)}_{0.2907} = 1.4586 \qquad E_{\text{est}} = 0.0201$$

and the error estimate is again within tolerance. Formula (10.5.4) suggests a stepsize of 0.1459; however, the distance to $t_f = 0.3$ is only 0.0093. Thus we use the stepsize $h_0 = 0.0093$.

Step 4. First try:

$$t = 0.2907 \qquad h_0 = 0.0093 \qquad t + h_0 = 0.3$$

The modified Euler approximations are

$$\xi^{(1)}_{0.3} = 1.4117 \qquad \xi^{(2)}_{0.3} = 1.4117 \qquad E_{\text{est}} = 0.0000$$

and we obtain the approximation $x(0.3) \approx 1.4117$, which is in error by 0.015.

When it is specified that an appropriate solution is desired with an error no more than 0.05, this is, of course, a statement concerning the *global* truncation error, while the choice (10.5.4) of h_{tol} is made to control the *local* truncation error at each step. Thus it is far from certain that we will obtain the specified accuracy. However, note that while h_{tol} is chosen so that the one-step approximation $\xi_{t+h_0}^{(1)}$ is accurate to within the given tolerance, the two-step approximation $\xi_{t+h_0}^{(2)}$ which is more accurate by a factor of 2^n, is actually used. This helps to compensate for the fact that we are controlling local, rather than global, error.

Algorithm 10.5.2 gives an implementation of the process described above. Observe that a minimum permissible stepsize h_{min} is specified. This is necessary since, as we saw in Sec. 10.2, there may be a point $t \in [t_0, t_f)$ at which the solution ceases to exist. At such a point the derivatives of x and hence the coefficient of h^n in the truncation error term approach infinity, and the stepsize necessary to attain a given tolerance falls to zero. The choice $h_{min} = 10^{-4}(t_f - t_0)$ is arbitrary. The only absolute stricture is that h_{min} be set far enough above the machine unit μ so that $t + h_{min} > t$ for any $t \in [t_0, t_f)$. The choice given simply has been found to strike a reasonable balance (on computers of about the speed of an IBM PC-AT) between the two undesirables of having the algorithm report failure on a problem when the solution in fact exists and having it grind on interminably when it does not. On faster computers a smaller value of h_{min} may be preferable, and, as will be discussed later, if the right-hand side of the differential equation has discontinuities, considerably smaller settings of h_{min} may be required.

Algorithm 10.5.2   Variable-step Runge-Kutta method

Input t_0, x_0 (Initial conditions)
             t_f (Point at which to find approximation)
             tol (Error tolerance)
             h_{min} (Minimum permissible stepsize)

Define $f(t, x)$ (Right-hand side of the differential equation)
             $RK(N_s, t_0, x_0, t, x)$ [A procedure that will calculate a
                          Runge-Kutta approximation x to $x(t)$ for the initial condition
                          $x(t_0) = x_0$ using N_s steps]

$h_0 := t_f - t_0$
$h_{min} := h_0 / 10^4$
$t := t_0, t_{next,} := t_0 + h_0$
$\xi_0 := x_0$
failed := FALSE

repeat
    done := FALSE
    repeat
        $RK(1, t, \xi_0, t_{next,} \xi_1)$

$$RK(2, t, \xi_0, t_{next}, \xi_2)$$
$$E_{est} := |\xi_1 - \xi_2|$$

if $E_{est} <$ tol then
$\qquad \{t := t_{next}$
$\qquad\quad \xi_0 := \xi_2$
$\qquad\quad$ done := TRUE$\}$

if $E_{est} > 0$ then
$$\{h_0 := h_0 \sqrt[n]{\frac{(1 - 2^{-n})\,\text{tol}}{E_{est}}}$$

$\qquad$ if (NOT done) and $(h_0 < h_{min})$ then failed := TRUE
$\qquad t_{next} := t + h_0$
$\qquad$ if $t_{next} > t_f$ then
$\qquad\quad \{h_0 := t_f - t$
$\qquad\qquad t_{next} := t_f\}\}$

else
$\qquad \{h_0 := t_f - t$
$\qquad\quad t_{next} := t_f\}$

until done or failed

until $(t \geq t_f)$ or failed

if $t \geq t_f$ then
$\qquad$ Output: ξ_0 [approximation to $x(t_f)$]

else
$\qquad$ Output: Message that method failed because stepsize fell below minimum, value of
$\qquad\qquad t, \xi_2,$ and E_{est} at point of failure.

Table 10.5.3 compares the results of applying the variable-step Algorithm 10.5.2 (with classical Runge-Kutta as the method) to the five problems given in Table 10.4.11 with the results given in Table 10.4.12 for the fixed-step classical Runge-Kutta (RK). The results shown are for a Turbo Pascal implementation of the algorithm running on an IBM 386 class PC. Also shown are the number of evaluations of the right-hand side (RHS) of the differential equation. For the fixed-step method this is simply 4 times the number of steps taken. For the variable-step method 12 evaluations are required for each try of each step. The number of such tries was ascertained by inserting a counter in the program.

TABLE 10.5.3
Comparison of fixed-step and variable-step classical Runge-Kutta methods

Prob.	Target error	Fixed-step RK 4			Variable-step RK 4		
		RHS evals.	Error	Time, s	RHS evals.	Error	Time, s
1	5E − 09	720	5.0E − 09	0.12	732	3.6E − 09	0.22
2	5E − 07	2120	4.8E − 07	0.32	852	7.2E − 06	0.24
3	5E − 09	1880	4.7E − 09	0.90	4596	2.5E − 09	2.77
4	5E − 09	440	4.7E − 09	0.58	996	3.4E − 09	1.44
5	5E − 06	3880	4.9E − 06	3.16	3672	5.9E − 05	3.50

As mentioned in Sec. 10.4 the number of steps required to compute an approximation to the given accuracy for the fixed-step method was determined by comparing approximations with the actual solution, and thus Table 10.5.3 does not represent a comparison of two practical methods. However, since the stepsize necessary for a fixed-step method to be accurate on some parts of the solution curve may be much smaller than is required on other parts of the curve, one would hope that a variable-step method would be competitive with a fixed-step method, even when the optimal number of steps was used for the fixed-step method.

In the three problems in which the target error was successfully attained by the variable-step Runge-Kutta method, it required 2 to 3 times longer than the fixed-step method. In problems 3 and 5 this is accounted for in part by the larger number of evaluations of the right-hand side of the differential equation that were required by the variable-step method. Another reason that more time was required by the variable-step method is that formula (10.5.4) was calculated as

$$h_{tol} = \exp\left[0.2 \ln \frac{(1 - 2^{-n})\,tol}{E_{est}}\right]$$

which adds two evaluations of transcendental functions to the computational overhead in the algorithm. For uncomplicated systems of low dimension this is a significant expense. In large or complicated systems of differential equations in which many calls to transcendental or other computationally expensive functions are made, it would be of less significance.

Observe that the variable-step method missed achieving the target accuracy by about a factor of 10 in problems 2 and 5. In the case of problem 2 this is not surprising; indeed this problem was placed in the list of test problems to illustrate the limitations of any scheme to control error in the approximation of solutions to differential equations. It turns out that the exact solutions to the initial value problem $x' = 1 + x^2$, $x(0) = x_0$ when $x_0 = 0$ and $x_0 = 5 \times 10^{-7}$ differ by about 10^{-4} at $t = 1.5$. This means that if an approximation scheme makes an error of 10^{-7} (the specified tolerance) on the first step and *no error thereafter*, it would make an error of about 10^{-4}. Under these circumstances, even the accuracy of 7×10^{-6} obtained by the variable-step method must be regarded as fortuitous.

Even on less difficult problems it is possible for a variable-step method such as that given by Algorithm 10.5.2 to fail to achieve its target accuracy, as was the case in problem 5. This is because, as mentioned earlier, the stepsize is chosen to keep the local, rather than global, truncation error within the target tolerance tol.

Concerning the efficiency of a variable-step method, the question of course is not so much "Is it acceptable?" as "Can we do better?" In the next subsection, in which we consider the Runge-Kutta-Felberg algorithms, we will see that we can indeed do better.

Runge-Kutta-Felberg Methods

Runge-Kutta-Felberg methods use the approach of the variable-step Runge-Kutta method given in the last subsection, but instead of using one and then two steps of a single Runge-Kutta method to estimate the local truncation error, they use two different Runge-Kutta methods, one with local truncation error of order n and the other with order $n + 1$. Again we assume that an approximation ξ_t to $x(t)$ has already been determined and that a candidate stepsize h_0 has been chosen. Let $\xi_{t+h_0}^{(1)}$ denote the approximation to $x(t + h_0)$ generated by the method with local order n and $\xi_{t+h_0}^{(2)}$ the approximation of the order $= (n + 1)$ method. Then an estimate of the local truncation error is $E_{est} := |\xi_{t+h_0}^{(1)} - \xi_{t+h_0}^{(2)}| \approx |C_1 h_0^n - C_2 h_0^{n+1}| \approx |C_1| h_0^n$, providing that h_0 is sufficiently small. This gives the estimate $|C| \approx E_{est}/h_0^n$, and hence the stepsize required to make the local truncation error about magnitude tol is

$$h_{tol} = \left(\frac{tol}{E_{est}}\right)^{1/n} h_0 \qquad (10.5.5)$$

To increase the likelihood that h_{tol} will produce a value of E_{est} less than tol, the formula used in practice is

$$h_{tol} = \sigma\left(\frac{tol}{E_{est}}\right)^{1/n} h_0 \qquad (10.5.6)$$

where σ is typically taken to be about 0.9.

The argument leading to (10.5.5) places no constraints on the two Runge-Kutta methods used other than that their local truncation errors should differ in order by one. Thus we are free to choose the two methods so as to make the algorithm as efficient as possible. Let us consider the case of choosing methods with local truncation errors of orders three and four. From Eqs. (10.3.16) and (10.3.21) the general forms of these methods are

$$m_1 = f(t, \xi_t)$$
$$m_2 = f(t + \alpha_2^{(1)}h, \xi_t + h\beta_{21}^{(1)}m_1) \qquad (10.5.7)$$
$$\xi_{t+h} = \xi_t + h(\omega_1^{(1)}m_1 + \omega_2^{(1)}m_2)$$

and

$$m_1 = f(t, \xi_t)$$
$$m_2 = f(t + \alpha_2^{(2)}h, \xi_t + h\beta_{21}^{(2)}m_1)$$
$$m_3 = f(t + \alpha_3^{(2)}h, \xi_t + h\beta_{31}^{(2)}m_1 + h\beta_{32}^{(2)}m_2) \qquad (10.5.8)$$
$$\xi_{t+h} = \xi_t + h(\omega_1^{(2)}m_1 + \omega_2^{(2)}m_2 + \omega_3^{(2)}m_3)$$

for the locally order-three and order-four methods, respectively. The computational expense of evaluating the right-hand-side $f(t, x)$ of the differential equation is expected to be a substantial part of the computational expense, and thus any reduction in the number of such evaluations will improve the efficiency of the method significantly.

Observe that the value of m_1 is the same in both (10.5.7) and (10.5.8); thus it need only be calculated once. The question then arises as to whether the value of m_2 can also be made the same for both methods. This is not obvious since as we saw in Sec. 10.3 the requirement that the method have a given order for the local truncation error implies relations among the α_i and β_{ij}. For the method (10.5.8) these are given by Eqs. (10.3.23) and (10.3.24). Let us choose the method (10.5.7) to be the modified Euler method whence $\alpha_2^{(1)} = \beta_2^{(1)} = \beta_{21}^{(1)} = 1$. That m_2 is the same in both methods then requires that $\alpha_2^{(2)} = \beta_{21}^{(2)} = 1$. With these choices of $\alpha_2^{(2)}$ and $\beta_{21}^{(2)}$, Eqs. (10.3.24) become

$$\omega_1^{(2)} + \omega_2^{(2)} + \omega_3^{(3)} = 1$$

$$\omega_2^{(2)} + (\beta_{31}^{(2)} + \beta_{32}^{(2)})\omega_3^{(2)} = \tfrac{1}{2}$$

$$\omega_2^{(2)} + (\beta_{31}^{(2)} + \beta_{32}^{(2)})^2\omega_3^{(2)} = \tfrac{1}{3} \qquad (10.5.9)$$

$$\beta_{31}^{(2)}\omega_3^{(2)} = \tfrac{1}{6}$$

Since from Eqs. (10.3.23) we have $\alpha_3^{(2)} = \beta_{31}^{(2)} + \beta_{32}^{(2)}$ and we want $0 \le \alpha_3^{(2)} \le 1$, we choose arbitrarily $\beta_{31}^{(2)} + \beta_{32}^{(2)} = \tfrac{1}{2}$. This in turn implies from the second and third of Eqs. (10.5.9) that $\omega_2^{(2)} = \tfrac{1}{6}$, $\omega_3^{(2)} = \tfrac{2}{3}$, and hence the fourth equation yields $\beta_{31}^{(2)} = \beta_{32}^{(2)} = \tfrac{1}{4}$.

The procedure that is followed for a Runge-Kutta-Felberg method is to calculate the difference

$$E_{\text{est}} = |\xi_{t+h}^{(1)} - \xi_{t+h}^{(2)}| = \left\| \left[\xi_t + \frac{h}{2}(m_1 + m_2) \right] - \left[\xi_t + \frac{h}{6}(m_1 + m_2 + 4m_3) \right] \right\|$$

$$= \left| \frac{h}{3}(2m_3 - m_1 - m_2) \right| \qquad (10.5.10)$$

of the approximations given by the two methods and, if it is less than tol, then to accept the approximation $\xi_{t+h}^{(2)}$ of the higher-order method. Thus it is never necessary to calculate explicitly the lower-order approximation $\xi_{t+h}^{(1)}$. Hence a *Runge-Kutta-Felberg 2(3)* method is given by

$$m_1 = f(t, \xi_t)$$

$$m_2 = f(t + h, \xi_t + hm_1)$$

$$m_3 = f\left(t + \frac{h}{2}, \xi_t + \frac{h}{4}(m_1 + m_2) \right) \qquad (10.5.11)$$

$$E_{\text{est}} = \left| \frac{h}{3}(2m_3 - m_1 - m_2) \right|$$

$$\xi_{t+h} = \xi_t + \frac{h}{6}(m_1 + m_2 + 4m_3)$$

By convention Runge-Kutta-Felberg methods are specified by the global orders of the two methods rather than the local orders, hence the designation 2(3). Clearly there are infinitely many 2(3) methods.

When a locally $O(h^3)$ Runge-Kutta method is used, the variable-step Runge-Kutta method given by Algorithm 10.5.2, as presented, requires six evaluations of the right-hand side of the differential equation for each try of each step. This can be reduced to five since the slope $m_1 = f(t, \xi_t)$ is used in both the one- and two-step calculations. Thus the Runge-Kutta-Felberg 2(3) method (10.5.11) saves two evaluations. When the approach used to obtain the 2(3) method is applied to higher-order methods, the savings are more substantial. Perhaps the most commonly used are the *Runge-Kutta-Felberg 4(5) methods*, one of which is

$$m_1 = f(t, \xi_t)$$
$$m_2 = f(t + \tfrac{2}{9}h, \xi_t + \tfrac{2}{9}hm_1)$$
$$m_3 = f(t + \tfrac{1}{3}h, \xi_t + \tfrac{1}{12}hm_1 + \tfrac{1}{4}hm_2)$$
$$m_4 = f(t + \tfrac{3}{4}h, \xi_t + \tfrac{69}{128}hm_1 - \tfrac{243}{128}hm_2 + \tfrac{135}{64}hm_3)$$
$$m_5 = f(t + h, \xi_t - \tfrac{17}{12}hm_1 + \tfrac{27}{4}hm_2 - \tfrac{27}{5}hm_3 + \tfrac{16}{15}hm_4)$$
$$m_6 = f(t + \tfrac{5}{6}h, \xi_t + \tfrac{65}{432}hm_1 - \tfrac{5}{16}hm_2 + \tfrac{13}{16}hm_3 + \tfrac{4}{27}hm_4 + \tfrac{5}{144}hm_5)$$
$$E_{est} = h\left| -\tfrac{1}{150}m_1 + \tfrac{3}{100}m_3 - \tfrac{16}{75}m_4 - \tfrac{1}{20}m_5 + \tfrac{6}{25}m_6 \right|$$
$$\xi_{t+h} = \xi_t + h\left(\tfrac{47}{450}m_1 + \tfrac{12}{25}m_3 + \tfrac{32}{225}m_4 + \tfrac{1}{30}m_5 + \tfrac{6}{25}m_6 \right)$$

(10.5.12)

This method requires only six evaluations of the right-hand side of the differential equation, whereas Algorithm 10.5.2, when using a fourth-order Runge-Kutta method, requires eleven.

Algorithm 10.5.4 gives an implementation of a Runge-Kutta-Felberg 4(5) method such as (10.5.12). The basic control structure is similar to that of Algorithm 10.5.2.

Algorithm 10.5.4   Runge-Kutta-Felberg 4(5) method

Input　　　t_0, x_0 (Initial conditions)
　　　　　　t_f (Point at which to find approximation)
　　　　　　tol (Error tolerance)
　　　　　　h_{min} (Minimum permissible stepsize)

Define　　 $f(t, x)$ (Right-hand side of the differential equation)

Declare　　Arrays m_i, α_i, β_{ij}, e_i, and ω_i, where m_i will hold the six slopes in (10.5.12), α_i the coefficients used in the first argument of f, β_{ij} the coefficients used in the second argument of f, e_i the coefficients used in the calculation of E_{est}, and ω_i the coefficients used in the calculation of ξ_{t+h}.

Initialize　Constant arrays α_i, β_{ij}, e_i, and ω_i.

$$h_0 := t_f - t_0$$
$$h_{\min} := h_0/10^4$$
$$t := t_0, t_{\text{next}} := t_0 + h_0$$
$$\xi_0 := x_0$$
failed $:=$ FALSE

repeat
    done $:=$ FALSE
    repeat
        $m_1 := f(t, \xi_0)$
        for $j = 2$ to 6
            $\{w := 0$
                for $k = 1$ to $j - 1$ $\{w := w + \beta_{jk}m_k\}$
                $m_j := f(t + \alpha_j h_0, \xi_0 + wh_0)\}$
        $E_{\text{est}} := 0$
        for $j = 1$ to 6 $\{E_{\text{est}} := E_{\text{est}} + he_j m_j\}$
        $E_{\text{est}} := |E_{\text{est}}|$
        if $E_{\text{est}} <$ tol then
            $\{w := 0$
                for $j = 1$ to 6 $\{w := w + \omega_j m_j\}$
                $\xi := \xi_0 + wh_0$
                $t := t + h_0$
                $\xi_0 := \xi$
                done $:=$ TRUE$\}$
        if $E_{\text{est}} > 0$ then
            $\{h_0 := 0.9h_0\sqrt[5]{\dfrac{\text{tol}}{E_{\text{est}}}}$
            if (NOT done) and $(h_0 < h_{\min})$ then failed $:=$ TRUE
            if $t + h_0 > t_f$ then $h_0 := t_f - t\}$
        else $h_0 := t_f - t$
    until done or failed
  until $(t \geq t_f)$ or failed

if $t \geq t_f$ then
    output: ξ [approximation to $x(t_f)$]
else
    output: Message that stepsize fell below permissible limit, values of t and
        E_{est} at point of failure.

Figures 10.5.5 to 10.5.8 indicate the width of the steps required by the Runge-Kutta-Felberg 4(5) method given by (10.5.12) on problems 1, 3, 4, and 5 in Table 10.4.11 when the target error was 0.005. As can be seen in the figures the width of steps varied considerably in problems 1 and 3 but not so much in problems 4 and 5.

Table 10.5.9 shows the outcome of applying the Runge-Kutta-Felberg 4(5) method given by (10.5.12) to the five test problems of Table 10.4.11. It fails to meet the target accuracy in two of the five problems. As noted earlier, failure on problem 2 is likely, and the shortfall in problem 5 is very small. In

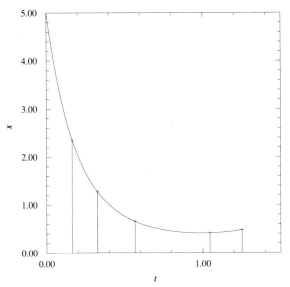

FIGURE 10.5.5

the four problems where the specified tolerance was attained or almost attained, we see that the Runge-Kutta-Felberg routine required no more, and sometimes many fewer, evaluations of the right-hand side of the differential equation than the fixed-step method. However, because the overhead in the Runge-Kutta-Felberg algorithm is much greater, execution times were about 50 percent higher in those problems in which a comparable number of evaluations

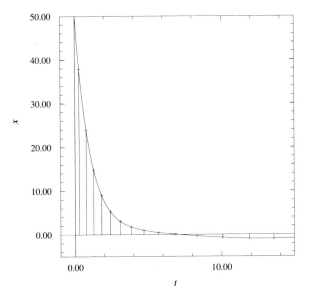

FIGURE 10.5.6

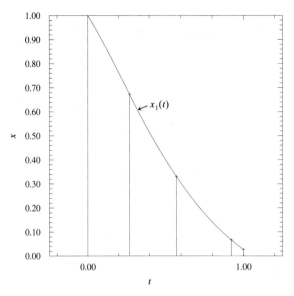

FIGURE 10.5.7

were required. Given that the execution times for the fixed-step method are based upon the unrealistic premise that the number of steps to be used is known beforehand, a loss of efficiency of this magnitude is not serious. In any event the gains in percentage terms over the variable-step Runge-Kutta Algorithm 10.5.2 are considerable.

In connection with the variable-step Runge-Kutta method, it was noted that while the stepsize was chosen to control the local truncation error of the

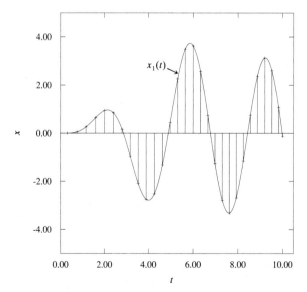

FIGURE 10.5.8

TABLE 10.5.9
Comparison of fixed-step classical Runge-Kutta and a variable-step Runge-Kutta-Felberg method

Prob.	Target error	Fixed-step RK 4			RKF 4(5)		
		RHS evals.	Error	Time, s	RHS evals.	Error	Time, s
1	5E − 09	720	5.0E − 09	0.12	324	4.9E − 09	0.15
2	5E − 07	2120	4.8E − 07	0.32	276	1.2E − 04	0.12
3	5E − 09	1880	4.7E − 09	0.90	1878	2.4E − 10	1.43
4	5E − 09	440	4.7E − 09	0.58	420	4.3E − 10	0.79
5	5E − 06	3880	4.9E − 06	3.16	1242	6.9E − 06	1.91

method when one step was used, that the approximation accepted was a two-step approximation helped compensate for the fact that local, rather than global, error was being controlled. Similarly, with Runge-Kutta-Felberg methods the stepsize is predominantly determined by the magnitude of the error made by the lower-order method, but the approximation used is that of the higher-order method. This means that the error should be smaller than predicted by about a factor of h, the stepsize. This is roughly the amount required to keep the global error within the specified tolerance.

Further Discussion of Runge-Kutta-Felberg Methods

In the discussion of variable-step methods thus far we have considered solely the problem of finding an approximation at a single point t_f. This poses a problem when one also wants to know the values of the solution at a set of points $\mathscr{P} = \{t_0, t_1, \ldots, t_n = t_f\}$ between t_0 and t_f, say for purposes of tabular or graphical representation of the solution. One possibility is simply to apply the variable-step method sequentially to the intervals (t_0, t_1), $(t_1, t_2), \ldots$, (t_{n-1}, t_f), with the approximation at the end of the interval (t_{i-1}, t_i) becoming the initial condition on the interval (t_i, t_{i+1}). Since there is a considerable amount of computation required in each step of a variable-step method, this will degrade the performance of the method somewhat if the number of intermediate points is large. A more efficient approach is to store the intermediate values $t_0, \bar{t}_1, \ldots, \bar{t}_{m-1}, t_m = t_f$ at which the solution is computed naturally during the process of finding the solution at t_f and then to approximate the solution at the points of $\mathscr{P}$ by some means of interpolation, Lagrange or Hermite, for instance. Sufficiently many points should be used in the interpolation so that the interpolation error is of the same order as that made by the variable-step method. For instance, if Lagrange interpolation is used in approximating the value of the solution at a point t, we might use the points $\bar{t}_{i-2}, \bar{t}_{i-1}, \bar{t}_i, \bar{t}_{i+1}, \bar{t}_{i+2}$ produced by a Runge-Kutta-Felberg 4(5) method in the interpolation formula

$$x(t) = x[\bar{t}_i] + x[\bar{t}_i, \bar{t}_{i+1}](t - \bar{t}_i)$$
$$+ x[\bar{t}_{i-1}, \bar{t}_i, \bar{t}_{i+1}](t - \bar{t}_i)(t - \bar{t}_{i+1})$$
$$+ x[\bar{t}_{i-1}, \bar{t}_i, \bar{t}_{i+1}, \bar{t}_{i+2}](t - \bar{t}_{i-1})(t - \bar{t}_i)(t - \bar{t}_{i+1})$$
$$+ x[\bar{t}_{i-2}, \bar{t}_{i-1}, \bar{t}_i, \bar{t}_{i+1}, \bar{t}_{i+2}](t - \bar{t}_{i-1})(t - \bar{t}_i)(t - \bar{t}_{i+1})(t - \bar{t}_{i+2})$$

$$(10.5.13)$$

which is $O(h^5)$ with $h := \max_{j=-1\ldots2} |\bar{t}_{i+j} - \bar{t}_{i+j-1}|$. This and other enhancements to the Runge-Kutta-Felberg algorithm are the topic of Project 1.

Another problem with the Runge-Kutta-Felberg method is that it will terminate in failure at points at which the right-hand side of the differential equation is discontinuous unless the minimum stepsize $h_{\min}$ in Algorithm 10.5.4 is set to be no more than one or two orders of magnitude greater than tol. For instance, consider the simple case of the initial value problem

$$x' = \begin{cases} x'_1 & t \le t_d \\ x'_2 & t > t_d \end{cases} \qquad x(t_0) = x_0 \qquad (10.5.14)$$

where x'_1 and x'_2 are constants. When the Runge-Kutta-Felberg method given by (10.5.12) has progressed to the point where $t_d \in (t, t + h_0)$, then E_{est} will take one of the following values:

$$E_{est} = \tfrac{2}{300} h_0 |x'_2 - x'_1| \qquad t_d \in \left(0, t + \frac{h_0}{3}\right)$$

$$E_{est} = \tfrac{7}{300} h_0 |x'_2 - x'_1| \qquad t_d \in \left(t + \frac{h_0}{3}, t + \frac{3h_0}{4}\right)$$

$$(10.5.15)$$

$$E_{est} = \tfrac{57}{300} h_0 |x'_2 - x'_1| \qquad t_d \in \left(t + \frac{3h_0}{4}, t + \frac{5h_0}{6}\right)$$

$$E_{est} = \tfrac{72}{300} h_0 |x'_2 - x'_1| \qquad t_d \in \left(t + \frac{5h_0}{6}, t + h_0\right)$$

Thus a successful crossing of the discontinuity requires a stepsize h_0 in the range $150/|x'_2 - x'_1|$ tol to $75/(18|x'_2 - x'_1|)$ tol whence $h_{\min}$ must be about as small as tol. This is in contrast to the situation when $f(t, x)$ is sufficiently differentiable for which we expect success if $h_{\min} < \sqrt[5]{\text{tol}}$.

Example 10.5.10. In Example 10.4.17 we saw that the classical Runge-Kutta method requires 700 steps, that is, 2800 evaluations of the right-hand side of the differential equation, to determine a solution to the initial value problem

$$x' = \begin{cases} 1 - x & t < \pi \\ -5x & t > \pi \end{cases} \qquad x(0) = 0 \qquad (10.5.16)$$

to within an accuracy of 5×10^{-5}. The solution to this problem is

$$x(t) = \begin{cases} 1 - e^{-t} & t < \pi \\ (1 - e^{-\pi})e^{-5(t-\pi)} & t > \pi \end{cases} \qquad (10.5.17)$$

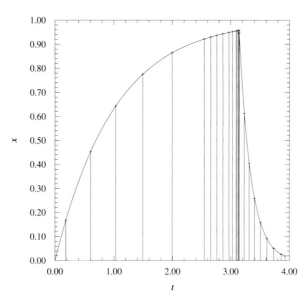

FIGURE 10.5.11

and thus $x(\pi) = 1 - e^{-\pi} \simeq 0.9568$. This gives $\lim_{t \to \pi^-} x'(t) \simeq 0.0432$ and $\lim_{t \to \pi^+} x'(t) \simeq -4.7389$ whence $x'(\pi^+) - x'(\pi^-) \simeq -4.8721$. Thus a stepsize reduction down to the range of $h_0 = 1.5 \times 10^{-3}$ to 5×10^{-5} will be necessary to cross the discontinuity at $t = \pi$. In fact, with tol $= 5 \times 10^{-5}$ reduction to a stepsize of about 6×10^{-4} was required to cross the discontinuity. This is only slightly above the value of $h_{min} = (5\pi/4) \times 10^{-4}$ used in Algorithm 10.5.4. Thus achieving stricter tolerances would necessitate reducing h_{min}. Such a reduction is not without costs, as it means that in problems in which the solution x or its derivatives are becoming infinite and the method necessarily must fail, termination may take substantially longer.

As Fig. 10.5.11 indicates, the stepsizes required to achieve an accuracy of 5×10^{-5} are in keeping with the prediction $h_0 \sim \sqrt[5]{5 \times 10^{-5}} \simeq 0.14$ except near $t = \pi$. It may be noted that approximating $x(5\pi/4)$ required only 408 evaluations of the right-hand side of the differential equation in (10.5.16). Thus compared to the fixed-step classical Runge-Kutta method, the Runge-Kutta-Felberg method was quite efficient in solving this problem.

Variable-Step Adams Predictor-Corrector Methods

Devising a variable-step method based upon an Adams predictor-corrector method poses an obvious difficulty. These methods derive their efficiency from saving and reusing evaluations of the right-hand side of the differential equation from previous steps. If the stepsize is changed, then these values become obsolete and must be recalculated. Thus frequent changes in the stepsize would severely degrade the efficiency of such a method. Consequently, care needs to be taken to avoid changing the stepsize too often and, when such a change is required, to do it in such a way that as few new evaluations as possible are required.

A natural error estimate E_{est} for a predictor-corrector method is the magnitude of difference between the predicted and corrected values. The new stepsize is taken to be proportional to the quantity $\sqrt[n]{tol/E_{est}}$, where n is the order of the local truncation error of the methods involved.

The decision procedure is to halve the stepsize h_0 when $E_{est} >$ tol for the current stepsize and to double it when it is predicted that this can be done without E_{est} exceeding tol. Halving the stepsize ensures that half the stored values of the right-hand $f(t, x)$ can be used for the new stepsize. The other values may be generated using some explicit method such as a Runge-Kutta method or by interpolation. The doubling of the stepsize can be accomplished without any new evaluations of the right-hand-side $f(t, x)$, providing that sufficiently many previous values of f have been stored. Implementing a variable-step predictor-corrector method along the lines described above is Project 4.

EXERCISES 10.5

1. Following the format of Example 10.5.1, use a variable-step Runge-Kutta method based upon Euler's method to find an approximation accurate to 0.05 to $x(0.5)$ for the initial value problem $x' = 5t - 2x$, $x(0) = 1$. Compare your answers with the actual solution after each step.

2. (a) Find a Runge-Kutta-Felberg 2(3) method for which the second-order method is the midpoint method.
   (b) Find a Runge-Kutta-Felberg 2(3) method for which the second-order method is Heun's method (see Exercise 7 in Sec. 10.3).

3. Write a computer program implementing the Runge-Kutta-Felberg 2(3) method given by (10.5.11) for a single first-order differential equation. Apply the method to approximate $x(2)$ for the initial value problem $x' = 5t - 2x$, $x(0) = 1$ to within an accuracy of 5×10^{-5}.

4. Write a computer program implementing the Runge-Kutta-Felberg 4(5) method given by (10.5.12) for a single first-order differential equation. Apply the method to approximate $x(2)$ for the initial value problem $x' = 5t - 2x$, $x(0) = 1$ to within an accuracy of 5×10^{-5}.

5. Compare the maximum, minimum, and average stepsizes required by the Runge-Kutta-Felberg 2(3) method for Exercise 3 to those of the 4(5) method of Exercise 4.

6. Use the Runge-Kutta-Felberg 4(5) method (10.5.12) to find and graph the temperature of an air-conditioned building [see Eq. (10.1.4)] when $k = 0.5\,\text{hour}^{-1}$, $A(t) = 70°$, $U_0 = 4°/\text{hour}$, $T_{crit} = 65°$, and $T(0) = 50°$. Use a tolerance that ensures that each point on the graph will be correctly plotted to the nearest pixel.

7. Repeat Exercise 6 when the ambient temperature is $A(t) = 70° + 20 \sin(\pi t/12)$.

8. Use the Runge-Kutta-Felberg 4(5) method (10.5.12) for the following initial value problems.
   (a) $x' = -t/x$, $x(0) = 1$, $0 \le t \le 2$
   (b) $x' = \dfrac{x^2 + x}{t}$, $x(1) = 1$, $1 \le t \le 3$

   Explain what happened in each case.

9. For the Runge-Kutta-Felberg 2(3) method (10.5.11), find how small h_0 would have to be for the method to successfully cross the discontinuity in the initial value problem (10.5.14). Confirm your analysis by finding how small h_0 became when crossing the discontinuity in the initial value problem (10.5.16).

10. Write a program implementing the Runge-Kutta-Felberg 4(5) method (10.5.12) for a system of differential equations. Apply it to find approximations to within an accuracy of 5×10^{-5} for $x_1(1)$ and $x_2(1)$ for the initial value problem

$$x_1' + x_2 = t^2 \qquad x_1(0) = 1$$
$$x_2' - x_1 = 1 \qquad x_2(0) = 0 \qquad (10.5.18)$$

11. Use the Runge-Kutta-Felberg 4(5) method (10.5.12) to graph solutions to the Lotka-Volterra model (10.1.17). Take $a = 2$, $b = 1$, $c = 1$, $d = 1$ and use the initial conditions $x_1(0) = 1$, $x_2(0) = 0.1$. Graph solutions over the time interval $0 \le t \le 10$.

12. Compare the execution times and accuracies given in Table 10.5.9 with those for your computer system. Try to account for any differences that you observe.

Use the Runge-Kutta-Felberg 4(5) method (10.5.12) for the exercises given below.

13. Exercise 18 in Sec. 10.3.
14. Exercise 19 in Sec. 10.3.
15. Exercise 20 in Sec. 10.3.
16. Exercise 21 in Sec. 10.3.
17. Exercise 22 in Sec. 10.3.
18. Apply the Runge-Kutta-Felberg 4(5) method (10.5.12) to Exercise 23 in Sec. 10.3. Use tolerances of 0.05 and 0.005 in lieu of the given number of steps.

10.6 VARIABLE-ORDER METHODS

In Sec. 9.5 we saw that Romberg integration, which is Richardson extrapolation applied to the trapezoid rule for numerical integration, provided a sequence of approximations of increasing order for definite integrals. In this section we develop methods for approximating the solution to the initial value problem

$$x' = f(t, x) \qquad x(t_0) = x_0 \qquad (10.6.1)$$

by means of Richardson extrapolation.

Let $A(h)$ be a scheme for approximating some quantity $A(0)$; that is, $\lim_{h \to 0} A(h) = A(0)$. The parameter h is typically an interval width. It is assumed that the error of approximation $A(0) - A(h)$ has an expansion in powers of h whence

$$A(0) = A(h) + a_1 h + a_2 h^2 + a_3 h^3 + a_4 h^4 + \cdots \qquad (10.6.2)$$

Recall that Richardson extrapolation entails using approximations $A(h_0)$, $A(h_1)$, $A(h_2)$, ... with $h_0 > h_1 > h_2 > \cdots$ to successively eliminate the terms in the error expansion, thereby producing approximations of higher and higher order. In Sec. 9.5 the sequence used was $h_j := h/2^j$, $j = 0, 1, 2, \ldots$, where h is some starting interval width; however, for our present purposes other sequences $\{h_j\}$ may be more advantageous. If a_1 in (10.6.2) is not zero, then the approximation scheme $A(h)$ is only $O(h)$. To obtain an $O(h^2)$ approximation we note that

$$A(0) = A(h_0) + a_1 h_0 + a_2 h_0^2 + a_3 h_0^3 + a_4 h_0^4 + \cdots$$
$$A(0) = A(h_1) + a_1 h_1 + a_2 h_1^2 + a_3 h_1^3 + a_4 h_1^4 + \cdots \qquad (10.6.3)$$

Upon subtracting h_0 times the second equation from h_1 times the first and solving for $A(0)$, we obtain

$$A(0) = \frac{h_1 A(h_0) - h_0 A(h_1)}{h_1 - h_0} - a_2 h_0 h_1 - a_3 h_0 h_1 (h_0 + h_1)$$
$$- a_4 (h_0^2 + h_0 h_1 + h_1^2) - \cdots$$

$$= A(h_1) + \frac{A(h_1) - A(h_0)}{h_0/h_1 - 1} - a_2 h_0 h_1 - a_3 h_0 h_1 (h_0 + h_1)$$
$$- a_4(h_0^2 + h_0 h_1 + h_1^2) - \cdots \tag{10.6.4}$$

Thus

$$A_1(h_0) := A(h_1) + \frac{A(h_1) - A(h_0)}{h_0/h_1 - 1} \tag{10.6.5}$$

is an $O(h_0^2)$ approximation to $A(0)$ since $h_1 < h_0$. Since any pair h_j, h_{j+1} could have been used in the elimination process above, we see that

$$A_1(h_j) := A(h_{j+1}) + \frac{A(h_{j+1}) - A(h_j)}{h_j/h_{j+1} - 1} \tag{10.6.6}$$

is an $O(h_j^2)$ approximation to $A(0)$.

We now have

$$A(0) = A_1(h_0) - a_2 h_0 h_1 - a_3 h_0 h_1 (h_0 + h_1) - a_4 h_0 h_1 (h_0^2 + h_0 h_1 + h_1^2) - \cdots \tag{10.6.7}$$

$$A(0) = A_1(h_1) - a_2 h_1 h_2 - a_3 h_1 h_2 (h_1 + h_2) - a_4 h_1 h_2 (h_1^2 + h_1 h_2 + h_2^2) - \cdots$$

Upon eliminating the terms involving a_2, we obtain (Exercise 1)

$$A(0) = A_2(h_0) + a_3 h_0 h_1 h_2 + a_4 h_0 h_1 h_2 (h_0 + h_1 + h_2) + \cdots \tag{10.6.8}$$

where

$$A_2(h_0) := A_1(h_1) + \frac{A_1(h_1) - A_1(h_0)}{h_0/h_2 - 1} \tag{10.6.9}$$

is an $O(h_0^3)$ approximation to $A(0)$. More generally

$$A_2(h_j) := A_1(h_{j+1}) + \frac{A_1(h_{j+1}) - A_1(h_j)}{h_j/h_{j+2} - 1} \tag{10.6.10}$$

is an $O(h_j^3)$ approximation to $A(0)$. Continuing in this manner, the recursively defined sequence

$$A_0(h_j) := A(h_j)$$
$$A_n(h_j) = A_{n-1}(h_{j+1}) + \frac{A_{n-1}(h_{j+1}) - A_{n-1}(h_j)}{h_j/h_{j+n} - 1} \tag{10.6.11}$$

is obtained.

On the basis of the results for $A(h_j)$ and $A_2(h_j)$, it seems that $A_n(h_j)$ provides an $O(h_j^{n+1})$ approximation to $A(0)$. This may be verified directly by following the evolution of the general term $a_n h^n$ in the error expansion but is perhaps obtained more easily by an alternative approach. The Lagrange interpolating polynomial for the points $(h_0, A(h_0))$, $(h_1, A(h_1))$, ..., $(h_n, A(h_n))$ is

TABLE 10.6.1

Level	$O(h_j)$		$O(h_j^2)$		$O(h_j^3)$		$O(h_j^4)$
0	$A_0(h_0)$						
1	$A_0(h_1)$	$\rightarrow$	$A_1(h_0)$				
2	$A_0(h_2)$	$\leftarrow$	$A_1(h_1)$	$\rightarrow$	$A_2(h_0)$		
3	$A_0(h_3)$	$\rightarrow$	$A_1(h_2)$	$\rightarrow$	$A_2(h_1)$	$\rightarrow$	$A_3(h_0)$

$$P_n(h) := A[h_n] + A[h_n, h_{n-1}](h - h_n) + A[h_n, h_{n-1}, h_{n-2}](h - h_n)(h - h_{n-1})$$
$$+ \cdots + A[h_n, h_{n-1}, \ldots, h_0, h](h - h_n)(h - h_{n-1}) \cdots (h - h_1)(h - h_0)$$
$$(10.6.12)$$

It is easily seen that $P_n(0) = A_n(h_0)$ (Exercise 2), which is why the process is referred to as extrapolation. From the remainder term of the Lagrange formula (4.3.13), it follows that the error made in approximating $A(0)$ by $P_n(0)$ is

$$\frac{(-1)^{(n+1)} A^{(n+1)}(\eta)}{(n+1)!} h_0 h_1 \cdots h_n$$

with $0 \le \eta \le h_0$ whence it is an $O(h_0^{n+1})$ approximation.

Table 10.6.1 shows the extrapolation sequence generated by the formulas (10.6.11). Only the first column requires application of the method $A(h)$. The higher-order refinements are generated by simple arithmetic computations and thus are inexpensive in terms of computing time.

Euler-Romberg Method

Let $\xi(t; h)$ denote the approximation to the solution $x(t)$ to (10.6.1) generated by Euler's method when a stepsize h is used. To apply Richardson extrapolation to Euler's method, it needs to be established that the global truncation error has an expansion in terms of h of

$$x(t) - \xi(t; h) = a_1 h + a_2 h^2 + \cdots \qquad (10.6.13)$$

This is shown to be the case for Euler's method by Gragg (1965), but we will not go into the details here. Richardson extrapolation applied to Euler's method is referred to as the Euler-Romberg method.

Example 10.6.2. Let us find an approximation to $x(0.1)$ for the solution to the initial value problem $x' = 5(t - 1)x$, $x(0) = 5$ using Euler's method with stepsizes of $h_0 = 0.1$, $h_1 = 0.05$, and $h_2 = 0.025$. The Euler method approximation to $x(0.1)$ with a stepsize of $h_0 = 0.1$ is $\xi(0.1; 0.1) = 5 + 0.1(5)(0 - 1)(5) = 2.50000$. For a stepsize of $h_1 = 0.05$ we have

TABLE 10.6.3
Errors made by the Euler-Romberg method for Example 10.6.2

Level	$O(h_j)$	$O(h_j^2)$	$O(h_j^3)$	$O(h_j^4)$	$O(h_j^5)$
0	6.1E − 01				
1	2.5E − 01	1.1E − 01			
2	1.2E − 01	1.9E − 02	1.1E − 02		
3	5.6E − 02	4.2E − 03	8.6E − 04	5.2E − 04	
4	2.7E − 02	9.9E − 04	9.0E − 05	2.0E − 05	1.3E − 05

$$\xi_1 = 5 + 0.05(5)(0 - 1)(5) = 3.75000$$

$$\xi_2 = 3.75 + 0.05(5)(0.1 - 1)(3.75) \approx 2.85938$$

Thus $\xi(0.1; 0.05) \approx 2.85938$, and from (10.6.11)

$$A_1(0.1) = \xi(0.1; 0.05) + \frac{\xi(0.1; 0.05) - \xi(0.1; 0.1)}{0.1/0.05 - 1} \approx 3.21876$$

The four-step Euler approximation $\xi(0.1; 0.025)$ turns out to be 2.99413 to five decimal places, and thus

$$A_1(0.05) = \xi(0.1; 0.025) + \frac{\xi(0.1; 0.025) - \xi(0.1; 0.05)}{0.05/0.025 - 1} \approx 3.12888$$

$$A_2(0.1) = A_1(0.05) + \frac{A_1(0.05) - A_1(0.1)}{0.1/0.025 - 1} \approx 3.09892$$

The actual solution is $x(t) = 5e^{(5/2)t^2 - 5t}$ whence $x(0.1) \approx 3.10943$. Thus the Euler-Romberg method makes an error of 0.011 despite the fact that the most accurate of the three Euler method approximations employed made an error of 0.12. Table 10.6.3 shows the errors made by the Euler-Romberg method through the fourth level of the extrapolation process using $h_j = 0.25/2^j$, $j = 0, 1, 2, 3, 4$.

Algorithm 10.6.4 gives an implementation of a fixed-step Euler-Romberg method. It is not expected that a single extrapolation sequence will be effective in approximating $x(t_f)$ when the interval (t_0, t_f) is large; hence the interval (t_0, t_f) is first subdivided into smaller intervals of width h, and the Euler-Romberg scheme is applied to each interval. As with any fixed-step method, the appropriate number of steps can be determined only by prior experimentation. A variable-step extrapolation method will be presented later in this section.

Algorithm 10.6.4 requires specification of a sequence $N_0, N_1, \ldots,$ N_{MaxLevel} in which N_j is the number of steps to use in obtaining the Euler method approximation at the jth level. For instance, in Table 10.6.3 the sequence $N_j = 2^j$, $j = 0, 1, 2, 3$ was used. Since the stepsize used by the Euler method is $h_j = h/N_j$, we have $h_j/h_{j+n} = N_{j+n}/N_j$ in (10.6.11) and it is this latter form that is used in the algorithm.

Algorithm 10.6.4   Fixed-step Euler-Romberg method

Input t_0, x_0 (initial conditions)
              t_f (value at which approximation is sought)
              N_{steps} (number of times to apply basic algorithm)
              tol (error tolerance for each application of the Euler-Romberg method)

Constant   *MaxLevel* (maximum permissible table depth)

Define   Function $f(t, x)$ (right-hand side of the differential equation)
              Function $Euler(t_0, x_0, h, N)$ which returns the Euler approximation to $x(t_0 + h)$ using N steps.

Declare   Arrays $CurrRow_k$ and $LastRow_k$ of dimension $MaxLevel + 1$. Array N_j containing the number of steps to use at the jth level.

Initialize N_j (step number sequence to be used)

$h := (t_f - t_0)/N_{\text{steps}}$
$t := t_0, \ x := x_0$
$i := 1$

repeat
    $CurrRow_0 := Euler(t, x, h, N_0)$
    $Level := 0$
    Success := FALSE
    Failure := FALSE
    repeat
        $LastRow := CurrRow$
        $Level := Level + 1$
        $CurrRow_0 := Euler(t, x, h, N_{Level})$
        for $k = 0$ to $Level - 1$

$$\left\{ CurrRow_{k+1} := CurrRow_k + \frac{CurrRow_k - LastRow_k}{N_{Level}/N_{Level-k-1} - 1} \right\}$$

        if $|CurrRow_{Level} - LastRow_{Level-1}| < $ tol then Success := TRUE
        else if $(Level = MaxLevel)$ then Failure := TRUE
    until Success or Failure
    if Success then
        $\{ t := t_0 + ih$
        $x := CurrRow_{Level}$
        Output: x (Approximation to the solution at t_f)
        $i := i + 1 \}$
until $(i > N_{\text{steps}})$ or Failure

if Failure then
    Output: Message of failure,
            Value of t and $CurrRow_{Level}$ at time of failure,
            Estimated error $CurrRow_{Level} - LastRow_{Level-1}$

Gragg's Method

Recall that the trapezoid rule was selected as the base method for Romberg integration because, from the Euler-Maclaurin summation formula, its error expansion has only even powers of the subinterval width h. The result of this is that as we move column by column to the right in the extrapolation table, the orders of the approximations increase by two each time, rather than one as they do in Table 10.6.1. Thus it is desirable to find a method $\xi(t; h)$ of approximating the solution to a differential equation that has the property

$$x(t) = \xi(t; h) + a_2 h^2 + a_4 h^4 + \cdots \qquad (10.6.14)$$

It turns out that the explicit Runge-Kutta methods do not possess such an expansion but that the multistep midpoint method [Eq. (10.4.2)]

$$\xi_{i+1} = \xi_{i-1} + 2hf(t_i, \xi_i) \qquad (10.6.15)$$

does. This is shown by Gragg (1965). We saw in Fig. 10.4.2 that this second-order method is not a particularly accurate one, having a tendency to oscillate. Fortunately this tendency can be suppressed by an averaging process

$$\bar{\xi}_i = \tfrac{1}{4}(\xi_{i-1} + 2\xi_i + \xi_{i+1}) \qquad (10.6.16)$$

Figure 10.6.5 shows graphs of the approximations generated by (10.6.15) with a stepsize of $h = \tfrac{3}{8}$ when (1) no averaging is used, (2) the average approximation $\bar{\xi}_i$ given by (10.6.16) is plotted at each step in lieu of the original approximation, and (3) the averaging process is applied only as correction at the last step. The gain in accuracy achieved by correcting only at the final step is important as this is what is done in *Gragg's method*, which is Richardson

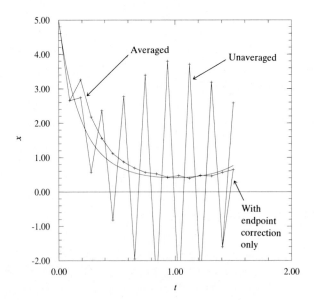

FIGURE 10.6.5

extrapolation applied to the multistep midpoint method (10.6.15) with the correction (10.6.16) applied at the last step. Algorithm 10.6.6 gives the details of the implementation of this modified midpoint method.

Algorithm 10.6.6 Midpoint method with endpoint correction

> *Input* t_0, x_0 (initial conditions)
> h (basic stepsize)
> N (≥ 2, number of steps into which to subdivide h)

> $h_s := h/N$
> $t := t_0, \, \xi_0 := x_0$
> $\xi_1 := \xi_0 + h_s f(t, \xi_0)$ (use Euler's method at first step)

> for $i = 1$ to $N - 1$ (middle steps)
> $\quad \{ t := t_0 + ih_s$
> $\quad \xi_2 := \xi_0 + 2h_s f(t, \xi_1)$ (multistep midpoint method)
> $\quad \xi_0 := \xi_1$
> $\quad \xi_1 := \xi_2 \}$

> $t := t_0 + Nh_s$ [last step, we are now approximating $x(t + h + h_s)$]
> $\xi_2 := \xi_0 + 2h_s f(t, \xi_1)$

> *Output* $(\xi_0 + 2\xi_1 + \xi_2)/4$ [smoothed approximation to $x(t + h)$]

When the process of eliminating higher-order terms in the error expansion that led to the recursive sequence (10.6.11) is applied to a method of numerical approximation having the form (10.6.14), the result is

$$A_0(h_j) = A(h_j)$$

$$A_n(h_j) = A_{n-1}(h_{j+1}) + \frac{A_{n-1}(h_{j+1}) - A_{n-1}(h_j)}{h_j^2/h_{j+n}^2 - 1} \tag{10.6.17}$$

A fixed-step Gragg's method may be obtained from Algorithm 10.6.4 by replacing Euler's method by the modified midpoint method of Algorithm 10.6.6 and using (10.6.17) in lieu of (10.6.11).

TABLE 10.6.7
Errors made by Gragg's method with $h_j = 0.25/2^j$, $j = 1, \ldots, 5$

Level	$O(h_j^2)$	$O(h_j^4)$	$O(h_j^6)$	$O(h_j^8)$	$O(h_j^{10})$
0	4.6E − 02				
1	2.0E − 02	1.1E − 02			
2	5.6E − 03	8.7E − 04	2.0E − 04		
3	1.4E − 03	5.9E − 05	4.3E − 06	1.1E − 06	
4	3.6E − 04	3.7E − 06	7.4E − 08	6.4E − 09	2.0E − 09

TABLE 10.6.8
Errors made by Gragg's method with $h_j = 0.25/N_j$, $N_j = 2, 4, 6, 8, 12$

Level	$O(h_j^2)$	$O(h_j^4)$	$O(h_j^6)$	$O(h_j^8)$	$O(h_j^{10})$
0	4.6E − 02				
1	2.0E − 02	1.1E − 02			
2	9.6E − 03	1.5E − 03	3.6E − 04		
3	5.6E − 03	4.1E − 04	3.0E − 05	7.8E − 06	
4	2.5E − 03	1.0E − 04	3.6E − 06	3.1E − 07	9.3E − 08

Tables 10.6.7 and 10.6.8 show extrapolation tables for Gragg's method for the problem of Example 10.6.2 using two different choices of the step sequence $h_j = h/N_j$, $j = 0, 1, 2, 3, 4$. It can be seen that in both cases the convergence along the main diagonal of the table is much quicker than that shown for the Euler-Romberg method in Table 10.6.3. While the convergence along the main diagonal in Table 10.6.7 is faster than in Table 10.6.8, the sequence of stepsizes used in the latter table, which is part of the longer sequence $h_j = h/N_j$ with $N_j = 2, 4, 6, 8, 12, 16, 24$, is preferred. For instance, suppose that an error tolerance tol $= 5 \times 10^{-5}$ had been specified. With either step sequence Gragg's method would have terminated successfully after level 4. However, since the modified midpoint method of Algorithm 10.6.6 requires $N + 1$ evaluations of the right-hand side of the differential equation each time it executes, completion of Table 10.6.7 requires 67 evaluations against 37 for Table 10.6.8. Even were the error tolerance such that the sequence used in Table 10.6.7 led to successful termination after level 3, 34 evaluations would have been required.

Since with a fixed-step Euler-Romberg or Gragg method we have control over the error, at least for each step, the choice of the stepsize for the most part affects the efficiency rather than the accuracy of the method. Table 10.6.9 shows the result of applying the fixed-step Gragg method to problem 5 of Table 10.4.11 for 500, 100, and 20 steps with tol $= 5 \times 10^{-6}$. The step sequence used was $h_j := h/N_j$ with $N_j = 2, 4, 6, 8, 12, 16, 24$. As can be seen the efficiency for the Gragg method with 20 steps is much better than the classical Runge-Kutta

TABLE 10.6.9

Method and steps	RHS evals.	Error	Time, s
Classical RK			
500 steps	3880	4.9E − 06	3.16
Gragg			
500 steps	7500	3.5E − 09	8.23
Gragg			
100 steps	2400	9.0E − 08	2.71
Gragg			
20 steps	1580	2.2E − 08	1.73

with 500 steps, but the Gragg method with 500 steps is far less efficient. The difficulty is that when the stepsize is small, Gragg's method only needs to go down a few levels in the extrapolation table to achieve the tolerance. If the extrapolation terminates on level 2, for instance, then 15 evaluations of the right-hand side of the differential equation have been used to obtain an order-six approximation; however, from Table 10.3.14 a Runge-Kutta method only requires seven evaluations to attain this order. An extrapolation method derives its efficiency from obtaining very high order approximations using relatively large stepsizes and is not particularly competitive otherwise. This indicates the need for a mechanism for varying the stepsize in the Gragg algorithm. This is the concern of the next subsection.

Variable-Step Gragg's Method

Algorithm 10.6.10 gives a variable-step Gragg's method. The stepsize is reduced when two consecutive elements on the main diagonal of the extrapolation table fail to agree to within the given tolerance when the last row MaxLevel of the table has been reached. At the outset the stepsize used is $h = t_f - t_0$. The method will often fail to achieve tolerance at this stepsize. Since the cost of such a failure is high, 72 wasted evaluations of the right-hand side of the differential equation, the stepsize is reduced by a factor of 10 at each instance of failure. After the first successful attainment of the tolerance, reduction by a factor of 10 is too severe, and from then on reductions in stepsize are by a factor of 2.

To decide when an increase in the stepsize is possible, a target level TargLevel, which is near MaxLevel, is specified. As noted previously, Gragg's method is not efficient when success is attained after only two or three levels of the extrapolation table have been calculated. If success is attained before the level of extrapolation reaches TargLevel, then the stepsize is doubled.

Algorithm 10.6.10   Variable-step Gragg method

Input t_0, x_0 (initial conditions)
              t_f (value at which approximation is sought)
              tol (error tolerance)

Constants   MaxLevel (maximum permissible table depth)
              TargLevel (target table depth)

Define   Function $f(t, x)$ (right-hand side of the differential equation)
              Function $MultiMid(t_0, x_0, h, N)$ [returns the approximation of Algorithm 10.6.6 to $x(t_0 + h)$ using N steps]

Declare   Arrays $CurrRow_k$ and $LastRow_k$ of dimension $MaxLevel + 1$. Array N_j containing the number of steps to use at the jth level.

Initialize N_j (step number sequence)

$h := t_f - t_0$
$h_{\min} := h/10^4$
ReductionFactor:=10
TotalFailure:=FALSE
$t := t_0, x := x_0$

repeat
    $Level := 0$
    $CurrRow_0 := MultiMid(t_0, x_0, h, N_0)$
    Success:=FALSE
    Failure:=FALSE
    repeat
        $LastRow := CurrRow$
        $Level := Level + 1$
        $CurrRow_0 := MultiMid(t_0, x_0, h, N_{Level})$
        for $k = 0$ to $Level - 1$

$$\left\{ CurrRow_{k+1} := CurrRow_k + \frac{CurrRow_k - LastRow_k}{\left(N_{Level}/N_{Level-k-1}\right)^2 - 1} \right\}$$

        $Error := |CurrRow_{Level} - LastRow_{Level-1}|$
        if $Error <$ tol then Success:=TRUE
        else if $Level = MaxLevel$ then Failure:=TRUE
    until Success or Failure

    if Success then
        $\{t := t + h, x := CurrRow_{Level}$
        ReductionFactor:=2
        if $Level < TargetLevel$ then $h := 2h$
        if $t + h > t_f$ then $h := t_f - t\}$
    else
        $\{h := h/$ReductionFactor
        if $h < h_{\min}$ then TotalFailure:=TRUE$\}$
until $(t \geq t_f)$ or TotalFailure

if NOT TotalFailure then
    Output: $CurrRow_{Level}$ [Approximation to $x(t_f)$]
else
    Output: Report that tolerance could not be achieved, values of t,
        $CurrRow_{Level}$, and $Error$ at time of failure.

Table 10.6.11 compares the results of applying a Turbo Pascal implementation of Algorithm 10.6.10 to the results for the Runge-Kutta-Felberg 4(5) method given by (10.5.12) on the test problems of Table 10.4.11. The step number sequence used was $N_j = 2, 4, 6, 8, 12, 16, 24$. MaxLevel was 6 and TargLevel was 5. As can be seen the variable-step Gragg method typically requires as many or more evaluations of the right-hand side than the Runge-Kutta-Felberg method, but the program overhead is sufficiently lower so that

TABLE 10.6.11

		RKF 4(5)			Variable-step Gragg		
Prob.	Target error	RHS evals.	Error	Time, s	RHS evals.	Error	Time, s
1	5E − 09	324	4.9E − 09	0.15	364	9.1E − 12	0.09
2	5E − 07	276	1.2E − 04	0.12	658	2.2E − 08	0.16
3	5E − 09	1878	2.4E − 10	1.43	1793	3.9E − 12	1.03
4	5E − 09	420	4.3E − 10	0.79	519	2.5E − 11	0.79
5	5E − 06	1242	6.9E − 06	1.91	1664	2.0E − 07	1.86

TABLE 10.6.12
Stepsizes required by Runge-Kutta-Felberg and Gragg methods for five test problems

	RKF 4(5)			Variable-step Gragg		
Prob.	Min	Max	Av	Min	Max	Av
1	0.010	0.059	0.026	0.125	0.250	0.179
2	0.005	0.207	0.054	0.056	0.600	0.214
3	0.018	0.208	0.097	0.300	1.200	0.968
4	0.016	0.075	0.043	0.300	0.600	0.375
5	0.017	0.131	0.097	0.200	0.800	0.741

the overall execution time is less. Thus on the whole it appears that the Gragg method is the most efficient for solving the problems of Table 10.4.11. However, it should be noted that these problems involve finding an approximation at a single point $t = t_f$. If approximations are also desired at points between t_0 and t_f, then these can be obtained by imposing a limitation on the maximum stepsize h that can be taken. As indicated in Table 10.6.9 this could severely diminish the efficiency of the Gragg method, and the Runge-Kutta-Felberg might be more appropriate in this situation.

Table 10.6.12 shows the least, greatest, and average stepsize that the Runge-Kutta-Felberg and Gragg methods used in the course of approximating the solutions to each of the test problems. As can be seen there the Gragg method can use a stepsize from 5 to 10 times larger than the Runge-Kutta-Felberg method.

EXERCISES 10.6

1. Go through the steps of eliminating the term $a_2 h_0 h_1$ in (10.6.7) to obtain (10.6.8).
2. Show that for $P_n(h)$ defined by (10.6.12), $P_n(0) = A_n(h_0)$.
3. Use the Euler-Romberg method to find an ap-

proximation accurate to 0.05 to $x(0.25)$ for the initial value problem $x' = 5t − 2x, x(0) = 1$. Compare your answers with the actual solution.
4. Write a computer program implementing the fixed-step Euler-Romberg method for a single

first-order differential equation. Apply the method to approximate $x(2)$ for the initial value problem $x' = 5t - 2x$, $x(0) = 1$ to within an accuracy of 5×10^{-5}. Compare the efficiency of the two-step number sequences $N_j = 2, 4, 6, 8, 12, 16, 24$ and $N_j = 1, 2, 4, 8, 16, 32, 64$.

5. Write a computer program implementing the fixed-step Gragg method for a single first-order differential equation. Apply the method to approximate $x(2)$ for the initial value problem $x' = 5t - 2x$, $x(0) = 1$ to within an accuracy of 5×10^{-5}.

6. Write a computer program implementing the variable-step Gragg method for a single first-order differential equation. Apply the method to approximate $x(2)$ for the initial value problem $x' = 5t - 2x$, $x(0) = 1$ to within an accuracy of 5×10^{-5}. Compare the efficiency of the two-step number sequences $N_j = 2, 4, 6, 8, 12, 16, 24$ and $N_j = 1, 2, 4, 8, 16, 32, 64$.

7. Use the variable-step Gragg method to find and graph the temperature of an air-conditioned building [see Eq. (10.1.4)] when $k = 0.5\,\text{hour}^{-1}$, $A(t) = 70°$, $U_0 = 4°/\text{hour}$, $T_{\text{crit}} = 65°$, and $T(0) = 50°$. Use a tolerance that ensures that each point on the graph will be correctly plotted to the nearest pixel.

8. Repeat Exercise 7 when the ambient temperature is $A(t) = 70° + 20° \sin(\pi t/12)$.

9. Write a program implementing the variable-step Gragg method for a system of differential equations. Apply it to find approximations to within an accuracy of 5×10^{-5} for $x_1(1)$ and $x_2(1)$ for the initial value problem

$$x_1' + x_2 = t^2 \qquad x_1(0) = 1$$
$$x_2' - x_1 = 1 \qquad x_2(0) = 0 \qquad (10.6.18)$$

10. Use the variable-step Gragg method to graph solutions to the Lotka-Volterra model (10.1.17). Take $a = 2$, $b = 1$, $c = 1$, $d = 1$ and use the initial conditions $x_1(0) = 1$, $x_2(0) = 0.1$. Graph solutions over the time interval $0 \le t \le 10$.

11. Compare the execution times and accuracies given in Table 10.6.11 with those for your computer system. Try to account for any differences that you observe.

12. Apply the variable-step Gragg method to the problem of Example 10.5.10 with a tolerance of 5×10^{-5}.

Use the variable-step Gragg method for the exercises given below.

13. Exercise 18 in Sec. 10.3
14. Exercise 19 in Sec. 10.3
15. Exercise 20 in Sec. 10.3
16. Exercise 21 in Sec. 10.3
17. Exercise 22 in Sec. 10.3
18. Apply the variable-step Gragg method to Exercise 23 in Sec. 10.3. Use tolerances of 0.05 and 0.005 in lieu of the number of steps.

10.7 STABILITY AND STIFF DIFFERENTIAL EQUATIONS

In many applications the solution to the initial value problem

$$x' = f(t, x) \qquad x(t_0) = x_0 \qquad (10.7.1)$$

approaches an equilibrium point $x = \bar{x}$, that is, a point for which $f(t, \bar{x}) \equiv 0$ for all t. Approximating solutions by numerical means as they approach an equilibrium point might be thought to be routine, but this is not always the case, as the following example shows.

Example 10.7.1. The solution $x(t) = e^{-t^2}$ to the initial value problem

$$x' = -2tx \qquad x(0) = 1 \qquad (10.7.2)$$

approaches the equilibrium point $x = 0$. Figure 10.7.2 shows the Euler method approximation for a stepsize of $h = 0.2$. It illustrates a difficulty, *instability*, that is encountered with many fixed-step numerical methods for solving differential equations.

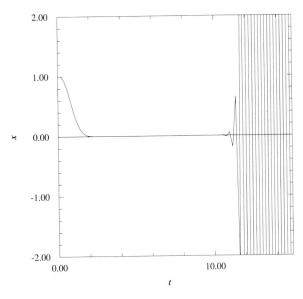

FIGURE 10.7.2

Stability of Euler's Method

To see more about the origin of such instability, let us consider a simpler problem, the application of Euler's method to the initial value problem

$$x' = \lambda x \qquad x(0) = x_0 \tag{10.7.3}$$

where λ is a constant. For Euler's method $\xi_{i+1} = \xi_i + h\lambda\xi_i = (1 + h\lambda)\xi_i$. Thus

$$\xi_0 = x_0$$
$$\xi_1 = (1 + h\lambda)\xi_0 = (1 + h\lambda)x_0$$
$$\xi_2 = (1 + h\lambda)\xi_1 = (1 + h\lambda)^2 x_0 \tag{10.7.4}$$

$$\cdots\cdots\cdots\cdots\cdots\cdots$$

$$\xi_i = (1 + h\lambda)\xi_{i-1} = (1 + h\lambda)^i x_0$$

When $\lambda < 0$, the actual solution $x(t) = x_0 e^{\lambda t}$ approaches the equilibrium point $x = 0$. The Euler method approximation $\xi_i = (1 + \lambda h)^i x_0$ also approaches $x = 0$ only if $|1 + \lambda h| < 1$, that is, $h\lambda \in (-2, 0)$. When $1 + h\lambda < -1$, the Euler method approximation will exhibit growing oscillations about $x = 0$. Figure 10.7.3 shows the Euler method approximation to the solutions of $x' = -5x$, $x(0) = 1$ when $h = 0.25$ and when $h = 0.5$.

The analysis of the stability of Euler's method for the initial value problem (10.7.3) leads us to expect that for problem (10.7.2), Euler's method will be stable so long as $-2ht \in (-2, 0)$. With $h = 0.2$ this gives $t \in (0, 5)$.

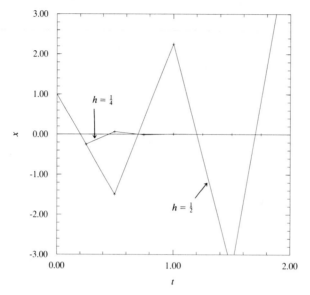

FIGURE 10.7.3

Table 10.7.4 shows the Euler approximations to (10.7.2) between $t = 3$ and $t = 7$. Observe that the approximations begin a steady oscillatory increase after $t = 5$, though this is not manifest in Fig. 10.7.2 until after $t = 10$.

More generally, suppose that the solution to the initial value problem (10.7.1) approaches the equilibrium point $x = 0$. Then from Taylor's theorem

$$f(t, x) = f(t, 0) + \frac{\partial f}{\partial x} (t, 0)x + O(|x|^2)$$

$$= \frac{\partial f}{\partial x} (t, 0)x + O(|x|^2)$$

(10.7.5)

TABLE 10.7.4

t_i	x_i	t_i	x_i
3.00000E + 00	2.54665E − 08	5.20000E + 00	−2.81486E − 11
3.20000E + 00	−5.09331E − 09	5.40000E + 00	3.04005E − 11
3.40000E + 00	1.42613E − 09	5.60000E + 00	−3.52646E − 11
3.60000E + 00	−5.13405E − 10	5.80000E + 00	4.37281E − 11
3.80000E + 00	2.25898E − 10	6.00000E + 00	−5.77211E − 11
4.00000E + 00	−1.17467E − 10	6.20000E + 00	8.08096E − 11
4.20000E + 00	7.04803E − 11	6.40000E + 00	−1.19598E − 10
4.40000E + 00	−4.79266E − 11	6.60000E + 00	1.86573E − 10
4.60000E + 00	3.64242E − 11	6.80000E + 00	−3.05980E − 10
4.80000E + 00	−3.05963E − 11	7.00000E + 00	5.26285E − 10
5.00000E + 00	2.81486E − 11		

since $f(t, 0) = 0$. The initial value problem

$$x' = \frac{\partial f}{\partial x}(t, 0)x \qquad x(t_0) = x_0 \qquad (10.7.6)$$

is referred to as the *linearized* form of 10.7.1. When $(\partial f/\partial x)(t, 0) \neq 0$, we expect that, at least roughly, Euler's method will be stable when $h(\partial f/\partial x)(t, 0)$ is in the interval of stability $(-2, 0)$ for the problem (10.7.3).

Based upon the above considerations, we make the following definition.

Definition 10.7.5. The *interval of absolute stability* of a numerical method is the interval of $h\lambda$ in which the approximation ξ_i to the solution $x(t_i) = x_0 e^{\lambda t_i}$ to (10.7.3) approaches zero as $i \to \infty$.

As the qualifier *absolute* indicates, other notions of stability pertain to numerical solutions to differential equations. We will mention another later.

Stability of Runge-Kutta Methods

The interval of absolute stability of other Runge-Kutta methods can be found in a similar manner to that used for Euler's method.

Example 10.7.6. Let us find the interval of absolute stability for the modified Euler method

$$\xi_{i+1} = \xi_i + \frac{h}{2}[f(t_i, \xi_i) + f(t_i + h, \xi_i + hf(t_i, \xi_i))]$$

With $f(t_i, \xi_i) = \lambda\xi_i$, we have $\xi_{i+1} = (1 + h\lambda + \frac{1}{2}h^2\lambda^2)\xi_i$ whence $\xi_i = (1 + h\lambda + h^2\lambda^2)^i x_0$. Thus the approximations approach $x = 0$ when $|1 + h\lambda + \frac{1}{2}h^2\lambda^2| < 1$. Since $1 + h\lambda + \frac{1}{2}h^2\lambda^2 = \frac{1}{2}(h\lambda + 1)^2 + \frac{1}{2}$, this implies that the interval of absolute stability for the modified Euler method is $h\lambda \in (-2, 0)$, as it was for Euler's method.

From the general formula (10.3.20) for a Runge-Kutta method, it is clear that application of the method to the problem (10.7.3) will lead to an equation of the form $\xi_{i+1} = P(\bar{h})\xi_i$, where $\bar{h} := h\lambda$. For the Euler and modified Euler methods we have found $P(\bar{h}) = 1 + \bar{h}$ and $P(\bar{h}) = 1 + \bar{h} + \frac{1}{2}\bar{h}^2$, respectively. Since $\xi_{i+1} = P(\bar{h})\xi_i$ implies $\xi_i = P(\bar{h})^i x_0$, the condition for absolute stability is $|P(\bar{h})| < 1$.

TABLE 10.7.7
Intervals of absolute stability for Runge-Kutta methods

Order	$P(\bar{h})$	Interval of absolute stability
1	$1 + \bar{h}$	$(-2, 0)$
2	$1 + \bar{h} + \frac{1}{2}\bar{h}^2$	$(-2, 0)$
3	$1 + \bar{h} + \frac{1}{2}\bar{h}^2 + \frac{1}{6}\bar{h}^3$	$(-2.51, 0)$
4	$1 + \bar{h} + \frac{1}{2}\bar{h}^2 + \frac{1}{6}\bar{h}^3 + \frac{1}{24}\bar{h}^4$	$(-2.78, 0)$

Table 10.7.7 shows the interval of absolute stability for Runge-Kutta methods with global truncation errors of order one through four. It turns out for these four orders that the polynomial $P(\bar{h})$ is the same for all methods of each order. Establishing this for orders two and three is Exercise 2a and b.

Linear Difference Equations

An nth-order linear difference equation has the form

$$a_n \xi_{i+n} + a_{n-1} \xi_{i+n-1} + \cdots + a_0 \xi_i = b_i \tag{10.7.7}$$

The coefficients a_j, $j = 0, \ldots, n$ and the sequence b_i, $i = 0, 1, \ldots$ are known; the unknown sequence ξ_i is called the *solution* to the difference equation. The process for solving linear difference equations parallels that for solving linear differential equations. Specifically, the general solution has the form $\xi_i = \xi_i^h + \xi_i^p$, where ξ_i^p is any particular solution of (10.7.7) and ξ_i^h is the general solution to the *associated homogeneous* equation

$$a_n \xi_{i+n} + a_{n-1} \xi_{i+n-1} + \cdots + a_0 \xi_i = 0 \tag{10.7.8}$$

It turns out that for the purpose of analyzing the stability of a numerical method, it is the solution to a homogeneous difference equation which is of interest. For instance, as we saw above, application of a Runge-Kutta method to the initial value problem (10.7.3) leads to the first-order difference equation $\xi_{i+1} - P(\bar{h})\xi_i = 0$ which has solution $\xi_i = P(\bar{h})^i x_0$ when $\xi_0 = x_0$. Let us seek solutions to (10.7.8) of the form $\xi_i = r^i$. This guess is suggested by the fact that we know linear homogeneous first-order difference equations have solutions of this form. Substituting $\xi_i = r^i$ into (10.7.8) and dividing by r^i yields the *characteristic equation* of (10.7.8),

$$a_n r^n + a_{n-1} r^{n-1} + \cdots + a_0 = 0 \tag{10.7.9}$$

Equation (10.7.9) has n solutions called the *characteristic roots*, when repeated roots are counted according to their multiplicity. For each distinct solution r of (10.7.9), r^i is a solution to (10.7.8). If r is a root of multiplicity m, then it turns out that $\{r^i, ir^i, \ldots, i^{m-1}r^i\}$ are solutions (Exercise 3b).

It is easily seen that solutions to (10.7.8) obey the *superposition principle*; that is, constant multiples of solutions and sums of solutions are again solutions. Thus, for instance, in the case when the n roots are distinct,

$$\xi_i = c_1 r_1^i + \cdots + c_n r_n^i \tag{10.7.10}$$

is a solution to (10.7.8). Indeed (10.7.10) is the general solution; that is, all solutions to (10.7.8) must have this form (Exercise 4b).

Consistency and Convergence of Multistep Methods

The general form of a multistep method is

$$\xi_{i+1} = \sum_{j=1}^{n} \alpha_j \xi_{i-j+1} + h \sum_{j=0}^{n} \beta_j f(t_{i-j+1}, \xi_{i-j+1}) \tag{10.7.11}$$

In Sec. 10.3 two basic attributes, *consistency* and *convergence*, were required of one-step approximation methods. Consistency is the requirement that the difference equation used by the method be equivalent to the differential equation as the stepsize h goes to zero. Convergence is the requirement that the approximations generated by the method approach the actual solution as the stepsize goes to zero.

If in (10.7.11) we fix upon a point $t \, (=t_i)$ and replace the approximations ξ_{i+j} by the actual solution $x(t + jh)$, then we obtain

$$x(t + h) = \sum_{j=1}^{n} \alpha_j x(t - (j-1)h) + h \sum_{j=0}^{n} \beta_j f(t - (j-1)h, x(t - (j-1)h))$$
$$(10.7.12)$$

Expanding the functions $x(t + jh) = x(t) + jhx'(t) + O(h^2)$ in Taylor series about t and dividing by h yields

$$\frac{\left(1 - \sum_{j=1}^{n} \alpha_j\right) x(t)}{h} + \left(1 + \sum_{j=1}^{n} (j-1)\alpha_j\right) x'(t)$$

$$= \sum_{j=0}^{n} \beta_j f(t - (j-1)h, x(t - (j-1)h)) + O(h) \qquad (10.7.13)$$

A multistep method is *consistent* if upon letting h approach zero in (10.7.13) we recover the differential equation $x' = f(t, x)$. Clearly the necessary and sufficient conditions for the consistency of the method are

$$\sum_{j=1}^{n} \alpha_j = 1$$
$$(10.7.14)$$
$$1 + \sum_{j=1}^{n} (j-1)\alpha_j = \sum_{j=0}^{n} \beta_j$$

For a multistep method to be convergent it must be consistent and also *Dalquist*, or *zero*, *stable*.

Definition 10.7.8. The multistep method (10.7.11) is *Dalquist stable* if the roots r_j of the polynomial $\rho(r) := r^n - \sum_{j=1}^{n} \alpha_j r^{n-j}$ satisfy $|r_j| \le 1$ and all roots for which $|r_j| = 1$ are simple.

Theorem 10.7.9. For the multistep method (10.7.11) to be convergent, it is necessary and sufficient that it be consistent and Dalquist stable.

The proof of this theorem is difficult. The reader is referred to Henrici (1962). While the requirements of consistency and convergence are fundamental, it is likely, assuming that a multistep method was derived from a finite difference or numerical quadrature formula, that they will be fulfilled. What often distinguishes effective multistep methods from ineffective ones is their interval of absolute stability.

Stability of Multistep Methods

To find the interval of absolute stability for a multistep method (10.7.11) we apply it to problem (10.7.3). Setting $f(t_i, \xi_i) = \lambda\xi_i$, we find that the approximations ξ_i generated by the multistep method satisfy the linear homogeneous difference equation

$$(1 - \beta_0\bar{h})\xi_{i+1} - \sum_{j=1}^{n} (\alpha_j + \beta_j\bar{h})\xi_{i-j+1} = 0 \qquad (10.7.15)$$

where as before $\bar{h} := h\lambda$. In accordance with Definition 10.7.5 we seek the values of $\bar{h} := h\lambda$ for which solutions to (10.7.15) approach zero as $i \to \infty$. The characteristic equation for the method is

$$\pi(r; \bar{h}) := (1 - \beta_0\bar{h})r^n - \sum_{j=1}^{n} (\alpha_j + \beta_j\bar{h})r^{n-j} = 0 \qquad (10.7.16)$$

The roots may be real or complex and may be of multiplicity greater than one. In all cases, however, the general solution is the sum of n terms of the form $ci^k r^i$, c a constant, and thus the condition that the approximations generated by the difference equation approach zero is $|r| < 1$ for all roots. In the case of a complex root, $r = a + ib$, $|r| = \sqrt{a^2 + b^2}$.

Example 10.7.10. Let us find the interval of absolute stability for the multistep midpoint method

$$\xi_{i+1} = \xi_{i-1} + 2hf(t_i, \xi_i) \qquad (10.7.17)$$

which is used in Gragg's method (Sec. 10.6). In (10.7.11) this corresponds to $n = 2$, $\alpha_1 = 0$, $\alpha_2 = 1$, $\beta_0 = \beta_2 = 0$, and $\beta_1 = 2$. The polynomial $\rho(r)$ in Definition 10.7.8 is $\rho(r) := r^2 - 1$. Thus the multistep midpoint rule is Dalquist stable, and as it is easily seen to be consistent as well, it converges by Theorem 10.7.9.

The characteristic polynomial (10.7.16) is $\pi(r; \bar{h}) = r^2 - 2\bar{h}r - 1$. The roots of this polynomial are $r_1 = \bar{h} + \sqrt{\bar{h}^2 + 1}$ and $r_2 = \bar{h} - \sqrt{\bar{h}^2 + 1}$, and thus the general solution to the difference equation $\xi_{i+1} = \xi_{i-1} + 2\bar{h}\xi_i$ is $\xi_i = c_1 r_1^i + c_2 r_2^i$. Hence both roots must be less than 1 in magnitude for the method to be absolutely stable. As can be seen from Fig. 10.7.11 the root r_1 decreases to 0 as $|\bar{h}|$ becomes large; however, the root r_2 is always less than -1 for $\bar{h} < 0$. Consequently the multistep midpoint method has no interval of absolute stability. Did it not possess an error expansion that makes it efficient for use in the Richardson extrapolation process used in Gragg's method, it would not otherwise merit much consideration as a method of approximating solutions to initial value problems.

Let us investigate further the convergence of the approximate solution $\xi_i = c_1 r_1^i + c_2 r_2^i$ given by the multistep midpoint rule to the actual solution $x(t) = x_0 e^{\lambda t}$. We have $\xi_0 = x_0$. Since the midpoint rule is a two-step method, ξ_1 must be approximated by another method. If Euler's method is used, then $\xi_1 = x_0 e^{\bar{h}} + O(\bar{h}^2)$. The constants c_1 and c_2 in the general solution $\xi_i = c_1 r_1^i + c_2 r_2^i$ to the difference equation then must satisfy the conditions

$$c_1 + c_2 = x_0$$
$$r_1 c_1 + r_2 c_2 = e^{\lambda\bar{h}} + O(\bar{h}^2) \qquad (10.7.18)$$

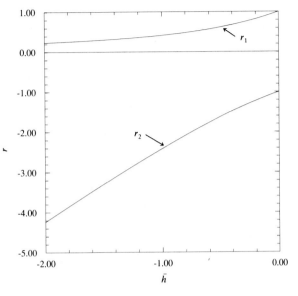

FIGURE 10.7.11

Solving for c_1 and c_2 gives

$$c_1 = \frac{x_0 r_2 - (x_0 e^{\lambda \bar{h}} + O(\bar{h}^2))}{r_2 - r_1}$$

$$c_2 = \frac{(x_0 e^{\lambda \bar{h}} + O(\bar{h}^2)) - x_0 r_1}{r_2 - r_1} \tag{10.7.19}$$

Upon computing Taylor expansions of r_1 and r_2, we see that

$$r_1 = \bar{h} + \sqrt{\bar{h}^2 + 1} = 1 + \bar{h} + O(\bar{h}^2) = e^{\bar{h}} + O(\bar{h}^2)$$

$$r_2 = \bar{h} - \sqrt{\bar{h}^2 + 1} = -1 + \bar{h} + O(\bar{h}^2) = -e^{-\bar{h}} + O(\bar{h}^2)$$

$$\frac{1}{r_2 - r_1} = \frac{1}{-2\sqrt{\bar{h}^2 + 1}} = -\frac{1}{2} + O(\bar{h}^2) \tag{10.7.20}$$

Substituting the expressions in (10.7.20) into (10.7.19) yields

$$c_1 = x_0 + O(\bar{h}^2)$$

$$c_2 = O(\bar{h}^2) \tag{10.7.21}$$

Let us fix a point $t = t_f$. With $h_N := t_f / N$ and $\bar{h}_N := \lambda h_N$, it can be shown that

$$r_1 = (\bar{h}_N + \sqrt{\bar{h}_N^2 + 1})^N = e^{\lambda t_f} + O(\bar{h}_N^2)$$

$$r_2 = (\bar{h}_N - \sqrt{\bar{h}_N^2 + 1})^N = -e^{-\lambda t_f} + O(\bar{h}_N^2) \tag{10.7.22}$$

(Exercise 9). Consequently

$$\xi_i = c_1 r_1^i + c_2 r_2^i = x_0 e^{i\bar{h}} + O(\bar{h}^2)(-1)^i e^{-i\bar{h}} + O(\bar{h}^2) \tag{10.7.23}$$

It is noteworthy here that the approximation to the actual solution $x(t) = x_0 e^{\lambda t}$ derives solely from the term $c_1 r_1^i$. The term $c_2 r_2^i$ has no beneficial effect on the accuracy of the solution. Rather, for $\lambda < 0$, it is an exponentially diverging term that acts to destroy the accuracy of the solution.

The situation for the multistep midpoint rule is illustrative of the general situation for multistep methods. The term $c_1 r_1^i$ corresponding to one root, called the *principal root*, of the characteristic polynomial $\pi(r; \bar{h})$ provides the approximation to the solution, while the terms $c_j r_j^i$, $j = 2, \ldots, n$, corresponding to the remaining roots either go to zero if $\bar{h}$ is within the interval of absolute stability or grow without bound if not. In neither instance do they enhance the accuracy of the approximation. For this reason the roots other than the principal one are variously called *extraneous, parasitic, or spurious* roots. Observe that the characteristic polynomial $\pi(r; \bar{h})$ approaches the polynomial $\rho(r)$ (see Definition 10.7.8) as $\bar{h}$ approaches zero. Thus the n roots of $\pi(r; \bar{h})$ approach those of $\rho(r)$. Providing that the method is consistent, the first condition in (10.7.14) implies that $r = 1$ is a root of $\rho(r)$. If the method is Dalquist stable, then $r = 1$ is a simple root, and thus only one root of $\pi(r; \bar{h})$ approaches it as $\bar{h}$ approaches zero. This root turns out to be the principal root, and its coefficient approaches x_0 while the coefficients of the spurious roots approach zero as $\bar{h} \to 0$.

Stability of the Adams Methods

Example 10.7.12. Let us examine the absolute stability of the second-order Adams-Bashforth and Adams-Moulton multistep formulas

$$\xi_{i+1} = \xi_i + h[\tfrac{3}{2} f(t_i, \xi_i) - \tfrac{1}{2} f(t_{i-1}, \xi_{i-1})]$$

$$\xi_{i+1} = \xi_i + \frac{h}{2} [f(t_i, \xi_i) + f(t_{i+1}, \xi_{i+1})]$$

(10.7.24)

These formulas were derived in Sec. 10.4.

For the Adams-Bashforth formula the characteristic polynomial $\pi(r; \bar{h}) = r^2 - (\tfrac{3}{2}\bar{h} + 1)r + \tfrac{1}{2}\bar{h}$ has roots

$$r_1 = \frac{(1 + \tfrac{3}{2}\bar{h}) + \sqrt{(1 + \tfrac{3}{2}\bar{h})^2 - 2\bar{h}}}{2}$$

$$r_2 = \frac{(1 + \tfrac{3}{2}\bar{h}) - \sqrt{(1 + \tfrac{3}{2}\bar{h})^2 - 2\bar{h}}}{2}$$

(10.7.25)

Figure 10.7.13 shows graphs of the two roots as functions of $\bar{h}$. The extraneous root r_2 eventually becomes less than -1. By solving the equation $r_2 = -1$, it is found that this happens at $\bar{h} = -1$. Thus the interval of absolute stability of the second-order Adams-Bashforth method is $(-1, 0)$.

For the order-two Adams-Moulton formula, also called the *implicit trapezoid rule*, the characteristic polynomial

$$\pi(r; \bar{h}) = \left(1 - \frac{\bar{h}}{2}\right)r^2 - \left(1 + \frac{\bar{h}}{2}\right)r$$

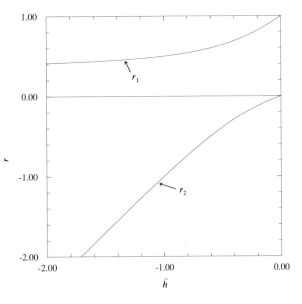

FIGURE 10.7.13

has roots

$$r_1 = \frac{2 + \bar{h}}{2 - \bar{h}} \qquad r_2 = 0 \tag{10.7.26}$$

Both these roots satisfy $|r| < 1$ for all $\bar{h} < 0$, and thus the interval of absolute stability of the implicit trapezoid method is $(-\infty, 0)$. When this is the case, a method is said to be *absolutely stable*.

While the implicit trapezoid method has the desirable property of absolute stability, its direct use is somewhat inefficient since Newton's method must be used at each step to solve for ξ_{i+1} (see Algorithm 10.4.7). As was noted in Sec. 10.4 an efficient use of the explicit and implicit Adams methods is to pair them in *predictor-corrector* schemes. For the second-order Adams methods the predictor-corrector scheme is

$$\xi_{i+1}^{\text{pred}} = \xi_i + \frac{h}{2} [3f(t_i, \xi_i) - f(t_{i-1}, \xi_{i-1})]$$

$$\xi_{i+1} = \xi_i + \frac{h}{2} [f(t_i, \xi_i) + f(t_{i+1}, \xi_{i+1}^{\text{pred}})] \tag{10.7.27}$$

The characteristic polynomial for the predictor-corrector scheme is

$$\pi(r; \bar{h}) = r^2 - (1 + \bar{h} + \tfrac{3}{4}\bar{h}^2)r + \tfrac{1}{4}\bar{h}^2 \tag{10.7.28}$$

(Exercise 8), and hence the roots are

$$r_1 = \frac{1 + \bar{h} + \tfrac{3}{4}\bar{h}^2 + \sqrt{(1 + \bar{h} + \tfrac{3}{4}\bar{h}^2)^2 - \bar{h}^2}}{2}$$

$$r_2 = \frac{1 + \bar{h} + \tfrac{3}{4}\bar{h}^2 - \sqrt{(1 + \bar{h} + \tfrac{3}{4}\bar{h}^2)^2 - \bar{h}^2}}{2} \tag{10.7.29}$$

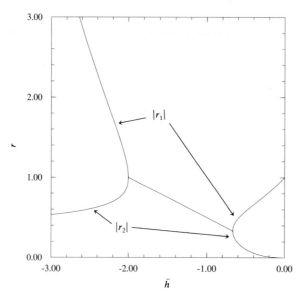

FIGURE 10.7.14

The roots are real in the interval $(-\infty, -2] \cup [-\frac{2}{3}, 0)$ and complex in the interval $(-2, -\frac{2}{3})$. Figure 10.7.14 shows the graphs of $|r_1|$ and $|r_2|$ for $\bar{h} \in [-3, 0]$. The modulus of the root r_1 exceeds 1 for $\bar{h} < -2$, and thus the interval of absolute stability is $(-2, 0)$. Hence unfortunately the second-order predictor-corrector method does not inherit the absolute stability of the corrector.

Table 10.7.15 shows the intervals of stability for the Adams methods of orders two through five, as well as the stability interval of the predictor-corrector method which pairs the explicit and implicit method of the same order. Observe that except for the order-two Adams-Moulton rule (the implicit trapezoid rule), none of the implicit Adams-Moulton methods are absolutely stable on the entire negative $\bar{h}$ axis. Note also that in contrast to the stability results shown in Table 10.7.7 for Runge-Kutta methods, the intervals of stability for Adams predictor-corrector methods narrow as the order of the method increases. Were this not the case, it would otherwise be most efficient to use the highest-order method available since Adams predictor-corrector methods require two evaluations of the right-hand side of the differential equation irrespective of the order.

TABLE 10.7.15
Intervals of absolute stability for the Adams methods

Steps	Bashforth	Moulton	Predictor-corrector
2	$(-1, 0)$	$(-\infty, 0)$	$(-2, 0)$
3	$(-0.55, 0)$	$(-6.0, 0)$	$(-1.8, 0)$
4	$(-0.3, 0)$	$(-3.0, 0)$	$(-1.3, 0)$
5	$(-0.2, 0)$	$(-1.8, 0)$	$(-0.95, 0)$

Stability of Systems of Differential Equations

As with the case of a single equation, let us commence the study of the stability of systems by considering the simple system

$$\mathbf{x}' = A\mathbf{x} \qquad \mathbf{x} = \mathbf{x}_0 \tag{10.7.30}$$

where A is an $n \times n$ constant matrix. Again we will be primarily interested in the behavior of the numerical solution under conditions in which the actual solution approaches the equilibrium point $\mathbf{x} = \mathbf{0}$.

Recall that the *eigenvalues* of the matrix A are the numbers λ for which the equation

$$A\mathbf{v} = \lambda\mathbf{v} \tag{10.7.31}$$

has solutions other than the trivial one $\mathbf{v} = \mathbf{0}$. Note that for (10.7.31) to have nontrivial solutions the eigenvalues must satisfy the polynomial equation $\det(A - \lambda I) = 0$, and hence there are at most n distinct eigenvalues for an $n \times n$ matrix. The *multiplicity* of an eigenvalue is its multiplicity as a root of the polynomial equation. The nontrivial solutions corresponding to a given eigenvalue λ are the *eigenvectors*. If the eigenvalue is of multiplicity one, then the eigenvector is unique up to multiplication by a constant. If all the eigenvalues are distinct, then the general solution to the differential equation (10.7.30) is

$$\mathbf{x}(t) = c_1\mathbf{v}_1 e^{\lambda_1 t} + \cdots + c_n\mathbf{v}_n e^{\lambda_n t} \tag{10.7.32}$$

where $\mathbf{v}_i$ is an eigenvector corresponding to the eigenvalue λ_i. The structure of the solution is more complicated when A has eigenvalues of higher multiplicity. For the sake of simplicity we will consider only the case in which all eigenvalues are distinct. The conclusions drawn will be valid when eigenvalues of higher multiplicity are present, however.

The solution to (10.7.30) approaches the equilibrium point $\mathbf{x} = \mathbf{0}$ when the real parts of all eigenvalues are negative. If we apply Euler's method to (10.7.30), we obtain the system of difference equations

$$\boldsymbol{\xi}_{i+1} = \boldsymbol{\xi}_i + hA\boldsymbol{\xi}_i \tag{10.7.33}$$

Let us look for solutions to this system of the form $\boldsymbol{\xi}_i = \mathbf{w}r^i$. Substituting this guess into (10.7.33) gives

$$\mathbf{w}r^{i+1} = \mathbf{w}r^i + hA\mathbf{w}r^i \tag{10.7.34}$$

which upon division by r^i and rearrangement gives

$$\left(\frac{r-1}{h}\right)\mathbf{w} = A\mathbf{w} \tag{10.7.35}$$

Thus the system of difference equations has a solution of the form $\mathbf{w}r^i$ providing $r = 1 + \lambda_i h$, where λ_i is an eigenvalue of A and $\mathbf{w} = \mathbf{v}_i$, the corresponding eigenvector. The general solution to the system (10.7.33) is then

$$\boldsymbol{\xi}_i = c_1\mathbf{v}_1(1 + \lambda_1 h)^i + \cdots + c_n\mathbf{v}_n(1 + \lambda_n h)^i \tag{10.7.36}$$

Clearly this solution will approach $\mathbf{0}$ providing $\max_{1 \leq i \leq n}|1 + h\lambda_i| < 1$.

When all eigenvalues are real, this is essentially the same result that was obtained in the case of a single equation, and hence we conclude that Euler's method is absolutely stable provided $h\lambda_i \in (-2, 0)$, $i = 1, \ldots, n$. For a complex eigenvalue, $\lambda = a + ib$, $|1 + h\lambda| = \sqrt{(1 + ha)^2 + (hb)^2}$. Thus $|1 + h\lambda| = 1$ is a circle in the complex plane, and Euler's method for systems is absolutely stable if $h\lambda$ lies within this circle for every eigenvalue of the system (10.7.30).

A similar analysis can be performed for Runge-Kutta methods of higher order to conclude that the solution to the difference equation arising from the application of the method to the system (10.7.30) is

$$\xi_i = c_1 \mathbf{v}_1 P(\bar{h}_1)^i + \cdots + c_n \mathbf{v}_n P(\bar{h}_n)^i \tag{10.7.37}$$

where $\bar{h}_j := h\lambda_j$ and $P(\bar{h})$ is as given in Table 10.7.7. For real eigenvalues the condition for absolute stability is that all $h\lambda_i$ lie in the intervals given in Table 10.7.7. For complex eigenvalues, $h\lambda$ must lie in the interior of a region in the complex plane. These regions are shown in Fig. 10.7.16 for Runge-Kutta methods of orders one through four.

In applying a multistep method to the system (10.7.30) the region of absolute stability is again a region in the complex plane. It can be shown that the implicit trapezoid rule remains absolutely stable; that is, it is stable for all $h\lambda$ when the real part of λ is negative. Unfortunately it can be shown [Dalquist (1963)] that no multistep method of global truncation order greater than two has this property. Indeed the implicit trapezoid rule is the optimal absolutely stable method in the sense that it has the smallest local truncation error among such methods.

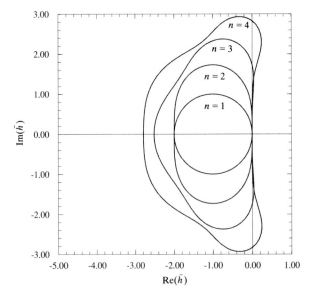

FIGURE 10.7.16

Stiff Systems of Differential Equations

To see the implications of the above analysis let us consider the problem of approximating the solution to the initial value problem

$$x'' + (b + 1)x' + bx = 0 \qquad x(0) = 1 \qquad x'(0) = 0 \qquad (10.7.38)$$

where $b > 0$. The solution to this problem is easily seen to be $x(t) = [b/(b - 1)]e^{-t} - [1/(b - 1)]e^{-bt}$. Problem (10.7.38) is equivalent to the initial value problem

$$
\begin{aligned}
x_1' &= x_2 & x_1(0) &= 1 \\
x_2' &= -bx_1 - (b + 1)x_2 & x_2(0) &= 0
\end{aligned}
\qquad (10.7.39)
$$

for first-order systems. The solution to (10.7.39) is

$$
\begin{aligned}
x_1(t) &= \frac{b}{b - 1} e^{-t} - \frac{1}{b - 1} e^{-bt} \\
x_2(t) &= -\frac{b}{b - 1} e^{-t} + \frac{b}{b - 1} e^{-bt}
\end{aligned}
\qquad (10.7.40)
$$

Now suppose that b is much larger than 1. Then from Table 10.7.7 a fourth-order Runge-Kutta method is stable for stepsizes for which $h < 2.78/b$. If $b = 1000$, for instance, then a stepsize of $h < 0.00278$ is required. This is far smaller than is required to keep the local truncation error of the method within all but the most stringent tolerances. Indeed, an implementation of the classical Runge-Kutta method in Turbo Pascal with a stepsize of $h = 0.0025$ produces approximations to the solution (10.7.40) that are about the size of the machine unit $\mu = 2 \times 10^{-12}$. However, if the stepsize is increased to $h = 0.003$, then the method is unstable and floating-point overflow quickly results.

A system of differential equations such as (10.7.38) with large b, in which the stepsize is constrained by the interval of absolute stability rather than the need to keep the truncation error small, is said to be *stiff*. Methods of numerical approximation such as the Runge-Kutta-Felberg and Gragg methods that efficiently control the local truncation error tend to do poorly on stiff problems. Table 10.7.17 shows the results of applying the Runge-Kutta-Felberg 4(5) method (10.5.12) and the variable-step Gragg method of Sec. 10.6 to problem (10.7.39) for different values of b. While for $b = 10$ the time required is in keeping with the results that we have obtained in applying these two methods to other nonstiff problems (Table 10.6.11), both methods become increasingly inefficient as the size of b is increased. This is somewhat ironic since for $b = 100$, 1000, or 10,000 the value of the actual solution at $t = 1$ is for all practical purposes $x_1(1) = [b/(b - 1)]e^{-1}$, $x_2(1) = -[b/(b - 1)]e^{-1}$, and thus as the value of b gets larger, it has a greater effect on the efficiency of numerical solution but a smaller effect on the actual solution.

To obtain the Runge-Kutta-Felberg approximation for $b = 10,000$ the parameter h_{min} in Algorithm 10.5.4 had to be reduced to $(t_f - t_0)/10^6$. As indicated in Table 10.7.17, Gragg's method failed for $b = 1000$ and $b = 10,000$.

TABLE 10.7.17
Times to approximate $x_1(1)$ and $x_2(1)$ for problem (10.7.39) to within an accuracy of 5×10^{-5}

	RKF 4(5)		Variable step Gragg	
b	RHS evals.	Time, s	RHS evals.	Time, s
10	90	0.06	294	0.11
100	246	0.16	1459	0.56
1,000	1812	1.16	Overflow	Overflow
10,000	17286	11.04	Overflow	Overflow

The difficulty was that the base method, the multistep midpoint rule, is so unstable that overflow resulted for the initial stepsize of $t_f - t_0$. This problem could have been avoided by specifying a smaller initial stepsize, but there is little reason to do so. Gragg's method simply is not suitable for stiff differential equations.

Obviously the notion of stiffness is very imprecise, depending as it does upon the interplay of the truncation error and the interval of absolute stability of a numerical method. About all that can be said generally is that it arises when the time intervals over which different "components" of a solution undergo changes of comparable magnitude are very different. For instance, in the solution (10.7.40) one component decays as e^{-t} while the other decays as e^{-bt}, which will be much faster if b is large. The solution to the differential equation $x' + bx = 10 \cos t$ has the form $x(t) = ce^{-bt} + A \cos t + B \sin t$. It too can be considered stiff for large b since the transient part of the solution ce^{-bt} changes over a much shorter time interval than does the steady state part $A \cos t + B \sin t$. A nonhomogeneous linear system $\mathbf{x}'(t) = A\mathbf{x} + \mathbf{f}(t)$ could be stiff either because the eigenvalues of A differ considerably from each other or because the transient solution decays on a time scale very different from that appropriate for measuring changes in the steady state solution.

Though the above discussion of stiffness has concerned linear differential equations, it is also applicable to nonlinear equations when changes in the solution occur on very different time scales on different parts of the domain of interest. For further discussion of stiff differential equations, see Shampine and Gear (1979).

Approximation of Solutions to Stiff Differential Equations

As indicated above, the methods that do well in approximating the solution to nonstiff differential equations may be quite ineffective when applied to stiff equations. A good approach to solving stiff equations is to use a method that is absolutely stable and thus whose interval of stability cannot become a constraint. As indicated before, the implicit trapezoid rule is such a method. Applying the implicit trapezoid rule to the initial value problem

$$\mathbf{x}' = \mathbf{f}(t, \mathbf{x}) \qquad \mathbf{x}(t_0) = \mathbf{x}_0 \qquad (10.7.41)$$

gives

$$\mathbf{G}(t_i, \boldsymbol{\xi}_i, \boldsymbol{\xi}_{i+1}; h) := \boldsymbol{\xi}_{i+1} - \boldsymbol{\xi}_i - \frac{h}{2} [\mathbf{f}(t_i, \boldsymbol{\xi}_i) + \mathbf{f}(t_i + h, \boldsymbol{\xi}_{i+1})] = \mathbf{0}$$

$$(10.7.42)$$

For known t_i and ξ_i this equation must be solved for the approximation ξ_{i+1} to $\mathbf{x}(t_{i+1})$.

It turns out that for stiff systems simple iteration will not always converge, and thus Newton's method for systems must be used. Recall from Sec. 2.6 that the Newton iteration for solving the system $\mathbf{F}(\mathbf{x}) = 0$ is $\mathbf{x}_{i+1} = \mathbf{x}_i - \{J[\mathbf{F}](\mathbf{x}_i)\}^{-1} F(\mathbf{x}_i)$, where $J[\mathbf{F}](\mathbf{x})$ is the jacobian matrix of $\mathbf{F}$,

$$J[\mathbf{F}](\mathbf{x}) := \begin{bmatrix} \dfrac{\partial F_1}{\partial x_1}(\mathbf{x}) \cdots \dfrac{\partial F_1}{\partial x_n}(\mathbf{x}) \\ \cdots\cdots\cdots\cdots\cdots\cdots \\ \dfrac{\partial F_n}{\partial x_1}(\mathbf{x}) \cdots \dfrac{\partial F_n}{\partial x_n}(\mathbf{x}) \end{bmatrix} \qquad (10.7.43)$$

The Jacobian matrix of the function $\mathbf{G}(t_i, \boldsymbol{\xi}_i, \boldsymbol{\xi}_{i+1}; h)$ is easily seen to be $J[\mathbf{G}](\xi_{i+1}) = I - (h/2)J[\mathbf{f}(t_i + h, \cdot)](\xi_{i+1})$. Algorithm 10.7.18 outlines a procedure for approximating $\mathbf{x}(t_f)$ using N steps of the implicit trapezoid rule.

Algorithm 10.7.18   Fixed-step implicit trapezoid method

Input $t_0, \mathbf{x}_0$ (initial conditions)
t_f (point at which to find approximation)
N (number of subintervals)
tol (tolerance for Newton's method)

$t := t_0, \ \xi_0 := x_0$
$h := (t_f - t_0)/N$
for $k = 1$ to N
   $\{\boldsymbol{\xi}_1 := \boldsymbol{\xi}_0 + h\mathbf{f}(t, \boldsymbol{\xi}_0)$ (use Euler's method to find initial guess for Newton's method)
   repeat (Newton's method)
      Solve the linear system $\{I - \dfrac{h}{2} J[\mathbf{f}(t_{i+1}, \cdot)] (\boldsymbol{\xi}_1)\}\Delta\boldsymbol{\xi} = -\mathbf{G}(t, \boldsymbol{\xi}_0, \boldsymbol{\xi}_1; h)$
      (see Sec. 2.6)
      $\boldsymbol{\xi}_1 := \boldsymbol{\xi}_1 + \Delta\boldsymbol{\xi}$
   until $\|\Delta\boldsymbol{\xi}\| < \text{tol}$
   $t := t_0 + kh$
   $\boldsymbol{\xi}_0 := \boldsymbol{\xi}_1\}$

Output ξ_1, approximation to $\mathbf{x}(t_f)$

As we have seen, variable-step methods are preferable to fixed-step methods. As might be expected from the fact that it is derived from the trapezoid rule for integration (Exercise 8a in Sec. 10.4), the truncation error for the implicit trapezoid method has an expansion in even powers of the stepsize h, and consequently Richardson extrapolation can be applied efficiently. Implementing this is Project 3.

TABLE 10.7.19
Times of execution for the implicit trapezoid rule
for (10.7.40) with a tolerance of 5×10^{-5}

b	RHS evals.	Time, s
10	144	0.15
100	480	0.52
1000	724	0.79
10,000	957	1.04

Table 10.7.19 shows the results of applying an extrapolation method based upon the implicit trapezoid rule to problem (10.7.39). As can be seen the implicit trapezoid rule is less efficient than the Runge-Kutta-Felberg 4(5) method (10.5.12) (see Table 10.7.17) for $b = 100$, but more efficient when $b = 1000$, and substantially more efficient for $b = 10,000$.

The Method of Choice

The first consideration in choosing a method for numerically solving a differential equation is whether or not a specific differential equation, or at least an equation from a very specific family of differential equations, is being solved. For instance, a program designed to illustrate planetary motions in a solar system need only be concerned with the most efficient means of solving the equations for gravitational attraction. In such a situation it may be ascertainable whether the equations are stiff or not. If they are not, then among the methods that we have considered, Gragg's method is likely to be the most efficient. If stiffness is a possibility, at least for some choices of parameters, then a Richardson extrapolation scheme based upon the implicit trapezoid rule is suggested.

A program designed to solve an arbitrary differential equation is more problematic. Here stiff equations are possible, but not necessarily likely. Gragg's method should probably be eliminated because its performance is so poor on stiff equations. Table 10.7.20 compares the results of applying the implicit trapezoid rule to the five test problems of Table 10.4.11 with those for

TABLE 10.7.20

Prob.	Target error	RKF 4(5)			Implicit trapezoid		
		RHS evals.	Error	Time, s	RHS evals.	Error	Time, s
1	5E − 09	324	4.9E − 09	0.15	658	1.8E − 11	0.35
2	5E − 07	276	1.2E − 04	0.12	941	2.0E − 08	0.50
3	5E − 09	1878	2.4E − 10	1.43	1536	5.0E − 12	1.17
4	5E − 09	420	4.3E − 10	0.79	756	4.5E − 13	2.23
5	5E − 06	1242	6.9E − 06	1.91	2376	1.3E − 08	8.20

the Runge-Kutta-Felberg 4(5) method (10.5.12). On problem 2 the implicit trapezoid takes longer, but it also achieves the specified tolerance whereas the Runge-Kutta-Felberg method does not. It also is faster on problem 3, which is somewhat stiff, but is 3 or 4 times slower for the remaining three problems. There is no clear way to choose between these two methods, since which one is more effective overall depends upon the mixture of stiff and nonstiff problems that the program is called upon to solve. It should be pointed out, however, that if the program permits the solution of systems of large dimension, the implicit trapezoid method may be preferable, since the larger the dimension of the system, the greater the likelihood that there will be a wide dispersion in the time scales over which various components of the solution change.

It is of course not necessary that a program confine itself to a single method of approximation. Suppose that an approximation ξ_i to $\mathbf{x}(t_i)$ has already been determined. Note that the system (10.7.41) can be written as

$$\mathbf{x}' = J[\mathbf{f}(t, \cdot)](\xi_i)(\mathbf{x} - \xi_i) + O(\|\mathbf{x} - \xi_i\|^2) \tag{10.7.44}$$

This means that the eigenvalues of the jacobian matrix $J[\mathbf{f}(t_i, \cdot)]\xi_i$ can be used to determine whether, at least locally, the solution is contending with a wide range of time scales. We will see in the next chapter that the largest and smallest eigenvalues of a matrix can be found fairly efficiently, and thus it is possible for the program itself to make an assessment of the stiffness of the differential equation. One widely used program that makes use of several different methods is *Gear's method*. This program is based upon the *backward differentiation formulas*, a set of implicit multistep methods having the form

$$\xi_{i+1} = \sum_{j=1}^{n} \alpha_j \xi_{i-j+1} + h\beta_0 \mathbf{f}(t_{i+1}, \xi_{i+1}) \tag{10.7.45}$$

See Exercise 7 in Sec. 10.4 for an example of a backward differentiation formula. The program is variable-order, as well as variable-step. It can switch to a higher-order formula if controlling the local truncation error is the dominant constraint upon the stepsize or to a lower-order formula if the region of stability is the constraining factor at the current order.

EXERCISES

1. Show that the midpoint rule (10.3.5) and Heun's rule both have $(-2, 0)$ as their intervals of absolute stability.

2. (a) Show that all second-order Runge-Kutta formulas lead to the polynomial $P(\bar{h})$ given in Table 10.7.7.

(b) Show that all third-order Runge-Kutta formulas lead to the polynomial $P(\bar{h})$ given in Table 10.7.7.

3. (a) Show that if a second-order homogeneous difference equation has a characteristic root r of multiplicity two, then $\xi_i^{(1)} = r^i$ and $\xi_i^{(2)} = ir^i$ are solutions to the difference equation.

(b) Show that if an nth-order homogeneous difference equation has a characteristic root r of multiplicity n, then $i^k r^i$ is a solution to the difference equation for $k = 0, \ldots, n - 1$. [Hint: In this case the difference equation can be written as $a_0(D_1 - r)^n \xi_i = 0$, where D_j is the shift operator $D_j \xi_i := \xi_{i+j}$.]

4. (a) Show that if ξ_i^h is a solution to (10.7.8) and ξ_i^p is any particular solution to (10.7.7), then $\xi_i^h + \xi_i^p$ is also a solution to (10.7.7).

(b) Let $\xi_i^{(1)}, \ldots, \xi_i^{(n)}$ be linearly independent solutions to the homogeneous difference equation (10.7.8). Show that $\Sigma_{j=1}^n c_j \xi_i^{(j)}$ is the general solution to (10.7.8); that is, for any solution $\bar{\xi}_i$ there is a choice of the constants c_j for which $\bar{\xi}_i = \Sigma_{j=1}^n c_j \xi_i^{(j)}$.

(c) Show that the solution set $\{r_1^i, \ldots, r_n^i\}$ is linearly independent when $r_j \neq r_k$ for $j \neq k$.

(d) Show that the solution set $\{r^i, ir^i, \ldots, i^k r^i\}$ is linearly independent.

5. Note that Euler's method with stepsize h when applied to the initial value problem $x' - \lambda x = e^{-t}$, $x(0) = x_0$ leads to the nonhomogeneous difference equation $\xi_{i+1} = (1 + \lambda h)\xi_i + he^{-ih}$. Solve this difference equation, and use the solution to analyze the stability of Euler's method for this problem.

6. Show that the
(a) Adams-Bashforth four-point formula
(b) Adams-Moulton four-point formula
is consistent and convergent.

7. (a) Show that the backward differentiation formula (10.4.28) is consistent and convergent.

(b) Show that both the explicit and implicit formulas in the Milne predictor-corrector scheme of Exercise 30 in Sec. 10.4 are consistent and convergent.

8. Verify the formula for the characteristic polynomial of the second-order Adams predictor-corrector method given by (10.7.28).

9. Verify the two assertions made about the characteristic roots of the multistep midpoint formula made in (10.7.22).

10. Find the interval of stability of the backward differentiation formula (10.4.29).

11. Find the interval of absolute stability for the implicit rule used in the Milne predictor-corrector scheme given by (10.4.30).

12. (a) Show that the explicit method

$$\xi_{i+1} = \tfrac{1}{2}\xi_i + \tfrac{1}{2}\xi_{i-1} + \tfrac{1}{2}hf(t_i, \xi_i) + hf(t_{i-1}, \xi_{i-1})$$
(10.7.46)

is consistent and convergent.

(b) Find the interval of absolute convergence of the method in part (a).

13. Show that when the eigenvalues of the matrix A in (10.7.30) are distinct, (10.7.32) is a solution.

14. On the basis of the fact that the interval of stability for the problem (10.7.3) is $(-2, 0)$, make predictions about the stability of the nonlinear problems

$$x' = -x - x^2 \qquad x(0) = 1 \qquad (10.7.47)$$

and

$$x' = -x^2 \qquad x(0) = 1 \qquad (10.7.48)$$

Confirm your conjectures with numerical experiments.

15. Euler's method when applied to the initial value problem (10.7.30) leads to the difference equation (10.7.33) which has solution (10.7.36).

(a) Find $c_1, c_2, v_1,$ and v_2 when the initial value problem is

$$\begin{bmatrix} x_1' \\ x_2' \end{bmatrix} = \begin{bmatrix} 0 & 1 \\ -b & -(b+1) \end{bmatrix} \begin{bmatrix} x_1 \\ x_2 \end{bmatrix}$$

$$\begin{bmatrix} x_1(0) \\ x_2(0) \end{bmatrix} = \begin{bmatrix} 1 \\ -1 \end{bmatrix} \qquad (10.7.49)$$

[see (10.7.39)] and thus conclude that the efficiency of Euler's method for this particular choice of initial conditions should not depend upon the value of b.

(b) Because of rounding error, Euler's method is actually the iteration

$$\xi_{i+1} = (I + hA)\xi_i + \mu_i \qquad (10.7.50)$$

where for simplicity let us take $\mu_i = \mu$, a constant. The components of μ will be small, on the order of the machine unit. Solve (10.7.50). Illustrate what you observe about the solution by making graphs of the Euler approximations to (10.7.50).

16. Show that the implicit trapezoid rule is absolutely stable for complex $\bar{h}$ with Re $\bar{h} < 0$.

17. Observe that in Fig. 10.7.16 Euler's method does not contain any part of the imaginary axis, other than the origin, in or on the boundary of its region of absolute stability. Discuss the implications of this. Illustrate by applying Euler's method for an initial value problem for the differential equation $x'' + x = 0$. Note that fourth-order Runge-Kutta methods contain a portion of the imaginary axis within their region of absolute stability. Again discuss the implications of this and illustrate with an example.

PROJECTS

1. Add the following features to the Runge-Kutta-Felberg Algorithm 10.5.4.

   (a) Have the program take as an input a number N of evenly spaced points at which to output values of the solution. Have the program store the values of the solution at points intermediate to t_0 and t_f as they are computed, and then interpolate these points to produce the values of the solution at the desired points. Test the program by having it compute values at 100 evenly spaced points for each of the test problems in Table 10.4.11. Compare the time of execution for this program with that of the basic Runge-Kutta Felberg-algorithm.

   (b) It is often desirable to know at what values of the independent variable t the solution to a differential equation assumes certain values of the dependent variable x. Produce a variant of the program in part (a) that will output all values of t in the domain (t_0, t_f) at which the solution assumed a particular value of x.

2. Make the enhancements described in Project 1 for Gragg's method.

3. (a) Develop a variable-step method similar to the Gragg Algorithm 10.6.10 that uses the implicit trapezoid rule as the base method for Richardson extrapolation. Use the program to reproduce the results of Tables 10.7.19 and 10.7.20 for your system. Try to account for any differences that you observe. The program should allow for the possibility that Newton's method will converge slowly or not at all by restricting the number of iterations that Newton's method is permitted. If the specified tolerance has not been achieved after the maximum number of iterations, then the stepsize should be reduced. Experiment with different settings for the maximum number of iterations.

   (b) Make the enhancements described in Project 1 for the extrapolated implicit trapezoid rule program.

4. Develop a variable-step method that uses the fourth-order Adams predictor-corrector formulas. See the end of Sec. 10.5 for a brief description of a way to do this. Compare the execution times with those given in Tables 10.6.11 and 10.7.17 and the stepsizes with those of Table 10.6.12. Replace the fourth-order predictor-corrector pair with the sixth-order predictor-corrector pair

$$\xi_{i+1} = \xi_i + \frac{h}{1440}\,(4277f_i - 7923f_{i-1} + 9982f_{i-2}$$
$$- 7298f_{i-3} + 2877f_{i-4} - 475f_{i-5})$$

$$\xi_{i+1} = \xi_i + \frac{h}{1440}\,(475f_{i+1} + 1427f_i - 798f_{i-1}$$
$$+ 482f_{i-2} - 173f_{i-3} + 27f_{i-4})$$

and assess the results.

5. Use contour plotting to produce the regions of stability for Runge-Kutta methods of orders one through four shown in Fig. 10.7.16. If your computer language does not support complex arithmetic, then use Euler's identity $\bar{h}^n = r^n e^{in\theta} = r^n \cos n\theta + i \sin n\theta$ to decompose the problem into real and imaginary parts. Also produce graphs of the regions of stability for the Adams-Bashforth, Moulton, and predictor-corrector methods of orders two through four.

6. Write a graphics program that will show the motions of the planets about the sun. Calculate the positions of the planets by solving numerically the system of differential equations arising from the law of gravitational attraction (see Exercise 13 in Sec. 10.1). How many planets can be handled simultaneously will depend upon the speed of your computer system. More accurate results will be obtained if the *Gaussian gravitational constant* rather than the universal gravitational constant is used, since the former constant is known to several more digits. See Danby (1962).

7. In some species of organisms it is found that there is a time lag T in the reaction of the species to constraints such as overcrowding and food shortages. The population growth of such species can be modeled by the following variant of the logistic equation

$$P'(t) = (a - bP(t - T))P(t) \qquad (10.7.51)$$

Such equations are referred to as *differential-delay* equations. What are appropriate initial conditions for a differential-delay equation? Modify Euler's method so that it will calculate approximations to solutions of (10.7.51). Use your modified method to draw graphs of solutions for $a = 1$, $b = 0.01$, $T = 3$, and various initial conditions. How do solutions appear to differ from those illustrated in Fig. 10.1.3 for the logistic model? Investigate the applicability of more efficient methods of numerical solution.

8. Select a problem of interest to you that requires numerical solution of differential equations and use one of the methods in this chapter to solve it. Justify your choice of method. An interesting collection of problems requiring numerical solution of differential equations can be found in Danby (1985).

CHAPTER 11

THE EIGENVALUE PROBLEM

11.1 POWER AND INVERSE POWER METHODS

Definition of Eigenvalue and Examples

The *eigenvalues* of an $n \times n$ matrix A are values of λ for which the equation

$$A\mathbf{v} = \lambda\mathbf{v} \tag{11.1}$$

has nontrivial solutions, that is, solutions other than $\mathbf{v} = \mathbf{0}$. Eigenvalues may be either real or complex numbers. The solutions $\mathbf{v} \neq \mathbf{0}$ corresponding to an eigenvalue are the *eigenvectors* for λ.

Example 11.1.1. An important source of eigenvalue problems is the solution of linear homogeneous first-order systems of differential equations

$$\mathbf{x}' = A\mathbf{x} \qquad \mathbf{x}(0) = \mathbf{x}_0 \tag{11.1.2}$$

where A is an $n \times n$ matrix. Let us look for solutions to (11.1.2) of the form $\mathbf{x}(t) = e^{\lambda t}\mathbf{v}$. Substituting into the differential equation gives $\lambda e^{\lambda t}\mathbf{v} = Ave^{\lambda t}$, and thus $\mathbf{x}(t) = e^{\lambda t}\mathbf{v}$ is a solution if λ is an eigenvalue of the matrix A and $\mathbf{v}$ is an eigenvector corresponding to λ. If the n eigenvalues λ_k, $k = 1, \ldots, n$, are distinct, then it can be shown that the corresponding eigenvectors $\mathbf{v}_k$ are linearly independent (Exercise 2). From this it easily follows that the general solution to the differential equation (11.1.2) is

$$\mathbf{x}(t) = c_1\mathbf{v}_1 e^{\lambda_1 t} + \cdots + c_n\mathbf{v}_n e^{\lambda_n t} \tag{11.1.3}$$

By virtue of the linear independence of the eigenvectors $\mathbf{v}_k$, it follows that the initial condition $\mathbf{x}(0) = \mathbf{x}_0$ uniquely determines the constants c_k.

Clearly, if $\mathbf{v}$ is an eigenvector, then so also is $c\mathbf{v}$ for any scalar $c \neq 0$. Likewise if $\mathbf{v}^{(1)}$ and $\mathbf{v}^{(2)}$ are eigenvectors corresponding to the same eigenvalue, then $\mathbf{v}^{(1)} + \mathbf{v}^{(2)}$ is also an eigenvector for that eigenvalue. Since (11.1.1) is equivalent to $(A - \lambda I)\mathbf{v} = \mathbf{0}$, the matrix $A - \lambda I$ must be singular when λ is an eigenvalue. Hence $\det(A - \lambda I) = 0$. The expression $P(\lambda) := \det(A - \lambda I)$, called the *characteristic polynomial* of A, is easily seen to be an nth-degree polynomial in λ, and thus there are at most n distinct eigenvalues of A. There may be fewer if some of the roots of $P(\lambda)$ have multiplicity greater than one. The *multiplicity* of the eigenvalue λ is the multiplicity of λ as a root of $P(\lambda)$. In this chapter we develop ways of finding the eigenvalues and eigenvectors of a matrix numerically.

Let us consider three examples that illustrate different situations which may arise in eigenvalue problems.

Example 11.1.2. To find the eigenvalues of

$$A = \begin{bmatrix} -21 & -9 & 12 \\ 0 & 6 & 0 \\ -24 & -8 & 15 \end{bmatrix} \tag{11.1.4}$$

we note that

$$A - \lambda I = \begin{bmatrix} -21 - \lambda & -9 & 12 \\ 0 & 6 - \lambda & 0 \\ -24 & -8 & 15 - \lambda \end{bmatrix} \tag{11.1.5}$$

Expanding the determinant $\det(A - \lambda I)$ gives

$$\det(A - \lambda I) = (6 - \lambda)[-(21 + \lambda)(15 - \lambda) + 288]$$
$$= (6 - \lambda)[\lambda^2 + 6\lambda - 27]$$
$$= -(\lambda - 6)(\lambda - 3)(\lambda + 9) \tag{11.1.6}$$

and thus the eigenvalues are $\lambda_1 = -9$, $\lambda_2 = 3$, $\lambda_3 = 6$. To find the eigenvectors corresponding to each eigenvalue, that is, all nontrivial vectors satisfying $(A - \lambda I)\mathbf{v} = \mathbf{0}$, we substitute each of the three eigenvalues into (11.1.5) and row-reduce the resulting matrix. This gives for $\lambda = -9$,

$$A + 9I = \begin{bmatrix} -12 & -9 & 12 \\ 0 & 15 & 0 \\ -24 & -8 & 24 \end{bmatrix} \xrightarrow[\text{row-reduce}]{} \begin{bmatrix} 1 & 0 & -1 \\ 0 & 1 & 0 \\ 0 & 0 & 0 \end{bmatrix} \tag{11.1.7}$$

Thus the components of an eigenvalue must satisfy the relations $v_1 - v_3 = 0$, $v_2 = 0$. Setting $v_3 = c_1$, this gives $\mathbf{v}^{(1)} = c_1[1, 0, 1]^T$ as the most general eigenvector corresponding to $\lambda_1 = -9$. A similar procedure can be applied to the other two eigenvalues. The result is that we have the eigenvectors

$$\mathbf{v}^{(1)} = \begin{bmatrix} 1 \\ 0 \\ 1 \end{bmatrix} \qquad \mathbf{v}^{(2)} = \begin{bmatrix} -1 \\ 3 \\ 0 \end{bmatrix} \qquad \mathbf{v}^{(3)} = \begin{bmatrix} 1 \\ 0 \\ 2 \end{bmatrix} \tag{11.1.8}$$

corresponding to the eigenvalues $\lambda_1 = -9$, $\lambda_2 = 3$, and $\lambda_3 = 6$, respectively. Observe that these three vectors form a basis for $\mathbb{R}^3$. As mentioned above, this is always the case when the eigenvalues are all of multiplicity one.

Example 11.1.3. For the matrix

$$A = \begin{bmatrix} 5 & 0 & -1 \\ 0 & 8 & 0 \\ -3 & 0 & 7 \end{bmatrix} \tag{11.1.9}$$

we have $\det(A - \lambda I) = -(\lambda - 4)(\lambda - 8)^2$, and thus we have one eigenvalue $\lambda_1 = 4$ of multiplicity one and another, $\lambda = 8$, of multiplicity two. Corresponding to $\lambda = 4$ we find in the same way as in Example 11.1.2 that an eigenvector is $\mathbf{v}^{(1)} = [1, 0, 1]^T$. For $\lambda = 8$, upon row-reducing the matrix $A - 8I$ we obtain

$$\begin{bmatrix} 3 & 0 & 1 \\ 0 & 0 & 0 \\ 0 & 0 & 0 \end{bmatrix} \tag{11.1.10}$$

Thus the only relation required of the components of an eigenvector is $3v_1 + v_3 = 0$. This means that any vector of the form

$$\mathbf{v} = c_2 \begin{bmatrix} 1 \\ 0 \\ -3 \end{bmatrix} + c_3 \begin{bmatrix} 1 \\ 1 \\ -3 \end{bmatrix} \tag{11.1.11}$$

is an eigenvector corresponding to $\lambda = 8$. Note that in this example, as in Example 11.1.2, the eigenvectors span $\mathbb{R}^3$.

Example 11.1.4. The matrix

$$A = \begin{bmatrix} 0 & 1 & 3 \\ 0 & 6 & 0 \\ -6 & 2 & 9 \end{bmatrix} \tag{11.1.12}$$

has an eigenvalue $\lambda = 2$ of multiplicity one and an eigenvalue $\lambda = 6$ of multiplicity two. The eigenvector corresponding to $\lambda = 3$ is $\mathbf{v}^{(1)} = [1, 0, 1]^T$. Upon row-reducing the matrix $A - 6I$ we obtain

$$\begin{bmatrix} 2 & 0 & -1 \\ 0 & 1 & 0 \\ 0 & 0 & 0 \end{bmatrix} \tag{11.1.13}$$

This gives only the eigenvector $\mathbf{v}^{(2)} = [1, 0, 2]^T$, and thus the eigenvectors of A span only a two-dimensional subspace of $\mathbb{R}^3$.

Matrices such as that of Example 11.1.4 for which the set of all eigenvectors does not span all of $\mathbb{R}^n$ are termed *defective*. As we will see, it is defective matrices whose eigenvalues are the most difficult to compute numerically.

Before discussing methods of approximating eigenvalues, we first mention an approach that should *not* be used, which is application of a root-finding technique such as Newton's method to the characteristic polynomial $P(\lambda) = \det(A - \lambda I)$. Computation of an $n \times n$ determinant requires $n!$ operations, and thus such an approach would be extremely inefficient. Moreover, as indicated by the example of Wilkinson discussed in Sec. 1.3, the roots of polynomials can be very sensitive to perturbations in their coefficients. If the coefficients were obtained by a process involving $n!$ computations, there likely would be such perturbations due to rounding error.

Indeed, one feasible approach to finding the roots of a polynomial is to convert it to an eigenvalue problem and then apply one of the techniques that will be developed in this chapter. The *companion matrix* of a polynomial

$$Q(x) = a_n x^n + a_{n-1} x^{n-1} + \cdots + a_1 x + a_0 \tag{11.1.14}$$

is

$$A = \begin{bmatrix} 0 & 1 & 0 & 0 & \cdots & 0 & 0 \\ 0 & 0 & 1 & 0 & \cdots & 0 & 0 \\ 0 & 0 & 0 & 1 & \cdots & 0 & 0 \\ \hdashline 0 & 0 & 0 & 0 & \cdots & 0 & 1 \\ -\dfrac{a_1}{a_n} & -\dfrac{a_2}{a_n} & -\dfrac{a_3}{a_n} & -\dfrac{a_4}{a_n} & \cdots & -\dfrac{a_{n-2}}{a_n} & -\dfrac{a_{n-1}}{a_n} \end{bmatrix} \tag{11.1.15}$$

and it is easily shown that the characteristic polynomial of A is $\det(A - \lambda I) = (1/a_n)Q(\lambda)$, whence the eigenvalues of A are the roots of Q (Exercise 5).

Power Method

If for a matrix A there is a unique eigenvalue, call it λ_1, of largest magnitude, that is, $|\lambda_1| > |\lambda_k|$ when $\lambda_k \neq \lambda_1$, then λ_1 is referred to as the *dominant* eigenvalue. The *power method* provides a simple means of determining the dominant eigenvalue and an associated eigenvector.

Let us for the moment assume that the matrix A is nondefective, and thus it has a set of eigenvectors $\{\mathbf{v}^{(1)}, \ldots, \mathbf{v}^{(n)}\}$ that span $\mathbb{R}^n$. Then for an arbitrarily chosen vector $\mathbf{x}^{(0)}$ we have $\mathbf{x}^{(0)} = \sum_{k=1}^{n} c_k \mathbf{v}^{(k)}$. Thus

$$\mathbf{x}^{(1)} := A\mathbf{x}^{(0)} = \sum_{k=1}^{n} c_k A\mathbf{v}^{(k)} = \sum_{k=1}^{n} c_k \lambda_k \mathbf{v}^{(k)}$$

$$\mathbf{x}^{(2)} := A^2 \mathbf{x}^{(0)} = A\mathbf{x}^{(1)} = \sum_{k=1}^{n} c_k \lambda_k^2 \mathbf{v}^{(k)} \tag{11.1.16}$$

$$\cdots\cdots\cdots\cdots\cdots\cdots\cdots\cdots\cdots\cdots\cdots$$

$$\mathbf{x}^{(j)} := A^j \mathbf{x}^{(0)} = A\mathbf{x}^{(n-1)} = \sum_{k=1}^{n} c_k \lambda_k^j \mathbf{v}^{(k)}$$

$$\cdots\cdots\cdots\cdots\cdots\cdots\cdots\cdots\cdots\cdots\cdots$$

Assume now that the multiplicity of the dominant eigenvalue is p, whence $\lambda_1 = \lambda_2 = \cdots = \lambda_p$, and that $|\lambda_1| > |\lambda_{p+1}| \geq \cdots \geq |\lambda_n|$. Then

$$\mathbf{x}^{(j)} = \lambda_1^j \left[\sum_{k=1}^{p} c_k \mathbf{v}^{(k)} + \sum_{k=p+1}^{n} c_k \left(\frac{\lambda_k}{\lambda_1} \right)^j \mathbf{v}^{(k)} \right] \tag{11.1.17}$$

and since $|\lambda_1| > |\lambda_k|$, $k = p+1, \ldots, n$, we have for large j, $\mathbf{x}^{(j)} \simeq \lambda_1^j \Sigma_{k=1}^p c_k \mathbf{v}^{(k)}$. This means that the ratio $x_i^{(j+1)}/x_i^{(j)}$ of any two corresponding components of $\mathbf{x}^{(j)}$ and $\mathbf{x}^{(j+1)}$ is an approximation to λ_1 and that $(1/\lambda_1^j)\mathbf{x}^{(j)}$ is an approximation to an eigenvector corresponding to λ_1.

Observe that the approach above fails if the selected initial vector $\mathbf{x}^{(0)}$ is such that $c_1 = \cdots = c_p = 0$. Without knowledge of the eigenvalues and eigenvectors, there is no way to exclude this possibility with certainty; however, as a practical matter it is unlikely that this will occur. Should it happen, it is clear that the power method would then find the eigenvalue with the next largest magnitude, assuming that it is unique and that its coefficient is not also zero.

In deriving the power method we assumed that there was a unique eigenvalue of largest magnitude and that the matrix A was nondefective. The latter assumption was made solely for simplicity. The power method will succeed at finding the dominant eigenvalue of a defective matrix, though it is more difficult to prove this, and the convergence of the method tends to be substantially slower.

The assumption that there is a dominant eigenvalue is essential, however. If there are two distinct eigenvalues of greatest magnitude, λ_1 and λ_2, then from (11.1.17)

$$\mathbf{x}^{(j)} \simeq \lambda_1^j \left[c_1 \mathbf{v}^{(1)} + \left(\frac{\lambda_2}{\lambda_1} \right)^j c_2 \mathbf{v}^{(2)} \right]$$

Consequently, as j becomes large, $\mathbf{x}^{(j)}$ does not approach any fixed vector and the power method will fail. For instance, if a real matrix A has a complex eigenvalue λ, then the complex conjugate $\bar{\lambda}$ is also an eigenvalue (Exercise 6). Thus if the eigenvalue of largest magnitude of a real matrix is complex, then the power method will fail.

From (11.1.17) it is clear that how large j must be for $\mathbf{x}^{(j)} \simeq c_1 \lambda_1^j \mathbf{v}^{(j)}$ depends upon the magnitude of the next largest eigenvalue λ_{p+1}. If the next largest eigenvalue is not too much smaller, then it is possible that the computation of the powers $A^j \mathbf{x}^{(0)}$ could lead to floating-point overflow or underflow, depending upon whether $|\lambda_1|$ is greater or less than 1, before the desired accuracy is attained. This necessitates a modification of the approach in which the vector $\mathbf{x}^{(j)}$ is normalized to be of unit magnitude after each multiplication by A.

Recall that the supremum norm of a vector is $\|\mathbf{x}\|_\infty := \max_{1 \leq i \leq n} |x_i|$. We define two sequences of vectors, $\mathbf{x}^{(j)}$ and $\mathbf{y}^{(j)}$, as follows. The vector $\mathbf{x}^{(0)}$ is chosen arbitrarily as before. Let m_0 be one of the components of $\mathbf{x}^{(0)}$ of greatest magnitude. Then $\mathbf{y}^{(0)} := (1/m_0)\mathbf{x}^{(0)}$, whence $\|\mathbf{y}^{(0)}\|_\infty = 1$. Now let $\mathbf{x}^{(1)} := A\mathbf{y}^{(1)}$, let m_1 be one of the components of $\mathbf{x}^{(1)}$ of largest magnitude, and let $\mathbf{y}^{(1)} := (1/m_1)\mathbf{x}^{(1)}$. Thus starting from $\mathbf{x}^{(0)}$ we compute the sequence

$$\mathbf{x}^{(0)}, \qquad\qquad \mathbf{y}^{(0)} := \frac{1}{m_0}\,\mathbf{x}^{(0)}$$

$$\mathbf{x}^{(1)} := A\mathbf{y}^{(0)} \qquad \mathbf{y}^{(1)} := \frac{1}{m_1}\,\mathbf{x}^{(1)} \qquad\qquad (11.1.18)$$

$$\cdots\cdots\cdots\cdots\cdots\cdots\cdots$$

$$\mathbf{x}^{(j)} := A\mathbf{y}^{(j-1)} \qquad \mathbf{y}^{(j)} := \frac{1}{m_j}\,\mathbf{x}^{(j)}$$

where at each stage m_j is a component of $\mathbf{x}^{(j)}$ of maximal magnitude.

To see the effect of this modification of the power method, let us assume that the eigenvectors $\mathbf{v}^{(k)}$ have been chosen so that $\|\mathbf{v}^{(k)}\|_\infty = 1$. For simplicity assume that the dominant eigenvalue λ_1 is of multiplicity one. Again assuming that the vectors $\{\mathbf{v}^{(k)}\}$ are a basis for $\mathbb{R}^n$, $\mathbf{y}^{(j)} = \sum_{k=1}^n c_k^{(j)}\mathbf{v}^{(k)}$ for some scalars $c_k^{(j)}$. Multiplication by A gives $\mathbf{x}^{(j+1)} = \sum_{k=1}^n \lambda_k c_k^{(j)}\mathbf{v}^{(k)}$ and thus pulls $\mathbf{y}^{(j)}$ in the direction of $\lambda_1\mathbf{v}^{(1)}$. Figure 11.1.5 illustrates the process of computing $\mathbf{x}^{(j+1)}$ from $\mathbf{x}^{(j)}$ in two dimensions when $\lambda_1 = -2$ and $\lambda_2 = \frac{1}{2}$. Clearly $\lim_{j\to\infty} \mathbf{x}^{(j)} = \lambda_1\mathbf{v}_1$. Thus the component of $\mathbf{x}^{(j)}$ of largest magnitude m_j approaches λ_1 and $\mathbf{y}^{(j)}$ approaches $\mathbf{v}^{(1)}$.

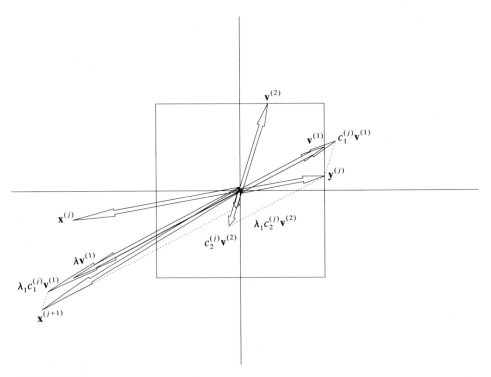

FIGURE 11.1.5

This normalization process can lead to difficulties when the eigenvector being sought has largest components of equal but opposite magnitude. In this case the sequence m_j of maximal components may converge to $-\lambda_1$ rather than λ_1, and the sequence of normalized eigenvectors $\mathbf{v}_j$ will not converge. See Exercise 8.

Algorithm 11.1.6 gives an implementation of the power method. As we have noted, there are circumstances under which it will not converge at all and other circumstances where it will converge very slowly; thus a limitation on the number of iterations is necessary. Note that the stopping criterion is based upon the convergence of the normalized eigenvectors $\mathbf{x}$ rather than the scale factors m. This prevents obtaining an incorrect answer in the situation illustrated by Exercise 8. Developing an alternative approach that will work in this situation is Exercise 9.

Algorithm 11.1.6   The power method

> *Input* A (matrix)
> tol (maximum error for eigenvalue)
> $\mathbf{x}^{(0)}$ (initial vector)
> MaxIter (maximum permissible number of iterations)
>
> Find an element m of $\mathbf{x}^{(0)}$ of largest magnitude
> $$\mathbf{x}:=\frac{1}{m}\mathbf{x}^{(0)}$$
> Iter:=0
>
> repeat
> > $\mathbf{x}_{\text{last}}:=\mathbf{x}$
> > $\mathbf{x}:=A\mathbf{x}$
> > Find an element m of $\mathbf{x}$ of largest magnitude
> > $$\mathbf{x}:=\frac{1}{m}\mathbf{x}$$
> > Iter:=Iter + 1
> until $\|\mathbf{x}-\mathbf{x}_{\text{last}}\|_\infty <$ tol or Iter > MaxIter
>
> if Iter $\leq$ MaxIter then
> > Output: m (approximation to the dominant eigenvalue)
> > $\mathbf{x}$ (eigenvector)

From Equation (11.1.17) the rate at which the power method converges is controlled by the ratio λ_k/λ_1, where λ_k is the next largest eigenvalue in magnitude. Table 11.1.7 shows the results of applying Algorithm 11.1.6 to the matrices of Examples 11.1.2 to 11.1.4. The tolerance specified was 5×10^{-6}. As can be seen the power method converges fairly rapidly to accurate answers for the nondefective matrices of Examples 11.1.2 and 11.1.3. The faster rate of convergence for Example 11.1.3 is due to the fact that the ratio of the two

TABLE 11.1.7

Example	Approx eigenvalue	Error	Iterations
11.1.2	$-8.999\,992\,0$	$8.0E-06$	42
11.1.3	$8.000\,000\,0$	0.0	17
11.1.4	$6.007\,749\,3$	$7.7E-03$	776

largest eigenvalues is $1/2$ as opposed to $2/3$ for Example 11.1.2. Convergence is very slow for the defective matrix in Example 11.1.4, and the error in the approximation is substantially less than the specified tolerance. As mentioned before, the analysis of convergence of the power method is more complicated for defective matrices, and, evidently, the rate of convergence is slower than would be indicated by the ratio of the two largest eigenvalues.

Shift of Origin

Let λ be an eigenvalue of A with associated eigenvector $\mathbf{v}$. Then for any number q

$$(A - qI)\mathbf{v} = A\mathbf{v} - q\mathbf{v} = (\lambda - q)\mathbf{v} \tag{11.1.19}$$

Thus $\lambda - q$ is an eigenvalue of $A - qI$ with associated eigenvector $\mathbf{v}$. This observation can be used to find the eigenvalue λ_n of A that is farthest from the dominant eigenvalue, assuming that it is unique. The power method is applied to A to determine an approximate value for the dominant eigenvalue λ_1. The power method is then applied to the matrix $A - \lambda_1 I$ which has eigenvalues 0, $\lambda_2 - \lambda_1, \ldots, \lambda_n - \lambda_1$ with $\lambda_n - \lambda_1$ being the largest in magnitude. By adding λ_1 to the result of this second application of the power method, λ_n is found.

Inverse Power Method

Note that if λ is a nonzero eigenvalue of a nonsingular matrix A with eigenvector $\mathbf{v}$, then $A\mathbf{v} = \lambda\mathbf{v}$ implies $A^{-1}\mathbf{v} = (1/\lambda)\mathbf{v}$. Thus λ^{-1} is an eigenvalue of A^{-1} with eigenvector $\mathbf{v}$. Combining this with an origin shift we see that if A is nondefective with eigenvalues $\{\lambda_k\}_{k=1}^{n}$ and corresponding linearly independent eigenvectors $\{\mathbf{v}_k\}_{k=1}^{n}$, then the eigenvalues of $(A - qI)^{-1}$ are $\{(\lambda_k - q)^{-1}\}_{k=1}^{n}$ while the eigenvectors remain $\{\mathbf{v}_k\}_{k=1}^{n}$.

The *inverse power method* is to apply the power method to the matrix $(A - qI)^{-1}$. To see the advantage of this, note that under the assumption that A is nondefective, the initial vector $\mathbf{x}^{(0)}$ in the power method has a representation of the form $\mathbf{x}^{(0)} = \sum_{k=1}^{n} c_k \mathbf{v}^{(k)}$. Successive iterations of the power method then produce

$$\mathbf{x}^{(1)} = (A - qI)^{-1}\mathbf{x}^{(0)} = \sum_{k=1}^{n} \frac{c_k}{\lambda_k - q} \mathbf{v}^{(k)}$$

$$\mathbf{x}^{(2)} = [(A - qI)^{-1}]^2\mathbf{x}^{(0)} = (A - qI)^{-1}\mathbf{x}^{(1)} = \sum_{k=1}^{n} \frac{c_k}{(\lambda_k - q)^2} \mathbf{v}^{(k)} \quad (11.1.20)$$

....

$$\mathbf{x}^{(j)} = [(A - qI)^{-1}]^j\mathbf{x}^{(0)} = (A - qI)^{-1}\mathbf{x}^{(j-1)} = \sum_{k=1}^{n} \frac{c_k}{(\lambda_k - q)^j} \mathbf{v}^{(k)}$$

....

Observe that if the parameter q is chosen sufficiently near to some eigenvalue λ_m, then $(\lambda_m - q)^{-1} > (\lambda_k - q)^{-1}$ for all $\lambda_k \neq \lambda_m$. Thus, for instance, if λ_m is of multiplicity one, then $\mathbf{x}^{(j)} \simeq c_m(\lambda_m - q)^{-j}\mathbf{v}^{(m)}$. The error made by the inverse power method after j iterations is proportional to

$$\left[\max_{k \neq m} \frac{(\lambda_k - q)^{-1}}{(\lambda_m - q)^{-1}} \right]^j = \left[\max_{k \neq m} \frac{\lambda_m - q}{\lambda_k - q} \right]^j \quad (11.1.21)$$

and consequently if the initial estimate q is relatively close to λ_m, convergence will be quite rapid.

As with the power method it is best to normalize the vectors produced at each iteration to be of norm 1. Algorithm 11.1.8 gives more detail. The computation of the iterates $\mathbf{x}^{(j)} = (A - qI)^{-1}\mathbf{x}^{(j-1)}$ is most efficiently accomplished by using LU decomposition with partial pivoting (Algorithm 6.4.4) to solve the system $(A - qI)\mathbf{x}^{(j)} = \mathbf{x}^{(j-1)}$. Since the matrix $A - qI$ does not change, the LU decomposition need be done only once. During the iteration only the forward elimination and backward substitution parts of Algorithm 6.4.4 need to be executed. If the guess q is an eigenvalue, then $A - qI$ is singular and the inverse power method is not applicable. Thus the LU decomposition algorithm must be able to detect this condition.

Algorithm 11.1.8 The inverse power method

Input A (matrix)
           q (estimate of eigenvalue)
           tol (maximum error of eigenvalue)
           $\mathbf{x}^{(0)}$ (initial vector)
           MaxIter (maximum permissible number of iterations)

Find the LU decomposition of $A - qI$
If A is singular then
      {Output: q (eigenvalue, we cannot return an eigenvector in this
                             case)
      Halt program}

Find an element m of $\mathbf{x}^{(0)}$ of largest magnitude

$$\mathbf{y} := \frac{1}{m} \mathbf{x}^{(0)}$$

Iter:=0

repeat

$\mathbf{y}_{\text{last}}:=\mathbf{y}$

Solve $(A - qI)\mathbf{x}:=\mathbf{y}$ by forward elimination and then backward
substitution

Find an element m of $\mathbf{x}$ of largest magnitude

$\mathbf{y}:=\dfrac{1}{m}\,\mathbf{x}$

Iter:=Iter + 1

until $\|\mathbf{y} - \mathbf{y}_{\text{last}}\|_{\infty} < \text{tol}$ or Iter $>$ MaxIter

if Iter $\le$ MaxIter then

Output $q + 1/m$ (approximation to the eigenvalue nearest q)
    $\mathbf{y}$ (eigenvector)

The inverse power method is often used to find eigenvectors using approximate eigenvalues determined by one of the methods that will be discussed in the next sections. Observe that it fails to find the eigenvector when the estimate q is exact. In this case entering a slightly less accurate guess will usually result in an eigenvector being found. Indeed the algorithm could be modified to add tol to q in this case and recompute the LU decomposition.

As mentioned earlier the power method cannot find complex eigenvalues of a real matrix. Providing that an implementation of the inverse power method supports complex arithmetic, it can find complex eigenvalues of real matrices. Clearly, however, the initial guess q must be complex to find a complex eigenvalue.

Gerschgorin Disk Theorem

The inverse power method requires the input of an initial guess q for the eigenvalue. The *Gerschgorin disk theorem* is useful in making such guesses.

Theorem 11.1.9. Every eigenvalue of a matrix A lies in the union of the disks

$$|z - a_{ii}| \le \sum_{j=1, j \ne i}^{n} |a_{ij}| \qquad i = 1, \dots, n \tag{11.1.22}$$

Remark. The variable z is complex-valued; hence each inequality in (11.1.22) defines a disk in the complex plane.

Proof. If λ is an eigenvalue of A and $\mathbf{v}$ a corresponding eigenvector, then

$$\sum_{j=1}^{n} a_{ij} v_j = \lambda v_i \qquad i = 1, \dots, n \tag{11.1.23}$$

We may assume that $\mathbf{v}$ has been normalized so that its maximum component, v_r say, is 1. Then the rth equation in (11.1.23) is

$$\lambda - a_{rr} = \sum_{j=1, j \neq r}^{n} a_{rj} v_j \tag{11.1.24}$$

whence

$$|\lambda - a_{rr}| \leq \sum_{j=1, j \neq r}^{n} |a_{rj}| \tag{11.1.25}$$

A related theorem supplies more detailed information about the location of the eigenvalues. Its proof is Exercise 7.

Theorem 11.1.10. If the union of any m Gerschgorin disks forms a connected set that is isolated from the remaining disks, then the union contains exactly m eigenvalues, when counted according to multiplicity.

Example 11.1.11. The matrix

$$A = \begin{bmatrix} 10 & 1 & 1 & 2 \\ 1 & 5 & 1 & 0 \\ 1 & 1 & -5 & 0 \\ 2 & 0 & 0 & -10 \end{bmatrix} \tag{11.1.26}$$

is symmetric and as a consequence has only real eigenvalues (Exercise 6). The Gerschgorin disks are

$$|z - 10| \leq 4$$
$$|z - 5| \leq 2$$
$$|z + 5| \leq 2 \tag{11.1.27}$$
$$|z + 10| \leq 2$$

(Fig. 11.1.12). From Theorem 11.1.10 the disks about -10 and -5 each must contain an eigenvalue. The other two eigenvalues must lie in the interval $[3, 14]$. Application of the inverse power method with a tolerance of 5×10^{-6} with initial guesses of $10, 5, -5$, and -10 leads to approximations of $\lambda_1 = 10.46975$, $\lambda_2 = 4.88031$, $\lambda_3 = -5.14971$, and $\lambda_4 = -10.20035$, respectively. The number of iterations required ranged from 9 to 13.

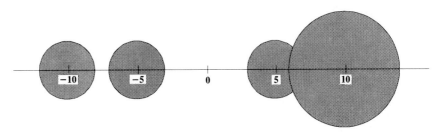

FIGURE 11.1.12

EXERCISES 11.1

1. Find all eigenvalues and eigenvectors. State whether the matrix is defective.

(a) $\begin{bmatrix} 3 & 3 & 2 \\ 1 & 1 & -2 \\ -1 & -3 & 0 \end{bmatrix}$

(b) $\begin{bmatrix} 1 & -1 & 0 \\ 1 & 3 & 0 \\ 1 & 1 & 2 \end{bmatrix}$

(c) $\begin{bmatrix} 2 & -1 & -1 \\ 0 & 3 & 1 \\ 0 & 1 & 3 \end{bmatrix}$

(d) $\begin{bmatrix} 1 & 1 & -1 \\ -2 & 2 & -3 \\ 0 & 1 & 0 \end{bmatrix}$

2. (a) Show that eigenvectors corresponding to distinct eigenvalues are linearly independent.
   (b) Show that if an $n \times n$ matrix A has n distinct eigenvalues, then the initial value problem (11.1.2) has a unique solution.

3. Write a program that uses the power method to find the eigenvalues which are largest and smallest in magnitude and the associated eigenvectors. Have the program report the number of iterations required in each case.
   (a) Test your program on Examples 11.1.2 to 11.1.4.
   (b) Use your program to find the largest and smallest eigenvalues and associated eigenvectors for each matrix in Exercise 1.
   (c) Use your program to find the largest and smallest eigenvalues and associated eigenvectors for the matrix

$$\begin{bmatrix} 10 & 1 & 0 & 2 & 0 \\ 1 & -5 & 0 & 1 & 1 \\ 0 & 0 & -7 & 0 & 1 \\ 2 & 1 & 0 & 8 & 1 \\ 0 & 1 & 1 & 1 & 12 \end{bmatrix} \quad (11.1.28)$$

4. Write a program that implements the inverse power method. Have it take an estimate of eigenvalue as input and report out the eigenvalue, an eigenvector, and the number of iterations required. Apply your program to find all eigenvalues and eigenvectors for the matrix in Exercise $3c$ to within an accuracy of 5×10^{-4}.

5. (a) Show for the matrix A given by (11.1.14) that $\det(A - \lambda I) = Q(\lambda)/a_n$.
   (b) Use your inverse power method program (Exercise 4) to find all roots of the polynomial $4x^5 - 77x^3 + 6x^2 + 208x - 96$.

6. (a) Show that if A is a real matrix, then if λ is an eigenvalue with eigenvector $\mathbf{v}$, then its complex conjugate $\bar{\lambda}$ is an eigenvalue with eigenvector $\bar{\mathbf{v}}$.
   (b) A matrix A is symmetric if it is equal to its transpose A^T. Show that a real symmetric matrix can have only real eigenvalues. (*Hint:* Show that $\lambda \mathbf{v}^T \bar{\mathbf{v}} = \bar{\lambda} \mathbf{v}^T \bar{\mathbf{v}}$.)

7. Prove Theorem 11.1.10. [*Hint:* Let $D := \text{diag}\{a_{11}, \ldots, a_{nn}\}$ and $B := A - D$. Consider the evolution of the roots of the characteristic polynomial $P_\epsilon(\lambda)$ of $C_\epsilon := D + \epsilon B$ as $\epsilon: 0 \to 1$.]

8. (a) Find the eigenvalues and eigenvectors of the matrix

$$\begin{bmatrix} 1 & 3 \\ 0 & -2 \end{bmatrix} \quad (11.1.29)$$

   (b) For the power method [Eq. (11.1.18), Algorithm 11.1.6] find $\mathbf{x}^{(j+1)}$ when $\mathbf{x}^{(j)} = [1, -1 + \epsilon]^T$ and also when $\mathbf{x}^{(j)} = [-1 + \epsilon, 1]^T$. Take $0 < \epsilon < \frac{1}{2}$. To what value does the sequence m_j of scale factors converge? What is the behavior of the normalized eigenvectors? Verify by having your power method program print out the sequence of scale factors for this example. Illustrate what is occurring with a diagram similar to Fig. 11.1.5.
   (c) Prove that *if* the sequence of normalized eigenvectors $\mathbf{y}^{(j)}$ in (11.1.18) converges, then the sequence m_j converges to λ_1.

9. Modify the power method (Algorithm 11.1.6) so that at each step the approximate eigenvector is normalized by $\|\mathbf{x}\|_\infty$ rather than the maximal component. Take as the approximation to the eigenvalue the ratio of some component of two successive $\mathbf{x}$'s. Show that this formulation will obtain the correct answer for the problem in Exercise 8.

10. Modify the inverse power method algorithm so that it will take complex numbers as initial estimates and return complex eigenvalues and eigenvectors. Test your program by finding all eigen-

values and eigenvectors of the matrices

(a) $\begin{bmatrix} 1 & 1 & -1 \\ -2 & 2 & -3 \\ 0 & 1 & 0 \end{bmatrix}$

(b) $\begin{bmatrix} 1 & 2 & 0 & 0 & 2 \\ -4 & 3 & -2 & -2 & 0 \\ -3 & 4 & -1 & 0 & 0 \\ 2 & -2 & 0 & 3 & -2 \\ 3 & -4 & 2 & 2 & -1 \end{bmatrix}$

11.2 EIGENVALUES OF SYMMETRIC MATRICES

In Sec. 11.1 we developed methods of finding individual eigenvalues. The power method can find the dominant eigenvalue, and the inverse power method can find the eigenvalue nearest to an initial estimate. In this and the next section we develop methods to find all eigenvalues of a given matrix. The basic approach is to find a sequence of *similarity transformations* that carry a given matrix into another matrix which has eigenvalues which are easily found.

Similar Matrices

Definition 11.2.1. Matrices A and B are *similar* if there is a nonsingular matrix S such that $B = S^{-1}AS$.

Observe that if $B = S^{-1}AS$ and λ is an eigenvalue of A with eigenvector $\mathbf{v}$, then since $A\mathbf{v} = \lambda\mathbf{v}$ we have

$$SS^{-1}ASS^{-1}\mathbf{v} = \lambda\mathbf{v}$$
$$BS^{-1}\mathbf{v} = \lambda S^{-1}\mathbf{v} \tag{11.2.1}$$

and thus λ is also an eigenvalue of the matrix B with eigenvector $S^{-1}\mathbf{v}$. Thus similar matrices have the same set of eigenvalues.

Theorem 11.2.2. For any matrix A there is a nonsingular matrix S such that $S^{-1}AS = \Lambda$, where Λ is a block diagonal matrix of the form

$$\Lambda = \begin{bmatrix} \Lambda_0 & 0 & 0 & \cdots & 0 \\ 0 & \Lambda_1 & 0 & \cdots & 0 \\ 0 & 0 & \Lambda_2 & \cdots & 0 \\ 0 & 0 & 0 & \cdots & 0 \\ 0 & 0 & 0 & \cdots & \Lambda_m \end{bmatrix} \tag{11.2.2}$$

where

$$\Lambda_0 = \begin{bmatrix} \lambda_1 & 0 & \cdots & 0 & 0 \\ 0 & \lambda_2 & \cdots & 0 & 0 \\ \multicolumn{5}{c}{\dotfill} \\ 0 & 0 & \cdots & 0 & \lambda_r \end{bmatrix} \qquad \Lambda_k = \begin{bmatrix} \lambda_{r+k} & 0 & \cdots & 0 & 0 \\ 1 & \lambda_{r+k} & \cdots & 0 & 0 \\ \multicolumn{5}{c}{\dotfill} \\ 0 & 0 & \cdots & 1 & \lambda_{r+k} \end{bmatrix} \tag{11.2.3}$$

for $k = 1, \ldots, m$.

Remark. The matrix S may be complex-valued, even if the matrix A is real.

The matrix Λ is said to be the *Jordan canonical form* of the matrix A. The proof of this theorem can be found in any comprehensive text on linear algebra. If the matrix A is nondefective, then $\Lambda = \Lambda_0$; that is, A is similar to a diagonal matrix. Subblocks with 1s below the main diagonal occur only for those eigenvalues whose number of linearly independent eigenvectors is less than the multiplicity of the eigenvalue. For this reason nondefective matrices are also called *diagonalizable*.

Since the eigenvalues of a matrix in Jordan canonical form are obvious, it would be a simple matter to find the eigenvalues of a matrix if it was possible to find directly, either analytically or numerically, the similarity matrix S that carries the matrix A in the matrix Λ. In general this is not feasible; however, schemes for finding all eigenvalues of a matrix do consist of sequences of similarity transformations each of which carries the original matrix A into a matrix that is "nearly" diagonal.

Symmetric Matrices

A matrix A is *symmetric* if $A^T = A$, that is, if $a_{ij} = a_{ji}$ for each element of A. As we will see in Chap. 12, boundary value problems for differential equations sometimes lead to eigenvalue problems for symmetric matrices. The next theorem gives some facts about symmetric matrices.

Theorem 11.2.3. If A is symmetric, then:
(a) Eigenvectors corresponding to distinct eigenvalues of A are orthogonal.
(b) All eigenvalues of A are real.
(c) A is diagonalizable.

Proof
(a) Let λ and μ be two eigenvalues of A with eigenvectors $\mathbf{u}$ and $\mathbf{v}$, respectively. Then

$$\lambda \mathbf{u}^T\mathbf{v} = (\lambda\mathbf{u})^T\mathbf{v} = (A\mathbf{u})^T\mathbf{v} = \mathbf{u}^TA^T\mathbf{v} = \mathbf{u}^T(A\mathbf{v}) = \mathbf{u}^T(\mu\mathbf{v}) = \mu\mathbf{u}^T\mathbf{v}$$

$$(11.2.4)$$

Thus either $\lambda = \mu$ or $\mathbf{u}^T\mathbf{v} = 0$; that is, $\mathbf{u}$ and $\mathbf{v}$ are orthogonal.
(b) This is very similar to (a). See Exercises in Sec. 11.1.
(c) This is more difficult. The reader is again referred to a comprehensive text on linear algebra.

Definition 11.2.4. A matrix S is *orthogonal* if $S^{-1} = S^T$.

Note that if A is symmetric and $B := S^{-1}AS$, where S is orthogonal, then

$$B^T = (S^TAS)^T = S^TA^T(S^T)^T = S^TAS = B \qquad (11.2.5)$$

and thus similarity transformations on symmetric matrices that use orthogonal matrices produce matrices which are again symmetric.

Let us find the most general 2×2 orthogonal matrix R. We have

$$R^{-1} = \begin{bmatrix} a & b \\ c & d \end{bmatrix}^{-1} = \frac{1}{\det(R)} \begin{bmatrix} d & -b \\ -c & a \end{bmatrix} \qquad R^{\mathrm{T}} = \begin{bmatrix} a & c \\ b & d \end{bmatrix} \qquad (11.2.6)$$

Then $R^{-1} = R^{\mathrm{T}}$ implies that $a = d$, $b = -c$, and $a^2 + c^2 = 1$ (Exercise 2). Consequently the most general orthogonal 2×2 matrix is

$$R = \begin{bmatrix} \cos\theta & -\sin\theta \\ \sin\theta & \cos\theta \end{bmatrix} \qquad (11.2.7)$$

Such matrices are referred to as *rotation matrices*, since the product $R[x, y]^{\mathrm{T}}$ represents a clockwise rotation of the vector $[x, y]^{\mathrm{T}}$ through an angle θ.

Rotation matrices can be used to solve the eigenvalue problem for 2×2 symmetric matrices. Let

$$A = \begin{bmatrix} a_{11} & a_{12} \\ a_{21} & a_{22} \end{bmatrix} \qquad (11.2.8)$$

with $a_{12} = a_{21}$. Then

$$\begin{bmatrix} a'_{11} & a'_{12} \\ a'_{21} & a'_{22} \end{bmatrix} := \begin{bmatrix} \cos\theta & \sin\theta \\ -\sin\theta & \cos\theta \end{bmatrix} \begin{bmatrix} a_{11} & a_{12} \\ a_{21} & a_{22} \end{bmatrix} \begin{bmatrix} \cos\theta & -\sin\theta \\ \sin\theta & \cos\theta \end{bmatrix} \qquad (11.2.9)$$

implies that

$$a'_{11} = a_{11} \cos^2\theta + a_{12} \sin 2\theta + a_{22} \sin^2\theta$$

$$a'_{21} = a'_{12} = \tfrac{1}{2}(a_{22} - a_{11}) \sin 2\theta + a_{12} \cos 2\theta \qquad (11.2.10)$$

$$a'_{22} = a_{11} \sin\theta - a_{12} \sin 2\theta + a_{22} \cos\theta$$

If θ is selected so that $a'_{12} = a'_{21} = 0$, then a'_{11} and a'_{22} will become the eigenvalues of the matrix A. Thus we choose θ to satisfy

$$\cot 2\theta = \frac{a_{11} - a_{22}}{2a_{12}} \qquad (11.2.11)$$

Jacobi Method

It is not feasible to construct orthogonal matrices that carry an $n \times n$ symmetric matrix A into a diagonal matrix; however, it is possible to carry A toward diagonal form by a sequence of similarity transformations. The matrix

$$\Omega(p, q) = \begin{bmatrix} 1 & 0 & \cdots & 0 & 0 & 0 & \cdots & 0 & 0 & 0 & \cdots & 0 \\ & & \ddots & & & & & & & & & \\ 0 & 0 & \cdots & 1 & 0 & 0 & \cdots & 0 & 0 & 0 & \cdots & 0 \\ 0 & 0 & \cdots & 0 & \cos\theta & 0 & \cdots & 0 & -\sin\theta & 0 & \cdots & 0 \\ 0 & 0 & \cdots & 0 & 0 & 1 & \cdots & 0 & 0 & 0 & \cdots & 0 \\ & & & & & & \ddots & & & & & \\ 0 & 0 & \cdots & 0 & 0 & 0 & \cdots & 1 & 0 & 0 & \cdots & 0 \\ 0 & 0 & \cdots & 0 & \sin\theta & 0 & \cdots & 0 & \cos\theta & 0 & \cdots & 0 \\ 0 & 0 & \cdots & 0 & 0 & 0 & \cdots & 0 & 0 & 1 & \cdots & 0 \\ & & & & & & & & & & \ddots & \\ 0 & 0 & \cdots & 0 & 0 & 0 & \cdots & 0 & 0 & 0 & \cdots & 1 \end{bmatrix}$$

$$(11.2.12)$$

where the two cosines are in the (p, p) and (q, q) positions and the sines are in the (p, q) and (q, p) positions, is called the *rotation matrix in the* (p, q) *plane*. If $A' := \Omega(p, q)^T A \Omega(p, q)$, where A is symmetric, then the components of A' are

$$a'_{pp} = a_{qq} \sin^2 \theta - a_{pq} \sin 2\theta + a_{pp} \cos^2 2\theta$$

$$a'_{qq} = a_{qq} \cos^2 \theta + a_{pq} \sin 2\theta + a_{pp} \sin^2 \theta$$

$$a'_{pq} = a_{pq} \cos 2\theta + \frac{1}{2} (a_{pp} - a_{qq}) \sin 2\theta \qquad (11.2.13)$$

$$a'_{pj} = -a_{qj} \sin \theta + a_{pj} \cos \theta \qquad (j \neq p, q)$$

$$a'_{qj} = a_{qj} \cos \theta + a_{pj} \sin \theta \qquad (j \neq p, q)$$

If θ is chosen so that

$$\cot 2\theta = \frac{a_{pp} - a_{qq}}{2a_{pq}} \qquad (11.2.14)$$

[see (11.2.11)], then $a'_{pq} = a'_{qp} = 0$. Moreover, the transformation $A' = \Omega(p, q)^T A \Omega(p, q)$ has the effect of reducing the sum of squares of the off-diagonal elements. Thus each application of the transformation carries the matrix toward diagonal form.

Theorem 11.2.5. With a'_{ij} defined by (11.2.13) where θ is defined by (11.2.11),

$$\sum_{i=1}^{n} \sum_{j=1, j \neq i}^{n} a'^2_{ij} = \sum_{i=1}^{n} \sum_{j=1, j \neq i}^{n} a^2_{ij} - 2a^2_{pq} \qquad (11.2.15)$$

The proof of this theorem follows from (11.2.13). It is Exercise 3.

The *Jacobi method* of finding the eigenvalues of a symmetric matrix is to apply repeatedly plane rotations until the sum of squares of the off-diagonal elements is smaller than some specified tolerance. At that point the elements along the main diagonal are taken as the approximations to the eigenvalues. Algorithm 11.2.6 gives the *classical Jacobi method* in which rotations are applied to annihilate the largest off-diagonal element. Since the search for this largest element can be expensive in terms of computer time, an often more efficient approach is one of the *cyclic Jacobi methods*, in which the off-diagonal elements are annihilated in some fixed order. Implementation of a cyclic Jacobi method is Project 1.

Algorithm 11.2.6 Classical Jacobi method

Input A_0 (symmetric matrix)
    tol (error tolerance)

$A = A_0$

$S := \sum_{i=2}^{n} \sum_{j=1}^{i-1} a^2_{ij}$ (half of sum of off-diagonal elements)

```
repeat
    Find the element a_pq, p ≠ q of A for which |a_pq| is largest
    S:=S − a²_pq
    A = Ω(p, q)ᵀAΩ(p, q) [calculated from (11.2.13)]
until S < tol
```

Output   Elements of the main diagonal of A as approximations to the eigenvalues of A_0.

Though we see that the application of the similarity transformation $\Omega(p, q)^T A \Omega(p, q)$ sets the elements a'_{pq} and a'_{qp} of the resulting matrix A' to zero, subsequent transformations intended to annihilate other elements in the pth or qth row or column will usually make the (p, q) element nonzero again. Thus after any finite number of steps most off-diagonal elements will be nonzero. It can be shown that in the limit the classical Jacobi process does lead to a diagonal matrix, which is to say that the elements along the diagonal converge to the eigenvalues of the matrix A [see Goldstine, Murray, and Von Neumann (1959)].

For reasons of efficiency it is preferable to avoid the calculation of θ, $\cos \theta$, and $\sin \theta$ using trigonometric and inverse trigonometric functions, as these are computationally expensive. In (11.2.14) let $\alpha := (a_{pp} - a_{qq})/(2a_{pq})$. Whatever the value of α, there is a θ in $[-\pi/4, \pi/4]$ for which $\cot 2\theta = \alpha$. For θ lying in this range, $\cos \theta$ will always be nonnegative and $\sin \theta$ will have the same sign as α. Applying the double-angle formula for the cotangent function gives $\cot^2 \theta - 1 = 2\alpha \cot \theta$. This is a quadratic equation in $\cot \theta$, and thus $\cot \theta = \beta := \alpha \pm \sqrt{\alpha^2 + 1}$. The quantities $\sin \theta$ and $\cos \theta$ needed for the matrix $R(p, q)$ can now be calculated as

$$\sin \theta = \frac{\text{sgn}(\beta)}{\sqrt{\beta^2 + 1}} \qquad \cos \theta = \beta \sin \theta \qquad (11.2.16)$$

where $\text{sgn}(\beta)$, the *signum* function, is 1 or −1 as β is positive or negative.

Householder Transformations

An alternative approach to the Jacobi methods is to first apply *Householder transformations* to transform the matrix A into a matrix in *upper Hessenberg form*

$$\begin{bmatrix} x & x & x & x & \cdots & x & x & x \\ x & x & x & x & \cdots & x & x & x \\ 0 & x & x & x & \cdots & x & x & x \\ 0 & 0 & x & x & \cdots & x & x & x \\ 0 & 0 & 0 & x & \cdots & x & x & x \\ \multicolumn{8}{c}{\dotfill} \\ 0 & 0 & 0 & 0 & \cdots & 0 & x & x \end{bmatrix} \qquad (11.2.17)$$

that is, a matrix in which all elements below the principal subdiagonal are zero.

The eigenvalues of a matrix in Hessenberg form are not obvious; however, there are methods, the *Householder-Given's algorithm* for symmetric matrices and the *QR algorithm* for general matrices, that can be efficiently applied to matrices in Hessenberg form to approximate the eigenvalues.

Definition 11.2.7. A *Householder matrix* is a matrix of the form

$$P = I - 2\mathbf{w}\mathbf{w}^{\mathrm{T}} \qquad (11.2.18)$$

where the vector $\mathbf{w}$ satisfies $\mathbf{w}^{\mathrm{T}}\mathbf{w} = \Sigma_{k=1}^{n} w_k^2 = 1$.

Theorem 11.2.8. A Householder matrix is symmetric and orthogonal.

Proof. That a Householder matrix is symmetric can be seen simply by writing it in component form

$$P = \begin{bmatrix} 1 - 2w_1^2 & -2w_1w_2 & -2w_1w_3 & \cdots & -2w_1w_n \\ -2w_2w_1 & 1 - 2w_2^2 & -2w_2w_3 & \cdots & -2w_2w_n \\ \vdots & & & & \vdots \\ -2w_nw_1 & -2w_nw_2 & -2w_nw_3 & \cdots & 1 - 2w_n^2 \end{bmatrix} \qquad (11.2.19)$$

That Householder matrices are orthogonal follows from

$$P^2 = (I - 2\mathbf{w}\mathbf{w}^{\mathrm{T}})(I - 2\mathbf{w}\mathbf{w}^{\mathrm{T}})$$

$$= I - 4\mathbf{w}\mathbf{w}^{\mathrm{T}} + 4\mathbf{w}\mathbf{w}^{\mathrm{T}}\mathbf{w}\mathbf{w}^{\mathrm{T}}$$

$$= I - 4\mathbf{w}\mathbf{w}^{\mathrm{T}} + 4\mathbf{w}\mathbf{w}^{\mathrm{T}} \qquad \text{(since } \mathbf{w}^{\mathrm{T}}\mathbf{w} = 1\text{)}$$

$$= I \qquad (11.2.20)$$

Thus $P^{-1} = P = P^{\mathrm{T}}$ since P is symmetric.

We now show that a set of Householder matrices $\{P_j\}_{j=1}^{n-2}$ can be constructed so that similarity transformation P_jAP_j results in a matrix which has the same elements as A below the principal subdiagonal in columns $1, \ldots, j-1$ and zeros below the principal subdiagonal in column j. It then follows that the sequence of transformations

$$A_0 = A$$
$$A_j = P_jA_{j-1}P_j \qquad j = 1, \ldots, n-2 \qquad (11.2.21)$$

will carry the matrix A into Hessenberg form. Note that if A is symmetric, then since the matrices P_j are orthogonal, the result will be a symmetric tridiagonal matrix.

The vector $\mathbf{w}^{(j)}$ used in the Householder matrix P_j will be taken to have the form $\mathbf{w}^{(j)} = [0, \ldots, 0, w_{j+1}^{(j)}, \ldots, w_n^{(j)}]^{\mathrm{T}}$. From (11.2.19) this gives the matrix P_j in the block form

$$P_j = \begin{bmatrix} I_{j\times j} & 0_{(n-j)\times j} \\ 0_{j\times(n-j)} & P_{22} \end{bmatrix} \qquad (11.2.22)$$

where $I_{j\times j}$ is the $j \times j$ identity matrix and $0_{i\times j}$ is an $i \times j$ matrix of zeros. The reason for choosing $\mathbf{w}^{(j)}$ in this way is that we then have

$$P_j A_{j-1} P_j = \begin{bmatrix} I_{j \times j} & 0_{j \times (n-j)} \\ 0_{(n-j) \times j} & P_{22} \end{bmatrix} \begin{bmatrix} A_{11} & A_{22} \\ A_{21} & A_{22} \end{bmatrix} \begin{bmatrix} I_{j \times j} & 0_{j \times (n-j)} \\ 0_{(n-j) \times j} & P_{22} \end{bmatrix}$$

$$= \begin{bmatrix} A_{11} & A_{12} P_{22} \\ P_{22} A_{21} & P_{22} A_{22} P_{22} \end{bmatrix} \tag{11.2.23}$$

The matrix A_{j-1} has zeros below the principal subdiagonal in columns $1, \ldots, j-1$ since it is the result of previous Householder transformations. Consequently we have

$$P_{22} A_{21} = \begin{bmatrix} 1 - 2(w_{j+1}^{(j)})^2 & \cdots & -2w_{j+1}^{(j)} w_n^{(j)} \\ \cdots\cdots\cdots\cdots\cdots\cdots\cdots\cdots \\ -2w_n^{(j)} w_{j+1}^{(j)} & \cdots & 1 - 2(w_n^{(j)})^2 \end{bmatrix} \begin{bmatrix} 0 & \cdots & 0 & a_{j+1,j}^{(j-1)} \\ \cdots\cdots\cdots\cdots\cdots \\ 0 & \cdots & 0 & a_{nj}^{(j-1)} \end{bmatrix}$$

$$= \begin{bmatrix} 0 & \cdots & 0 & a_{j+1,j}^{(j-1)} - 2w_{j+1}^{(j)} \sum_{i=j+1}^{n} a_{ij}^{(j-1)} w_i^{(j)} \\ \cdots\cdots\cdots\cdots\cdots\cdots\cdots\cdots\cdots\cdots\cdots\cdots\cdots\cdots \\ 0 & \cdots & 0 & a_{nj}^{(j-1)} - 2w_n^{(j)} \sum_{i=j+1}^{n} a_{ij}^{(j-1)} w_i^{(j)} \end{bmatrix} \tag{11.2.24}$$

Thus if the Householder matrices are chosen to have the form (11.2.22), then the zeros below the principal subdiagonal in the first $j-1$ columns are preserved, and zeros can be created below the principal subdiagonal in the jth column by choosing $w_{j+1}^{(j)}, \ldots, w_n^{(j)}$ so that

$$a_{kj}^{(j-1)} - 2w_k^{(j)} \sum_{i=j+1}^{n} a_{ij}^{(j-1)} w_i^{(j)} = 0 \qquad k = j+2, \ldots, n \tag{11.2.25}$$

Since there are $n - j$ unknowns $w_k^{(j)}$, it may appear that (11.2.25) could be required for $k = j + 1$ also. However the $w_k^{(j)}$'s are subject to the condition $\sum_{k=j+1}^{n} w_k^{(j)} = 1$. As a consequence of this we have, with $\sigma := \sum_{k=j+1}^{n} a_{kj}^{(j-1)} w_k^{(j)}$,

$$\sum_{k=j+1}^{n} (a_{kj}^{(j-1)} - 2w_k^{(j)} \sigma)^2 = \sum_{k=j+1}^{n} (a_{kj}^{(j-1)})^2 - 4\sigma \sum_{k=j+1}^{n} a_{kj}^{(j-1)} w_k^{(j)}$$

$$+ 4\sigma^2 \sum_{k=j+1}^{n} (w_k^{(j)})^2$$

$$= \sum_{k=j+1}^{n} (a_{kj}^{(j-1)})^2 - 4\sigma^2 + 4\sigma^2$$

$$= \sum_{k=j+1}^{n} (a_{kj}^{(j-1)})^2 \tag{11.2.26}$$

If the $w_k^{(j)}$ have been chosen to satisfy Eqs. (11.2.25), then all terms but the first in the sum on the left in (11.2.26) are zero and we are left with

$$a_{j+1,j}^{(j-1)} - w_{j+1}^{(j)} \sigma = \pm s_j \tag{11.2.27}$$

where the quantity

$$s_j = \sqrt{\sum_{k=j+1}^{n} (a_{kj}^{(j-1)})^2}$$ (11.2.28)

is known by the time the $w_k^{(j)}$'s are being calculated. Thus we see that in general (11.2.25) cannot be satisfied for $k = j + 1$.

Now observe that from (11.2.25) and (11.2.27)

$$\sigma = \sum_{k=j+1}^{n} a_{kj}^{(j-1)} w_k^{(j)}$$

$$= (\pm s_j + 2w_{j+1}^{(j)}\sigma)w_{j+1}^{(j)} + \sum_{k=j+2}^{n} (2w_k^{(j)}\sigma)w_k^{(j)}$$

$$= \pm s_j w_{j+1}^{(j)} + 2\sigma$$ (11.2.29)

and thus

$$\sigma = \mp s_j w_{j+1}^{(j)}$$ (11.2.30)

This gives

$$(w_{j+1}^{(j)})^2 = \frac{s_j \mp a_{j+1,j}^{(j-1)}}{2s_j}$$

$$w_k^{(j)} = \mp \frac{a_{kj}^{(j-1)}}{2w_{j+1}^{(j)}s_j} \qquad k = j+2, \ldots, n$$ (11.2.31)

Since the quantity $w_{j+1}^{(j)}$ occurs in the denominators of the $w_k^{(j)}$ in (11.2.31), the $\mp$ choice in (11.2.31) should be made so as to preclude the possibility of the cancellation of the most significant digits. Hence we make the determination of sign $\mp a_{j+1,j}^{(j-1)} = \text{sgn}(a_{j+1,j}^{(j-1)})a_{j+1,j}^{(j-1)} = |a_{j+1,j}^{(j-1)}|$. This gives finally

$$w_{j+1}^{(j)} = \frac{s_j + |a_{j+1,j}^{(j-1)}|}{2s_j}$$

$$w_k^{(j)} = \text{sgn}(a_{j+1,j}^{(j-1)}) \frac{a_{kj}^{(j-1)}}{2w_{j+1}^{(j)}s_j} \qquad k = j+2, \ldots, n$$ (11.2.32)

It is easily verified that the choice of the $w_k^{(j)}$'s given by (11.2.32) satisfies the condition $\sum_{k=j+1}^{n} (w_k^{(j)})^2 = 1$ (Exercise 4).

Our objective is to calculate the matrix $A_j = P_j A_{j-1} P_j$. Because of the special form of the matrices P_j, calculating A_j as a product of three matrices, which would entail $2n^3$ multiplications, is unnecessarily inefficient. First note that with

$$
\mathbf{v}^{(j)} := \begin{bmatrix} 0 \\ \vdots \\ 0 \\ \operatorname{sgn}(a_{(j+1,j)}^{(j-1)})s_j + a_{j+1,j}^{(j-1)} \\ a_{j+2,j}^{(j-1)} \\ \vdots \\ a_{nj}^{(j-1)} \end{bmatrix} \tag{11.2.33}
$$

we have

$$
\mathbf{w}^{(j)} = \frac{\operatorname{sgn}(a_{j+1,j}^{(j-1)})}{2w_{j+1}^{(j)}s_j}\,\mathbf{v}^{(j)} \tag{11.2.34}
$$

It is now easy to verify that the Householder matrix is

$$
P_j = I - \alpha_j \mathbf{v}^{(j)}\mathbf{v}^{(j)\mathrm{T}} \tag{11.2.35}
$$

where

$$
\alpha_j = \frac{1}{(s_j + |a_{j+1,j}^{(j)}|)s_j} \tag{11.2.36}
$$

Thus

$$
\begin{aligned}
P_j A_{j-1} P_j &= (I - \alpha_j \mathbf{v}^{(j)}\mathbf{v}^{(j)\mathrm{T}})A_{j-1}(I - \alpha_j \mathbf{v}^{(j)}\mathbf{v}^{(j)\mathrm{T}}) \\
&= A_{j-1} - \alpha_j \mathbf{v}^{(j)}\mathbf{v}^{(j)\mathrm{T}}A_{j-1} - \alpha_j A_{j-1}\mathbf{v}^{(j)}\mathbf{v}^{(j)\mathrm{T}} + \alpha_j^2 \mathbf{v}^{(j)}\mathbf{v}^{(j)\mathrm{T}}A_{j-1}\mathbf{v}^{(j)}\mathbf{v}^{(j)\mathrm{T}}
\end{aligned} \tag{11.2.37}
$$

With the definitions

$$
\mathbf{x}^{(j)} := \alpha_j A_{j-1}\mathbf{v}^{(j)} \qquad \mathbf{y}^{(j)} := \alpha_j A_{j-1}^{\mathrm{T}}\mathbf{v}^{(j)}
$$

$$
\mu_j := \frac{1}{2}\,\alpha \mathbf{v}^{(j)\mathrm{T}}\mathbf{x}^{(j)} = \frac{1}{2}\,\alpha^2\mathbf{v}^{(j)\mathrm{T}}A_{j-1}\mathbf{v}^{(j)} \tag{11.2.38}
$$

$$
\mathbf{p}^{(j)} := \mathbf{y}^{(j)} - \mu_j \mathbf{v}^{(j)} \qquad \mathbf{q}^{(j)} := \mathbf{x}^{(j)} - \mu_j \mathbf{v}^{(j)}
$$

we then see that

$$
P_j A_{j-1} P_j = A_{j-1} - \mathbf{v}^{(j)}\mathbf{p}^{(j)\mathrm{T}} - \mathbf{q}^{(j)}\mathbf{v}^{(j)\mathrm{T}} \tag{11.2.39}
$$

Algorithm 11.2.9   Householder reduction to Hessenberg form

Input A_0 (matrix to reduce)

$A := A_0$
for $j = 1$ to $n - 2$
    Find s_j from (11.2.28)
    if $s_j \neq 0$ then
        {Find α, $\mathbf{v}$ [formulas (11.2.36) and (11.2.35)]
        $\mathbf{z} := \alpha \mathbf{v}$

$$\mathbf{x} := A\mathbf{z}, \quad \mathbf{y} := A^{\mathrm{T}}\mathbf{z}$$
$$\mu := \tfrac{1}{2}\alpha\mathbf{v}^{\mathrm{T}}\mathbf{x}$$
$$\mathbf{z} := \mu\mathbf{v}$$
$$\mathbf{p} := \mathbf{y} - \mathbf{z}, \quad \mathbf{q} := \mathbf{x} - \mathbf{z}$$
$$A := A - \mathbf{v}\mathbf{p}^{\mathrm{T}} - \mathbf{q}\mathbf{v}^{\mathrm{T}}\}$$

Output   *A* (matrix in Hessenberg form)

Algorithm 11.2.9 requires $O(4n^2)$ multiplications to calculate A_j from A_{j-1} as opposed to the $O(2n^3)$ multiplications required by direct multiplication of $P_j A_{j-1} P_j$. Thus the entire $n-2$ steps of the reduction process require $O(4n^3)$ and $O(2n^4)$ multiplications, respectively. This does not make any use, however, of the fact that some of the elements of the vectors $\mathbf{v}^{(j)}$ are zero. If this is taken into account, then the number of multiplications can be reduced to $O(\tfrac{5}{3}n^3)$ (Exercise 6).

Algorithm 11.2.9 can be applied to any matrix, symmetric or nonsymmetric. The algorithm can be made more efficient in the symmetric case, since if A is symmetric, then $\mathbf{x} = \mathbf{y}$ and consequently $\mathbf{p} = \mathbf{q}$.

Example 11.2.10. Table 11.2.11 shows the result of applying the Householder reduction algorithm to the 5×5 symmetric matrix defined by $A_{ij} = ij/(i+j)$. The machine unit for this computation was $\mu = 2 \times 10^{-12}$, and thus the elements above and below the upper and lower principal subdiagonals are zero to machine accuracy. While in theory the result is a symmetric tridiagonal matrix, this symmetry is not completely maintained because of the nonassociativity of computer arithmetic.

Table 11.2.12 shows the reduction to Hessenberg form of the 5×5 symmetric matrix $A_{ij} = ij$. Since the machine unit for the arithmetic is about $\mu = 2 \times 10^{-12}$, it is not clear that all the elements of the last two or three rows are different from zero. In fact it is seen easily that the matrix has $\lambda = 0$ as an eigenvalue of multiplicity four.

TABLE 11.2.11

5.000E − 01	−1.530E + 00	−1.245E − 12	−9.197E − 13	−2.942E − 13
−1.530E + 00	6.781E + 00	6.205E − 01	−9.800E − 13	2.712E − 13
−1.245E − 12	6.205E − 01	2.150E − 01	−9.663E − 03	4.974E − 13
−9.197E − 13	0.000E + 00	−9.663E − 03	3.698E − 03	−1.389E − 04
−2.942E − 13	0.000E + 00	0.000E + 00	−1.389E − 04	3.513E − 05

TABLE 11.2.12

1.000E + 00	−7.348E + 00	−8.192E − 12	3.162E − 12	1.519E − 12
−7.348E + 00	5.400E + 01	1.176E − 10	7.819E − 12	−9.549E − 12
−8.192E − 12	1.099E − 10	−6.485E − 11	−1.498E − 11	0.000E + 00
3.162E − 12	−2.211E − 23	−1.498E − 11	3.841E − 12	6.647E − 12
1.519E − 12	−7.153E − 23	−1.323E − 23	6.647E − 12	2.796E − 12

Eigenvalues of a Tridiagonal Symmetric Matrix

The application of Householder transformations to a symmetric matrix leads to a symmetric tridiagonal matrix. Since the Householder transformations are similarity transformations, the eigenvalues of the tridiagonal matrix are the same as those of the original matrix. Thus once we have a scheme for solving symmetric tridiagonal matrices, we then have a means of finding the eigenvalues of any symmetric matrix.

Let T be a symmetric tridiagonal matrix with entries $a_{11}, \ldots, a_{nn}$ along the main diagonal and $b_1, \ldots, b_{n-1}$ along the upper and lower principal subdiagonals. The characteristic polynomial is then

$$\det(T - \lambda I) =$$

$$\det \begin{bmatrix} a_{11} - \lambda & b_1 & 0 & \cdots & 0 & 0 & 0 \\ b_1 & a_{22} - \lambda & b_2 & \cdots & 0 & 0 & 0 \\ 0 & b_2 & a_{33} - \lambda & \cdots & 0 & 0 & 0 \\ \cdots & \cdots & \cdots & \cdots & \cdots & \cdots & \cdots \\ 0 & 0 & 0 & \cdots & a_{n-2,n-2} - \lambda & b_{n-2} & 0 \\ 0 & 0 & 0 & \cdots & b_{n-2} & a_{n-1,n-1} - \lambda & b_{n-1} \\ 0 & 0 & 0 & \cdots & 0 & b_{n-1} & a_{nn} - \lambda \end{bmatrix}$$

$$(11.2.40)$$

Let

$$M_1 := [a_{11} - \lambda], \quad M_2 := \begin{bmatrix} a_{11} - \lambda & b_1 \\ b_1 & a_{22} - \lambda \end{bmatrix},$$

$$M_3 := \begin{bmatrix} a_{11} - \lambda & b_1 & 0 \\ b_1 & a_{22} - \lambda & b_2 \\ 0 & b_2 & a_{33} - \lambda \end{bmatrix}, \ldots \qquad (11.2.41)$$

be the leading principal submatrices of $T - \lambda I$. It is then easily seen by expanding about the last row of M_i that

$$\det(M_i) = (a_{ii} - \lambda)\det(M_{i-1}) - b_{i-1}^2 \det(M_{i-2}) \qquad (11.2.42)$$

With the definitions $P_0 \equiv 1$, $P_j(\lambda) := \det(M_j)$, $j = 1, \ldots, n$, we then have the recurrence relation

$$P_i(\lambda) = (a_{ii} - \lambda)P_{i-1} - b_{i-1}^2 P_{i-2}(\lambda) \qquad i = 2, \ldots, n \qquad (11.2.43)$$

which can be used to calculate $P_n(\lambda) = \det(T - \lambda I)$, whose roots are the eigenvalues of T. Computing $P_n(\lambda)$ in this way requires only $2n - 4$ multiplications and $2n - 2$ squares. Thus, unlike the situation for a general matrix where calculating the characteristic polynomial requires $n!$ multiplications, it is feasible to find the eigenvalues of a tridiagonal matrix by applying a root-finding scheme to its characteristic polynomial.

The *Householder-Given's* algorithm for finding the eigenvalues of a symmetric matrix consists of the following steps:

1. Reduce the matrix to a symmetric tridiagonal matrix by Householder transformations.
2. Find a set of intervals $[a_i, b_i]$ each of which contains a single root of $P_n(\lambda)$.
3. Solve for the root in each interval $[a_i, b_i]$ by the bisection method (see Sec. 2.1).

Accomplishing step 2 requires use of theorems about the relations among the polynomials $P_i(\lambda)$.

Theorem 11.2.13. If the coefficients b_i in $T - \lambda I$ are all nonzero, then the roots of the polynomial $P_i(\lambda)$ are distinct and separated from each other by the roots of $P_{i-1}(\lambda)$.

Proof. We first establish that two successive polynomials P_{i-1}, P_i cannot have a common root. Suppose that $P_i(c) = P_{i-1}(c) = 0$ for some c. Then (11.2.43) implies that $-b_{i-1}^2 P_{i-2}(c) = 0$ which in view of the hypothesis $b_i \neq 0$, $i = 1, \ldots, n-1$, implies that $P_{i-2}(c) = 0$. This then permits us to argue that $P_{i-3}(c) = \cdots = P_1(c) = 0$, but clearly P_1 and P_2 have no common zero when $b_1 \neq 0$. Therefore, once we have demonstrated the root separation property, it follows that the roots of each P_i must be distinct.

    The root separation property is established by induction. Let $\lambda_j^{(i)}$, $j = 1, \ldots, i$, with $\lambda_j^{(i)} < \lambda_{j+1}^{(i)}$, denote the roots of P_i. It is seen easily that the root $\lambda_1^{(1)} = a_1$ of P_1 separates the two roots of P_2. Assume as the induction hypothesis that P_{i-2} and P_{i-1} have the root separation property. For any root $\lambda_j^{(i-1)}$, $j = 1, \ldots, i-1$, of P_{i-1} we have $P_i(\lambda_j^{(i-1)}) = -b_{i-1}^2 P_{i-2}(\lambda_i^{(i-1)})$, and thus P_i and P_{i-2} have opposite signs at the roots of P_{i-1}. It is seen easily that $P_i(\lambda) = (-1)^i \lambda^i + O(\lambda^{i-1})$. Thus as $\lambda \to +\infty$, both P_{i-2} and P_i approach either $+\infty$ if i is even or $-\infty$ if i is odd. Since P_{i-2} has no roots greater than the largest root $\lambda_{i-1}^{(i-1)}$ of P_{i-1}, it follows that $P_{i-2}(\lambda_{i-1}^{(i-1)})$ is positive if i is even and negative if i is odd. Therefore, since $P_i(\lambda_{i-1}^{(i-1)})$ has the opposite sign, it must have at least one root greater than $\lambda_{i-1}^{(i-1)}$. A similar argument establishes that P_i must have at least one root less than the smallest root $\lambda_1^{(i-1)}$ of P_{i-1}. Now consider two consecutive roots $\lambda_j^{(i-1)}$, $\lambda_{j+1}^{(i-1)}$ of P_{i-1}. The polynomial P_{i-2} has opposite signs at these two roots, since by the induction hypothesis its roots separate those of P_{i-1}. Consequently P_i must have opposite signs at these roots, and hence P_i has at least one root in $(\lambda_j^{(i-1)}, \lambda_{j+1}^{(i-1)})$. Since the $i - 1$ roots of P_{i-1} partition the real line into i intervals $(-\infty, \lambda_1^{(i-1)})$, $(\lambda_1^{(i-1)}, \lambda_2^{(i-1)})$, $\ldots$, $(\lambda_{i-1}^{(i-1)}, \infty)$ and we have established that P_i has at least one root in each, it follows that P_i has exactly one root in each. This establishes the root separation property for P_i.

Example 11.2.14. Figure 11.2.15 shows $P_1(\lambda)$, $P_2(\lambda)$, and $P_3(\lambda)$ for the symmetric tridiagonal matrix

$$T = \begin{bmatrix} -1 & -1 & 0 \\ -1 & 1 & 2 \\ 0 & 2 & 2 \end{bmatrix} \tag{11.2.44}$$

Note that the two roots $\lambda_1^{(2)}$, $\lambda_2^{(2)}$ of $P_2(\lambda)$ separate the three roots $\lambda_1^{(3)}$, $\lambda_2^{(3)}$, $\lambda_3^{(3)}$ of $P_3(\lambda)$ and that the root $\lambda_1^{(1)} = -1$ of $P_1(\lambda)$ separates those of $P_2(\lambda)$.

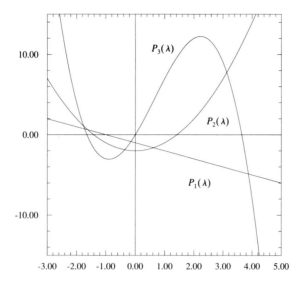

10.00

$P_3(\lambda)$

0.00

$P_2(\lambda)$

$P_1(\lambda)$

-10.00

-3.00 -2.00 -1.00 0.00 1.00 2.00 3.00 4.00 5.00 **FIGURE 11.2.15**

For some λ consider the sequence of signs of the polynomials 1, $P_1(\lambda)$, $P_2(\lambda), \ldots, P_n(\lambda)$. For instance, in Fig. 11.2.15 the sequence of signs is $+ + + -$ for any $\lambda \in (\lambda_1^{(3)}, \lambda_1^{(2)})$. We say there is a *sign agreement* if the sign of $P_i(\lambda)$ is the same as that of $P_{i-1}(\lambda)$. When $P_i(\lambda) = 0$, this is considered a sign agreement with $P_{i-1}(\lambda)$. Thus for any λ between the first roots of $P_3(\lambda)$ and $P_2(\lambda)$ there are two sign agreements in the sequence $+ + + -$. Let $N(\lambda)$ denote the number of sign agreements at λ.

Theorem 11.2.16. For any λ, $N(\lambda)$ is equal to the number of roots of $P_n(\lambda)$ that are greater than λ.

The proof of this theorem follows readily from Theorem 11.2.13 and is left as Exercise 7. Theorem 11.2.16 remains true even if some of the subdiagonal elements b_i are zero since this situation may be regarded as the limiting case of the general situation in which they are nonzero.

Algorithm 11.2.17 indicates the Householder-Given's method which uses Theorem 11.2.16 to find the eigenvalues of a symmetric tridiagonal matrix. It attempts to use a bisection procedure to find intervals containing a single eigenvalue. First, an interval $[a_{max}, b_{max}]$ is found which contains all eigenvalues. This may be done by using the Gerschgorin disk Theorem 11.1.9. Thus we have $N(a_{max}) = n$ and $N(b_{max}) = 0$. If at the midpoint m of $[a_{max}, b_{max}]$, $N(m) = n - 1$, then the interval $[a_{max}, m]$ contains only the leftmost eigenvalue. If $N(m) = n$, then the interval $[m, b_{max}]$ contains all eigenvalues, and we can repeat the bisection process on this interval. Finally, if $N(m) < n - 1$, then the interval $[a_{max}, m]$ contains more than one eigenvalue, and we repeat the bisection process on it. When an interval is found that contains exactly one root, the bisection method for root finding is applied to it to locate the root to within the specified tolerance.

The process could fail to find an interval with one root if there are eigenvalues of higher multiplicity or eigenvalues that differ by not much more than the machine unit of the arithmetic used. Thus an alternative to finding an interval with a single root is necessary. When the bisection process continues until the endpoints of the candidate interval $[a, b]$ differ by less than the error tolerance, then the midpoint of the interval is reported as an eigenvalue $N(a) - N(b)$ times.

Algorithm 11.2.17   Eigenvalues of a symmetric tridiagonal matrix

> *Input* T ($n \times n$ symmetric tridiagonal matrix)
> tol (error tolerance)

> Determine an interval $[a_{max}, b_{max}]$ containing all eigenvalues of T
> $k := n$
> $a := a_{max}, \ b := b_{max}$
> repeat
> $m := (a + b)/2$
>   if $N(m) = k - 1$ then Done:=TRUE
>   while NOT Done
>   {if $N(m) < k - 1$ then $b := m$ else $a := m$
> $m := (a + b)/2$
>   if $b - a < $ tol then
>   {for $i = 1$ to $k - N(b)$
>   {Output: m}
> $k := N(b)$
>   Done := TRUE}
>   else
>   if $N(m) = k - 1$ then Done:=TRUE}
>   if $N(m) = k - 1$ then
>   {$b := m$
>   Apply bisection method to find the root λ in $[a, b]$
>   Output: λ
> $k := k - 1$}
> $a := b, \ b := b_{max}$
> until $k = 0$

The combination of Algorithms 11.2.9 and 11.2.17 now gives us a means of finding the eigenvalues of any symmetric matrix.

Example 11.2.18. Tables 11.2.19 and 11.2.20 show the results of applying Algorithm 11.2.17 to the matrices shown in Tables 11.2.11 and 11.2.12, which were obtained by application of Householder transformations to the symmetric matrices $A_{ij} = ij/(i + j)$ and $A_{ij} = ij$, respectively. The error tolerance was 5×10^{-8}. For the former matrix the algorithm was able to find intervals containing a single eigenvalue for each of the five eigenvalues. The endpoints of these intervals are

TABLE 11.2.19

a_i	b_i	λ_i
$-7.32946E-05$	$5.46950E-04$	$8.45708E-06$
$5.46950E-04$	$1.81864E-02$	$6.44974E-04$
$1.81864E-02$	$1.59026E-01$	$1.85607E-02$
$1.59026E-01$	$4.59548E+00$	$2.94099E-01$
$6.81371E+00$	$7.92282E+00$	$7.18669E+00$

TABLE 11.2.20

a_i	b_i	λ_i
$-6.45945E-09$	$2.51575E-08$	$9.34904E-09$
$-6.45945E-09$	$2.51575E-08$	$9.34904E-09$
$-6.45945E-09$	$2.51575E-08$	$9.34904E-09$
$-6.45945E-09$	$2.51575E-08$	$9.34904E-09$
$5.37674E+01$	$5.76079E+01$	$5.50000E+01$

given in the first two columns of the table as a_i and b_i. For the latter matrix it found an interval about the origin of width less than the error tolerance that still contained four eigenvalues. As noted in Example 11.2.10 this matrix has $\lambda = 0$ as an eigenvalue of multiplicity four.

Once the eigenvalues of the matrix have been found, a corresponding eigenvector can be found by using each approximation in the inverse power method (Algorithm 11.1.8). Since the inverse power method itself refines the estimate of the eigenvalue and converges rapidly when the estimate is near to the actual eigenvalue, it is better to use rather large tolerances in the Householder-Givens algorithm and more stringent tolerances, if required, in the inverse power method.

EXERCISES 11.2

1. Apply Householder transformations to find a tri-diagonal matrix that is similar to

$$\begin{bmatrix} 1 & 2 & 1 & -2 \\ 2 & 2 & 0 & 3 \\ 1 & 0 & 3 & -3 \\ -2 & 3 & -3 & 4 \end{bmatrix} \quad (11.2.45)$$

2. (a) Show that for a 2×2 matrix the most general orthogonal transformation is given by (11.2.7).

   (b) Justify the contention that (11.2.7) rotates any vector clockwise through an angle θ.

3. Prove Theorem 11.2.5.

4. Show that for $\mathbf{w}^{(j)}$ given by (11.2.32), $\sum_{k=1}^{n} (w_k^{(j)})^2 = 1$.

5. Write a computer program that will use Householder transformations to carry a matrix into upper Hessenberg form. Apply your program to the two matrices in Example 11.2.10, the matrix in Exercise 1, and the matrix

$$\begin{bmatrix} 1 & 2 & 0 & 0 & 2 \\ -4 & 3 & -2 & -2 & 0 \\ -3 & 4 & -1 & 0 & 0 \\ 2 & -2 & 0 & 3 & -2 \\ 3 & -4 & 2 & 2 & -1 \end{bmatrix} \quad (11.2.46)$$

6. (a) Count the number of multiplications and divisions required to calculate A_j from A_{j-1} in Algorithm 11.2.9 for a general (nonsymmetric) matrix. What is the corresponding figure for a symmetric matrix?

(*b*) Rewrite Algorithm 11.2.9 so that the fact that the first $j - 1$ components of each vector $\mathbf{w}^{(j)}$ are zero is exploited. Show that the number of multiplications required for the entire reduction to Hessenberg form is now $O(5n^3/3)$.

7. Prove Theorem 11.2.16.

8. (*a*) Write a program that implements the classical Jacobi method for finding eigenvalues. Test the program on the matrix of Exercise 1 and on the matrix of Exercise 3 in Sec. 11.1.

   (*b*) Extend the program so that it uses the inverse power method to find eigenvectors associated with each eigenvalue. Test your programs on the matrices of part (*a*).

9. (*a*) Write a program that uses Algorithms 11.2.9 and 11.2.17 to find the eigenvalues of a sym-

metric matrix. Test the program on the matrix of Exercise 1 and on the matrix of Exercise 3 in Sec. 11.1.

   (*b*) Extend the program so that it uses the inverse power method to find eigenvectors associated with each eigenvalue. Test your programs on the matrices of part (*a*).

10. Compare the time required by the classical Jacobi method (Exercise 8*a*) to find the eigenvalues of a symmetric matrix to those required by the Householder-Givens approach (Exercise 9). Use symmetric matrices with elements produced by a random number generator. Also compare the performance of the two algorithms on the matrix $[ij/(i + j)]_{i,j=1}^n$.

11.3   THE QR ALGORITHM

The Jacobi and Householder-Givens methods of finding all eigenvalues considered in the previous section are applicable only to symmetric matrices. In this section we describe the *QR algorithm* which can find all eigenvalues of a general matrix.

Starting with $A_1 := A$, the QR algorithm iteratively computes similar matrices A_i, $i = 2, \ldots$, in two stages:

1. Factor A_i into $Q_i R_i$, where Q_i is an orthogonal matrix and R_i is an upper triangular matrix.
2. Define $A_{i+1} := R_i Q_i$.

Note that $R_i := Q_i^{-1} A_i$ from stage 1, and thus from stage 2 $A_{i+1} = Q^{-1} A_i Q = Q_i^T A_i Q_i$, whence all A_i are similar to A and thus have the same eigenvalues. It turns out that in the case where the eigenvalues of A all have different magnitudes, $|\lambda_1| > |\lambda_2| > \cdots > |\lambda_n|$, the QR iterates A_i approach an upper triangular matrix, and thus elements of the main diagonal approach the eigenvalues. When there are distinct eigenvalues of the same size, the iterates A_i may not approach an upper triangular matrix; however, they do approach a matrix that is near enough to an upper triangular matrix to allow us to find the eigenvalues. The convergence of the QR algorithm will be discussed in more detail later.

Factorization of a Matrix into QR Form

Let us first consider the problem of factoring a matrix A into QR form. Let the unknown columns of Q be denoted by $\mathbf{q}^{(j)}$, $j = 1, \ldots, n$. For Q to be orthogonal we need $Q^{-1} = Q^T$, that is, $Q^TQ = I$. For this condition to be met it is necessary and sufficient that

$$\mathbf{q}^{(i)T}\mathbf{q}^{(j)} = \mathbf{q}^{(i)} \cdot \mathbf{q}^{(j)} = \begin{cases} 1 & i = j \\ 0 & i \neq j \end{cases} \tag{11.3.1}$$

A set of vectors $\{\mathbf{q}^{(i)}\}$ satisfying (11.3.1) is said to be *orthonormal*.

The matrix A is to have the factorization

$$\begin{bmatrix} a_{11} & \cdots & a_{1n} \\ a_{21} & \cdots & a_{2n} \\ \cdots & & \cdots \\ a_{n1} & \cdots & a_{nn} \end{bmatrix} = \begin{bmatrix} q_{11} & \cdots & q_{1n} \\ q_{21} & \cdots & q_{2n} \\ \cdots & & \cdots \\ q_{n1} & \cdots & q_{nn} \end{bmatrix}\begin{bmatrix} r_{11} & r_{12} & \cdots & r_{1n} \\ 0 & r_{22} & \cdots & r_{2n} \\ \cdots & & & \cdots \\ 0 & 0 & \cdots & r_{nn} \end{bmatrix} \tag{11.3.2}$$

This requires

$$\mathbf{a}^{(1)} = r_{11}\mathbf{q}^{(1)} \tag{11.3.3}$$

where $\mathbf{a}^{(1)}$ denotes the first column of the matrix A. Upon taking the 2-norm of each side of (11.3.3) we obtain $\|\mathbf{a}^{(1)}\|_2 = \|r_{11}\mathbf{q}^{(1)}\|_2 = r_{11}$ from (11.3.1), assuming that we choose $r_{11} > 0$. We then have

$$\mathbf{q}^{(1)} = \frac{1}{\|\mathbf{a}^{(1)}\|_2}\mathbf{a}^{(1)} \tag{11.3.4}$$

From (11.3.2)

$$\mathbf{a}^{(2)} = r_{12}\mathbf{q}^{(1)} + r_{22}\mathbf{q}^{(2)} \tag{11.3.5}$$

Upon taking the dot product of this equation with $\mathbf{q}^{(1)}$ we obtain

$$\mathbf{a}^{(2)} \cdot \mathbf{q}^{(1)} = r_{12}\mathbf{q}^{(1)} \cdot \mathbf{q}^{(1)} + r_{22}\mathbf{q}^{(2)} \cdot \mathbf{q}^{(1)} = r_{12} \tag{11.3.6}$$

from (11.3.1). From (11.3.5)

$$\|\mathbf{a}^{(2)} - r_{12}\mathbf{q}^{(1)}\|_2 = \|r_{22}\mathbf{q}^{(2)}\|_2 = r_{22} \tag{11.3.7}$$

if we choose $r_{22} > 0$, and hence

$$\mathbf{q}^{(2)} = \frac{1}{r_{22}}(\mathbf{a}^{(2)} - r_{12}\mathbf{q}^{(1)}) \tag{11.3.8}$$

where r_{12}, r_{22}, and $\mathbf{q}^{(1)}$ are given by (11.3.6), (11.3.7), and (11.3.4).

In general assume that $\mathbf{q}^{(1)}, \ldots, \mathbf{q}^{(j)}$ have been determined. From (11.3.2)

$$\mathbf{a}^{(j+1)} = \sum_{i=1}^{j} r_{i,j+1}\mathbf{q}^{(i)} \tag{11.3.9}$$

Proceeding as above we obtain

$$r_{i,j+1} = \mathbf{a}^{(j+1)} \cdot \mathbf{q}^{(i)} \qquad i = 1, \ldots, j$$

$$r_{j+1,j+1} = \left\| \mathbf{a}^{(j+1)} - \sum_{i=1}^{j} r_{i,j+1}\mathbf{q}^{(i)} \right\|_2 \tag{11.3.10}$$

$$\mathbf{q}^{(j+1)} = \frac{1}{r_{j+1,j+1}} \left(\mathbf{a}^{(j+1)} - \sum_{i=1}^{j} r_{i,j+1}\mathbf{q}^{(i)} \right)$$

(Exercise 2).

In deriving (11.3.10) we have assumed at each stage that $r_{ii} \neq 0$, and the construction of Q fails without this. From (11.3.10) we see that $r_{j+1,j+1} = 0$ implies that $\mathbf{a}^{(j+1)}$ is a linear combination of the vectors $\mathbf{q}^{(i)}$, $i = 1, \ldots, j$, each of which is a linear combination of the vectors $\mathbf{a}^{(1)}, \ldots, \mathbf{a}^{(j)}$. Thus the factorization procedure will succeed as long as no column vector of A is a linear combination of other column vectors, that is, when A is nonsingular. Note that the only latitude in the construction of Q and R was in the choice of the signs of the elements along the main diagonal of R. Hence we have shown the following theorem.

> **Theorem 11.3.1.** Any nonsingular matrix A can be factored into a product $A = QR$ of an orthogonal matrix Q and an upper triangular matrix R. If it is required that the diagonal elements of R be positive, then the matrix Q is uniquely determined.

Gramm-Schmidt Orthonormalization

The process that was used above to obtain the orthogonal matrix Q for the factorization $A = QR$ was, in fact, an application of the *Gramm-Schmidt* process for constructing an orthonormal basis spanning the same vector space as some given nonorthogonal basis. In this case the nonorthogonal basis was the columns of the matrix A, and the resulting orthonormal basis became the columns of Q.

The Gramm-Schmidt process has a geometric interpretation that is useful in understanding the QR method. The vector $r_{12}\mathbf{q}^{(1)}$ in (11.3.8) is the component of the vector $\mathbf{a}^{(2)}$ in the direction of $\mathbf{q}^{(1)}$, which has the same direction as $\mathbf{a}^{(1)}$. Clearly, subtracting this component from $\mathbf{a}^{(2)}$ leaves a vector that is perpendicular to $\mathbf{a}^{(1)}$. An equivalent way of looking at this is that $\mathbf{a}^{(2)} - r_{12}\mathbf{q}^{(1)}$ is the projection of the vector $\mathbf{a}^{(2)}$ into the $(n-1)$-dimensional plane perpendicular to $\mathbf{a}^{(1)}$. See Fig. 11.3.2. Similarly, the vector $\mathbf{q}^{(3)}$ is obtained by first projecting the vector $\mathbf{a}^{(3)}$ into the plane perpendicular to $\mathbf{a}^{(1)}$ and then projecting this projection into the $(n-2)$-dimensional plane perpendicular to the plane spanned by the vectors $\mathbf{q}^{(1)}$ and $\mathbf{q}^{(2)}$.

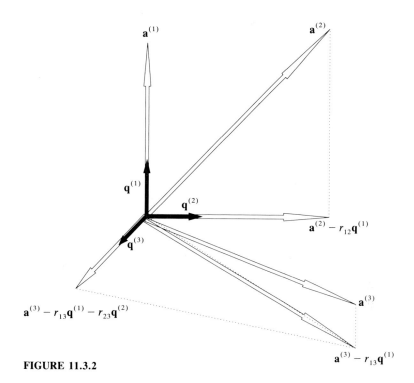

FIGURE 11.3.2

Convergence of the QR Method

As mentioned previously it is hoped that the QR iterates A_m approach a matrix of upper triangular form. We now want to see why this is the case. In doing so we will establish that the QR method is in some sense an extension of the power method discussed in Sec. 11.1. Our approach, which follows that of Watkins (1982), will be geometric and intuitive rather than analytic.

The QR method generates a sequence of iterates $A_{i+1} = Q_i^T A_i Q_i$. Consequently we have

$$A_{m+1} = Q_m^T A_m Q_m = Q_m^T Q_{m-1}^T A_{m-1} Q_{m-1} Q_m = \cdots = Q_m^T \cdots Q_1^T A Q_1 \cdots Q_m \tag{11.3.11}$$

With $Q_m' := Q_1 \cdots Q_m$, we see that A_{m+1} is related to the original matrix A by the similarity transformation

$$A_{m+1} = Q_m'^T A Q_m' \tag{11.3.12}$$

where the matrix Q_m' is orthogonal.

Now observe that

$$
\begin{aligned}
A^2 &= AA \\
&= Q_1 R_1 Q_1 R_1 \\
&= Q_1 A_2 R_1 \\
&= Q_1 Q_2 R_2 R_1 \\
&=: Q_2' R_2'
\end{aligned}
\tag{11.3.13}
$$

With $R'_m := R_m \cdots R_1$, it is easy to show by a similar argument that the mth power of the matrix A is

$$A^m = Q'_m R'_m \qquad (11.3.14)$$

(Exercise 3). Thus the matrix Q'_m occurring in (11.3.12) is the orthogonal matrix that occurs in the QR factorization of the power matrix A^m. Consequently from Theorem 11.3.1 the columns of Q'_m are the orthonormalization of the columns of the matrix A^m.

The columns of A^m are $A^m \mathbf{e}^{(1)}, \ldots, A^m \mathbf{e}^{(n)}$, where $\mathbf{e}^{(1)} = [1, 0, \ldots, 0]^T, \ldots, \mathbf{e}^{(n)} = [0, \ldots, 0, 1]^T$ is the standard basis for $\mathbb{R}^n$. Let us assume first that the eigenvalues of A satisfy $|\lambda_1| > \cdots > |\lambda_n|$. In this case the eigenvectors $\mathbf{v}^{(1)}, \ldots, \mathbf{v}^{(n)}$ of A span $\mathbb{R}^n$, and thus each standard basis vector has the representation $e^{(j)} = \Sigma_{k=1}^n c_k^{(j)} \mathbf{v}^{(k)}$, whence

$$A^m \mathbf{e}^{(1)} = \lambda_1^m c_1^{(1)} \mathbf{v}^{(1)} + \lambda_2^m c_2^{(1)} \mathbf{v}^{(2)} + \cdots + \lambda_n^m c_n^{(1)} \mathbf{v}^{(n)}$$

$$A^m \mathbf{e}^{(2)} = \lambda_1^m c_1^{(2)} \mathbf{v}^{(1)} + \lambda_2^m c_2^{(2)} \mathbf{v}^{(2)} + \cdots + \lambda_n^m c_n^{(2)} \mathbf{v}^{(n)} \qquad (11.3.15)$$

$$\cdots \cdots \cdots \cdots \cdots \cdots \cdots \cdots \cdots \cdots \cdots \cdots \cdots \cdots \cdots \cdots$$

$$A^m \mathbf{e}^{(n)} = \lambda_1^m c_1^{(n)} \mathbf{v}^{(1)} + \lambda_2^m c_2^{(n)} \mathbf{v}^{(2)} + \cdots + \lambda_n^m c_n^{(n)} \mathbf{v}^{(n)}$$

For m sufficiently large, all the column vectors $A^m \mathbf{e}^{(j)}$ of A^m will point approximately in the direction of $\mathbf{v}^{(1)}$ since λ_1^m is much larger than any other λ_k^m. Let us consider what happens when they are orthonormalized to produce the columns $\mathbf{q}^{(1)'}, \ldots, \mathbf{q}^{(n)'}$ of Q'_m. Upon dividing $A^m \mathbf{e}^{(1)}$ by its length, which is approximately $|\lambda_1^m c_1^{(1)}|$, we obtain $\mathbf{q}^{(1)'} \simeq \pm \mathbf{v}^{(1)}$. The second column $\mathbf{q}^{(2)'}$ of Q'_m is obtained by projecting $A^m \mathbf{e}^{(2)}$ into the plane perpendicular to $\mathbf{q}^{(1)'}$. Thus $\mathbf{q}^{(1)'}$ and $\mathbf{q}^{(2)'}$ lie in the same plane as $A^m \mathbf{e}^{(1)}$ and $A^m \mathbf{e}^{(2)}$. Since λ_2^m is much larger than λ_k^m, $k \geq 3$, we see from (11.3.15) that the plane of $\mathbf{q}^{(1)'}$ and $\mathbf{q}^{(2)'}$ is approximately the same as that of $\mathbf{v}^{(1)}$ and $\mathbf{v}^{(2)}$. If we let $\langle \mathbf{v}^{(1)}, \mathbf{v}^{(2)} \rangle$ denote the subspace spanned by $\mathbf{v}^{(1)}$ and $\mathbf{v}^{(2)}$, then this can be expressed as $\langle \mathbf{q}^{(1)'}, \mathbf{q}^{(2)'} \rangle \simeq \langle \mathbf{v}^{(1)}, \mathbf{v}^{(2)} \rangle$. Because it has been assumed that all the eigenvalues have distinct magnitudes, this sort of reasoning can be continued to conclude that

$$\langle \mathbf{q}^{(1)'} \rangle \simeq \langle \mathbf{v}^{(1)} \rangle$$

$$\langle \mathbf{q}^{(1)'}, \mathbf{q}^{(2)'} \rangle \simeq \langle \mathbf{v}^{(1)}, \mathbf{v}^{(2)} \rangle \qquad (11.3.16)$$

$$\cdots \cdots \cdots \cdots \cdots \cdots \cdots \cdots \cdots \cdots \cdots \cdots \cdots \cdots$$

$$\langle \mathbf{q}^{(1)'}, \ldots, \mathbf{q}^{(n-1)'} \rangle \simeq \langle \mathbf{v}^{(1)}, \ldots, \mathbf{v}^{(n-1)} \rangle$$

A vector subspace S is *invariant* under the transformation T if $\mathbf{v} \in S$ implies that $T\mathbf{v} \in S$. Clearly, any subspace spanned by the eigenvectors of a matrix A is invariant under multiplication by A. From (11.3.16) we conclude that the spaces $\langle \mathbf{q}^{(1)'} \rangle$, $\langle \mathbf{q}^{(1)'}, \mathbf{q}^{(2)'} \rangle$, . . . are approximately invariant under multiplication by A.

The similarity relation (11.3.12) can be written

$$
A_{m+1} =
\begin{bmatrix}
\mathbf{q}^{(1)'\mathrm{T}} \\
\mathbf{q}^{(2)'\mathrm{T}} \\
\cdots \\
\mathbf{q}^{(n)'\mathrm{T}}
\end{bmatrix}
A[\mathbf{q}^{(1)'}, \mathbf{q}^{(2)'}, \ldots, \mathbf{q}^{(n)'}]
$$

$$
=
\begin{bmatrix}
\mathbf{q}^{(1)'\mathrm{T}} A\mathbf{q}^{(1)'} & \mathbf{q}^{(1)'\mathrm{T}} A\mathbf{q}^{(2)'} & \cdots & \mathbf{q}^{(1)'\mathrm{T}} A\mathbf{q}^{(n)'} \\
\mathbf{q}^{(2)'\mathrm{T}} A\mathbf{q}^{(1)'} & \mathbf{q}^{(2)'\mathrm{T}} A\mathbf{q}^{(2)'} & \cdots & \mathbf{q}^{(2)'\mathrm{T}} A\mathbf{q}^{(n)'} \\
\cdots\cdots\cdots\cdots\cdots\cdots\cdots\cdots\cdots\cdots\cdots\cdots\cdots\cdots \\
\mathbf{q}^{(n)'\mathrm{T}} A\mathbf{q}^{(1)'} & \mathbf{q}^{(n)'\mathrm{T}} A\mathbf{q}^{(2)'} & \cdots & \mathbf{q}^{(n)'\mathrm{T}} A\mathbf{q}^{(n)'}
\end{bmatrix}
\quad (11.3.17)
$$

Since, approximately for large m, $A\mathbf{q}^{(1)'} \in \langle \mathbf{q}^{(1)'} \rangle$ and each $\mathbf{q}^{(j)'}$, $j \geq 2$, is orthogonal to $\mathbf{q}^{(1)'}$, the first column of A_{m+1} is approximately zero below the main diagonal. Likewise, $A\mathbf{q}^{(2)'} \in \langle \mathbf{q}^{(1)'}, \mathbf{q}^{(2)'} \rangle$, and since $\mathbf{q}^{(1)'}$ and $\mathbf{q}^{(2)'}$ are orthogonal to $\mathbf{q}^{(j)'}$, $j \geq 3$, the second column of A_{m+1} is approximately zero below the main diagonal. Continuing in this manner we conclude that all entries of A_{m+1} below the main diagonal are approximately zero for m sufficiently large; that is, A_m is approximately upper triangular. Observe that the ratios $(\lambda_{k+1}/\lambda_k)^m$ determine how large m must be for the various approximations (11.3.16) to be accurate. Consequently, the convergence of the QR method could be slow in the case where there are two eigenvalues of nearly equal magnitude.

Example 11.3.3. Table 11.3.4 shows the QR iterates A_9, A_{10}, A_{11}, and A_{12} of the matrix

$$
A =
\begin{bmatrix}
-21 & -9 & 12 \\
0 & 6 & 0 \\
-24 & -8 & 15
\end{bmatrix}
\quad (11.3.18)
$$

which has eigenvalues $\lambda_1 = -9$, $\lambda_2 = 3$, and $\lambda_3 = 6$ (Example 11.1.2). As indicated by the preceding analysis, the elements below the main diagonal approach zero. Notice that above the main diagonal not all elements are converging. This, however, is irrelevant. So long as the elements below the main diagonal converge to zero, the elements of the main diagonal will approach the eigenvalues.

The analysis above applied to the situation in which all eigenvalues had different magnitudes. Let us now consider the situation in which $|\lambda_1| = |\lambda_2| > |\lambda_3| > \cdots > |\lambda_n|$. We continue to assume that the matrix A is nondefective and thus that its eigenvectors span $\mathbb{R}^n$. In this case we cannot conclude that $\langle \mathbf{q}^{(1)'} \rangle \simeq \langle \mathbf{v}^{(1)} \rangle$ in (11.3.16); however, the remaining relations are still valid.

TABLE 11.3.4

A_9		
$-8.99954E + 00$	$3.44109E + 00$	$-3.77977E + 01$
$-3.49643E - 05$	$6.00001E + 00$	$-7.07225E - 01$
$-1.48339E - 04$	$3.40309E - 05$	$2.99953E + 00$

A_{10}		
$-9.00015E + 00$	$-3.44129E + 00$	$3.77975E + 01$
$-1.16558E - 05$	$6.00000E + 00$	$-7.07067E - 01$
$-4.94515E - 05$	$-1.13450E - 05$	$3.00016E + 00$

A_{11}		
$-8.99995E + 00$	$3.44122E + 00$	$-3.77976E + 01$
$-3.88515E - 06$	$6.00000E + 00$	$-7.07120E - 01$
$-1.64833E - 05$	$3.78151E - 06$	$2.99995E + 00$

A_{12}		
$-9.00002E + 00$	$-3.44124E + 00$	$3.77976E + 01$
$-1.29506E - 06$	$6.00000E + 00$	$-7.07102E - 01$
$-5.49448E - 06$	$-1.26052E - 06$	$3.00002E + 00$

Thus for large m the matrix A_{m+1} in (11.3.17) has the form (approximately)

$$A_{m+1} = \begin{bmatrix} \mathbf{q}^{(1)'^T}A\mathbf{q}^{(1)'} & \mathbf{q}^{(1)'^T}A\mathbf{q}^{(2)} & \mathbf{q}^{(1)'^T}A\mathbf{q}^{(3)'} & \cdots & \mathbf{q}^{(1)'^T}A\mathbf{q}^{(n)'} \\ \mathbf{q}^{(2)'^T}A\mathbf{q}^{(1)'} & \mathbf{q}^{(2)'^T}A\mathbf{q}^{(2)} & \mathbf{q}^{(2)'^T}A\mathbf{q}^{(3)'} & \cdots & \mathbf{q}^{(2)'^T}A\mathbf{q}^{(n)'} \\ 0 & 0 & \mathbf{q}^{(3)'^T}A\mathbf{q}^{(3)'} & \cdots & \mathbf{q}^{(3)'^T}A\mathbf{q}^{(n)'} \\ \cdots & \cdots & \cdots & \cdots & \cdots \\ 0 & 0 & 0 & \cdots & \mathbf{q}^{(n)'^T}A\mathbf{q}^{(n)'} \end{bmatrix}$$

$$(11.3.19)$$

While in this case the eigenvalues cannot simply be read from the diagonal, observe that the characteristic polynomial of (11.3.19) is

$$\det \begin{bmatrix} \mathbf{q}^{(1)'^T}A\mathbf{q}^{(1)'} - \lambda & \mathbf{q}^{(1)'^T}A\mathbf{q}^{(2)'} \\ \mathbf{q}^{(2)'^T}A\mathbf{q}^{(1)'} & \mathbf{q}^{(2)'^T}A\mathbf{q}^{(2)'} - \lambda \end{bmatrix} (\mathbf{q}^{(3)'^T}A\mathbf{q}^{(3)'} - \lambda) \cdots (\mathbf{q}^{(N)'^T}A\mathbf{q}^{(N)'} - \lambda)$$

$$(11.3.20)$$

The eigenvalues of the first 2×2 block can be found by applying the quadratic formula to its characteristic polynomial, and hence the situation in which two eigenvalues have the same magnitude poses little difficulty.

Example 11.3.5. Table 11.3.6 shows the QR iterates A_9, A_{10}, A_{11}, and A_{12} for the matrix

$$A = \begin{bmatrix} 1 & 1 & -1 \\ -2 & 2 & -3 \\ 0 & 1 & 0 \end{bmatrix} \qquad (11.3.21)$$

TABLE 11.3.6

A_9		
3.26904E − 01	1.18159E + 00	8.92247E − 01
−3.76898E + 00	1.67248E + 00	8.34995E − 01
0.00000E + 00	8.76859E − 04	1.00061E + 00

A_{10}		
1.88518E + 00	3.63333E + 00	−7.57183E − 01
−1.31674E + 00	1.14846E − 01	9.61650E − 01
0.00000E + 00	6.63423E − 04	9.99976E − 01

A_{11}		
2.17184E − 01	3.70445E + 00	−1.17254E + 00
−1.24526E + 00	1.78295E + 00	3.54562E − 01
0.00000E + 00	3.05028E − 04	9.99867E − 01

A_{12}		
1.32049E + 00	1.05280E + 00	−5.50831E − 01
−3.89687E + 00	6.79563E − 01	−1.09416E + 00
0.00000E + 00	7.71073E − 05	9.99952E − 01

which has eigenvalues $\lambda_1 = 1 + 2i$, $\lambda_2 = 1 - 2i$, $\lambda_3 = 1$. The elements of the first 2×2 subblock are not converging; however, note that for the subblock in the iterate A_9,

$$\det \begin{bmatrix} 1.88518 - \lambda & 3.63333 \\ -1.31674 & 0.114\,846 - \lambda \end{bmatrix} = \lambda^2 - 2.000\,026\lambda + 5.000\,656 = 0 \tag{11.3.22}$$

has solutions $\lambda = 1.000\,013 \pm 2.000\,158i$. Even though the corresponding sub-blocks for A_{10}, A_{11}, and A_{12} have different elements, they have approximately the same eigenvalues. Under the assumption $\langle \mathbf{q}^{(1)'}, \mathbf{q}^{(2)'} \rangle = \langle \mathbf{v}^{(1)}, \mathbf{v}^{(2)} \rangle$ this can be shown to be the case independent of the particular choice of the orthonormal vectors $\mathbf{q}^{(1)'}$ and $\mathbf{q}^{(2)'}$ (Exercise 4).

While we assumed for definiteness that there were two eigenvalues having the dominant magnitude, it is clear that a situation such as $|\lambda_1| > \cdots > |\lambda_k| = |\lambda_{k+1}| > \cdots > |\lambda_n|$ would simply locate the 2×2 subblock farther down the principal diagonal. The presence of more than one pair of eigenvalues of the same magnitude similarly would pose no difficulty as it would simply mean the presence of several 2×2 subblocks along the diagonal. However, the presence of three or more eigenvalues of equal magnitude might result in subblocks of dimension three or greater along the main diagonal, and these would pose a problem since it is not easy to find the eigenvalues of such subblocks.

We will not pursue discussion of the convergence of the QR method for general $n \times n$ matrices. The reason is that the calculation of the matrices Q_i and R_i from A_i using (11.3.10) requires $O(n^3/2)$ multiplications. The calculation of $A_{i+1} = R_i Q_i$ requires another $O(n^3/6)$ multiplications. Thus $O(2n^3/3)$ multiplications are required per iteration. For large matrices this would make the QR algorithm prohibitively slow. What is done in practice is first to transform the original matrix A to a matrix H that is in upper Hessenberg form [Eq. (11.2.17)]. How to do this using Householder transformations is indicated in Algorithm 11.2.9. While this transformation requires $O(5n^3/3)$ multiplications (Exercise 6 in Sec. 11.2), we will see that it permits the number of multiplications to be reduced to $O(4n^2)$ per iteration.

It is easily seen from (11.3.10) that the similarity transforms $A_j = Q_{j-1}^T A_{j-1} Q_{j-1}$ preserve Hessenberg form. Thus, starting from Hessenberg form, it is easy to characterize the situation that we hope will arise.

> **Definition 11.3.7.** Let $H_1 := H$ be a matrix in upper Hessenberg form. The sequence of iterates $H_i = Q_{i-1}^T H_{i-1} Q_{i-1}$, $i = 1, 2, \ldots$, generated by the QR method *converges* if at least one of each pair, $h_{k+1,k}$, $h_{k+2,k+1}$, $k = 1, \ldots, n-2$, converges to zero.

Convergence defined in this way implies that the QR iterates will approach a block upper triangular matrix with blocks of dimension one or two along the main diagonal, and as we have seen, it is easy to compute the eigenvalues of a matrix in this form. The next theorem, due to Parlett (1968), gives conditions for the QR iterates to converge.

> **Theorem 11.3.8.** When applied to a matrix H in Hessenberg form, the QR iterates converge if and only if among each set of eigenvalues of equal magnitude at most two have even multiplicity and at most two have odd multiplicity.

Thus for a matrix having no more than two eigenvalues of any given magnitude, the QR iterates converge. This is consistent with the analysis given previously, but the theorem indicates additionally that the multiplicities of the two eigenvalues are immaterial. The theorem also indicates that convergence cannot occur when more than four distinct eigenvalues of the same magnitude are present.

Implementation of the QR Method

Once the matrix is in Hessenberg form, rather than using (11.3.10) it turns out to be more efficient to find Q and R by use of *Given's rotations*. Recall from Eq. (11.2.12) that the matrix

$$\Omega(j, j+1)^T = \begin{bmatrix} 1 & 0 & \cdots & 0 & 0 & \cdots & 0 \\ 0 & 1 & \cdots & 0 & 0 & \cdots & 0 \\ & & \ddots & & & & \\ 0 & 0 & \cdots & \cos\theta & \sin\theta & \cdots & 0 \\ 0 & 0 & \cdots & -\sin\theta & \cos\theta & \cdots & 0 \\ & & & & & \ddots & \\ 0 & 0 & \cdots & 0 & 0 & \cdots & 1 \end{bmatrix} \qquad (11.3.23)$$

is referred to as a rotation in the j, $j+1$ plane. For a Given's rotation, θ is chosen so that the transformation $\Omega(j, j+1)^{\mathrm{T}}H$ annihilates the $(j+1, j)$ element of the matrix H. Clearly the transformation $\Omega(j, j+1)^{\mathrm{T}}H$ preserves any pairs of zeros in jth and $(j+1)$st rows and first through $(j-1)$st columns. Thus if H has upper Hessenberg form, then the transformation

$$R = \Omega(n-1, n)^{\mathrm{T}} \cdots \Omega(2, 3)^{\mathrm{T}}\Omega(1, 2)^{\mathrm{T}}H \qquad (11.3.24)$$

results in an upper triangular matrix. Each matrix $\Omega(j, j+1)$ is clearly orthogonal, and thus so is the matrix

$$Q = \Omega(1, 2)\Omega(2, 3) \cdots \Omega(n-1, n) \qquad (11.3.25)$$

The next QR iterate is then

$$H' = RQ = \Omega(n-1, n)^{\mathrm{T}} \cdots \Omega(1, 2)^{\mathrm{T}}H\Omega(1, 2) \cdots \Omega(n-1, n) \quad (11.3.26)$$

Thus we need not ever explicitly compute the factors Q and R; rather we can find the next iterate by means of the subiteration

$$H := \Omega(j, j+1)^{\mathrm{T}}H\Omega(j, j+1) \qquad j = 1, \ldots, n-1 \qquad (11.3.27)$$

The angle θ to be used in the Given's rotation need never be computed explicitly. The $(j+1, j)$ element of the transformation $H' = \Omega(j, j+1)^{\mathrm{T}}H$ is $h'_{j+1,j} = -\sin\theta\, h_{jj} + \cos\theta\, h_{j+1,j}$. Consequently, if θ is restricted to be in the interval $[0, \pi]$, then

$$\sin\theta = \frac{1}{\sqrt{1+\beta^2}} \qquad \cos\theta = \frac{\beta}{\sqrt{1+\beta^2}} \qquad \text{with } \beta := \frac{h_{jj}}{h_{j+1,j}}$$

$$(11.3.28)$$

Algorithm 11.3.9 gives an implementation of the QR method. The basic steps are conversion to Hessenberg form, computation of the QR iterates via the subiteration $H = \Omega(j, j+1)^{\mathrm{T}}H\Omega(j, j+1)$, testing for convergence according to the criterion set out in Definition 11.3.7, and computation of the eigenvalues after convergence. Observe that after each QR iterate is calculated, the elements of the Hessenberg matrix below the main subdiagonal are set to zero. The reason this is done is that while Hessenberg matrices are invariant under the QR transformation, they are not invariant under the transformations $\Omega(j, j+1)^{\mathrm{T}}H\Omega(j, j+1)$ involved in the subiteration. Thus typically there will be small nonzero numbers below the main subdiagonal after applying the sequence of Given's rotations because of rounding error. If these elements are not reset to zero, their cumulative effect over many iterations can be substantial.

If the matrix H is singular, there is no QR decomposition. However, it turns out that the rounding error entailed in the transformation to Hessenberg form is often enough to make the Hessenberg matrix nonsingular when the original matrix is singular. In this case the QR algorithm will find eigenvalues, but convergence may be slow and the accuracy less than the specific tolerance. See Exercise 9.

Algorithm 11.3.9 QR method

Input A ($n \times n$ matrix)
tol (error tolerance)
MaxIter (maximum permissible number of iterations)

Find an upper Hessenberg matrix H similar to A (Algorithm 11.2.9)

iter:$=0$
repeat
    iter:$=$iter$+1$
    for $j = 1$ to $n - 1$
        {if $h_{j+1,j} \neq 0$ then [find $\Omega(j, j+1)^{\mathrm{T}} H \Omega(j, j+1)$]

$$\{\beta := \frac{h_{jj}}{h_{j+1,j}}$$

$$s := \frac{1}{\sqrt{1+\beta^2}}, \; c := \beta s$$

        $\boldsymbol{\xi} := H(j, \cdot), \; \boldsymbol{\eta} := H(j+1, \cdot)$ [jth and $(j+1)$st rows of H]
        for $k := \max(1, j-2)$ to n
            $\{h_{jk} := c\xi_k + s\eta_k$
             $h_{j+1,k} := -s\xi_k + c\eta_k\}$
        $h_{j+1,j} := 0$
        $\boldsymbol{\xi} := H(\cdot, j), \; \boldsymbol{\eta} := H(\cdot, j+1)$ [jth and $(j+1)$st columns of H]
        for $k = 1$ to $\min(j+2, n)$
            $\{h_{jk} := c\xi_k + s\eta_k$
             $h_{k,j+1} := -s\xi_k + c\eta_k\}$
    }}
    for $j = 1$ to $n - 2$
        for $i = j + 2$ to n
            $h_{ij} := 0$

    Done:$=$TRUE
    if iter $\leq$ MaxIter then
        for $k = 1$ to $n - 2$
            {if Done and $(|h_{k+1,k}| <$ tol or $|h_{k+2,k+1}| <$ tol) then Done:$=$TRUE
            else Done:$=$FALSE}
until Done

if iter $<$ MaxIter then
    $\{k:=1$
    repeat
        if $|h_{k+1,k}| <$ tol then
            {Output: h_{kk} and 0 (real and imaginary parts of an
                eigenvalue)
            $k := k + 1\}$
        else
            {Solve the equation $\lambda^2 - (h_{k,k} + h_{k+1,k+1})\lambda + h_{kk}h_{k+1,k+1}$

$$-h_{k+1,k}h_{k,k+1} = 0$$
for a pair of eigenvalues
Output: real and imaginary part of eigenvalue
$k := k + 2\}$
until $k \geq n$
if $k = n$ then
Output: h_{nn} and 0 (real and imaginary parts of last
eigenvalue)$\}$

else

Output: Maximum number of iterations exceeded

If eigenvectors are needed also, then the eigenvalues found by the QR method can be used as guesses in the inverse power method (Algorithm 11.1.8). In this case it is not necessary to use a stringent tolerance in the QR method since the inverse power method will also refine the initial estimate of each eigenvalue to a specified tolerance.

Example 11.3.10. Algorithm 11.3.9 was applied to find the eigenvalues of the matrix

$$A = \begin{bmatrix} 1 & 2 & 0 & 0 & 2 \\ -4 & 3 & -2 & -2 & 0 \\ -3 & 4 & -1 & 0 & 0 \\ 2 & -2 & 0 & 3 & -2 \\ 3 & -4 & 2 & 2 & -1 \end{bmatrix} \tag{11.3.29}$$

to within a tolerance of 5×10^{-6}. The actual eigenvalues of this matrix are $\lambda_1 = 3$, $\lambda_{2,3} = 1 \pm 2i$, $\lambda_{4,5} = \pm i$. Table 11.3.11 shows the forty-third QR iterate of the matrix after it had been changed to upper Hessenberg form. This table was printed before the step in the algorithm that sets elements below the main subdiagonal to zero. Note that it meets the convergence criterion of Definition 11.3.7 by virtue of the $(2,1)$ and $(4,3)$ entries being less than 5×10^{-6}. Table 11.3.12 shows the computed eigenvalues, which are seen to be within the specified tolerance.

TABLE 11.3.11

3.00000E + 00	4.67348E + 00	3.98224E − 01	7.54247E + 00	−3.33333E + 00
4.33834E − 06	5.38544E − 01	−2.65709E + 00	−2.96814E − 01	−1.04945E − 01
3.46945E − 18	1.58555E + 00	1.46146E + 00	1.11591E + 00	3.94527E − 01
5.43247E − 33	−1.07706E − 27	4.69741E − 15	1.11111E − 01	3.22126E + 00
0.00000E + 00	3.04638E − 27	−1.61559E − 27	−3.14270E − 01	−1.11111E − 01

TABLE 11.3.12

Re(λ_i)	Im(λ_i)
2.999 995 758 2E + 00	0.000 000 000 0E + 00
1.000 002 120 8E + 00	2.000 002 947 8E + 00
1.000 002 120 8E + 00	−2.000 002 947 8E + 00
−6.082 245 818 1E − 12	9.999 999 999 6E − 01
−6.082 245 818 1E − 12	−9.999 999 999 6E − 01

TABLE 11.3.13

Dimension	Most	Least	Average
10×10	711	103	404
20×20	9026	332	2271

As indicated before, the QR method as implemented in Algorithm 11.3.9 requires $O(4n^2)$ multiplications per iteration. For matrices of large dimension this still may mean rather long execution times. Table 11.3.13 shows the results of applying Algorithm 11.3.9 to 10 matrices of dimension ten and 10 matrices of dimension twenty with elements between -1 and 1 assigned by a random number generator. The most, least, and average number of iterations for the 10 matrices of each dimension are shown. The error tolerance was 5×10^{-6}. As can be seen the QR method may require many iterations to converge. Recall that the rate of convergence of the method depends upon the ratios $|\lambda_k/\lambda_{k+1}|$, where the eigenvalues are assumed to be enumerated in descending order. This accounts for the wide disparity in the number of iterations necessary.

Origin Shifts

As Table 11.3.13 suggests it is desirable to accelerate the convergence of the QR method, especially for large matrices. As with the power methods discussed in Sec. 11.1, this may be accomplished with shifts of origin. The modified QR algorithm for a matrix $H_1 = H$ in Hessenberg form is then:

1. Factor $H_i - qI$ into a product $Q_i R_i$ of an orthogonal matrix Q_i and an upper triangular matrix R_i.
2. Define $H_{i+1} := R_i Q_i + qI$.

It is easily shown that $H_{i+1} = Q_i^T H_i Q_i$, and thus the iterates H_i are all similar and consequently have the same eigenvalues.

Let us suppose that the eigenvalues of H are all of different magnitudes. Recall that in this case the QR iterates converge to an upper triangular matrix. To accelerate the convergence of the method, the shift q should be chosen to be an approximation to some eigenvalue of H. According to the Gerschgorin disk Theorem 11.1.9, there is an eigenvalue in the interval $[h_{nn} - |h_{n,n-1}|, h_{nn} + |h_{n,n-1}|]$, and unless $|h_{n,n-1}|$ is much larger than the other elements of the matrix, h_{nn} will be the best available estimate of an eigenvalue. Thus we start with $q = h_{nn}$ and after each iteration set q equal to the new value of h_{nn}. When $|h_{n,n-1}| < \text{tol}$, h_{nn} is accepted as the approximation to an eigenvalue. At this point the nth row and column of the matrix can be eliminated and the process repeated for the resulting smaller matrix. The implementation of the process described here is Project 2.

The use of origin shifts described above will become more complicated if eigenvalues of equal magnitude are present, and especially if complex conjugate pairs of eigenvalues are present since this would entail choices of the estimate q that are complex numbers. In this situation a variation of the QR method called *double QR* is commonly used. In the double QR method the QR iterations are grouped in pairs:

$$H_{2i-1} - q_{2i-1}I = Q_{2i-1}R_{2i-1}$$
$$H_{2i} = R_iQ_i + q_{2i-1}I$$

$$(11.3.30)$$

$$H_{2i} - q_{2i}I = Q_{2i}R_{2i}$$
$$H_{2i+1} = R_{2i}Q_{2i}$$

It can be shown that if the shifts are chosen so that $q_{2i} = \bar{q}_{2i-1}$, then the iterates $H_1 = H, H_3, H_5, \ldots$ will be real-valued (Exercise 12), while the iterates $H_2, H_4, \ldots$ generally will be complex-valued. It is possible, however, to avoid explicit computation of the complex-valued iterates and thereby avoid any use of complex arithmetic. Efficient implementation of the double QR method is very involved however, and we do not pursue it further. Interested readers may consult Golub and Van Loan (1989), Wilkinson (1965), and Young and Gregory (1972) for more about this method.

EXERCISES 11.3

1. (*a*) Use Gramm-Schmidt orthonormalization to find the QR factorization for

$$\begin{bmatrix} 1 & 0 & 2 \\ -2 & 1 & 1 \\ -2 & -5 & 1 \end{bmatrix}$$

(*b*) Use Given's rotations to find the QR factorization for

$$\begin{bmatrix} 3 & 4 & 0 & 1 \\ 1 & -3 & 2 & 1 \\ 0 & -2 & 1 & 0 \\ 0 & 0 & 1 & 2 \end{bmatrix}$$

2. Derive the general formulas (11.3.10) for the QR factorization.

3. Show formula (11.3.14).

4. Suppose that an $n \times n$ matrix A has eigenvalues λ_1 and λ_2 with corresponding eigenvectors $\mathbf{v}^{(1)}$ and $\mathbf{v}^{(2)}$. Show that if $\mathbf{q}^{(1)}$ and $\mathbf{q}^{(2)}$ are orthonormal vectors such that $\langle \mathbf{q}^{(1)}, \mathbf{q}^{(2)} \rangle = \langle \mathbf{v}^{(1)}, \mathbf{v}^{(2)} \rangle$, then the matrix

$$\begin{bmatrix} \mathbf{q}^{(1)^T}A\mathbf{q}^{(1)} & \mathbf{q}^{(1)^T}A\mathbf{q}^{(2)} \\ \mathbf{q}^{(2)^T}A\mathbf{q}^{(1)} & \mathbf{q}^{(2)^T}A\mathbf{q}^{(2)} \end{bmatrix} \quad (11.3.31)$$

also has λ_1 and λ_2 as eigenvalues. (*Hint*: We can write $\mathbf{v}^{(1)} = \mathbf{v}^{(1)} \cdot \mathbf{q}^{(1)}\mathbf{q}^{(1)} + \mathbf{v}^{(1)} \cdot \mathbf{q}^{(2)}\mathbf{q}^{(2)}$.)

5. Show that if $H_1 = H$ has upper Hessenberg form, then all its QR iterates $H_{i+1} = Q_i^T H_i Q_i$ have upper Hessenberg form.

6. Explain why Theorem 11.3.8 implies that the QR method will not converge for an upper Hessenberg matrix having more than four distinct eigenvalues of the same magnitude.

7. Write a computer program implementing the QR Algorithm 11.3.9. Use it to find the eigenvalues of the following matrices:

(*a*) $$\begin{bmatrix} 4 & -1 & 0 & 0 & -1 \\ 1 & 1 & 0 & -1 & -1 \\ -2 & -3 & 1 & -1 & -4 \\ -1 & 1 & 0 & 3 & 1 \\ -1 & 4 & 0 & 1 & 6 \end{bmatrix}$$

(*b*) $$\begin{bmatrix} 2 & 2 & 0 & 0 & 2 \\ 0 & 6 & 0 & 2 & 2 \\ -3 & 4 & -2 & 2 & -2 \\ 1 & -3 & 0 & 2 & -2 \\ -1 & -6 & 0 & -2 & -2 \end{bmatrix}$$

(*c*) $$\begin{bmatrix} 1 & -1 & 0 & 0 & -1 \\ -3 & -2 & 0 & -5 & -1 \\ -6 & -3 & -2 & -5 & -4 \\ 3 & 1 & 0 & 4 & 1 \\ 3 & 4 & 0 & 5 & 3 \end{bmatrix}$$

8. Extend the program of Exercise 7 so that it will use inverse iteration to compute an eigenvector corresponding to each eigenvalue.

9. Test the QR program of Exercise 7 on the following singular matrices:

   (a) $(ij)_{i,j=1}^5$

   (b) $$\begin{bmatrix} 4 & -1 & 0 & 0 & 1 \\ 1 & 1 & 0 & -1 & -1 \\ -2 & -3 & 1 & -1 & 4 \\ -1 & 1 & 0 & 3 & 1 \\ 2 & -2 & 1 & 1 & 5 \end{bmatrix}$$

10. (a) For symmetric matrices the QR method can be simplified somewhat. Write a computer program that takes advantage of the simplifications which derive from specializing to the symmetric case. Test your program on the matrix $[ij/(i+j)]_{i,j=1}^5$ and on the matrix of Exercise 1 in Sec. 11.2.

   (b) Using randomly generated symmetric matrices, compare the time of execution of your QR program for symmetric matrices with
      i. The QR method for general matrices
      ii. The classical Jacobi method
      iii. The Householder-Givens method

11. (a) Show that computation of each QR iterate from the formulas (11.3.10) requires $O(2n^3/3)$ multiplications.

   (b) Show that computation of each QR iterate from the formulas (11.3.26) requires $O(4n^2)$ multiplications.

12. (a) For the double QR method show that $H_{2j+1} = (Q_{2j-1}Q_{2i})^T H_{2i-1}(Q_{2i-1}Q_{2i})$.

   (b) Show that $(H_{2i-1} - q_{2i-1}I)(H_{2i} - q_{2i}I) = Q_{2i-1}Q_{2i}R_{2i-1}R_{2i}$.

   (c) Show that if $q_{2i} = \bar{q}_{2i-1}$ and H_1 is real-valued, then all the odd QR iterates H_{2i+1} are real-valued.

PROJECTS

1. In the classical Jacobi method finding the maximal off-diagonal element in order to annihilate it can be time-consuming. An alternative is simply to proceed through the off-diagonal elements of a symmetric matrix in some systematic fashion, annihilating each in turn until the sum of the squares of all off-diagonal elements is less than the specified tolerance. A commonly used sequence of elements is $(2, 1)$, $(3, 1)$, $(3, 2)$, $(4, 1)$, $(4, 2)$, $(4, 3), \ldots, (n, 1), \ldots, (n, n-1)$. The Jacobi method that annihilates the elements in this order is sometimes referred to as the *row-cyclic* Jacobi method. At the completion of each cycle the sum of squares of all the off-diagonal elements is computed, and, if it is not less than the tolerance, the cycle is repeated. Write a computer program implementing the row-cyclic Jacobi method. By recording times of execution for finding the eigenvalues of various symmetric matrices, compare its efficiency to the classical Jacobi method and, if possible, to the Householder-Givens and QR methods. As the annihilation process proceeds, some of the elements will become quite small, and it may not be worth the computational expense to annihilate them. Experiment with threshold values beneath

which the annihilation of the current element in the cyclic sequence is skipped to determine if this improves the efficiency of the program.

2. Implement the acceleration procedure described in the subsection "*Origin Shifts*" in Sec. 11.3. Compare execution times of this method with the QR Algorithm 11.3.9 for randomly generated matrices.

3. In Sec. 10.7 *stiff* systems of differential equations were considered. For the initial value problem

$$\mathbf{x}' = A\mathbf{x} \qquad \mathbf{x}(0) = x_0 \qquad (1)$$

for a linear system of differential equations, the system is stiff if the eigenvalues of A are negative and vary widely in magnitude. For stiff systems it may be more efficient to use a method such as Richardson extrapolation applied to the implicit trapezoid rule rather than a scheme such as Gragg's method.

   (a) Write a program that will solve the initial value problem (1) when the matrix A is constant. Have the program find the greatest and least eigenvalue of A, and, based upon the disparity of these two eigenvalues, determine whether to use the Gragg method or the implicit trapezoid rule.

(b) If the matrix A depends upon time t, then the stiffness of the system may change as the solution evolves. Modify the program in (a) so that the greatest and least eigenvalues of A are computed at regular intervals and the method of solution is changed if an increase or decrease in the stiffness of the system warrants it. Test the program on matrices of the forms $A(t) = (1 - t)B + tC$ and $A(t) = tB + (1 - t)C$, where the eigenvalues of B are widely dispersed but those of C are not.

CHAPTER
12

BOUNDARY VALUE PROBLEMS FOR DIFFERENTIAL EQUATIONS

12.1 INTRODUCTION: SHOOTING METHOD

In Chap. 10 we studied methods of approximating solutions to *initial value problems* for ordinary differential equations. In an initial value problem the values of the solution and lower-order derivatives up to order one less than the order of the differential equation are specified at an initial time. Another type of problem that arises frequently in applications is a *boundary value problem* in which, in addition to the differential equation, information about the solution and perhaps some derivatives is specified at two different values of the independent variable. The general form of a boundary value problem for a second-order differential equation is

$$y''(x) = F(x, y, y')$$

$$\alpha_a y(a) + \beta_a y'(a) = \gamma_a \qquad \alpha_b y(b) + \beta_b y'(b) = \gamma_b \tag{12.1.1}$$

Example 12.1.1 Temperature distribution on a bar. The *heat flux* ϕ across a surface is the rate per unit area that heat energy crosses the surface. *Fourier's law* is the empirical observation that heat flux is proportional to the temperature gradient. Consider a thin bar of constant cross-sectional area that is insulated along its lateral surface, and let the temperature at time t at position x along the bar be denoted by $u(x, t)$. Then the temperature gradient is simply $\partial u/\partial x$, and Fourier's law states

$$\phi(x, t) = -k(x) \frac{\partial u}{\partial x}(x, t) \tag{12.1.2}$$

where $k(x) > 0$ is the *thermal conductivity* of the bar's material. This together with conservation of energy permits us to conclude (Exercise 14) that temperature as a function of position and time is governed by the partial differential equation

$$c(x)\rho(x)\frac{\partial u}{\partial t} - \frac{\partial}{\partial x}\left(k(x)\frac{\partial u}{\partial x}\right) = q(x, t) \tag{12.1.3}$$

where $\rho(x)$ and $c(x)$ are, respectively, the density and the specific heat of the bar at position x. The term $q(x, t)$ is the *source* term, and it is the heat energy per unit volume per unit time that is being generated within the bar due to mechanical stress or chemical reaction.

Suppose that the bar is of length L and that its left end is at $x = 0$. The evolution of the temperature u is not completely determined until the nature of heat transfer at the ends $x = 0$ and $x = L$ has been specified. One possibility is that the temperature at the end of the bar is maintained at some fixed value, for example, $u(0, t) = u_1$ if the left end is under consideration. Maintaining a fixed temperature at an end usually requires that heat energy either be supplied or removed at the end. Another possibility is that the bar might be insulated at an end. This means that there is no flux across the plane of the end of the bar, and thus from Fourier's law an insulated left end satisfies the condition $(\partial u/\partial x)(0, t) = 0$. A third possibility is that an end is in contact with the surrounding medium. If the temperature of the medium is u_A, then it is often assumed that the heat flux across the end is proportional to the difference between u_A and the temperature at the end of the bar. This gives the conditions $(\partial u/\partial x)(0, t) = \kappa[u(0, t) - u_A]$ and $(\partial u/\partial x)(L, t) = -\kappa[u(L, t) - u_A]$, where $\kappa > 0$ is a constant when, respectively, the left and right ends are under consideration. Note that if the proportionality constant κ is very small, then $\partial u/\partial x \approx 0$, while if it is very large, then $u(0, t) - u_A = (1/\kappa)\,\partial u/\partial x \approx 0$. Thus the insulated and fixed-temperature end conditions can be regarded as limiting cases of the more general case of heat exchange with the surrounding medium.

In addition to specifying the nature of heat transfer at the ends, the initial temperature distribution $u(x, 0) = g(x)$ along the bar must be given. With specification of heat exchange conditions for each end of the bar and an initial temperature distribution, the temperature of the bar is uniquely determined for all time $t > 0$.

If the source term q is independent of time t, then the temperature of the bar will approach a *steady state* temperature distribution $u = u_{ss}(x)$ that is independent of time. In this case $\partial u/\partial t = 0$ in (12.1.3), and we have the boundary value problem

$$[k(x)u_{ss}'(x)]' = -q(x)$$
$$\alpha_0 u_{ss}(0) + \beta_0 u_{ss}'(0) = \gamma_0 \qquad \alpha_L u_{ss}(L) + \beta_L u_{ss}'(L) = \gamma_L \tag{12.1.4}$$

which has the form (12.1.1). Note that the assumed form of the boundary conditions encompasses all three types of heat exchange at the ends of the bar which we discussed.

Boundary value problems of the form (12.1.4) occur in many other applications, among them transverse deflection of a cable, axial deformation of a bar, fluid flow through pipes, laminar flow in a channel, flow through porous media, and electrostatics.

Shooting Method for a Linear Differential Equation

If the differential equation in the boundary value problem (12.1.1) is linear and the boundary conditions consist of simply specifying the value of the unknown function $y(x)$ at each endpoint, then (12.1.1) has the form

$$y''(x) + p(x)y'(x) + q(x)y(x) = r(x)$$
$$y(a) = \gamma_a \qquad y(b) = \gamma_b$$

$$(12.1.5)$$

Recall from Chap. 10 that the solution to the initial value problem

$$Y''(x) + p(x)Y'(x) + q(x)Y(x) = r(x)$$
$$Y(a) = Y_1 \qquad Y'(a) = m$$

$$(12.1.6)$$

can be approximated by first transforming (12.1.6) to an initial value problem for a first-order system of differential equations

$$v_1' = v_2 \qquad v_1(a) = Y_1$$

$$(12.1.7)$$

$$v_2' = -p(x)v_2 - q(x)v_1 + r(x) \qquad v_2(a) = m$$

with $v_1 := Y$ and $v_2 := Y'$ and then applying a numerical method such as the classical Runge-Kutta method to (12.1.7). The *shooting method* entails finding the value of the slope m for which the solution to (12.1.6) becomes the same as the solution to (12.1.5). The left boundary condition in (12.1.5) dictates that we take $Y_1 = \gamma_a$ in (12.1.7), and thus we want to adjust m so that the solution Y passes through the point (b, γ_b) and hence satisfies the right boundary condition.

Example 12.1.2. Consider the initial value problem

$$y'' - 3y' + 2y = x \qquad y(0) = 1 \qquad y(1) = 2 \qquad (12.1.8)$$

Figure 12.1.3 shows graphs of solutions to the initial value problem

$$Y'' - 3Y' + 2Y = x \qquad Y(0) = 1 \qquad Y'(0) = m \qquad (12.1.9)$$

generated by the classical Runge-Kutta method for several choices of the slope m. Clearly some slope between $m = 0.75$ and $m = 0.8$ will cause the solution to the initial value problem (12.1.9) to pass through the point $(1, 2)$ and thus be the solution to the boundary value problem (12.1.8).

For linear differential equations the slope m that makes the solution of the initial value problem (12.1.6) the same as the solution to the boundary value problem (12.1.5) can be calculated explicitly. The general solution to the second-order linear differential equation in (12.1.5) has the form $y(x) = c_1 y_1(x) + c_2 y_2(x) + y_p(x)$, where y_1 and y_2 are linearly independent solutions to the associated homogeneous equation [(12.1.5) with $f(x) \equiv 0$] and y_p is any particular solution to the full equation. Let us choose y_1 to satisfy the initial conditions $y_1(a) = 1$, $y_1'(a) = 0$ and let y_2 be chosen so that $y_2(a) = 0$, $y_2'(a) = 1$.

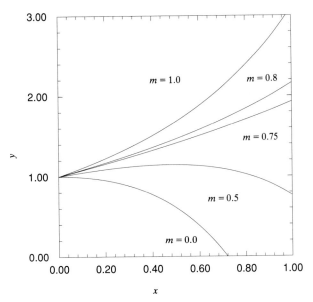

FIGURE 12.1.3

It is easily seen that the solutions y_1 and y_2 chosen in this way are linearly independent. Even if we cannot find these solutions analytically, their values can be approximated at any point by using one of the numerical methods of Chap. 10. Also let y_p be chosen so that $y_p(a) = 0$, $y'_p(a) = 0$. Note that with these choices of y_1, y_2, and y_p, $y'(a) = c_1 y'_1(a) + c_2 y'_2(a) + y'_p(a) = c_2$, and thus c_2 is the unknown slope $y'(a)$. We want the solution y to satisfy the conditions

$$y(a) = c_1 y_1(a) + c_2 y_2(a) + y_p(a) = \gamma_a$$
$$y(b) = c_1 y_1(b) + c_2 y_2(b) + y_p(b) = \gamma_b \qquad (12.1.10)$$

Because of the way we have chosen y_1, y_2, and y_p, the first equation in (12.1.10) is simply $c_1 = \gamma_a$. The second equation can be solved to obtain

$$c_2 = m = \frac{\gamma_2 - \gamma_1 y_1(b) - y_p(b)}{y_2(b)} \qquad (12.1.11)$$

provided that $y_2(b) \neq 0$.

Algorithm 12.1.4 summarizes the discussion above. Its extension to the more general boundary conditions used in (12.1.1) is Exercise 11.

Algorithm 12.1.4   Shooting method for a linear equation
Input a, b (endpoints of interval)
          γ_a, γ_b [values of $y(a)$, $y(b)$]

Use a numerical method to find $y_1(b)$ for the initial value problem

$$y''(x) + p(x)y'(x) + q(x)y(x) = 0, \; y(a) = 1, \; y'(a) = 0$$

Use a numerical method to find $y_2(b)$ for the initial value problem

$$y''(x) + p(x)y'(x) + q(x)y(x) = 0, \ y(a) = 0, \ y'(a) = 1$$

Use a numerical method to find $y_p(b)$ for the initial value problem

$$y''(x) + p(x)y'(x) + q(x)y(x) = r(x), \ y(a) = 0, \ y'(a) = 0$$

if $y_2(b) \neq 0$ then $m := \dfrac{\gamma_b - \gamma_a y_1(b) - y_p(b)}{y_2(b)}$

else

    {Output: Message that algorithm has failed
    Halt program}

Use a numerical method to find $y(x)$ at whatever values of x the solution is required by solving the initial value problem

$$y''(x) + p(x)y'(x) + q(x)y(x) = r(x), \ y(a) = \gamma_a, \ y'(a) = m$$

Remark 1. If the numerical method of solving initial value problems is $O(h^n)$, then (12.1.11) determines m to order n in the stepsize h (Exercise 7). The shooting method itself then results in an $O(h^n)$ approximation. This follows from Theorem 10.2.13 for Euler's method and from corresponding results for other numerical solution methods of initial value problems.

Remark 2. When $y_2(b)$ is small, then its numerical approximation may be subject to large relative errors, and the value of the slope m resulting from (12.1.11) is then inaccurate. In this case it is sometimes better to use a *reverse* shooting method that determines the slope at the right endpoint which makes the solution to an initial value problem the same as the given boundary value problem.

Example 12.1.5. The solution to the boundary value problem (12.1.8) of Example 12.1.2 is

$$y(x) = \frac{(e^2 - 3)e^x + (3 - e)e^{2x}}{4e(e - 1)} + \frac{3}{4} + \frac{1}{2}x \qquad (12.1.12)$$

TABLE 12.1.6

x	$y(x)$	$Actual - y(x)$
0.0000	1.000 000 0	0.0000E + 00
0.1000	1.078 045 4	−1.3300E − 08
0.2000	1.159 428 3	−2.8418E − 08
0.3000	1.244 585 8	−4.4824E − 08
0.4000	1.334 019 8	−6.1536E − 08
0.5000	1.428 308 1	−7.6980E − 08
0.6000	1.528 117 8	−8.8727E − 08
0.7000	1.634 220 8	−9.3190E − 08
0.8000	1.747 512 5	−8.5214E − 08
0.9000	1.869 034 3	−5.7516E − 08
1.0000	2.000 000 0	0.0000E + 00

Table 12.1.6 shows the accuracy obtained when Algorithm 12.1.4 is applied. The numerical method used for all calculations was the classical Runge-Kutta method with a stepsize of $h = 0.1$. Even though this is not a particularly small stepsize, the resulting answers are quite accurate. The classical Runge-Kutta method is order four in the stepsize h. Consequently it is expected that a stepsize of $h = 0.02$ should result in a reduction in the error by a factor of $5^4 = 625$. Indeed, when a stepsize of 0.02 is used, the error at $x = \frac{1}{2}$ falls from 7.7×10^{-8} to 1.4×10^{-10}, which is an improvement by a factor of 550.

Sturm-Liouville Boundary Value Problems

Formula (12.1.11) fails to determine a slope for the initial value problem (12.1.6) if $y_2(b) = 0$. In this case the actual solution to the boundary value problem either has no solution or it has infinitely many solutions. To see this note that if $y(x)$ is a solution to the boundary value problem (12.1.5), then so is $y(x) + cy_2(x)$ for any constant c since y_2 is a solution to the associated homogeneous equation for which $y_2(a) = y_2(b) = 0$. Thus if the problem (12.1.5) has any solutions at all, it has infinitely many.

Sturm-Liouville theory provides some more information on the existence and uniqueness of solutions to boundary value problems for linear differential equations.

A *homogeneous Sturm-Liouville boundary value problem* has the form

$$[\rho(x)y'(x)]' + \lambda\sigma(x)y(x) = 0$$
$$\alpha_a y(a) + \beta_a y'(a) = 0 \qquad \alpha_b y(b) + \beta_b y'(b) = 0 \tag{12.1.13}$$

where $\rho(x), \sigma(x) > 0$ are continuous functions. The boundary value problem

$$y''(x) + p(x)y'(x) + q(x)y(x) = 0$$
$$\alpha_a y(a) + \beta_a y'(a) = 0 \qquad \alpha_b y(b) + \beta_b y'(b) = 0 \tag{12.1.14}$$

for a general second-order homogeneous differential equation can be put in Sturm-Liouville form by multiplying the differential equation by

$$\mu(x) := \exp\left[\int_a^x p(u)\,du\right] \tag{12.1.15}$$

to obtain

$$[\mu(x)y'(x)]' + \mu(x)q(x)y(x) = 0 \tag{12.1.16}$$

Since $\mu(x) > 0$ for all x, (12.1.16) is a Sturm-Liouville differential equation providing that $q(x)$ does not change sign in the interval $[a, b]$; that is, if $q(x) < 0$ in $[a, b]$, then we can take $\sigma(x) = -\mu(x)q(x)$ and $\lambda = -1$, whereas if $q(x) > 0$, then we can take $\sigma(x) = \mu(x)q(x)$ and $\lambda = 1$. Whether the nonhomogeneous Sturm-Liouville problem

$$[\rho(x)y'(x)]' + \lambda\sigma(x)y(x) = \omega(x)$$
$$\alpha_a y(x_a) + \beta_a y'(a) = \gamma_a \qquad \alpha_b y(b) + \beta_b y'(b) = \gamma_b \tag{12.1.17}$$

has a unique solution depends upon whether the homogeneous problem (12.1.13) has solutions other than the trivial solution $y \equiv 0$.

Definition 12.1.7. A value λ for which (12.1.13) has nontrivial solutions is called an *eigenvalue* of the problem. The nontrivial solutions are called the *eigenfunctions* corresponding to that eigenvalue.

Theorem 12.1.8. The problem (12.1.13) has an enumerable set of eigenvalues $\lambda_1 < \lambda_2 < \cdots$, all of which are real. Moreover $\lim_{k \to \infty} \lambda_k = \infty$. If $y_k(x)$ is an eigenfunction corresponding to λ_k, then all eigenfunctions for λ_k have the form $cy_k(x)$ for some constant c.

A proof of this theorem may be found in Haberman (1983), for instance.

Example 12.1.9. The homogeneous Sturm-Liouville problem

$$y'' + \lambda y = 0 \qquad y(0) = 0 \qquad y(1) = 0 \qquad (12.1.18)$$

has eigenvalues $\lambda_k = k^2 \pi^2$, $k = 1, 2, \ldots$. The eigenfunctions corresponding to λ_k are $y_k(x) = c_k \sin k\pi x$ (Exercise 3).

Theorem 12.1.10. If all solutions to (12.1.13) satisfy

$$p(a)y(a)y'(a) - p(b)y(b)y'(b) \geq 0 \qquad (12.1.19)$$

then all eigenvalues are nonnegative.

A proof of this result can again be found in Haberman (1983). Note that in the case of simple boundary conditions such as $y(a) = 0$, $y(b) = 0$, Theorem 12.1.10 implies that all the eigenvalues are nonnegative.

Theorem 12.1.11. The nonhomogeneous Sturm-Liouville problem (12.1.17) has a unique solution if and only if λ is not an eigenvalue of (12.1.13).

This theorem is a particularization of a very general result about linear problems called the *Fredholm alternative*. Again, see Haberman (1983) for further discussion.

Example 12.1.12. The boundary value problem

$$y'' - (1 + x^2)y = \sin x \qquad y(0) = 1 \qquad y'(1) = -2 \qquad (12.1.20)$$

has a unique solution since $\lambda = -1$ is not an eigenvalue of the homogeneous problem

$$y'' + \lambda(1 + x^2)y = 0 \qquad y(0) = 0 \qquad y(1) = 0 \qquad (12.1.21)$$

by virtue of Theorem 12.1.10.
    The boundary value problem

$$y'' + y = e^{-x} \qquad y(0) = 1 \qquad y(1) = -2 \qquad (12.1.22)$$

has a unique solution since from Example 12.1.9 we know that $\lambda = 1$ is not an eigenvalue of the homogeneous problem.

The foregoing considerations are summarized in the next theorem.

Theorem 12.1.13. For the boundary value problem for a general second-order linear differential equation

$$y'' + p(x)y' + q(x)y = r(x)$$

$$\alpha_a y(a) + \beta_a y'(a) = \gamma_a \qquad \alpha_b y(b) + \beta_b y'(b) = \gamma_b$$

(12.1.23)

has a unique solution if either of the following sets of conditions is met:
(a) $q(x) < 0$ for all $x \in [a, b]$, and $\mu(a)y(a)y'(a) - \mu(b)y(b)y'(b) \geq 0$ for all solutions to

$$y''(x) + p(x)y'(x) + q(x)y(x) = 0$$

$$\alpha_a y(a) + \beta_a y'(a) = 0 \qquad \alpha_b y(b) + \beta_b y'(b) = 0$$

(12.1.24)

(b) $q(x) > 0$ for $x \in [a, b]$, and (12.1.24) has no solutions other than $y(x) \equiv 0$.

Example 12.1.14. The partial differential equation (12.1.3) governs the temperature distribution on a bar when the lateral surface of the bar is insulated. If instead the surface of the bar is in contact with a surrounding medium which has constant temperature u_A, and the flow of heat energy is assumed to be proportional to the difference of the temperature of the bar and u_A, then the governing differential equation is

$$c(x)\rho(x) \frac{\partial u}{\partial t} = \frac{\partial}{\partial x}\left(k(x)\frac{\partial u}{\partial x}\right) - \frac{PH}{A}[u(x, t) - u_a]$$

(12.1.25)

when there are no other heat sources (Exercise 15). Here $H > 0$ is the constant of proportionality and A and P are, respectively, the area and perimeter of the cross section of the bar.

Assuming that the heat transfer conditions at the ends of the bar are among the three cases discussed in Example 12.1.1, the steady state temperature on the bar is a solution to a Sturm-Liouville problem of the form

$$[k(x)u'_{ss}(x)]' - \frac{PH}{A}u_{ss}(x) = \frac{PH}{A}u_A$$

$$\alpha_0 u_{ss}(0) + \beta_0 u'_{ss}(0) = \gamma_0 \qquad \alpha_L u_{ss}(L) + \beta_L u'_{ss}(L) = \gamma_L$$

(12.1.26)

Since the quantity PH/A is positive, Theorem 12.1.13 guarantees that there is a unique equilibrium temperature (Exercise 8).

The eigenvalues of a Sturm-Liouville problem are often themselves of physical interest, as the next example indicates. Solution of such problems is considered further in Project 3.

Example 12.1.15. Small transverse displacements $y(x, t)$ of an elastic string are governed by the partial differential equation

$$\rho(x) \frac{\partial^2 y}{\partial t^2} = T_0 \frac{\partial^2 y}{\partial x^2}$$

(12.1.27)

where T_0 is the tension in the string, assumed constant, and $\rho(x)$ is the linear density at position x along the string. Let us suppose that the string is of length L

and that its ends are fastened to the x axis; that is,

$$y(0, t) = y(L, t) = 0 \qquad (12.1.28)$$

for $t \geq 0$. The *fundamental modes of vibration* are solutions of the form

$$y(x, t) = Y(x) \cos(\omega t + \theta) \qquad (12.1.29)$$

Substituting (12.1.29) into (12.1.27) and (12.1.28) gives the eigenvalue problem

$$Y'' + \omega^2 \frac{\rho(x)}{T_0} Y(x) = 0 \qquad Y(0) = 0 \qquad Y(L) = 0 \qquad (12.1.30)$$

The eigenvalues ω_n^2 of this problem are the squares of the *fundamental angular frequencies of vibration* ω_n. It turns out that any motion of the string satisfying the boundary conditions (12.1.28) can be written as a superposition of the fundamental modes of vibration

$$y(x, t) = \sum_{n=1}^{\infty} c_n Y_n(x) \cos(\omega_n t + \theta_n) \qquad (12.1.31)$$

where Y_n is an eigenfunction corresponding to ω_n. For most initial displacements and velocities of the string, the first few fundamental modes of vibration are of greatest amplitude and hence dominate the frequencies of the sound waves that arise from the motion of the string.

Shooting Method for Nonlinear Differential Equations

The structure of solutions to linear differential equations permitted us to find the explicit formula (12.1.11) for the unknown slope of the solution $y(x)$ at $x = a$ in the case of the boundary value problem (12.1.5) for a linear differential equation and thereby to find the solution by use of a numerical scheme for solving initial value problems. Now let us consider the corresponding problem for a nonlinear differential equation

$$y'' = F(x, y, y')$$
$$y(a) = \gamma_a \qquad y(b) = \gamma_b \qquad (12.1.32)$$

We adopt the same strategy of choosing the slope $m := y'(a)$ that produces an initial value problem which has the same solution as the given boundary value problem. Let $y(m; x)$ denote the solution to the initial value problem

$$y'' = F(x, y, y') \qquad y(a) = \gamma_a \qquad y'(a) = m \qquad (12.1.33)$$

The problem is to solve the equation

$$y(m; b) = \gamma_b \qquad (12.1.34)$$

for m. This may be done approximately using Newton's method (Sec. 2.3). The Newton iterates for (12.1.34) are

$$m_{i+1} = m_i - \frac{y(m_i; b) - \gamma_b}{(\partial y / \partial m)(m_i; b)} \qquad (12.1.35)$$

To apply Newton's method we need a means of calculating $(\partial y/\partial m)(m; b)$. Upon taking the partial derivative with respect to m of each equation in (12.1.33) we obtain

$$\left(\frac{\partial y}{\partial m}\right)''(m; x) = \frac{\partial F}{\partial y}(x, y(m; x), y'(m; x))\frac{\partial y}{\partial m}(m; x)$$

$$+ \frac{\partial F}{\partial y'}(x, y(m; x), y'(m; x))\left(\frac{\partial y}{\partial m}\right)'(m; x) \quad (12.1.36)$$

$$\frac{\partial y}{\partial m}(m; a) = 0 \qquad \left(\frac{\partial y}{\partial m}\right)'(m; a) = 1$$

Here a prime continues to denote differentiation with respect to x. Since the system (12.1.36) depends upon the solution of (12.1.33), we must regard (12.1.33) and (12.1.36) as an initial value problem for a system of two second-order differential equations. For the purpose of numerical solution this is converted to an equivalent initial value problem for a system of four first-order differential equations by letting $v_1 := y(m; x)$, $v_2 := y'(m; x)$, $v_3 := (\partial y/\partial m)(m; x)$, and $v_4 := (\partial y/\partial m)'(m; x)$. This results in the initial value problem

$$v_1' = v_2 \qquad v_1(a) = \gamma_a$$
$$v_2' = F(x, v_1, v_2) \qquad v_2(a) = m$$
$$v_3' = v_4 \qquad v_3(a) = 0 \qquad\qquad (12.1.37)$$
$$v_4' = \frac{\partial F}{\partial y}(x, v_1, v_2)v_3 + \frac{\partial F}{\partial y'}(x, v_1, v_2)v_4 \qquad v_4(a) = 1$$

The solutions $v_1(b)$ and $v_3(b)$ are the quantities required in the Newton iteration (12.1.35). Algorithm 12.1.16 summarizes the preceding considerations.

Algorithm 12.1.16 Shooting method for nonlinear equations
Input a, b (endpoints of the interval)
        γ_a, γ_b [values of $y(a), y(b)$]
        tol (tolerance for Newton's method)
        $F(x, y, y')$, $\dfrac{\partial F}{\partial y}(x, y, y')$, $\dfrac{\partial F}{\partial y'}(x, y, y')$ (right-hand side of
            differential equation and its partial derivatives)
        m (initial guess for slope at left endpoint)
        MaxIter (maximum permissible number of iterations of
            Newton's method)

    Iter:=0
    repeat
        Use a numerical method to solve the initial value problem (12.1.37).
        $m_{\text{last}} := m$

if $v_3(b) \neq 0$ then $m := m - \dfrac{v_1(b) - \gamma_b}{v_3(b)}$

else

      {Output: Method failed for initial guess m

      Halt program}

   Iter:=Iter + 1

until $|m - m_{last}| <$ tol or Iter $>$ MaxIter

If Iter $\leq$ MaxIter then

Use a numerical method to solve the initial value problem (12.1.33) to find $y(x)$ at those values of x of interest.

Example 12.1.17. The boundary value problem

$$y'' = 2yy' \qquad y(0) = -1 \qquad y\left(\frac{\pi}{2}\right) = 1 \qquad (12.1.38)$$

has solution $y(x) = \tan(x - \pi/4)$. For this problem

$$F(x, y, y') = 2yy' \qquad \frac{\partial F}{\partial y} = 2y' \qquad \frac{\partial F}{\partial y'} = 2y \qquad (12.1.39)$$

Table 12.1.18 shows the results of applying Algorithm 12.1.16 to this problem. The classical Runge-Kutta method was used with a stepsize of $h = 0.1$ and a tolerance for Newton's method of tol $= 5 \times 10^{-4}$. Though the stepsize is not particularly small, the answers are fairly accurate.

TABLE 12.1.18

x	$y(x)$	Actual $- y(x)$
0.0000	−1.000 000 0	0.0000E + 00
0.1571	−0.726 600 0	5.7503E − 05
0.3142	−0.509 582 4	5.6924E − 05
0.4712	−0.324 963 8	4.4084E − 05
0.6283	−0.158 413 7	2.9215E − 05
0.7854	−0.000 014 5	1.4520E − 05
0.9425	0.158 384 2	2.8485E − 07
1.0996	0.324 933 0	−1.3349E − 05
1.2566	0.509 550 5	−2.5056E − 05
1.4137	0.726 571 4	−2.8830E − 05
1.5708	1.000 000 0	−2.3683E − 09

EXERCISES 12.1

1. Solve the given boundary value problems analytically.

(a) $y'' - 5y' + 6y = 1 + x^2$, $y(0) = 2$, $y(2) = -1$
(b) $y'' + y = \sin x$, $y(0) = 0$, $y(1) = 1$

2. In the boundary value problem (12.1.4) for the steady state temperature distribution on a bar, let $k(x) \equiv 1$ and $q(x) = 1 + x^2$. Find analytically solutions for each of the following sets of boundary conditions.

(a) $u_{ss}(0) = 0$, $u_{ss}(L) = 1$
(b) $u'_{ss}(0) = 0$, $u'_{ss}(L) = 1$
(c) $u_{ss}(0) = 0$, $u_{ss}(L) = 1$

3. (a) Verify that the eigenvalues and eigenfunctions of the problem

$$y'' + \lambda y = 0 \qquad y(0) = 0 \qquad y(1) = 0$$

are as stated in Example 12.1.9.
   (b) Find the eigenvalues and eigenfunctions for the following Sturm-Liouville problems
      i. $y'' + \lambda y = 0$, $y'(0) = 0$, $y'(1) = 0$
      ii. $y'' + \lambda y = 0$, $y(0) = 0$, $y'(1) = 0$
      iii. $(xy')' + \lambda y/x = 0$, $y(1) = 0$, $y(2) = 0$
      (*Hint*: The differential equation is a Cauchy-Euler equation.)

4. (a) Derive an implicit relation for the eigenvalues of the problem

$$y'' + \lambda y = 0 \qquad y'(0) = 0 \qquad y'(1) + y(1) = 0$$

   (b) Use Newton's method to find the first three eigenvalues of the problem in part (a).

5. Solve the given boundary value problems analytically.
   (a) $y'' + y = 1 - 2x$, $y'(0) = 1$, $y'(1) = 0$
   (b) $y'' + y = 0$, $y'(0) = 1$, $y'(1) + y(1) = 0$

6. Derive a formula for the solution to the boundary value problem

$$y'' + \lambda y = 1 \qquad y(0) = 1 \qquad y(1) = 0$$

Why is this formula invalid when λ is an eigenvalue of the homogeneous problem?

7. Show that if $y_1(b)$, $y_2(b)$, and $y_p(b)$ are $O(h^n)$ approximations, then the slope m given by (12.1.11) is also an $O(h^n)$ approximation.

8. Use Theorem 12.1.13 to establish that (12.1.26) has a unique solution when the ends of the bar are subject to any combination of the three heat transfer conditions discussed in Example 12.1.1.

9. Write a computer program implementing the shooting method for second-order linear differential equations, Algorithm 12.1.4. Test it on the boundary value problems of Example 12.1.5 and Exercise 1.

10. For small positive ϵ the boundary value problem

$$y'' + (\pi^2 + \epsilon)y = 0 \qquad y(0) = 0 \qquad y(1) = 0$$
$$(12.1.40)$$

has a unique solution; however, there are infinitely many solutions when $\epsilon = 0$ (see Example 12.1.9). Solve (12.1.40) analytically, and apply the shooting method program of Exercise 9 to find the effect of being near an eigenvalue on the accuracy of the shooting method.

11. Write a computer program implementing the shooting method for a second-order linear differential equation with the boundary conditions

$$\alpha_a y(a) + \beta_a y'(a) = \gamma_a \qquad \alpha_b y(b) + \beta_b y'(b) = \gamma_b$$

where β_a, $\beta_b \neq 0$. Test the program on the boundary value problems of Exercise 5.

12. Write a computer program implementing the shooting method for second-order nonlinear differential equations, Algorithm 12.1.16. Test it on the boundary value problem of Example 12.1.17.

13. Write a computer program implementing the shooting method for a second-order nonlinear differential equation with the boundary conditions

$$\alpha_a y(a) + \beta_a y'(a) = \gamma_a \qquad \alpha_b y(b) + \beta_b y'(b) = \gamma_b$$

where β_a, $\beta_b \neq 0$. Test the program on the boundary value problem

$$y'' = 2yy' \qquad y'(0) = 4 \qquad y'\left(\frac{\pi}{8}\right) = 2$$

which has exact solution $y(x) = 2 \tan 2x$.

14. (a) If the convention is adopted that heat flux $\phi(x, t)$ on a bar is positive when the flux is from left to right, then argue from conservation of energy that for the segment of the bar between positions x and Δx

$$\frac{\partial}{\partial t} \int_x^{x + \Delta x} c(\xi)\rho(\xi)u(\xi, t) \, d\xi$$
$$= -\phi(x + \Delta x, t) + \phi(x, t)$$
$$+ \int_x^{x + \Delta x} q(\xi, t) \, d\xi$$

   (b) Derive the partial differential equation (12.1.3) governing temperature distribution on a bar by letting $\Delta x \to 0$.

15. Derive Eq. (12.1.25).

12.2 FINITE DIFFERENCE METHOD

In this section we examine another approach to solving the boundary value problem

$$y''(x) + p(x)y'(x) + q(x)y(x) = r(x)$$

$$y(a) = \gamma_a \qquad y(b) = \gamma_b \tag{12.2.1}$$

for a linear differential equation. Extensions of the method to other boundary conditions and to nonlinear differential equations are considered in the exercises.

Let $x_i := a + ih$, $i = 1, \ldots, N$, with $h := (b - a)/N$ be a uniform partition of the interval $[a, b]$ into N subintervals. The *finite difference method* consists of replacing the derivatives in (12.2.1) by one of the approximations for the derivative found in Sec. 9.2. From Tables 9.2.7 and 9.2.10 we have the equations

$$y'(x) = \frac{1}{2h}[y(x+h) - y(x-h)] - \frac{h^2}{6} y^{(3)}(\xi)$$

$$y''(x) = \frac{1}{h^2}[y(x+h) - 2y(x) + y(x-h)] - \frac{h^2}{12} y^{(4)}(\bar{\xi}) \tag{12.2.2}$$

for some $\xi, \bar{\xi} \in [x - h, x + h]$. Let $y_i := y(x_i)$. Substituting (12.2.2) into the differential equation in (12.2.1) and multiplying by h^2 gives

$$y_{i+1} - 2y_i + y_{i-1} + \frac{h}{2} p(x_i)(y_{i+1} - y_{i-1}) + h^2 q(x_i)y_i$$

$$= h^2 r(x_i) + \frac{h^4}{12}[y^{(4)}(\bar{\xi}_i) + p(x_i)y^{(3)}(\xi_i)]$$

or $\tag{12.2.3}$

$$\left[1 - \frac{h}{2} p(x_i)\right]y_{i-1} + [-2 + h^2 q(x_i)]y_i + \left[1 + \frac{h}{2} p(x_i)\right]y_{i+1}$$

$$= h^2 r(x_i) + \frac{h^4}{12}[y^{(4)}(\bar{\xi}_i) + p(x_i)y^{(3)}(\xi_i)]$$

for $i = 1, \ldots, N - 1$. If we set $y_0 = \gamma_a$ and $y_N = \gamma_b$ and neglect the $O(h^4)$ term, then (12.2.3) constitutes $N - 1$ linear equations in $N - 1$ unknowns; that is, the solutions $\eta_1, \ldots, \eta_{N-1}$ to the system

$$[-2 + h^2 q(x_1)]\eta_1 + \left[1 + \frac{h}{2} p(x_1)\right]\eta_2 = h^2 r(x_1) - \left[1 - \frac{h}{2} p(x_1)\right]\gamma_a$$

$$\left[1 - \frac{h}{2} p(x_i)\right]\eta_{i-1} + [-2 + h^2 q(x_i)]\eta_i \tag{12.2.4}$$

$$+ \left[1 + \frac{h}{2} p(x_i)\right]\eta_{i+1} = h^2 r(x_i) \qquad \text{for } i = 2, \ldots, N - 2$$

$$\left[1 - \frac{h}{2}\, p(x_{N-1})\right]\eta_{N-2} + [-2 + h^2 q(x_{N-1})]\eta_{N-1}$$

$$= h^2 r(x_{N-1}) - \left[1 + \frac{h}{2}\, p(x_{N-1})\right]\gamma_b$$

give approximations $\eta_i \simeq y_i$ to the unknown solution at the nodal points x_i.

Implementation of the Finite Difference Method

With

$$t_i := -2 + h^2 q(x_i) \qquad v_i := 1 + \frac{h}{2}\, p(x_i) \qquad w_i := 1 - \frac{h}{2}\, p(x_i) \qquad i = 1, \ldots, N-1$$

$$b_1 := h^2 r(x_1) - \left[1 - \frac{h}{2}\, p(x_1)\right]\gamma_a \qquad b_{N-1} := h^2 r(x_{N-1}) - \left[1 + \frac{h}{2}\, p(x_{N-1})\right]\gamma_b$$

$$b_i := h^2 r(x_i) \qquad i = 2, \ldots, N-2 \tag{12.2.5}$$

the system (12.2.4) has the tridiagonal matrix form

$$A\boldsymbol{\eta} := \begin{bmatrix} t_1 & v_1 & 0 & \cdots & 0 & 0 & 0 \\ w_2 & t_2 & v_2 & \cdots & 0 & 0 & 0 \\ \multicolumn{7}{c}{\dotfill} \\ 0 & 0 & 0 & \cdots & w_{N-2} & t_{N-2} & v_{N-2} \\ 0 & 0 & 0 & \cdots & 0 & w_{N-1} & t_{N-1} \end{bmatrix} \begin{bmatrix} \eta_1 \\ \eta_2 \\ \vdots \\ \eta_{N-2} \\ \eta_{N-1} \end{bmatrix} = \mathbf{b} := \begin{bmatrix} b_1 \\ b_2 \\ \vdots \\ b_{N-2} \\ b_{N-1} \end{bmatrix} \tag{12.2.6}$$

This system may be solved by finding an LU decomposition of the matrix A and then solving the system $L\mathbf{z} = \mathbf{b}$ by forward elimination and the system $U\boldsymbol{\eta} = \mathbf{z}$ by backward substitution. See Sec. 6.4. By virtue of the tridiagonal form of A, the computations involved in finding the LU factorization may be abbreviated. Indeed, observe that

$$LU := \begin{bmatrix} 1 & 0 & 0 & \cdots & 0 & 0 & 0 \\ l_1 & 1 & 0 & \cdots & 0 & 0 & 0 \\ 0 & l_2 & 1 & \cdots & 0 & 0 & 0 \\ \multicolumn{7}{c}{\dotfill} \\ 0 & 0 & 0 & \cdots & l_{N-3} & 1 & 0 \\ 0 & 0 & 0 & \cdots & 0 & l_{N-2} & 1 \end{bmatrix} \begin{bmatrix} d_1 & u_1 & 0 & \cdots & 0 & 0 & 0 \\ 0 & d_2 & u_2 & \cdots & 0 & 0 & 0 \\ 0 & 0 & d_3 & \cdots & 0 & 0 & 0 \\ \multicolumn{7}{c}{\dotfill} \\ 0 & 0 & 0 & \cdots & 0 & d_{N-2} & u_{N-2} \\ 0 & 0 & 0 & \cdots & 0 & 0 & d_{N-1} \end{bmatrix}$$

$$= \begin{bmatrix} d_1 & u_1 & 0 & \cdots & 0 & 0 & 0 \\ l_1 d_1 & l_1 u_1 + d_2 & u_2 & \cdots & 0 & 0 & 0 \\ 0 & l_2 d_2 & l_2 u_2 + d_3 & \cdots & 0 & 0 & 0 \\ \multicolumn{7}{c}{\dotfill} \\ 0 & 0 & 0 & \cdots & l_{N-3} d_{N-3} & l_{N-3} u_{N-3} + d_{N-2} & u_{N-2} \\ 0 & 0 & 0 & \cdots & 0 & l_{N-2} d_{N-2} & l_{N-2} u_{N-2} + d_{N-1} \end{bmatrix}$$

$$\tag{12.2.7}$$

Thus we have

$$d_1 = t_1 \qquad u_1 = v_1$$

(12.2.8)

$$l_i = \frac{w_{i+1}}{d_i} \qquad d_{i+1} = t_{i+1} - l_i u_i \qquad u_i = v_i \qquad \text{for } i = 1, \ldots, N-2$$

Algorithm 12.2.1 gives more detail concerning the implementation of the finite difference method. In the event that one of the diagonal elements d_i is zero, the algorithm terminates in failure. Conditions under which this possibility can be precluded will be given later.

Algorithm 12.2.1   Finite difference scheme for linear equations

Input a, b (Endpoints of interval)
             γ_a, γ_b [Values of $y(a)$ and $y(b)$]
             N (Number of subintervals)

Define $p(x), q(x), r(x)$ (Coefficients of differential equation)

$h := (b - a)/N$
for $i = 0$ to N
$\quad x_i := a + ih$

Definition of system
$d_1 := -2 + h^2 q(x_1), u_1 := 1 + \dfrac{h}{2} p(x_1)$
$b_1 := h^2 r(x_1) - \left[1 - \dfrac{h}{2} p(x_1) \right] \gamma_a$

for $i = 2$ to $N - 2$

$\quad \{ l_{i-1} := 1 - \dfrac{h}{2} p(x_i), d_i := -2 + h^2 q(x_i)$

$\quad\quad u_i := 1 + \dfrac{h}{2} p(x_i)$

$\quad\quad b_i := h^2 r(x_i) \}$

$l_{N-2} := 1 - \dfrac{h}{2} p(x_{N-1}), d_{N-1} := -2 + h^2 q(x_{N-1})$

$b_{N-1} := h^2 r(x_{N-1}) - \left[1 + \dfrac{h}{2} p(x_{N-1}) \right] \gamma_b$

Factorization
for $i = 1$ to $N - 2$
$\quad \{ \text{if } d_i \neq 0 \text{ then}$

$\quad\quad \{ l_i := \dfrac{l_i}{d_i}, d_{i+1} := d_{i+1} - l_i u_i \}$

$\quad$ else
$\quad\quad$ {Output: Method failed
$\quad\quad$ Halt program}}

Forward elimination

$z_1 := b_1$

for $i = 1$ to $N - 2$

$\qquad z_{i+1} := b_{i+1} - l_i z_i$

Backward substitution

$$\eta_{N-1} := \frac{z_{N-1}}{d_{N-1}}$$

for $i = N - 2$ downto 1

$$\eta_i := \frac{z_i - u_i \eta_{i+1}}{d_i}$$

Output $\qquad \eta_i, i = 1, \ldots, N - 1$ [Approximations to $y(\xi_i)$]

Example 12.2.2. Table 12.2.3 shows the result of applying Algorithm 12.2.1 to the boundary value problem

$$y'' - 3y' + 2y = x \qquad y(0) = 1 \qquad y'(1) = 2 \qquad (12.2.9)$$

with $N = 10$ (see Example 12.1.5). The results are less accurate than those of the shooting method for the same problem (see Table 12.1.6); however, Table 12.2.4 shows that the times of execution for the problem on an IBM class 386 PC without numeric coprocessor are shorter. Consequently, the finite difference approach may be preferable when it is desired to find the approximate value to a solution to a boundary value problem at a large number of points between a and b.

TABLE 12.2.3

x_i	$y(x_i)$	*Actual* $- y(x_i)$
0.00	1.0000E + 00	0.0000E + 00
0.10	1.0780E + 00	6.7315E − 05
0.20	1.1593E + 00	1.3828E − 04
0.30	1.2444E + 00	2.0989E − 04
0.40	1.3337E + 00	2.7764E − 04
0.50	1.4280E + 00	3.3503E − 04
0.60	1.5277E + 00	3.7291E − 04
0.70	1.6338E + 00	3.7864E − 04
0.80	1.7472E + 00	3.3506E − 04
0.90	1.8688E + 00	2.1909E − 04
1.00	2.0000E + 00	0.0000E + 00

TABLE 12.2.4

	Shooting method		Finite difference	
N	Time	Error at $x = 0.5$	Time	Error at $x = 0.5$
10	0.040	7.7E − 08	0.006	3.4E − 04
50	0.198	1.4E − 10	0.033	1.3E − 05

Convergence of the Finite Difference Method

From Sec. 12.1 there is a unique solution to the boundary value problem when p and q are continuous and $q(x) < 0$ for $x \in [a, b]$. In this case if the interval width is chosen so that

$$\frac{h}{2} \max_{[a,b]} |p(x)| < 1 \qquad (12.2.10)$$

then we have

$$|t_1| > 2 > v_1$$
$$|t_i| > 2 = |w_i| + |v_i| \qquad (12.2.11)$$
$$|t_{N-1}| > 2 > |w_{N-2}|$$

The tridiagonal matrix A is thus diagonally dominant and therefore has a unique solution by Theorem 6.6.4. Moreover it can be shown that Algorithm 12.2.1 will successfully find an approximation when $q(x) < 0$ under these conditions (Exercise 1).

In Table 12.2.4 a decrease in the interval width by a factor of 5 caused the error made by the finite difference method to decrease by a factor of 26. This suggests that the finite difference method is of order two in the interval width h. In the next theorem we establish this under certain conditions.

Theorem 12.2.5. For any h satisfying (12.2.10) the solution $\{\eta_i\}_{i=1}^{N-1}$ to the linear system (12.2.4) satisfies

$$|\eta_i - y(x_i)| \le h^2 \frac{m_4 + 2p^* m_3}{12 q_*} \qquad (12.2.12)$$

where $q_* > 0$ is the minimum on $[a, b]$ of $-q(x)$, and p^*, m_3, and m_4 are the maximums on $[a, b]$ of $p(x)$, $y^{(3)}(x)$, and $y^{(4)}(x)$, respectively.

Proof. In (12.2.3) we let $\epsilon_i := \frac{1}{12}[y^{(4)}(\bar{\xi}_i) - y^{(3)}(\xi_i)]$. With $e_i := y(x_i) - \eta_i$ and the notation (12.2.5), it follows upon subtracting (12.2.4) from (12.2.3) that

$$t_1 e_1 + v_1 e_2 = h^4 \epsilon_1$$
$$w_i e_{i-1} + t_i e_i + v_i e_{i+1} = h^4 \epsilon_i \qquad (12.2.13)$$
$$w_{N-2} e_{N-2} + t_{N-1} e_{N-1} = h^4 \epsilon_{N-1}$$

With $e := \max_{1 \le i \le N-1} |e_i|$ and $\epsilon := \max_{1 \le i \le N-1} |\epsilon_i|$, we have upon employing the inequalities (12.2.11)

$$|t_1||e_1| \le |v_1||e_2| + |\epsilon_1| h^4 \le 2e + \epsilon h^4$$
$$|t_i||e_i| \le |w_i||e_{i-1}| + |v_i||e_{i+1}| + |\epsilon_i| h^4 \le 2e + \epsilon h^4 \qquad (12.2.14)$$
$$|t_{N-1}||e_{N-1}| \le |w_{N-2}||e_{N-2}| + |\epsilon_{N-1}| h^4 \le 2e + \epsilon h^4$$

Since $|t_i| = |-2 + h^2 q(x_i)| \ge 2 + h^2 q_*$, we conclude that $(2 + h^2 q_*)|e_i| \le 2e + \epsilon h^4$ and thus, since there is an i for which $|e_i| = e$, that $q_* e \le \epsilon h^2$. Hence (12.2.12) follows upon noting that $|\epsilon_i| \le \frac{1}{12}(m_4 + p^* m_3)$.

EXERCISES 12.2

1. Show that when the matrix A in (12.2.6) is diagonally dominant, then none of the diagonal elements d_i will be zero and hence Algorithm 12.2.1 will successfully find an approximation. (*Hint*: Show by induction that $(d_i + v_i < 0.)$

2. Write a program implementing the finite difference Algorithm 12.2.1. Test the program on the boundary value problem of Example 12.1.5 and the boundary value problem of Exercise 1 in Sec. 12.1.

3. For small positive ϵ the boundary value problem

$$y'' + (\pi^2 + \epsilon)y = 0 \qquad y(0) = 0 \qquad y(1) = 0$$
$$(12.2.15)$$

has a unique solution; however, there are infinitely many solutions when $\epsilon = 0$ (see Example 12.1.9). Solve (12.2.15) analytically, and apply the finite difference program of Exercise 2 to find the effect of being near an eigenvalue on the accuracy of the finite difference method.

4. Write a program implementing the finite difference method for second-order linear differential equations with the boundary conditions

$$\alpha_a y(a) + \beta_a y'(a) = \gamma_a \qquad \alpha_b y(b) + \beta_b y'(b) = \gamma_b$$

where β_a, $\beta_b \neq 0$. Test the program on the boundary value problems of Exercise 5 in Sec. 12.1.

5. Use the approximations (12.2.2) to develop a program for solving the boundary value problem

$$y'' = F(x, y, y') \qquad y(a) = \gamma_a \qquad y(b) = \gamma_b$$

for a general second-order nonlinear differential equation. Test the program on the boundary value problem of Example 12.1.17.

6. Use the approximations (12.2.2) to develop a program for solving the boundary value problem

$$y'' = F(x, y, y') \qquad \alpha_a y(a) + \beta_a y'(a) = \gamma_a$$
$$\alpha_b y(b) + \beta_b y'(b) = \gamma_b$$

with β_a, $\beta_b \neq 0$. Test the program on the boundary value problem of Exercise 13 in Sec. 12.1.

PROJECTS

1. (*a*) Develop a program to solve the boundary value problem

$$y''''(x) + a_3(x)y'''(x) + a_2(x)y''(x) + a_1(x)y'(x)$$
$$+ a_0(x)y(x) = f(x)$$
$$(1)$$
$$y(a) = \gamma_a \qquad y'(a) = \gamma_a' \qquad y(b) = y_b$$
$$y'(b) = b_b'$$

using a shooting method.

(*b*) The transverse deflection of a beam subject to a transverse load $f(x)$ is governed by the fourth-order differential equation

$$\frac{d^2}{dx^2}\left(r(x)\frac{d^2 w}{dx^2}\right) + f(x) = 0 \qquad (2)$$

where x is a position along the beam and $r(x)$ is the *flexural rigidity* of the beam at position x.
  i. Interpret the boundary conditions in (1) in terms of a beam.
  ii. Solve the problem
$$w'''' = -(1 + x^2)$$
$$w(0) = 0 \qquad w'(0) = 0 \qquad w(1) = 1 \qquad (3)$$
$$w'(1) = -1$$

analytically, and compare the exact answers with the approximations generated in part (*a*).
  iii. Solve (2) with $r(x) = 1 + x^2$ and $f(x) = \sin \pi x$ and the boundary conditions $w(0) = w'(0) = w(1) = w'(1) = 0$.

2. (*a*) Use a finite difference method on the problem (1).
   (*b*) Do part (*b*) of Project 1.

3. Develop a finite difference scheme that will convert the Sturm-Liouville eigenvalue problem

$$[\rho(x)y'(x)]' + \lambda\sigma(x)y(x) = 0$$
$$y(0) = 0 \qquad y(L) = 0 \qquad (4)$$

into an eigenvalue problem for a matrix.
   (*a*) Write a program based on the power or inverse power method (Sec. 11.1) that finds the smallest eigenvalue of (4). Use this program to find the lowest fundamental frequency for the vibrating string problem (12.1.30) when $T_0 = 1$ and
    i. $\rho(x) \equiv 1$
    ii. $\rho(x) = 1 + x^2$

For (i) compare the approximate answer with the exact answer.

(b) Write a program that will find all eigenvalues of the matrix eigenvalue problem arising from (4) using the QR algorithm (Sec. 11.3). Apply the program to problems (i) and (ii) of part (a). For (i) compare the approximate eigenvalues with the exact eigenvalues.

4. The partial differential equation governing the temperature $u(x, y, t)$ at a position (x, y) in a plate is

$$c(x, y)\rho(x, y)\frac{\partial u}{\partial t} - \nabla \cdot [k(x, y)\nabla u] = q(x, y, t) \quad (5)$$

(see Example 12.1.1). When q is a function of position alone, then the temperature in the plate will approach a steady state distribution which is a solution to the partial differential equation

$$\nabla \cdot (k(x, y)\nabla u) = -q(x, y) \quad (6)$$

For a rectangular plate $a \leq x \leq b$, $c \leq y \leq d$ and the boundary conditions

$$\begin{align} u(a, y) &= u_a(y) & u(x, c) &= u_c(x) \\ u(b, y) &= u_b(y) & u(x, d) &= u_d(x) \end{align} \quad (7)$$

where the temperature functions u_a, u_b, u_c, and u_d are given, use a finite difference scheme to find a linear system whose solution approximates the temperature $u(x_i, y_j)$ at regularly spaced nodes in the interior of the plate.

(a) Test the program on the problem

$$\frac{\partial^2 u}{\partial x^2} + \frac{\partial^2 u}{\partial y^2} = 6 + 8(x^2 + y^2)$$

$$\begin{align} u(0, y) &= 2y^2 & u(x, 0) &= x^2 \\ u(1, y) &= 1 + 6y^2 & u(x, 1) &= 1 + 5x^2 \end{align} \quad (8)$$

which has exact solution $u(x, y) = x^2 + 2y^2 + 4x^2y^2$.

(b) Find approximations to steady state temperature on the plate $0 \leq x \leq 2$, $0 \leq y \leq 1$ when $k(x, y) = 1 + x^2 + y^2$, $q(x, y) = xy$, and the left and lower sides of the plate are held at temperature $u = 1$ while the right and upper sides are held at temperature $u = 0$.

CHAPTER
13

THE IMPACT
OF PARALLEL
COMPUTERS

13.1 INTRODUCTION: PARALLEL ARCHITECTURES

In this final chapter, we consider one of the most exciting challenges to face numerical analysts and software designers in recent years: the advent of parallel computers. Throughout the "computer age," the major developments in technology have been spurred by the need for ever-greater computer power to solve bigger and bigger problems faster and faster. Indeed, since the end of World War II, the development of numerical analysis and computer technology have been inextricably bound up together with significant advances in one being spawned by or spawning corresponding advances in the other.

The fact that the newer computer systems are faster than their predecessors and use different architectures may not at first seem to require anything special of the numerical software industry other than providing the opportunity for solving bigger problems and therefore, perhaps, more realistic models of the physical or economic situation. However, as we will soon see, these new architectures have sufficiently different characteristics that the preferred algorithm for a particular task may be completely different from that which has been found optimal on serial machines.

One simple example of this is that no one would seriously advocate the Gauss-Jordan elimination for solving a linear system on a serial machine since it is so much more expensive than, say, Gauss elimination with partial pivoting. On a massively parallel array processor, however, Gauss-Jordan would be the most efficient direct method of solving systems which fit into the processor array. We will discuss this in more detail in Sec. 13.3.

First, we need to describe some of the different types of parallel computers and their architectural classification. What do we mean by a parallel computer? At its simplest level, a parallel computer is just a computer which can perform more than one task simultaneously. We have to be a bit more careful with such a definition, since, taken too literally, this would include almost all computers. After all, memory buses fetch more than 1 bit from memory at one time, multipliers often work on the mantissas and exponents of the operands simultaneously, and modern fast arithmetic and logic units frequently operate on all bits of the operands simultaneously. Indeed, much research has been devoted to efficient low-level parallel algorithms and hardware for these sorts of tasks. Several such units even use different number representations *internally* to maximize their performance.

When we talk of parallel computers, we are thinking generally of higher-level operations being performed simultaneously, or *in parallel*. This statement remains (intentionally) somewhat vague; the ideas of parallelism will become clearer as we proceed.

Three main types of parallel processors are covered by our definition: *vector*, *array*, and *hypercube* processors.

Many special-purpose parallel architectures have also been designed for performing particular tasks. These include "butterfly networks" for fast Fourier transforms (FFTs—see Sec. 5.6) and signal processing. The term *systolic array* is used for arrays of simple one-operation processors linked together to perform a specific overall task. Such arrays can be designed specifically for many often-performed tasks such as matrix transposition. We are concerned here with more general-purpose parallel processors.

According to most definitions of parallel processors the vector processors should not really be included in our list. However, this class includes nearly all the so-called supercomputers—Cray, Cyber, Convex, for example. The vector computers are essentially *very* fast serial machines in which vector operations run especially fast because they are *pipelined*. This means that in the addition of two vectors a, b, for example, the elements of the vectors are reaching the processor in a steady stream. Thus, for example, once the "pipe" is full, we may have the result $a_{i-1} + b_{i-1}$ being stored while $a_i + b_i$ is being computed and a_{i+1}, b_{i+1} are being fetched from memory.

Such a description fits within our vague definition but does not satisfy most definitions of parallel processing, because there are no two data-processing operations being performed at the same time. This pipelining is usually much more powerful than our example suggests and can result in enormous speedup compared to serial machines. It should also be observed here that many vector processors do have a low level of genuine parallelism in that often there will be a small number of separate vector processors available.

From the point of view of the numerical software designer, these architectures present less of a new challenge. The programmer does not determine which operations are performed on which processor; this is done by the compiler. The program used is often essentially the same as would be adopted for a serial machine. The comparison between different algorithms for performing a task on such a machine is essentially the same as for serial architectures. Of course, some adjustments can be made by the experienced user to maximize the vectorization of his or her program.

We will return briefly to vector processors in discussing efficiency and speedup later in this section and will discuss some of the improvements available for vector processors in the next few sections. Apart from that, we will concentrate for most of this chapter on array and hypercube architectures. These typify the two main classes of parallel machines:

SIMD   Single-instruction multiple-data stream
MIMD   Multiple-instruction multiple-data stream

This classification was originally proposed by Flynn (1966) and also includes MISD and SISD. The last characterizes a serial machine, of course.

An array processor typically consists of a large rectangular mesh of processors controlled by a master control unit which sends the same instruction to all the processors. Figure 13.1.1 illustrates the case of a 4×4 array. Typical array sizes for such architectures vary from 32×32, or 1024 processors, up to 65,536 processors. Examples of such machines are the AMT Distributed Array Processor (DAP), the MasPar MP-1, and the Connection Machine. The connections between processors are not always in a simple grid, but this model is sufficient for our purposes.

Each processor has its own memory. Thus, for example, two matrices

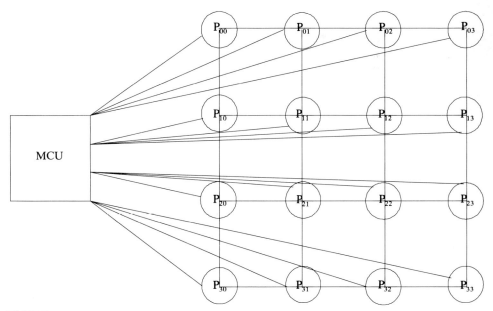

FIGURE 13.1.1
A 4×4 array processor.

A, B could be stored in such a way that their (i, j)th elements are stored in processor$_{ij}$'s memory. The single instruction $C = A + B$ would then have the effect that *simultaneously* each processor adds its element of A to its element of B and stores the result as its element of C. The addition of two matrices is therefore reduced to just a single operation.

The significance of the interprocessor connections is that two connected processors can "talk to each other." This means, for example, that processor P_{21} in Fig. 13.1.1 can only send information to (or receive information from) any of processors P_{11}, P_{20}, P_{31}, or P_{22}. It follows that communication of data between distant processors in the array can be very slow. Communication between P_{01} and P_{32} in the array of Fig. 13.1.1 would require four stages of nearest-neighbor communication. Some of these arrays also have a secondary lattice of connections which allows very fast transmission of data between the processors by allowing a processor to "broadcast" data to every processor in its row or its column.

The *topology* of the basic interprocessor connections is often slightly more complex than Fig. 13.1.1 suggests, also. There are often connections "round the back" connecting processors on opposite edges of the array. If there is just one set of such connections, the processor array resembles a cylinder, while if both sets of edge connections are present, the array is equivalent to a torus.

It is important to observe here that the individual processors in these "massively parallel" arrays are often quite simple, usually only either 1-bit or 4-bit processors. This has the effect that *individual* arithmetic operations will usually be significantly *slower* than on conventional machines. The great speedup that can be achieved is therefore derived entirely from the fact that large numbers of such operations are performed simultaneously. The performance of such machines will clearly vary considerably with different types of problem. For problems which fit the architecture well, performance can be amazingly good. Not only the problem but also the choice of algorithm used must be appropriate for optimal performance.

MIMD architectures can have a wide variety of interconnection topologies including meshes, rings, trees, and hypercubes. The individual processors are usually much more powerful than in SIMD machines, and usually there are fewer of them—anything from 4 to about 64 being typical. This model can also be used to describe what might be called "network parallelism" in which a program running on one node of a computer network looks for processors that are currently inactive and "farms" parts of the overall task out to them. Such an arrangement would typically behave like a tree architecture. Examples of tree and hypercube topologies are illustrated in Figs. 13.1.2 and 13.1.3.

These architectures typically fall into the *distributed memory* classification. That is, each processor has its own local memory, although there is usually a host processor (for the hypercube) or root processor (for a tree) which may well provide the main memory for a particular program while the

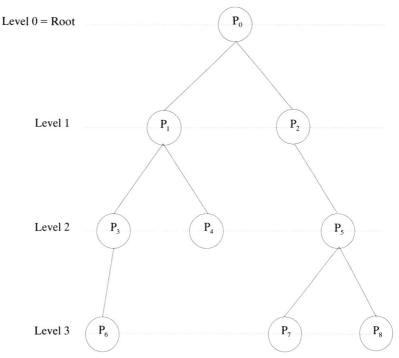

Level 0 = Root P_0

Level 1 P_1 P_2

Level 2 P_3 P_4 P_5

Level 3 P_6 P_7 P_8

FIGURE 13.1.2
An MIMD tree architecture.

individual processor nodes store only the data needed for their part of the overall operation.

We observe that communication times for a tree structure such as that in Fig. 13.1.2 may be high if much interprocessor communication of data is necessary. For example, communication between processors P_6 and P_7 has to be passed through the root and requires the use of six interprocessor links. This is the worst case for this particular tree which thus has a *communications diameter* of 6. This is a large diameter for a system of nine processors.

The importance of the communications diameter is, of course, problem-dependent. If very little communication through long links is needed, then the fact that the diameter is large is of little concern. Clearly the speed of these links relative to other operations affects the acceptability of a particular communications diameter for a given task.

On a regular $n \times n$ mesh of processors, the worst communications link would be over a full diagonal which yields a diameter of $2(n-1)$. With connections joining the edges so that the topology resembles a torus, this is reduced to the distance between the "center" and a corner which for even n is just n.

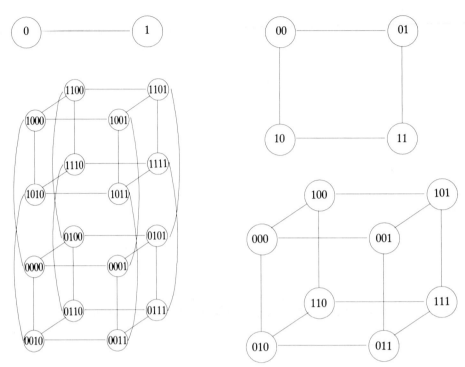

FIGURE 13.1.3
Hypercubes of dimensions one through four.

Conceptually, hypercube architectures have processors at each vertex of an *n*-dimensional cube with interprocessor connections along each edge of the hypercube. A two-node hypercube therefore consists of just two connected processors, while a four-node one is equivalent to having processors at the vertices of a square. Three- and four-dimensional hypercubes are illustrated in Fig. 13.1.3. It is apparent from these examples that the communications diameter of the *n*-dimensional hypercube is *n*. The binary encoding of the processor labels is used in Fig. 13.1.3 to illustrate the organization of interprocessor communications on a hypercube.

Each successive higher-dimensional hypercube consists of two copies of the previous one with the corresponding nodes connected. The nodes are labeled by putting a 0 at the begining of each node number on one copy and a 1 at the beginning for nodes on the other copy. By labeling the nodes in this way, it follows that neighboring nodes have labels which differ in exactly 1 bit. The number of bits in which two node labels differ represents their distance apart. This coding is also used to determine the routing of any interprocessor communications.

There are many good sources of further detail on the hardware and communication aspects of the various parallel processor architectures, and we do not pursue these matters any further here. Our interest in these architectures lies in their impact on numerical analysis and scientific computing.

The primary impact is, of course, in terms of speed. It is necessary to be able to measure the improvement that is being achieved to decide on efficient techniques for solving problems. It is also important to design an algorithm to take full advantage of the architecture—the same algorithm is unlikely to be optimal on a PC, a Cray, a DAP, and a hypercube, for example.

It is also important to bear in mind that there is often a trade-off between speedup and numerical stability which must be considered when selecting algorithms for particular tasks. We will return to this issue later. The rest of this section provides an introduction to the ideas of *efficiency* and *speedup*.

We begin by considering the effects of pipelining in a vector processor. Consider first the simplest of vector operations—addition of two vectors—on a vector processor, that is, the implementation of a loop such as

$$\text{for } i := 1 \text{ to } n \qquad c[i] := a[i] + b[i]$$

Now each component of the vector a must be fetched from memory, at the correctly synchronized times elements of b must be fetched and added to the corresponding elements of a, and the result is output as a component of c. The operation of the pipeline is such that, once the process is started, $a[i + 1]$ is being fetched from memory *at the same time as* $b[i]$ is being fetched and added to $a[i]$ *at the same time as* $c[i - 1]$ is being output to memory. (Actually, the operation is more complex than this since there are several subtasks to be performed in these different stages.) The process is summarized in Table 13.1.4.

Let us examine the time required for such a vector addition. Suppose for simplicity that each of the operations takes approximately the same time, τ say. (In fact, the second operation would be the slowest in this example, but the necessary synchronization demands that the other operations have the same *effective* speed.) The total time for this operation would then be $(n + 2)\tau$. The 2τ for steps 1 and 2 for which there is no output is referred to as the *start-up* time, or the *delay*. For a serial architecture in which each of these steps requires the same time τ, the overall time would be $3n\tau$, so the vector processor has speeded the overall operation by the factor $3n/(n + 2)$.

For a vector or parallel processor, the speedup factor is defined by

$$S = \frac{\text{serial processor time}}{\text{vector or parallel processor time}} \qquad (13.1.1)$$

TABLE 13.1.4
Pipelined addition of two vectors

Step	1	2	3	$\cdots$	n	$n + 1$	$n + 2$
Fetch	a_1	a_2	a_3	$\cdots$	a_n		
Fetch and add		$a_1 + b_1$	$a_2 + b_2$	$\cdots$	$a_{n-1} + b_{n-1}$	$a_n + b_n$	
Output			c_1	$\cdots$	c_{n-2}	c_{n-1}	c_n

Occasionally it is convenient to indicate the number of processors being used, in which case we use the notation S_p, where p is the number of processors. The efficiency of a parallel algorithm is defined by the ratio of the actual speedup to the optimal speedup factor p. That is,

$$E = \frac{S_p}{p} \tag{13.1.2}$$

In the case of our vector addition example, we see that the speedup is small for short vectors: $S(3) = 1.8$, while for long vectors it approaches 3: $S(128) \approx 2.95$ and $S(1024) \approx 2.994$. (Here we have used $S(n)$ to denote the speedup for vector length n.) The speedup in this example is bounded by 3 since the overall operation was separated into three parts. The more complex the operations which the pipeline can handle the better is the potential speedup.

As a second example we consider the formation of the scalar product of two vectors. The typical serial loop for this is

$$s := 0$$
$$\text{for } i = 1 \text{ to } n \qquad s := s + a[i] * b[i]$$

and most vector processors can handle the multiply-and-add operation in this loop without the need to store intermediate results. This would therefore yield a pipeline operation such as that depicted in Table 13.1.5.

In this case there is just one output step at the end of the computation, whereas in the serial version output of all partial sums is also necessary. The timing for this operation, again using τ for the time of one step, is therefore $(n + 3)\tau$ which must be compared with $4n\tau$ for the serial algorithm. The asymptotic speedup factor of 4 is approached for long vectors. Here $S(n) = 4n/(n + 3)$, so $S(3) = 2$, $S(128) \approx 3.91$, $S(1024) \approx 3.99$.

We should observe here that if all the partial sums were wanted, they could be stored at each step at no additional cost. Although this may seem unlikely for a scalar product calculation, there are other similar computations in which the partial sums would be required. Some discrete probability distributions, for example, entail computations of this sort.

Note here that the loops we have vectorized in this brief account are especially simple. Most vectorizing compilers will handle loops which are much more complicated than these. However, as the loops become more complex,

TABLE 13.1.5
Pipelined scalar product

Step	1	2	3	$\cdots$	n	$n+1$	$n+2$	$n+3$
Fetch	a_1	a_2	a_3	$\cdots$	a_n			
Fetch, multiply		$a_1 b_1$	$a_2 b_2$	$\cdots$	$a_{n-1}b_{n-1}$	$a_n b_n$		
Accumulate		$s = 0$	$s + a_1 b_1$	$\cdots$	$s + a_{n-2}b_{n-2}$	$s + a_{n-1}b_{n-1}$	$s + a_n b_n$	
Output								s

especially if branching or conditional statements are used in the loop, the vector speedup will be eroded. It is critical that the vectors be processed in a regularly ordered way. For the loops above, the vectors are handled in their natural order so that contiguous blocks of memory are fed into the pipelines. Taking regularly spaced components from the arrays would also be handled efficiently.

As a final comment on this topic, we remark that vector pipelines have a maximum capacity which is usually of the order of 128 for the vector length. This has an obvious effect on the choice of program parameters. For example, we saw that for vectors of length 128, the addition loop obtained a speedup of 2.95. A vector of length 129 would be treated as one of length 128, followed by one of length 1. The total time for this addition would thus be 133τ, which represents a speedup of approximately 2.91. For the scalar product, operation speedup of 3.91 for $n = 128$ is reduced to $S(129) \approx 3.82$. The longer the start-up time, the greater the erosion of vector speed for inconvenient vector lengths.

For the remainder of the section, we consider briefly the question of speedup on an array processor with p processors. The addition of two vectors of length no more than p can be achieved in just a single operation time by using processor P_i to obtain $c_i = a_i + b_i$ simultaneously for every i. Since p is typically 1024 or more for such machines, we obtain 100 percent efficiency for vectors of this length. Vectors of greater length, n say, must be separated into pieces of length p. There will be $\lceil n/p \rceil$ such pieces, and the operation will take this number of operation times. (Recall that $\lceil x \rceil$ denotes the least integer greater than, or equal to, x.) The speedup is thus given by

$$S_p(n) = \frac{n}{\lceil n/p \rceil}$$

and the efficiency is

$$E_p(n) = \frac{n}{p \lceil n/p \rceil}$$

However, this figure can be misleading in that it compares the parallel time with the serial time for the same operation *using the same type of processor*, but, as was remarked earlier, the individual processors in a massively parallel array are typically significantly slower than those of powerful serial machines.

What does the array processor do for the scalar product computation? Here the basic calculation would be rearranged in a form which can be summarized as

$$\mathbf{c} := \mathbf{a} * \mathbf{b}$$
$$s := \text{Sum}(\mathbf{c})$$

where the multiplication of the two vectors is interpreted as being element by element and the Sum operation sums the elements of the vector $\mathbf{c}$. (Such operations are common "built-ins" for array processors and form part of the usual array extensions of high-level languages.) Now all the products can be formed entirely in parallel in a single operation time (assuming $n \leq p$). The summation of the elements of **c** can then be performed using a reduction algorithm like those considered in the next section. This will take $\lceil \log_2 n \rceil$ addition times plus a communication time penalty. Given the presence of the rapid communication highways described above and the fact that the individual arithmetic operations are usually slower on such a machine, this communication time is almost negligible.

For a vector of length 1024, assuming at least that many processors, the total time is therefore $1 + \log_2 1024 = 11$ array floating-point operations, or *paraflops*. This compares with 2048 flops for the serial algorithm. The speedup is thus

$$S_{1024}(1024) = \tfrac{2048}{11} \simeq 186.2$$

which represents an efficiency of only about 18 percent. Even with the use of the reduction algorithm, the essentially serial nature of the summation reduces the efficiency greatly.

In the next section, we will consider how some of the fundamental operations of scientific computing can be reorganized for efficient parallel computation. In the remaining sections we apply some of these ideas to fundamental numerical problems such as solution of linear systems and partial differential equations.

13.2 BASIC ALGORITHMS

In this section, we consider some of the basic operations of scientific computing and how they can be performed efficiently on parallel computers.

We begin with the reduction algorithms alluded to in the last section for summation. Suppose then that we wish to evaluate the sum

$$S = \sum_{i=1}^{N} x_i \tag{13.2.1}$$

where, for simplicity, we will assume that the number of terms N is a power of 2,

$$N = 2^n \tag{13.2.2}$$

say. Note that sums of other numbers of terms can be extended with the required number of zeros. For a parallel machine with at least 2^n processors, extending the sum with additional zeros costs nothing. For a vector processor, it might be more efficient to adjust the first step of Algorithm 13.2.2 below.

The reduction process consists of adding pairs of terms together and so halving the number of terms to be summed at the next stage. The process is then repeated until there are just two terms to be added which yield the final sum. All the additions at any one stage can be performed in parallel—or by using a vector pipeline. Clearly n stages of this *reduction* algorithm will be required. The algorithm is summarized in Fig. 13.2.1.

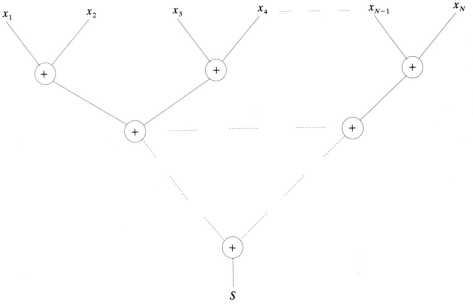

FIGURE 13.2.1
Reduction process for summation.

Algorithm 13.2.2. Reduction algorithm for summation

$$Input \qquad x_1^{(0)}, x_2^{(0)}, \ldots, x_N^{(0)}, \text{ where } N = 2^n$$

Loop   For $k = 1$ to n
          $N := N/2$
          for $i = 1$ to N
              $x_i^{(k)} := x_{2i-1}^{(k-1)} + x_{2i}^{(k-1)}$

Output $S := x_1^{(n)}$

It is apparent that the operation count for an array processor with sufficient processors is reduced to just $n = \lceil \log_2 N \rceil$ paraflops as stated in the previous section. Each stage of the algorithm is well-suited to vectorization, except that the last few would have short vector lengths, so serial processing may be preferable for the final two or three stages.

Similar algorithms could obviously be used for any binary operation such as extended products or finding the maximal element of a vector. Algorithms such as this are often called *recursive doubling* algorithms since the number of terms that have been added together is doubled at each stage.

Example 13.2.3. Use Algorithm 13.2.2 to sum the first 32 terms of the harmonic series. Each result will be stored to two significant figures.

Solution. We begin with

$$x_1^{(0)} = 1.0 \qquad x_2^{(0)} = 0.50 \qquad x_3^{(0)} = 0.33 \qquad x_4^{(0)} = 0.25$$
$$x_5^{(0)} = 0.20 \qquad x_6^{(0)} = 0.17 \qquad x_7^{(0)} = 0.14 \qquad x_8^{(0)} = 0.13$$
$$x_9^{(0)} = 0.11 \qquad x_{10}^{(0)} = 0.10 \qquad x_{11}^{(0)} = 0.091 \qquad x_{12}^{(0)} = 0.083$$
$$x_{13}^{(0)} = 0.077 \qquad x_{14}^{(0)} = 0.071 \qquad x_{15}^{(0)} = 0.067 \qquad x_{16}^{(0)} = 0.063$$
$$x_{17}^{(0)} = 0.059 \qquad x_{18}^{(0)} = 0.056 \qquad x_{19}^{(0)} = 0.053 \qquad x_{20}^{(0)} = 0.050$$
$$x_{21}^{(0)} = 0.048 \qquad x_{22}^{(0)} = 0.045 \qquad x_{23}^{(0)} = 0.043 \qquad x_{24}^{(0)} = 0.042$$
$$x_{25}^{(0)} = 0.040 \qquad x_{26}^{(0)} = 0.038 \qquad x_{27}^{(0)} = 0.037 \qquad x_{28}^{(0)} = 0.036$$
$$x_{29}^{(0)} = 0.034 \qquad x_{30}^{(0)} = 0.033 \qquad x_{31}^{(0)} = 0.032 \qquad x_{32}^{(0)} = 0.031$$

We therefore obtain

$$x_1^{(1)} = 1.5 \qquad x_2^{(1)} = 0.58 \qquad x_3^{(1)} = 0.37 \qquad x_4^{(1)} = 0.27$$
$$x_5^{(1)} = 0.21 \qquad x_6^{(1)} = 0.17 \qquad x_7^{(1)} = 0.15 \qquad x_8^{(1)} = 0.13$$
$$x_9^{(1)} = 0.12 \qquad x_{10}^{(1)} = 0.10 \qquad x_{11}^{(1)} = 0.093 \qquad x_{12}^{(1)} = 0.085$$
$$x_{13}^{(1)} = 0.078 \qquad x_{14}^{(1)} = 0.073 \qquad x_{15}^{(1)} = 0.067 \qquad x_{16}^{(1)} = 0.063$$

which in turn yields

$$x_1^{(2)} = 2.1 \qquad x_2^{(2)} = 0.64 \qquad x_3^{(2)} = 0.38 \qquad x_4^{(2)} = 0.28$$
$$x_5^{(2)} = 0.22 \qquad x_6^{(2)} = 0.18 \qquad x_7^{(2)} = 0.15 \qquad x_8^{(2)} = 0.13$$

The next stage reduces this to just four terms:

$$x_1^{(3)} = 2.7 \qquad x_2^{(3)} = 0.66 \qquad x_3^{(3)} = 0.40 \qquad x_4^{(3)} = 0.28$$

from which we obtain

$$x_1^{(4)} = 3.4 \qquad \text{and} \qquad x_2^{(4)} = 0.68$$

which, finally, yields $S = x_1^{(5)} = 4.1$, which is correct to two significant figures.

We commented in the previous section that there are situations where all the partial sums are required, and it is perhaps not immediately clear how the above algorithm could be modified to produce such output efficiently. There is a simple modification of the basic algorithm which produces in the same number of stages the complete array of partial sums. During each stage more additions are required, but for a parallel machine this carries no penalty. On a vector processor, the cost is greater since the vector length for each stage is considerably increased. The shortest vector length that could be required is $N/2$. For the implementation discussed here vectors of length N are used throughout. (On the other hand, this means that every stage makes efficient use of the pipeline.)

In describing this version of the algorithm it is convenient to introduce two "vector shift" operations which are frequently used in parallel and vector processing. These are included in the extended instruction sets of some parallel programming languages.

The operators SHLP and SHRP (shift left planar and shift right planar) are defined so that, for example, SHRP $(\mathbf{v}, k)$ shifts the vector $\mathbf{v}$ to the right by k places, with *planar* meaning that zeros are shifted into the vacated locations.

(There are corresponding *cyclic* operators SHLC and SHRC which have the effect of "wrapping the vector around" so that elements which are shifted out at one end are brought back in at the other.)

The instruction SHRP($\mathbf{v}, k$) is therefore equivalent to the *temporary* use of the loops

$$\text{for } i := N \text{ downto } k + 1 \; v[i] := v[i - k]$$
$$\text{for } i := k \text{ downto } 1 \; v[i] := 0$$

where N is the vector length. For example, if $\mathbf{v}$ is a vector with five components,

$$\text{SHRP}(\mathbf{v}, 2) = (0, 0, v_1, v_2, v_3)$$

For this same vector, the other shift operators yield

$$\text{SHLP}(\mathbf{v}, 2) = (v_3, v_4, v_5, 0, 0)$$
$$\text{SHRC}(\mathbf{v}, 2) = (v_4, v_5, v_1, v_2, v_3)$$
$$\text{SHLC}(\mathbf{v}, 2) = (v_3, v_4, v_5, v_1, v_2)$$

Note that use of these shift operators does not affect the stored arrays but simply which elements are used.

The modified recursive doubling algorithm for summation is based on adding shifted copies of the original vector together and then repeating the process with the resulting vectors with increasing shifts in such a way that after $n = \log_2 N$ steps the complete array of partial sums has been computed.

Suppose then that we wish to evaluate all partial sums of S given by (13.2.1). We denote these partial sums by

$$s_k = \sum_{i=1}^{k} x_i \tag{13.2.3}$$

and, more generally,

$$s_{jk} = \sum_{i=j}^{k} x_i \tag{13.2.4}$$

so that

$$s_k = s_{1k}$$

Now, the first step of the algorithm is the addition of the two vectors $\mathbf{x}$ and SHRP($\mathbf{x}, 1$) which generates the vector $\mathbf{x}^{(1)}$ whose components are given by

$$(x_1^{(1)}, x_2^{(1)}, \ldots, x_N^{(1)}) = (x_1 + 0, x_2 + x_1, \ldots, x_N + x_{N-1})$$
$$= (s_{11}, s_{12}, \ldots, s_{N-1,N}) \tag{13.2.5}$$

At the next stage, we set

$$\mathbf{x}^{(2)} = \mathbf{x}^{(1)} + \text{SHRP}(\mathbf{x}^{(1)}, 2)$$

which yields the vector $(s_{11}, s_{12}, s_{13}, s_{14}, s_{25}, \ldots, s_{N-3,N})$. The process continues with the shift being doubled at each iteration. After k stages, the vector $\mathbf{x}^{(k)}$ has its first 2^k components of the form s_{1i}, while the remaining ones are of the form $s_{i+1,i+2^k}$.

Algorithm 13.2.4 Recursive doubling for partial sums

$\quad$ *Input* $\qquad x_1^{(0)}, x_2^{(0)}, \ldots, x_N^{(0)}$, where $N = 2^n$; sh:$=1$;

$\quad$ *Loop* $\qquad$ for $k = 1$ to n
$\qquad\qquad\qquad \mathbf{x}^{(k)}:=\mathbf{x}^{(k-1)} + \text{SHRP}(\mathbf{x}^{(k-1)}, \text{sh})$
$\qquad\qquad\qquad$ sh:$=2*$sh

$\quad$ *Output* $\qquad$ s:$=\mathbf{x}^{(n)}$

Example 13.2.5. We illustrate this algorithm by finding the partial sums of the first eight positive integers.

Solution. Initially,

$$\mathbf{x}^{(0)} = (1, 2, 3, 4, 5, 6, 7, 8) \qquad \text{and} \qquad \text{SHRP}(\mathbf{x}^{(0)}, 1) = (0, 1, 2, 3, 4, 5, 6, 7)$$

so that

$$\mathbf{x}^{(1)} = (1, 3, 5, 7, 9, 11, 13, 15) \qquad \text{and} \qquad \text{SHRP}(\mathbf{x}^{(1)}, 2) = (0, 0, 1, 3, 5, 7, 9, 11)$$

from which we obtain at the next stage

$$\mathbf{x}^{(2)} = (1, 3, 6, 10, 14, 18, 22, 26) \qquad \text{and} \qquad \text{SHRP}(\mathbf{x}^{(2)}, 4) = (0, 0, 0, 0, 1, 3, 6, 10)$$

and, finally,

$$\mathbf{s} = \mathbf{x}^{(3)} = (1, 3, 6, 10, 15, 21, 28, 36)$$

Algorithm 13.2.4 requires just n paraflops and n vector shift operations. For even moderate vector lengths, this will represent a substantial saving over the conventional serial algorithm. Notice that for an array processor the longest shift which is required is by half of one side of the array. For example, if $N = 1024$ and we are operating on a 32×32 processor array, shifts of length $1, 2, 4, 8$, and 16 are required after which all shifts are by multiples of 32 which will correspond to a shift by a small number in the other "coordinate" direction.

We will see further examples of the use of recursive doubling later in this section. First, though, we turn to some of the fundamental matrix operations. We have seen throughout this book that linear algebraic problems and their solutions lie at the heart of much numerical analysis, so efficient algorithms for performing the basic arithmetic of vectors and matrices are essential. In the last section, we considered the addition of two vectors and the formation of scalar products.

Addition of two matrices is completely straightforward on an array processor, while on a vector machine it is only necessary to make sure that the loops are organized so that contiguous blocks of memory are fed to the pipeline. If matrices are stored rowwise, then they should be added rowwise, too. The array extensions of programming languages will often do this automatically.

Matrix multiplication is more complicated. Consider first the matrix-vector product $A\mathbf{x}$. This product can be written in two equivalent forms:

$$(A\mathbf{x})_i = \sum_j \mathbf{a}_{ij}x_j = \mathbf{a}_{i*}\mathbf{x} \qquad (13.2.6)$$

where $\mathbf{a}_{i*}$ is the (row) vector formed from the ith row of A, or

$$A\mathbf{x} = \sum_j x_j\mathbf{a}_{*j} \qquad (13.2.7)$$

where $\mathbf{a}_{*j}$ is the (column) vector formed from the jth column of A. These two expressions correspond to the two possible orderings of the loops in the conventional serial algorithms for this product: (13.2.6) is equivalent to

> for $i = 1$ to N
> $b_i := 0$
>   for $j = 1$ to N
> $b_i := b_i + a_{ij}x_j$

while (13.2.7) corresponds to

> for $i = 1$ to N $b_i := 0$
> for $j = 1$ to N
>   for $i = 1$ to N
> $b_i := b_i + a_{ij}x_j$

where $\mathbf{b}$ denotes $A\mathbf{x}$.

The first of these algorithms consists of the formation of a scalar product for each component of the product. It is therefore well-suited to vector processors in which this operation is efficiently handled. The second algorithm has more obvious parallelism when it is rewritten in vector form:

> $\mathbf{b} := \mathbf{0}$
> for $j = 1$ to N $\mathbf{b} := \mathbf{b} + x_j\mathbf{a}_{*j}$

in which a constant multiple of one vector is being added to another. This operation is often available within vector processors, so this version of the algorithm could be expected to be implemented very efficiently by the compiler.

The slight preference for this vector addition form of the algorithm becomes much stronger for implementation on array processors since we may again use a built-in recursive doubling summation to reduce the operation count. Again, for illustrative purposes, we will assume that our arrays fit the processor array. (The extension to other situations is essentially straight-forward.)

Suppose therefore that there are N^2 processors and that the matrix A is $N \times N$. The first step is to "broadcast" the vector across the "rows" of the processor array to produce a matrix MATR(**x**) each of whose rows is just the vector **x**. Yet again, instructions such as MATR, and the corresponding MATC (C for columns), are typically available on such array processors. Also, recall that such broadcasting operations can be readily achieved using the fast data highways.

The (elementwise) array product of A and MATR(**x**) is now a single paraflop and produces the matrix whose jth column is the product of the jth column of A and x_j. That is,

$$A * \text{MATR}(\mathbf{x}) = \begin{bmatrix} a_{11}x_1 & a_{12}x_2 & \cdots & a_{1n}x_n \\ a_{21}x_1 & a_{22}x_2 & \cdots & a_{2n}x_n \\ \cdots\cdots\cdots\cdots\cdots\cdots\cdots\cdots \\ a_{n1}x_1 & a_{n2}x_2 & \cdots & a_{nn}x_n \end{bmatrix}$$

The required vector **b** can now be obtained by summation across each of the rows of this matrix. This can be achieved simultaneously for the rows using the built-in reduction algorithm for summation with an instruction like RSUM. Thus the overall algorithms reduces to the single instruction

$$\mathbf{b} := \text{RSUM}(A * \text{MATR}(\mathbf{x}));$$

that is, it becomes just $1 + \log_2 N$ paraflops. For vectors of length 64 on a 64×64 processor array, this is just 7 paraflops.

The serial operation count for the formation of this product, not counting the initialization of **b** or any of the loop control overhead, is $2N^2$, which for this example is 8192. Clearly the individual arithmetic operations on the array processor can be very much slower than their serial counterparts and still yield signficant overall speedup.

We turn next to matrix-matrix products. On a vector processor, these can be accomplished by a serial loop of matrix-vector products which in turn can be performed in either of the ways outlined above. Plainly, a similar approach can be used on our array processor, the only additional complication being the allocation of the resulting vectors to the correct rows or columns of the product. Depending on the relative dimensions of the matrices and of the processor array, this approach may indeed be preferred. However, in our idealized situation in which the matrix dimension fits the processor dimensions, there is an alternative which may be preferable.

Suppose then that we wish to find $C = AB$, where A, B are $N \times N$ matrices. Of course, we have

$$c_{ij} = \sum_k a_{ik} b_{kj}$$

Following the notation used above, we will write $\mathbf{a}_{*k}$ and $\mathbf{b}_{k*}$ for the kth column of A and the kth row of B. It follows that MATC($\mathbf{a}_{*k}$) is the matrix with all its columns equal to the kth column of A. Similarly $\text{MATR}(\mathbf{b}_{k*})$ has the kth row of B for all its rows. The array product of these two matrices $\text{MATC}(\mathbf{a}_{*k}) * \text{MATR}(\mathbf{b}_{k*})$ therefore has as its ijth element the quantity $a_{ik}b_{kj}$. Summing over k yields the final form

$$C = \sum_k \text{MATC}(\mathbf{a}_{*k}) * \text{MATR}(\mathbf{b}_{k*}) \tag{13.2.8}$$

The summation here would necessarily be serial since the full parallelism is already in use in forming the array products. Thus the overall cost of this algorithm is $2N$ paraflops. For the same dimensions as before, $N = 64$, this implies that the matrix-matrix product takes 128 paraflops as compared with 7 for the matrix-vector product. This algorithm represents a considerable saving relative to a serial application of matrix-vector products.

The corresponding serial processor operation count, again neglecting initialization and overhead, is $2N^3$ flops. For our example, that is 524,288 floating-point operations. The operation count has been reduced by a factor of 4096—exactly the number of processors—so the algorithm has 100 percent efficiency according to the measure introduced in (13.1.2). Equally important-ly, even if the arithmetic is 100 times slower than with our serial processor, the array processor would have provided a speedup factor better than 40 over that processor.

Even in the case of simple linear algebraic arithmetic operations, we see that the choice and design of the algorithm can have significant effect on the speed and efficiency of computation. We continue this theme with a brief look at some other basic operations of scientific computing: evaluating polynomials, computing terms of linear and nonlinear recurrence relations, and solving tridiagonal systems. We will consider more general linear systems in more detail in the remaining sections of the chapter.

In the case of polynomial evaluation, which we have seen used in several settings—notably interpolation and approximation of functions—we presented Horner's algorithm (Sec. 3.1) as being both efficient and numerically stable. A similar idea can be incorporated into a recursive-doubling-style algorithm for vector and parallel processors.

Recall that Horner's rule for evaluation of a polynomial

$$p(x) = a_N x^N + a_{N-1} x^{N-1} + \cdots + a_1 x + a_0 \tag{13.2.9}$$

is equivalent to rewriting it in the form

$$p(x) = \{ \cdots [(a_N x + a_{N-1}) x + a_{N-2}] x + \cdots + a_1 \} x + a_0 \tag{13.2.10}$$

For simplicity in the description of our recursive doubling approach we will again assume that N has a convenient form, namely,

$$N = 2^n - 1$$

so that there is a total of 2^n terms to be summed. Since practical situations do not usually demand the evaluation of polynomials of very high degree, we also assume that the number of processors available, or the maximum vector length, is greater than N.

The recursive doubling algorithm is based on rewriting our polynomial as

$$p_1(x) = (a_N x + a_{N-1})x^{N-1} + (a_{N-2}x + a_{N-3})x^{N-3} + \cdots + (a_3 x + a_2)x^2$$
$$+ (a_1 x + a_0) \tag{13.2.11}$$

which can be viewed as a polynomial in x^2 so that the process can be repeated, combining pairs of terms of p_1, and so on, until we have p_n which will be the required value $p(x)$.

Example 13.2.6. For a polynomial of degree seven, the first step generates the terms

$$(a_7 x + a_6) \quad (a_5 x + a_4) \quad (a_3 x + a_2) \quad (a_1 x + a_0)$$

which we denote $a_3^{(1)}, a_2^{(1)}, a_1^{(1)}, a_0^{(1)}$, respectively, and $x^{(1)} = x^2$. The second stage applies the same procedure to this new array to yield

$$a_1^{(2)} = a_3^{(1)}x^{(1)} + a_2^{(1)} \qquad a_0^{(2)} = a_1^{(1)}x^{(1)} + a_0^{(1)} \qquad \text{and} \qquad x^{(2)} = (x^{(1)})^2 = x^4$$

from which one final step gives

$$a_0^{(3)} = a_1^{(2)}x^{(2)} + a_0^{(2)}$$

which is the required value.

For example, if the polynomial

$$p(x) = x - \frac{x^2}{2} + \frac{x^3}{3} - \frac{x^4}{4} + \frac{x^5}{5} - \frac{x^6}{6} + \frac{x^7}{7}$$

is to be evaluated at $x = \frac{1}{2}$, then we get, working to three decimal places,

$$a_3^{(1)} = \frac{x}{7} - \frac{1}{6} = -0.095 \qquad a_2^{(1)} = \frac{x}{5} - \frac{1}{4} = -0.150 \qquad a_1^{(1)} = \frac{x}{3} - \frac{1}{2} = -0.333$$

$$a_0^{(1)} = x - 0 = 0.500 \qquad \text{and} \qquad x^{(1)} = 0.25$$

The next stage produces

$$a_1^{(2)} = -0.174 \qquad a_0^{(2)} = 0.417 \qquad \text{and} \qquad x^{(2)} = 0.0625$$

Finally, therefore,

$$p(0.5) = a_0^{(3)} = 0.406$$

which is indeed correct to this accuracy.

We observe that the repeated squaring of the argument x can be achieved within the same structure by setting $a_{-1} = x$, $a_{-2} = 0$. With this notation, the process is summarized as Algorithm 13.2.7.

Algorithm 13.2.7   Reduction algorithm for polynomial evaluation

   Input   Coefficients $a_0^{(0)}, a_1^{(0)}, \ldots, a_N^{(0)}$ where $N = 2^n - 1$;
                argument $x =: a_{-1}^{(0)}$, $a_{-2}^{(0)} := 0$

Loop   for $k = 1$ to n
$\qquad a_{-2}^{(k)} := 0, \; N := \lceil N/2 \rceil$
$\qquad$ for $i = -1$ to N
$\qquad\qquad a_i^{(k)} := a_{2i+1}^{(k-1)} a_{-1}^{(k-1)} + a_{2i}^{(k-1)}$

Output $p(x) := a_0^{(n)}$.

Neglecting any overhead and initialization, this entails $2n = 2\lceil \log_2 N \rceil$ paraflops on an array processor. The serial operation count for Horner's rule or the above algorithm is $2N$. Horner's rule does not vectorize as efficiently since the result of each step is needed before computation of the next step can begin.

Many numerical processes involve the computation of terms of a recurrence relation. Indeed, Horner's rule is one example since it can be written in the form

$$b_N := a_N$$
$$\text{for } i = N - 1 \text{ downto } 0$$
$$b_i := b_{i+1} x + a_i$$

Euler's method for solving initial value problems is another instance.

Consider a general second-order linear recurrence relation

$$x_{i+1} = a_i x_i + b_i x_{i-1} \qquad x_0, x_1 \text{ given} \qquad (13.2.12)$$

and suppose that we wish to evaluate the terms $x_2, x_3, \dots, x_{N+1}$. It may seem that this is an inescapably serial operation and that there is no opportunity for exploiting parallelism. However, we have already seen in the special case of Horner's rule that the computation can be rearranged for parallel computers. Horner's rule is, of course, just a first-order relation.

The trick for (linear) recurrence relations of order k is to consider the generation of vectors of length k consisting of successive terms of the sequence. It turns out that there is then a first-order linear relation between successive vectors. That is, each subsequent vector is obtained by multiplying its predecessor by a $k \times k$ matrix.

For our second-order relation, denote by $\mathbf{x}_i$ the vector $\begin{bmatrix} x_{i+1} \\ x_i \end{bmatrix}$. The initial conditions amount to specifying the vector $\mathbf{x}_0$. Now

$$x_2 = a_1 x_1 + b_1 x_0$$
$$x_1 = 1 x_1 + 0 x_0$$

so that we can write

$$\mathbf{x}_1 = A_1 \mathbf{x}_0$$

where $A_1 = \begin{bmatrix} a_1 & b_1 \\ 1 & 0 \end{bmatrix}$.

In general, we have

$$\mathbf{x}_i = A_i \mathbf{x}_{i-1} \qquad (13.2.13)$$

where

$$A_i = \begin{bmatrix} a_i & b_i \\ 1 & 0 \end{bmatrix} \qquad (13.2.14)$$

It follows that

$$\mathbf{x}_N = A_N A_{N-1} \cdots A_1 \mathbf{x}_0 \qquad (13.2.15)$$

This extended matrix product can now be computed using a typical reduction algorithm, provided that sufficiently many 2×2 matrix products can be performed in parallel. The vector $\mathbf{x}_0$ could be treated as a further matrix A_0 whose second column is zero.

For a 64×64 processor array, we can use four processors for each matrix multiplication, in which case 3 paraflops are required. This would be sufficient for $N = 2047$ so that 2048 matrices are to be multiplied including $\mathbf{x}_0$. With eight processors per matrix multiply, the operation count can be reduced to 2 paraflops by computing all the relevant products simultaneously. Since the number of matrix pairs to be multiplied is halved on each iteration, this will eventually be the case.

For $N = 1023$, every stage can be performed in 2 paraflops; there are $\log_2 1024 = 10$ such stages of matrix multiplications. The complete operation count is therefore 20, or $2\lceil \log_2 N \rceil$. This technique also vectorizes efficiently. The generalization of these ideas to higher-order relations is straightforward.

The same basic approach can be used for certain nonlinear recurrences, too. For example, the relation

$$x_{i+1} = a_i + \frac{b_i}{x_i} \qquad x_0 \text{ given}$$

is transformed into a second-order linear recurrence by setting

$$X_i = \prod_{j=0}^{i} x_j \qquad x_{i+1} = \frac{X_{i+1}}{X_i}$$

so that

$$X_{i+1} = a_i X_i + b_i X_{i-1} \qquad X_0 = x_0 \qquad X_{-1} = 1$$

This last recurrence can then be solved in just the same way as above. However, we must observe that such transformations are not risk-free.

For example, consider the sequence defined by

$$x_{i+1} = 1 + \frac{1}{x_i} \qquad x_0 = 1$$

This is a well-behaved sequence:

$$x_1 = 2, x_2 = \tfrac{3}{2}, x_3 = \tfrac{5}{3}, \ldots$$

In general, $x_i = F_{i+1}/F_i$, where the F_i are the terms of the standard Fibonacci sequence (see Sec. 8.2). It is well-known that the limit of this sequence is the golden mean $(1 + \sqrt{5})/2 \approx 1.618$, which, incidentally, is therefore the limit of the sequence of convergents of the continued fraction (see Sec. 5.5)

$$1 + \cfrac{1}{1 + \cfrac{1}{1 + \cfrac{1}{1 + \cdots}}}$$

Now applying the above transformation to this example, it follows that $X_i = F_{i+1}$, which grows very rapidly. Overflow becomes a real danger.

Modifying this example slightly to

$$x_{i+1} = 10 + \frac{10}{x_i} \qquad x_0 = 0.1$$

produces a sequences X_i which overflows IEEE single-precision arithmetic in about 35 iterations, despite the fact that the values x_i that we really seek stay in the interval $[10, 11]$ from $i = 2$ onwards. In looking for efficient parallel algorithms, we must keep an eye on the numerical wisdom of the algorithms under consideration.

The final "basic task" that we consider in this section is the solution of tridiagonal systems of linear equations. Such a system can be written as

$$c_i x_{i-1} + a_i x_i + b_i x_{i+1} = d_i \qquad (13.2.16)$$

We will for convenience extend the coefficient vectors with 0s for the off-diagonal coefficients and 1s for the diagonal. Thus, for example, for $i \le 0$, we have $b_i = c_i = 0$, $a_i = 1$.

The technique adopted here is similar in nature to the recursive doubling approach except that the number of equations considered at each stage remains fixed. It is the separation between interacting equations which is doubled. This process is known as *cyclic odd-even reduction*; it was originally proposed by Hockney (1965).

The idea is to eliminate x_{i-1} and x_{i+1} in Eq. (13.2.16) using the corresponding equations for $i - 1$ and $i + 1$. Of course, this introduces coefficients of both x_{i-2} and x_{i+2}. This process is performed in parallel for each row and then repeated using equations separated by 2.

Consider then the three equations

$$
\begin{aligned}
c_{i-1} x_{i-2} + a_{i-1} x_{i-1} + b_{i-1} x_i & & & = d_{i-1} \\
c_i x_{i-1} + a_i x_i + b_i x_{i+1} & & & = d_i \\
c_{i+1} x_i + a_{i+1} x_{i+1} + b_{i+1} x_{i+2} & & & = d_{i+1}
\end{aligned}
$$

We use the multipliers

$$m_i^- = \frac{c_i}{a_{i-1}} \qquad m_i^+ = \frac{b_i}{a_{i+1}}$$

and then subtract m_i^- times the top equation and m_i^+ times the bottom one from the middle one. This yields

$$c_i^{(1)} x_{i-2} + a_i^{(1)} x_i + b_i^{(1)} x_{i+2} = d_i^{(1)}$$

where $c_i^{(1)} = -m_i^- c_{i-1}$, $b_i^{(1)} = -m_i^+ b_{i+1}$, $a_i^{(1)} = a_i - m_i^- b_{i-1} - m_i^+ c_{i+1}$, and $d_i^{(1)} = d_i - m_i^- d_{i-1} - m_i^+ d_{i+1}$.

Example 13.2.8. Use the cyclic odd-even reduction algorithm to solve

$$\begin{bmatrix} 4 & 1 & & & & & & \\ 1 & 4 & 1 & & & & & \\ & 1 & 4 & 1 & & & & \\ & & 1 & 4 & 1 & & & \\ & & & 1 & 4 & 1 & & \\ & & & & 1 & 4 & 1 & \\ & & & & & 1 & 4 & 1 \\ & & & & & & 1 & 4 \end{bmatrix} \begin{bmatrix} x_1 \\ x_2 \\ x_3 \\ x_4 \\ x_5 \\ x_6 \\ x_7 \\ x_8 \end{bmatrix} = \begin{bmatrix} 4.1 \\ 1.41 \\ 0.141 \\ 1.014 \\ 4.101 \\ 1.41 \\ 0.141 \\ 0.014 \end{bmatrix}$$

Solution. For the first iteration, all the multipliers are $\frac{1}{4}$, and so we obtain the partitioned matrix and right-hand side

$$\begin{bmatrix} 3.75 & 0 & -0.25 & 0 & 0 & 0 & 0 & 0 & \vdots & 3.7475 \\ 0 & 3.5 & 0 & -0.25 & 0 & 0 & 0 & 0 & \vdots & 0.34975 \\ -0.25 & 0 & 3.5 & 0 & -0.25 & 0 & 0 & 0 & \vdots & -0.465 \\ 0 & -0.25 & 0 & 3.5 & 0 & -0.25 & 0 & 0 & \vdots & -0.0465 \\ 0 & 0 & -0.25 & 0 & 3.5 & 0 & -0.25 & 0 & \vdots & 3.495 \\ 0 & 0 & 0 & -0.25 & 0 & 3.5 & 0 & -0.25 & \vdots & 0.3495 \\ 0 & 0 & 0 & 0 & -0.25 & 0 & 3.5 & 0 & \vdots & -0.215 \\ 0 & 0 & 0 & 0 & 0 & -0.25 & 0 & 3.75 & \vdots & -0.02125 \end{bmatrix}$$

This time most multipliers are $-0.25/3.5$, and the result is

$$\begin{bmatrix} 3.73 & 0 & 0 & 0 & -0.02 & 0 & 0 & 0 & \vdots & 3.71 \\ 0 & 3.48 & 0 & 0 & 0 & -0.02 & 0 & 0 & \vdots & 0.346 \\ 0 & 0 & 3.47 & 0 & 0 & 0 & -0.02 & 0 & \vdots & 0.0345 \\ 0 & 0 & 0 & 3.46 & 0 & 0 & 0 & -0.02 & \vdots & 0.00345 \\ -0.2 & 0 & 0 & 0 & 3.46 & 0 & 0 & 0 & \vdots & 3.45 \\ 0 & -0.02 & 0 & 0 & 0 & 3.47 & 0 & 0 & \vdots & 0.345 \\ 0 & 0 & -0.02 & 0 & 0 & 0 & 3.48 & 0 & \vdots & 0.0346 \\ 0 & 0 & 0 & -0.02 & 0 & 0 & 0 & 3.73 & \vdots & 0.00371 \end{bmatrix}$$

One further step reduces the system to the diagonal form

$$\begin{bmatrix} 3.73 & 0 & 0 & 0 & 0 & 0 & 0 & 0 & \vdots & 3.73 \\ 0 & 3.48 & 0 & 0 & 0 & 0 & 0 & 0 & \vdots & 0.348 \\ 0 & 0 & 3.47 & 0 & 0 & 0 & 0 & 0 & \vdots & 0.0347 \\ 0 & 0 & 0 & 3.46 & 0 & 0 & 0 & 0 & \vdots & 0.00347 \\ 0 & 0 & 0 & 0 & 3.46 & 0 & 0 & 0 & \vdots & 3.47 \\ 0 & 0 & 0 & 0 & 0 & 3.47 & 0 & 0 & \vdots & 0.347 \\ 0 & 0 & 0 & 0 & 0 & 0 & 3.48 & 0 & \vdots & 0.0348 \\ 0 & 0 & 0 & 0 & 0 & 0 & 0 & 3.73 & \vdots & 0.00373 \end{bmatrix}$$

which clearly has the correct solution vector

$$(1, 0.1, 0.01, 0.001, 1, 0.1, 0.01, 0.001)$$

to within a very small accumulated rounding error.

This method, which is summarized as Algorithm 13.2.9, is easily implemented using the various array operations for shifts. In Algorithm 13.2.9 we will assume that there are $N = 2^n$ equations. As we saw in the example, $n = \log_2 N$ steps are needed. Remember when reading this algorithm that arithmetic operations on vector arrays are elementwise array operations, so division of one "vector" by another is meaningful.

Algorithm 13.2.9 Reduction algorithm for tridiagonal systems

Input Dimension $N = 2^n$
Coefficient vectors **a**, **b**, **c**
Right-hand-side vector **d**

Initialize Shift interval, $i:=1$

Declare Multiplier vectors **mplus, mminus**

Loop for $k = 1$ to n
  mplus$:=$**b**$/\text{SHLP}(\mathbf{a}, i)$
  mminus$:=$**c**$/\text{SHRP}(\mathbf{a}, i)$
  a$:=$**a** $-$ **mminus** $*\text{SHRP}(\mathbf{b}, i) -$ **mplus** $*\text{SHLP}(\mathbf{c}, i)$
  b$:=-$**mplus** $*\text{SHLP}(\mathbf{b}, i)$
  c$:=-$**mminus** $*\text{SHRP}(\mathbf{c}, i)$
  d$:=$**d** $-$ **mminus** $*\text{SHRP}(\mathbf{d}, i) -$ **mplus** $*\text{SHLP}(\mathbf{d}, i)$
  $i:=2*i$

Output Solution vector **x**$:=$**d**$/$**a**.

Recall that:
The jth element of $\text{SHLP}(\mathbf{v}, i)$ is v_{i+j}.
The jth element of $\text{SHRP}(\mathbf{v}, i)$ is v_{j-1}.
with zeros used whenever the subscript range is exceeded.

Assuming that the processor array can accommodate the arrays of length N, the operation count for this algorithm is 12 paraflops per iteration of the loop plus one for the final division. For a system of 4096 such equations on a 64×64 array of processors, this represents a total of 145 paraflops.

The same algorithm would be most inefficient on a serial processor. The parallel algorithm is sufficiently well-designed that the efficiency measure used earlier would be 100 percent. For a more realistic measure of how well this algorithm might perform, let us compare it with an efficient serial algorithm.

A simple Gauss elimination procedure for the tridiagonal system (13.2.16) is

$$
\begin{aligned}
&\text{for } i = 2 \text{ to } N \\
&\quad m:=c_i/a_{i-1} \\
&\quad c_i:=0 \\
&\quad a_i:=a_i - mb_{i-1} \\
&\quad d_i:=d_i - md_{i-1}
\end{aligned}
$$

$$x_N := d_N / a_N$$

$$\text{for } i = N - 1 \text{ downto } 1$$
$$x_i := (d_i - b_i x_{i+1}) / a_i$$

The operation count for this algorithm is $8(N-1)+1$. For our system of dimension 4096, this represents a total of 32,761 flops which must be compared with the 145 paraflops for the above algorithm. The operations ratio is thus approximately $226:1$, or, for machines with similar operation times, this is equivalent to 6 minutes:1 day! Typically though, as we have commented previously, individual arithmetic times will be slower on the array processor, so the saving is unlikely to be this dramatic.

We have seen in this section that many of the fundamental operations of scientific computing can be reorganized for very efficient use of parallel or vector machines. In the next few sections, we look in greater detail at some more general problems and their solution on parallel machines. Hypercube architectures have been mentioned very little in this section since their impact is somewhat different. They are well-suited to parallelism of a much coarser nature in which the overall problem is separated into subproblems, and these are handled by the various processors which are typically serial in nature.

EXERCISES 13.2

1. Write a program to implement the recursive doubling summation Algorithms 13.2.2 and 13.2.4. Test these for summation of series, and compare your results with those obtained using conventional serial algorithms.

2. Implement the reduction Algorithm 13.2.7 for polynomial evaluation, and compare it with Horner's rule for both numerical performance and speed.

3. Write a program to evaluate Legendre and Chebyshev polynomials from their recurrence relations using the algorithm described by Eqs. (13.2.13) to (13.2.15). (See Chap. 7.)

4. Try to devise a modification of the recurrence relation algorithm which will generate *all* the terms of the sequence. (*Hint*: try to combine the ideas of Algorithm 13.2.4 for generating all partial sums with the algorithm for recurrence relations.)

5. Write a program to implement the cyclic reduction algorithm for the solution of tridiagonal systems. Test it and time it to illustrate that the choice of algorithm is critical to performance. This algorithm will be much slower than Gauss elimination on a serial machine, and yet we have seen that it is vastly superior for an array processor.

13.3 SOLUTION OF LINEAR EQUATIONS: ARRAY PROCESSORS

In this section we restrict our attention to just one problem and one type of architecture: the solution of linear systems of equations on array processors. We will begin with a look at direct methods such as LU factorization. Later we will also look at some of the iterative techniques and how they can be efficiently implemented on such a machine. When appropriate we will also comment on the suitability of the methods discussed for other computer architectures. For details of the underlying methods refer to Chap. 6.

In Sec. 6.1, we studied the operation count for basic Gaussian elimination and found that approximately $2N^3/3$ flops are required for the solution of an $N \times N$ system by this method. An obvious alternative to Gauss elimination is the Gauss-Jordan process, in which we eliminate both above and below the diagonal in the elimination phase. This has the effect of reducing the "back substitution" to the solution of a diagonal system which, of course, amounts to just N divisions.

Why was this method not even mentioned in Chap. 6? It appears to have much to commend it, *but* it turns out to have a significantly higher operation count than does the standard elimination method. The full serial algorithm can be summarized in a manner similar to Algorithm 6.1.2 as Algorithm 13.3.1.

Algorithm 13.3.1   Serial Gauss-Jordan algorithm

> *Input* N, an $N \times N$ matrix A, and an N vector $\mathbf{b}$
>
> *Elimination*
>
> > for $j = 1$ to N
> > > for $i = 1$ to N
> > > > if $i \neq j$ then
> > > > > $m_{ij} := a_{ij}/a_{ii}$; $a_{ij} := 0$; $b_i := b_i - m_{ij} b_j$
> > > > > for $k = j + 1$ to N $a_{ik} := a_{ik} - m_{ij} a_{jk}$
>
> *Output*   for $i = 1$ to N $x_i := b_i/a_{ii}$

The serial operation count for this algorithm is approximately N^3, which represents an increase of about 50 percent over Gauss elimination.

Let us now consider the parallel (or vector) implementation of these two algorithms. To simplify the comparison we reproduce below the serial Gauss elimination algorithm from Sec. 6.1.

Algorithm 13.3.2   Serial Gauss elimination algorithm

> *Input* N, an $N \times N$ matrix A, and an N vector $\mathbf{b}$
>
> *Forward elimination*
>
> > for $j = 1$ to $N - 1$
> > > for $i = j + 1$ to N
> > > > $m_{ij} := a_{ij}/a_{jj}$
> > > > $a_{ij} := 0$; $b_i := b_i - m_{ij} * b_j$
> > > > for $k = j + 1$ to N
> > > > > $a_{ik} := a_{ik} - m_{ij} * a_{jk}$
>
> *Back substitution*
>
> > for $i = N$ downto 1
> > > $x_i := b_i$
> > > for $j = i + 1$ to N
> > > > $x_i := x_i - a_{ij} * x_j$
> > > $x_i := x_i/a_{ii}$
>
> *Output*   Solution vector $\mathbf{x}$.

To reexpress these algorithms in a form suitable for array (or vector) processing, it is necessary to introduce the notion of *masking*.

A *logical mask* is an array of single-bit entries, interpreted as "true" or "false," which can be used to inhibit certain operations on particular parts of a vector or array. When programming with such masks, it is important to note that they do *not* inhibit the arithmetic itself but merely the assignment of the result to the appropriate memory location. This can be critical in trying to inhibit invalid operations. The use of masks is usually free on an array processor but carries a small operational cost on a vector processor.

The notation we use for such masks here is illustrated by the following examples. The statement

$$\mathbf{a}[i > 2] := \mathbf{a} - m * \mathbf{b}$$

is equivalent to the loops

$$\text{for } i = 3 \text{ to } N \qquad a_i := a_i - mb_i$$

or

$$\text{for } i = 1 \text{ to } N \qquad \text{if } i > 2 \text{ then } a_i := a_i - mb_i$$

Alternatively an explicit mask may be employed as in the statement

$$\mathbf{a}[\text{not}(\mathbf{used})] := \mathbf{a} - m * \mathbf{b}$$

using the bit-vector mask, **used**; this is equivalent to the loop

$$\text{for } i = 1 \text{ to } N$$
$$\text{if not}(\mathbf{used}[i]) \text{ then } a_i := a_i - mb_i$$

so that only those elements of **a** for which the corresponding entry in **used** is false are modified.

We will also make further use of some of the notation introduced in the last section:

MATR(**x**) is a matrix each of whose rows is **x**.
MATC(**x**) is a matrix each of whose columns is **x**.
$\mathbf{a}_{*j}$ is the vector formed from the jth column of A.
$\mathbf{a}_{i*}$ is the vector formed from the ith row of A.

The Gauss elimination algorithm can now be rewritten as Algorithm 13.3.3.

Algorithm 13.3.3 Parallel version of Gauss elimination

Input	N, an $N \times N$ matrix A, and an N vector **b**
Elimination	for $j = 1$ to $N - 1$
	$\quad \mathbf{m}[i > j] := \mathbf{a}_{*j} / a_{jj}$
	$\quad \mathbf{a}_{*j}[i > j] := 0$
	$\quad \mathbf{b}[i > j] := \mathbf{b} - b_j * \mathbf{m}$
	$\quad A[i > j, k > j] := A - \text{MATC}(\mathbf{m}) * \text{MATR}(\mathbf{a}_{j*})$

Back substitution	for $i = N$ downto 1
	$x_i := (b_i - \text{SUM}(\mathbf{a}_{i*} * \mathbf{x}))/a_{ii}$
Output	Solution vector $\mathbf{x}$.

Note that in the back substitution phase we have taken advantage of the fact that $a_{ij} = 0$ for $i > j$. This back substitution can be improved, but that is not of present concern. It is easy to see that the forward elimination requires $5(N - 1)$ paraflops. Two broadcasting operations are also needed for each iteration. The back substitution needs a further $3N$ paraflops together with N vector summations. The summation operations can be avoided as we will see later so that the total operation count for Gauss elimination becomes approximately $8N$.

We also note for future reference that the LU factorization of A can be achieved using the forward elimination phase of the above algorithm, without the operations on the right-hand-side vector $\mathbf{b}$, in just $3N$ paraflops.

The main point of this discussion is that, on our array processor, elimination above the diagonal can be achieved at zero additional cost by simply replacing the masks "$i > j$" with "$i \neq j$." This yields the following algorithm for parallel Gauss-Jordan.

Algorithm 13.3.4 Parallel Gauss-Jordan

Input	N, an $N \times N$ matrix A, and an N vector $\mathbf{b}$
Elimination	for $j = 1$ to $N - 1$
	$\mathbf{m}[i \neq j] := \mathbf{a}_{*j}/a_{jj}$
	$\mathbf{a}_{*j}[i \neq j] := 0$
	$\mathbf{b}[i \neq j] := \mathbf{b} - b_j * \mathbf{m}$
	$A[i \neq j, k > j] := A - \text{MATC}(\mathbf{m}) * \text{MATR}(\mathbf{a}_{j*})$
Output	Solution vector $\mathbf{x} := \mathbf{b}/\text{DIAG}(A)$

This algorithm requires just $5N$ paraflops. We see that instead of being some 50 percent more expensive, as it is in serial form, the parallel Gauss-Jordan algorithm is some 50 percent cheaper than the parallel version of Gauss elimination.

Of course, Gauss-Jordan still has some of the other drawbacks of its serial version. It is not readily suited to solving several systems using the same coefficient matrix since it does not correspond to a simple factorization of A just using the multipliers. Multiple right-hand sides do not present a problem if they are all to be solved simultaneously, but this is not typically the case.

Especially in the absence of pivoting, Gauss-Jordan is also numerically less stable than is Gauss elimination. Nonetheless, we can see from this discussion that for well-conditioned "one-off" problems *which fit the processor array*, Gauss-Jordan becomes a very efficient technique.

To improve the numerical stability of either of these techniques, pivoting is known to be desirable. We discuss briefly the use of partial (or column) pivoting within the Gauss-Jordan algorithm.

The only real extra task is the location of the pivot element within the current row. This task is also made easy by many array processors which have efficient procedures for identifying the maximum element of an array and/or the position of that maximum. We will use MAX and PMAX for these two operations, respectively. Thus PMAX returns the position of the maximal element of a vector—or, strictly, the position of its first occurrence—so that

$$\text{PMAX}(\mathbf{x}) = \text{MIN}\{i: x_i = \text{MAX}(\mathbf{x})\} \qquad (13.3.1)$$

There is a slight complication in that the pivot must be chosen from among those rows which have not yet been used in the elimination process. This is achieved by using a vector mask to remember which rows have been used. We must also store a vector of pivots used which will be the required divisors for the final solution phase. This is all summarized in Algorithm 13.3.5.

Algorithm 13.3.5   Gauss-Jordan with pivoting

Input	N, an $N \times N$ matrix A, and an N vector $\mathbf{b}$
Initialize	**used**:=false; **pivot**:=0
Elimination	for $j = 1$ to $N - 1$
	$\quad p := \text{PMAX}(\text{ABS}(\mathbf{a}_{*j}[\text{not}(\mathbf{used})]))$
	$\quad \mathbf{pivot}[j] := a_{pj}$
	$\quad \mathbf{used}[p] := \text{true}$
	$\quad \mathbf{m}[i \neq p] := \mathbf{a}_{*j}/\mathbf{pivot}[j]$
	$\quad \mathbf{a}_{*j}[i \neq p] := 0$
	$\quad \mathbf{b}[i \neq p] := \mathbf{b} - b_p * \mathbf{m}$
	$\quad A[i \neq p, k > j] := A - \text{MATC}(\mathbf{m}) * \text{MATR}(\mathbf{a}_{p*})$
Output	Solution vector $\mathbf{x} := \mathbf{b}/\mathbf{pivot}$

It is seen that the only additional operations compared with a step of the standard Gauss-Jordan algorithm are two assignment statements and location of the pivots. This is a small price for the additional stability. The algorithm remains very efficient.

Unfortunately, as we observed earlier, the Gauss-Jordan algorithm does not lend itself easily to repeated solution using the same coefficient matrix. For this task, one of the various matrix factorization schemes is preferred. There are many such algorithms available including some specifically designed for parallel architectures.

One of the simplest is, of course, LU factorization with partial pivoting. The pivoting can be achieved in much the same way as was used here for Gauss-Jordan. The factorization is therefore just a blend of this pivoting routine with the forward elimination phase of the Gauss elimination Algorithm 13.3.3 (omitting the step modifying **b**). Of course, we must also store the multipliers in the matrix array, but that presents no difficulty (see Exercise 2). The solution process is then completed using a forward and backward solve as usual. We will discuss the efficient parallel solution of such triangular systems shortly.

Modifications for the Cholesky factorization of a symmetric positive definite matrix are straightforward. Indeed, since pivoting is not required, this becomes a simplification.

One of the factorization approaches developed specifically for parallel processors is the WZ factorization which gets its name from the shapes of the two factor matrices—plus a bit of imagination! In this algorithm, the coefficient matrix A is factorized as the product WZ, where these two matrices have shapes illustrated by Fig. 13.3.6 in which nonzero entries only occur in shaded regions.

The elements of these two factors are computed in a recursive manner. Two rows of Z are computed, in parallel, using relatively small matrix-matrix multiplications, and then two columns of W are computed by solving, in parallel, several 2×2 systems of equations. Such a scheme would also, therefore, be well-suited to implementation on a shared-memory architecture. For details of the factorization algorithm and the subsequent solution phase see, for example, Modi (1988).

The other principal factorization technique involves the use of the QR factorization using either Givens' rotations or Householder reflections as an equation solver. This factorization is discussed in its more usual context as an eigenvalue method in Chap. 11. Again the factorization can be vectorized or parallelized efficiently and leads to a final solution phase consisting of the solution of a triangular system. It is to this last problem that we now direct our attention.

In Algorithm 13.3.3, a *scalar product* form was used for the back substitution. This involved computing scalar products of rows of the matrix with the current solution vector. However, the summation that is required for this scalar product operation is potentially expensive, and so other processes are desirable.

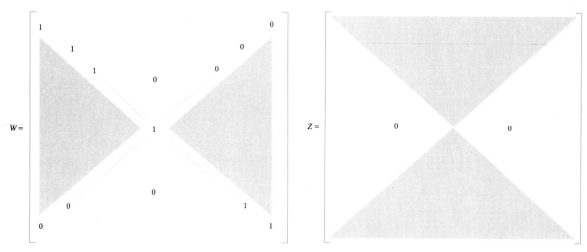

FIGURE 13.3.6
Form of the WZ factorization.

We will consider the solution of a lower triangular system such as that encountered in the forward solve of the LU technique. There is, of course, a completely analogous approach for upper triangular systems. Suppose then that we seek the solution of

$$L\mathbf{x} = \mathbf{b} \tag{13.3.2}$$

where, for convenience, we write L in the form

$$L = \begin{bmatrix} 1 & & & & \\ -m_{21} & 1 & & & \\ -m_{31} & -m_{32} & 1 & & \\ \vdots & \vdots & & \vdots & \\ -m_{N1} & -m_{N2} & \cdots & & 1 \end{bmatrix} \tag{13.3.3}$$

The main alternative to the scalar product form is the *column sweep* algorithm in which we adjust the complete right-hand-side vector for the contribution of x_i as soon as this quantity is known. Thus the first step is to set

$$x_1 = b_1$$

and then, for every $j > 1$,

$$b_j := b_j + m_{j1}x_1$$

which we can write in vector form as

$$\mathbf{b}[j > 1] := \mathbf{b} + x_1 * \mathbf{m}_{*1} \tag{13.3.4}$$

The obvious extension to the subsequent steps is included in the following algorithm.

Algorithm 13.3.7   Column sweep algorithm for triangular system

> *Input*   Unit lower triangular matrix L in the form (13.3.3)
> Right-hand-side vector $\mathbf{b}$
>
> *Loop*   for $k = 1$ to N
> $\quad x_k := b_k$
> $\quad \mathbf{b}[j > k] := \mathbf{b} + x_k * \mathbf{m}_{*k}$
>
> *Output*   Solution vector $\mathbf{x}$.

Note that the mask used in modifying the vector $\mathbf{b}$ is not strictly necessary here since the earlier components of this vector will not be used further in the solution process.

This routine has an operation count of just $2N$ paraflops. For the situation where the diagonal entries are not all 1s, there would be a division in the first step of the loop. Thus a typical forward and backward solve for LU factors would require a total of $5N$ paraflops, which is comparable with the Gauss-Jordan algorithm.

In the case of QR factorization, only one triangular system must be solved after a matrix transpose and a matrix-vector multiply. Transposing a matrix is communications-intensive, and this must be accounted for in overall timing considerations. The matrix-vector operation, which we saw in the previous section, is an $O(\log_2 N)$ paraflop task. Thus QR factorization is likely to be quicker and is known to have greater numerical stability.

There is a further alternative to the column sweep algorithm for triangular systems using the product form of the inverse of L and a recursive doubling algorithm for the matrix multiplications. For a full matrix, this requires a very large number, $O(N^3)$, of processors if it is to be efficient. However, for banded systems (such as often result from finite difference solutions of partial differential equations), this approach can be used to good effect since the early products are sparse. Thus for the first several steps many matrix products can be performed in parallel. As the density of the products increases, the number of matrix products needed declines. Such a scheme clearly needs careful organization. [Again, more detail of this algorithm is to be found in Modi (1988).]

It is also important to note that the error bound for the recursive algorithm grows rapidly compared with that for the column sweep or scalar product forms, making it unsuitable for ill-conditioned systems.

We finish this section with a brief look at iterative methods for general systems on array processors. We write

$$A = L + D + U \tag{13.3.5}$$

where L, D, U are the lower triangular, diagonal, and upper triangular parts of A. The Jacobi iteration (see Sec. 6.6) takes the form

$$\mathbf{x}_{k+1} = D^{-1}[\mathbf{b} - (L + U)\mathbf{x}_k] \tag{13.3.6}$$

which is clearly parallelizable for a large-enough array. Each iteration consists of a matrix-vector multiplication—an operation which we have already seen can be efficiently parallelized or vectorized—followed by a vector subtraction and division.

On an array processor, this results in a total of $3 + \log_2 N$ paraflops per iteration. For $N = 64$, this is 9 paraflops per iteration, which means that some 30 iterations can be used before the cost of Gauss-Jordan is reached. For full matrices this typically would be insufficient to obtain great accuracy.

We saw in Chap. 6 that the Gauss-Seidel iteration usually converges about twice as fast as the Jacobi iteration. The Gauss-Seidel iteration can be written as

$$\mathbf{x}_{k+1} = D^{-1}(\mathbf{b} - L\mathbf{x}_{k+1} - U\mathbf{x}_k) \tag{13.3.6}$$

so that the most recent estimates are used at every stage. At first sight, this appears to be inherently serial since the computation of $x_2^{(k+1)}$, for example, requires the value of $x_1^{(k+1)}$. It turns out that many sparse matrices have structures that admit modified Gauss-Seidel iterations which can be performed in parallel. This in turn allows parallel implementations of various SOR-type algorithms. We study this in detail in the next section in the context of finite difference solutions of partial differential equations.

EXERCISES 13.3

1. Write a program to implement the Gauss-Jordan algorithm both with and without pivoting. Compare its performance with that of the Gauss elimination, or LU factorization, algorithm in terms of speed and accuracy.

2. Write a pseudocode similar to Algorithms 13.3.3 and 13.3.5 for parallel implementation of LU factorization with partial pivoting.

3. Write a parallel algorithm for upper triangular systems with nonunit diagonals. Incorporate the column sweep algorithms into your LU solver. Compare the performance with the scalar product form. Is there any significant difference on a serial processor?

13.4 PARTIAL DIFFERENTIAL EQUATIONS

We begin this section by considering finite difference approaches to the solution of partial differential equations. In fact, we will concentrate, for illustrative purposes, on just one such equation. Almost all methods of solution for such equations begin with some discretization, either from finite difference approximations of the equation or from finite element triangulation of the domain. This will result in a, perhaps large, sparse system of linear equations to be solved either for the function values or for the coefficients of the finite element solution.

We use Poisson's equation

$$u_{xx} + u_{yy} = f \qquad (13.4.1)$$

to illustrate the ideas. For a finite difference solution of this equation on a rectangular domain, we obtain approximate values of the solution at points of a regular rectangular grid of points. Suppose then that the domain is $[0, a] \times [0, b]$ and that the mesh of grid points is such that

$$x_i = ih \qquad y_j = jk \qquad (13.4.2)$$

where the steplengths are given by

$$h = \frac{a}{m} \qquad k = \frac{b}{n} \qquad (13.4.3)$$

We denote by u_{ij} the approximation to the true value $u(x_i, y_j)$ of the solution at the grid point (x_i, y_j).

We can use the symmetric difference formula from Table 9.2.6 to approximate the second partial derivatives of u at each mesh point. Thus

$$u_{xx}(x_i, y_j) \simeq \frac{u_{i+1j} - 2u_{ij} + u_{i-1j}}{h^2}$$

$$u_{yy}(x_i, y_j) \simeq \frac{u_{ij+1} - 2u_{ij} + u_{ij-1}}{k^2}$$

$$(13.4.4)$$

and substituting these into the differential equation (13.4.1), we obtain, for each interior point, the equation

$$k^2(u_{i+1j} - 2u_{ij} + u_{i-1j}) + h^2(u_{ij+1} - 2u_{ij} + u_{ij-1}) = h^2k^2f_{ij} \quad (13.4.5)$$

where $f_{ij} = f(x_i, y_j)$. This can be rearranged to yield

$$u_{ij} = \frac{k^2(u_{i+1j} + u_{i-1j}) + h^2(u_{ij+1} + u_{ij-1}) - h^2k^2f_{ij}}{2(h^2 + k^2)} \quad (13.4.6)$$

which represents a linear system of equations for the values u_{ij} in a form suitable for iteration.

Boundary conditions will, of course, affect the equations relating to the boundary points. For example, Dirichlet boundary conditions, in which u is specified on the boundary, determine the values u_{ij} for $i = 0, m, j = 0, n$, and only the remaining values would be modified by the iterations. Neumann conditions specify the normal derivative of u on the boundary, and these are incorporated by modified equations using difference approximations to these derivatives. Our primary concern here is with the iterative solution of the linear systems that arise.

We will assume that the two steplengths h and k are equal. We can then rewrite (13.4.6) as though the meshsize parameters satisfy $h = k = 1$. That is,

$$v_{ij} = \frac{v_{i-1j} + v_{i+1j} + v_{ij-1} + v_{ij+1} - \phi_{ij}}{4} \quad (13.4.7)$$

where

$$v_{ij} \simeq v(i, j) = u(ih, jh) \qquad \phi_{ij} = \phi(i, j) = h^2f(ih, jh) \quad (13.4.8)$$

This leads in a natural way to a Jacobi iteration (see Sec. 6.6) in which the function values are updated according to

$$v_{ij}^{(p+1)} = \frac{v_{i-1j}^{(p)} + v_{i+1j}^{(p)} + v_{ij-1}^{(p)} + v_{ij+1}^{(p)} - \phi_{ij}}{4} \quad (13.4.9)$$

On an array processor, we can allocate v_{ij} to processor (i, j) and then update all entries simultaneously. This could be achieved using extensions of the shift operators defined earlier to allow shifts in the other coordinate direction.

We denote by SHUP and SHDP for shift-up-planar and shift-down-planar, respectively, the shift operators for the y direction. With this notation, the iteration (13.4.9) can be written

$$\mathbf{v}^+ := \{\text{SHRP}(\mathbf{v}, 1) + \text{SHLP}(\mathbf{v}, 1) + \text{SHUP}(\mathbf{v}, 1) + \text{SHDP}(\mathbf{v}, 1) - \phi\}/4 \quad (13.4.10)$$

where we have suppressed the iteration number. The use of (13.4.10) entails 5 paraflops per iteration. This can be improved upon by combining pairs of these shifts so that we set

$$\mathbf{w} := \text{SHRP}(\mathbf{v}, 1) + \text{SHDP}(\mathbf{v}, 1)$$
$$\mathbf{v}^+ := \{\mathbf{w} + \text{SHLP}(\text{SHUP}(\mathbf{w}, 1), 1) - \phi\}/4$$

which reduces the operation count to 4 paraflops per iteration at the cost of an additional double shift. Typically, division by a power of the base 2 is a very fast operation, and so this represents a saving of close to 33 percent in the arithmetic.

With appropriate ordering of the data points, it would be similarly straightforward to vectorize this iteration. We do not discuss the details of this vectorization, since we will see shortly that there are considerable improvements over this Jacobi scheme available for both vector and parallel processors. The numerical performance of the various iterations under discussion are compared later in the section.

In Sec. 6.6., we discovered that the Gauss-Seidel iteration is usually more rapidly convergent than the Jacobi iteration. In the Gauss-Seidel iteration, the most recently computed values are used on the right-hand side. This appears to be an essentially serial idea in that no component of the solution can be updated until its predecessor, in whatever order is being used for the mesh points, has been updated.

This is misleading, however. Supposing we order the points (i, j) in increasing order of $i + j$ with, say, j increasing within each such set of vertices. Thus, for the Dirichlet boundary conditions where v is known on the boundary, we update the approximate solution in the order v_{11}; v_{21}, v_{12}; v_{31}, v_{22}, v_{13}; .... However, once v_{21} and v_{12} have been updated, we have all the information needed for the *next* update of v_{11} as well as the initial update of v_{31}, v_{22}, v_{13}. This information is then sufficient to update v_{21} and v_{12} again and $v_{41}, v_{32}, v_{23}, v_{14}$ for the first time.

On our parallel processor, we can therefore update our solution in a staggered way according to Table 13.4.1. This implies that, at any one time, our parallel implementation of the Gauss-Seidel iteration has different mesh points at different iterations.

Eventually, this process reaches "saturation point" after which all mesh points with $i + j$ even are updated on one iteration and those with $i + j$ odd are updated in the next. This observation leads naturally to the suggestion that the updates could be performed in this manner from the start. This *red-black*, or *black-white*, or *checkerboard* ordering is equivalent to applying the update to all

TABLE 13.4.1
Parallel updating schedule for Gauss-Seidel iteration

Step	Values generated
1	$v_{11}^{(1)}$
2	$v_{21}^{(1)}, v_{12}^{(1)}$
3	$v_{11}^{(2)}, v_{31}^{(1)}, v_{22}^{(1)}, v_{13}^{(1)}$
4	$v_{21}^{(2)}, v_{12}^{(2)}, v_{41}^{(1)}, v_{32}^{(1)}, v_{23}^{(1)}, v_{14}^{(1)}$
5	$v_{11}^{(3)}, v_{31}^{(2)}, v_{22}^{(2)}, v_{13}^{(2)}, v_{51}^{(1)}, v_{42}^{(1)}, v_{33}^{(1)}, v_{24}^{(1)}, v_{15}^{(1)}$

points with $i + j$ even and odd alternately. This is equivalent to using the Jacobi iteration for all $i + j$ even and then the Gauss-Seidel iteration for mesh points with $i + j$ odd.

In Fig. 13.4.2, we update all the "black" (shaded) points on the odd iterations and all "white" points on the even iterations, always using the most recent values. (The confusion over the name derives from the different coloring schemes for checkerboards, or draughtboards, used on the two sides of the Atlantic.)

The iteration equations are therefore

$$v_{ij}^{(p+1)} = \begin{cases} \dfrac{v_{i-1j}^{(p)} + v_{i+1j}^{(p)} + v_{ij-1}^{(p)} + v_{ij+1}^{(p)} - \phi_{ij}}{4} & i + j \text{ even} \\[3mm] \dfrac{v_{i-1j}^{(p+1)} + v_{i+1j}^{(p+1)} + v_{ij-1}^{(p+1)} + v_{ij+1}^{(p+1)} - \phi_{ij}}{4} & i + j \text{ odd} \end{cases}$$

$$(13.4.11)$$

We observe that this red-black iteration is approximately twice as expensive as the Jacobi iteration since only half the points can be updated in a single parallel step. Experimental evidence suggests that the Gauss-Seidel iteration typically converges approximately twice as fast as Jacobi. These two factors appear to cancel one another out. Of course, the same-sized processor array can be used for a system with twice as many mesh points for the red-black iteration, which may provide enough additional accuracy to tip the balance in favor of this scheme.

The numerical performance of both the Gauss-Seidel and red-black schemes will be examined later along with Jacobi and the successive overrelaxation (SOR) schemes to which we turn our attention now. (See Sec. 6.7 for a description of the basic SOR technique.)

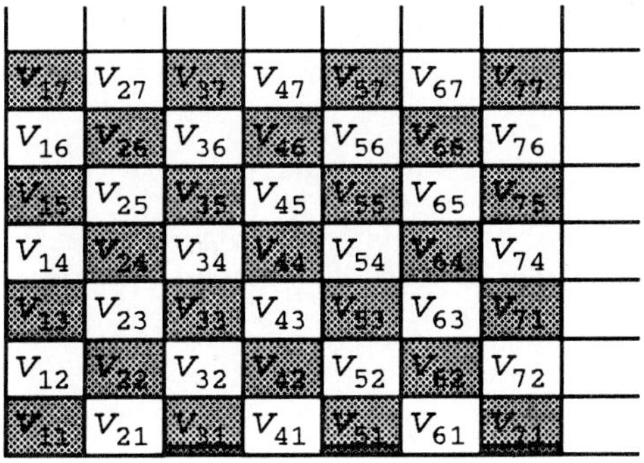

FIGURE 13.4.2
Checkerboard, or black-white or red-black ordering.

We will base our SOR iteration on the red-black ordering so that all points are updated every second iteration. The SOR method can be viewed as iterating according to

$$\mathbf{v}^+ := \omega \mathbf{v}_{\mathrm{RB}} + (1 - \omega)\mathbf{v} \tag{13.4.12}$$

where the subscript RB signifies the red-black update of $\mathbf{v}$, ω is the relaxation parameter, and the iteration number has been suppressed. We recall that the relaxation parameter is chosen from the interval $(0, 2)$ and, for successive *over*relaxation, $\omega > 1$. In fact, for the red-black ordering this updating, and the relaxation, would be performed for the red points and then for the black in turn.

The red-black SOR iteration is therefore given by the equations

$$w_{ij} = \frac{v^{(p)}_{i-1j} + v^{(p)}_{i+1j} + v^{(p)}_{ij-1} + v^{(p)}_{ij+1} - \phi_{ij}}{4} \qquad i + j \text{ even}$$

$$v^{(p+1)}_{ij} = \omega w_{ij} + (1 - \omega)v^{(p)}_{ij} \tag{13.4.13}$$

$$w_{ij} = \frac{v^{(p+1)}_{i-1j} + v^{(p+1)}_{i+1j} v^{(p+1)}_{ij-1} + v^{(p+1)}_{ij+1} - \phi_{ij}}{4} \qquad i + j \text{ odd}$$

$$v^{(p+1)}_{ij} = \omega w_{ij} + (1 - \omega)v^{(p)}_{ij}$$

We turn now to the examination of the numerical performance of these various algorithms which should properly be called, respectively, *point* Jacobi (13.4.9), *point* Gauss-Seidel, and so on, since we will see shortly that other orderings of the computation will lead to *line* and *block* forms of these algorithms too.

Throughout the study of these finite difference techniques, we will consider the solution of Poisson's equation with steplength 1 in each direction on the rectangle $[0, n + 1] \times [0, m + 1]$ for the special case where the true solution satisfies

$$v_{ij} = \sin \frac{2\pi i}{n + 1} \sin \frac{5\pi j}{m + 1} + \sin \frac{3\pi i}{n + 1} \sin \frac{11\pi j}{m + 1}$$

In every case we will take as our initial guess $v_{ij} = 0$ for all i, j and consider two separate cases

$$m = 5, n = 10 \qquad \text{and} \qquad m = 15, n = 15$$

The means of comparison is the maximum error at each iteration. That is, we tabulate $\|\mathbf{v}^{(p)} - \mathbf{v}\|_\infty$ at each iteration p for each of the methods. This error is tabulated for the first 10 iterations of each of six different approaches: Jacobi, Gauss-Seidel, red-black iterations, and red-black SOR with three values of the relaxation parameter. These are $\omega = 0.8$, 1.2, and 1.4.

The first of these is not really successive *over*relaxation since $\omega < 1$. This underrelaxation iteration is similar to the "damped Jacobi" iteration which is sometimes advocated as the underlying iteration of multigrid methods. [See, for example, McCormick (1987) for an introduction to multigrid methods.]

The results for these point iteration methods are presented in Table 13.4.3. It is apparent that the SOR method has yielded significant improvement over the Jacobi and Gauss-Seidel iterations. The ordinary red-black iteration also shows good convergence for these problems especially when we take into account the fact that it uses only 5 paraflops per iteration compared with 8 for the SOR methods.

We have already seen that the ordering of the mesh points can have a marked effect on the performance of iterative methods. The above examples all used *point ordering* in which individual points are updated. These orderings were chosen so that many points could be updated simultaneously on an array processor.

For a vector processor or if the mesh is large enough that it cannot all be updated in one paraflop, these same orderings could be used so that we apply the update to appropriately chosen subsets of the mesh. However, in such cases, other orderings may be more suitable. One natural vectorization is to consider the overall mesh as an array of "vectors," each of which consists of a line of the original mesh. This leads to the use of *line orderings*.

TABLE 13.4.3
Point iterative schemes for Poisson's equation

Iter.	Jacobi	Gauss-Seidel	Red-black	$\omega = 0.8$	SOR $\omega = 1.2$	$\omega = 1.4$
			(a) $m = 5, n = 10$			
1	0.76397	0.69088	0.74144	0.97304	0.65466	0.77241
2	0.57253	0.42312	0.43527	0.60938	0.30734	0.21901
3	0.43527	0.29152	0.25170	0.41822	0.07784	0.15058
4	0.33100	0.18943	0.14555	0.29540	0.02878	0.05041
5	0.25170	0.13096	0.08417	0.21032	0.00778	0.01474
6	0.19141	0.09354	0.04867	0.15007	0.00254	0.01104
7	0.14555	0.06714	0.02815	0.10714	0.00072	0.00317
8	0.11069	0.04857	0.01628	0.07650	0.00022	0.00105
9	0.08417	0.03541	0.00941	0.05463	0.00007	0.00076
10	0.06401	0.02602	0.00544	0.03901	0.00002	0.00019
			(b) $m = 15, n = 15$			
1	0.83727	0.80918	0.83727	1.01600	0.71204	0.78189
2	0.56065	0.47389	0.40663	0.61617	0.24599	0.25498
3	0.40663	0.30713	0.22152	0.40079	0.06861	0.12302
4	0.29968	0.20514	0.12120	0.27033	0.01665	0.05132
5	0.22152	0.13983	0.06632	0.18526	0.00368	0.02095
6	0.16385	0.09751	0.03629	0.12778	0.00075	0.01265
7	0.12120	0.07034	0.01986	0.08835	0.00014	0.00330
8	0.08965	0.05422	0.01087	0.06114	0.00003	0.00190
9	0.06632	0.04207	0.00595	0.04233	0.00000	0.00068
10	0.04906	0.03290	0.00325	0.02931	0.00000	0.00018

The idea here is to update a complete "line" of mesh points at each step. In what follows we will think of a line as a column $\mathbf{v}_{\cdot j}$. The Jacobi line iteration is then obtained by rearranging (13.4.7) to yield

$$v_{i-1j}^{(p+1)} - 4v_{ij}^{(p+1)} + v_{i+1j}^{(p+1)} = \phi_{ij} - v_{ij+1}^{(p+1)} - v_{ij-1}^{(p+1)} \qquad (13.4.14)$$

which is a diagonally dominant tridiagonal system of equations for the new line. We will denote the tridiagonal matrix by A.

This tridiagonal system can be solved using the cyclic reduction scheme discussed in Sec. 13.2. In a natural way, this iteration admits Gauss-Seidel and red-black modifications. The red-black line ordering is such that all odd-numbered lines are updated in parallel and then all even-numbered ones in another parallel step. This can be further modified to include relaxation. In Table 13.4.5, we present the results of these various approaches for the same example as was used above to test the point iterations.

The red-black line SOR algorithm is summarized as follows.

Algorithm 13.4.4   Red-black line SOR Poisson equation solver

> *Input*   Right-hand side ϕ_{ij}
> Initial guess $\mathbf{v}$, relaxation parameter ω
>
> *Repeat until convergence*
> For all odd j:
> $\mathbf{r} := \phi - \text{SHLP}(\mathbf{v}) - \text{SHRP}(\mathbf{v})$
> Solve the tridiagonal system $A\mathbf{w}_{\cdot j} = \mathbf{r}_{\cdot j}$
> Relax: $\mathbf{v}_{\cdot j} = \omega \mathbf{w}_{\cdot j} + (1 - \omega)\mathbf{v}_{\cdot j}$
>
> For all even j:
> $\mathbf{r} := \phi - \text{SDLP}(\mathbf{v}) - \text{SHRP}(\mathbf{v})$
> Solve the tridiagonal systems $A\mathbf{w}_{\cdot j} = \mathbf{r}_{\cdot j}$
> Relax: $\mathbf{v}_{\cdot j} = \omega \mathbf{w}_{\cdot j} + (1 - \omega)\mathbf{v}_{\cdot j}$
>
> *Output*   Approximate solution $\mathbf{v}$.

The results in Table 13.4.5 again show that the red-black SOR iteration with a relaxation parameter of 1.2 performs well. It is noticeable in the case of a 15×15 mesh that the plain red-black iteration is also highly effective. This iteration is slightly cheaper as a result of eliminating the relaxing steps of Algorithm 13.4.4 which represent 3 paraflops each. However, for these line iterations that is a small saving since the tridiagonal solve will usually require significantly more operations. As with the point iterations, the damped scheme using $\omega = 0.8$ appears not to justify the additional work.

In the serial implementation of the iterations used to obtain these results, the cyclic reduction scheme was not used for the tridiagonal matrix. The meshsize is small enough that a tridiagonal LU factorization algorithm could be used efficiently for this part of the process. In general, the overall system of equations is of block tridiagonal form with the diagonal blocks themselves

TABLE 13.4.5
Line iterative schemes for Poisson's equation

Iter.	Jacobi	Gauss-Seidel	Red-black	$\omega = 0.8$	SOR $\omega = 1.2$	$\omega = 1.4$
			(a) $m = 5, n = 10$			
1	1.31711	0.73326	1.31711	1.06972	1.56449	1.81187
2	0.91838	0.30797	0.64389	0.65012	0.51792	0.68561
3	0.64389	0.12371	0.32161	0.41919	0.15474	0.13453
4	0.45388	0.05602	0.16387	0.28157	0.04403	0.12439
5	0.32161	0.02503	0.08499	0.19411	0.01110	0.03457
6	0.22902	0.01103	0.04477	0.13592	0.00274	0.01347
7	0.16387	0.00488	0.02424	0.09605	0.00068	0.00849
8	0.11777	0.00219	0.01328	0.06822	0.00016	0.00123
9	0.08499	0.00099	0.00731	0.04860	0.00004	0.00128
10	0.06158	0.00046	0.00404	0.03468	0.00001	0.00040
			(b) $m = 15, n = 15$			
1	0.94328	0.80420	0.92516	0.74324	1.12135	1.31754
2	0.46795	0.34112	0.22807	0.39879	0.09459	0.43815
3	0.23254	0.22418	0.05660	0.23451	0.04292	0.10433
4	0.11575	0.13739	0.01414	0.14197	0.00472	0.09720
5	0.05771	0.08926	0.00356	0.08670	0.00148	0.02578
6	0.02882	0.05084	0.00090	0.05309	0.00027	0.00910
7	0.01442	0.03481	0.00023	0.03254	0.00004	0.00639
8	0.00722	0.02487	0.00006	0.01994	0.00001	0.00160
9	0.00362	0.01798	0.00002	0.01223	0.00000	0.00082
10	0.00182	0.01343	0.00000	0.00749	0.00000	0.00037

being tridiagonal and the off-diagonal blocks being just negative identity matrices.

For the particular boundary conditions used here, the diagonal block is constant and so can be factorized just once. Its factors can then be stored and used to perform the line iterations efficiently. Several copies of the factors of the diagonal block could be stored on a processor array to allow the corresponding red-black schemes to be implemented.

Many other orderings have been proposed for iterative solution of partial differential equations using finite differences employing a variety of blocks of mesh points. The interested reader is referred to the paper of O'Leary (1984) which is devoted to the study of different orderings for parallel processing.

We complete this section with an introduction to the finite element solution of a partial differential equation, again using Poisson's equation for illustrative purposes. We do not consider here the details of the finite element solution but just outline the process to illustrate the nature of the resulting computational problem. [More detail of the finite element method can be found in, for example, Prenter (1975), Schatz (1985), and Showalter (1985).]

Consider the solution of a differential equation of the form

$$-\nabla \cdot A^T \nabla u = f \qquad (13.4.15)$$

on a region $\Omega \subseteq \mathbb{R}^2$ subject to Dirichlet boundary conditions $u(x) = 0$ on the boundary $\partial\Omega$ of Ω. Poisson's equation corresponds to taking $A = I$ in (13.4.15).

The first step in the solution process is to obtain the *weak form* of the equation by multiplying (13.4.15) by an arbitrary test function v which satisfies the Dirichlet boundary conditions and then integrating over the region Ω. (That is, we form the inner product of each side of the equation with v.) Applying Green's theorem to the left-hand side, we get

$$\int_\Omega (\nabla u)^T A \nabla v \, dx = \langle f, v \rangle = \int_\Omega f(x)v(x) \, dx \qquad (13.4.16)$$

It follows therefore that the solution of Poisson's equation satisfies

$$\int_\Omega \nabla u \cdot \nabla v \, dx = \langle f, v \rangle \qquad (13.4.17)$$

for every v in our space of test functions. [These test functions satisfy the boundary conditions and have first partial derivatives that are square-integrable over Ω. They are in the *Sobolev* space $H_0^1(\Omega)$.]

The next step of this *Galerkin* method is to partition, or *triangulate*, the domain Ω into either triangles (or, for higher-dimensional problems, simplexes) or rectangles. This triangulated domain is usually denoted Ω^h, where h is a "meshsize" indicating the size of the subdomains. The approximate solution we obtain is formed as a piecewise function defined separately on each *element*. For triangular elements, piecewise linear functions are used, while for rectangular elements, piecewise bilinear functions form the basis of the solution.

The situation is illustrated in Fig. 13.4.6, in which we have a region subdivided into 12 triangles by placing 11 *nodes* in the interior and on the boundary of the original polygonal region. In this case, we would approximate our solution by a piecewise linear function which is continuous across each of the element boundaries.

We thus seek a "solution" u^h to (13.4.17) from this space of functions and denote the corresponding set of test functions by v^h. We therefore require u^h satisfying

$$\int_{\Omega^h} \nabla u^h \cdot \nabla v^h \, dx = \int_{\Omega^h} f(x)v^h(x) \, dx$$

The function spaces are now finite-dimensional. Let $\{\phi_1, \phi_2, \ldots, \phi_n\}$ be a basis for this space. [Such a basis could be formed by taking ϕ_i to be the piecewise linear function satisfying $\phi_i(V_j) = \delta_{ij}$, where V_j denotes the jth vertex or node and δ_{ij} is the Kronecker δ.]

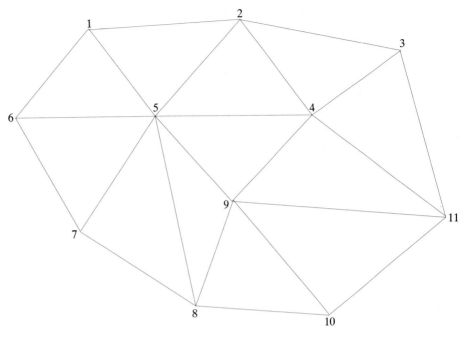

FIGURE 13.4.6
Triangulation of a domain.

In terms of this basis, we can write

$$u^h = \sum \alpha_i \phi_1 \qquad v^h = \sum \beta_i \phi_i \qquad (13.4.18)$$

and denote

$$\int_{\Omega^h} \nabla\phi_i \cdot \nabla\phi_j \, dx =: m_{ij} \qquad (13.4.19)$$

$$\int_{\Omega^h} f\phi_i \, dx =: f_i$$

It follows that we now require a solution u^h such that

$$\sum_{i,j} \beta_i m_{ij} \alpha_j = \sum_i \beta_i f_i$$

for every v^h, or, equivalently, for every choice of the coefficients $\beta_1, \beta_2, \ldots, \beta_n$. It follows that our approximate solution is given by

$$\sum_{j=1}^{n} m_{ij} \alpha_j = f_i \qquad (13.4.20)$$

or

$$M\alpha = f .$$

The choice of the basis functions and the ordering of the vertices will affect the sparsity structure of the matrix M. For the labeling of the vertices used in Fig. 13.4.6, and the piecewise linear basis defined by the requirement $\phi_i(V_j) = \delta_{ij}$ so that $m_{ij} = 0$ for any pair of vertices V_i, V_j which are not vertices of a common triangle, we would obtain a matrix M with the form

$$\begin{bmatrix} x & x & & & x & x & & & & & & \\ x & x & x & x & x & & & & & & & \\ & x & x & x & & & & & & & x & \\ & x & x & x & x & & & & x & & x & \\ x & x & & x & x & x & x & x & x & & & \\ x & & & & x & x & x & & & & & \\ & & & & x & x & x & x & & & & \\ & & & & x & & x & x & x & x & & \\ & & & x & x & & & x & x & x & x & \\ & & & & & & & x & x & x & x & \\ & & x & x & & & & & x & x & x & \end{bmatrix} \tag{13.4.21}$$

where x indicates a nonzero entry. We see that M has considerable sparsity and is symmetric. Its sparsity, however, does not have a simple pattern like the block multidiagonal forms typical of finite difference methods. Clearly the solution methods must be designed to retain as much sparsity as possible.

The structure of the matrix is not necessary well-suited to the regular nature of the array or vector processor iterations discussed above. For realistic problems, of course, the matrix dimensions would be very much larger than this. The next section is devoted to methods for solving large sparse systems of equations.

EXERCISES 13.4

1. Write programs to implement the various point iterative methods discussed in this section for Poisson's equation. Test them by choosing the right-hand side appropriate for a known solution. Experiment with the relaxation parameter to try to find the optimum value. Do you get the same value for other problems or other measures of the error?

2. Modify your programs from Exercise 1 to the case of Neumann boundary conditions and repeat your experiments.

3. Continue your experimentation with the line iterative schemes. Try both LU factorization and cyclic reduction algorithms for the tridiagonal system solver.

13.5 LARGE SPARSE SYSTEMS

This final section concerns itself with the solution of the large sparse linear systems which typically result, for example, from the finite element solution of partial differential equations. In much of what follows we will, in fact, refer for simplicity to fairly small—but still sparse—systems. For much of the time, we will use a system which would result from the triangulation of the domain shown in Fig. 13.4.6. This yields a matrix with the structure illustrated in (13.4.21).

We will also restrict our attention to symmetric matrices and assume further that the matrices under discussion are positive definite. There are extensions of all that is said here to more general settings, and the interested reader is referred, for example, to George (1989) for a more substantial survey of this topic. As with the rest of this chapter, we strive only to introduce some of the basic ideas and considerations of parallel algorithm design.

In this section, we consider just one solution technique: the Cholesky factorization $A = LL^T$ and solution of triangular systems. The structure of the matrix, the pivoting strategy, and the ordering of the rows or columns can have a very marked effect on the sparsity of the factors. As a simple, though extreme, example of this consider the factorization of a matrix which has its only nonzero entries in its top row, first column, and diagonal. Its Cholesky factors are easily seen to be dense. We summarize this statement with the "structure equation"

$$
\begin{bmatrix}
x & x & x & x & x & x & x \\
x & x &   &   &   &   &   \\
x &   & x &   &   &   &   \\
x &   &   & x &   &   &   \\
x &   &   &   & x &   &   \\
x &   &   &   &   & x &   \\
x &   &   &   &   &   & x
\end{bmatrix}
=
\begin{bmatrix}
x &   &   &   &   &   &   \\
x & x &   &   &   &   &   \\
x & x & x &   &   &   &   \\
x & x & x & x &   &   &   \\
x & x & x & x & x &   &   \\
x & x & x & x & x & x &   \\
x & x & x & x & x & x & x
\end{bmatrix}
$$

$$
\times
\begin{bmatrix}
x & x & x & x & x & x & x \\
  & x & x & x & x & x & x \\
  &   & x & x & x & x & x \\
  &   &   & x & x & x & x \\
  &   &   &   & x & x & x \\
  &   &   &   &   & x & x \\
  &   &   &   &   &   & x
\end{bmatrix}
\qquad (13.5.1)
$$

Here, and throughout the section, we will assume that no zeros are created during the elimination process since the algorithms under consideration can only take account of predictable sparsity. With the simple reordering resulting from writing the first equation last and the first unknown last, the original matrix has its final column and bottom row nonzero. The factorization then yields

$$
\begin{bmatrix}
x &   &   &   &   &   & x \\
  & x &   &   &   &   & x \\
  &   & x &   &   &   & x \\
  &   &   & x &   &   & x \\
  &   &   &   & x &   & x \\
  &   &   &   &   & x & x \\
x & x & x & x & x & x & x
\end{bmatrix}
=
\begin{bmatrix}
x &   &   &   &   &   &   \\
  & x &   &   &   &   &   \\
  &   & x &   &   &   &   \\
  &   &   & x &   &   &   \\
  &   &   &   & x &   &   \\
  &   &   &   &   & x &   \\
x & x & x & x & x & x & x
\end{bmatrix}
\begin{bmatrix}
x &   &   &   &   &   & x \\
  & x &   &   &   &   & x \\
  &   & x &   &   &   & x \\
  &   &   & x &   &   & x \\
  &   &   &   & x &   & x \\
  &   &   &   &   & x & x \\
  &   &   &   &   &   & x
\end{bmatrix}
$$

$$
(13.5.2)
$$

in which the sparsity has been preserved totally.

Our assumption that the original matrix is positive definite and symmetric implies that pivoting does not affect the numerical stability of the elimination process. This in turn implies that the system can be ordered so as to maximize the sparsity of the Cholesky factors.

We do not consider the choice of ordering here but concentrate instead on the sparsity structure of the factors for a given matrix. The idea is to be able to predict the sparsity structure of the Cholesky factors *in advance* of the computation so that the storage of the matrix can be economical and *static*. For a genuinely large system with perhaps several hundred (or even thousand) rows and columns, the storage requirement and efficient access to the nonzero entries is a major consideration. To have a static storage scheme for the triangular factors is a desirable aspect of this even if it results in the storage of some of the zero entries.

Much of the following discussion of the storage scheme is applicable to any computer architecture, including serial machines where economization of storage for large sparse systems is similarly important. The discussion of parallel implementation of the algorithms follows later. We begin with the *symbolic factorization* and storage of the lower triangular factor L.

The idea of the symbolic factorization stage is to obtain the sparsity structure of L to organize its storage and numerical factorization efficiently. Much of the work on the symbolic factorization and the subsequent algorithms is due to George and Liu, although the fundamental result below is credited to Parter; again see George (1989). The motivation for these results is to obtain conditions under which elements of L are nonzero so that we can compute the structure of this factor in advance of the numerical factorization.

Proposition 13.5.1 (Parter). The element $l_{ij} \neq 0$ $(i > j)$ if and only if either (a) $a_{ij} \neq 0$ or (b) there exists $k < j$ for which $l_{ik}, l_{jk} \neq 0$.

Proof. Recall that we assume no cancellation takes place, so if $a_{ij} \neq 0$, it is necessarily the case that $l_{ij} \neq 0$. We do not give a full proof of this result but merely indicate it by consideration of an example.

Suppose that $a_{75} = 0$ but that $l_{75} \neq 0$. How has this fill-in occurred? The simplest case is that in which for some $k < 5$, both a_{5k} and a_{7k} are nonzero. By the symmetry of A it follows that $a_{k5} \neq 0$ from which it follows that the elimination in column k assigns

$$l_{75} = a_{75} - \frac{a_{k5} a_{7k}}{a_{kk}}$$

where all these elements are interpreted as taking their values at the appropriate stage of the elimination. This entry will be further scaled at a later stage, but we are only interested here in the fact that it becomes nonzero. The mere fact that we use the values appropriate to this stage of the implementation demonstrates that nonzero entries l_{5k} and l_{7k} would have the same effect.

This basic proposition allows us to obtain the structure of L. First, we introduce some notation in which we follow George (1989). The symbol Ω is used to denote the set of subscripts of nonzero entries in a matrix or vector:

$$\Omega(\mathbf{v}) = \{i: v_i \neq 0\}$$
$$\Omega(M) = \{(i, j): m_{ij} \neq 0\} \tag{13.5.3}$$

As before, we denote by M_{i*}, M_{*j}, respectively, the ith row and jth column of a matrix M. The lower triangular part of a matrix M is denoted $M^\triangleright$ and its upper triangular part by $M^\triangleleft$, so

$$m_{ij}^\triangleright = \begin{cases} m_{ij} & i \geq j \\ 0 & i < j \end{cases} \qquad m_{ij}^\triangleleft = \begin{cases} 0 & i > j \\ m_{ij} & i \leq j \end{cases} \tag{13.5.4}$$

For example, $\Omega(A_{*j}^\triangleright)$ represents the rows in which there are nonzero entries on or below the diagonal in column j of A. Finally, we set

$$r_j = \min\{i > j: l_{ij} \neq 0\} = \min(\Omega(L_{*j}) - \{j\}) \tag{13.5.5}$$

provided this set is nonempty. In the event that there are no nonzero entries below the diagonal in column j of L, we set $r_j = j$.

Using Proposition 13.5.1, it is now fairly easy to establish that

$$\Omega(L_{*j}) - \{j\} \subseteq \Omega(L_{*r_j}) \tag{13.5.6}$$

The significance of this result is that if $i > r_j$ and $i \in \Omega(L_{*j})$, then, when considering column i, we need only look back to column r_j and not right back to column j since all the relevant information will be carried there. Repeated application of this result yields the conclusion

$$\Omega(L_{*j}) = \bigcup_{r_k=j} \Omega(L_{*k}) \cup \Omega(A_{*j}^\triangleright) - \{1, 2, \ldots, j-1\} \tag{13.5.7}$$

Example 13.5.2. We consider the Cholesky factorization of the matrix resulting from the finite-element triangulation illustrated in Fig. 13.4.6. This matrix has the sparsity structure shown in (13.4.21). Applying the above results, we see that the Cholesky factors have the structure shown below. The structure of column 5 of L is seen to be the union of column 5 of A and column 4 of L.

$$A = \begin{bmatrix} x & x & & & x & x & & & & & & \\ x & x & x & x & x & & & & & & & \\ & & x & x & x & & & & & & x & \\ & & x & x & x & x & & & & & x & x \\ x & x & & & x & x & x & x & x & x & & \\ x & & & & x & x & x & & & & & \\ & & & & & x & x & x & x & & & \\ & & & & x & & x & x & x & x & & \\ & & & x & x & & & x & x & x & x & \\ & & & & & & & & x & x & x & x \\ & & x & x & & & & & & x & x & x \end{bmatrix} \tag{13.5.8}$$

$$
L = \begin{bmatrix}
x & & & & & & & & & & & \\
x & x & & & & & & & & & & \\
& x & x & & & & & & & & & \\
& & x & x & x & & & & & & & \\
x & x & 2 & x & x & & & & & & & \\
x & 1 & 2 & 2 & x & x & & & & & & \\
& & & & x & x & x & & & & & \\
& & & & x & 5 & x & x & & & & \\
& & & x & x & 4 & 5 & x & x & & & \\
& & & & & & & x & x & x & & \\
& x & x & 3 & 3 & 5 & 5 & x & x & x &
\end{bmatrix}
\tag{13.5.9}
$$

The numerical entries in the matrix L represent the *fill-in* that occurs in the factorization. Thus, for example, a 3 indicates that fill occurs in that position during elimination in column 3. We should point out that this is different from the order in which we would discover the fill-in by applying (13.5.7). (See Exercise 2.)

For this particular matrix, we see that $r_j = j + 1$ in all cases except for $j = 11$. This will turn out to have severe disadvantages for the parallel implementation of the factorization and subsequent solution of the system.

The application of Eq. (13.5.7) is summarized (again following George) in Algorithm 13.5.3 below. The sets S_j consist of the column numbers of those columns of L whose structure will be used in computing $\Omega(L_{*j})$.

Algorithm 13.5.3

Input The sparsity structure $\Omega(A)$ of the $n \times n$ matrix A

Initialize for $j = 1$ to n $S_j := \emptyset$

Loop for $j = 1$ to n
$$\Omega(L_{*j}) := \Omega(A_{*j}^{\triangleright})$$
for $i \in S_j$ $\Omega(L_{*j}) := \Omega(L_{*j}) \cup \Omega(L_{*i}) - \{i\}$
if $\Omega(L_{*j}) = \{j\}$ then $r_j := j$
else $r_j := \min(\Omega(L_{*j}) - \{j\})$
$S_{r_j} := S_{r_j} \cup \{j\}$

Output The structure $\Omega(L)$ of the Cholesky factor L of A.

Before proceeding to the subsequent stages of the solution of our linear system, we discuss briefly a possible (static) storage scheme for the Cholesky factor. Four simple arrays, two for the numerical entries and two to indicate where these occur, suffice.

DIAG used to store the diagonal entries, l_{ii}
NZ stores the *off-diagonal nonzero* elements of L listed column by column
NZSUB stores the row subscripts of the elements of NZ in the corresponding order

PCOL   points to position in NZ of first member of each column

For the example used above, the contents of these arrays would be:

DIAG $l_{11}, l_{22}, l_{33}, l_{44}, l_{55}, l_{66}, l_{77}, l_{88}, l_{99}, l_{10,10}, l_{11,11}$

NZ $l_{21}, l_{51}, l_{61}, l_{32}, l_{42}, l_{52}, l_{62}, l_{43}, l_{53}, l_{63}, l_{11,3}, l_{54}, l_{64}, l_{94}, l_{11,4}, l_{65},$
$l_{75}, l_{85}, l_{95}, l_{11,5}, l_{76}, l_{86}, l_{96}, l_{11,6}, l_{87}, l_{97}, l_{11,7}, l_{98}, l_{10,8}, l_{11,8}, l_{10,9},$
$l_{11,9}, l_{11,10}$

NZSUB   2, 5, 6, 3, 4, 5, 6, 4, 5, 6, 11, 5, 6, 9, 11, 6, 7, 8, 9, 11, 7, 8, 9, 11,
8, 9, 11, 9, 10, 11, 10, 11, 11

PCOL   1, 4, 8, 12, 16, 21, 25, 28, 31, 33

It is possible to compress this storage scheme further by eliminating any repeated sequences of subscripts in NZSUB, but we do not concern ourselves with the details here. [Again, see George (1989).]

We turn now to the numerical factorization of A. Algorithm 13.3.3 shows one possible parallel implementation of the LU factorization in which the bottom-right $(n - j) \times (n - j)$ submatrix is modified as a result of the elimination in column j. The corresponding Cholesky algorithm is just one approach to its implementation. An alternative which often leads to a better balanced parallel implementation is to use the approach of the column sweep algorithm for solving triangular systems. (See Algorithm 13.3.7).

The *column Cholesky* algorithm for a full matrix can be written in the following form. This will, of course, require modification for efficient parallel application to sparse matrix problems.

Algorithm 13.5.4

> *Input* N, an $N \times N$ symmetric positive definite matrix A

> *Cholesky factorization*   for $j = 1$ to N
>   for $k = 1$ to $j - 1$
> $A[i \geq j, j] := A_{*j} - l_{jk} L_{*k}$
> $l_{jj} := \sqrt{a_{jj}}$
>   for $k = j + 1$ to N $l_{kj} := a_{kj}/l_{jj}$

> *Output*   Cholesky factor L

Recall that using the mask results in only the elements in rows j through N being modified by the first k loop in which multiples of the columns of L that are already known are subtracted from the current elimination column. The second k loop modifies the current column to take account of the scaling of the diagonal element computed using the square root.

For the sparse matrix case under current consideration, the first k loop is only needed for those k for which $l_{jk} \neq 0$, and similarly the second loop is only executed for nonzero positions in L. This results in the following amendment of the above algorithm for the numerical factorization phase for a sparse matrix.

Algorithm 13.5.5   Numerical phase of sparse Cholesky factorization

Input N, an $N \times N$ symmetric positive definite matrix A
Cholesky factorization   for $j = 1$ to N
$\qquad$ for $k \in \Omega(L_{j*}) - \{j\}$
$\qquad\qquad$ $A[i \in \Omega(L_{*j}), j] := A_{*j} - l_{jk}L_{*k}$
$\qquad$ $l_{jj} := \sqrt{a_{jj}}$
$\qquad$ for $k \in \Omega(L_{*j}) - \{j\}$ $\qquad$ $l_{kj} := a_{kj}/l_{jj}$

Output   Cholesky factor L

An efficient implementation of this algorithm requires the storage of a list of columns which affect column j.

In considering the parallel implementation of sparse Cholesky factorization algorithms, it is often useful to obtain the *elimination tree* for the matrix. The elimination tree has been found especially helpful in studying the effect of reordering the system. This has been discussed extensively by, among others, Duff and Liu; references to these works are included in George (1989). To obtain the elimination tree in a particular case, we use the *parent* vector which consists of the quantities r_j defined by (13.5.5). This vector contains the row index of the first nonzero off-diagonal entry in column j of L provided such an

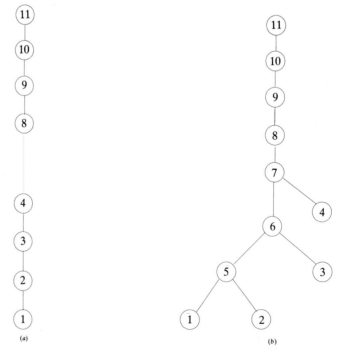

(a) $\qquad\qquad\qquad\qquad\qquad\qquad\qquad$ *(b)*

FIGURE 13.5.6
Elimination trees for two vertex orderings.

entry exists. It follows that the elimination in column j cannot be performed until that in column r_j has been completed.

For the example considered earlier, we saw that $r_j = j + 1$ for $j \le 10$. This means that elimination in column 10 is dependent on that in column 9, which in turn depends on that in column 8, and so on. The elimination tree for this case consists, therefore, of a single stem; it is illustrated in Fig. 13.5.6a. Such linear growth in the elimination tree is usually associated with a low level of available parallelism.

We stated above that the ordering of the system can have a significant effect on the parallel solution of a system of equations. The matrix in our example was derived from a finite-element triangulation of the domain of a differential equation. Reordering the labeling of the vertices of this triangulation will not affect the *number* of nonzero entries in the matrix, but it will affect their positions and therefore the amount of fill-in and the elimination tree. In Fig. 13.5.7 we reproduce the original triangulation of Fig. 13.4.6 together with a relabeled version of the same triangulation.

For the second of these orderings we see that the matrix A has the structure

$$
A = \begin{bmatrix}
x & & & & & x & & x & & & x \\
& x & & & & x & x & & & x & \\
& & x & & & x & x & & & x & \\
& & & x & & & x & x & & & x \\
x & x & & & x & & & & x & & x \\
& x & x & & & x & & & x & x & \\
& & x & x & & & x & & & x & x \\
x & & & & x & & & x & & & x \\
& x & & & x & x & & & x & x & x \\
& & x & & & x & x & & x & x & x \\
x & & & & x & x & & x & x & x & x
\end{bmatrix}
\tag{13.5.10}
$$

Applying Algorithm 13.5.3 to this matrix structure yields the following structure for the Cholesky factor:

$$
L = \begin{bmatrix}
x & & & & & & & & & & \\
& x & & & & & & & & & \\
& & x & & & & & & & & \\
& & & x & & & & & & & \\
x & x & & & x & & & & & & \\
& x & x & & & x & + & x & & & \\
& & x & x & & & x & + & x & & \\
x & & & & x & + & & + & x & & \\
& x & & & & x & x & & + & x & \\
& & x & & & & x & x & + & x & x \\
x & & & & x & x & & x & x & x & x & x
\end{bmatrix}
\tag{13.5.11}
$$

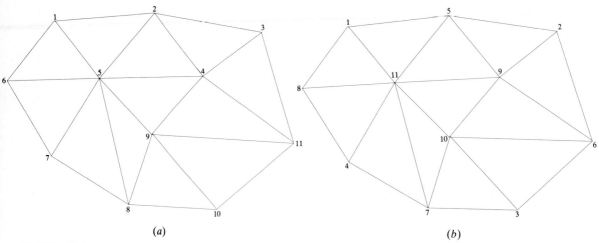

FIGURE 13.5.7
Two vertex orderings for the triangulation of Fig. 13.4.6.

where a + indicates a fill-in location. The components r_j of the parent vector can now be used to obtain the elimination tree for this ordering. This elimination tree is shown in Fig. 13.5.6b. We will see that this ordering has considerable parallelism. We also observe that there is significantly less fill-in created using this second ordering; in (13.5.9) 11 additional elements are required, whereas in (13.5.11) there are only 6 fill-in locations.

It is to be noted that the ordering used to get (13.5.10) results in considerably greater sparsity in the top-left corner of the matrix, while the densest row is pushed further down. It is only a slight oversimplification to say that minimal fill-in and maximal parallelism result from orderings with these properties. However, the above ordering is *not* claimed to be optimal for this particular problem. It is simply an improvement over the original.

The development of algorithms to achieve optimal ordering of the matrix is beyond the scope of this book. We turn our attention to the parallel implementation of the factorization algorithm. The solution of the triangular systems can be achieved using the column sweep algorithm, Algorithm 13.3.7, modified to take account of the sparsity in much the same way as the factorization stage.

For simplicity in the implementation, we will suppose that we are implementing the factorization algorithm on a shared-memory architecture so that all processors have access to all global data. The essential structure of the algorithm can be modified for shared-memory MIMD machines by the addition of a *map* that carries the information on which processor has which columns and then the various interprocessor communication instructions.

The procedure described below will be illustrated by following through the Cholesky factorization of the matrix (13.5.10) on four processors in some detail.

A typical simple model for a shared-memory parallel program consists of a *master* segment which consists of the initialization and scheduling of the various subtasks which are to be performed, the so-called slave processes. For the particular task of Cholesky factorization, the basic subtasks are the formation of the columns of L. We will denote by Col(j) the task of generating column j of L. The master segment for the Cholesky factorization would thus look like the following.

Master:
    Input   An $n \times n$ matrix A
                  The sparsity structure $\Omega(L)$ of the Cholesky factor L

    Initialize   (We discuss this initialization shortly.)

    for $j = 1$ to n schedule task Col(j)

To see what is needed in the initialization step, it is necessary to consider some of the details of the tasks Col(j) and their interrelations.

In Algorithm 13.5.5, we see that this subtask separates naturally into two pieces: the modification of column j by column k for all relevant columns to the left and then the scaling of column j by the square root of its diagonal element. Following George, we denote these subtasks by cmod(j, k) and cdiv(j), respectively. As to the interrelation of the tasks, we see that as soon as cdiv(k) has been completed, column k can be used to modify any (appropriate) columns to its right. Thus, for example, we may be able to begin Col(5) before Col(3) has been completed. We will see in the example that Col(11) is begun before Col(8) has finished and even before Col(9) or Col(10) has started.

It is clearly necessary to keep a record of which columns are available and for which ones they are required. Much of this information is carried in $\Omega(L)$, from which we can obtain the row subscript of the next (nonzero) entry in a column of L below a particular entry. That is, we will set

$$\text{next}[j, k] := \min\{i > j: l_{ik} \neq 0\}$$

provided such an entry exists. (If not, we set next $[j, k] := n + 1$.) For convenience in our algorithm, we will suppose that the array next$[j, k]$ is available, although the information would actually be obtained from the sparsity structure.

Once cmod(j, k), that is, the modification of column j by column k, is completed, next$[j, k]$ is the next column which will be modified by column k. Therefore, k is added to the (beginning of the) list of columns ready to modify column next$[j, k]$. Of course, k must also be removed from the list for column j. In this way the lists are necessarily disjoint and so can be stored in a single integer vector, the *link* vector whose entries we denote by l_j. Thus, for example, setting $l_{10} := 3$ implies that the next use of column 3 will be in cmod(10, 3). A similar updating of the link vector takes place after each cdiv(j), which is when column j becomes available to modify other columns.

To avoid conflicts, it is necessary to ensure that two processors do not try to update this link vector simultaneously. (In Example 13.5.10 below, we do allow apparently simultaneous adjustments to this vector where they do not conflict.) Setting an element $l_j := 0$ implies an inactive state for column j—either it is completely finished or it is waiting because no other column is yet available to modify column j. This is the initial status of all columns.

The other important information which must be stored is used to determine when a particular column has been modified by all columns to its left and so is ready to be scaled and then used to modify other columns to its right. This is achieved by simply storing an array nmod[j] that is initialized to the number of columns which must modify column j and is reduced by 1 after each cmod(j, k). Once cmod[j] = 0, cdiv(j) can be performed. It follows that the initial values for this array given by

$$\text{nmod}[j] := |\Omega(L_{j*})| - 1$$

which is the number of off-diagonal (nonzero) entries in *row* j of L.

The remaining aspect of the master segment is the scheduling of the tasks Col(j). The scheduling algorithm must assign to an available processor one of the Col(j) which has not yet been started—but which one? In Example 13.5.10 below we use the following decision process. We look first at those columns for which l_k is nonzero since these are precisely the ones for which Col(k) could be started immediately. From among these we choose the (smallest) one for which nmod[k] is maximized since this gives a good chance of balancing the load. If $l_k = 0$ for all remaining columns, then we simply choose the next one in numerical order.

We will denote by S the set of columns for which Col(j) is still to be started and $\Omega S := \{k \in S : l_k \neq 0\}$. The next column chosen is then Col(j) given by

$$\text{if } \Omega S \neq \emptyset \text{ then } j := \text{PMAX}(\text{nmod}[\Omega S]) \text{ else } j := \min S$$

where, we recall from (13.3.1), that PMAX yields the position of (the first occurrence of) the maximum.

The master segment of our implementation can now be summarized as follows.

Algorithm 13.5.8   Master segment of parallel Cholesky factorization for a large sparse matrix

$\quad\quad$ *Input* $\quad\quad$ An $n \times n$ matrix A

$\quad\quad\quad\quad\quad\quad\quad$ The sparsity structure $\Omega(L)$ of the Cholesky factor L

$\quad\quad$ *Initialize* $\quad$ $S := \{1, 2, \ldots, n\}$

$\quad\quad\quad\quad\quad\quad\quad$ $\Omega S := \emptyset$

$\quad\quad\quad\quad\quad\quad\quad$ for $j = 1$ to n

$\quad\quad\quad\quad\quad\quad\quad\quad$ $l_j := 0$

$\quad\quad\quad\quad\quad\quad\quad\quad$ $\text{nmod}[j] := |\Omega(L_{j*})| - 1$

Scheduling   for $i = 1$ to n
  wait until a processor is available
  if $\Omega S \neq \emptyset$ then $j:=\text{PMAX}(\text{nmod}[\Omega S])$ else $j:=\min S$
  assign Col(j) to available processor
  $S:=S-\{j\}$

It remains to consider the details of the slave process Col(j). This consists of Algorithm 13.5.5 augmented with the updating of the link vector and nmod.

Algorithm 13.5.9   Slave process Col(j)

  while nmod[j] > 0
    wait until $l_j > 0$
    $k:=l_j, l_j:=l_k$
    cmod(j, k)
    next:=next[j, k]
    if next $\leq n$ then $l_k:=$next, $l_{\text{next}}:=k$

  cdiv(j)
  next:=next[j, j]
  if next $\leq n$ then $l_j:=$next, $l_{\text{next}}:=j$

Example 13.5.10. Apply the Cholesky factorization of Algorithms 13.5.8 and 13.5.9 to the matrix given by (13.5.10) using four processors.

$\mathbf{P_1}$	$\mathbf{P_2}$	$\mathbf{P_3}$	$\mathbf{P_4}$
Col(1)	**Col(2)**	**Col(3)**	**Col(4)**
nmod[1] = 0	nmod[2] = 0	nmod[3] = 0	nmod[4] = 0
cdiv(1)	cdiv(2)	cdiv(3)	cdiv(4)
next:=5	next:=5	next:=6	next:=7
$l_1:=0, l_5:=1$			
Col(5)	$l_2:=1, l_5:=2$	$l_3:=0, l_6:=3$	$l_4:=0, l_7:=4$
nmod[5] = 2	**Col(6)**	**Col(7)**	**Col(8)**
$l_5 = 2$	nmod[6] = 3	nmod[7] = 3	nmod[8] = 4
$k:=2, l_5:=1$	$l_6 = 3$	$l_7 = 4$	$l_8 = 0$
cmod(5, 2)	$k:=3, l_6:=0$	$k:=4, l_7:=0$	WAIT
nmod[5]:=1	cmod(6, 3)	cmod(7, 4)	
next:=6	nmod[6]:=2	nmod[7]:=2	
$l_2:=0, l_6:=2$	next:=7	next:=8	
$l_5 = 1$	$l_3:=0, l_7:=3$	$l_4:=0, l_8:=4$	
$k:=1, l_5:=0$	$l_6 = 2$	$l_7 = 3$	$l_8 = 4$
cmod(5, 1)	$k:=2, l_6:=0$	$k:=3, l_7:=0$	$k:=4, l_8:=0$
nmod[5]:=0	cmod(6, 2)	cmod(7, 3)	cmod(8, 4)
next:=8	nmod[6]:=1	nmod[7]:=1	nmod[8]:=3
$l_1:=0, l_8:=1$	next:=9	next:=10	next:=11
cdiv(5)	$l_2:=0, l_9:=2$	$l_3:=0, l_{10}:=3$	$l_4:=0, l_{11}:=4$
next:=6	WAIT	WAIT	$l_8 = 1$
$l_5:=0, l_6:=5$			$k:=1, l_8:=0$

Col(11)
nmod[11] = 7
$l_{11} = 4$
$k:=4, l_{11}:=0$
cmod(11, 4)
nmod[11]:=6
next:=12
$l_{11} = 1$
$k:=1, l_{11}:=0$
cmod(11, 1)
nmod[11]:=5
next:=12
WAIT

$l_{11} = 7$
$k:=7, l_{11}:=0$
cmod(11, 7)
nmod[11]:=4
next:=12
WAIT

$l_{11} = 5$
$k:=5, l_{11}:=0$
cmod(11, 5)
nmod[11]:=3
next:=12
$l_{11} = 8$
$k:=8, l_{11}:=0$
cmod(11, 8)
nmod[11]:=2
next:=12

$l_6 = 5$
$k:=5, l_6:=0$
cmod(6, 5)
nmod[6]:=0
next:=8
$l_5:=0, l_8:=5$
cdiv(6)
next:=7
$l_6:=0, l_7:=6$
Col(10)
nmod[10] = 5
$l_{10} = 3$
$k:=3, l_{10}:=0$
cmod(10, 3)
nmod[10]:=4
next:=12
WAIT

$l_{10} = 7$
$k:=7, l_{10} = 0$
cmod(10, 7)
nmod[10]:=3
next:=11
$l_7:=0, l_{11}:=7$
$l_{10} = 6$
$k:=6, l_{10}:=0$
cmod(10, 6)
nmod[10]:=2
next:=12
$l_{10} = 8$
$k:=8, l_{10}:=0$
cmod(10, 8)
nmod[10]:=1
next:=11

$l_8:=0, l_{11}:=8$
WAIT

$l_{10} = 9$
$k:=9, l_{10}:=0$

$l_7 = 6$
$k:=6, l_7:=0$
cmod(7, 6)
nmod[7]:=0
next:=9
$l_6:=5, l_9:=6$
cdiv(7)
next:=8
$l_7:=0, l_8:=7$
Col(9)
nmod[9] = 4
$l_9 = 6$
$k:=6, l_9:=5$
cmod(9, 6)
nmod[9]:=3
next:=10

$l_6:=0, l_{10}:=6$
$l_9 = 8$
$k:=8, l_9:=5$
cmod(9, 8)
nmod[9]:=2
next:=10
$l_8:=0, l_{10}:=8$
$l_9 = 5$
$k:=5, l_9:=2$
cmod(9, 5)
nmod[9]:=1
next:=11
$l_5:=0, l_{11}:=5$
$l_9 = 2$
$k:=2, l_9:=0$
cmod(9, 2)
nmod[9]:=0
next:=12
cdiv(9)
next:=10
$l_9:=0, l_{10}:=9$
FINISHED

cmod(8, 1)
nmod[8]:=2
next:=11
$l_1:=0, l_{11}:=1$
WAIT

$l_8 = 5$
$k:=5, l_8:=0$
cmod(8, 5)
nmod[8]:=1
next:=9
$l_5:=2, l_9:=5$
WAIT

$l_8 = 7$
$k:=7, l_8:=0$
cmod(8, 7)
nmod[8]:=0
next:=10
$l_7:=0, l_{10}:=7$
cdiv(8)
next:=9
$l_8:=5, l_9:=8$
FINISHED

WAIT

$l_{11}:=9$
$k:=9,\ l_{11}:=0$
cmod$(11,9)$
nmod$[11]:=1$
next:$=12$
$l_{11}=10$
$k:=10,\ l_{11}:=0$
cmod$(11,10)$
nmod$[11]:=0$
next:$=12$
cdiv(11)
next:$=12$
FINISHED

cmod$(10,9)$
nmod$[10]:=0$
next:$=11$
$l_9:=0,\ l_{11}:=9$
cdiv(10)
next:$=11$
$l_{10}:=0,\ l_{11}:=10$
FINISHED

It is apparent that for much of the time all four processors are busy, and so the parallelism has been exploited quite efficiently. The scheduling procedure outlined in Algorithm 13.5.8 has the effect of initiating Col(11) before either Col(9) or Col(10), and this has the effect of balancing the computation so that processor 1 finishes Col(11) very quickly after processor 2 has completed Col(10).

This chapter has provided only a very cursory introduction to the ideas of parallel algorithms. This is an increasingly important aspect of scientific computing, and the interested reader is urged to look into it in greater detail. The books of Duff, Erisman, and Reid (1986), Modi (1988), and Ortega (1988) and the lecture notes of George (1989) are recommended.

EXERCISES 13.5

1. Find sets $\Omega(A)$, $\Omega(A_{*j}^{\triangleright})$ for each j for the matrix of (13.5.8).
2. Use the information from Exercises 1 and Eq. (13.5.7) to obtain the structure of the Cholesky factor L of A. Check that your answer agrees with (13.5.9). Work Algorithm 13.5.3 for this example, and confirm that it yields the same structure.
3. Apply Algorithm 13.5.3 to the second vertex ordering given in Fig. 13.5.7 to show that the Cholesky factor L has the structure given in (13.5.11). Verify that this yields the elimination tree given in Fig. 13.5.6b.
4. Create a triangulation of a region in the plane. Try to label the vertices in such a way as to maximize the sparsity of the Cholesky factor of the resulting matrix. Apply Algorithms 13.5.8 and 13.5.9 to the factorization of the matrix.

APPENDIX

A

BACKGROUND THEOREMS IN REAL ANALYSIS

A.1 SEQUENCES

Basic Definitions

1. A *sequence* (x_n) is said to *converge* or *tend* to the *limit* L as $n \to \infty$ if for every $\epsilon > 0$ there exists an integer N such that

$$|x_n - L| < \epsilon \qquad \text{whenever } n > N$$

We write $\lim_{n \to \infty} x_n = L$ or $x_n \to L$ as $n \to \infty$.

2. A sequence (x_n) is *bounded* if there exist constants m and M such that

$$m \leq x_n \leq M \qquad \text{for all } n$$

or, equivalently, there exists a number K such that

$$|x_n| \leq K \qquad \text{for all } n$$

3. A sequence (x_n) is *increasing* if

$$x_n \leq x_{n+1} \qquad \text{for all } n$$

It is *strictly increasing* if $x_n < x_{n+1}$ for all n. Similar definitions apply to the terms *decreasing* and *strictly decreasing*. (x_n) is called *monotone* if it has any of these properties.

Important examples of Convergent Sequences

$$x^n \to 0 \text{ as } n \to \infty \qquad \text{for every } |x| < 1$$

$$\frac{1}{n^p} \to 0 \text{ as } n \to \infty \qquad \text{for any positive } p$$

$$n^p x^n \to 0 \text{ as } n \to \infty \qquad \text{for every } |x| < 1 \text{ and every } p$$

Elementary Properties of Limits

Let $(a_n), (b_n)$, and (c_n) be convergent sequences with limits a, b, and c. Then

$$a_n \pm b_n \to a \pm b \qquad \text{as } n \to \infty$$

$$a_n b_n \to ab \qquad \text{as } n \to \infty$$

$$ca_n \to ca \qquad \text{as } n \to \infty \text{ for any constant } c$$

$$\frac{1}{a_n} \to \frac{1}{a} \qquad \text{as } n \to \infty \text{ provided } a_n, a \neq 0$$

$$a_n < b_n \text{ for every } n \text{ implies } a \leq b$$

The sandwich rule. If $a_n \leq c_n \leq b_n$ for every n and $a = b$, then $c = a$.

Theorem 1. If (a_n) is bounded and monotone, then it is convergent.

Specifically, if (a_n) is increasing and bounded, then

$$\lim_{n \to \infty} a_n = \sup_n a_n$$

while for (a_n) decreasing and bounded,

$$\lim_{n \to \infty} a_n = \inf_n a_n$$

Recall that the *supremum* of a bounded set S of real numbers is the least upper bound for S; that is,

$$\sup(s \in S) = \min\{x \in \mathbb{R}: x \geq s, \forall s \in S\}$$

and the *infimum* is the greatest lower bound

$$\inf(s \in S) = \max\{x \in \mathbb{R}: x \leq s, \forall s \in S\}$$

It follows, for example, that

$$b = \sup[a, b] = \sup[a, b) = \sup(a, b) = \sup(-\infty, b)$$

The supremum of S satisfies:

1. $\sup S \geq s$ for every $s \in S$.
2. For every $\epsilon > 0$ there exists $s \in S$ such that

$$s > \sup S - \epsilon$$

There is a corresponding characterization of inf S.

Theorem 2. Any bounded sequence has a convergent subsequence.

A.2   SERIES

The *series* $\Sigma\, a_n$ which has the *sequence of terms* (a_n) is defined by

$$\sum_{n=1}^{\infty} a_n = a_1 + a_2 + \cdots$$

This series is said to converge if the *sequence of partial sums* (s_n) where

$$s_n = \sum_{k=1}^{n} a_k = a_1 + a_2 + \cdots + a_n$$

converges. If $s_n \to s$, then we say the series has the *sum s* and write

$$\sum a_n = s$$

If $\Sigma\, |a_n|$ converges, then $\Sigma\, a_n$ is called *absolutely convergent*.

Important Examples

Geometric series $\Sigma\, x^n = \dfrac{1}{1-x}$ if $|x| < 1$

$\Sigma\, x^n$ diverges if $|x| \geq 1$

Harmonic series $\Sigma\, \dfrac{1}{n}$ diverges

p series $\Sigma\, \dfrac{1}{n^p}$ converges for $p > 1$, diverges for $p \leq 1$

Tests for Convergence

1. *Vanishing test*: If $\Sigma\, a_n$ converges, then $a_n \to 0$ as $n \to \infty$.
2. *Comparison test*: If $0 \leq a_n \leq b_n$ and $\Sigma\, b_n$ converges, then $\Sigma\, a_n$ converges.
3. *Ratio test*: If

$$\left| \frac{a_{n+1}}{a_n} \right| \to L < 1 \text{ as } n \to \infty$$

then $\Sigma\, a_n$ converges absolutely. If

$$\left| \frac{a_{n+1}}{a_n} \right| \to L > 1 \text{ as } n \to \infty$$

then $\Sigma\, a_n$ diverges.

4. *Alternating sign test*: If (a_n) is a decreasing sequence with limit 0, then $\Sigma (-1)^n a_n$ converges.

Power Series

The power series $\Sigma a_n x^n$ converges for all $|x| < R$, the radius of convergence, and diverges if $|x| > R$. The radius of convergence is given by

$$R = \frac{1}{L} \qquad \text{where } L = \lim \left| \frac{a_{n+1}}{a_n} \right|$$

If $f(x) = \Sigma a_n x^n$ for $|x| < R$, then $\Sigma na_n x^{n-1}$ has radius of convergence equal to R and $\Sigma na_n x^{n-1} = f'(x)$ for $|x| < R$. Similarly, we can integrate the power series term by term to get

$$\int_0^x f(t)\,dt = \Sigma \frac{a_n x^{n+1}}{n+1} \qquad |x| < R$$

A.3 REAL FUNCTIONS

Basic Definitions

1. A real-valued function f of a real variable is *continuous at a point a* if

$$\lim_{x \to a} f(x) = f(a)$$

Equivalently, f is continuous at a if $f(x_n) \to f(a)$ for every sequence (x_n) in the domain of f such that $x_n \to a$ as $n \to \infty$.

2. The function f is continuous on the interval $[a, b]$ if it is continuous at every point of (a, b) and

$$\lim_{x \to a+} f(x) = f(a) \qquad \lim_{x \to b-} f(x) = f(b)$$

3. The function f is *Lipschitz continuous*, or satisfies a *Lipschitz condition*, on $[a, b]$ if there exists a constant Λ, a *Lipschitz constant*, such that

$$|f(x_1) - f(x_2)| \le \Lambda |x_1 - x_2| \qquad \text{for all } x_1, x_2 \in [a, b]$$

4. The function f is *differentiable* at a point x_0 and has derivative $f'(x_0)$ there if

$$\lim_{h \to 0} \frac{f(x_0 + h) - f(x_0)}{h} = f'(x_0)$$

5. f is differentiable on $[a, b]$ if it is differentiable at every point of (a, b) and the appropriate one-sided limits exist at a and b.

Function Space Notation

Denote by $\mathbf{C}[a, b]$ the space of real-valued functions that are continuous on the interval $[a, b]$. There are similar definitions for open intervals, half-lines, etc.

Denote by $f^{(k)}$ the kth derivative of the function f and by $\mathbf{C}^n[a, b]$ the space of all real-valued functions for which:

1. f is n times differentiable on (a, b); that is, $f^{(n)}$ is defined on (a, b).
2. $f^{(n-1)}$ is continuous on the interval $[a, b]$.

Denote by $\mathbf{C}^\infty[a, b]$ the space of infinitely differentiable functions on $[a, b]$.

These spaces are examples of (infinite-dimensional) linear, or vector, spaces.

We denote by Π_n the (finite-dimensional linear) space of all real polynomials of degree at most n.

Important Theorems

(Corollary to) Unique Factorization Theorem. Let $p \in \Pi_n$. If there exist $n + 1$ distinct points $x_0, x_1, \ldots, x_n$ such that $p(x_i) = 0$ for each i, then $p(x) \equiv 0$.

Intermediate Value Theorem (IVT). Let $f \in \mathbf{C}[a, b]$. If $f(a) < \eta < f(b)$, then there exists $c \in (a, b)$ such that $f(c) = \eta$.

It follows that if $f \in \mathbf{C}[a, b]$ satisfies $f(a)f(b) < 0$, then the equation $f(x) = 0$ has a solution in (a, b).

Maximum Modulus Theorem. Let $f \in \mathbf{C}[a, b]$. Then f is bounded on $[a, b]$ and attains its bounds. That is, there exist real numbers m and M and corresponding $x_m, x_M \in [a, b]$ such that

$$m = f(x_m) = \min_{[a,b]} f(x) \qquad M = f(x_M) = \max_{[a,b]} f(x)$$

Proof. To see that f is bounded, suppose not and obtain a contradiction. If f is not bounded above, then there is a sequence of points (x_n) in $[a, b]$ such that $f(x_n) > n$ for each n. Since (x_n) is bounded sequence, it has a convergent subsequence. Denote the limit of this subsequence by c; it follows that f is undefined at $c \in [a, b]$, which is the desired contradiction.

It follows that there exists a number $M = \sup\{f(x): x \in [a, b]\}$. Again, there is a sequence (x_n) in $[a, b]$ such that $f(x_n) > M - 1/n$ for each n. As above there is a convergent subsequence with limit x_M, say. It follows from the continuity of f that $f(x_M) = M$.

Similar proofs hold for the minimum.

It now follows that if f' is continuous and so bounded by Λ, say, on $[a, b]$, then, by the mean value theorem below, f satisfies a Lipschitz condition with Lipschitz constant Λ.

Rolle's Theorem. Let $f \in \mathbf{C}^1[a, b]$. If $f(a) = f(b) = 0$, then there exists $c \in (a, b)$ such that $f'(c) = 0$.

From this it is easy to establish the following theorem.

Mean Value Theorem (MVT). Let $f \in C^1[a, b]$. There exists $c \in (a, b)$ such that

$$f'(c) = \frac{f(b) - f(a)}{b - a}$$

As a consequence of the maximum modulus and intermediate value theorems, we have the following theorem.

Integral form of MVT. Let $f, g \in C[a, b]$. If $g(x)$ does not change sign in $[a, b]$, then there exists $c \in (a, b)$ such that

$$\int_a^b f(x)g(x) \, dx = f(c) \int_a^b g(x) \, dx$$

Taylor's Theorem. Let $f \in C^{n+1}[a, b]$ and denote $b - a$ by h. There exists $c \in (a, b)$ such that

$$f(b) = f(a) + hf'(a) + \frac{h^2}{2} f''(a) + \cdots + \frac{h^n}{n!} f^{(n)}(a) + \frac{h^{n+1}}{(n+1)!} f^{(n+1)}(c)$$

Integral form of Taylor's theorem. Under the same hypotheses as above,

$$f(b) = f(a) + hf'(a) + \frac{h^2}{2} f''(a) + \cdots + \frac{h^n}{n!} f^{(n)}(a) + \int_a^b \frac{(b-t)^n}{n!} f^{(n+1)}(t) \, dt$$

In both the above, the terms

$$f(b) = f(a) + hf'(a) + \frac{h^2}{2} f''(a) + \cdots + \frac{h^n}{n!} f^{(n)}(a)$$

are called the *Taylor polynomial* of degree n for f at a; the final terms are different forms of the *remainder*.

Taylor series expansion. Let $f \in C^\infty[a, b]$. Then, with $h = b - a$,

$$f(b) = f(a) + hf'(a) + \frac{h^2}{2} f''(a) + \cdots = \sum_{k=1}^{\infty} \frac{h^k}{k!} f^{(k)}(a)$$

Theorems for Multivariate Functions

Consider now a real-valued function f of n variables,

$$f(x) = f(x_1, x_2, \ldots, x_n)$$

Denote by **D** the differential operator given by

$$\mathbf{D} = \left(\frac{\partial}{\partial x_1}, \frac{\partial}{\partial x_2}, \ldots, \frac{\partial}{\partial x_n} \right)^{\mathrm{T}}$$

Let $\mathbf{x}_0$ and $\mathbf{h}$ be given. By applying the corresponding single-variable results to the function $\phi(t) = f(\mathbf{x} + t\mathbf{h})$ on the interval $[0, 1]$, we may deduce, assuming the necessary smoothness in f and therefore ϕ, the following.

Mean Value Theorem. There exists $\tau \in (0, 1)$ such that

$$f(\mathbf{x}_0 + \mathbf{h}) - f(\mathbf{x}_0) = \phi(1) - \phi(0) = \phi'(t) = \mathbf{h}^{\mathrm{T}} Df(\mathbf{x}_0 + \tau\mathbf{h}) = \sum_{i=1}^{n} \frac{h_i \partial f(\mathbf{x}_0 + \tau\mathbf{h})}{\partial x_i}$$

For a function of two variables $f(x, y)$ with $\mathbf{h} = (h, k)^{\mathrm{T}}$, this reduces to

$$f(x + h, y + k) - f(x, y) = h \frac{\partial f(x + \tau h, y + \tau k)}{\partial x} + k \frac{\partial f(x + \tau h, y + \tau k)}{\partial y}$$

for some $0 < \tau < 1$.

Taylor's theorem. There exists $\tau \in (0, 1)$ such that

$$f(\mathbf{x}_0 + \mathbf{h}) = \sum_{k=0}^{N} (\mathbf{h}^{\mathrm{T}} D)^k \frac{f(\mathbf{x}_0)}{k!} + \frac{(\mathbf{h}^{\mathrm{T}} D)^{N+1} f(\mathbf{x}_0 + \tau\mathbf{h})}{(N + 1)!}$$

The second-degree Taylor polynomial for f at $\mathbf{x}_0$ is therefore given by

$$f(\mathbf{x}_0) + \mathbf{h}^{\mathrm{T}} Df(\mathbf{x}_0) + \frac{\mathbf{h}^{\mathrm{T}} J(\mathbf{x}_0) \mathbf{h}}{2}$$

where the Hessian matrix of f at $\mathbf{x}_0$, $J(\mathbf{x}_0)$, has i, jth element

$$J_{ij} = \frac{\partial^2 f}{\partial x_i \partial x_j}$$

evaluated at $\mathbf{x}_0$. Expanded out, this expression is

$$f(\mathbf{x}_0) + \sum_{i=1}^{n} h_i \frac{\partial f(\mathbf{x}_0)}{\partial x_i} + \frac{1}{2} \sum_{i=1}^{n} \sum_{j=1}^{n} h_i h_j \frac{\partial^2 f(\mathbf{x}_0)}{\partial x_i \partial x_j}$$

which for a function of two variables is

$$f(x, y) + h \frac{\partial f(x, y)}{\partial x} + k \frac{\partial f(x, y)}{\partial y}$$

$$+ \frac{1}{2} \left[h^2 \frac{\partial^2 f(x, y)}{\partial x^2} + 2hk \frac{\partial^2 f(x, y)}{\partial x \partial y} + k^2 \frac{\partial^2 fx, y}{\partial y^2} \right]$$

B.1 LINEAR SPACES

Throughout this appendix, we will deal with *linear*, or *vector*, *spaces* over the real numbers. In many instances similar statements and results to those quoted apply to complex linear spaces as well as to vector spaces over other fields.

Basic Definitions

By a *vector* or *linear* space over the real numbers, **V**, we mean a set of elements or vectors with the following properties:

1. For any pair of elements $\mathbf{u}, \mathbf{v} \in \mathbf{V}$, there is a vector $\mathbf{w} = \mathbf{u} + \mathbf{v}$.
2. The addition in property 1 is associative and commutative. There is a *zero element* **0** such that $\mathbf{0} + \mathbf{u} = \mathbf{u}$ for every $\mathbf{u} \in \mathbf{V}$, and each element **u** has a negative, denoted $-\mathbf{u}$, such that $\mathbf{u} + (-\mathbf{u}) = \mathbf{0}$.
3. For any element **u** and any *scalar* (or real number) α there is an element **v** such that $\mathbf{v} = \alpha \mathbf{u}$. Also this *scalar multiplication* is distributive over addition both of vectors and of scalars: thus

$$(\alpha + \beta)\mathbf{u} = \alpha\mathbf{u} + \beta\mathbf{u} \qquad \text{and} \qquad \alpha(\mathbf{u} + \mathbf{v}) = \alpha\mathbf{u} + \alpha\mathbf{v}$$

Important Examples

$\mathbb{R}^n$ Euclidean *n*-dimensional space with the usual vector addition and scalar multiplication. For example, $\mathbb{R}^2$ consists of pairs (x, y) of real numbers with addition and scalar multiplication defined by

$$(x_1, y_1) + (x_2, y_2) = (x_1 + x_2, y_1 + y_2) \qquad \alpha(x, y) = (\alpha x, \alpha y)$$

Π_N The space of all real polynomials of degree not more than N with the usual addition and multiplication by constants (scalars).

$C[a, b]$ The space of all continuous functions on the closed interval $[a, b]$ with the usual addition of functions and multiplication by constants. Similarly, the spaces $\mathbf{C}^n[a, b]$ and $\mathbf{C}^\infty[a, b]$ are vector spaces.

U is a linear *subspace* of **V** if it is a subset of **V** and is itself a linear space. Thus, for example, Π_N is a subspace of Π_{N+1} for each N; these are both subspaces of $\mathbf{C}^\infty[a, b]$, which in turn is a subspace of $\mathbf{C}^n[a, b]$ for every n. All these are subspaces of $C[a, b]$.

Further Definitions

V is called an inner product space if it is a linear space on which there is defined a *scalar* or *inner product*: that is a function often denoted by $(.\,,.)$ or $\langle .\,,.\rangle$ from $\mathbf{V} \times \mathbf{V}$ to the real numbers with the properties that it is symmetric, bilinear (that is, linear in each of the variables), and satisfies

$$(\mathbf{v}, \mathbf{v}) \geq 0 \qquad \text{for every } \mathbf{v} \in \mathbf{V}$$

with

$$(\mathbf{v}, \mathbf{v}) = 0 \qquad \text{(if and) only if } \mathbf{v} = \mathbf{0}$$

Two vectors $\mathbf{u}, \mathbf{v}$ are said to be *orthogonal*, or *normal*, if $(\mathbf{u}, \mathbf{v}) = 0$.
Any inner product space can also be made a *normed* linear space by defining the *norm function* $\|\cdot\|$ on **V** by

$$\|\mathbf{v}\| = \sqrt{(\mathbf{v}, \mathbf{v})}$$

It is easy to check that such a definition satisfies the definition of a norm, namely, that for $\mathbf{u}, \mathbf{v} \in \mathbf{V}$ and $\alpha \in \mathbb{R}$,

$$\|\mathbf{v}\| \geq 0 \qquad \text{with } \|\mathbf{v}\| = 0 \text{ if and only if } \mathbf{v} = \mathbf{0}$$
$$\|\alpha \mathbf{v}\| = |\alpha| \|\mathbf{v}\|$$

and the *triangle inequality* $\|\mathbf{u} + \mathbf{v}\| \leq \|\mathbf{u}\| + \|\mathbf{v}\|$.
All the above examples can be made normed linear spaces under several different norms. They can also have inner products defined on them in natural ways. In the case of $\mathbb{R}^2$, for example, we can define an inner product in the usual way by

$$(\mathbf{u}, \mathbf{v}) = \mathbf{u}^\mathsf{T} \mathbf{v} = \mathbf{u} \cdot \mathbf{v} = u_1 v_1 + u_2 v_2$$

This gives rise to the usual, *euclidean*, norm

$$\|\mathbf{v}\| = \sqrt{\mathbf{v}^T \mathbf{v}} = \sqrt{v_1^2 + v_2^2}$$

$\mathbb{R}^2$ is also a normed linear space under either of the norms $\|\mathbf{v}\|_1 = |v_1| + |v_2|$ or $\|\mathbf{v}\|_\infty = \max(|v_1|, |v_2|)$. The euclidean norm is often denoted by $\|\cdot\|_2$.

The norms $\|\mathbf{v}\|_1$ and $\|\mathbf{v}\|_\infty$ above are *not* derived from inner products, so a normed linear space is not necessarily an inner product space.

B.2 LINEAR INDEPENDENCE AND BASES

Basic Definitions

A set of elements $\{\mathbf{v}_1, \mathbf{v}_2, \ldots, \mathbf{v}_n\}$ of a vector space $\mathbf{V}$ is said to be *linearly independent* if there is no nontrivial linear combination of them which is $\mathbf{0}$; that is,

$$\text{If } \alpha_1 \mathbf{v}_1 + \alpha_2 \mathbf{v}_2 + \cdots + \alpha_n \mathbf{v}_n = \mathbf{0}, \text{ then } \alpha_1 = \alpha_2 = \cdots = \alpha_n = 0$$

The set of all elements of $\mathbf{V}$ which can be expressed as linear combinations of the vectors $\{\mathbf{v}_1, \mathbf{v}_2, \ldots, \mathbf{v}_n\}$ is called their *linear span*; it is a subspace of $\mathbf{V}$. If the set $\{\mathbf{v}_1, \mathbf{v}_2, \ldots, \mathbf{v}_n\}$ of linearly independent vectors spans the whole of $\mathbf{V}$, then it is called a *basis* for $\mathbf{V}$, and $\mathbf{V}$ has *dimension n*.

Important Examples

The vectors $(1, 0)$ and $(0, 1)$ form a basis for $\mathbb{R}^2$. So do the pairs of vectors $(1, 0)$ and $(1, 1)$, $(2, 1)$ and $(1, -2)$, and $(1, 1)$ and $(3, 2)$. The first of these bases, $(1, 0)$ and $(0, 1)$, is called the *standard basis* for $\mathbb{R}^2$.

The set of vectors $\mathbf{e}_i$ in $\mathbb{R}^n$, where $\mathbf{e}_1 = (1, 0, 0, \ldots, 0)$, $\mathbf{e}_2 = (0, 1, 0, \ldots, 0), \ldots, \mathbf{e}_n = (0, 0, \ldots, 0, 1)$, forms the standard basis for $\mathbb{R}^n$. This basis has the property that the vector

$$\mathbf{x} = (x_1, x_2, \ldots, x_n)$$

is represented in terms of this standard basis by

$$\mathbf{x} = x_1 \mathbf{e}_1 + x_2 \mathbf{e}_2 + \cdots + x_n \mathbf{e}_n$$

The standard basis for $\mathbb{R}^n$ has the additional property that it is an *orthonormal* basis for $\mathbb{R}^n$ with the euclidean inner product and norm; that is, each pair of basis elements is orthogonal and of norm (or magnitude) 1.

The standard basis for the polynomial space Π_N consists of the polynomials $1, x, x^2, \ldots, x^N$, so the polynomial $a_0 + a_1 x + \cdots + a_N x^N$ can be represented by the vector $(a_0, a_1, \ldots, a_N)$. This standard basis is not an orthogonal basis for any of the usual inner products on Π_N. Such orthogonal and orthonormal bases do exist for different inner products. Examples are the Legendre and Chebyshev polynomials.

The infinite-dimensional spaces $\mathbf{C}[a, b]$, $\mathbf{C}^n[a, b]$, $\mathbf{C}^\infty[a, b]$ have no finite—or even countable—bases.

Nonsingular square matrices provide a convenient way to describe a change of bases in a finite-dimensional vector space, that is, a simple mapping between representations of the same element in terms of different bases.

B.3   TRANSFORMATIONS AND MATRICES

A *linear mapping* or *transformation* between two vector spaces **U** and **V** is a linear function $f: \mathbf{U} \to \mathbf{V}$; that is, f satisfies

$$f(\alpha \mathbf{u}_1 + \beta \mathbf{u}_2) = \alpha f(\mathbf{u}_1) + \beta f(\mathbf{u}_2)$$

for every $\mathbf{u}_1, \mathbf{u}_2 \in \mathbf{U}$ and all scalars α, β.

If **U** and **V** are finite-dimensional spaces with bases $\{\mathbf{u}_1, \mathbf{u}_2, \ldots, \mathbf{u}_n\}$ and $\{\mathbf{v}_1, \mathbf{v}_2, \ldots, \mathbf{v}_m\}$, then such a transformation can be fully described by expressing $f(\mathbf{u}_j)$ as a linear combination of $\{\mathbf{v}_1, \mathbf{v}_2, \ldots, \mathbf{v}_m\}$ for each $j = 1, 2, \ldots, n$, that is, by a set of identities

$$f(\mathbf{u}_j) = a_{1j}\mathbf{v}_1 + a_{2j}\mathbf{v}_2 + \cdots + a_{mj}\mathbf{v}_m \qquad j = 1, 2, \ldots, n$$

We thus have a natural $m \times n$ matrix representation of this transformation as

$$A = \begin{bmatrix} a_{11} & a_{12} & \cdots & a_{1n} \\ a_{21} & a_{22} & \cdots & a_{2n} \\ \vdots & & & \vdots \\ a_{m1} & a_{m2} & \cdots & a_{mn} \end{bmatrix}$$

Notice here that we assume that elements of a finite-dimensional vector space are represented as a *column* vector of coefficients of the basis being used.

For example, the linear transformation between $\mathbb{R}^2$ and $\mathbb{R}^3$ which maps the standard basis $(1, 0)^T$ and $(0, 1)^T$ to the vectors $(1, 0, 0)^T$ and $(0, 1, 2)^T$ in $\mathbb{R}^3$ is characterized by the equations

$$f((1, 0)^T) = 1\mathbf{e}_1 + 0\mathbf{e}_2 + 0\mathbf{e}_3 \qquad f((0, 1)^T) = 0\mathbf{e}_1 + 1\mathbf{e}_2 + 2\mathbf{e}_3$$

where we use $\mathbf{e}_1, \mathbf{e}_2, \mathbf{e}_3$ to represent the standard basis of $\mathbb{R}^3$. This is represented by the 3×2 matrix

$$\begin{bmatrix} 1 & 0 \\ 0 & 1 \\ 0 & 2 \end{bmatrix}$$

from which we deduce that the image under this transformation of the vector $\mathbf{u} = (u_1, u_2)^T$ is

$$\begin{bmatrix} 1 & 0 \\ 0 & 1 \\ 0 & 2 \end{bmatrix} \begin{bmatrix} u_1 \\ u_2 \end{bmatrix} = \begin{bmatrix} 1u_1 + 0u_2 \\ 0u_1 + 1u_2 \\ 0u_1 + 2u_2 \end{bmatrix}$$

If the two spaces are the same, that is, if $\mathbf{U} = \mathbf{V}$, then the transformation is a mapping of **U** into itself. The set of vectors $\{A\mathbf{u}_1, A\mathbf{u}_2, \ldots, A\mathbf{u}_n\}$ spans the range of the transformation (which is necessarily a subspace of **U**). If these vectors are linearly independent, it follows that the range is **U** itself; that is, the mapping is one-to-one and onto. In that case the vectors $\mathbf{a}_j = A\mathbf{u}_j$ form another basis for **U**, and A represents a change of basis from the original $\{\mathbf{u}_1, \mathbf{u}_2, \ldots, \mathbf{u}_n\}$ to this new basis $\{\mathbf{a}_1, \mathbf{a}_2, \ldots, \mathbf{a}_n\}$.

Each $\mathbf{a}_j$ is represented in terms of the original basis by

$$\mathbf{a}_j = a_{1j}\mathbf{u}_1 + a_{2j}\mathbf{u}_2 + \cdots + a_{nj}\mathbf{u}_n$$

or, in other words, by the vector $(a_{1j}, a_{2j}, \ldots, a_{nj})^{\mathrm{T}}$, which is simply the jth column of the matrix A. For this reason, the range of the transformation represented by A is often called the *column* space of the matrix A.

The dimension of the column space of A is called the rank of A whether or not the matrix A is square.

For an $n \times n$ matrix A, if the rank is n, then the matrix is called *nonsingular*. In this case, we have seen that the transformation it represents is one-one and onto and therefore invertible. The inverse transformation also has a matrix representation, and its matrix A^{-1} is also the *inverse* of the matrix A; that is,

$$AA^{-1} = A^{-1}A = I$$

where I is the identity matrix. The $n \times n$ identity matrix is

$$I = \begin{bmatrix} 1 & 0 & 0 & \cdots & 0 \\ 0 & 1 & 0 & \cdots & 0 \\ \multicolumn{5}{c}{\cdots\cdots\cdots\cdots\cdots\cdots} \\ 0 & 0 & \cdots & 0 & 1 \end{bmatrix}$$

EXERCISE B.3

The first three Chebyshev polynomials are given by

$$T_0(x) = 1$$

$$T_1(x) = x$$

$$T_2(x) = 2x^2 - 1$$

Find the matrix for the transformation of a polynomial in Π_2 written in terms of this basis as $c_0 T_0 + c_1 T_1 + c_2 T_2$ into Π_2 with its standard basis. Find the matrix for the inverse transformation, and check that this is indeed the inverse of the first matrix.

B.4 SYSTEMS OF LINEAR EQUATIONS

We restrict our attention here to systems of n equations in n unknowns. Such a system can be written in matrix form as

$$Ax = b$$

where $\mathbf{b} \in \mathbb{R}^n$, or more fully as

$$\begin{bmatrix} a_{11} & a_{12} & \cdots & a_{1n} \\ a_{21} & a_{22} & \cdots & a_{2n} \\ \multicolumn{4}{c}{\cdots\cdots\cdots\cdots\cdots} \\ a_{n1} & a_{n2} & \cdots & a_{nn} \end{bmatrix} \begin{bmatrix} x_1 \\ x_2 \\ \cdots \\ x_n \end{bmatrix} = \begin{bmatrix} b_1 \\ b_2 \\ \cdots \\ b_n \end{bmatrix}$$

or in full as

$$a_{11}x_1 + a_{12}x_2 + \cdots + a_{1n}x_n = b_1$$
$$a_{21}x_1 + a_{22}x_2 + \cdots + a_{2n}x_n = b_2$$
$$\cdots \cdots \cdots \cdots \cdots \cdots \cdots \cdots$$
$$a_{n1}x_1 + a_{n2}x_2 + \cdots + a_{nn}x_n = b_n$$

If the matrix A is nonsingular (that is, has rank n, or nonzero determinant), then it has an inverse, and the system above has the unique solution $\mathbf{x} = A^{-1}\mathbf{b}$.

If, on the other hand, A is a singular matrix, then:

1. *Either* rank $A < $ rank $[A|\mathbf{b}]$, and there are no solutions.
2. *Or* rank $A = $ rank $[A|\mathbf{b}]$, and there are infinitely many solutions.

By $[A|\mathbf{b}]$ we denote the $n \times (n+1)$ matrix whose first n columns are those of A and whose $(n+1)$th column is $\mathbf{b}$. The alternatives 1 and 2 reflect the possibilities of the system of equations being respectively inconsistent or consistent but having redundancy. The first situation is described as being *overdetermined*, while the second is *underdetermined*.

The cases where the matrix is singular correspond to the transformation represented by A mapping the original vector space onto a proper subspace.

Among such transformations, a particularly important subclass are called projections. A *projection* from $\mathbf{U}$ onto a subspace $\mathbf{V}$ is a linear mapping P with the properties that:

1. $P(\mathbf{u}) \in \mathbf{V}$ for every $\mathbf{u} \in \mathbf{U}$.
2. $P(\mathbf{v}) = \mathbf{v}$ for every $\mathbf{v} \in \mathbf{V}$.

It follows that $P^2 = P$, and this property is inherited by its matrix representation; it is shared by some other matrices. An *orthogonal projection* on an inner product space has the additional property:

3. $P(\mathbf{u}) = \mathbf{0}$ for every $\mathbf{u} \in \mathbf{V}^{\perp}$

where the *orthogonal complement* $\mathbf{V}^{\perp} = \{\mathbf{u} \in \mathbf{U}: (\mathbf{u}, \mathbf{v}) = 0, \ \forall \mathbf{v} \in \mathbf{V}\}$. If the subspace $\mathbf{V}$ has a basis $\{\mathbf{v}_1, \mathbf{v}_2, \ldots, \mathbf{v}_m\}$ and we extend this to a basis for $\mathbf{U}$ by adding the vectors $\mathbf{u}_{m+1}, \mathbf{u}_{m+2}, \ldots, \mathbf{u}_n$, then P has the matrix representation *relative to this basis*

$$\left[\begin{array}{c|c} I_m & 0 \\ \hline 0 & 0_{n-m} \end{array}\right]$$

where 0_{n-m} denotes the $(n-m) \times (n-m)$ zero matrix and the other zeros denote the appropriate $m \times (n-m)$ and $(n-m) \times m$ zero matrices.

B.5 EIGENVALUES, MATRIX NORMS, AND CONDITION NUMBERS

The $n \times n$ matrix A has an *eigenvalue* λ if there exists a (nonzero) vector $\mathbf{x}$ for which

$$A\mathbf{x} = \lambda \mathbf{x}$$

The vector $\mathbf{x}$ is called an *eigenvector* corresponding to λ.

Eigenvalues and eigenelements for general linear operators on other linear spaces are defined similarly.

> **Theorem 1.** Eigenvectors corresponding to distinct eigenvalues of a symmetric matrix are orthogonal.

Important Properties

1. If A has n distinct eigenvalues, then the corresponding eigenvectors form a basis for $\mathbb{R}^n$.
2. If A is real symmetric $(A = A^T)$, then it has n real eigenvalues.
3. The eigenvalue λ has (geometric) multiplicity k if the linear subspace of all eigenvectors corresponding to λ is of dimension k.
4. The determinant of a matrix is equal to the product of all its eigenvalues, each taken with its correct multiplicity. If A has all its eigenvalues nonzero, then A is nonsingular.
5. If A is nonsingular, then the eigenvalues of A^{-1} are the reciprocals of those of A.
6. The spectral radius $\rho(A)$ is the largest eigenvalue (in absolute value) of A.

Matrix Norms

Corresponding to any vector norm on $\mathbb{R}^n$ there is a matrix norm on $n \times n$ matrices defined by

$$\|A\| = \max\left\{ \frac{\|A\mathbf{x}\|}{\|\mathbf{x}\|} : \mathbf{x} \in \mathbb{R}^n \right\}$$

It follows that

$$\|A\|_\infty = \max_i \sum_j |a_{ij}| \quad \text{and} \quad \|A\|_1 = \max_j \sum_i |a_{ij}|$$

Less easy to establish is that

$$\|A\|_2 = \sqrt{\rho(A^T A)}$$

which for a symmetric matrix reduces to simply $\|A\|_2 = \rho(A)$. One immediate consequence of the definition is that for any matrix norm

$$\|A\mathbf{x}\| \le \|A\| \|\mathbf{x}\|$$

Condition Number

The condition number of A for a given vector norm and its associated matrix norm, $\kappa(A)$, is given by

$$\kappa(A) = \|A\| \|A^{-1}\|$$

For any matrix norm it is easy to prove that $\kappa(A) \geq |\lambda_{max}/\lambda_{min}|$, where λ_{max} and λ_{min} are the largest and smallest eigenvalues of A in absolute value.

Generally speaking, the smaller the condition number (that is, the closer it is to unity) the better is the matrix A suited to numerical computation. Ill-conditioned matrices (those with large condition numbers) are more sensitive to roundoff errors or perturbations in the data.

EXERCISES 1.1

1. 0010, 1101, 1010
3. 4-bit: $I_{max} = 7$, $I_{min} = -8$

EXERCISES 1.2

1. $\pi \simeq (0.314\,159)_{10} \times 10^1$
   $\pi \simeq (0.11001001000011111110)_2 \times 2^2$
   $\pi \simeq (0.3243F)_{16} \times 16^1$

2. $1 = (0.100\,000)_2 \times 2^1 \qquad \frac{1}{1} = (0.100\,000)_2 \times 2^1$
   $2 = (0.100\,000)_2 \times 2^2 \qquad \frac{1}{2} = (0.100\,000)_2 \times 2^0$
   $3 = (0.110\,000)_2 \times 2^2 \qquad \frac{1}{3} = (0.101\,011)_2 \times 2^{-1}$
   $4 = (0.100\,000)_2 \times 2^3 \qquad \frac{1}{4} = (0.100\,000)_2 \times 2^{-1}$
   $5 = (0.101\,000)_2 \times 2^3 \qquad \frac{1}{5} = (0.110\,011)_2 \times 2^{-2}$

EXERCISES 1.3

1. $0.333\ldots \times 10^{-5}$, $0.333\,325 \le x < 0.333\,335$
2. 111.1104, 135.8016

EXERCISES 1.4

3. Consider numbers whose normalized form has the minimum exponent.

EXERCISES 1.5

1. $\delta(f(4)) \simeq 0.00125$, $\sqrt{3.995} = 1.99875$, $\sqrt{4.005} = 2.00125$ to five decimal places.
3. $\delta(\tau) \simeq 0.001$ second; $\rho(\tau) \simeq 0.0003$

EXERCISES 1.6

1. $N = 9$, $\delta(e^8) \simeq 2.89 \times 10^{-4}$

EXERCISES 1.7

1. $N = 8$

EXERCISES 2.1

1. Interval $\left[\dfrac{7\pi}{32}, \dfrac{\pi}{4}\right]$; 24 more iterations are needed

EXERCISES 2.2

2. $n \geq 30$
3. Iteration (iii), $n = 3$

EXERCISES 2.3

1. $x_0 = 1$, $x_1 = 0.750\,364$, $x_2 = 0.739\,113$, $x_3 = x_4 = x_5 = 0.739\,085\,133$
3. Iterates: 2.5, 2.709 273 170, 2.718 266 884, 2.718 281 828 repeated

EXERCISES 2.4

1. Using $x_0 = 3$, $x_1 = 4$: $x_2 = 3.57143$, $x_3 = 3.60377$, $x_4 = 3.60556$, $x_5 = 3.60555$, $x_6 = 3.60555$
2. $x_{n+1} = x_{n-1} + x_n - x_{n-1}x_n c$

EXERCISES 2.5

1. Using rearrangement: $x = \ln(1 + 2x)$.
Function iterates: 1.5, 1.38629, 1.32776, 1.29624, 1.27884, 1.26911.
Aitken iterates: 1.26566, 1.25945, 1.25742, 1.25675

EXERCISES 2.6

2. $\dfrac{\partial y}{\partial x} = \dfrac{2x + y}{y} > 1$ throughout the positive quadrant

EXERCISES 2.7

1. 3.97 to two decimal places

EXERCISES 3.1

1. 6.6512

EXERCISES 3.3

1. $y = 3.606\ 737\ 6,$ $Y = 3,$ $U = 0.606\ 737\ 6,$ $V = \frac{19}{32},$ $W = 0.012\ 987\ 6$
   12.182 494 0
2. $F = 1.2,\ N = 1,\ G = 0.2,\ H = 0.2/2.2 = 0.090\ 909 \cdots$

EXERCISES 3.4

1. $1.2345 = 1 + \frac{1}{2} - \frac{1}{4} - \frac{1}{8} + \frac{1}{16} + \frac{1}{32} + \frac{1}{64} + \frac{1}{128} - \frac{1}{256}$
2. For $a < \frac{1}{2},\ a^k > 2a^{k+1} > a^{k+1} + 2a^{k+2} > \cdots$
   Fix k and use induction on $n - k$.

EXERCISES 4.2

1. $p_3(0.15) = 0.26245,\ p_3(0.28) = 0.44460$
2. Either use induction or show that for $i \neq j$, $(x_i - x_j)$ is a factor of the determinant.

EXERCISES 4.3

1. $f[x_0, x_1] = -1,\ f[x_0, x_2] = 1,\ f[x_0, x_1, x_2] = 1$
3. Divided difference table:

```
−1   2
              −1
     0   1          1
              2          0
     2   5          1          p(x) = 2 − 1(x + 1) + 1(x + 1)x
              5          0               = x² + 1
     3  10          1
              7
     4  17
```

EXERCISES 4.4

1. $A_{2,2} = \dfrac{(x_1 - \bar{x})A_{1,2} - (x_2 - \bar{x})A_{1,1}}{x_1 - x_2}$

Expand and simplify to obtain
$f[x_0] + (\bar{x} - x_0)f[x_0, x_1] + (\bar{x} - x_0)(\bar{x} - x_1)f[x_0, x_1, x_2]$

EXERCISES 4.5

1. $\sin 1.23 \approx 0.8912 + (1.3)(0.0408) + \dfrac{(1.3)(0.3)(-0.0092)}{2}$

$+ \dfrac{(1.3)(0.3)(-0.7)(-0.0006)}{6}$

≈ 0.9425

$\sin 1.96 \approx 0.9093 + (-0.4)(-0.0370) + \dfrac{(-0.4)(0.6)(-0.0095)}{2}$

$+ \dfrac{(-0.4)(0.6)(1.6)(-0.0001)}{6}$

≈ 0.9252

EXERCISES 5.1

2. 0.0912, 0.2594, 0.4032, 0.5289, 0.6405

EXERCISES 5.2

1. 0.09353, 0.26286, 0.40531, 0.53079, 0.64139

EXERCISES 5.4

1. $$\begin{bmatrix} 2h_0 & h_0 & & & & \\ h_0 & 2(h_0 + h_1) & h_1 & & & \\ 0 & h_1 & 2(h_1 + h_2) & h_2 & & \\ \cdots\cdots\cdots\cdots\cdots\cdots\cdots\cdots\cdots\cdots\cdots\cdots\cdots\cdots\cdots\cdots\cdots\cdots \\ & & h_{N-2} & 2(h_{N-2} + h_{N-1}) & h_{N-1} \\ & & & h_{N-1} & 2h_{N-1} \end{bmatrix} \begin{bmatrix} A_0 \\ A_1 \\ \vdots \\ \\ A_N \end{bmatrix}$$

$$= 6 \begin{bmatrix} d_0 - f'_0 \\ \Delta d_0 \\ \vdots \\ \Delta d_{N-2} \\ f'_N - d_{N-1} \end{bmatrix}$$

EXERCISES 5.5

1. For example, $\dfrac{1}{x+1} - \dfrac{1}{x+2} \notin \mathcal{P}_{0,1}$

2. $\dfrac{7 - 2x - x^2}{1 + x/2}$

EXERCISES 5.6

1. $\sin 3x = 3 \sin x - 4 \sin^3 x$
   $\cos 3x = 4 \cos^3 x - 3 \cos x$
   $\sin 4x = 4 \sin x \cos x (\cos^2 x - \sin^2 x)$
   $\cos 4x = 8 \cos^4 x - 8 \cos^2 x + 1$

EXERCISES 5.7

1. $2 - x + (x - 1)(y^2 - y + 2), 3.375$

EXERCISES 6.1

1. $(10, 1, 0.2, 0.03, 0.004, 0.0005)$
3. No solution: $a = 4, b \neq 9$;
   Infinitely many: $a = 4, b = 9$; General solution: $x_1 = x_3 = \alpha$; $x_2 = 3 - 2\alpha$
   Unique solution: $a \neq 4$; for example, for $a = 5, b = 10, x_1 = x_2 = x_3 = 1$

EXERCISES 6.3

1. $\|A\|_1 = 32$, $\|\mathbf{b}\|_1 = 75$, $\|\mathbf{r}\|_1 = 3.1 \times 10^{-11}$, $\|A^{-1}\|_1 \geq 10^5$, $\kappa(A)_1 \geq 3.2 \times 10^6$,
   $\|\mathbf{x}^*\|_1 = 4$
   Relative error $\approx 1.3 \times 10^{-6}$; true relative error $\approx 4 \times 10^{-6}$

EXERCISES 6.4

1. $\begin{bmatrix} 1 & 2 & 3 & 4 \\ 2 & 2 & 3 & 4 \\ 3 & 3 & 3 & 4 \\ 4 & 4 & 4 & 4 \end{bmatrix} = \begin{bmatrix} \frac{1}{4} & 1 & 0 & 0 \\ \frac{1}{2} & 0 & 1 & 0 \\ \frac{3}{4} & 0 & 0 & 1 \\ 1 & 0 & 0 & 0 \end{bmatrix} \begin{bmatrix} 4 & 4 & 4 & 4 \\ 0 & 1 & 2 & 3 \\ 0 & 0 & 1 & 2 \\ 0 & 0 & 0 & 1 \end{bmatrix}$

EXERCISES 6.6

1. First three iterations:

	x_1	x_2	x_3	x_4	x_5
0	0	0	0	0	0
1	0.5	−0.2	0.4	−0.2	0.5
2	0.59	−0.55	0.58	−0.58	0.59
3	0.774	−0.666	0.738	−0.666	0.774

EXERCISES 7.1

1. Linear: $\quad 1.175 + 1.104x$
   Quadratic: $\quad 0.996 + 1.104x + 0.538x^2$
   Cubic: $\quad 0.996 + 0.994x + 0.538x^2 + 0.184x^3$

EXERCISES 7.2

1. $p(x) = \frac{1}{2}$ is best constant and linear approximation. Best quadratic is $0.972 - 0.405x^2$ to three decimal places.

EXERCISES 7.3

1. For $w(x) = 1$:
   $$f \approx 1.175 + 1.104x + 0.538(x^2 - \frac{1}{3}) + 0.184(x^3 - 3x/5)$$
   $$= 0.996 + 0.994x + 0.538x^2 + 0.184x^3$$

EXERCISES 7.4

3. $1, x - 1, x^2 - 4x + 2, x^3 - 9x^2 + 18x - 6$

EXERCISES 7.5

3. $b_k = 0$, $a_0 = \pi$, $a_k = -4/k^2\pi$ for k odd, $a_k = 0$ for k even

EXERCISES 7.6

3. $1, x^2 - 5/2, x^3 - 17x/20$;
   Coefficients $a_0 = 0.994$, $a_1 = 1.148$, $a_2 = 0.548$

EXERCISES 7.8

1. $B_{0,0}(x) = 1$, $B_{1,0}(x) = 1 - x$, $B_{1,1}(x) = x$, $B_{2,0}(x) = (1 - x)^2$,
   $B_{2,1}(x) = 2x(1 - x)$, $B_{2,2}(x) = x^2$, $B_{3,0}(x) = (1 - x)^3$,
   $B_{3,1}(x) = 3x(1 - x)^2$, $B_{3,2}(x) = 3x^2(1 - x)$, $B_{3,3}(x) = x^3$,
   $B_{4,0}(x) = (1 - x)^4$, $B_{4,1}(x) = 4x(1 - x)^3$, $B_{4,2}(x) = 6x^2(1 - x)^2$,
   $B_{4,3}(x) = 4x^3(1 - x)$, $B_{4,4}(x) = x^4$
   $B_4(\sin \pi x; x) = 2.828x(1 - x)^3 + 6x^2(1 - x)^2 + 2.828x^3(1 - x)$

EXERCISES 8.1

2. $x = 1$, $y = 1$; Minimum
4. $h = 2r$

EXERCISES 8.2

1. With derivatives: $[3.55, 6.11]$; without: $[3.55, 11.23]$
3. Bracket should contain minimum point 5.2945.

EXERCISES 8.3

1. Minima at $(\pm\sqrt{2}, 2)$
   Vertices for first few iterations:

   $\mathbf{G} = (1, 0), \quad \mathbf{H} = (0, 0), \quad \mathbf{S} = (0, 1), \quad \mathbf{R} = (-1, 1), \mathbf{E} = (-2, 1.5)$

   $\mathbf{G} = (0, 0), \quad \mathbf{H} = (0, 1), \quad \mathbf{S} = (-1, 1), \quad \mathbf{R} = (-1, 2)$

   $\mathbf{G} = (0, 1), \quad \mathbf{H} = (-1, 2), \mathbf{S} = (-1, 1), \quad \mathbf{R} = (-2, 2), \mathbf{C} = (-0.5, 1.25)$

   $\mathbf{G} = (-0.5, 1), \mathbf{H} = (-1, 1), \mathbf{S} = (-1, 1.5)$

EXERCISES 8.4

$(0, 0), \ (1, 0), \ (0.730, 0.135), \ (0.642, 0.248), \ (0.884, 0.517)$

EXERCISES 8.5

3. (a) Iteration 1: $\mathbf{x}_0 = (4, 2)$, $\mathbf{x}_1 = (2, 2)$, $\mathbf{x}_3 = (2, 1)$, $\mathbf{x}_4 = (0, 0)$
   (b) Iteration 1: $(3, 1)$, $(1, 1)$, $(1, 0.5)$, $(0.2, 0.3)$
       Iteration 2: $(0.2, 0.3)$, $(0.2, 0.1)$, $(0.04, 0.06)$, $(0, 0)$
   (c) Iteration 1: $(1, 1)$, $(1, 1)$, $(1, 0.5)$, $(1, 0.5)$
       Iteration 2 fails since $\mathbf{s}_0 = \mathbf{s}_1$.
5. Iterates: $(2.5, 2)$, $(1, 2)$, $(-0.5, -1)$

EXERCISES 8.6

1. For example, starting from $(0, -0.5, 1)$, next two iterates are $(0, -0.5, 0.5)$ and $(-0.2, -0.4, 0.4)$, the true minimum. From most starting points three iterations would be required.

EXERCISES 8.7

2. Starting from $(0, 0)$:
   Iteration 1: $\alpha = \frac{1}{4}$, $\mathbf{x}_1 = (0, -0.5)$
   Iteration 2: $\alpha = \frac{1}{4}$, $\mathbf{x}_2 = (-0.2, -0.9)$
   Iteration 3: $\alpha = \frac{1}{64}$, $\mathbf{x}_3 = (-0.239, -0.982)$
   Iteration 4: $\alpha = 1$, $\mathbf{x}_4 = (-0.280, -1.025)$

EXERCISES 8.8

1. $k = 1$: $F = (x + 1)^2 + x^2$; Minimum at $x = -\frac{1}{2}$
   $k = 10$: $F = (x + 1)^2 + 10x^2$; Minimum at $x = -\frac{1}{11}$
   In general: $x_k = -\dfrac{1}{1 + 10^{k-1}} \to 0$

EXERCISES 9.2

1. $f'(0) = 1.94 \pm 0.08$, $f'(0.25) = 1.02 \pm 0.02$, $f'(0.5) = 0.10 \pm 0.08$
2. $f''(0) = -2.208 \pm 0.024$, $f''(0.5) = -1.352 \pm 0.008$
4. $f[1, 1, 1] = 1.950 \pm 0.001$
   $f[0.9, 1, 1.1] = -0.500 \pm 0.01$
   $f[0.9, 1, 1.1, 1.2] = 0.333 \pm 0.067$
   $f[0.8, 0.9, 1, 1.1, 1.2] = -0.418 \pm 0.335$
   $f'(1) = 1.996 \pm 0.001$
5. With $\delta = 5 \times 10^{-5}$ (for instance)
   (a) With $M_3 = 64$, $h_{min} = 0.013$, $E(h_{min}) = 0.006$, $\Delta_{cent} f(2\pi/3) = -3.4615$, actual error $= 0.0025$
   (b) With $M_3 = 0.36$, $h_{min} = 0.075$, $E(h_{min}) = 0.001$, $\Delta_{cent} f(2) = 1.3600$, actual error $= 0.0009$
   (c) With $M_3 = 5$, $h_{min} = 0.031$, $E(h_{min}) = 0.002$, $\Delta_{cent} f(1) = -0.5000$, actual error $= 0$
12. $-2.25x(0.25) + x(0.5) = -1$, $x(0.25) - 2.25x(0.5) + x(0.75) = 0$,
    $x(0.5) - 2.25x(0.75) = 1 \to x(0.25) \approx -\frac{4}{9}$, $x(0.5) = 0$, $x(0.75) = \frac{4}{9}$

EXERCISES 9.3

1. (a) Trapezoid: $N = 135$; Simpson: $N = 42$
   (b) Trapezoid: $N = 637$; Simpson: $N = 40$
8. Actual $= \frac{46}{15}$
   (a) $n = 2$: 2.247, $|\text{error}| = 0.82$; $n = 3$, 2.8333, $|\text{error}| = 0.23$; $n = 4$: 2.9045, $|\text{error}| = 0.16$
   (b) (i) $n = 2$: 0.494 240 2, (ii) $n = 3$, 0.493 105 5, (iii) $n = 4$: 0.493 106 1
9. (a) $N = \sqrt{\dfrac{1}{480 \, \text{tol}}} \left(\sqrt{\dfrac{c\rho}{kt}} \, x \right)^5$

EXERCISES 9.4

1. (a) Level 1: $(0, 3)$, $\tau = 5 \times 10^{-4}$, Reject
      Level 2: $(0, \frac{3}{2})$, $\tau = 2.5 \times 10^{-4}$, Reject
      Level 3: $(0, \frac{3}{4})$, $\tau = 1.25 \times 10^{-4}$, Accept, $I_{test} = 0.22320$
      Level 3: $(\frac{3}{4}, \frac{3}{2})$: $\tau = 1.25 \times 10^{-4}$, Accept, $I_{test} = 0.36617$
      Level 2: $(\frac{3}{2}, 3)$, $\tau = 2.5 \times 10^{-4}$, Accept, $I_{test} = 0.56192$
      $I \approx 1.1513$

(b) Level 1: $(0, 3)$, $\tau = 10^{-4}$, Reject
    Level 2: $(0, \frac{3}{2})$, $\tau = 5 \times 10^{-5}$, Reject
    Level 3: $(0, \frac{3}{4})$, $\tau = 2.5 \times 10^{-5}$, Reject
    Level 4: $(0, \frac{3}{8})$, $\tau = 1.25 \times 10^{-5}$, Accept, $I_{test} = 0.06559$
    Level 4: $(\frac{3}{8}, \frac{3}{4})$, $\tau = 1.25 \times 10^{-5}$, Accept, $I_{test} = 0.14952$
    Level 3: $(\frac{3}{4}, \frac{3}{2})$, $\tau = 2.5 \times 10^{-5}$, Accept, $I_{test} = 0.23218$
    Level 3: $(\frac{3}{2}, \frac{9}{4})$, $\tau = 2.5 \times 10^{-5}$, Accept, $I_{test} = 0.49526$
    Level 3: $(\frac{9}{4}, 3)$, $\tau = 2.5 \times 10^{-5}$, Accept, $I_{test} = 0.00311$
    $I \approx 0.49992$

EXERCISES 9.5

1. 1.32768
3. (a) $A_i(h) = \dfrac{2^i A_{i-1}(h/2) - A_{i-1}(h)}{2^i - 1}$
   (b) 1.99708
6. 0.95050

EXERCISES 9.6

1. (a) 2-point: 1.50000; 3-point: 1.58333
   (b) 2-point: 0.64937; 3-point: 0.46397
2. (a) $N = 4$ (b) $N = 3$
4. $x = \pi/2$: 2-point: 0.78540; 3-point: 0.78540
   $x = \pi$: 2-point: 1.19283; 3-point: 1.60607
   $x = 3\pi/2$: 2-point: 2.35619; 3-point: 2.35619
   $x = 2\pi$: 2-point: 5.91941; 3-point: 1.47667
5. 2-point: -0.95143; 3-point: -0.43220
6. (a) 2-point: 1.48095; 3-point: 1.64560
   (b) 2-point: 0.60955; 3-point: 0.40080

EXERCISES 10.1

3. (a) $\quad x(t) = \left[C\mu(t)^{-1} + (n-1)\mu(t)^{-1} \int_0^t b(\tau)\mu(\tau)d\tau \right]^{-1/(n-1)}$

where

$$\mu(t) := \exp\left[(n-1) \int_0^t a(\tau)d\tau \right]$$

4. (a) $x(t) = \frac{5}{2}t - \frac{5}{4} + \frac{9}{4}e^{-2t}$; $h = 1$, $x \approx 6.00$ (error $= 2.2$), $h = 0.5$, $x \approx 3.75$ (error $= 0.041$).
   (b) $x(3) = 27$; however, the actual solution is discontinuous at $t = 2$, and thus this approximation has no validity.

10. $x_1(t) = 2 \cos t - 2 \sin t + 2t - 1$, $x_2(t) = 2 \cos t + 2 \sin t + t^2 - 2$
$h = 1$: $x_1(1) \approx 1$ (error $= 0.60$), $x_2(1) = 2$ (error $= 0.24$)
$h = 0.5$: $x_1(1) \approx 0.625$ (error ≈ 0.23), $x_2(1) \approx 2$ (error $= 0.23$).

11. $u_1' = u_2$, $u_1(0) = 1$, $u_2' = -u_1 + e^{-t}$, $u_2(0) = 0$
$u_1(t) = \frac{1}{2} \cos t + \frac{1}{2} \sin t + \frac{1}{2} e^{-t}$, $u_2(t) = -\frac{1}{2} \sin t + \frac{1}{2} \cos t - \frac{1}{2} e^{-t}$
$h = 0.5$: $x_1(0.5) \approx 1$ (error $= 0.018$), $x_2(0.5) \approx 0$ (error $= 0.10$)
$h = 0.25$: $x_1(0.5) \approx 1$ (error $= 0.018$), $x_2(0.5) \approx -0.55$ (error $= 0.049$)

12. (b) $mx'' + (k + 3\epsilon s^2)x + 3\epsilon sx^2 + \epsilon x^3 = 0$, where s is the unique real solution to $\epsilon s^3 + ks = mg$.

13. (a) (i) With $t^* = \sqrt{\dfrac{g}{R}}\, t$, $b^* = Rb$

(ii)
$$u_1' = u_2, \quad u_1(0) = \frac{h_0}{R}$$

$$u_2' = b^* u_2^2 - \frac{1}{1 + u_1^2}, \quad u_2(0) = \frac{v_0}{\sqrt{Rg}}$$

(b) $x_j' = v_j$, $v_j' = \dfrac{-Gm_2 x_j}{(x_1^2 + x_2^2 + x_3^2)^{(3/2)}}$, $j = 1, 2, 3$

19. With $t^* := \sqrt{\dfrac{g}{l}}\, t$, $\dfrac{d^2\theta^*}{dt^{*2}} + \dfrac{1}{\theta_0} \sin(\theta_0 \theta) = 0$, $\theta^*(0) = 1$, $\dfrac{d\theta^*}{dt^*}(0) = 0$.
$\frac{1}{6}\theta_0^2 \ll 1$.

22. With $x_1^* := \dfrac{d}{c} x_1$, $x_2^* := \dfrac{bd}{ad - ce} x_2$, $t^* := ct$, $a^* := \dfrac{a}{c}$, $e^* := \dfrac{e}{d}\dfrac{dx_1^*}{dt^*}$

$= [a^* - (a^* - e^*)x_2^* - e^* x_1^*] x_1^*$, $\dfrac{dx_2^*}{dt^*} = (-1 + x_1^*)x_2^*$

25. $\bar{x}_1 = \bar{x}_2 = \sqrt{b(r - 1)}$, $\bar{x}_3 = r - 1$

EXERCISES 10.2

2. (a) Yes, $L = 3$ (for instance). (b) No. (c) No.
6. Let $x \equiv 0$, $0 \le t < 2$ and $x \equiv 0$, $x = (\frac{1}{3}t^2 - \frac{2}{3}t)^{3/2}$, or $x = -(\frac{1}{3}t^2 - \frac{2}{3}t)^{3/2}$ on $2 \le t < \infty$.

EXERCISES 10.3

1. See Exercise 4, Sec. 10.1, for actual solution.

Method	h	$x(2)$	Error
(a) Mod. Euler	1.0	6.0000	2.2
	0.5	3.8906	0.099
(b) Midpoint	1.0	6.0000	2.2
	0.5	3.8906	0.099
(c) Class. RK	1.0	4.0000	0.21
	0.5	3.7945	0.003

8. Coefficient of h^3 term:

$$(\tfrac{1}{6} - \tfrac{1}{2}\omega_2\alpha^2)f_{i,tt} + (\tfrac{1}{3} - \omega_2\alpha\beta)f_i f_{i,tx} + (\tfrac{1}{6} - \tfrac{1}{2}\omega_2\beta^2)f_i^2 f_{i,xx} + \tfrac{1}{6}f_{i,t}f_{i,x} + \tfrac{1}{6}f_i f_{i,x}^2$$

14. See Exercise 10, Sec. 10.1, for actual solution.

Method	h	$x_1(1)$	Error	$x_2(1)$	Error
Mod. Euler	1.0	0.39844	8E − 4	1.81250	5E − 2
	0.5	0.39614	2E − 3	1.77375	1E − 2
Midpoint	1.0	0.33984	6E − 2	1.79688	3E − 2
	0.5	0.38218	2E − 2	1.76819	5E − 3
Class. RK	1.0	0.39788	2E − 4	1.76294	6E − 4
	0.5	0.39768	2E − 5	1.76352	3E − 5

EXERCISES 10.4

1. See Exercise 4, Sec. 10.1, for actual solution.

Method	h	$x(2)$	Error
(a) AB2	1	1.50000	2.2
	0.5	3.89063	0.10
(b) APC2	1	6.00000	2.2
	0.5	3.62695	0.16
(c) Impl.	1	3.75000	0.04
	0.5	3.77777	0.01

13. See Exercise 10, Sec. 10.1, for actual solution.

Method	h	$x_1(1)$	Error	$x_2(1)$	Error
(a) AB2	1	0.25000	0.15	1.85938	0.10
	0.5	0.33786	0.06	1.76753	0.004
(b) APC2	1	0.41016	0.012	1.76563	0.002
	0.5	0.40652	0.009	1.76280	0.0007
(c) Impl.	1	0.45329	0.056	1.77509	0.012
	0.5	0.41193	0.014	1.76662	0.003

EXERCISES 10.5

1. Step 1:
   First try: $t = 0$, $x = 1$, $h = 0.5$
   $E_{est} = 0.5625$ Reject: $h_{tol} = 0.1291$
   Second try: $t = 0$, $x = 1$, $h = 0.1291$
   $E_{est} = 0.0375$ Accept: $\xi_{t+h}^{(2)} = 0.7793$, $h_{tol} = 0.1291$
   Step 2:
   First try: $t = 0.1291$, $x = 0.7793$, $h = 0.1291$
   $E_{est} = 0.0284$ Accept: $\xi_{t+h}^{(2)} = 0.6899$, $h_{tol} = 0.1482$

Step 3:
First try: $t = 0.2582$, $x = 0.6899$, $h = 0.1482$
$E_{est} = 0.0284$ Accept: $\xi_{t+h} = 0.7052$, $h_{tol} = 0.1702$
Step 4:
First try: $t = 0.4064$, $x = 0.7052$, $h = 0.0936$
$E_{est} = 0.0082$ Accept: $\xi_{t+h}^{(2)} = 0.7716$ (error $= 0.056$)

EXERCISES 10.6

3. Extrapolation table:
0.500
0.6406   0.7813
0.6939   0.7472   0.7358   (Error $= 0.004$)

EXERCISES 10.7

5. $\xi_i = \left(x_0 + \dfrac{h}{1 + \lambda h - e^{-h}}\right)(1 + \lambda\eta)^i - \dfrac{he^{-ih}}{1 + \lambda h - e^{-h}}$

10. $-\frac{1}{6} \le \bar{h} < 0$, $|r_{1,2}| = \dfrac{\frac{4}{3} \pm \sqrt{\frac{4}{9} + \frac{8}{3}\bar{h}}}{2(1 - \frac{2}{3}\bar{h})}$

$\bar{h} < -\frac{1}{6}$, $|r_{1,2}| = \dfrac{\sqrt{\frac{4}{3} - \frac{8}{3}\bar{h}}}{2(1 - \frac{2}{3}\bar{h})}$; Interval of stability: $(-\infty, 0)$

11. $r = \dfrac{4\bar{h}/3 \pm \sqrt{4\bar{h}^2/3 + 4}}{2(1 - \bar{h}/3)}$. No interval of absolute stability.

EXERCISES 11.1

1. (a) $\lambda_1 = -2$, $v_1 = c[-1, 1, 0]^T$; $\lambda_2 = 2$, $v_2 = c[1, -1, 0]^T$; $\lambda_3 = 4$, $v_3 = c[1, 1, 0]^T$

(b) $\lambda_{1,2,3} = 2, 2, 2$, $v = c_1[-1, 1, 0]^T + c_2[0, 0, 1]^T$

(c) $\lambda_{1,2} = 2, 2$, $v = c_1[0, -1, 1]^T + c_2[1, 0, 0]^T$; $\lambda_3 = 4$, $v_3 = c[-1, 1, 1]$

3. (c) $\lambda_{max} \approx 12.550$, $v \approx [0.354, 0.010, 0.051, 0.398, 1.000]^T$ $\lambda_{min} \approx -7.053$, $v \approx [-0.001, 0.016, 1.000, 0.003, -0.053]^T$

4. $\lambda_1 \approx -7.054$, $v_1 \approx [-0.002, 0.026, 1.000, 0.002, -0.054]^T$

$\lambda_2 \approx -5.174$, $v_2 \approx [-0.058, 1.00, -0.029, -0.063, -0.053]^T$

$\lambda_3 \approx 6.632$, $v_3 \approx [-0.599, 0.018, -0.014, 1.000, -0.187]^T$

$\lambda_4 \approx 11.049$, $v_4 \approx [1.000, 0.059, -0.030, 0.495, -0.550]^T$

$\lambda_5 \approx 12.546$, $v_5 \approx [0.350, 0.099, 0.051, 0.396, 1.000]^T$.

5. Actual roots are -4, -2, $\frac{1}{2}$, $\frac{3}{2}$, 4.

10. Actual eigenvalues are (a) $1, 1 \pm 2i$ and (b) $3, \pm i, 1 \pm 2i$.

EXERCISES 11.2

1. $\begin{bmatrix} 1.00 & -3.00 & 0 & 0 \\ -3.00 & 1.67 & 3.07 & 0 \\ 0 & 3.07 & 4.19 & 2.82 \\ 0 & 0 & 2.82 & 3.15 \end{bmatrix}$

5. $\begin{bmatrix} 1.00 & -0.32 & 1.83 & 1.33 & 1.67 \\ 6.16 & 3.37 & -5.74 & -4.02 & 1.01 \\ 0.00 & 0.78 & -1.03 & -0.28 & -1.18 \\ 0.00 & 0.00 & -1.36 & 1.62 & 0.56 \\ 0.00 & 0.00 & 0.00 & -2.56 & 0.04 \end{bmatrix}$

6. (a) $(n-2)(3n^2 + \frac{7}{2}n + \frac{9}{2})$

EXERCISES 11.3

1. (a) $Q = \begin{bmatrix} 0.33 & -0.20 & 0.92 \\ -0.67 & 0.64 & 0.38 \\ -0.67 & -0.74 & 0.08 \end{bmatrix}$ $R = \begin{bmatrix} 3.00 & 2.67 & -0.67 \\ 0.00 & 4.35 & -0.51 \\ 0.00 & 0.00 & 2.30 \end{bmatrix}$

(b) $Q = \begin{bmatrix} 0.95 & 0.28 & 0.01 & -0.14 \\ 0.32 & -0.85 & 0.03 & 0.41 \\ 0.00 & -0.44 & 0.06 & -0.90 \\ 0.00 & 0.00 & 1.00 & 0.07 \end{bmatrix}$

$R = \begin{bmatrix} 3.16 & 2.85 & 0.63 & 1.26 \\ 0.00 & 4.57 & -2.14 & -0.57 \\ 0.00 & 0.00 & 1.00 & 1.98 \\ 0.00 & 0.00 & 0.00 & 0.41 \end{bmatrix}$

7. Actual eigenvalues are (a) 1, 2, 3, 4, 5, (b) $-2, 3 \pm i, 1 \pm i$, (c) $\pm 1, \pm 2, 4$.

EXERCISES 12.1

1. (a) $y(x) = \dfrac{179e^6 + 277}{108(e^6 - e^4)} e^{2x} - \dfrac{277 + 179e^4}{108(e^6 - e^4)} e^{4x} + \frac{37}{108} + \frac{5}{18}x + \frac{1}{6}x^2$

(b) $y(x) = \dfrac{2 + \cos 1}{2 \sin 1} \sin x - \frac{1}{2}x \cos x$

2. (a) $u(x) = \left(\dfrac{12 + 6L^2 + L^4}{12L} \right)x - \frac{1}{2}x^2 - \frac{1}{12}x^4$

(b) No solution

(c) $u(x) = 1 + \frac{1}{2}L^2 + \frac{1}{12}L^4 - \frac{1}{2}x^2 - \frac{1}{12}x^4$

3. (b) (i) $\lambda_k = \dfrac{k^2 \pi^2}{L^2}$, $y_k(x) = C_k \cos\left(\dfrac{k\pi}{L} x \right)$

(ii) $\lambda_k = \dfrac{(2k-1)^2\pi^2}{4}$, $y_k(x) = C_k \sin \dfrac{(2k-1)\pi}{2} x$

(iii) $\lambda_k = \dfrac{k^2\pi^2}{(\ln 2)^2}$, $y_k(x) = C_k \sin \dfrac{k\pi}{\ln 2} x$

4. (a) $\sqrt{\lambda} \sin\sqrt{\lambda} + \cos\sqrt{\lambda} = 0$
   (b) $\lambda_1 = 7.83$, $\lambda_2 = 37.47$, $\lambda_3 = 86.82$

5. (a) $y(x) = \dfrac{2 - \cos 1}{\sin 1} \cos x + 3 \sin x + 1 - 2x$

   (b) $y(x) = \dfrac{\sin 1 + \cos 1}{\sin 1 - \cos 1} \cos x + \sin x$

6. $y(x) = \dfrac{1}{\lambda} + \dfrac{\lambda - 1}{\lambda} \cos\sqrt{\lambda}x - \left(\dfrac{1}{\lambda} + \dfrac{\lambda - 1}{\lambda} \cos\sqrt{\lambda}\right) \dfrac{\sin\sqrt{\lambda}x}{\sin\sqrt{\lambda}}$

BIBLIOGRAPHY

Abramowitz, M., and Stegun, I. A. (1965): *Handbook of Mathematical Functions*, Dover, New York.

Barlow, J. L., and Bareiss, E. H. (1985): "On Roundoff Error Distributions in Floating-Point and Logarithmic Arithmetic," *Computing*, vol. 24, pp. 325–341.

Benford, F. (1937): "The Law of Anomalous Numbers," *Proc. Am. Philos. Soc.*, vol. 78, pp. 551–572.

Brent, R. P. (1973): *Algorithms for Minimization without Derivatives*, Prentice-Hall, Englewood Cliffs, N.J.

—— (1978): "A Fortran Multiple-Precision Arithmetic Package," *ACM Trans. Math. Software*, vol. 4, pp. 57–70.

Broyden, C. G. (1967): "Quasi-Newton Methods and Their Application to Function Minimization," *Math. Comput.*, vol. 21, pp. 368–381.

—— (1970): "The Convergence of a Class of Double-Rank Minimization Algorithms," *J. IMA*, vol. 6, pp. 222–231.

Chvatal, V. (1983): *Linear Programming*, Freeman, New York.

Conte, S. D., and de Boor, C. (1980): *Elementary Numerical Analysis*: *An Algorithmic Approach*, 3d ed., McGraw-Hill, New York.

Cooley, J. W., and Tukey, J. W. (1965): "An Algorithm for the Machine Calculation of Complex Fourier Series," *Math. Comput.*, vol. 19, pp. 297–301.

Dalquist, G. G. (1963): "Stability Questions for Some Numerical Methods for Ordinary Differential Equations," *Proc. Symp. Appl. Math.*, vol. 15, Amer. Math. Soc., Providence, R.I.

Danby, J. M. A (1962): *Fundamentals of Celestial Mechanics*, Macmillan, New York.

—— (1985): *Computing Applications to Differential Equations*, Reston, Reston, Va.

Davidon, W. C. (1959): "Variable Metric Methods of Minimization, *Argonne Natl. Lab. Rep.* ANL-5990.

de Boor, C. (1978): *A Practical Guide to Splines*, Springer-Verlag, New York.

Dixon, L. C. W. (1972): "Variable Metric Methods: Necessary and Sufficient Conditions for Identical Behaviour on Non-quadratic Functions," *J. Optim. Theory Appl.* vol. 10, pp. 34–40.

—— (1973): "Conjugate Directions without Linear Searches," *J. IMA*, vol. 11, pp. 317–328.

Duff, I. S., Erisman, A. M., and Reid, J. K. (1986): *Direct Methods for Sparse Matrices*, Oxford University Press, Oxford.

Fiacco, A. V., and McCormick, G. P. (1968): *Nonlinear Programming*, Wiley, New York.

Fletcher, R. (1970): "A New Approach to Variable Metric Algorithms," *Comput. J.*, vol. 13, pp. 317–322.

—— and Powell, M. J. D. (1964): "A Rapidly Convergent Descent Method for Minimization," *Comput. J.*, vol. 6, pp. 163–168.

—— and Reeves, C. M. (1964): "Function Minimization by Conjugate Gradients," *Comput. J.*, vol. 7, pp. 149–154.

Flynn, M. J. (1966): "Very High Speed Computing Systems," *Proc. IEEE*, vol. 54, pp. 1901–1909.

George, J. A. (1989): "Solution of Sparse Systems of Equations on Multiprocessor Architectures," in P. R. Turner (ed.), *Numerical Analysis and Parallel Processing*, Lecture Notes in Mathematics, Springer-Verlag, Heidelberg.

Goldfarb, D. (1970): "A Family of Variable Metric Methods Derived by Variational Means," *Math. Comput.*, vol. 24, pp. 23–26.

Goldstine, H. H, Murray, F. J., and von Neumann, J. (1959): "The Jacobi Method for Real Symmetric Matrices," *J. Assoc. Comp. Mach.*, vol. 6, pp. 59–96.

Golub, G. H., and Van Loan, C. F. (1989): *Matrix Computations*, 2d ed., Johns Hopkins, Baltimore.

Gragg, W. B. (1965): "On Extrapolation Algorithms for Ordinary Initial Value Problems," *J. SIAM Numer. Anal.*, ser. B, vol. 2, no. 3, pp. 384–403.

Haberman, R. (1983): *Elementary Applied Partial Differential Equations*, Prentice-Hall, Englewood Cliffs, N.J.

Henrici, P. (1962): *Discrete Variable Methods in Ordinary Differential Equations*, Wiley, New York.

—— (1974): *Applied and Computational Complex Analysis*, vol. 1, Wiley, New York.

—— (1986): *Applied and Computational Complex Analysis*, vol. 3, Wiley, New York.

Hockney, R. W. (1965). "A Fast Direct Solver of Poisson's Equation Using Fourier Series," *J. ACM*, vol. 12, pp. 95–113.

Huang, H. Y. (1970): "Unified Approach to Quadratically Convergent Algorithms for Function Minimization," *J. Opt. Theory Appl.*, vol. 5, pp. 405–423.

Isaacson, E., and Keller, H. B. (1966): *Analysis of Numerical Methods*, Wiley, New York.

Lanczos, C. (1988): *Applied Analysis*, Dover, New York.

Lin, C. C., and Segal, L. A. (1974): *Mathematics Applied to Deterministic Problems in the Natural Sciences*, Macmillan, New York.

Lorenz, E. (1963): "Deterministic Non-Periodic Flows," *J. Atmos. Sci.*, vol. 20, pp. 130–141.

McCormick, S. F. (ed.) (1987): *Multigrid Methods*, SIAM, Philadelphia.

Modi, J. J. (1988): *Parallel Algorithms and Matrix Computation*, Oxford University Press, Oxford.

Nelder, J. A., and Mead, R. (1965): "A Simplex Method for Function Minimization," *Comput. J.*, vol. 7, pp. 308–313.

Nemytskii, V. V., and Stepanov, V. V. (1989): *Qualitative Theory of Differential Equations*, Dover, New York.

O'Leary, D. P. (1984): "Ordering Schemes for Parallel Processing of Certain Mesh Problems," *SIAM J. Sci. Stat. Comput.*, vol. 5, pp. 620–632.

Ortega, J. M. (1988): *Introduction to Parallel and Vector Solution of Linear Systems*, Plenum, New York.

Parlett, B. N. (1968): "Global Convergence of the Basic QR Algorithm on Hessenberg Matrices," *Math. Comput.*, vol. 22, pp. 803–817.

Polak, E., and Ribiere, G. (1969): "Note sur la Convergence de Methodes de Directions Conjugees," *Rev. Fr. Rech. Oper.*, vol. 16-R1, pp. 35–43.

Powell, M. J. D. (1964): "An Efficient Method for Finding the Minimum of a Function of Several Variables without Calculating Derivatives," *Comput. J.*, vol. 7, pp. 155–162.

Prenter, P. M. (1975): *Splines and Variational Methods*, Wiley, New York.

Rutishauser, H. (1954): "Der Quotienten-Differenzen-Algorithmus," *Z. Angew. Math. Phys.*, vol. 7, pp. 562–570.

Schatz, A. (1985): *An Introduction to the Analysis of the Error in the Finite Element Method for Second-Order Elliptic Boundary Value Problems*, in P. R. Turner ed., *Numerical Analysis, Lancaster 1984*, Lecture Notes in Mathematics, Springer-Verlag, Heidelberg.

Shampine, L. F., and Gear, C. W. (1979): "A User's View of Solving Stiff Ordinary Differential Equations," *SIAM Rev.*, vol. 21, no. 1, pp. 1–17.

Shanno, D. F. (1970): "Conditioning of Quasi-Newton Methods for Function Minimization," *Math. Comput.*, vol. 24, pp. 647–656.

Showalter, R. E. (1985): *Variational Theory and Approximation of Boundary Value Problems*, in P.

R. Turner (ed.), *Numerical Analysis, Lancaster 1984*, Lecture Notes in Mathematics, Springer-Verlag, Heidelberg.

Stoer, J., and Bulirsch, R. (1980): *Introduction to Numerical Analysis*, Springer-Verlag, New York.

Swann, W. H. (1964): "Report on the Development of a New Direct Search Method of Optimization," ICI Industries, Central Instruments Laboratory Research Note 64/3.

Turner, P. R. (1982): "The Distribution of Leading Significant Digits," *IMA J. Numerical Anal.*, vol. 2, pp. 407–412.

Volder, J. (1959): "The CORDIC Computing Technique," *IRE Trans. Comput.*, vol. EC8, pp. 330–334.

Watkins, D. S. (1982): "Understanding the QR Algorithm," *SIAM Rev.*, vol. 24, no. 4, pp. 427–440.

Wilkinson, J. H. (1963): *Rounding Errors in Algebraic Processes*, H.M. Stationery Office, London.

——— (1965): *The Algebraic Eigenvalue Problem*, Oxford University Press, Oxford.

Young, D. M., and Gregory, R. T. (1972): *A Survey of Numerical Mathematics*, vol. II, Dover, New York.

Zangwill, W. J. (1969): *Nonlinear Programming*: *A Unified Approach*, Prentice-Hall, Englewood Cliffs, N.J.

INDEX